T0251401

SI BASE UNITS

Unit	Symbol
meter	m
kilogram	kg
second	s
Kelvin	K
mole	mol

SI DERIVED UNITS

Unit	Symbol	Formula
Hertz	Hz	1/s
Newton	N	kg m/s^2
Pascal	Pa	N/m^2
Joule	J	N m
Watt	W	J/s

COMMON CONVERSION FACTORS (BY ROWS)

MASS

kg	lbmass
1	2.205
0.4536	1

LENGTH

m	cm	in	ft	mi
1	100	39.37	3.281	
0.01	1	0.3937		
	2.54	1		
	30.48	12	1	
1609			5280	1

AREA

m^2	cm^2	in^2	ft^2
1	1E4	1550	10.76
1E-4	1	0.1550	1.076E-3
6.452E-4	6.452	1	6.944E-3
0.0929	929	144	1

VOLUME

m^3	cm^3	in^3	ft^3	gal (US)	gal (UK)	bbl (petrol)
1	1E6		35.31	264.2	220	6.29
1E-6	1	0.06102	3.531E-5			
1.639E-5	16.39	1	5.787E-4			
0.02832	2.832E4		1	7.481	6.229	0.1781
3.785E-3	3785	231	0.1337	1	0.8327	0.02381
4.546E-3	4546	277.4	0.1605	1.201	1	0.02859
0.1590	1.590E5	9702	5.615	42	34.97	1

DENSITY

kg/m^3	gm/cm^3	lb/ft^3
1	1E-3	0.06243
1E3	1	62.43
16.02	0.01602	1

FORCE

N	lbf
1	0.2248
4.448	1

PRESSURE

kPa	psi	in HOH 60°F	in Hg 60°F	atm	bar
1	0.1450	4.019	0.2961	9.869E-3	0.01
6.895	1	27.71	2.042	0.06805	0.06895
0.2488	0.03609	1	0.07369	2.456E-3	2.488E-3
3.377	0.4898	13.57	1	0.0333	0.03377
101.3	14.7	407.2	30.01	1	1.013
100	14.5	401.9	29.61	0.9869	1

VELOCITY

m/s	m/h	ft/s	ft/min	in/s	mi/h
1		3.281	196.9	39.37	2.237
	1		0.05648	0.01094	
0.3048	1097	1	60	12	0.6818
5.080E-3	18.29	0.01667	1	0.2000	0.01136
0.05682	91.44	0.0833	5.000	1	0.05682
0.4470	1609	1.467	88.00	17.60	1

ACCELERATION

m/s^2	ft/s^2
1	3.281
0.3048	1

ENERGY

J	kW h	cal	ft lbf	Btu
1	2.778E-7	0.239	0.7376	9.484E-4
3.600E6	1	8.604E5	2.655E6	3414
4.184	1.162E-6	1	3.086	3.968E-3
1.356	3.766E-7	0.324	1	1.286E-3
1054	2.929E-4	252	777.6	1

POWER

W	ft lbf/s	ft lbf/hr	hp
1	0.7376	2655	1.341E-3
1.356	1	3600	1.818E-3
3.766E-4	2.778E-4	1	5.051E-7
745.7	550.0	1.980E6	1

ENERGY FLUX

kW/m^2	Btu/(h ft^2)
1	317
3.155E-3	1

HEAT CAPACITY

kj/(kg K)	Btu/(lb °F)
1	0.2388
4.187	1

CONDUCTIVITY

W/(m K)	Btu/(h ft °F)
1	0.5778
1.731	1

DIFFUSIVITY

m^2/s	ft^2/s
1	10.76
0.09290	1

VISCOSITY

N s/m^2	lb/(ft s)	lb/(ft h)	cP
1	0.6720	2419	1000
1.488	1	3600	1488
4.134E-4	2.778E-4	1	0.4134
0.001	6.720E-4	2.419	1

Momentum, Heat, and Mass Transfer Fundamentals

Momentum, Heat, and Mass Transfer Fundamentals

David P. Kessler
Robert A. Greenkorn

Purdue University
West Lafayette, Indiana

CRC Press
Taylor & Francis Group
Boca Raton London New York

CRC Press is an imprint of the
Taylor & Francis Group, an **informa** business

To our grandchildren:

Merrideth, Matthew, Mitchell (D.P.K.)

Megan, Samuel, Symone (R.A.G.)

PREFACE

This text springs from our experience over the past 30+ years teaching the momentum, heat, and mass transfer/transport sequence in the School of Chemical Engineering at Purdue University. As faculty members in a state land-grant institution, we encounter students with a wide variety of backgrounds planning for a wide variety of ultimate careers. We believe that with a firm grasp of engineering fundamentals, our graduates can readily progress to careers that involve either highly technical functions or broader responsibilities in management.

Our objective with this volume is to provide a foundation in basic momentum, heat, and mass transfer/transport sufficient to permit the student to do elementary design and analysis, and adequate as a base from which to learn more advanced concepts. We present the fundamentals of both microscopic and macroscopic processes. The text is built around a large number of examples which are worked in detail. Many of the examples are, of course, idealized, because their objective is to illustrate elementary principles, but we have kept them as realistic as possible.

Since the book is intended as a textbook, we have incorporated a high level of detail and redundancy in an effort to make the text readable for those just being introduced to the area. At the expense of conciseness in many places, we have attempted to avoid those gaps in derivations that are obvious to one familiar to the area, but utterly opaque to the novice.

We have included many references to more advanced material, or simply to other approaches to the material herein. Age and, we hope, wisdom has disabused us of any conceit that we possesses the only valid approach to the subject. To make access to other material easier for the student, we have included page numbers for most references to avoid the necessity of an index search in the citation. In the same vein, we have attempted to make our nomenclature conform to the most common usage in the area and have incorporated an extensive nomenclature table. An abbreviated thumb index permits rapid access to chapters and the more commonly accessed tables and figures.

Problems in momentum, heat, and mass transfer in fluids are profoundly difficult because the most practical application of the area is to the turbulent flow regime. To date, very little in the way of rigorous solution to even the most elementary turbulent flow problems is possible, although the exponential increase in computing power with time holds great promise for the future.

We feel that dimensional analysis is, and for the foreseeable future will be, still crucial for design of experiments, scale-up of equipment, and simplifying differential equations and associated boundary conditions.

Closed-form analytical solutions are still important for the insight that they bring; however, most applied problems seem clearly destined to be solved numerically. For this reason we have included a fair amount of numerical solution techniques.

Emphasis is mostly on finite difference algorithms, which are often easily implemented on a spreadsheet. Finite element analysis is heavily reliant on software, which requires too heavy a time investment to permit other than the abbreviated treatment here, although the area is certainly of growing importance. We have attempted to give an overview sufficient to enable the student to read further on his/her own.

At Purdue we cover the material in this text in two three-credit required undergraduate courses. In general, depending on the background of the students, the material in Chapter 5 on systems of units will receive more or less emphasis. Similarly, the transfer of heat by radiation stands alone and can be tailored to the desires of the particular professor. Perry's *Chemical Engineer's Handbook*[1] remains a useful supplemental reference for physical properties and empirical correlations. One of the commercial design packages can give help with physical property estimation, pumps and pipe networks, and more detailed heat exchanger design.

David P. Kessler

Robert A. Greenkorn

[1] Perry, R. H. and D. W. Green, Eds. (1984). *Chemical Engineer's Handbook.* New York, NY, McGraw-Hill.

THUMB INDEX

TABLE OF CONTENTS

experienced had the real atmosphere been in the box. As we have indicated by lighter density shading in Figure 1.1-1 (b), **all the features of the real atmosphere will not be present in the model, hence, the outputs of the model will not be precisely those of the real weather.** If one went to a more sophisticated (complicated) model, more features would be represented (usually) more accurately. The price of a more sophisticated model is usually more effort and expense to employ the model.

Notice that the model could be anything. At one extreme it could, to take a very poor[1] example of a model, incorporate a dart board with combinations of the wind, precipitation, and temperature, as shown in Figure 1.1-2. We could randomly throw a dart at the board and use as the model output the content of the label where the dart lands.

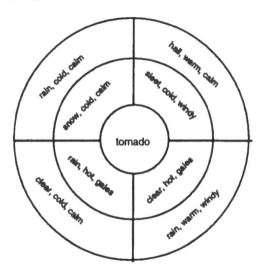

Figure 1.1-2 A poor model of the weather

Several things are immediately obvious about this model. First, it does not include anywhere near all possible weather patterns (outputs) - as just one example, there is no label for hail on a windy, cool day. Second, it does not include all (input) parameters. For example, it incorporates effects of humidity, barometric pressure, prevailing winds, etc., only insofar as they might affect the

---

[1] Probably the worst model ever used is the frontier method (hopefully apocryphal) for weighing hogs. First, they would laboriously tie the squirming hog to the end of a long sturdy log, then place the log across a stump for a fulcrum. Next they would search for a rock which, when tied to the other end of the log, balanced the weight of the hog. Then they would guess the weight of the rock.

person throwing the dart (and he/she is probably indoors). In fact, the output of the model bears no systematic relationship to the inputs. Third, it is not very precise in its outputs - it predicts only "windy" or "calm" without giving a wind velocity.

It will, however, give the correct prediction occasionally, if only by chance. **One should always be alert to the possibility that bad models can fortuitously give correct results,** even though the probability of this occurring is usually vanishingly small. For this reason, statistical methods should be used to test models, even though we will not have room here to explore this subject thoroughly.

At the other end of the spectrum are weather prediction models resident on large computers, containing hundreds of variables and large systems of partial differential equations. If we use such a very sophisticated model, we still need to look inside the black box and ask the same questions that we ask of the dart board model. (For example, some of the results of chaos theory pose the question of whether accurate long-term modeling of the weather is even possible.)

Although engineers and scientists model many of the same physical situations, the scientist is interested in the model that gives maximum accuracy and best conformity with physical reality, while the engineer is interested in the model which offers the precision and accuracy sufficient for an adequate solution of the immediate problem without unnecessary expenditure of time and money.

It should be noted that the best model of a system is the system itself. One would like to put the system to be modeled within the black box and use the system itself to generate the outputs. One would then obtain the most representative outputs.[2]

These outputs would be the solution of the true differential equations that describe the system behavior, even though from the outside of the box one would not know the form of these equations. The system, through its response to the inputs, "solves" the equations and presents the solution as the outputs. (We will see later, in Chapter 5, that through dimensional analysis of the inputs to the system it is possible to obtain the coefficients in the equations describing the system, even absent knowledge of the form of the equations themselves.)

---

[2] Even if we conduct measurements on the actual system under the same input conditions we still must face questions of reproducibility and experimental error.

If we cannot make measurements on the system itself[3] under the desired inputs, we must replace the system with a more convenient model. One of two avenues is usually chosen: either a scaled laboratory model or a mathematical model. The choice of an appropriately scaled laboratory model and conditions can be accomplished with the techniques presented in Chapter 5. Using a mathematical model presents the problem of including all the important effects while retaining sufficient simplicity to permit the model to be solved.

Originally, engineering mathematical models could be formulated solely in terms of intuitive quantities such as size, shape, number, etc. For example, it was intuitively obvious to the designer of a battering ram that if he made it twice as heavy and/or had twice as many men swing it, the gates to the city would yield somewhat sooner. This happy but unsophisticated state of affairs has evolved into one aesthetically more pleasing (more elegant and more accurate) but intuitively less satisfying.

Nature, though simple in detail, has proved to be very complex in aggregate. Describing the effect of a change in crude oil composition on the product distribution of a refinery is a far cry from deducing the effect of adding another man to the battering ram. This leaves us with a choice - we may model the processes of nature either by complex manipulations of simple relationships, or we may avoid complexity in the manipulations by making the relationships more abstract.

For example, in designing a pipeline to carry a given flow rate of a liquid, we might wish to know the pressure drop required to force the liquid through the pipeline at a given rate. One approach to this problem would be to take a series of pipes of varying roughness, diameter and length, apply various pressure drops, and measure the resulting flow rates using various liquids. The information could be collected in an encyclopedic set of tables or graphs, and we could design pipeline flow by looking up the exact set of conditions for each design. The work involved would be extreme and the information would be hard to store and retrieve, but one would need only such intuitively satisfying concepts as diameter, length and flow rate.[4] If, however, we are willing to give up these elementary concepts in favor of two abstractions called the **Reynolds number** and the **friction factor**, we can gather all the information needed in twenty or

---

[3] For example, the system may be inaccessible, as in the case of a reservoir of crude oil far beneath the ground; a hostile environment, as in the case of the interior of a nuclear reactor; or the act of measurement may destroy the system, as in explosives research.

[4] To run all combinations of only 10 different pipe diameters, 10 pressure drops and 100 fluids would require 10 x 10 x 100 = 10,000 experiments - and we still would have completely neglected important variables such as pipe roughness.

thirty experiments and summarize it on a single page. (We will learn how this is done in Chapter 6.)

1.1.1   Mathematical models and the real world

As the quantities we manipulate grow more abstract, our rules for manipulating these quantities grow more formal; that is, more prescribed by that system of logic called mathematics, and less by that accumulation of experience and application of analogy which we call physical intuition. We are thus driven to methods whereby we replace the system by a mathematical model, operate on the model with minimal use of physical intuition, and then apply the result of the model to the system. Obviously, the better the model incorporates the desired features of the system, the better our conclusions will be, and conversely. We must always be aware of the fact that the model is not the real world, a fact all too easy to ignore (to the detriment of our conclusions and recommendations) as the model becomes more and more abstract.

This procedure is exemplified by the way one balances a checkbook. The symbols in the checkbook do not come from observing the physical flow of money to and from the bank, but, if the manipulations are performed correctly, the checkbook balance (the output of the model) will agree with the amount in the bank at the end of the month. In this case the mathematical model represents **one** aspect of the real world *exactly* - that is, the account balance. This is not true in general for models (and is seldom true for engineering models). The outputs of the model are useful only in the areas where the predictions of the model and the results of the real world are approximately the same.

For example, a model of fluid flow which neglects fluid viscosity may give excellent results for real problems at locations where the velocity *gradients* are small, since the viscosity operates via such gradients, but the same model is useless to apply to problems involving sizable gradients in velocity. We frequently use just such a model for flow around airfoils by neglecting the presence of velocity gradients "far from" the airfoil. In the case of a plane flying at, say, 550 miles per hour, "far from" may be only the thickness of a sheet of paper above the wing surface. In this case, the relative velocity between the wing and the air goes from 0 to almost 550 miles per hour in this very short distance and then remains relatively constant outside this *boundary layer*. Obviously, **within** this layer the velocity gradient cannot be neglected: a model that neglects viscosity would give erroneous results.

Approximate models that are easily manipulated without external aids (e.g., pencil and paper, computer, etc.) are sometimes called "rules of thumb."[5] Rules of thumb are also referred to as *heuristics*.[6]

A practical engineering rule of thumb is that an economic velocity in pipes for fluids with properties not far from those of water is perhaps 10-15 ft/s. This model gives a crude optimum between the increased capital cost of pipe to keep pumping costs low (low velocity) and the increased pumping cost for the same flow rate to use smaller diameter pipe (high velocity). This rule of thumb should obviously not be extrapolated to fluids such as air, molasses, or molten polymers.

Practicing engineers apply models at a variety of levels of sophistication. Some engineers work at a purely experimental level, treating variables in their physically most intuitive form and determining simple correlations of the behavior of these variables by exhaustive experimental studies in the laboratory. Other engineers work with mathematical models of relatively low sophistication but based on physical or chemical laws, and fit these models to experimental data taken in the laboratory by adjustment of the various parameters that appear in the models. Still other engineers work almost completely in abstractions, often with mathematical models which describe processes for which data cannot be obtained in the laboratory - perhaps because sensing devices to acquire this data are not available or because the environment in which the data must be acquired is too corrosive, too hot, etc.

The most effective **engineer** is one who chooses the modeling process most economical in obtaining the answer **required** - that is, an answer

---

[5] An example of a much more approximate rule of thumb is that an alligator's length in feet is the same as the distance between its eyes in inches [attributed to Joan Isbell, horticulturist, Ithaca, New York, in Parker, T. (1987). *Rules of Thumb 2*. Boston, MA, Houghton Mifflin, p. 2]. There are several compendia of such rules of thumb, both humorous and otherwise: for example, also see Parker, T. (1983). *Rules of Thumb*. Boston, MA, Houghton Mifflin. and Parker, T. (1990). *Never Trust a Calm Dog and Other Rules of Thumb*. New York, NY, Harper Perennial Division of Harper Collins Publishers..

[6] An **heuristic** is defined as (Gove, P. B., Ed. (1981). *Third New International Dictionary of the English Language - Unabridged*. Springfield, MA, G and C Merriam Co.): "providing aid or direction in the solution of a problem but otherwise unjustified or incapable of justification ... of or relating to exploratory problem-solving techniques that utilize self educating techniques (as the evaluation of feedback) to improve performance." Rules of thumb are seldom unjustified, but usually rely on long experience (feedback) which has not been quantified in the form of rigorous mathematical models.

sufficiently accurate but of no higher order of accuracy than is necessary to draw conclusions appropriate to the problem.[7]

For example, in a model to predict the appropriate time for soft-boiling an egg, it might be desired to incorporate the effect of barometric pressure on the boiling temperature of water, since going from sea level to the top of a mountain can produce a significant change in this temperature, and a lower boiling temperature would require a longer time for the necessary heat transfer. It would be useless, however, to determine the barometric pressure effect to six significant figures, because the change in cooking time produced by uncertainty in other factors (e.g., egg-to-egg size variation, ambient temperature when the egg is removed from the water, how long it is until the egg is broken after removing it from the water, shell thickness, and so on) far exceeds a change produced by altering the boiling temperature in the sixth digit.

It is very embarrassing and not uncommon for engineers to find themselves refining their estimates of the effect of a factor to the point of irrelevance, while neglecting some far more important variable. The reader is cautioned to attempt **always** to work on the **limiting step** in the process.

As stated above, a model is useful only insofar as it gives useful predictions. A bad or incomplete model will guarantee bad or incomplete predictions **regardless of how elegantly it is manipulated**. Unfortunately, as models become more complex it becomes progressively easier to lose sight of this elementary fact.

For example, the checkbook model gives us the amounts of the checks but no information about **where** the checks were cashed - this simply is not a feature included in the model, and it would be absurd to try to predict where a check would be cashed by looking at the checkbook. Even a feature of the model (the checkbook balance) that supposedly perfectly reproduces a feature of the system (the account balance) must be checked by periodic reconciliation with the bank account, because human error (checkbook entries omitted or incorrectly entered, subtracted, or added), computer error (by the bank) or malice (embezzlement) may interfere with an otherwise perfect model. Similarly, in engineering, even models that are thought to be very accurate must still be checked against real-world data to ensure that the assumed accuracy is, in fact, true.

---

[7] For an excellent discussion of experiment vis-a-vis theory, see S. W. Churchill, *Chem. Eng. Progr.*, **66**(7):86(1970).

1.1.2   Scale of the model

A system can be described at different *scales*. In this text we will make most frequent use of the *macroscopic* and the *microscopic* scales. Macroscopic and microscopic are merely ways of saying "less detail" and "more detail."[8]

We define the *macroscopic* scale as one at which the model predicts only what is happening at points in **time,** not points in **space**. We may **use** information about space effects (initial and boundary conditions) in evaluating terms (integrals) in the model, but we do not **predict** space effects. In each term of a macroscopic model we integrate out the space effects individually across each input or output area and over the volume of the system itself ("lumping"). As a consequence of these integrations, macroscopic models do not predict what is happening at individual points, but only the gross effect over the total input area, the total output area, or the total internal volume.

For example, consider modeling energy for the case of a fluid flowing through a single inlet into a tank which contains a heating coil and out through a single outlet. A macroscopic model of this situation would be concerned only with the total amount of energy flowing into the inlet as a function of time, the total amount of energy flowing out of the outlet as a function of time, and the total energy contained within the tank as a function of time. The model might use the entering and exiting velocity and/or temperature profiles (space variations) if **known** as a function of time, but these would be buried in the integrands of terms, and the model would not **predict** what these profiles were at unknown points, nor would it predict the energy content of the fluid at individual points within the tank. Macroscopic models therefore tend to produce ordinary differential equations (with the independent variable being time) for unsteady-state systems, or algebraic equations for steady-state systems. Process control theory classifies such models as *lumped-parameter* models.

A *microscopic* model, by contrast, **does** include prediction of what happens at a points in space as well as time. Consider the case of fluid in laminar flow in a single pipe: a microscopic model would predict velocity and temperature at individual points in **both space and time** across the inlet, across the outlet,

---

[8] The microscopic scale as defined here refers to a scale that, although small, is still large enough that continuum assumptions about materials are valid. Below this scale lies the *molecular* scale. Later we will also make brief reference to the *megascopic* scale, a scale even larger than the macroscopic scale in the sense that large regional anomalies at the macroscopic scale level must be taken into account - for example, large-scale heterogeneities in flow in porous media. An example of the megascopic scale would be a primarily sandstone underground reservoir containing isolated large pockets of shale.

and within the pipe. This means that, as well as the value of the integral, we must concern ourselves with what is happening within the integrands that appear in terms in the macroscopic balances. As a consequence, microscopic models usually lead to partial differential equations, whose independent variables are the three space coordinates in addition to time.

We will apply three different transfer coefficients in modeling at the **macroscopic** scale: By *freestream* conditions, we refer to conditions far enough from the surface in an external flow that they are essentially constant.

• The first, C_D, is for momentum flux or shear stress:

$$\tau \sim \frac{F}{A} = C_D \left(\frac{\rho v_\infty^2}{2} - \frac{\rho v_s^2}{2} \right)$$

                                                                (1.1.2-1)

where τ is the magnitude of the shear stress, F is the magnitude of the drag force (skin friction), A is a characteristic area, ρ is the density, v_s is the velocity in the x-direction at the surface (equal to zero for a coordinate system fixed to the surface), and v_∞ is the freestream velocity. The shear stress relationship is shown as proportional rather than equal because the area on which the drag coefficient is based is usually not the area on which the force acts directly, but rather some typical area such as the cross-sectional area normal to flow. **The second term in parenthesis is usually zero, and is shown only to indicate the parallel to the definitions of the other coefficients.**

C_D as written above is primarily used for external flows - that is, flow **around** bodies. A modified form, called the friction factor, is usually used when the flow is **through** (inside) things - e.g., pipes - and in its definition, since there is no constant freestream velocity, an area-averaged velocity is usually used.

• The second, h, is for energy, specifically, for heat:

$$q_s = h \left(T - T_s \right)$$

                                                                (1.1.2-2)

where q_s is the heat flux at the surface; h is the single-phase heat transfer coefficient; T is either the freestream temperature or an average temperature, depending on whether external or internal flows are being modeled; and T_s is the temperature at the surface.

• The third, k_x, is for mass transfer:

$$\boxed{N_A = k_z \left(z_A - z_{As} \right)}$$
(1.1.2-3)

where N_A is the mass flux of species A, k_z is the single-phase mass transfer coefficient, z_{As} is the concentration of A at the surface, and z_A is either the freestream concentration or an average concentration of A depending on whether external or internal flows are being modeled.

The above equations describe three **transfer** coefficients. From Equation (1.1.2-1), C_D is the momentum **transfer** coefficient; from Equation (1.1.2-2), h is the heat **transfer** coefficient; and from Equation (1.1.2-3), k_z is the mass **transfer** coefficient. These coefficients describe the respective fluxes in terms of specific driving forces **within a single phase**. We can define other driving forces as well, which must be paired with different numerical coefficient values.

We will later also define, in terms of appropriate **overall** driving forces, **overall** heat transfer coefficients (denoted by U with a subscript indicating the particular area on which they are based) and **overall** mass transfer coefficients (denoted by K with a subscript denoting the phase on which they are based), but we will not apply a corresponding overall coefficient to momentum transfer because such a coefficient is not presently in engineering use.

At the **microscopic** level we define the transfer in terms of point values using differential equations. The resulting coefficients are called *transport coefficients*. We illustrate using one-dimensional forms.

• For momentum:

$$\boxed{\tau_{yx} = -\mu \frac{\partial v_x}{\partial y}}$$
(1.1.2-4)

where τ_{yx} is both (both quantities have the same units)

a) the flux of x-momentum in the y-direction, and

b) the shear stress in the x-direction on a surface of constant y exerted by the fluid in the region of lesser y,

and μ is the (Newtonian) viscosity, v_x is the x-component of velocity, and y is the y-coordinate.

• For energy:

$$q_x = -k \frac{\partial T}{\partial x}$$

(1.1.2-5)

where q is the heat flux in the x-direction, k is the thermal conductivity, T is the temperature, and x is the x-coordinate.

• For mass (using for illustration the special case of equimolar counterdiffusion):

$$N_{Az} = -\mathcal{D}_{AB} \frac{\partial x_A}{\partial z}$$

(1.1.2-6)

where N_{Az} is mass flux of species A in the z-direction, $\mathcal{D}_{AB}$ is the diffusivity of A through B (also called the diffusion coefficient), x_A is the mole fraction of A, and z is the z-coordinate.

The viscosity, μ, is the momentum **transport** coefficient, the thermal conductivity, k, is the heat **transport** coefficient, and the diffusivity, $\mathcal{D}_{AB}$, is the mass **transport** coefficient.

1.2 The Entity Balance[9]

In our study we will be concerned with the movement of mass, energy, and momentum much in the same way that a business is concerned with the transport of goods and services via manufacturing and sales efforts. Just as a business needs an accounting procedure to permit description of the inventory and movement of goods, we need an accounting procedure to help us describe the storage and movement of mass, energy, and momentum. This latter accounting procedure, coupled with some physical laws (for mass and energy, conservation laws, and for momentum, Newton's second law) permits us to develop macroscopic balances which tell us **how much** mass, energy, and momentum has been transferred, but not how the **rate** of this transfer relates to the physical variables in the system (that is, to velocity, temperature, and concentration differences and to the physical properties of the fluid). The macroscopic balances are analogous to the part of a company's annual report, which tells the amount of sales over the past year and how much the inventory has changed, but not, for example, how market conditions affected the sales rate.

In our course of study, we shall be faced with the manipulation of many abstractions: "lost work," and "enthalpy," to name just two. Many of these entities will be handled by the simple accounting procedure which we will now describe - in particular, those relating to mass, energy, and momentum. Our accounting procedure (the entity balance) must be applied to a **system** or **control volume**. This accounting procedure will lead us in a natural way to the mathematical model describing a process.

By *system* or *control volume* we mean some unambiguously defined region of space. For example, if we fill a balloon with air and release it, We might take as our system the space interior to the balloon. This would provide a perfectly acceptable system even though it moves about and changes size in a peculiar way as it flies about the room. Although **acceptable**, it might or might not be the most **convenient** system, and one of our problems in many situations is the selection of the most convenient system.

One sometimes uses as a system a specific quantity of matter, all parts of which remain in proximity; for example, a "clump" of fluid. This is consistent with the concept of a system since this matter unambiguously defines a region in space. The idea of a system can be illustrated by considering fluid flow in a pipe. We could choose as our system the region bounded by two planes normal

---

[9] An entity is a thing which has reality and distinctness of being, either in fact or for thought. (*Webster's Third New International Dictionary of the English Language*, G & C Merriam Company, Springfield, MA, 1969.)

to the pipe axis and the **inner** wall of the pipe. We could equally well use the region between the two planes and the **outer** wall of the pipe. Or, to take yet a third system, we could select a certain mass of fluid and follow the mass as it moves in space.

The *entity balance* used for our accounting system is

$$\boxed{\text{Input} + \text{Generation} = \text{Output} + \text{Accumulation}}$$

(1.2-1)

By **input** we mean that which crosses the system boundary from outside to inside in time Δt; by output, the converse. (There is no real need to define the second term, output. One could instead speak of positive and negative inputs - it is simply traditional.[10])

By **accumulation** we mean the result of subtracting that which was in the system at the **beginning** of some time interval from that which was there at the end of the interval. As opposed to the input/output case, we do **not** define a term called "depletion" (although we could); we speak instead of a "negative accumulation."

By **generation** we mean that which appears within the system without either being present initially or being transferred in across the boundary. It materializes, somewhat as the ghost of Hamlet's father, but in a far more predictable fashion. Similarly to the case of accumulation, we do not refer to "consumption," but rather to "negative accumulation."

You will note that the above is full of "that which"; we have carefully avoided saying just what it is for which we are accounting. This is deliberate, and is done to stress the generality of the procedure, which, as we shall see below, is applicable to people and money as well as to mass, energy, and momentum.

Note that our entity balance is equally valid when applied to **rates**.[11] By considering smaller and smaller time intervals we obtain

---

[10] As Tevye says in *Fiddler on the Roof* before singing "Tradition": "You may ask, 'How did this tradition start?' I'll tell you - I don't know! But it's a tradition. Because of our traditions, everyone knows who he is and what God expects him to do." Even engineering and science are bound by tradition in their methods of communication.

[11] The definition of a **instantaneous rate** for an entity is

$$\lim_{\Delta t \to 0} \left[\frac{\text{change in entity in } \Delta t}{\Delta t} \right]$$

$$\lim_{\Delta t \to 0} \left[\frac{\text{Input} + \text{Generation}}{\Delta t} = \frac{\text{Output} + \text{Accumulation}}{\Delta t} \right]$$

$$\lim_{\Delta t \to 0} \left[\frac{\text{Input}}{\Delta t} \right] + \lim_{\Delta t \to 0} \left[\frac{\text{Generation}}{\Delta t} \right] = \qquad\qquad (1.2\text{-}2)$$

$$\lim_{\Delta t \to 0} \left[\frac{\text{Output}}{\Delta t} \right] + \lim_{\Delta t \to 0} \left[\frac{\text{Accumulation}}{\Delta t} \right]$$

$$\boxed{\begin{array}{l} \text{Input rate} + \text{Generation rate} = \\[1em] \qquad \text{Output rate} + \text{Accumulation rate} \end{array}} \qquad (1.2\text{-}3)$$

so the entity balance applies to rates as well as to amounts.

Example 1.2-1   An entity balance

The entity balance can be applied to very general sorts of quantities so long as they are quantifiable, either on a discrete scale (e.g., particles) or a continuous scale (e.g., energy). When applied to discrete entities it is frequently called a **population balance**. One such entity, of course, is **people**, although automobiles, tornadoes, buildings, trees, bolts, marriages, etc., can also be described using the entity balance approach.

If the entity balance is applied to people with a political unit (e.g., a city, county, state, etc.) as the system, the input term is calculable from immigration statistics and the output from emigration statistics. The accumulation term is calculable from census figures (the differences in the number within the system over some prescribed time interval). The generation term is made up of births (positive) and deaths (negative).

For example, suppose that a political unit (the system) had a population of 1,000,000 people at the beginning of the year $(t = 0)$. At the end of the year $(t + \Delta t)$ suppose the population was 1,010,000. Thus the **accumulation** over a year (Δt) was

$$1,010,000 - 1,000,000 = 10,000 \qquad\qquad (1.2\text{-}4)$$

During this time suppose that 18,000 people died and 30,000 were born, giving a net **generation** of

$$30,000 - 18,000 = 12,000 \text{ people} \qquad (1.2\text{-}5)$$

If 40,000 people immigrated (input), how many emigrated (output)?

Solution

Application of the entity balance shows

$$40,000 + 12,000 = \text{Output} + 10,000 \qquad (1.2\text{-}6)$$

Solving, we see that 42,000 people emigrated.

An edifying exercise is to check published statistics on population, birth/death, and immigration/emigration for consistency by using the entity balance. Such a calculation performed on the world population reveals a large accumulation term, often referred to as the "population explosion."

1.2.1 Conserved quantities

It is interesting that when our equation is applied to some entities, there is never any **generation**. Quantities which do not exhibit generation - that is, quantities which are neither created nor destroyed - we term *conserved* quantities. For these quantities the entity balance contains only three terms

$$\text{Input} = \text{Output} + \text{Accumulation} \qquad (1.2.1\text{-}1)$$

This balance can also be written in terms of rates

$$\text{Input Rate} = \text{Output Rate} + \text{Accumulation Rate} \qquad (1.2.1\text{-}2)$$

For example, if I apply the entity balance to the amount in my checking account as the system, I never find money spontaneously appearing. (If I didn't put it in, it isn't there to remove - the first great law of personal finance. We note in passing that an embezzler may be regarded as either an output or a negative accumulation term but the total amount of money is still conserved.)

Quantities which are conserved under some assumptions or approximations may not be conserved under other assumptions. Banks and the government can create and destroy money, but this requires a model of a different level, much as including nuclear reactions in our energy balance models would destroy the assumption of conservation of energy.

We know that matter and energy are **not** conserved in processes where nuclear reactions are taking place or where things move at speeds approaching the speed of light. Matter and energy can, instead, be transmuted into one another, and only their totality is conserved.

In the course of this book, we will assume that **total** mass and **total** energy are conserved quantities, as they are for all practical purposes in the processes we will consider. On the other hand, momentum (which can be created or destroyed by applied forces), mass of an individual species (which can be created or destroyed by chemical reaction), and mechanical energy and thermal energy (which can each be transformed into the other) are not generally conserved in our processes.

Another difference in applying the entity balance to momentum is that we will generate a **vector** equation rather than a **scalar** equation.

We will, therefore, utilize **balance** equations, not all of which are **conservation** equations, and **conservation** equations, **all** of which are **balance** equations. Conservation equations are a sub-set of balance equations.

1.2.2   Steady-state processes

Another special case of the entity balance which is often of interest is for a **steady-state** process. By a steady-state process we mean one which does not change in time - if we look at the process initially, then look at it again after some time has elapsed, the process still looks the same. This means that we are concerned with the entity balance in the form of **rates** as expressed in Equation (1.2-3). Note that this does **not** imply that all rates are zero. It **does** imply that **accumulation** rate is zero, because the very nature of accumulation means that content of the system is changing in time. Input and output rates, however, can be different from zero so long as they are constant in time, as can generation rates - it is only necessary that these rates balance one another so that there is no accumulation.

For a steady-state process, therefore, the entity balance is written as

Input rate + Generation rate = Output rate

or

$$\boxed{\text{Generation rate} = \text{Output rate} - \text{Input rate}}$$

(1.2.2-1)

1.3 The Continuum Assumption

We will treat only models in which each fluid property (for example, the density, the temperature, the velocity, etc.) is assumed to have a **unique** value at a given point at a given time. Further, we assume these values to be **continuous** in space and time. Even though we know that fluids are composed of molecules, and therefore, within volumes approaching molecular size, density fluctuates (and temperature is hard even to define), in order to use the concepts of calculus and differential equations we model such properties as continuous.

For example, we define the density at a point as the mass per unit volume; but, since a point has no volume, it is not clear what we mean. By our definition we **might** mean

$$\rho = \lim_{\Delta V \to 0} \left(\frac{\Delta m}{\Delta V} \right)$$

(1.3-1)

where ρ is the density, and Δm is the mass in an elemental volume ΔV.

Consider applying this definition to a gas at uniform temperature and pressure. Assume for the moment that we can observe Δm and ΔV instantaneously, that is, that we can "freeze" the molecular motion as we make any one observation. For a range of values of ΔV suppose we make many observations of $\Delta m/\Delta V$ at various times but at the same point.

At large values of ΔV, because the total number of molecules in our volume would fluctuate very little, the density from successive observations would be constant for all practical purposes. As we approach smaller volumes, however, our observed value would no longer be **unique**, because a gas is not a continuum, but is made up of small regions of large density (the molecules) connected by regions of zero density (empty space). In fact, we would see a situation somewhat as depicted in Figure 1.3-1, where, as our volume shrinks, we begin to see fluctuation in the observed density even in a gas at constant temperature and pressure.

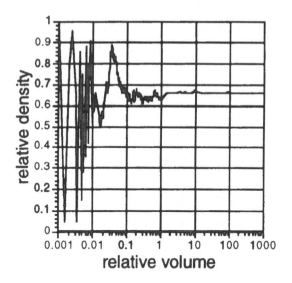

Figure 1.3-1 Breakdown of continuum assumption

For example, if our ΔV were of the same order of volume as an atomic nucleus, the observed density would vary from zero to the tremendous density of the nucleus itself, depending on whether or not a nucleus (or, perhaps, some extra-nuclear particle) were actually present in ΔV at the time of the observation. To circumvent such problems we make the arbitrary restriction that we will treat only problems where we may assume that the material behavior we wish to predict can be described adequately by a model based on a continuum assumption. In practice, this usually means that we do not attempt to model problems where distances of the order of the mean free paths of molecules are important.

The continuum assumption does fail under some circumstances of interest. For example, the size of the interconnected pores of a porous medium may be of the order of the mean free path of gas molecules flowing through the pores. In this situation, "slip flow" (where the velocity at the gas-solid interface is not zero) can result. Another example is in the upper atmosphere where the mean free path of the molecules is very large - in fact, it may become of the order of a hundred miles. In this situation, slip flow can occur at the surface of vehicles traversing the space.

1.4 Fluid Behavior

The reader is encouraged to make use of the resources available in flow visualization, e.g., Japan, S. o. M. E., Ed. (1988). *Visualized flow: Fluid motion in basic and engineering situations revealed by flow visualization.* Thermodynamics and Fluid Mechanics. New York, NY, Pergamon, Japan, T. V. S. o., Ed. (1992). *Atlas of Visualization.* Progress in Visualization. New York, NY, Pergamon Press, and Van Dyke, M. (1982). *An Album of Fluid Motion.* Stanford, CA, The Parabolic Press, which contain numerous photos, both color and black and white. There are also many films and video tapes available. There is nothing quite so convincing as seeing the actual phenomena that the basic models in fluid mechanics attempt to describe - for example, that the fluid really does stick to the wall.

1.4.1 Laminar and turbulent flow

We normally classify continuous fluid motion in several different ways: for example, as **inviscid** vs. **viscous***;* **compressible** vs. **incompressible**. Viscosity is that property of a fluid that makes honey and molasses "thicken" with a decrease in temperature. Inviscid flow is modeled by assuming that there is no viscous effect. What we sometimes call ideal fluids or ideal fluid flow is modeled by not including in the equations of motion the property called viscosity. Viscous flow is modeled including these viscous effects.

The viscosity of a fluid can range over many orders of magnitude. For example, in units of μPa s, air has a viscosity of about 10^{-5}; water, about 10^{-3}; honey, about 10; and molten polymers, perhaps 10^{4}; a ratio of largest to smallest of 1,000,000,000. (This is by no means the total range of viscous behavior, merely a sample of commonly encountered values in engineering problems.)

In viscous flow we can have two further sub-classifications. The first is *laminar* flow, a flow dominated by viscous forces. In steady laminar flow, the fluid "particles" march along smoothly in files as if on a railroad track. If one releases dye in the middle of a pipe in which fluid is flowing at steady-state, as shown in Figure 1.4.1-1, at low velocities the dye will trace out a straight line. At higher velocities the fluid moves chaotically as shown and quickly mixes across the entire pipe cross-section. The low velocity behavior is *laminar* and the higher velocity is *turbulent*.

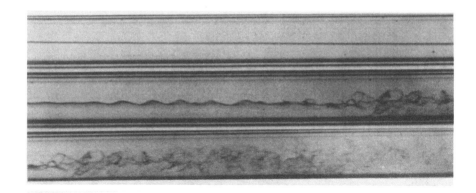

Figure 1.4.1-1 Injection of dye in pipe flow[12] laminar (top) to turbulent (bottom)

Laminar flow is dominated by viscous forces and turbulent flow is dominated by inertial forces. In a plot of point velocity versus time we would see a constant velocity in steady laminar flow, but in "steady" turbulent flow we would see instead a velocity varying chaotically about the time-average velocity at the point.

Later in the text we will discuss dimensionless numbers and their meaning. At this point we merely state that value of the Reynolds number, a dimensionless group, lets us differentiate between laminar and turbulent flow. The Reynolds number is composed of the density and the viscosity of the fluid flowing in the pipe, the pipe diameter, and the area-averaged velocity, and represents the ratio of inertial forces to viscous forces.

$$\text{Reynolds Number} = \text{Re} = \frac{D\,v\,\rho}{\mu} = \frac{\text{inertial forces}}{\text{viscous forces}} \qquad (1.4.1\text{-}1)$$

In pipe flow if the value of the Reynolds number is less than about 2100, the flow is laminar, and as the value of the Reynolds number increases beyond 2100 the flow undergoes a transition to turbulent. This transitional value is different for different geometrical configurations, for example, flow over a flat plate.

---

[12] Japan, S. o. M. E., Ed. (1988). *Visualized flow: Fluid motion in basic and engineering situations revealed by flow visualization*. Thermodynamics and Fluid Mechanics. New York, NY, Pergamon, p. 25.

1.4.2   Newtonian fluids

When two "layers" of fluid flow past each other as sketched in Figure 1.4.2-1, where the upper layer is moving faster than the lower layer, the deformation produces a shear stress t_{yx} proportional to velocity **gradient**. (The profile in Figure 1.4.2-1 is shown as linear for ease of illustration.) The faster-moving layer exerts a force which tends to accelerate the slower-moving layer, and the slower-moving layer exerts a force which tends to decelerate the faster-moving layer, much as if the layers rubbed against one another like the successive cards in a deck which is being spread (even though this is not what actually happens, it is a convenient **model** for visualization purposes).

This differs from a solid, in which the shear stress can exist simply because of relative **displacement** of adjacent layers without relative **motion** of the adjacent layers.

The shear stress here is normal to the y direction. Different fluids exhibit different resistance to shear.

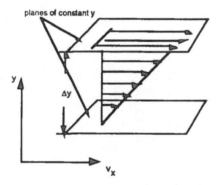

Figure 1.4.2-1   Shear between layers of fluid

A **Newtonian fluid** follows Newton's law of viscosity.[13] Newton's law of viscosity relates the shear stress to the velocity gradient as

---

[13] All fluids that do not behave according to this equation are called non-Newtonian fluids. Since the latter is an enormously larger class of fluids, this is much like classifying mammals into platypuses and non-platypuses; however, it is traditional.
A complete definition of a Newtonian fluid comprises

$$\tau_{xx} = -2\mu \frac{\partial v_x}{\partial x} + \frac{2}{3}(\mu - \kappa)(\nabla \cdot v)$$

(footnote continued on following page)

$$\boxed{\tau_{yx} = -\mu \frac{dv_x}{dy}}$$

$$(1.4.2-1)$$

where μ is the coefficient of viscosity, v_x is the x-component of the velocity vector, y is the y-coordinate, and τ_{yx} can be regarded as either (both quantities have the same units)

   a) the flux of x-momentum in the y-direction, or

   b) the shear stress in the x-direction exerted on a surface of constant y by the fluid in the region of lesser[14] y

Units of shear stress are the same as units of momentum flux.

$$\tau_{yy} = -2\mu \frac{\partial v_y}{\partial y} + \frac{2}{3}(\mu - \kappa)(\nabla \cdot v)$$

$$\tau_{zz} = -2\mu \frac{\partial v_z}{\partial z} + \frac{2}{3}(\mu - \kappa)(\nabla \cdot v)$$

$$\tau_{xy} = \tau_{yx} = -\mu \left(\frac{\partial v_x}{\partial y} + \frac{\partial v_y}{\partial x}\right)$$

$$\tau_{yz} = \tau_{zy} = -\mu \left(\frac{\partial v_y}{\partial z} + \frac{\partial v_z}{\partial y}\right)$$

$$\tau_{zx} = \tau_{xz} = -\mu \left(\frac{\partial v_z}{\partial x} + \frac{\partial v_x}{\partial z}\right)$$

where k is the **bulk viscosity**. The one-dimensional form shown will suffice for our purposes. For further details, see Bird, R. B., W. E. Stewart, et al. (1960). *Transport Phenomena*. New York, Wiley; Longwell, P. A. (1966). *Mechanics of Fluid Flow*, McGraw-Hill; Prager, W. (1961). *Introduction to Mechanics of Continua*. Boston, MA, Ginn and Company; Bird, R. B., R. C. Armstrong, et al. (1977). *Dynamics of Polymeric Liquids: Volume 1 Fluid Mechanics*. New York, NY, John Wiley and Sons; and Bird, R. B., R. C. Armstrong, et al. (1987). *Dynamics of Polymeric Liquids: Volume 1 Fluid Mechanics*. New York, NY, Wiley.

[14] Note the **actual, not the absolute** sense is implied; i.e., y = -10 is less than y = -5.

$$\text{shear stress} = \frac{\text{force}}{\text{area}} = \frac{\text{mass} \times \text{acceleration}}{\left[\text{length}\right]^2}$$

$$= [M]\frac{[L]}{[t]^2}\frac{1}{[L]^2} = \frac{[M]}{[L][t]^2}$$

$$\text{momentum flux} = \frac{\text{momentum}}{\left(\text{area}\right)\left(\text{time}\right)} = \frac{\text{mass} \times \text{velocity}}{\left(\text{length}\right)^2 \text{time}} \qquad (1.4.2\text{-}2)$$

$$= \frac{[M]\frac{[L]}{[t]}}{[L]^2[t]} = \frac{[M]}{[L][t]^2}$$

Equation (1.4.2-1) is often written without the minus sign.[15] We choose to include the minus sign herein because the shear stress can be interpreted (and, in the case of a fluid, perhaps more logically be interpreted) as a momentum flux.[16] Momentum "runs down the velocity gradient" as a consequence of the second law of thermodynamics much as water runs downhill; i.e., flows from regions of higher velocity to regions of lower velocity. It therefore seems more logical to associate a negative sign with τ_{yx} in the above illustrated case. Since the Newtonian viscosity is by convention a positive number, and the velocity gradient is shown as positive, hence the minus sign to ensure the negative direction for the momentum flux τ_{yx}.

---

[15] For example, see McCabe, W. L., J. C. Smith, et al. (1993). *Unit Operations of Chemical Engineering*, McGraw-Hill, p. 46; Bennett, C. O. and J. E. Myers (1974). *Momentum, Heat, and Mass Transfer*. New York, NY, McGraw-Hill, p. 18; Fox, R. W. and A. T. McDonald (1992). *Introduction to Fluid Mechanics*. New York, NY, John Wiley and Sons, Inc., p. 27; Welty, J. R., C. E. Wicks, et al. (1969). *Fundamentals of Momentum, Heat, and Mass Transfer*, New York, NY, Wiley, p. 92; Ahmed, N. (1987). *Fluid Mechanics*, Engineering Press, p. 15; Potter, M. C. and D. C. Wiggert (1991). *Mechanics of Fluids*. Englewood Cliffs, NJ, Prentice Hall, p. 14; Li, W.-S. and S.-H. Lam (1964). *Principles of Fluid Mechanics*. Reading, MA, Addison-Wesley, p. 3; Roberson, J. A. and C. T. Crowe (1993). *Engineering Fluid Mechanics*. Boston, MA, Houghton Mifflin Company, p. 16; Kay, J. M. and R. M. Nedderman (1985). *Fluid Mechanics and Transfer Processes*. Cambridge, England, Cambridge University Press, p. 10.

[16] This follows the convention used in Bird, R. B., W. E. Stewart, et al. (1960). *Transport Phenomena*. New York, Wiley, p. 5; Geankoplis, C. J. (1993). *Transport Processes and Unit Operations*. Englewood Cliffs, NJ, Prentice Hall, p. 43.

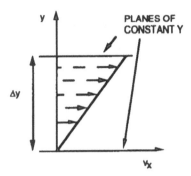

Figure 1.4.2-2 Momentum transfer between layers of fluid

Figure 1.4.2-3 shows the various cases, Quadrant 1 being the example above.

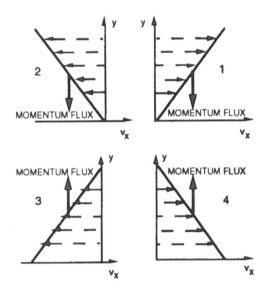

Figure 1.4.2-3 Sign convention for momentum flux between layers of fluid

A similar interpretation can be made for regarding τ_{yx} as a shear stress.

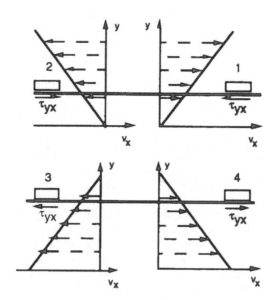

Figure 1.4.2-4 Sign convention for shear stress on surface layers of fluid

Note that

> • in Quadrant 1 the fluid in the region of lesser y has **lower** velocity in the **positive** x-direction than the fluid above it; therefore, the shear stress exerted is to the **left** (negative)

> • in Quadrant 2 the fluid in the region of lesser y has **lower** velocity in the **negative** x-direction than the fluid above it; therefore, the shear stress exerted is to the **right** (positive)

> • in Quadrant 3 the fluid in the region of lesser y has **greater** velocity in the **negative** x-direction than the fluid above it; therefore, the shear stress exerted is to the **left** (negative)

> • in Quadrant 4 the fluid in the region of lesser y has **greater** velocity in the **positive** x-direction than the fluid above it; therefore, the shear stress exerted is to the **right** (positive)

all consistent with (b) above. By Newton's first law the shear stresses on the corresponding opposite (upper) faces of the planes of constant y are the negative of the forces shown in Figure 1.4.2-4.

The cases are summarized in Table 1.4.2-1.

Table 1.4.2-1 Summary of sign convention for stress/momentum flux tensor

Quadrant	Sign of Derivative	Direction of Momentum flux or Shear stress
1	positive	negative
2	negative	positive
3	positive	negative
4	negative	positive

We frequently refer to the action of viscosity as being "fluid friction." In fact, viscosity is not exhibited in this manner. Rather than friction from layers of fluid rubbing against one another (much as consecutive cards in a deck of cards sliding past each other), momentum is transferred in a fluid by molecules in a slower-moving layer of fluid migrating to a faster-moving region of fluid and vice versa, as shown schematically in Figure 1.4.2-2, where some molecules from the A layer are shown to have migrated into the B layer, displacing in turn molecules from this layer.

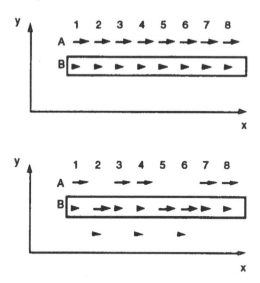

Figure 1.4.2-5 Migration of momentum by molecular motion

In the top sketch of Figure 1.4.2-2 we have shown the x-components of momentum of molecules at a particular instant in two layers A and B of fluid, with a larger x-component in the upper layer. We have assumed for illustrative purposes that the x-component of momentum is uniform in each layer. If the molecules are all of the same type (mass), the momentum vectors would then be proportional to the velocity vectors. Each molecule could, of course, also have y- and z-components of velocity (not shown).

Assume that during some small time increment Δt, molecules at x-locations 2, 5, and 6 are displaced in the negative y-direction as shown (because of the y-component of their velocity). As molecules from layer A at 2, 5, and 6 mix with layer B, they will increase the momentum of layer B. (Ultimately, of course, they collide with and exchange x-momentum with the remaining molecules in layer B). Since molecules in layer B possessing less momentum have been replaced with fluid from layer A with higher momentum, layer B will have gained momentum and thus positive momentum transfer will have been effected from layer A to layer B. (Molecules displaced from layer B similarly transfer momentum to the layer below.) This is a geometrically very simplified version of the ultimate momentum transfer mechanism in a real fluid.

An analogy might be as follows: Suppose that one day you are standing quietly on roller skates, and a friend of yours skates past at 20 mph and hands to you an anvil which he happens to be holding. Needless to say, the jump of this anvil from a region of higher velocity (him) to a region of lower velocity (you) will induce a velocity in the lower-velocity region (you). This is much like the situation as it happens in a fluid, except in a fluid the random motion of the molecules causes them to jump from one velocity region to another.

It is often useful to divide the fluid viscosity by the fluid density to give a quantity called the *kinematic viscosity*, ν, ($\nu = \mu/\rho$). The kinematic viscosity is a measure of the inherent resistance of a particular fluid to flow, or, more particularly, to transition from laminar to turbulent flow. This can be seen from the Reynolds number, which may be written $Re = Dv/\nu$, where ν summarizes the fluid properties and the other two variables the ambient flow conditions. Kinematic viscosity measures the rate of diffusion of momentum: fluids with low kinematic viscosity support turbulence more readily than those fluids with high kinematic viscosity.

For a Newtonian fluid, viscosity is a state property; that is, it is a function of pressure, temperature, and composition.[17] Figure 1.4.2-6 gives values of viscosity for common fluids.

---

[17] Normally the viscosity of a gas increases with temperature; normally the viscosity of a liquid decreases with temperature.

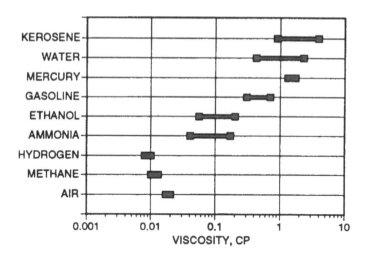

Figure 1.4.2-6 Viscosity of common fluids. (Temperature range approximately 20°F-150°F.) Not intended for design purposes.

Example 1.4.2-1   Flow of fluids between fixed parallel plates

For flow of a fluid between two fixed parallel plates of a length L, the velocity distribution in the x-direction as a function of plate separation is

$$v_x = \frac{(p_1 - p_2)}{2\mu L}\left[1 - \left(\frac{y}{B}\right)^2\right]$$

(1.4.2-3)

where $(p_1 - p_2)$ is the pressure drop over the length of the plate, 2B is the plate separation, and y is the distance from the center line.

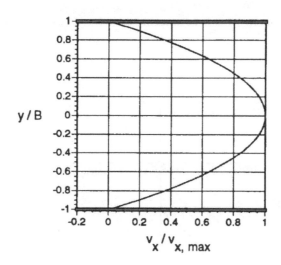

Determine the shear stress at the plate walls caused by water flowing between the plates under a pressure drop of 1 psi, if B is 0.1 ft and L is 10 ft.

Solution

$$\frac{dv_x}{dy} = -\frac{(p_1 - p_2)}{\mu L} y \tag{1.4.2-4}$$

$$-\mu \frac{dv_x}{dy} = \tau_{yx}\Big|_{y=B} = \frac{(p_1 - p_2)}{L} B \tag{1.4.2-5}$$

$$\tau_{yx}\Big|_{y=B} = \frac{\left(1\frac{lbf}{in^2}\right)}{(10\,ft)} (0.1\,ft) = 0.01 \frac{lbf}{in^2} \tag{1.4.2-6}$$

1.4.3   Complex fluids

A plot of shear stress versus velocity gradient for gases and most homogeneous non-polymeric liquids is a straight line, indicating that they may be modeled by Newton's law of viscosity. (The slope of the straight line is the viscosity.) Complex fluids require more complex models than Newton's law of viscosity.   In Figure 1.4.3-1 the straight line indicated by (a) is Newtonian fluid behavior.

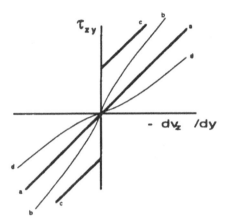

Figure 1.4.3-1 Complex fluids

A *Bingham plastic,* represented by curve (c), does not flow until the applied shear stress reaches a yield point τ_0, and then for higher shear stress follows Newtonian behavior. The relationship between shear stress and velocity gradient for a Bingham plastic may be represented by the model

$$\tau_{yz} = \tau_0 - \mu_p \frac{dv_z}{dy} \tag{1.4.3-1}$$

Toothpaste typically tends to follow a Bingham model.

A *pseudoplastic* fluid, for example, some polymer melts or polymer solutions, has behavior such that the apparent viscosity **decreases** as the velocity gradient increases. In other words, the harder these fluids are "pushed," the easier they flow. Curve (b) in Figure 1.4.3-1 represents a pseudoplastic fluid. A *power law model* may be used to represent the relation between shear stress and velocity gradient for these fluids.

$$\tau_{yz} = -m \left| \frac{dv_z}{dy} \right|^{n-1} \frac{dv_z}{dy} \tag{1.4.3-2}$$

A pseudoplastic fluid is modeled with an exponent $n < 1$. For $n = 1$, $m = \mu$; and Equation (1.4.3-2) reduces to Newton's law of viscosity.

A *dilatant* fluid, such as some suspensions, has a behavior such that the apparent viscosity **increases** as the velocity gradient increases. The harder you "push," the more the fluid "resists" (not unlike many people - but this is engineering, not psychology). Curve (d) in Fig. 1.4.3-1 represents a dilatant fluid. This type of fluid is modeled with the power law model, Equation (1.4.3-2), with an exponent n >1.

A *viscoelastic* fluid such as rubber cement has **strain recovery** or **recoil** when stress is removed. A simple mechanical analog of a viscoelastic fluid is that of a dashpot and a spring connected in series as in Figure 1.4.3-2.

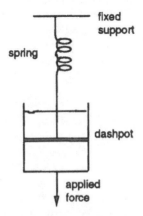

Figure 1.4.3-2 Mechanical analog of viscoelasticity

The dashpot (analogous to an automotive shock absorber) taken in isolation is analogous to a viscous fluid - the spring in isolation is analogous to the elastic behavior of a solid. When force is applied, the spring stretches (proportional to the displacement) and the dashpot moves (proportional to the **rate** of displacement). When force is released, the dashpot remains in place but the spring recoils and/or oscillates depending on the degree of damping. Viscoelastic fluid models include a "spring constant" and time-dependent behavior

$$\tau_{yz} + \lambda \frac{d\tau_{yz}}{dt} = -\mu \frac{dv_z}{dy} \qquad (1.4.3\text{-}3)$$

where λ is the elastic term.

Other types of non-Newtonian fluids exhibit time-dependent behavior. These fluids often said to exhibit "memory." This is because the fluid appears to remember what happened to it in the immediate past; that is, how **long** it has been sheared and/or stretched. Thixotropic fluids show a **decrease** in apparent viscosity with time under constant rate of shear. Rheopectic fluids show an **increase** in apparent viscosity with time under constant rate of shear.

We have used simplified examples to introduce complex fluid behavior. Although we will not discuss complex fluids in great detail, they are extremely important in industry. For example, many textile fibers are made by extrusion of a polymer melt, a highly non-Newtonian fluid. In pipeline flow we can sometimes cause useful drag reduction by adding a polymer which induces non-Newtonian behavior in the fluid. Many mixing processes in the synthetic rubber industry involve mixing of fluids which behave in a non-Newtonian manner. Production of such simple cosmetic articles as toothpaste, face cream, etc. involves flow of complex fluids - frequently with the accompanying problems of heat transfer and sometimes mass transfer.

This large and very interesting area of fluid mechanics is still developing rapidly. For further information there are a wide range of multiple resources, from the applied to the theoretical. The reader might find a starting point in one of the references cited below.[18]

1.4.4   Compressible vs. incompressible flows

Flows in which the density variation in the fluid can be neglected may be modeled as incompressible, even fluids usually thought of as compressible, such as gases. Only if the density variation must be considered are the flows modeled as compressible.

Most liquid flows can be modeled as incompressible. (Cavitation is an exception.) Gas flows at low speeds can frequently be modeled as incompressible. For values of the Mach number (another dimensionless group, which represents the ratio of the local velocity to the speed of sound) less than 0.3, the error is less than 5% in assuming incompressible behavior. For Mach

[18] Tadmor, Z. and C. Gogos (1979). *Principles of Polymer Processing*. New York, NY, John Wiley and Sons; Dealy, J. M. and K. F. Weissbrun (1990). *Melt Rheology and its Role in Plastics Processing*. New York, NY, Van Nostrand Reinhold; Mashelkar, R. A., A. S. Mujumdar, et al., Eds. (1989). *Transport Phenomena in Polymeric Systems*. Chichester, W. Sussex, England, Ellis Horwood, Ltd.; Joseph, D. D. (1990). *Fluid Dynamics of Viscoelastic Liquids*. New York, NY, Springer-Verlag; Bird, R. B., R. C. Armstrong, et al. (1987). *Dynamics of Polymeric Liquids: Volume 1 Fluid Mechanics*. New York, NY, Wiley; Bird, R. B., C. F. Curtis, et al. (1987). *Dynamics of Polymeric Liquids: Volume 2 Kinetic Theory*. New York, NY, Wiley.

numbers greater than 0.3, gases normally should be modeled as compressible fluids.

1.5 Averages

We make extensive use of the concept of average, particularly in conjunction with area-averaged velocity, bulk or mixing-cup temperature, and bulk or mixing-cup concentration, all of which are attributes of flowing streams. All averages we will use may be subsumed under the generalized concept of average discussed below.

Among other uses, averages link microscopic models to macroscopic models, the variables in the macroscopic models being space and/or time averages of the variables in the microscopic models. Time averages are used as models for velocity fields in turbulent flow, where the rapid local fluctuations in velocity are often either not mathematically tractable or not of interest.

1.5.1   General concept of average

In one dimension, consider a function

$$y = f(x) \tag{1.5.1-1}$$

By the *average value of y with respect to x* we mean a number such that, if we multiply it by the range on x, we get the same result as would be obtained by integrating y over the same range. We write this definition as

$$y_{average}\left(b-a\right) = \bar{y}_x\left(b-a\right) \equiv \int_{x=a}^{x=b} f(x)\, dx \tag{1.5.1-2}$$

We use the overbar to indicate an average and subscript(s) to indicate with respect to which variable(s) the average is taken.[19] If it is clear with respect to which variable the average is taken, we will omit the subscript.

This definition can be rewritten as

---

[19] There is obviously overlap between this notation and the notation sometimes used for partial derivatives. The context will always make the distinction clear in cases treated here. There are also a number of notations used for special averages, such as the area average and bulk quantities. Treatment of these follows.

$$y_{average} = \bar{y} \equiv \frac{\int_{x=a}^{x=b} f(x)\, dx}{\int_{x=a}^{x=b} dx}$$

(1.5.1-3)

where we have omitted the subscript because there is only one independent variable with respect to which the average can be taken.

This same definition can be extended to the case where y is a function of many variables. We define *the average of y with respect to* x_n as

$$\bar{y}_{x_n} \equiv \frac{\int_{x_n=a}^{x_n=b} f(x_1, x_2, x_3, \ldots x_n, \ldots x_N)\, dx_n}{\int_{x_n=a}^{x_n=b} dx_n}$$

(1.5.1-4)

Sometimes we take the average with respect to more than one independent variable; to do this we integrate over the second, third, etc., variables as well. For example, the average of y with respect to x_1, x_6, and x_{47} is

$$\bar{y}_{x_1, x_6, x_{47}} = \frac{\int_{x_{47}=a}^{x_{47}=b} \int_{x_6=a}^{x_6=b} \int_{x_1=a}^{x_1=b} f(x_1, x_2, x_3, \ldots x_n, \ldots x_N)\, dx_1\, dx_6\, dx_{47}}{\int_{x_{47}=a}^{x_{47}=b} \int_{x_6=a}^{x_6=b} \int_{x_1=a}^{x_1=b} dx_1\, dx_6\, dx_{47}}$$

(1.5.1-5)

Notation for averages does not always distinguish with respect to which variables they are taken. This information must be known in order to apply an average properly, as must the range over which the average is taken.

We will in this course of study frequently use three particular averages: the area average, the time average, and the bulk average.

Example 1.5.1-1   Time-average vs distance-average speed

Suppose I drive from Detroit to Chicago (300 miles), and upon arrival am asked "What was your average speed?" If I answer, "60 mph," I am probably thinking somewhat as follows: My speed vs. time (neglecting short periods of acceleration and deceleration) might be approximated by the following:

- First 45 minutes - speed about 40 mph during time to get out of Detroit

- Stopped for 15-minute coffee break

- Resumed driving, drove at 70 mph (exceeding speed limit by 5 mph) for 3.5 hours

- Passed highway accident, reduced speed to 50 mph for last 30 minutes to Chicago city limits

A plot of my approximate speed versus time, s = f(t), is shown in Figure 1.5.1-1.

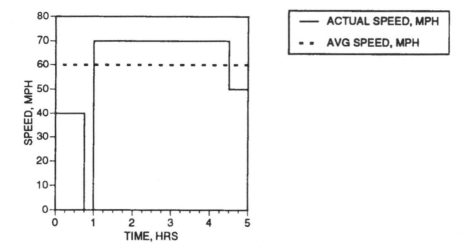

Figure 1.5.1-1 Time-average speed for travel between two points

Here speed is the dependent variable and time the independent variable:

$$s = f(t)$$

$(1.5.1-6)$

The distance traveled is the integral of speed by time.

By declaring my average speed to be 60 mph, I mean that multiplying 60 mph, the average speed, by the elapsed time of 5 hours will yield the total distance, 300 mi. In Figure 1.5.1-1 this means that the area beneath the line at the average speed is the same as the integral of the actual speed vs. time curve. Writing this in the form of our definition of average we have:

$$S_{\text{time average}} = \frac{\displaystyle\int_0^5 f(t)\, dt}{\displaystyle\int_0^5 dt} =$$

$$\frac{\left[(0.75-0)(40)+(1-0.75)(0)+(4.5-1)(70)+(5-4.5)(50)\right]\text{hr}\,\frac{\text{mi}}{\text{hr}}}{(5-0)\,\text{hr}}$$

$$= 60\,\frac{\text{mi}}{\text{hr}}$$

$$(1.5.1\text{-}7)$$

Note that we could, however, equally well regard speed as a function of distance, x, not time, t, and get another function s = g(x) as shown in Figure 1.5.1-2.

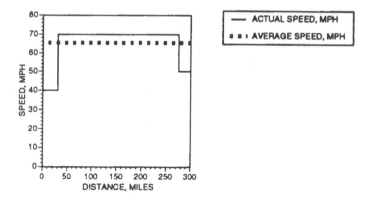

Figure 1.5.1-2 Distance-average speed for travel between two points

This new function tells me how fast I was going as a function of **where** I was rather than **what time** it was. I can equally well define an average speed based on this function

$$S_{\text{distance average}} = \frac{\int_0^{300} g(x)\, dx}{\int_0^{300} dx} =$$

$$\frac{\left[(30-0)(40) + (275-30)(70) + (300-275)(50)\right] \text{mi} \frac{\text{mi}}{\text{hr}}}{(300-0)\, \text{mi}} = \qquad (1.5.1\text{-}8)$$

$$\frac{19{,}600 \frac{(\text{mi})^2}{\text{hr}}}{300} \frac{}{\frac{\text{mi}}{\text{mi}}} = 65.3 \frac{\text{mi}}{\text{hr}}$$

This average speed is different, since it is intended to be used differently. Instead of multiplying by elapsed **time** we multiply by elapsed **distance** to get the integral:

$$\int_0^{300} g(x)\, dx = (300)\, \text{mi}\, (65.3)\, \frac{\text{mi}}{\text{hr}} = 19{,}590 \frac{(\text{mi})^2}{\text{hr}} \qquad (1.5.1\text{-}9)$$

The integral of $g(x)$, however, does not have the intuitive appeal of the integral of $f(t)$ in the numerator of the right-hand side of Equation (1.5.1-7), which is the total distance traveled.

1.5.2 Velocity averages

There are two primary averages that are used in conjunction with fluid velocity, although many are possible:

• area-averaged velocities, and

• time-averaged velocities

Area-averaged velocity

A special average that is used frequently enough to deserve a unique notation is the area-averaged velocity, <v>, which is defined as a number such that, if it is multiplied by the basis area through which the fluid is flowing, yields the volumetric flow rate.

$$\langle v \rangle \equiv \frac{Q}{A} = \frac{\int_A \left(\vec{v} \cdot \vec{n} \right) dA}{\int_A dA} \tag{1.5.2-1}$$

This is clearly just a special case of our definition of general average above, where the average here is with respect to **area**.[20]

We must be careful to get the signs correct when using area-averaged velocities or volumetric flow rates. The equation above will give the correct sign (positive for output, negative for input) for **substituting** in equations applied to a particular system or control volume; however, when written standing alone (not in a balance equation) the absolute value of the area-averaged velocity or volumetric flow rate is usually implied, without explicitly writing the absolute value sign, even when it might be appropriate. For example, when simply giving values for a system where the subscript 1 denotes input conditions (cos α is negative) it is usual to write, for example

$$\langle v \rangle_1 = \frac{Q_1}{A_1} = \frac{100 \frac{ft^3}{s}}{10 \, ft^2} \tag{1.5.2-2}$$

or

$$\langle v \rangle_1 = 10 \frac{ft}{s} \tag{1.5.2-3}$$

and the reader is expected to insert the sign that is appropriate when substituting in the balance for the particular system, rather than writing

$$\left| Q_1 \right| = \left| -50 \, gpm \right| \tag{1.5.2-4}$$

or

$$\left| \langle v_1 \rangle \right| = \left| -10 \frac{ft}{s} \right| \tag{1.5.2-5}$$

Any area could be chosen for the basis area for the area-averaged velocity; however, in practice the basis is usually either

---

[20] In doing the actual integration, it is of course necessary to express A in terms of the appropriate coordinates - e.g., $dA = dx \, dy$ - and insert the appropriate limits.

1) the area normal to flow, or

2) the surface area (of the control volume) through which the flow is passing

Example 1.5.2-1   Area-averaged velocity for laminar pipe flow

The velocity profile for laminar flow of liquid in a horizontal pipe of radius R is given by the Hagen-Poiseuille "law" - although it is not a law in the strict sense of the word (we will derive this relationship in Chapter 6)

$$v_z = \frac{(p_1 - p_2) R^2}{4 \mu L} \left[1 - \left(\frac{r}{R} \right)^2 \right] \qquad (1.5.2\text{-}6)$$

which is sketched in Figure 1.5.2-1. At $z = 0$, $p = p_1$; at $z = L$, $p = p_2$.

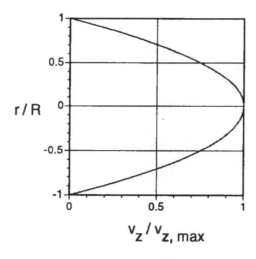

Figure 1.5.2-1 Velocity profile

a) Find the area-averaged velocity in terms of the maximum (centerline) velocity

b) A pressure drop of 10^{-4} psi per foot is imposed on water at $70°F$ flowing in a tube with 1 inch inside diameter. Calculate the area-averaged velocity. Viscosity of water is 1 cP.[21]

Solution

a) The maximum (centerline) velocity is

$$v_{z,\,max} = \frac{(p_1 - p_2)R^2}{4\mu L}\left[1 - \left(\frac{r}{R}\right)^2\right]\Bigg|^?_{r=0} = \frac{(p_1 - p_2)R^2}{4\mu L} \qquad (1.5.2\text{-}7)$$

Calculating the area-averaged velocity using the positive sign for the dot product

$$\langle v_z \rangle = \frac{\int_A (v \cdot n)\, dA}{\int_A dA} = \frac{\int_A v_z\, dA}{\int_A dA}$$

$$= \frac{\int_0^R \dfrac{(p_1 - p_2)R^2}{4\mu L}\left[1 - \left(\dfrac{r}{R}\right)^2\right] 2\pi r\, dr}{\int_0^R 2\pi r\, dr}$$

$$= \frac{(p_1 - p_2)R^2}{4\mu L} \cdot \frac{\int_0^R \left[1 - \left(\dfrac{r}{R}\right)^2\right] r\, dr}{\int_0^R r\, dr}$$

$$= \frac{(p_1 - p_2)R^2}{4\mu L} \cdot \frac{\left[\dfrac{r^2}{2} - \left(\dfrac{1}{R}\right)^2 \dfrac{r^4}{4}\right]_0^R}{\left[\dfrac{r^2}{2}\right]_0^R}$$

[21] Perry, R. H. and D. W. Green, Eds. (1984). *Chemical Engineer's Handbook*. New York, McGraw-Hill, pp. 3-252.

$$= \frac{(p_1 - p_2) R^2}{4 \mu L} \frac{\left[\frac{R^2}{2} - \frac{R^2}{4}\right]}{\frac{R^2}{2}}$$

$$\langle v_z \rangle = \frac{(p_1 - p_2) R^2}{8 \mu L} \qquad (1.5.2\text{-}8)$$

Taking the ratio of the two velocities

$$\frac{\langle v_z \rangle}{v_{z,\,max}} = \frac{\left[\frac{(p_1 - p_2) R^2}{8 \mu L}\right]}{\left[\frac{(p_1 - p_2) R^2}{4 \mu L}\right]} = \frac{1}{2} \qquad (1.5.2\text{-}9)$$

so **for laminar tube flow the area-averaged velocity is half the centerline velocity**

b) For the given conditions

$$\langle v_z \rangle = \frac{(p_1 - p_2) R^2}{8 \mu L}$$

$$= \frac{\left(10^{-4}\right) \frac{lbf}{in^2} \left(0.5\right)^2 in^2}{(8)(1) \, cp \frac{\left(6.72 \times 10^{-4}\right) lbm}{cp \; ft \; s} (1) \, ft} \frac{(32.2) \; lbm \; ft}{lbf \; s^2}$$

$$\qquad\qquad\qquad\qquad\qquad\qquad\qquad (1.5.2\text{-}10)$$

$$\langle v_z \rangle = 0.15 \frac{ft}{s}$$

Time-averaged velocity

A second velocity average that will be of considerable utility to us (e.g., in our discussion of turbulence) is the **time-averaged point velocity**. This velocity is a direct application of our definition of general average, and is illustrated by the following example.

Example 1.5.2-2   Time-averaged velocity for turbulent flow

Turbulent flow involves rapid fluctuations in time of the velocity at a point. We can represent the velocity in time at any point as the sum of a time-averaged component and a fluctuation

$$\boxed{\mathbf{v} = \bar{\mathbf{v}} + \mathbf{v}'}$$

(1.5.2-11)

where

$$\mathbf{v} = \mathbf{v}(x_i, t)$$

$$\bar{\mathbf{v}} = \frac{\displaystyle\int_0^{t_0} \mathbf{v}\, dt}{\displaystyle\int_0^{t_0} dt}$$

$$\mathbf{v}' = \mathbf{v}'(x_i, t)$$

(1.5.2-12)

Notice that the time-averaged velocity may depend on the time t_0 over which the average is taken if velocity is not constant in time (if the average velocity is increasing or decreasing with time). The average magnitude of fluctuations can also vary with time. We assume that our time interval over which the time-average is taken is long compared to the period of fluctuations, but short with respect to changes in the time average.

The velocity profile for time-averaged velocity for turbulent flow of liquid in a horizontal pipe of radius R is modeled reasonably well by the 1/7 power distribution originally proposed by Prandtl.[22]

---

[22] Schlichting, H. (1960). *Boundary Layer Theory*, McGraw-Hill, p. 506.

$$\mathbf{v}_z = \mathbf{v}_{z,\,max}\left[1-\left(\frac{r}{R}\right)\right]^{\frac{1}{7}}$$

(1.5.2-13)

Calculate the area-averaged, time-averaged velocity in terms of the maximum (centerline) velocity.

Solution

Using the definition

$$\langle \mathbf{v}_z \rangle = \frac{\int_A \mathbf{v}_z\,dA}{\int_A dA}$$

$$= \frac{\int_0^R \mathbf{v}_{z,\,max}\left[1-\left(\frac{r}{R}\right)\right]^{\frac{1}{7}} 2\,\pi\,r\,dr}{\int_0^R 2\,\pi\,r\,dr}$$

$$= \frac{2\,\pi\,R^2\,\mathbf{v}_{z,\,max}\int_0^R\left[1-\left(\frac{r}{R}\right)\right]^{\frac{1}{7}}\frac{r}{R}\,d\!\left(\frac{r}{R}\right)}{\pi\,R^2}$$

$$= 2\,\mathbf{v}_{z,\,max}\int_0^R\left[1-\left(\frac{r}{R}\right)\right]^{\frac{1}{7}}\frac{r}{R}\,d\!\left(\frac{r}{R}\right)$$

(1.5.2-14)

substituting $\tau = (1 - r/R)$

$$\langle \mathbf{v}_z \rangle = 2\,\mathbf{v}_{z,\,max}\int_0^1 (\tau)^{\frac{1}{7}}\,(1-\tau)\,d(1-\tau)$$

$$= 2\,\bar{\mathbf{v}}_{z,\,max}\int_1^0\left((\tau)^{\frac{8}{7}}-(\tau)^{\frac{1}{7}}\right)d\tau$$

$$= 2\,\bar{\mathbf{v}}_{z,\,max}\left[\left(\frac{7}{15}\,(\tau)^{\frac{15}{7}}-\frac{7}{8}\,(\tau)^{\frac{8}{7}}\right)\right]_1^0$$

$$= 2\,\mathbf{v}_{z,\,max}\left(-\frac{7}{15}+\frac{7}{8}\right) = 2\left[\frac{-56+105}{120}\right]\mathbf{v}_{z,\,max}$$

$$\langle v_z \rangle = \frac{49}{60} v_{z,\,max}$$

$$(1.5.2\text{-}15)$$

1.5.3   Temperature averages

The **bulk temperature, T_b**, is defined as the temperature that would result if the fluid flowing were permitted to exit the system through the basis area and were caught in a cup and mixed perfectly. For this reason it is sometimes called the **mixing cup** temperature. Using the energy balance this implies

$$\int_A C_p (T_b - T_{ref})\, \rho \,(v \cdot n)\, dA \equiv \int_A C_p (T - T_{ref})\, \rho \,(v \cdot n)\, dA$$

$$T_b \int_A C_p \rho \,(v \cdot n)\, dA \equiv \int_A T\, C_p \rho \,(v \cdot n)\, dA$$

$$- T_{ref} \int_A C_p \rho \,(v \cdot n)\, dA$$

$$+ T_{ref} \int_A C_p \rho \,(v \cdot n)\, dA$$

$$T_b \int_A C_p \rho \,(v \cdot n)\, dA \equiv \int_A T\, C_p \rho \,(v \cdot n)\, dA$$

$$\boxed{T_b \equiv \frac{\displaystyle\int_A T\, C_p \rho \,(v \cdot n)\, dA}{\displaystyle\int_A C_p \rho \,(v \cdot n)\, dA}}$$

$$(1.5.3\text{-}1)$$

Notice that the weighting here is not simply with respect to area, but with respect to **energy flow rate**.

We usually apply bulk temperatures in models for which either or both of heat capacity and density may be taken as constant across the flow cross-section.[23] If the heat capacity is constant across the flow cross-section and can be canceled in the numerator and denominator, the weighting becomes with respect to **mass flow rate**.

---

[23] Otherwise we would have to define an average heat capacity or density in addition to the bulk temperature, which would result in the definition of one average depending upon how another average was defined - a situation ripe for ambiguity.

$$T_b = \frac{\int_A T \rho \left(\mathbf{v} \cdot \mathbf{n} \right) dA}{\int_A \rho \left(\mathbf{v} \cdot \mathbf{n} \right) dA} \tag{1.5.3-2}$$

If, in addition, the density is constant across the cross-section, the weighting is with respect to **volumetric flow rate**.

$$T_b = \frac{\int_A T \left(\mathbf{v} \cdot \mathbf{n} \right) dA}{\int_A \left(\mathbf{v} \cdot \mathbf{n} \right) dA} \tag{1.5.3-3}$$

At times we also make the assumption that the velocity is constant across the flow cross-section (sometimes called *plug flow* or *rodlike flow*), and then the weighting becomes with respect to **area** (the same weighting as we used above for the velocity).

$$T_b = \frac{\int_A T \, dA}{\int_A dA} = \langle T \rangle \tag{1.5.3-4}$$

It is only under the additional constraints of constant density, heat capacity, and velocity that the area-average temperature is equivalent to the bulk temperature.

The definition of a bulk enthalpy or mixing cup enthalpy follows directly from assuming constant heat capacity in the definition of bulk temperature and inserting the definition of enthalpy

$$T_b \int_A C_p \rho \left(\mathbf{v} \cdot \mathbf{n} \right) dA \equiv \int_A C_p T \rho \left(\mathbf{v} \cdot \mathbf{n} \right) dA$$
$$\int_A T_b C_p \rho \left(\mathbf{v} \cdot \mathbf{n} \right) dA \equiv \int_A C_p T \rho \left(\mathbf{v} \cdot \mathbf{n} \right) dA \tag{1.5.3-5}$$

subtracting

$$\int_A C_p T_{ref} \rho \left(\mathbf{v} \cdot \mathbf{n} \right) dA \tag{1.5.3-6}$$

from each side gives

$$\int_A C_p \left(T_b - T_{ref}\right) \rho \left(v \cdot n\right) dA \; \equiv \; \int_A C_p \left(T - T_{ref}\right) \rho \left(v \cdot n\right) dA \quad (1.5.3\text{-}7)$$

Using the definition of enthalpy

$$\hat{H} \equiv C_p \left(T - T_{ref}\right) \qquad\qquad\qquad (1.5.3\text{-}8)$$

and defining

$$\hat{H}_b \equiv C_p \left(T_b - T_{ref}\right) \qquad\qquad\qquad (1.5.3\text{-}9)$$

gives, upon substitution

$$\int_A \hat{H}_b \left(v \cdot n\right) dA \; \equiv \; \int_A \hat{H} \left(v \cdot n\right) dA$$

$$\hat{H}_b \int_A \left(v \cdot n\right) dA \; \equiv \; \int_A \hat{H} \left(v \cdot n\right) dA$$

$$\boxed{\; \hat{H}_b \equiv \dfrac{\displaystyle\int_A \hat{H} \left(v \cdot n\right) dA}{\displaystyle\int_A \left(v \cdot n\right) dA} \;} \qquad\qquad (1.5.3\text{-}10)$$

Example 1.5.3-1   Area-average temperature vs. bulk temperature

Assume that we model the axial velocity in a pipe as laminar flow (cf. Example 1.5.2-1)

$$v_z = \frac{\left(p_1 - p_2\right) R^2}{4 \mu L} \left[1 - \left(\frac{r}{R}\right)^2\right] \qquad\qquad (1.5.3\text{-}11)$$

Assume that the temperature distribution across the cross-section may be modeled as the simple linear profile

$$\frac{T - T_R}{T_0 - T_R} = 1 - \frac{r}{R} \qquad\qquad\qquad (1.5.3\text{-}12)$$

where

T_R = temperature at r = R (at the pipe wall) = 50°F
T_0 = temperature at r = 0 (at the center line) = 100°F

Calculate the area-average temperature and the bulk temperature.

Solution

Rearranging the temperature profile

$$T = \left(1 - \frac{r}{R}\right)\left(T_0 - T_R\right) + T_R$$

(1.5.3-13)

Using the area-average temperature (applicable for plug flow with constant heat capacity and density across the flow cross-section) gives

$$\langle T \rangle = \frac{\int_A T \, dA}{\int_A dA}$$

$$= \frac{\int_0^R \left[\left(1 - \frac{r}{R}\right)\left(T_0 - T_R\right) + T_R\right] 2\pi r \, dr}{\int_0^R 2\pi r \, dr}$$

$$= \frac{\left(T_0 - T_R\right)\int_0^R \left(r - \frac{r^2}{R}\right) dr}{\int_0^R r \, dr} + \frac{\left(T_R\right)\int_0^R r \, dr}{\int_0^R r \, dr}$$

$$= \frac{\left(T_0 - T_R\right)\left[\frac{r^2}{2} - \frac{r^3}{3R}\right]_0^R}{\left[\frac{r^2}{2}\right]_0^R} + \frac{\left(T_R\right)\left[\frac{r^2}{2}\right]_0^R}{\left[\frac{r^2}{2}\right]_0^R}$$

$$= \frac{\left(T_0 - T_R\right)\left[\frac{1}{2} - \frac{1}{3}\right]}{\left[\frac{1}{2}\right]} + \left(T_R\right)$$

$$= \frac{1}{3}\left(T_0 - T_R\right) + \left(T_R\right)$$

$$\langle T \rangle = \frac{1}{3}\left(100 - 50\right) \, °F + \left(50\right) \, °F = 67 \, °F$$

(1.5.3-14)

In contrast, using the definition of bulk temperature and the definition of area-averaged velocity

$$T_b = \frac{\int_A T\, v_z\, dA}{\int_A v_z\, dA} = \frac{\int_0^R T\, v_z\, 2\pi r\, dr}{\int_0^R v_z\, 2\pi r\, dr} = \frac{\int_0^R T\, v_z\, 2\pi r\, dr}{\langle v_z \rangle\, \pi R^2} \qquad (1.5.3\text{-}15)$$

Substituting the temperature and velocity profiles and grouping the constants

$$
\begin{aligned}
T_b &= \frac{[T_o - T_s]}{\langle v_z \rangle}\left[\frac{(p_1 - p_2)R^2}{4\mu L}\right]\int_0^R \left[1 - \left(\tfrac{r}{R}\right)\right]\left[1 - \left(\tfrac{r}{R}\right)^2\right]\frac{2r}{R^2}\, dr \\
&\quad + \frac{T_s}{\langle v_z \rangle}\left[\frac{(p_1 - p_2)R^2}{4\mu L}\right]\int_0^R \left[1 - \left(\tfrac{r}{R}\right)^2\right]\frac{2r}{R^2}\, dr \\
&= 2\frac{[T_o - T_s]}{\langle v_z \rangle}\left[\frac{(p_1 - p_2)R^2}{4\mu L}\right] \\
&\quad \times \int_{\frac{r}{R}=0}^{\frac{r}{R}=1}\left[1 - \left(\tfrac{r}{R}\right)\right]\left[1 - \left(\tfrac{r}{R}\right)^2\right]\left(\tfrac{r}{R}\right)d\!\left(\tfrac{r}{R}\right) \qquad (1.5.3\text{-}16) \\
&\quad + \frac{T_s}{\langle v_z \rangle}\left[\frac{(p_1 - p_2)R^2}{4\mu L}\right]\int_{\frac{r}{R}=0}^{\frac{r}{R}=1}\left[1 - \left(\tfrac{r}{R}\right)^2\right](2)\left(\tfrac{r}{R}\right)d\!\left(\tfrac{r}{R}\right)
\end{aligned}
$$

Using the relationship developed in Example 1.5.1-1 to simplify the constants

$$\langle v_z \rangle = \frac{(p_1 - p_2)R^2}{8\mu L} \qquad (1.5.3\text{-}17)$$

$$
\begin{aligned}
\left[\frac{[T_o - T_s]}{\langle v_z \rangle}\right]\left[\frac{(p_y - p_2)R^2}{4\mu L}\right] &= \left[\frac{[T_o - T_s]}{\langle v_z \rangle}\right]\left[2\langle v_z \rangle\right] \\
&= 2[T_o - T_s]
\end{aligned}
$$

$$\left[\frac{T_s}{\langle v_z \rangle}\right]\left[\frac{(p_1 - p_2)R^2}{4\mu L}\right] = \left[\frac{T_s}{\langle v_z \rangle}\right]\left[2\langle v_z \rangle\right] = 2\,T_s \qquad (1.5.3\text{-}18)$$

Substituting

$$T_b = 4\left[T_o - T_s\right]\int_{(\frac{r}{R})=0}^{(\frac{r}{R})=1}\left[1 - \left(\frac{r}{R}\right) - \left(\frac{r}{R}\right)^2 + \left(\frac{r}{R}\right)^3\right]\left(\frac{r}{R}\right)d\left(\frac{r}{R}\right)$$

$$+ 2\,T_s\int_{(\frac{r}{R})^2=1}^{(\frac{r}{R})^2=1}\left[1 - \left(\frac{r}{R}\right)^2\right]d\left[\left(\frac{r}{R}\right)^2\right] \qquad (1.5.3\text{-}19)$$

Substituting for the dummy variables of integration to simplify notation

$$T_b = 4\left[T_o - T_s\right]\int_0^1\left[x - x^2 - x^3 + x^4\right]dx$$

$$+ 2\,T_s\int_0^1\left[1 - y\right]dy \qquad (1.5.3\text{-}20)$$

$$T_b = 4\left[T_o - T_s\right]\left[\frac{x^2}{2} - \frac{x^3}{3} - \frac{x^4}{4} + \frac{x^5}{5}\right]_0^1$$

$$+ 2\,T_s\left[y - \frac{y^2}{2}\right]_0^1$$

$$= 0.467\left[T_o - T_s\right] + T_s \qquad (1.5.3\text{-}21)$$

$$T_b = (0.467)(100 - 50)\,°F + (50)\,°F = 73.4\,°F \qquad (1.5.3\text{-}22)$$

The bulk temperature is higher than the area-averaged temperature for this case because the flow velocity is higher toward the center of the tube where the temperature is higher, and this weights the average more heavily with the higher temperatures.

Example 1.5.3-2   Bulk temperature for quadratic temperature profile, laminar pipe flow

For the flow situation described in Example 1.5.1-1, suppose the temperature distribution in the water to be given by:

$$\frac{T - T_s}{T_o - T_s} = 1 - \frac{4}{3}\left(\frac{r}{R}\right)^2 + \frac{1}{3}\left(\frac{r}{R}\right)^4 \tag{1.5.3-23}$$

where

$$T_0 = 140^\circ F = \text{centerline temperature}$$
$$T_s = 50^\circ F = \text{wall temperature}$$

Calculate the bulk temperature for the given conditions.

Solution

Solving the profile for T

$$T = \left[T_o - T_s\right]\left[1 - \frac{4}{3}\left(\frac{r}{R}\right)^2 + \frac{1}{3}\left(\frac{r}{R}\right)^4\right] + T_s \tag{1.5.3-24}$$

Using the definition of bulk temperature and the definition of area-averaged velocity

$$T_b = \frac{\int_A T v_z \, dA}{\int_A v_z \, dA} = \frac{\int_0^R T v_z \, 2\pi r \, dr}{\int_0^R v_z \, 2\pi r \, dr} = \frac{\int_0^R T v_z \, 2\pi r \, dr}{\langle v_z \rangle \, \pi R^2} \tag{1.5.3-25}$$

Substituting the temperature and velocity profiles and grouping the constants

$$T_b = \frac{[T_o - T_s]}{\langle v_z \rangle} \left[\frac{(p_1 - p_2) R^2}{4 \mu L} \right]$$

$$\times \int_0^R \left[1 - \frac{4}{3} \left(\frac{r}{R} \right)^2 + \frac{1}{3} \left(\frac{r}{R} \right)^4 \right] \left[1 - \left(\frac{r}{R} \right)^2 \right] \frac{2r}{R^2} dr \qquad (1.5.3\text{-}26)$$

$$+ \frac{T_s}{\langle v_z \rangle} \left[\frac{(p_1 - p_2) R^2}{4 \mu L} \right] \int_0^R \left[1 - \left(\frac{r}{R} \right)^2 \right] \frac{2r}{R^2} dr$$

Defining new constants for compactness in notation

$$\beta_1 \equiv \left[\frac{[T_o - T_s]}{\langle v_z \rangle} \right] \left[\frac{(p_1 - p_2) R^2}{4 \mu L} \right]$$

$$= \left[\frac{[T_o - T_s]}{\langle v_z \rangle} \right] \left[2 \langle v_z \rangle \right] = 2 \left[T_o - T_s \right]$$

$$\beta_2 \equiv \left[\frac{T_s}{\langle v_z \rangle} \right] \left[\frac{(p_1 - p_2) R^2}{4 \mu L} \right]$$

$$= \left[\frac{T_s}{\langle v_z \rangle} \right] \left[2 \langle v_z \rangle \right] = 2 T_s \qquad (1.5.3\text{-}27)$$

$$T_b = \beta_1 \int_{\left(\frac{r}{R} \right)^2 = 0}^{\left(\frac{r}{R} \right)^2 = 1} \left[1 - \frac{4}{3} \left(\frac{r}{R} \right)^2 + \frac{1}{3} \left(\frac{r}{R} \right)^4 \right.$$

$$\left. - \left(\frac{r}{R} \right)^2 + \frac{4}{3} \left(\frac{r}{R} \right)^4 - \frac{1}{3} \left(\frac{r}{R} \right)^6 \right] d \left[\left(\frac{r}{R} \right)^2 \right] \qquad (1.5.3\text{-}28)$$

$$+ \beta_2 \int_{\left(\frac{r}{R} \right)^2 = 1}^{\left(\frac{r}{R} \right)^2 = 1} \left[1 - \left(\frac{r}{R} \right)^2 \right] d \left[\left(\frac{r}{R} \right)^2 \right]$$

Substituting $x = (r/R)^2$ to simplify notation

$$T_b = \beta_1 \int_0^1 \left[1 - \frac{4}{3}x + \frac{1}{3}x^2 - x + \frac{4}{3}x^2 - \frac{1}{3}x^3 \right] dx$$

$$+ \beta_2 \int_0 [1 - x] dx \tag{1.5.3-29}$$

$$T_b = \beta_1 \left[x - \frac{7}{3}\frac{x^2}{2} + \frac{5}{3}\frac{x^3}{3} - \frac{1}{3}\frac{x^4}{4} \right]_0^1$$

$$+ \beta_2 \left[x - \frac{x^2}{2} \right]_0^1 = 0.306\,\beta_1 + 0.5\,\beta_2 \tag{1.5.3-30}$$

$$T_b = 0.306\,\beta_1 + 0.5\,\beta_2 = (0.306)(2)\left[T_o - T_s\right] + (0.5)(2)\left[T_s\right]$$

$$= 0.612\,T_o + 0.388\,T_s = (0.612)(140) + (0.388)(50)$$

$$T_b = 105\ ^\circ F$$

$$\tag{1.5.3-31}$$

1.5.4   Concentration averages

The **bulk concentration** is defined in a way similar to the bulk temperature, but with concentration substituted for enthalpy. The **bulk mass concentration of species A**, ω_{Ab}, is defined as the concentration that would result if the fluid flowing were permitted to exit the system through the basis area and were caught in a cup and mixed perfectly. For this reason it is sometimes called the **mixing cup** concentration. We define the bulk concentration only for models which assume total mass density and total molar density to be constant across the flow cross-section, for the same reason as for assuming constant properties in the case of bulk temperature.

Letting ω_A represent the instantaneous point mass concentration of species A we define the bulk mass concentration using the condition

$$\left(\omega_A\right)_b \int_A \rho\,(v \cdot n)\,dA \equiv \int_A \omega_A\,\rho\,(v \cdot n)\,dA \tag{1.5.4-1}$$

For the special but common case that $\rho = $ constant

$$(\omega_A)_b \int_A (v \cdot n) \, dA = \int_A \omega_A (v \cdot n) \, dA$$

$$(\omega_A)_b = \frac{\int_A \omega_A (v \cdot n) \, dA}{\int_A (v \cdot n) \, dA} \qquad (1.5.4\text{-}2)$$

Since the definition is analogous to that of bulk temperature, we have again used the subscript b to denote the bulk condition.

Since

$$\omega_A \left[\frac{\text{mass A}}{\text{total mass}}\right] \rho \left[\frac{\text{total mass}}{\text{vol}}\right] = \rho_A \left[\frac{\text{mass A}}{\text{vol}}\right]$$

$$= c_A \left[\frac{\text{moles A}}{\text{vol}}\right] M \left[\frac{\text{mass A}}{\text{mol A}}\right]$$

$$= x_A \left[\frac{\text{mol A}}{\text{total mol}}\right] c \left[\frac{\text{total mol}}{\text{vol}}\right] M \left[\frac{\text{mass A}}{\text{mol A}}\right] \qquad (1.5.4\text{-}3)$$

$$\omega_A = x_A \, c \, M$$

(where ρ and c are, respectively, the total mass concentration and the total molar concentration) by substituting in the definition for bulk mass concentration we obtain

$$(x_A)_b \, c \, M \equiv \frac{\int_A x_A \, c \, M \, (v \cdot n) \, dA}{\int_A (v \cdot n) \, dA}$$

$$(1.5.4\text{-}4)$$

For the special but common case that c = constant

$$\left(x_A\right)_b \equiv \frac{\int_A x_A \left(v \cdot n\right) dA}{\int_A \left(v \cdot n\right) dA}$$

$$(1.5.4\text{-}5)$$

Similarly, for constant ρ and c, using

$$x_A \frac{\text{mol A}}{\text{total mol}} \quad c \frac{\text{total mol}}{\text{vol}} = c_A \frac{\text{mol A}}{\text{vol}}$$
$$\omega_A \frac{\text{mass A}}{\text{total mass}} \quad \rho \frac{\text{total mass}}{\text{vol}} = \rho_A \frac{\text{mass A}}{\text{vol}}$$

$$(1.5.4\text{-}6)$$

gives

$$\left(c_A\right)_b \equiv \frac{\int_A c_A \left(v \cdot n\right) dA}{\int_A \left(v \cdot n\right) dA}$$

$$\left(\rho_A\right)_b \equiv \frac{\int_A \rho_A \left(v \cdot n\right) dA}{\int_A \left(v \cdot n\right) dA}$$

$$(1.5.4\text{-}7)$$

Example 1.5.4-1   Bulk concentration

Suppose that for the same pipe flow problem the concentration profile is represented by

$$\frac{c_A - c_{AR}}{c_{A0} - c_{AR}} = 1 - \frac{r}{R}$$

$$(1.5.4\text{-}8)$$

where

$$c_A = c_{AR} = 0.01 \text{ moles/ft}^3 \text{ at } r = R$$
$$c_A = c_{A0} = 0.02 \text{ moles/ft}^3 \text{ at } r = 0$$

Calculate the bulk concentration.

Solution

Solving for the concentration profile explicitly in terms of c_A

$$c_A = \left(c_{A0} - c_{AR} \right) \left(1 - \frac{r}{R} \right) + c_{AR} \qquad (1.5.4\text{-}9)$$

Since here

$$\mathbf{v} \cdot \mathbf{n} = v_z \qquad (1.5.4\text{-}10)$$

the appropriate relationship for bulk concentration is

$$\left(c_A \right)_b \equiv \frac{\int_A c_A \, v_z \, dA}{\int_A v_z \, dA} \qquad (1.5.4\text{-}11)$$

Using the concentration profile above, the Hagen-Poiseuille velocity profile

$$v_z = \left[\frac{\left(p_1 - p_2 \right) R^2}{4 \, \mu \, L} \right] \left[1 - \left(\frac{r}{R} \right)^2 \right] \qquad (1.5.4\text{-}12)$$

and the corresponding area-averaged velocity

$$\langle v_z \rangle = \left[\frac{\left(p_1 - p_2 \right) R^2}{8 \, \mu \, L} \right] \qquad (1.5.4\text{-}13)$$

to substitute in the equation for bulk concentration gives

$$\left(c_A \right)_b = \left\{ \int_0^R \left[\left(c_{A0} - c_{AR} \right) \left(1 - \frac{r}{R} \right) + c_{AR} \right] \left[\frac{\left(p_1 - p_2 \right) R^2}{4 \, \mu \, L} \right] \left[1 - \left(\frac{r}{R} \right)^2 \right] 2 \pi r \, dr \right\}$$

$$+ \left\{ \int_0^R \left[\frac{(p_1 - p_2) \, R^2}{4 \mu L} \right] \left[1 - \left(\frac{r}{R} \right)^2 \right] 2 \pi r \, dr \right\} \tag{1.5.4-14}$$

$$\left(c_A \right)_b = \frac{\int_0^R \left[\left(c_{A0} - c_{AR} \right) \left(1 - \frac{r}{R} \right) + c_{AR} \right] \left[1 - \left(\frac{r}{R} \right)^2 \right] 2 \pi r \, dr}{\int_0^R \left[1 - \left(\frac{r}{R} \right)^2 \right] 2 \pi r \, dr} \tag{1.5.4-15}$$

$$\left(c_A \right)_b = \left(c_{A0} - c_{AR} \right) \left\{ \int_0^R \left(1 - \frac{r}{R} \right) \left[1 - \left(\frac{r}{R} \right)^2 \right] 2 \pi r \, dr \right.$$

$$\left. + c_{AR} \int_0^R \left[1 - \left(\frac{r}{R} \right)^2 \right] 2 \pi r \, dr \right\}$$

$$+ \int_0^R \left[1 - \left(\frac{r}{R} \right)^2 \right] 2 \pi r \, dr \tag{1.5.4-16}$$

$$\left(c_A \right)_b = \frac{\left(c_{A0} - c_{AR} \right) \int_0^1 \left(1 - \frac{r}{R} \right) \left[1 - \left(\frac{r}{R} \right)^2 \right] 2 \left[\frac{r}{R} \right] d \left(\frac{r}{R} \right)}{\int_0^R \left[1 - \left(\frac{r}{R} \right)^2 \right] 2 \left(\frac{r}{R} \right) d \left(\frac{r}{R} \right)} + c_{AR}$$

$$\tag{1.5.4-17}$$

letting $\gamma = r/R$

$$\left(c_A \right)_b = \left(c_{A0} - c_{AR} \right) \frac{\int_0^1 \left(1 - \gamma \right) \left[1 - (\gamma)^2 \right] \left[\gamma \right] d(\gamma)}{\int_0^1 \left[1 - (\gamma)^2 \right] \left[\gamma \right] d(\gamma)} + c_{AR} \tag{1.5.4-18}$$

$$\left(c_A \right)_b = \left(c_{A0} - c_{AR} \right) \frac{\int_0^1 \left[\gamma - \gamma^2 - \gamma^3 + \gamma^4 \right] d(\gamma)}{\int_0^1 \left[\gamma - (\gamma)^3 \right] d(\gamma)} + c_{AR} \tag{1.5.4-19}$$

$$(c_A)_b = (c_{A0} - c_{AR}) \frac{\left[\left[\frac{\gamma^2}{2} - \frac{\gamma^3}{3} - \frac{\gamma^4}{4} + \frac{\gamma^5}{5}\right]\right]_0^1}{\left[\left[\frac{\gamma^2}{2} - \frac{\gamma^4}{4}\right]\right]_0^1} + c_{AR} \qquad (1.5.4\text{-}20)$$

$$(c_A)_b = (c_{A0} - c_{AR}) \frac{\frac{7}{60}}{\frac{15}{60}} + c_{AR} = \frac{7}{15}(c_{A0} - c_{AR}) + c_{AR} \qquad (1.5.4\text{-}21)$$

$$(c_A)_b = \frac{7}{15}(0.02 - 0.01) + (0.01)$$
$$(c_A)_b = 0.0147 \qquad (1.5.4\text{-}22)$$

1.5.5 Arithmetic, logarithmic, and geometric means

We will later have many occasions to use three different averages[24] of the value of a given quantity at two conditions.

• The arithmetic mean of θ evaluated at condition 1 and condition 2 is defined as

$$\boxed{\text{Arithmetic mean} \equiv \frac{\theta_1 + \theta_2}{2}} \qquad (1.5.5\text{-}1)$$

The arithmetic mean is often referred to simply as the mean without qualifying adjective, and conventionally one assumes the arithmetic mean is the one intended if no qualifying adjective is appended.

• The logarithmic mean of θ evaluated at condition 1 and condition 2 is defined as

$$\boxed{\text{Logarithmic mean} \equiv \frac{\theta_1 - \theta_2}{\ln\left|\frac{\theta_1}{\theta_2}\right|}} \qquad (1.5.5\text{-}2)$$

---

[24] *Mean* and *average* denote the same quantity.

• The geometric mean of θ evaluated at condition 1 and condition 2 is defined as

$$\boxed{\text{Geometric mean} \equiv \sqrt{\theta_1 \, \theta_2}} \qquad (1.5.5\text{-}3)$$

The geometric mean frequently has utility in problems whose natural description is in spherical coordinates.

The logarithmic mean is often abbreviated to log mean. It frequently has utility in problems whose natural description is in cylindrical coordinates. Notice that natural logarithms are required, not logarithms to the base 10, which would yield a different number.

The logarithmic mean is symmetrical

$$\frac{\theta_1 - \theta_2}{\ln\left[\dfrac{\theta_1}{\theta_2}\right]} = \frac{\theta_1 - \theta_2}{-\ln\left[\dfrac{\theta_2}{\theta_1}\right]} = \frac{\theta_2 - \theta_1}{\ln\left[\dfrac{\theta_2}{\theta_1}\right]} \qquad (1.5.5\text{-}4)$$

Two cases of the log mean are worth examining.

Case I: $\theta_1 > \theta_2 > 0$ \qquad\qquad\qquad\qquad\qquad (1.5.5-5)

In this case the numerator is positive, the fraction in the denominator is positive and greater than 1, so the logarithm is positive, and the resulting log mean is positive.

Case II : $\theta_1 < \theta_2 < 0$ \qquad\qquad\qquad\qquad\qquad (1.5.5-6)

In this case the numerator is negative, the fraction in the denominator is positive and greater than 1, so the logarithm is positive, and the resulting log mean is negative.

This shows that the log mean preserves the sign of the original quantities if they are both positive or both negative. Notice that the log mean of quantities which differ in sign is not defined, because the logarithm of a negative number is undefined.

Example 1.5.5-1   Case examples of logarithmic mean

Compute the logarithmic mean of the following quantities:

Case	θ_1	θ_2
I	10	5
II	-10	-5

Solution

$$\text{Log mean} \equiv \frac{\theta_1 - \theta_2}{\ln\left[\dfrac{\theta_1}{\theta_2}\right]} \qquad (1.5.5\text{-}7)$$

Case I

$$\text{Log mean} = \frac{10 - 5}{\ln\left[\dfrac{10}{5}\right]} = 7.21 \qquad (1.5.5\text{-}8)$$

Case II

$$\text{Log mean} = \frac{-10 - (-5)}{\ln\left[\dfrac{(-10)}{(-5)}\right]} = -7.21 \qquad (1.5.5\text{-}9)$$

Example 1.5.5-2   Approximation of logarithmic mean by arithmetic mean

For ease in calculation, it is sometimes desirable to model the logarithmic mean of two quantities, A and B, by the arithmetic mean. To determine the error introduced in this approximation, consider the function

$$\frac{\text{Arithmetic mean}\left[A, B\right]}{\text{Logarithmic mean}\left[A, B\right]} = \frac{\left[\dfrac{A+B}{2}\right]}{\left[\dfrac{A-B}{\ln\left(\dfrac{A}{B}\right)}\right]} = \frac{1}{2}\frac{\left(\dfrac{A}{B}+1\right)}{\left(\dfrac{A}{B}+1\right)}\ln\left(\dfrac{A}{B}\right) \qquad (1.5.5\text{-}10)$$

The following plot shows this function.

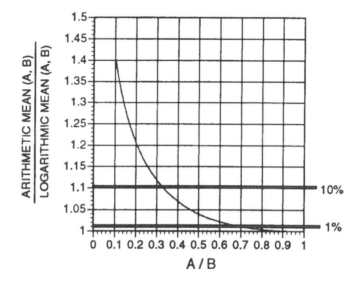

A discrepancy of less than 1% is obtained between ratios of 0.7 to 1.0. At a ratio of 0.5 the error is still less than 5%. The discrepancy is less than 10% clear down to the value of 0.33. Outside of these limits the error grows rapidly, as can be seen.

1.6 Scalars, Vectors, Tensors and Coordinate Systems

1.6.1   The viscous stress tensor

In the same way that a vector, which is a first-order tensor, associates a scalar (its component), which is a zero-order tensor, with each direction in space, higher-order tensors associate tensors of one lower order with each direction in space. A *second-order tensor* associates a *vector* with each direction in space.

A second-order tensor that will be of much use to us is the *viscous stress tensor*

$$\tau = \begin{bmatrix} \tau_{11} & \tau_{12} & \tau_{13} \\ \tau_{21} & \tau_{22} & \tau_{23} \\ \tau_{31} & \tau_{32} & \tau_{33} \end{bmatrix}$$

(1.6.1-1)

This tensor returns, for any direction in space, a vector that represents the viscous stress on a surface whose outward normal coincides with this direction in space.

Another second-order tensor that will be of use to us is the *total stress tensor*

$$\mathbf{T} = \begin{bmatrix} \mathbf{T}_{11} & \mathbf{T}_{12} & \mathbf{T}_{13} \\ \mathbf{T}_{21} & \mathbf{T}_{22} & \mathbf{T}_{23} \\ \mathbf{T}_{31} & \mathbf{T}_{32} & \mathbf{T}_{33} \end{bmatrix} = \tau + \mathrm{p}\,\boldsymbol{\delta}$$

(1.6.1-2)

This tensor returns, for any direction in space, a vector that represents the viscous stress (momentum flux) plus the pressure stress on a surface whose outward normal coincides with this direction in space.

Components of the viscous stress tensor

If we take the dot product of the viscous stress tensor successively with the rectangular Cartesian coordinate directions, we find first that for the 1- or x-direction

$$\begin{aligned}
\tau \cdot \delta_1 \Rightarrow \delta_i \tau_{ij} \delta_{1j} &= \delta_1 \tau_{11} \delta_{11} + \delta_1 \tau_{12} \delta_{12} + \delta_1 \tau_{13} \delta_{13} \\
&\quad + \delta_2 \tau_{21} \delta_{11} + \delta_2 \tau_{22} \delta_{12} + \delta_2 \tau_{23} \delta_{13} \\
&\quad + \delta_3 \tau_{31} \delta_{11} + \delta_3 \tau_{32} \delta_{12} + \delta_3 \tau_{33} \delta_{13} \\
&= \delta_1 \tau_{11}\,(1) + \delta_1 \tau_{12}\,(0) + \delta_1 \tau_{13}\,(0) \\
&\quad + \delta_2 \tau_{21}\,(1) + \delta_2 \tau_{22}\,(0) + \delta_2 \tau_{23}\,(0) \\
&\quad + \delta_3 \tau_{31}\,(1) + \delta_3 \tau_{32}\,(0) + \delta_3 \tau_{33}\,(0) \\
&= \delta_1 \tau_{11} + \delta_2 \tau_{21} + \delta_3 \tau_{31}
\end{aligned}$$

(1.6.1-3)

Similarly, for the 2- (y-) and 3- (z-) directions

$$\tau \cdot \delta_2 \Rightarrow \delta_i \tau_{ij} \delta_{2j} = \delta_1 \tau_{12} + \delta_2 \tau_{22} + \delta_3 \tau_{32}$$
$$\tau \cdot \delta_3 \Rightarrow \delta_i \tau_{ij} \delta_{3j} = \delta_1 \tau_{13} + \delta_2 \tau_{23} + \delta_3 \tau_{33} \qquad (1.6.1\text{-}4)$$

This shows that the columns (or rows, since the viscous stress tensor is symmetric) furnish the components of the vectors that represent the viscous stresses in the respective coordinate directions.

We now illustrate the geometric significance of the above equations. In Figure 1.6.1-1(a) we have shown an arbitrarily oriented surface, A, with its outwardly directed unit normal n, and the surface traction vector, τ, that some particular stress tensor associates with the direction in space indicated by n. The vector τ can be decomposed into a vector in the normal direction, τ_n, and a vector lying in the plane of the surface, τ_s, as indicated. Such a procedure can be carried out for any surface at an arbitrary orientation in space.

Consider then the particular surfaces B, C, and D, whose outward unit normals are the basis vectors for rectangular Cartesian coordinates, δ_1, δ_2, and δ_3. The stress tensor will return an appropriate surface traction for each of these directions in space, indicated in the figure by τ_1, τ_2, and τ_3. Each of these vectors may be decomposed into a vector in the direction of the outward normal and a vector lying in the plane of the surface. (Remember that we deal with only one of these vectors at a time - we choose one direction, the tensor gives us back one vector - despite the appearance from the figure that all three vectors appear simultaneously.)

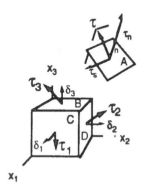

Figure 1.6.1-1 (a) Vectors associated by a particular viscous stress tensor with the direction of the rectangular Cartesian axes

Because of the orientation of the surfaces chosen, the vector lying in the plane of the surface can be further decomposed into its components along the coordinate axes that were used to determine that particular surface. We show in Figure 1.6.1-1 (a) the vectors associated with each face (coordinate direction in space), and, in addition, in Figure 1.6.1-1 (b) show the 3-direction vector decomposed into its component vectors along the rectangular Cartesian axes. These component vectors can also be written in terms of the scalar components multiplied by the appropriate basis vector.

$$\boldsymbol{\tau} \cdot \boldsymbol{\delta}_3 \;\Rightarrow\; \boldsymbol{\delta}_i\, \tau_{ij}\, \boldsymbol{\delta}_{3j} = \boldsymbol{\delta}_1\, \tau_{13} + \boldsymbol{\delta}_2\, \tau_{23} + \boldsymbol{\delta}_3\, \tau_{33} \qquad (1.6.1\text{-}5)$$

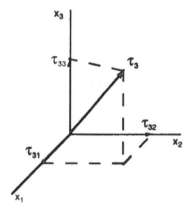

Figure 1.6.1-1 (b) Vector associated with the 3-direction decomposed into its components

1.6.2   Types of derivatives

There are three types of derivatives that we shall meet in our consideration of fluid mechanics.

Partial derivative

If a scalar function $\Phi(x_1, x_2, x_3, t)$ has a *total differential*, it may be written as

$$d\Phi = \left[\frac{\partial \Phi}{\partial t}\right]_{x_1, x_2, x_3} dt$$

$$+ \left[\frac{\partial \Phi}{\partial x_1}\right]_{x_2, x_3, t} dx_1 + \left[\frac{\partial \Phi}{\partial x_2}\right]_{x_1, x_3, t} dx_2 + \left[\frac{\partial \Phi}{\partial x_3}\right]_{x_1, x_2, t} dx_3 \qquad (1.6.2\text{-}1)$$

which is most often written omitting the indication of what is held constant in each partial derivative as

$$d\Phi = \frac{\partial \Phi}{\partial t} dt + \frac{\partial \Phi}{\partial x_1} dx_1 + \frac{\partial \Phi}{\partial x_2} dx_2 + \frac{\partial \Phi}{\partial x_3} dx_3 \qquad (1.6.2\text{-}2)$$

The coefficient of the first term is the partial derivative with respect to time

$$\left[\frac{\partial \Phi}{\partial t}\right]_{x_1, x_2, x_3} \qquad (1.6.2\text{-}3)$$

which gives the time rate of change of the function Φ at a point in space.

Total derivative

By dividing by dt, the total differential can be written as the *total time derivative*

$$\begin{aligned}
\frac{d\Phi}{dt} &= \frac{\partial \Phi}{\partial t} + \frac{\partial \Phi}{\partial x_1} \frac{dx_1}{dt} + \frac{\partial \Phi}{\partial x_2} \frac{dx_2}{dt} + \frac{\partial \Phi}{\partial x_3} \frac{dx_3}{dt} \\
&= \partial_0 \Phi + \partial_i \Phi \frac{dx_i}{dt}
\end{aligned} \qquad (1.6.2\text{-}4)$$

This expression represents the change in time of the function Φ ($\partial \Phi / \partial t$) as we move about with arbitrary velocities in the coordinate directions (dx_1/dt, dx_2/dt, dx_3/dt).

Substantial derivative, material derivative, derivative following the motion

If we constrain the motion to follow the motion of the individual fluid particles, we obtain the *substantial* derivative or *material* derivative or derivative *following the motion*, given by

$$\frac{D\Phi}{Dt} = \frac{\partial \Phi}{\partial t} + \frac{\partial \Phi}{\partial x_1} v_1 + \frac{\partial \Phi}{\partial x_2} v_2 + \frac{\partial \Phi}{\partial x_3} v_3$$

$$= \partial_0 \Phi + v_i \partial_i \Phi$$

(1.6.2-5)

where ∂_0 refers to the partial derivative with respect to time. The substantial derivative can be written in operator form as

$$\frac{D}{Dt} = \frac{\partial}{\partial t} + (v \cdot \nabla) \stackrel{\text{rect cart}}{\Rightarrow} \partial_0 + v_i \partial_i$$

(1.6.2-6)

Example 1.6.2-1   Rate of change of pollen density

Suppose that you are a sufferer with hay fever who decides to seek relief by going aloft in a blimp. Let Φ represent the grains of pollen per unit volume at any point at any time (pollen count). Let our reference coordinate axis system be rectangular Cartesian coordinates fixed to the surface of the earth.

Before you take off, while the blimp is still stationary but with the wind blowing past, if you put your head out of the window your nose will experience a **rate of change** in pollen count which corresponds to the partial derivative of Φ with respect to time, $\partial \Phi / \partial t$ (because the velocity in each of the coordinate directions is zero).

After the blimp takes off, and is ascending to altitude at some arbitrary velocity (speed and direction), suppose you again open the window and put out your head. Your nose will experience a rate of change in pollen count described by $d\Phi / dt$, because the blimp now has non zero velocity components with respect to the coordinate axes.

Suppose now that after some time aloft the pilot shuts off the engine, so that the blimp (assumed to be neutrally buoyant, or weightless) acts like a free balloon and is carried by the air at the velocity of the wind. Now if you put your head out of the window your nose will experience a rate of change in pollen count described by the substantial derivative, $D\Phi / Dt$, because your velocity now corresponds to the velocity of the fluid - you are *following the motion of the fluid* (air)

$$\left\{ \begin{array}{l} \dfrac{dx_1}{dt} = v_1 \\[2mm] \dfrac{dx_2}{dt} = v_2 \\[2mm] \dfrac{dx_3}{dt} = v_3 \end{array} \right\} \qquad\qquad (1.6.2\text{-}7)$$

1.6.3   Transport theorem

Since our study here is fluids, which are frequently continuously deforming, it is often convenient to adopt the Lagrangian viewpoint, where we ride on a fluid particle to observe changes in our variables rather than taking the Eulerian viewpoint of an observer fixed to the ground. The substantial derivative for a general Cartesian tensor is sometimes denoted by a dot in the superscript position

$$\Omega^{\bullet}_{kl...} = \partial_0\, \Omega_{kl...} + v_j\, \partial_j \Omega_{kl...} \qquad\qquad (1.6.3\text{-}1)$$

Consider now how the volume integral of some tensorial property of the continuum, $T_{kl...}$, changes as we move with the fluid.

$$I = \int_V \Omega_{kl...}\, dV \qquad\qquad (1.6.3\text{-}2)$$

where the integral is taken at some time t.

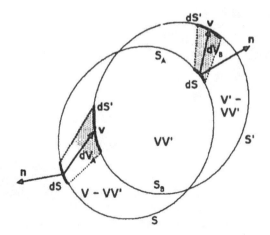

Figure 1.6.3-1 Motion of continuum

In Figure 1.6.3-1, consider the points that at time t are in V, the volume interior to S. In some time Δt, these points will move to the volume V', interior to S'. Points which are on the elemental surface dS will move to dS'. By the definition of the material derivative which declares it to be the time rate of change following the motion of the continuum, we then have that

$$I^\bullet = \lim_{\Delta t \to 0} \left\{ \frac{\left[\int_V \Omega_{kl...} \, dV \right]_{V', \, t+\Delta t} - \left[\int_V \Omega_{kl...} \, dV \right]_{V, \, t}}{\Delta t} \right\} \qquad (1.6.3\text{-}3)$$

However, we see that those points common to both V and V' will generate terms in the numerator which involve only the change in time, not in space. Denoting this common volume as VV', we can write, since V = VV' + (V - VV') and V' = VV' + (V' - VV')

$$I^* = \lim_{\Delta t \to 0} \left\{ \frac{\left[\int_V \Omega_{kl...} \, dV \right]_{VV', t+\Delta t} - \left[\int_V \Omega_{kl...} \, dV \right]_{VV', t}}{\Delta t} \right.$$

$$\left. + \frac{\left[\int_V \Omega_{kl...} \, dV \right]_{(V-VV'), t+\Delta t} - \left[\int_V \Omega_{kl...} \, dV \right]_{(V'-VV'), t}}{\Delta t} \right\} \qquad (1.6.3-4)$$

The first term in the limit gives the partial derivative of the integral with respect to time. For convenience we denote V - VV' as V_A and V' - VV' as V_B. By recognizing that the volumes not common to V and V' can be written in terms of the elemental volumes swept out by the elemental surfaces as indicated

$$dV_A = \left(v \cdot n \right) dS$$
$$dV_B = \left(v \cdot n \right) dS \qquad\qquad (1.6.3-5)$$

the second term can be written as

$$\lim_{\Delta t \to 0} \left\{ \frac{\left[\int_S \Omega_{kl...} v_i n_i \, \Delta t \, dS \right]_{S_A} + \left[\int_S \Omega_{kl...} v_i n_i \, \Delta t \, dS \right]_{S_B}}{\Delta t} \right\} = \int_S \Omega_{kl...} v_i n_i \, dS$$

$$(1.6.3-6)$$

giving the result

$$\boxed{I^* = \int_V \partial_0 \Omega_{kl...} \, dV + \int_S \Omega_{kl...} v_j n_j \, dS} \qquad (1.6.3-7)$$

We can apply the Gauss theorem to the second term

$$I^*_{kl...} = \int_V \left[\partial_0 \Omega_{kl...} + \partial_j \left(v_j \Omega_{kl...} \right) \right] dV$$

$$\boxed{I^{\bullet}_{kl...} = \int_V \left[\Omega^{\bullet}_{kl...} + \Omega_{kl...} \, \partial_j v_j \right] dV}$$

(1.6.3-8)

If $I_{kl..}$ has a constant value for the region of the continuum under consideration (a **conserved** quantity), then $I^{\bullet}_{kl..} = 0$. If this is to be true for arbitrary values of the volume, the only way it can be true is for the integrand above to be zero. We will use this argument several times in our course of study to obtain microscopic equations from macroscopic balances.

Chapter 1   Problems

1.1 At the end of June your checkbook balance is $356. During July you wrote $503 in checks and deposited $120. What is your balance (accumulation) at the end of July? What is your accumulation rate per day in July?

1.2 At the end of January your parts warehouse contains 35,133 parts. During February you sold 21,316 parts from the warehouse and manufactured 3,153 which were added to the warehouse. How many parts are in the warehouse at the end of February?

1.3 What are the conversion factors for:

> Newtons (N) to pounds force (lbf)
> Pascals (Pa) to pounds force per square inch (lbf/in^2)
> Watts (W) to horsepower (hp)
> meters cubed (m^3) to quarts (qt)

1.4 To convert between English Units and SI units calculate conversion factors for the following:

> 1 lbf to Newtons (N)
> 1 lbf/in^2 to Pascal (Pa)
> 1 BTU to Joules (J)
> 1 hp to Watts (W)
> 1 quart to m^3

1.5 Two infinite plates are separated 0.01 in. by a liquid with a viscosity of 1.1 centipoise (cP). The top plate moves at a velocity of 0.3 ft/s. The bottom plate is stationary. If the velocity decreases linearly from the top plate through the fluid to the bottom plate, what is the shear stress on each plate?

1.6 Two infinite plates are separated 0.2 mm by a liquid with viscosity of 0.9 cP. The top plate moves at a velocity of 0.1 m/s; the bottom plate is stationary. Assume velocity decreases linearly through the fluid to 0 at the bottom plate. What is the shear stress on each plate?

1.7 The expression for velocity v_x, for fluid flow between two parallel plates is

$$v_x = \frac{(p_1 - p_2)B^2}{2\mu L}\left[1\left(\frac{y}{B}\right)^2\right]$$

2B is plate separation, L is length of the two plates, (p_1-p_2) is the pressure drop causing flow and μ is the fluid viscosity. Determine the expression for the average velocity, $<v_x>$.

1.8 If the coordinate system in problem 1.7 is changed to the bottom plate of the slit and the expression for velocity is

$$v_x = \frac{(p_1 - p_2)B^2}{\mu L}\left[\frac{1}{2}\left(\frac{y}{B}\right)^2 - \left(\frac{y}{B}\right)\right]$$

Determine the average velocity, $<v_x>$.

1.9 The relationship

$$\mathbf{v} - \mathbf{w} = \sum_{i=1}^{3}\mathbf{i}\,v_i - \sum_{i=1}^{3}\mathbf{i}\,w_i = \sum_{i=1}^{3}\mathbf{i}(v_i - w_i) = v_i - w_i$$

shows how to transform symbolic vector notation to index vector notation. Do the same for

      a. The product of a scalar and a vector s **v**
      b. The scalar product of two vectors **v** • **w**
      c. The cross product of two vectors **v** × **w**

1.10 Show that

$$\mathbf{a} \times (\mathbf{b} \times \mathbf{c}) = \mathbf{a}\,(\mathbf{b} \cdot \mathbf{c}) - \mathbf{c}(\mathbf{a} \cdot \mathbf{b})$$

1.11 Show that

$$\bar{\tau} : \nabla v = \nabla \bullet (\bar{\tau} \bullet v) - v \bullet (\nabla \bullet \bar{\tau})$$

1.12 The sketch shows rectangular and spherical coordinates where

$-\infty \le x \le \infty, \ -\infty \le y \le \infty, \ -\infty < z < \infty,$

$0 < r \le \infty, \ 0 \le \theta \le \pi, \text{ and } 0 \le \phi \le 2\pi$

Show the relations for two coordinate systems, i.e., $x = x(r,\theta, \phi)$, $y = y(r, \theta, \phi)$, $z = z(r,\theta, \phi)$.

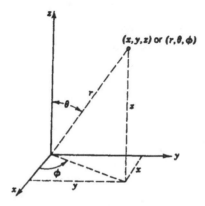

1.13 The sketch shows rectangular and cylindrical coordinates, where

$-\infty \le x \le \infty, \ -\infty \le y \le \infty, \ -\infty < z < \infty, \ 0 < r \le \infty, \ 0 \le \phi \le 2\pi$.

The coordinates are related by $x = r \cos \theta$, $y = r \sin \theta$ and $z = z$. Apply the chain rule and show the relationships between the derivatives.

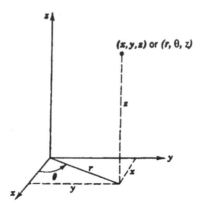

2

THE MASS BALANCES

We are concerned in this course of study with mass, energy, and momentum. We will proceed to apply the entity balance to each in turn. There also are sub-classifications of each of these quantities that are of interest. In terms of mass, we are often concerned with accounting for both total mass and mass of a species.

2.1 The Macroscopic Mass Balances

Consider a system of an arbitrary shape fixed in space as shown in Figure 2.1-1. V refers to the volume of the system, A to the surface area. ΔA is a small increment of surface area; ΔV is a small incremental volume inside the system; v is the velocity vector, here shown as an input; n is the outward normal; α is the angle between the velocity vector and the outward normal, here greater than π radians or 180° because the velocity vector and the outward normal fall on opposite sides of ΔA.

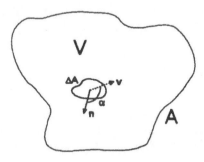

Figure 2.1-1 System for mass balances

Using the continuum assumption, the density can be modeled functionally by

$$\rho = \rho\,(x_1, x_2, x_3, t) \tag{2.1-1}$$

2.1.1 The macroscopic total mass balance

We now apply the entity balance, written in terms of rates, term by term to total mass. Since total mass will be conserved in processes we consider, the generation term will be zero.

$$\begin{bmatrix} \text{output} \\ \text{rate} \\ \text{of total mass} \end{bmatrix} - \begin{bmatrix} \text{input} \\ \text{rate} \\ \text{of total mass} \end{bmatrix} + \begin{bmatrix} \text{accumulation} \\ \text{rate} \\ \text{of total mass} \end{bmatrix} = 0 \tag{2.1.1-1}$$

The total mass in ΔV is approximately

$$\rho\,\Delta V \tag{2.1.1-2}$$

where ρ is evaluated at any point in ΔV and Δt.

The total mass in V can be obtained by summing over all of the elemental volumes in V and taking the limit as ΔV approaches zero

$$\begin{bmatrix} \text{total mass} \\ \text{in} \\ \text{V} \end{bmatrix} \cong \sum [\rho\,\Delta V] = \lim_{\Delta V \to 0} \left\{ \sum [\rho\,\Delta V] \right\} = \int_V \rho\,dV \tag{2.1.1-3}$$

Accumulation of mass

First examine the rate of accumulation term. The accumulation (not rate of accumulation) over time Δt is the difference between the mass in the system at some initial time t and the mass in the system at the later time $t + \Delta t$.

$$\begin{bmatrix} \text{accumulation} \\ \text{of total mass} \\ \text{in V} \\ \text{during } \Delta t \end{bmatrix} = m_{t+\Delta t} - m_t = \int_V \rho\,dV \bigg|_{t+\Delta t} - \int_V \rho\,dV \bigg|_t$$

$$[M] \Leftrightarrow \left[\frac{M}{L^3}\right]\left[L^3\right] \qquad (2.1.1\text{-}4)$$

The rate of accumulation is obtained by taking the limit

$$\begin{bmatrix} \text{rate of} \\ \text{accumulation} \\ \text{of total mass} \\ \text{in V} \end{bmatrix} = \lim_{\Delta t \to 0} \left[\frac{\left[\int_V \rho \, dV\right]_{t+\Delta t} - \left[\int_V \rho \, dV\right]_t}{\Delta t} \right] = \frac{d}{dt} \int_V \rho \, dV]$$

$$\left[\frac{M}{t}\right] \Leftrightarrow \left[\frac{1}{t}\right]\left[\frac{M}{L^3}\right]\left[L^3\right] \qquad (2.1.1\text{-}5)$$

Input and output of mass

The rate of input and rate of output terms may be evaluated by considering an arbitrary small area ΔA on the surface of the control volume. The velocity vector will not necessarily be normal to the surface.

The approximate volumetric flow rate[1] through the elemental area ΔA can be written as the product of the velocity normal to the area evaluated at some point within ΔA multiplied by the area.

$$\begin{bmatrix} \text{volumetric} \\ \text{flow rate} \\ \text{through} \, \Delta A \end{bmatrix} \equiv (v \cdot n) \, \Delta A \qquad (2.1.1\text{-}6)$$

$$\left[\frac{L^3}{t}\right] \Leftrightarrow \left[\frac{L}{t}\right]\left[L^2\right]$$

Notice that the volumetric flow rate written above will have a negative sign for inputs (where $\alpha > \pi$) and a positive sign for outputs (where $\alpha < \pi$) and so the appropriate sign will automatically be associated with either input or output terms, making it no longer necessary to distinguish between input and output.

The approximate mass flow rate through ΔA can be written by multiplying the volumetric flow rate by the density at some point in ΔA

---

[1] We frequently need to refer to the volumetric flow rate independent of a control volume. In doing so we use the symbol Q and restrict it to the absolute value

$$Q = \int_A |(v \cdot n)| \, dA$$

$$\begin{bmatrix} \text{mass} \\ \text{flow rate} \\ \text{through } \Delta A \end{bmatrix} \cong \rho \left(\mathbf{v} \cdot \mathbf{n} \right) \Delta A \qquad (2.1.1\text{-}7)$$

$$\left[\frac{M}{t} \right] \Leftrightarrow \left[\frac{M}{L^3} \right] \left[\frac{L^3}{t} \right]$$

The mass flow rate through the total external surface area A can then be written as the limit of the sum of the flows through all the elemental ΔAs as the elemental areas approach zero. Notice that in the limit it no longer matters where in the individual ΔA we evaluate velocity or density, since a unique point is approached for each elemental area.[2]

$$\begin{bmatrix} \text{mass} \\ \text{flow rate} \\ \text{through } A \end{bmatrix} \cong \sum_A \left[\rho \left(\mathbf{v} \cdot \mathbf{n} \right) \Delta A \right]$$

$$= \lim_{\Delta A \to 0} \left\{ \sum_A \left[\rho \left(\mathbf{v} \cdot \mathbf{n} \right) \Delta A \right] \right\}$$

$$= \int_A \rho \left(\mathbf{v} \cdot \mathbf{n} \right) dA \qquad (2.1.1\text{-}8)$$

Substitution in the entity balance then gives the *macroscopic total mass balance*

$$\left\{ \begin{bmatrix} \text{Total mass} \\ \text{output} \\ \text{rate} \end{bmatrix} - \begin{bmatrix} \text{Total mass} \\ \text{input} \\ \text{rate} \end{bmatrix} \right\} + \begin{bmatrix} \text{Total mass} \\ \text{accumulation} \\ \text{rate} \end{bmatrix} = 0$$

---

[2] We frequently need to refer to the mass flow rate independent of a control volume. In doing so we use the symbol w and restrict it to the absolute value

$$w = \int_A \rho \left| \left(\mathbf{v} \cdot \mathbf{n} \right) \right| dA$$

For constant density systems $w = \rho Q$

$$\int_A \rho \left(v \cdot n\right) dA + \frac{d}{dt} \int_V \rho \, dV = 0$$

$$w_{out} - w_{in} + \frac{dm}{dt} = 0$$

(2.1.1-9)

We can use the latter form of the balance if we know total flow rates and the change in mass of the system with time; otherwise, we must evaluate the integrals. In some cases, where density is constant or area-averaged velocities are known, this is simple, as detailed below. Otherwise, we must take into account the variation of velocity and density across inlets and outlets and of density within the system

Simplified forms of the macroscopic total mass balance

Many applications involve constant density systems. Further, our interest is often only in total flow into or out of the system rather than the manner in which flow velocity varies across inlet and outlet cross-sections.

Such problems are usually discussed in terms of area-average velocities. The area-average velocity is simply a number that, if multiplied by the flow cross-section, gives the same result as integrating the velocity profile across the flow cross-section, i.e., the total volumetric flow rate. (So it is really the area average of the normal component of velocity.)

In such cases

$$\int_A \rho \left(v \cdot n\right) dA + \frac{d}{dt} \int_V \rho \, dV = 0$$

$$\rho \int_A \left(v \cdot n\right) dA + \rho \frac{d}{dt} \int_V dV = 0$$

(2.1.1-10)

which can be written in terms of mass flow rate as

$$\rho \langle v \rangle A_{out} - \rho \langle v \rangle A_{in} + \rho \frac{dV}{dt} = 0$$

(2.1.1-11)

or in volumetric flow rate

$$\langle v \rangle A_{out} - \langle v \rangle A_{in} + \frac{dV}{dt} = 0$$

$$Q_{out} - Q_{in} + \frac{dV}{dt} = 0 \qquad\qquad (2.1.1\text{-}12)$$

Example 2.1.1-1   Mass balance on a surge tank

Tanks[3] are used for several purposes in processes. The most obvious is for storage. They are also used for mixing and as chemical reactors. Another purpose of tankage is to supply surge capacity to smooth out variations in process flows; e.g., to hold material produced at a varying rate upstream until units downstream are ready to receive it. Consider the surge tank shown in Figure 2.1.1-1.

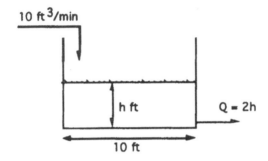

Figure 2.1.1-1 Surge tank

Suppose water is being pumped into a 10-ft diameter tank at the rate of 10 ft^3/min. If the tank were initially empty, and if water were to leave the tank at a rate dependent on the liquid level according to the relationship $Q_{out} = 2h$ (where the units of the constant 2 are [ft^2/min], of Q are [ft^3/min], and of h are [ft]), find the height of liquid in the tank as a function of time.

Solution

This situation is adequately modeled by constant density for water, so

$$Q_{out} - Q_{in} + \frac{dV}{dt} = 0 \qquad\qquad (2.1.1\text{-}13)$$

---

[3] "Tank" is here used in the generic sense as a volume used to hold liquid enclosed or partially enclosed by a shell of some sort of solid material. In process language, a distinction is often made among holding tanks, mixers, and reactors.

$$2 h \left[\frac{ft^3}{min}\right] - 10\left[\frac{ft^3}{min}\right] + \frac{d}{dt}\left[\frac{1}{min}\right]\left(\frac{\pi \, 10^2}{4} h\right)\left[ft^3\right] = 0 \qquad (2.1.1\text{-}14)$$

$$h - 5 + 39.3 \frac{dh}{dt} = 0 \qquad (2.1.1\text{-}15)$$

$$-39.3 \int_0^h \frac{(-dh)}{5-h} = \int_0^t dt \qquad (2.1.1\text{-}16)$$

$$t = -39.3 \ln\left[\frac{5-h}{5}\right] \qquad (2.1.1\text{-}17)$$

$$\exp\left(-\frac{t}{39.3}\right) = \left[\frac{5-h}{5}\right] \qquad (2.1.1\text{-}18)$$

$$h = 5\left[1 - \exp\left(-\frac{t}{39.3}\right)\right] ft \qquad (2.1.1\text{-}19)$$

Note that as time approaches infinity, the tank level stabilizes at 5 ft. It is usually desirable that surge tank levels stabilize before the tank overflows for reasons of safety if nothing else. The inlet flow should not be capable of being driven to a larger value than the outlet flow.

From a safety standpoint it would be important to examine the input flow rate to make sure it is the maximum to be expected, and to make sure that the exit flow rate is realistic - for example, could a valve be closed or the exit piping be obstructed in some other way, e.g., fouling or plugging by debris (perhaps a rag left in the tank after cleaning), such that the exit flow becomes proportional to some constant with numerical value less than 2. Overflow could be serious even with cool water because of damage to records or expensive equipment; with hot, toxic, and/or corrosive liquids the hazard could be severe.

Example 2.1.1-2   Volumetric flow rate of fluid in laminar flow in circular pipe

The (axially symmetric) velocity profile for fully developed laminar flow in a smooth tube of constant circular cross-section is

$$v = v_{max}\left[1 - \left(\frac{r}{R}\right)^2\right]$$

(2.1.1-20)

Calculate the volumetric flowrate.

Solution

Choose the flow direction to be the z-direction.

a) Since we are not referring the volumetric flow to any particular system or control volume, we use the absolute value

$$Q = \left|\int_A (v \cdot n) \, dA\right| = \left|\int_0^R v_{max}\left[1 - \left(\frac{r}{R}\right)^2\right](k \cdot n) \, 2\pi r \, dr\right|$$

(2.1.1-21)

$$Q = 2\pi v_{max}\int_0^R \left[1 - \left(\frac{r}{R}\right)^2\right]\left|(k \cdot n)\right| r \, dr$$

(2.1.1-22)

Noting that

$$\left|(k \cdot n)\right| = |k||n||\cos(\alpha)| = |1||1|\left|\cos\begin{pmatrix}0\\ \text{or}\\ \pi\end{pmatrix}\right| = 1$$

(2.1.1-23)

$$Q = 2\pi v_{max}\int_0^R \left[1 - \left(\frac{r}{R}\right)^2\right] r \, dr$$

(2.1.1-24)

$$Q = 2\pi v_{max}\int_0^R \left[r - \left(\frac{r^3}{R^2}\right)\right] dr$$

(2.1.1-25)

$$Q = 2\pi v_{max}\left[\frac{r^2}{2} - \frac{r^4}{4R^2}\right]_0^R$$

(2.1.1-26)

$$Q = 2\pi v_{max}\left[\frac{R^2}{2} - \frac{R^2}{4}\right]$$

(2.1.1-27)

$$Q = \tfrac{1}{2} v_{max} \left(\pi R^2 \right) \tag{2.1.1-28}$$

Example 2.1.1-3   Air storage tank

A tank of volume $V = 0.1 \ m^3$ contains air at 1000 kPa and (uniform) density of 6 kg/m^3. At time $t = 0$, a valve at the top of the tank is opened to give a free area of 100 mm^2 across which the air escape velocity can be assumed to be constant at 350 m/s. What is the initial rate of change of the density of air in the tank?

Solution

$$\int_A \rho \left(v \cdot n \right) dA + \frac{d}{dt} \int_V \rho \ dV = 0 \tag{2.1.1-29}$$

Examining the second term

$$\frac{d}{dt} \int_V \rho \ dV = \frac{d}{dt} \left(\rho \int_V dV \right) = \frac{d}{dt} \left(\rho \ V \right) = \rho \frac{dV}{dt} + V \frac{d\rho}{dt} \tag{2.1.1-30}$$

But, since the tank volume is constant, dV/dt = 0 and

$$\frac{d}{dt} \int_V \rho \ dV = V \frac{d\rho}{dt} \tag{2.1.1-31}$$

The first term yields, since the only term is an output

$$\int_A \rho \left(v \cdot n \right) dA = \rho \left(v \cdot n \right) \int_A dA = \rho \ v \ A \tag{2.1.1-32}$$

Substituting in the mass balance

$$\rho \ v \ A + V \frac{d\rho}{dt} = 0 \tag{2.1.1-33}$$

$$\frac{d\rho}{dt} = -\frac{\rho \ v \ A}{V}$$

$$= -(6)\left[\frac{kg}{m^3}\right](350)\left[\frac{m}{s}\right](100)\left[mm^2\right]\frac{1}{(0.1)\left[m^3\right]}\left[\frac{m^2}{10^6\,mm^2}\right]$$

$$= -2.1\,\frac{kg}{m^2\,s}$$

$$(2.1.1\text{-}34)$$

Example 2.1.1-4   Water manifold

Water is in steady flow through the manifold[4] shown below, with the indicated flow directions known.

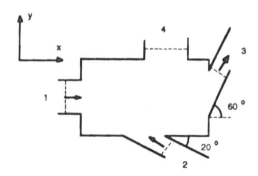

The following information is also known.

---

[4] A **manifold** is a term loosely used in engineering to indicate a collection of pipes connected to a common region: often, a single inlet which supplies a number of outlets; e.g., the burner on a gas stove or the fuel/air intake manifold for an multi-cylinder internal combustion engine. If the volume of the region is large compared to the volume of the pipes, one does not ordinarily use the term manifold - for example, a number of pipes feeding into or out of a single large tank could be called but would not ordinarily be called a manifold.

ρ	$62.4 \frac{\text{lbmass}}{\text{ft}^3}$
A^1	0.3 ft^2
A^2	0.5 ft^2
A^3	0.5 ft^2
A^4	0.4 ft^2
v_1	$81 \frac{\text{ft}}{\text{s}}$
Q^2	$0.9 \frac{\text{ft}^3}{\text{s}}$
w^3	$250 \frac{\text{lbmass}}{\text{s}}$

Assuming that velocities are constant across each individual cross-sectional area, determine v_4.

Solution

We apply the macroscopic total mass balance to a system which is bounded by the solid surfaces of the manifold and the dotted lines at the fluid surfaces.

$$\int_A \rho \left(v \cdot n\right) dA + \frac{d}{dt} \int_V \rho \, dV = 0 \qquad (2.1.1\text{-}35)$$

At steady state

$$\frac{d}{dt} \int_V \rho \, dV = 0 \qquad (2.1.1\text{-}36)$$

so

$$\int_A \rho \left(v \cdot n\right) dA = 0 \qquad (2.1.1\text{-}37)$$

and, recognizing that at the solid surfaces

$$\left(v \cdot n\right) = 0 \qquad (2.1.1\text{-}38)$$

the macroscopic total mass balance reduces to

$$\int_{A_1} \rho \left(\mathbf{v} \cdot \mathbf{n} \right) dA + \int_{A_2} \rho \left(\mathbf{v} \cdot \mathbf{n} \right) dA$$
$$+ \int_{A_3} \rho \left(\mathbf{v} \cdot \mathbf{n} \right) dA \qquad\qquad (2.1.1\text{-}39)$$
$$+ \int_{A_4} \rho \left(\mathbf{v} \cdot \mathbf{n} \right) dA = 0$$

Examining the first term,

$$\int_{A_1} \rho \left(\mathbf{v} \cdot \mathbf{n} \right) dA = \int_{A_1} (62.4) \left[\frac{\text{lbmass}}{\text{ft}^3} \right] \left(8\, \mathbf{i} \left[\tfrac{\text{ft}}{\text{s}} \right] \cdot \mathbf{n} \right) dA$$
$$= (62.4) \left[\frac{\text{lbmass}}{\text{ft}^3} \right] (8) \left[\tfrac{\text{ft}}{\text{s}} \right] \int_{A_1} \left(\mathbf{i} \cdot \mathbf{n} \right) dA \qquad (2.1.1\text{-}40)$$

but the outward normal at A_1 is in the negative x-direction, so the velocity vector at A_1 and the outward normal are at an angle of $180°$, giving

$$\left(\mathbf{i} \cdot \mathbf{n} \right) = |1| \, |1| \cos \left(\pi \text{ radians} \right) = -1 \qquad\qquad (2.1.1\text{-}41)$$

and therefore

$$\int_{A_1} \rho \left(\mathbf{v} \cdot \mathbf{n} \right) dA = -(62.4) \left[\frac{\text{lbmass}}{\text{ft}^3} \right] (8) \left[\tfrac{\text{ft}}{\text{s}} \right] \int_{A_1} dA$$
$$= -(62.4) \left[\frac{\text{lbmass}}{\text{ft}^3} \right] (8) \left[\tfrac{\text{ft}}{\text{s}} \right] (A_1)$$
$$= -(62.4) \left[\frac{\text{lbmass}}{\text{ft}^3} \right] (8) \left[\tfrac{\text{ft}}{\text{s}} \right] (0.3) \left[\text{ft}^2 \right]$$
$$= -150 \, \frac{\text{lbmass}}{\text{s}} \qquad\qquad (2.1.1\text{-}42)$$

Proceeding to the second term, and recognizing that this also is an input term so will have a negative sign associated with it as in the first term

$$\int_{A_2} \rho \left(\mathbf{v} \cdot \mathbf{n} \right) dA = \rho \int_{A_2} \left(\mathbf{v} \cdot \mathbf{n} \right) dA = -\rho\, Q_2$$

$$= -(62.4)\left[\frac{\text{lbmass}}{\text{ft}^3}\right](0.9)\left[\frac{\text{ft}^3}{\text{s}}\right]$$

$$= -56\,\frac{\text{lbmass}}{\text{s}} \tag{2.1.1-43}$$

The third term, which is an output (therefore the dot product gives a positive sign), then gives

$$\int_{A_3} \rho\,(v \cdot n)\,dA = \rho \int_{A_3} (v \cdot n)\,dA$$

$$= \rho\,Q_3 = w_3 = 250\,\frac{\text{lbmass}}{\text{s}} \tag{2.1.1-44}$$

The fourth term simplifies to

$$\int_{A_4} \rho\,(v \cdot n)\,dA = \rho\,(v \cdot n)\int_{A_4} dA = \rho\,(v_4 \cdot n)\,A_4$$

$$\int_{A_4} \rho\,(v \cdot n)\,dA = (62.4)\left[\frac{\text{lbmass}}{\text{ft}^3}\right](v_4 \cdot n)\left[\frac{\text{ft}}{\text{s}}\right](0.4)\left[\text{ft}^2\right]$$

$$= (25)\left[\frac{\text{lbmass}}{\text{ft}}\right](v_4 \cdot n)\left[\frac{\text{ft}}{\text{s}}\right]$$

$$= 25\,(v_4 \cdot n)\,\frac{\text{lbmass}}{\text{s}} \tag{2.1.1-45}$$

Substituting these results into the simplified macroscopic total mass balance

$$-(150)\left[\frac{\text{lbmass}}{\text{s}}\right] - (56)\left[\frac{\text{lbmass}}{\text{s}}\right] + (250)\left[\frac{\text{lbmass}}{\text{s}}\right]$$

$$+\left\{(25)\left[\frac{\text{lbmass}}{\text{ft}}\right](v_4 \cdot n)\left[\frac{\text{ft}}{\text{s}}\right]\right\} = 0 \tag{2.1.1-46}$$

Solving

$$v_4 \cdot n = -1.76\,\frac{\text{ft}}{\text{s}} \tag{2.1.1-47}$$

The negative sign indicates that the velocity vector is in the opposite direction to the outward normal. At A_4 the outward normal is in the positive y-direction; therefore, the velocity vector is in the negative y-direction, or

$$v_4 = -1.76 \frac{ft}{s} \, j$$

$$(2.1.1-48)$$

Water flows into the system at A_4.

Notice that if A_4 had emerged from the bottom of the manifold rather than the top, the outward normal would have been in the negative rather than the positive y-direction, giving a velocity vector in the positive y-direction, so the resultant flow would still have been into the system - we cannot change the mass balance by simply changing the orientation of the inlet or outlet to a system (we can, of course, change the momentum balance by such a change). The reader is encouraged to change the orientation of A_4 to horizontally to the left or right and prove that the mass balance is unaffected.

2.1.2   The macroscopic species mass balance

As noted earlier when we discussed conserved quantities, **total mass can be modeled as conserved** under processes of interest to us here. However, **individual species are not, in general, conserved**, especially in processes where chemical reactions occur. Chemical reaction can give a positive or negative generation of an individual species.

For example, if we feed a propane-air mixture into a combustion furnace there will be a negative generation of propane (since propane disappears) and there will be a positive generation of CO_2.[5] In such a case, to apply the entity balance in a useful way we need a way to describe the rate of generation of a species by chemical reaction.

Let us consider for the moment applying, to a control volume fixed in space, the entity balance (again, as for total mass, written in terms of rates) now to **mass of a chemical species**, which is assumed to be involved in processes for which the generation is by chemical reaction.

---

[5] We can, of course, at the same time generate elemental carbon (in soot, etc.), CO, and other byproducts of combustion. There will also be a negative generation of O_2.

Generation of mass of a species

Consider the generation rate term. Call the **rate** of species mass generation per **unit volume** of the ith chemical species r_i, which will be a function of position in space and of time

$$r_i = r_i(x_1, x_2, x_3, t) \qquad (2.1.2\text{-}1)$$
$$\left[\frac{M}{L^3 t}\right]$$

The rate of generation of mass of species i in a volume ΔV is then approximately equal to

$$\begin{bmatrix} \text{rate of generation} \\ \text{of mass of species i} \\ \text{in volume } \Delta V \end{bmatrix} = r_i \Delta V \qquad (2.1.2\text{-}2)$$

$$\left[\frac{M}{t}\right] \Leftrightarrow \left[\frac{M}{L^3 t}\right]\left[L^3\right]$$

To get the rate of generation throughout our control volume, we add the contributions of all the constituent elemental volumes in V. Taking the limit as ΔV approaches 0, we obtain the rate of generation term in the entity balance

$$\begin{bmatrix} \text{rate of generation} \\ \text{of mass of species i} \\ \text{in volume V} \end{bmatrix} = \lim_{\Delta V \to 0} \sum r_i \Delta V = \int_V r_i \, dV \qquad (2.1.2\text{-}3)$$

Accumulation of mass of a species

In the same manner as for total mass, the **amount** of accumulation of mass of species i (not rate of accumulation) over time Δt is the difference between the mass of species i in the system at some initial time t and the mass of species i in the system at the later time $t + \Delta t$. The mass of species i in the system at a given time is determined by taking the sum of the masses of species i in all of the elemental volumes that make up the system and then letting the elemental volumes approach zero, in a similar way to that shown in Equation 2.1.1-5 for total mass.

$$\begin{bmatrix} \text{accumulation} \\ \text{of mass of species i} \\ \text{in volume V} \end{bmatrix} = \left(m_i \right)_{t+\Delta t} - \left(m_i \right)_t$$

$$= \int_V \rho_i \, dV \bigg|_{t+\Delta t} - \int_V \rho_i \, dV \bigg|_t \qquad (2.1.2\text{-}4)$$

$$[M] \Leftrightarrow \left[\frac{M}{L^3} \right] \left[L^3 \right]$$

The **rate** of accumulation of mass of species i is obtained by taking the limit as Δt approaches 0

$$\begin{bmatrix} \text{rate of accumulation} \\ \text{of mass of species i} \\ \text{in volume V} \end{bmatrix} = \lim_{\Delta t \to 0} \left[\frac{\int_V \rho_i \, dV \big|_{t+\Delta t} - \int_V \rho_i \, dV \big|_t}{\Delta t} \right]$$

$$= \frac{d}{dt} \int_V \rho_i \, dV \qquad (2.1.2\text{-}5)$$

$$\left[\frac{M}{t} \right] \Leftrightarrow \left[\frac{1}{t} \right] \left[\frac{M}{L^3} \right] \left[L^3 \right]$$

Input and output of mass of a species

The input and output for mass of a species are readily obtained in the same manner as for total mass. The approximate mass flow rate of species i through an elemental area ΔA can be written by multiplying the volumetric flow rate[6] by the mass concentration of species i, ρ_i, (rather than the total density, as was done for total mass) at some point in ΔA

$$\begin{bmatrix} \text{mass flow rate} \\ \text{of species i} \\ \text{through } \Delta A \end{bmatrix} \equiv \rho_i \left(v \cdot n \right) \Delta A \qquad (2.1.2\text{-}6)$$

---

[6] Strictly speaking, we must replace v by v_i, since the individual species may not move at the velocity v in the presence of concentration gradients. The difference between v and v_i is negligible, however, for most problems involving flowing streams. We discuss the difference between v and v_i in our treatment of mass transfer.

$$\left[\frac{M}{t}\right] \Leftrightarrow \left[\frac{M}{L^3}\right]\left[\frac{L^3}{t}\right]$$

The mass flow rate through the total external surface area A can then be written as the limit of the sum of the flows through all the elemental ΔAs as the elemental areas approach zero.

$$\begin{bmatrix} \text{mass flow rate} \\ \text{of species i} \\ \text{through A} \end{bmatrix} \equiv \sum_A \left[\rho_i\,(v \cdot n)\,\Delta C\right]$$

$$= \lim_{\Delta A \to 0} \left\{\sum_A \left[\rho_i\,(v \cdot n)\,\Delta A\right]\right\}$$

$$= \int_A \rho_i\,(v \cdot n)\,dA \qquad (2.1.2\text{-}7)$$

Rearranging the entity balance as

$$\begin{bmatrix} \text{rate of output} \\ \text{of mass} \\ \text{of species i} \end{bmatrix} - \begin{bmatrix} \text{rate of input} \\ \text{of mass} \\ \text{of species i} \end{bmatrix}$$

$$+ \begin{bmatrix} \text{rate of accumulation} \\ \text{of mass} \\ \text{of species i} \end{bmatrix} = \begin{bmatrix} \text{rate of generation} \\ \text{of mass} \\ \text{of species i} \end{bmatrix} \qquad (2.1.2\text{-}8)$$

entering the various terms as just developed

$$\boxed{\int_A \rho_i\,(v \cdot n)\,dA + \frac{d}{dt}\int_V \rho_i\,dV = \int_V r_i\,dV} \qquad (2.1.2\text{-}9)$$

which is the **macroscopic mass balance for an individual species, i.**

If there is no generation term, the macroscopic total mass balance and the macroscopic mass balance for an individual species look very much the same except for the density term.

Water (whose density ρ may be assumed to be independent of the concentration of A) with $\rho = 1000$ kg/m^3 is flowing at 0.02 m^3/min through a pipe of inside diameter 0.05 m into a perfectly mixed tank containing 15 m^3 solution and out at the same rate through a pipe of diameter 0.02 m as shown in the following illustration.

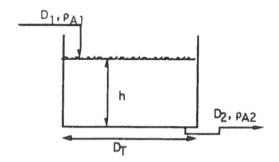

The entering liquid contains 20 kg/m^3 of A. The zero-order reaction

$$2\,A \rightarrow B \tag{2.1.2-10}$$

takes place with rate $r_B = 0.08$ kg B/(min m^3).

If the initial concentration of A in the tank is 50 kg/m^3, find the outlet concentration of A at t = 100 min.

Solution

Applying the macroscopic mass balance for an individual species

$$\int_A \rho_A (v \cdot n)\, dA + \frac{d}{dt}\int_V \rho_A\, dV = \int_V r_A\, dV \tag{2.1.2-11}$$

Evaluating term by term, starting with the first term on the left-hand side

$$\int_A \rho_A (v \cdot n) \, dA = (\rho_A)_{out} \int_{A_{out}} (v \cdot n) \, dA$$

$$+ (\rho_A)_{in} \int_{A_{in}} (v \cdot n) \, dA$$

$$= (\rho_A)_{out} Q_{out} - (\rho_A)_{in} Q_{in} \qquad (2.1.2\text{-}12)$$

But, since the volumetric flow rates in and out are the same in absolute value

$$\int_A \rho_A (v \cdot n) \, dA = [Q] \left[(\rho_A)_{out} - (\rho_A)_{in} \right] \qquad (2.1.2\text{-}13)$$

The second term on the left-hand side is

$$\frac{d}{dt} \int_V \rho_A \, dV = \frac{d}{dt} \left[(\rho_A)_{tank} \int_{V_{tank}} dV \right] = \frac{d}{dt} \left[(\rho_A)_{tank} V_{tank} \right]$$

$$= (\rho_A)_{tank} \frac{dV_{tank}}{dt} + V_{tank} \frac{d}{dt} (\rho_A)_{tank} \qquad (2.1.2\text{-}14)$$

But a total mass balance combined with constant density and the fact that inlet and outlet volumetric flow rates are the same gives

$$\rho Q_2 - \rho Q_1 + \rho \frac{dV_{tank}}{dt} = 0$$

$$\frac{dV_{tank}}{dt} = 0 \qquad (2.1.2\text{-}15)$$

so

$$\frac{d}{dt} \int_V \rho_A \, dV = V_{tank} \frac{d}{dt} (\rho_A)_{tank} \qquad (2.1.2\text{-}16)$$

Examining the right-hand side, we note that the reaction rate of A is twice the negative of that of B by the stoichiometry

$$r_A = -2 r_B \qquad (2.1.2\text{-}17)$$

$$\int_V r_A \, dV = r_A \int_{V_{tank}} dV = r_A V_{tank}$$

$$= -2 r_B V_{tank} \frac{mol\ A}{min} \qquad (2.1.2\text{-}18)$$

Substituting in the balance

$$Q\left[\left(\rho_A\right)_{out} - \left(\rho_A\right)_{in}\right] + V_{tank} \frac{d}{dt}\left(\rho_A\right)_{tank} = -2 r_B V_{tank} \qquad (2.1.2\text{-}19)$$

but, since the tank is perfectly stirred

$$\left(\rho_A\right)_{out} = \left(\rho_A\right)_{tank} = \rho_A \qquad (2.1.2\text{-}20)$$

which gives

$$Q\left(\rho_A - \rho_{Ain}\right) + V_{tank} \frac{d\rho_A}{dt} = -2 r_B V_{tank}$$

$$V_{tank} \frac{d\rho_A}{dt} = -Q\left(\rho_A - \rho_{Ain}\right) - 2 r_B V_{tank}$$

$$V_{tank}\left[\frac{d\rho_A}{-Q\left[\rho_A - \rho_{Ain}\right] - 2 r_B V_{tank}}\right] = dt \qquad (2.1.2\text{-}21)$$

Integrating and applying the initial condition

$$V_{tank} \int_{\rho_{A0}}^{\rho_A} \frac{d\rho_A}{\left(-Q\left[\rho_A - \rho_{Ain}\right] - 2 r_B V_{tank}\right)} = \int_0^t dt \qquad (2.1.2\text{-}22)$$

$$V_{tank} \int_{\rho_{A0}}^{\rho_A} \frac{d\rho_A}{\left(-Q\rho_A + \left[Q\rho_{Ain} - 2 r_B V_{tank}\right]\right)} = \int_0^t dt$$

$$-\frac{V_{tank}}{Q} \int_{\rho_{A0}}^{\rho_A} \frac{-Q \, d\rho_A}{\left(-Q\rho_A + \left[Q\rho_{Ain} - 2r_B V_{tank}\right]\right)} = t \qquad (2.1.2-23)$$

$$-\frac{V_{tank}}{Q} \ln \left\{ \frac{\left[Q\rho_{Ain} - 2r_B V_{tank}\right] - Q\rho_A}{\left[Q\rho_{Ain} - 2r_B V_{tank}\right] - Q\rho_{A0}} \right\} = t \qquad (2.1.2-24)$$

$$\frac{\left[Q\rho_{Ain} - 2r_B V_{tank}\right] - Q\rho_A}{\left[Q\rho_{Ain} - 2r_B V_{tank}\right] - Q\rho_{A0}} = \exp\left[-\frac{Q}{V_{tank}} t\right] \qquad (2.1.2-25)$$

Evaluating the constant term

$$Q\rho_{Ain} - 2r_B V_{tank} = (0.02)\left[\frac{m^3}{min}\right](20)\left[\frac{kg}{m^3}\right]$$

$$- (2)(0.08)\left[\frac{kg}{min \ m^3}\right](15)[m^3]$$

$$Q\rho_{Ain} - 2r_B V_{tank} = -2\frac{kg}{min} \qquad (2.1.2-26)$$

Introducing the required time

$$\frac{-(2)\left[\frac{kg}{min}\right] - (0.02)\left[\frac{m^3}{min}\right](\rho_A)\left[\frac{kg}{m^3}\right]}{-(2)\left[\frac{kg}{min}\right] - (0.02)\left[\frac{m^3}{min}\right](50)\left[\frac{kg}{m^3}\right]}$$

$$= \exp\left[-\frac{0.02\frac{m^3}{min}}{15 \ m^3} 100 \ min\right]$$

$$(0.02)(\rho_A) = \left[(2) + (0.02)(50)\right]\exp\left[-\left(\frac{0.02}{15}\right)(100)\right] - 2$$

$$(\rho_A) = 31.3\frac{kg}{m^3} \qquad (2.1.2-27)$$

Example 2.1.2-2   Macroscopic species mass balance with first-order irreversible reaction

Suppose that a reaction B → C takes place in a perfectly mixed tank. (By **perfectly mixed** we mean that ρ_i is independent of location in the tank, although it can depend on time. Although no real vessel is perfectly mixed, this is often a good model.)

Further suppose that the initial concentration of B in the tank is 5 lbm/ft³, that the inlet concentration of B is 15 lbm/ft³, that the volume of liquid in the tank is constant at 100 ft³, and that flow in and out of the tank is constant at 10 ft³/min.

If $r_B = (- k \, \rho_B)$ ($k = 0.1$ min⁻¹) what is ρ_B as a function of time?

Solution

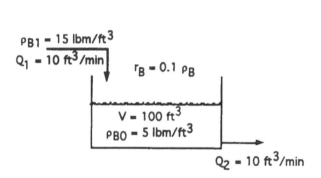

Figure 2.1.2-1 Perfectly mixed tank with reaction

Applying the macroscopic mass balance for an individual species

$$\int_A \rho_B \left(v \cdot n \right) dA + \frac{d}{dt} \int_V \rho_B \, dV = \int_V r_B \, dV \qquad (2.1.2\text{-}28)$$

Evaluating term by term

$$\int_A \rho_B \left(v \cdot n \right) dA = \left(\rho_B \right)_{out} \int_{A_{out}} \left(v \cdot n \right) dA + \left(\rho_B \right)_{in} \int_{A_{in}} \left(v \cdot n \right) dA$$

$$= \left(\rho_B \right)_{out} Q_{out} - \left(\rho_B \right)_{in} Q_{in}$$

$$= \left(\rho_B\right)_{out}\left[\frac{\text{lbmass B}}{\text{ft}^3}\right](10)\left[\frac{\text{ft}^3}{\text{min}}\right]$$

$$- (15)\left[\frac{\text{lbmass B}}{\text{ft}^3}\right](10)\left[\frac{\text{ft}^3}{\text{min}}\right]$$

$$= (10)\left[\left(\rho_B\right)_{out} - (15)\right]\frac{\text{lbmass B}}{\text{min}} \tag{2.1.2-29}$$

$$\frac{d}{dt}\int_V \rho_B\, dV = \frac{d}{dt}\left[\left(\rho_B\right)_{tank}\int_{V_{tank}} dV\right] = \frac{d}{dt}\left[\left(\rho_B\right)_{tank} V_{tank}\right]$$

$$= \left(\rho_B\right)_{tank}\frac{dV_{tank}}{dt} + V_{tank}\frac{d}{dt}\left[\left(\rho_B\right)_{tank}\right]$$

$$= V_{tank}\frac{d}{dt}\left[\left(\rho_B\right)_{tank}\right]$$

$$= (100)\,\text{ft}^3\left(\frac{d}{dt}\left[\left(\rho_B\right)_{tank}\right]\right)\frac{\text{lbm B}}{\text{ft}^3\,\text{min}}$$

$$= (100)\left(\frac{d}{dt}\left[\left(\rho_B\right)_{tank}\right]\right)\frac{\text{lbm B}}{\text{min}} \tag{2.1.2-30}$$

$$\int_V r_B\, dV = r_B\int_{V_{tank}} dV$$

$$= r_B V_{tank} = \left(-0.1\right)\left[\rho_B\right]_{tank}\frac{\text{lbm B}}{\text{ft}^3\,\text{min}}(100)\,\text{ft}^3$$

$$= \left(-0.1\right)\left[\rho_B\right]_{tank}(100)\frac{\text{lbm B}}{\text{min}} \tag{2.1.2-31}$$

Substituting in the balance

$$[10]\left[\left(\rho_B\right)_{out} - 15\right]\frac{\text{lbm B}}{\text{min}} + (100)\left(\frac{d}{dt}\left[\left(\rho_B\right)_{tank}\right]\right)\frac{\text{lbm B}}{\text{min}}$$

$$= \left(-0.1\right)\left[\rho_B\right]_{tank}(100)\frac{\text{lbm B}}{\text{min}} \tag{2.1.2-32}$$

but, since the tank is perfectly stirred

$$\left(\rho_B\right)_{out} = \left(\rho_B\right)_{tank} = \rho_B \tag{2.1.2-33}$$

which gives

$$(10)\left(\rho_B - 15\right) + (100)\left(\frac{d\rho_B}{dt}\right) = (-0.1)\left(\rho_B\right)(100)$$

$$\left(\rho_B - 15\right) + 10\frac{d\rho_B}{dt} = \rho_B$$

$$10\frac{d\rho_B}{dt} = 15 - 2\rho_B$$

$$\frac{10\,d\rho_B}{\left(15 - 2\rho_B\right)} = dt$$

$$\frac{10}{(-2)}\int_5^{\rho_B}\frac{(-2)\,d\rho_B}{\left(15 - 2\rho_B\right)} = \int_0^t dt$$

$$(-5)\left[\ln\left(15 - 2\rho_B\right)\right]_5^{\rho_B} = t$$

$$\frac{t}{(-5)} = \left[\ln\left(\frac{15 - 2\rho_B}{15 - 10}\right)\right] = \left[\ln\left(\frac{15 - 2\rho_B}{5}\right)\right]$$

$$\left(\frac{15 - 2\rho_B}{5}\right) = e^{(-0.2\,t)}$$

$$\rho_B = \frac{\left(15 - 5\,e^{(-0.2\,t)}\right)}{2} = 2.5\left(3 - e^{(-0.2\,t)}\right) \tag{2.1.2-34}$$

2.2   The Microscopic Mass Balances

The microscopic mass balance is expressed by a differential equation rather than by an equation in integrals. In contrast to the macroscopic balances, which are usually applied to determine input/output relationships, microscopic balances permit calculation of profiles (velocity, temperature, concentration) at individual points within systems.

2.2.1   The microscopic total mass balance (continuity equation)

The macroscopic total mass balance was determined (Equation 2.1.1-9) to be

$$\int_A \rho \left(\mathbf{v} \cdot \mathbf{n} \right) dA + \frac{d}{dt} \int_V \rho \, dV = 0 \qquad (2.2.1\text{-}1)$$

which can be written as

$$\int_A \left(\mathbf{n} \cdot \rho \, \mathbf{v} \right) dA + \frac{d}{dt} \int_V \rho \, dV = 0 \qquad (2.2.1\text{-}2)$$

In Chapter 1 from the Gauss theorem we showed

$$\int_V \left(\nabla \cdot \mathbf{u} \right) dV = \int_S \left(\mathbf{n} \cdot \mathbf{u} \right) dS \qquad (2.2.1\text{-}3)$$

where $\mathbf{u}$ was an arbitrary vector (we have changed the symbol from $\mathbf{v}$ to $\mathbf{u}$ to make clear the distinction between this arbitrary vector and the velocity vector).

The macroscopic mass balance was derived for a stationary control volume, and so the limits in the integrals therein are not functions of time. Application of the *Leibnitz rule* [7] therefore yields

$$\frac{d}{dt} \int_V \rho \, dV = \int_V \frac{\partial \rho}{\partial t} dV \qquad (2.2.1\text{-}4)$$

Applying these two results to the macroscopic total mass balance gives (note that multiplying the velocity vector by the scalar ρ simply gives another vector)

$$\int_V \left(\nabla \cdot \rho \, \mathbf{v} \right) dV + \int_V \frac{\partial \rho}{\partial t} dV = 0 \qquad (2.2.1\text{-}5)$$

Combining the integrands, since the range of integration of each is identical

---

[7] The Leibnitz rule is useful in interchanging differentiation and integration operations. It states that for f a continuous function with continuous derivative

$$\frac{d}{dt} \int_{a(t)}^{b(t)} f(x, t) \, dx = f[b(t), t] \, b'(t) - f[a(t), t] \, a'(t) + \int_{a(t)}^{b(t)} \frac{\partial f(x, t)}{\partial t} dx$$

Kaplan, W. (1952). *Advanced Calculus*. Reading, MA, Addison-Wesley, p. 220.

$$\int_V \left[\left(\nabla \cdot \rho \, \mathbf{v} \right) + \frac{\partial \rho}{\partial t} \right] dV = 0 \qquad\qquad (2.2.1\text{-}6)$$

Since the limits of integration are arbitrary, the only way that the left-hand side can vanish (to maintain the equality to zero) is for the integrand to be identically zero.

$$\boxed{\left(\nabla \cdot \rho \, \mathbf{v} \right) + \frac{\partial \rho}{\partial t} = 0} \qquad\qquad (2.2.1\text{-}7)$$

This is the **microscopic total mass balance** or the **equation of continuity**.

Special cases of the continuity equation[8]

For **incompressible** fluids, the time derivative vanishes, and by factoring the constant density the continuity equation may be written as

$$\rho \left(\nabla \cdot \mathbf{v} \right) = 0 \qquad\qquad (2.2.1\text{-}8)$$

but we can divide by the density since it is non-zero for cases of interest, giving

$$\boxed{\nabla \cdot \mathbf{v} = 0} \qquad\qquad (2.2.1\text{-}9)$$

For **steady flow**, the density does not vary in time, so the continuity equation becomes

$$\boxed{\nabla \cdot \rho \, \mathbf{v} = 0} \qquad\qquad (2.2.1\text{-}10)$$

These relationships often prove convenient in simplifying other equations.

---

[8] These forms, as well as being used in and of themselves, are often used to simplify other equations by eliminating the corresponding terms.

Continuity equation in different coordinate systems

For computation of numerical results, it is necessary to write the continuity equation in a particular coordinate system. Table 2.2.1-1 gives the component form in rectangular, cylindrical, and spherical coordinates.

Table 2.2.1-1 Continuity equation (microscopic total mass balance) in rectangular, cylindrical, and spherical coordinate frames

RECTANGULAR COORDINATES

$$\frac{\partial \rho}{\partial t} + \frac{\partial}{\partial x}\left(\rho\, v_x\right) + \frac{\partial}{\partial y}\left(\rho\, v_y\right) + \frac{\partial}{\partial z}\left(\rho\, v_z\right) = 0$$

CYLINDRICAL COORDINATES

$$\frac{\partial \rho}{\partial t} + \frac{1}{r}\frac{\partial}{\partial r}\left(\rho\, r\, v_r\right) + \frac{1}{r}\frac{\partial}{\partial \theta}\left(\rho\, v_\theta\right) + \frac{\partial}{\partial z}\left(\rho\, v_z\right) = 0$$

SPHERICAL COORDINATES

$$\frac{\partial \rho}{\partial t} + \frac{1}{r^2}\frac{\partial}{\partial r}\left(\rho\, r^2\, v_r\right) + \frac{1}{r\sin\theta}\frac{\partial}{\partial \theta}\left(\rho\, v_\theta \sin\theta\right) + \frac{1}{r\sin\theta}\frac{\partial}{\partial \varphi}\left(\rho\, v_\varphi\right) = 0$$

Example 2.2.1-1   Velocity components in two-dimensional steady incompressible flow, rectangular coordinates

The x-component of velocity in a two-dimensional steady incompressible flow is given by

$$v_x = c\, x \tag{2.2.1-11}$$

where c is a constant.

What is the form of the y-component of velocity?

Solution

The continuity equation reduces to

$$\nabla \cdot \mathbf{v} = 0 \tag{2.2.1-12}$$

which is, for a two-dimensional flow

$$\frac{\partial v_x}{\partial x} + \frac{\partial v_y}{\partial y} = 0 \tag{2.2.1-13}$$

$$\frac{\partial v_x}{\partial x} = -\frac{\partial v_y}{\partial y} \tag{2.2.1-14}$$

But from the given x-component

$$\frac{\partial v_x}{\partial x} = \frac{\partial (c\,x)}{\partial x} = c = -\frac{\partial v_y}{\partial y} \tag{2.2.1-15}$$

Integrating with respect to y

$$\int \left(\frac{\partial v_y}{\partial y}\right) dy = -\int c\,dy = -c\,y + f(x) = v_y \tag{2.2.1-16}$$

Notice that since we are integrating a **partial** derivative of a function of x and y with respect to y, we can determine the result of such an integration only to within a function of x rather than a constant, because any such function disappears if we take the partial derivative with respect to y. Consequently, the result shown would, as is required, satisfy the inverse operation

$$-\left(\frac{\partial v_y}{\partial y}\right)_x = -\left(\frac{\partial \left[-c\,y + f(x)\right]}{\partial y}\right)_x = c \tag{2.2.1-17}$$

We have added the subscript x to the partial derivative to emphasize that x is being held constant during the operation. We should write such a subscript in conjunction with all partial derivatives; however, the subscripts are usually obvious so we customarily omit them for conciseness.

Therefore, the most general form is

$$v_y = -cy + f(x)$$

(2.2.1-18)

This form includes, of course

$$f(x) = \text{constant}$$

(2.2.1-19)

and

$$f(x) = 0$$

(2.2.1-20)

Example 2.2.1-2   Velocity components in two-dimensional steady incompressible flow, cylindrical coordinates

Water is in steady one-dimensional axial flow in a uniform-diameter cylindrical pipe. The velocity is a function of the radial coordinate only. What is the form of the radial velocity profile?

Solution

The continuity equation in cylindrical coordinates is

$$\frac{\partial \rho}{\partial t} + \frac{1}{r}\frac{\partial}{\partial r}\left(\rho \, r \, v_r\right) + \frac{1}{r}\frac{\partial}{\partial \theta}\left(\rho \, v_\theta\right) + \frac{\partial}{\partial z}\left(\rho \, v_z\right) = 0$$

(2.2.1-21)

which, for steady flow reduces to

$$\frac{1}{r}\frac{\partial}{\partial r}\left(\rho \, r \, v_r\right) + \frac{1}{r}\frac{\partial}{\partial \theta}\left(\rho \, v_\theta\right) + \frac{\partial}{\partial z}\left(\rho \, v_z\right) = 0$$

(2.2.1-22)

which then becomes, for velocity, a function of r only

$$\frac{1}{r}\frac{d}{dr}\left(\rho \, r \, v_r\right) = 0$$

(2.2.1-23)

where we have replaced the partial derivative symbols with ordinary derivatives because we now have an unknown function of only one independent variable. This then becomes, for an incompressible fluid

$$\rho \, \frac{1}{r}\frac{d}{dr}\left(r \, v_r\right) = 0$$

(2.2.1-24)

We multiply both sides by the radius, and, since the density is not equal to zero, we can divide both sides by the density

$$\frac{d}{dr}(r\,v_r) = 0 \qquad (2.2.1\text{-}25)$$

Integrating

$$\int \frac{d}{dr}(r\,v_r)\,dr = \int 0\,dr \qquad (2.2.1\text{-}26)$$

$$r\,v_r = \text{constant}$$

which gives as the form of the radial velocity profile

$$v_r = \frac{\text{constant}}{r} \qquad (2.2.1\text{-}27)$$

Example 2.2.1-3   Compression of air

Air is slowly compressed in a cylinder by a piston. The motion of the air is one-dimensional. Since the motion is slow, the density remains constant in the z-direction even though changing in time. As the piston moves toward the cylinder head at speed v, what is the time rate of change of the density of the air?

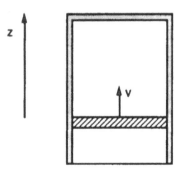

Figure 2.2.1-1 Air compression by piston

Solution

We begin with the continuity equation

$$\left(\nabla \cdot \rho \, \mathbf{v}\right) + \frac{\partial \rho}{\partial t} = 0 \tag{2.2.1-28}$$

In cylindrical coordinates

$$\frac{\partial \rho}{\partial t} + \frac{1}{r}\frac{\partial}{\partial r}\left(\rho \, r \, v_r\right) + \frac{1}{r}\frac{\partial}{\partial \theta}\left(\rho \, v_\theta\right) + \frac{\partial}{\partial z}\left(\rho \, v_z\right) = 0 \tag{2.2.1-29}$$

In one dimension (the z-direction)

$$\frac{\partial \rho}{\partial t} + \frac{\partial}{\partial z}\left(\rho \, v_z\right) = 0 \tag{2.2.1-30}$$

Expanding the derivative of the product

$$\frac{\partial \rho}{\partial t} + \rho \frac{\partial v_z}{\partial z} + v_z \frac{\partial \rho}{\partial z} = 0 \tag{2.2.1-31}$$

However, density is constant in the z-direction, so

$$\frac{\partial \rho}{\partial t} + \rho \frac{\partial v_z}{\partial z} = 0 \tag{2.2.1-32}$$

2.2.2 The microscopic species mass balance

We can develop a microscopic mass balance for an individual species in the same general way that we developed the microscopic total mass balance, but, as in the macroscopic case, individual species are not conserved but may be created or destroyed (e.g., by chemical reactions), so a generation term must be included in the model.

The **macroscopic** mass balance for an individual species took the form

$$\int_A \rho_i \left(\mathbf{v} \cdot \mathbf{n}\right) dA + \frac{d}{dt}\int_V \rho_i \, dV = \int_V r_i \, dV \tag{2.2.2-1}$$

Models based on the above equation (which are adequate for many real problems) are restricted to cases where the velocity of the bulk fluid, $\mathbf{v}$, and the velocity of the individual species, v_i, are assumed to be identical.

As was just noted, individual species may not move at the same velocity as the bulk fluid, for example, because of concentration gradients in the fluid. This becomes of much more a concern in certain classes of problems we wish to model at the microscopic level than it was at the macroscopic level.

We begin with the more general equation, which includes the species velocity rather than the bulk fluid velocity

$$\int_A \rho_i \left(v_i \cdot n \right) dA + \frac{d}{dt} \int_V \rho_i \, dV = \int_V r_i \, dV \qquad (2.2.2\text{-}2)$$

By incorporating the species velocity v_i instead of the velocity of the bulk fluid v we can model cases where v_i and v are significantly different. This is the whole point of mass transfer operations.

The Leibnitz rule and the Gauss theorem can be applied as in the case of the microscopic total mass balance, giving

$$\int_V \left(\nabla \cdot \rho_i v_i + \frac{\partial \rho_i}{\partial t} - r_i \right) dV = 0 \qquad (2.2.2\text{-}3)$$

Again the limits of integration are arbitrary, so it follows that

$$\boxed{\nabla \cdot \rho_i v_i + \frac{\partial \rho_i}{\partial t} = r_i} \qquad (2.2.2\text{-}4)$$

which is the *microscopic mass balance for an individual species*. There is one such equation for each species. In a mixture of n constituents, the sum of all n of the species equations must yield the total mass balance, and therefore only n of the n+1 equations are independent.

The microscopic mass balance in the form above is not particularly useful for two reasons:

> • First, it does not relate the concentration to the properties of
> the fluid. This is done via a constitutive equation (such an
> equation relates the flow variables to the way the fluid is
> constituted, or made up).

• Second, it is written in terms of mass concentration, while most mass transfer process operate via molar concentrations (diffusive processes tend to be proportional to the relative number of molecules, not the relative amount of mass).

Diffusion

An example of a constitutive equation relating fluxes and concentrations is Fick's law of diffusion

$$N_A - x_A \left[N_A + N_B \right] = -c \, \mathcal{D}_{AB} \, \nabla x_A \qquad (2.2.2\text{-}5)$$

where N_i denotes the molar flux of the ith species with respect to axes fixed in space, x_i is the mole fraction of species i, c is the total molar concentration, and $\mathcal{D}_{AB}$ is the diffusion coefficient.

The molar flux can be written in terms of the velocity of the ith species, v_i, which is defined as the *number average velocity*

$$N_i = c \, v_i \qquad (2.2.2\text{-}6)$$

It is not difficult to see that substituting this relationship in Fick's law to obtain velocity as a function of concentration and then putting the result in the microscopic species mass balances (adjusting the units appropriately) is not a trivial task. We will return to this later in Section 8.2.

Chapter 2   Problems

2.1 For the following situation:
      a. Calculate $(v \cdot n)$
      b. Calculate v_x
      c. Calculate w

Use coordinates shown.

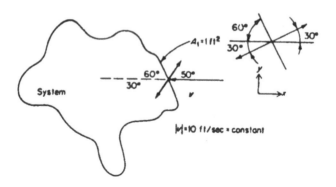

$$|v| = 10 \text{ ft/sec} = \text{constant}$$

2.2 For the following

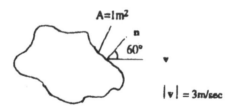

$A = 1 \text{ m}^2$

$|v| = 3 \text{m/sec}$

a. Calculate $(v \cdot n)$
b. Calculate w $(\rho = 999 \text{ kg/m}^3)$
c. Calculate $\int_A \rho (v \cdot n) \, dA$ if $|v| = \left(1 - \frac{r}{R}\right) v_{max}$

   where $v_{max} = 4.5$ m/s.

2.3 Calculate $\int_A \rho (v \cdot n) \, dA$ for the following situation

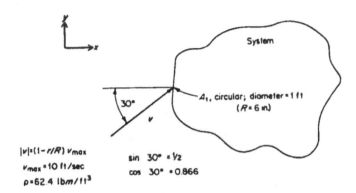

$|v| = (1 - r/R) \, v_{max}$

$v_{max} = 10 \, ft/sec$

$\rho = 62.4 \, lbm/ft^3$

sin $30° = \frac{1}{2}$

cos $30° = 0.866$

Use coordinates shown.

2.4 A pipe system sketched below is carrying water through section 1 at a velocity of 3 ft/s. The diameter at section 1 is 2 ft. The same flow passes through section 2 where the diameter is 3 ft. Find the discharge and bulk velocity at section 2.

$3 \, ft/sec \, v_1 \longrightarrow$

2.5 Consider the steady flow of water at 70°F through the piping system shown below where the velocity distribution at station 1 may be expressed by

$$v(r) = 9.0 \left(1 - \frac{r^2}{R_1^2} \right) \qquad ft/hr$$

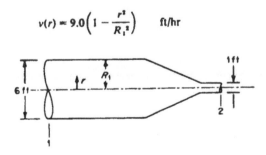

Find the average velocity at 2.

2.6 Water is flowing in the reducing elbow as shown. What is the velocity and mass flow rate at 2?

$$\rho = 62.4 \frac{lbm}{ft^3}$$

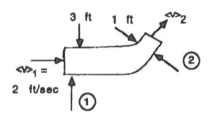

2.7 The tee shown in the sketch is rectangular and has 3 exit-entrances. The width is w. The height h is shown for each exit-entrance on the sketch. You may assume steady, uniform, incompressible flow.

     a. Write the total macroscopic mass balance and use the
        assumptions to simplify the balance for this situation.
     b. Is flow into or out of the bottom arm (2) of the tee.
     c. What is the magnitude of the velocity into or out of the
        bottom arm (2) of the tee?

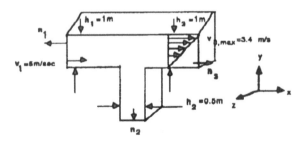

2.8 The velocity profile for turbulent flow of a fluid in a circular smooth tube follows a power law such that

$$|v| = \left(1 - \frac{r}{R}\right)^{\frac{1}{7}} v_{max}$$

Find the value of $<v>/v_{max}$ for this situation. If $n = 7$ this is the result from the Blasius resistance formula.

(Hint: let $y = R - r$)

2.9 Oil of specific gravity 0.8 is pumped to a large open tank with a 1.0-ft hole in the bottom. Find the maximum steady flow rate Q_{in}, ft^3/s, which can be pumped to the tank without its overflowing.

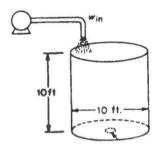

2.10 Oil at a specific gravity of 0.75 is pumped into an open tank with diameter 3 m and height 3 m. Oil flows out of a pipe at the bottom of the tank that has a diameter of 0.2 m. What is the volumetric flow rate that can be pumped in the tank such that it will not overflow?

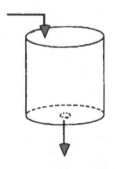

2.11 In the previous problem assume water is pumped in the tank at a rate of 3 m^3/min. If the water leaves the tank at a rate dependent on the liquid level such that $Q_{out} = 0.3h$ (where 0.3 is in m^2/min and Q is in m^3/min, h is in m. What is the height of liquid in the tank as a function of time?

2.12 Shown is a perfectly mixed tank containing 100 ft^3 of an aqueous solution of 0.1 salt by weight. At time $t = 0$, pumps 1 and 2 are turned on. Pump 1 is pumping pure water into the tank at a rate of 10 ft³/hr and pump 2 pumps solution from the tank at 15 ft³/hr. Calculate the concentration in the outlet line as an explicit function of time.

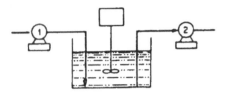

2.13 Water is flowing into an open tank at the rate of 50 ft³/min. There is an opening at the bottom where water flows out at a rate proportional to

$$w = 0.1 \, \rho \, \sqrt{2 \, g \, h} \, \frac{lbm}{s}$$

where h = height of the liquid above the bottom of the tank. Set up the equation for height versus time; separate the variables and integrate. The area of the bottom of the tank is 100 ft².

2.14 Salt solution is flowing into and out of a stirred tank at the rate of 5 gal solution/min. The input solution contains 2 lbm salt/gal solution. The volume of the tank is 100 gal. If the initial salt concentration in the tank was 1 lbm salt/gal, what is the outlet concentration as a function of time?

2.15 A batch of fuel-cell electrolyte is to be made by mixing two streams in large stirred tank. Stream 1 contains a water solution of H_2SO_4 and K_2SO_4 in mass fractions x_A and x_{B1} lbm/lbm of solution, respectively (A = H_2SO_4, B = K_2SO_4). Stream 2 is a solution of H_2SO_4 in water having a mass fraction x_{A2}.

If the tank is initially empty, the drain is closed, and the streams are fed at steady rates w_1 and w_2 lbm of solution/hr, show that x_A is independent of time and find the total mass m at any time.

(Hint: Do not expand the derivative of the product in the species balance, but integrate directly.)

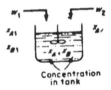

Concentration
in tank

2.16 You have been asked to design a mixing tank for salt solutions. The tank volume is 126 gal and a salt solution flows into and out of the tank at 7 gal of solution/min. The input salt concentration contains (3 lbm salt)/gal of solution. If the initial salt concentration in the tank is (1/2 lbm salt)/gal, find:

  a. The output concentration of salt as a function of time assuming that the output concentration is the same as that within the tank.
  b. Plot the output salt concentration as a function of time and find the asymptotic value of the concentration $m(t)$ as $t \to \infty$.
  c. Find the response time of the stirring tank, i.e., the time at which the output concentration $m(t)$ will equal 62.3 percent of the change in the concentration from time $t = 0$ to time $t \to \infty$.

3

THE ENERGY BALANCES

3.1 The Macroscopic Energy Balances

The entity balance can be applied to forms of energy to obtain, for example, macroscopic energy balances. **Total** energy, like **total** mass, can be modeled as a conserved quantity for processes of interest here, which do not involve interconversion of energy and mass by either nuclear reactions or mass moving at speeds approaching that of light. On the other hand, neither **thermal** energy nor **mechanical** energy is conserved in general, but each can be created or destroyed by conversion to the opposite form - thermal to mechanical (for example, via combustion processes in engines), or, more frequently, mechanical to thermal (through the action of friction and other irreversible processes).

3.1.1   Forms of energy

Just as we did for mass balances, to develop the macroscopic energy balances we consider an arbitrary system fixed in space. Since mass (because of its position, motion or physical state) carries with it associated energy, each of the macroscopic energy balances will therefore contain a term which corresponds to each term in the corresponding macroscopic mass balance.

Energy associated with mass can be grouped within three classifications.

1. Energy present because of position of the mass in a field (e.g., gravitational, magnetic, electrostatic). This energy is called potential energy, Φ.

2. Energy present because of translational or rotational motion of the mass. This energy is called kinetic energy, K.

3. All other energy associated with mass (for example, rotational and vibrational energy in chemical bonds, energy of

Brownian motion[1], etc.). This energy is called internal energy, U.

In general, U, Φ, and K are functions of position and time.

In addition, however, energy is also transported across the boundaries of a system in forms **not associated with mass**. This fact introduces terms in the macroscopic (and microscopic) energy balances that have no counterpart in the overall mass balances.

Although energy **not** associated with any mass may cross the boundary of the system, **within** the system we will consider only energy that is associated with mass (we will not, for example, attempt to model systems within which a large part of the energy is in the form of free photons).

As we did with mass, we assume that we continue to deal with *continua*, that is, that energy, like mass, is "smeared out" locally. A consequence of this is that we group the energy of random molecular translation into internal energy, not kinetic energy, even though at a microscopic level it represents kinetic energy. The continuum assumption implies that we cannot see the level of detail required to perceive kinetic energy of molecular motion, but rather simply see a local region of space containing internal energy.

We shall denote the amount of a quantity associated with a unit mass of material by placing a caret, (^) above the quantity. Thus, for example, $\hat{U}$ is the internal energy per unit mass (F L/M) and $\hat{W}$ is the rate of doing work per unit mass (F L/[M t]).

3.1.2   The macroscopic total energy balance

We now apply the entity balance, written in terms of rates, term by term to **total energy**. Since total energy will be conserved in processes we consider, the generation term will be zero.

---

[1] Even though Brownian motion represents kinetic energy of the molecules, this kinetic energy is below the scale of our continuum assumption. It is furthermore, random and so not recoverable as work (at least, until someone discovers a real-life Maxwell demon).

$$\left[\begin{array}{c} \text{rate of output of} \\ \text{total energy} \end{array}\right] - \left[\begin{array}{c} \text{rate of input of} \\ \text{total energy} \end{array}\right]$$
$$+ \left[\begin{array}{c} \text{rate of accumulation of} \\ \text{total energy} \end{array}\right] = 0 \qquad (3.1.2\text{-}1)$$

Rate of accumulation of energy

The energy in ΔV is approximately[2]

$$\left(\hat{U} + \hat{\Phi} + \hat{K}\right) \rho \, \Delta V \qquad (3.1.2\text{-}2)$$

where $\hat{U}$, $\hat{\Phi}$, $\hat{K}$, and ρ are each evaluated at arbitrary points in ΔV and Δt.

The total energy in V can be obtained by summing over all of the elemental volumes in V and taking the limit as ΔV approaches zero

$$\left[\begin{array}{c} \text{total energy} \\ \text{in} \\ V \end{array}\right] \equiv \sum_{V} \left[\left(\hat{U} + \hat{\Phi} + \hat{K}\right) \rho \, \Delta V\right] \qquad (3.1.2\text{-}3)$$

$$= \lim_{\Delta V \to 0} \left\{ \sum_{V} \left[\left(\hat{U} + \hat{\Phi} + \hat{K}\right) \rho \, \Delta V\right] \right\} \qquad (3.1.2\text{-}4)$$

$$= \int_{V} \left(\hat{U} + \hat{\Phi} + \hat{K}\right) \rho \, dV \qquad (3.1.2\text{-}5)$$

Applying the definition of accumulation yields

---

[2] This is a model which neglects systems that contain energy not associated with mass that changes significantly with time, as well as neglecting any conversion of mass into energy or vice versa, as we have stipulated earlier - it would not be suitable, for example, in making an energy balance on the fireball of a nuclear explosion where both large, changing amounts of radiation are involved as well as processes that convert mass into energy.

$$\begin{bmatrix} \text{accumulation} \\ \text{of energy} \\ \text{in V} \\ \text{during } \Delta t \end{bmatrix} = \int_V \left(\hat{U} + \hat{\Phi} + \hat{K} \right) \rho \, dV \bigg|_{t+\Delta t}$$

$$- \int_V \left(\hat{U} + \hat{\Phi} + \hat{K} \right) \rho \, dV \bigg|_t \qquad (3.1.2\text{-}6)$$

$$[FL] \Leftrightarrow \left[\frac{FL}{M} \right] \left[\frac{M}{L^3} \right] \left[L^3 \right]$$

We then obtain the energy accumulation **rate** by dividing by Δt and taking the limit as Δt approaches zero

$$\begin{bmatrix} \text{energy} \\ \text{accumulation} \\ \text{rate} \end{bmatrix} = \lim_{\Delta t \to 0} \left[\frac{\int_V \left(\hat{U} + \hat{\Phi} + \hat{K} \right) \rho \, dV \big|_{t+\Delta t} - \int_V \left(\hat{U} + \hat{\Phi} + \hat{K} \right) \rho \, dV \big|_t}{\Delta t} \right]$$

$$\boxed{\begin{bmatrix} \text{energy} \\ \text{accumulation} \\ \text{rate} \end{bmatrix} = \frac{d}{dt} \int_V \left(\hat{U} + \hat{\Phi} + \hat{K} \right) \rho \, dV} \qquad (3.1.2\text{-}7)$$

$$\left[\frac{FL}{t} \right] \Leftrightarrow \left[\frac{FL}{M} \right] \left[\frac{M}{L^3} \right] \left[L^3 \right] \left[\frac{1}{t} \right]$$

Rates of input and output of energy

We will first consider input and output of energy **during addition or removal of mass**. As each increment of mass is added to or removed from the system, energy also crosses the boundary in the form of

- kinetic, potential and internal energy **associated** with the mass

- energy transferred in the **process** of adding or removing the mass.

Consider for the moment the system shown in Figure 3.1.2-1. Illustrated is the addition to a system of material which is contained in a closed rigid container (a beverage can, if you wish).

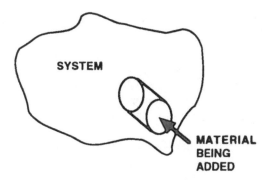

Figure 3.1.2-1 Flow work

Note that to move the additional mass inside the system we must **push** - that is, we must compress some of the material already in the system to make room for the added material (thereby doing work on the mass already in the system and increasing its energy).

In the typical process shown in Figure 3.1.2-1, we do not change the energy associated with the **added** mass, but we cannot either add or remove mass without performing this *flow work* on the mass already in the system.

In general, the flow work necessary to add a unit mass to a system is the pressure at the point where the mass crosses the boundary multiplied by the volume added

$$\left[\begin{array}{c} \text{flow work} \\ \text{per} \\ \text{unit mass} \end{array} \right] = p\,\hat{V} \qquad (3.1.2\text{-}8)$$

Therefore, the total energy associated with transferring a unit mass into a system is the sum of the internal, potential, and kinetic energy associated with the mass plus the flow work (whether or not the material is contained in a rigid container)

$$\left(\hat{U} + \hat{\Phi} + \hat{K} \right) + p\,\hat{V} \qquad (3.1.2\text{-}9)$$

Thermodynamics teaches that it is convenient to combine the flow work with the energy associated with the mass added or removed by defining the abstraction *enthalpy, H.*

$$\hat{H} = \hat{U} + p\,\hat{V} \tag{3.1.2-10}$$

Notice that the enthalpy combines some energy **associated** with mass with some energy **not associated** with mass (the flow work). Definition of enthalpy is not made for any compelling physical reason but rather for convenience - the two terms always occur simultaneously and so it is convenient to group them under the same symbol.

This makes the energy associated with transferring a unit mass into the system

$$\left(\hat{H} + \hat{\Phi} + \hat{K}\right) \tag{3.1.2-11}$$

Multiplying the energy associated with transferring a unit mass into the system by the mass flow rate gives the net rate (output minus input) of energy transfer

- associated with mass and

- from flow work

$$\begin{vmatrix} \text{energy} \\ \text{input and output} \\ \text{rate} \end{vmatrix} = \int_A \left(\hat{H} + \hat{\Phi} + \hat{K}\right) \rho\,(\mathbf{v} \cdot \mathbf{n})\,dA \tag{3.1.2-12}$$

Now consider the remainder of the energy that crosses the boundary **not associated with mass**. We arbitrarily (but for sound reasons) divide this energy into two classes - *heat* and *work*.

By **heat** we mean the amount of energy crossing the boundary

- not associated with mass

- which flows as a result of a temperature gradient.

By **work** we mean energy crossing the boundary

- not associated with mass

   • which does **not** transfer as a result of the temperature gradient **except for** flow work.

Since this is definition by exclusion, or by using a complementary set, what we classify as work needs some clarification.

We normally think of work as a force acting through a distance - for example, as transmitted by a rotating shaft passing through the boundary, so-called *shaft work*. Note that even though such a shaft has mass, its mass is not **crossing** the boundary - each part of the mass remains either **within** or **outside** the system - so the energy is not transmitted by being **associated with mass crossing the boundary.**

The definition of work also leads to some things being classified as work which we might not usually think of as work - for example, energy transferred across the boundary via electrical leads. Even though electrical current is associated with electrons, which can be assigned a minuscule mass (although they also have a wave nature), at scales of interest to us our models do not "see" electrons. Therefore, such energy is lumped into the work classification by default because it is not flow work, is not associated with mass, and does not flow as a result of a temperature gradient.

By convention, **heat into** the system and **work out of** the system have traditionally been regarded as **positive** (probably because the originators of the convention worked with steam power plants where one puts heat in and gets work out - a good memory device by which to remember the convention).[3]

We will denote the **rate** of transfer of heat or work across a boundary as $\dot{Q}$ [4] or $\dot{W}$ (units: [F L/t]). If we wish to refer only to the **amount** of heat or work (units: [F L]), we will use Q or W.

From thermodynamics we know that heat and work are not exact differentials, i.e., they are not independent of the path, although their

---

[3] There is currently movement away from this convention to one in which both heat and work entering the system are regarded as positive. In this text we adhere to the older convention.

[4] We use $\dot{Q}$ both for volumetric flow rate and heat transfer. The distinction is usually clear from the context; however, it would be clearer to use a different symbol. We are faced here with the two persistent problems with engineering notation: a) traditional use, and b) the scarcity of symbols that have not already been appropriated. Since heat is almost universally designated by the letter Q or q, we will bow to tradition and not attempt to find another symbol.

difference (the internal energy - from first law) is independent of path. Therefore, we usually write small amounts of heat or work as δQ or δW rather than dQ or dW to remind ourselves of this fact.

If we add the rates that heat and work cross the boundary for each differential area on the surface of the system we obtain

$$\begin{bmatrix} \text{energy input rate} \\ \text{from heat in} \\ \text{across } \Delta A \end{bmatrix} = \delta Q$$

$$\begin{bmatrix} \text{energy output rate} \\ \text{from work out} \\ \text{across } \Delta A \end{bmatrix} = \delta W \tag{3.1.2-13}$$

with units in each case of

$$\left[\frac{F L}{t} \right] \tag{3.1.2-14}$$

Adding the contributions of each elemental area and letting the elemental area approach zero we have

$$\begin{bmatrix} \text{energy input rate} \\ \text{from heat in} \\ \text{across } A \end{bmatrix} = \lim_{\Delta A \to 0} \sum_A \delta Q = Q$$

$$\begin{bmatrix} \text{energy output rate} \\ \text{from work out} \\ \text{across } A \end{bmatrix} = \lim_{\Delta A \to 0} \sum_A \delta W = W \tag{3.1.2-15}$$

Substituting in the entity balance both the terms for energy associated with mass and not associated with mass, remembering our sign convention for heat and work, and moving the heat and work terms to the right-hand-side of the equation

$$\int_A \left(\hat{H} + \hat{\Phi} + \hat{K} \right) \rho \left(\mathbf{v} \cdot \mathbf{n} \right) dA$$
$$+ \frac{d}{dt} \int_V \left(\hat{U} + \hat{\Phi} + \hat{K} \right) \rho \, dV = \dot{Q} - \dot{W}$$

(3.1.2-16)

which is the *macroscopic total energy balance.* Be sure to remember that the **enthalpy** is associated only with the input/output term, and the **internal energy** with the accumulation term.

Simplified forms of the macroscopic total energy balance

The macroscopic total energy balance often can be simplified for particular situations. Many of these simplifications are so common that it is worth examining them in some detail.

For example, we know from calculus that we always can rewrite the first term as

$$\int_A \hat{H} \rho \left(\mathbf{v} \cdot \mathbf{n} \right) dA + \int_A \hat{\Phi} \rho \left(\mathbf{v} \cdot \mathbf{n} \right) dA + \int_A \hat{K} \rho \left(\mathbf{v} \cdot \mathbf{n} \right) dA \qquad \text{(3.1.2-17)}$$

We now proceed to examine ways of simplifying each of the above terms, initially examining the second term, then the third term, and then the first term.

The potential energy term

We observe that the imposed force fields of most interest to us as engineers are **conservative**; that is, the force may be expressed as the gradient of a *potential function,* defined up to an additive constant. This function gives the value of **the potential energy per unit mass** referred to some datum point (the location of which determines the value of the additive constant).

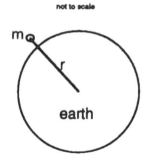

Figure 3.1.2-2 Gravitational field of earth

For example, in the case illustrated in Figure 3.1.2-2, we show a mass, m, in the gravitational potential field, Φ, of the earth, with reference point of zero energy at the center of the earth. In spherical coordinates the potential function Φ may be written as[5]

$$\Phi = \frac{G\,m\,m_e}{r}$$

$$[F][L] \Leftrightarrow \frac{\frac{[F][L]^2}{[M]^2}[M][M]}{[L]}$$

(3.1.2-18)

where typically

Φ = potential energy, $[N][m]$

G = a universal constant which has
       the same value for any pair of particles

   $= -6.6726 \times 10^{-11} \left[\dfrac{N\,m^2}{kg^2}\right]$

m = mass of object, $[kg]$

m_e = mass of earth, $[kg]$

   $= 5.97 \times 10^{24} [kg]$

───────────────

[5] Resnick, R. and D. Halliday (1977). *Physics: Part One.* New York, NY, John Wiley and Sons, p. 338 ff.

r = distance from the center of gravity of the earth
to center of gravity of object $[m]$

This relation is valid only outside the earth's surface.

The force produced by such a field is the **negative** of the **gradient** of the potential, and in this case lies in the r-direction because the gradient of the potential is zero in the $\theta-$ and $\phi-$directions (derivatives with respect to these two coordinate directions are zero).

$$\left(\mathbf{F} \right)_r = -\nabla \Phi \cdot \frac{\mathbf{r}}{|\mathbf{r}|} = -\nabla_r \Phi = -\frac{\partial \Phi}{\partial r} = \frac{G m m_e}{r^2} = m g \qquad (3.1.2\text{-}19)$$

$(G m_e/r^2)$ is commonly called g, the acceleration of gravity $[L]/\{t\}^2$.

This yields the expected result; that is, the force is inversely proportional to the square of the distance from the center of the earth and directed toward the center. At sea level g is approximately 32.2 ft/sec^2 or 9.81 m/s^2.

We can now rewrite our potential function as

$$\Phi = \frac{G m m_e}{r} = \frac{G m_e}{r^2} m r = m g r$$
$$\hat{\Phi} = \frac{\Phi}{m} = \frac{m g r}{m} = g r \qquad (3.1.2\text{-}20)$$

In practice, most processes take place over small **changes** in r, at large **values** of r, that is, at the surface of the earth, where $(r_2 - r_1)$ is of perhaps the order of 100 meters, while r_2 and r_1 themselves are of the order of 6.37 x 10^6 meters (the radius of the earth). Under such a circumstance the acceleration of gravity $(G m_e/r^2)$ is nearly constant, as we now demonstrate.

Examining the value of $(G m_e/r^2)$ for such a case we find that

$$\frac{G m_e}{r^2} = \frac{\left(-6.6726 \times 10^{-11} \right) \frac{N\,m^2}{kg^2} \left(5.97 \times 10^{24} \right) kg}{\left(6.37 \times 10^6 \right)^2 m^2} = -9.81727 \frac{N}{kg}$$

$$\frac{G\,m_e}{r^2} = \frac{\left(-6.6726\times10^{-11}\right)\frac{N\,m^2}{kg^2}\left(5.97\times10^{24}\right)kg}{\left(6.37\times10^6+100\right)^2 m^2} = -9.81696\frac{N}{kg}$$

$$(3.1.2\text{-}21)$$

(Note that N/kg = (kg m/s^2)/kg = m/s^2.) The difference does not appear until the fifth significant digit.[6] Consequently we usually assume in our models that $(G\,m_e/r^2) = g$ is constant (such problems as placing items into orbit from the earth surface excepted, obviously).

Assumption of constant acceleration of gravity reduces the potential energy term to

$$\int_A \Phi\rho\left(v\cdot n\right)dA = \int_A g\,r\,\rho\left(v\cdot n\right)dA = g\int_A r\,\rho\left(v\cdot n\right)dA \quad (3.1.2\text{-}22)$$

We usually use one of the rectangular coordinates rather than the spherical coordinate r, since the geometry at the earth's surface can be adequately treated in rectangular Cartesian coordinates - for example

$$\boxed{\int_A \Phi\rho\left(v\cdot n\right)dA = g\int_A z\,\rho\left(v\cdot n\right)dA}$$

$$(3.1.2\text{-}23)$$

Consider now the case of a system with a **single inlet denoted by 1 and a single outlet denoted by 2**, with the z-direction being parallel to and opposite in direction to the gravity vector. Assume that **change** in potential energy is negligible across the cross-section of the inlet and across the cross-section of the outlet, **although potential energy is, in general, different at the two locations** (for example, a four-inch pipe inlet 100 feet vertically below a six-inch pipe outlet).

$$\int_A \Phi\rho\left(v\cdot n\right)dA = g\int_A z\,\rho\left(v\cdot n\right)dA$$

$$= g\,z_1\int_{A_1}\rho\left(v\cdot n\right)dA + g\,z_2\int_{A_2}\rho\left(v\cdot n\right)dA$$

---

[6] We obtained 9.82 m/s^2 rather than 9.81 m/s^2 for our value because of the approximate nature of the numbers we used. This did not matter for the illustration because we sought differences, not absolute values.

$$\boxed{\int_A \Phi \rho \left(v \cdot n \right) dA = g\, z_2\, w_2 - g\, z_1\, w_1} \tag{3.1.2-24}$$

If **density is constant across each cross-section individually**

$$\int_A \Phi \rho \left(v \cdot n \right) dA = g \int_A z \rho \left(v \cdot n \right) dA \tag{3.1.2-25}$$

$$= g \left[z_1 \rho_1 \int_{A_1} \left(v \cdot n \right) dA + z_2 \rho_2 \int_{A_2} \left(v \cdot n \right) dA \right]$$

$$\boxed{\int_A \Phi \rho \left(v \cdot n \right) dA = g \left[z_2 \rho_2 \langle v \rangle_2 A_2 - z_1 \rho_1 \langle v \rangle_1 A_1 \right]} \tag{3.1.2-26}$$

If **density in addition is the same at 1 and 2**

$$\int_A \Phi \rho \left(v \cdot n \right) dA = g \rho \left[z_2 \langle v \rangle_2 A_2 - z_1 \langle v \rangle_1 A_1 \right] \tag{3.1.2-27}$$

$$\boxed{\int_A \Phi \rho \left(v \cdot n \right) dA = g \rho \left[z_2 Q_2 - z_1 Q_1 \right]} \tag{3.1.2-28}$$

If we further assume **steady-state** conditions, so that $w_1 = w_2 = w$; $Q_1 = Q_2 = Q$

$$\boxed{\begin{aligned} \int_A \Phi \rho \left(v \cdot n \right) dA &= g \rho\, Q \left[z_2 - z_1 \right] \\ &= g\, w \left[z_2 - z_1 \right] \end{aligned}} \tag{3.1.2-29}$$

The kinetic energy term

From physics we know that the kinetic energy of a unit mass is expressible as

$$\hat{K} = \frac{\left| v \right|^2}{2} = \frac{v^2}{2} \tag{3.1.2-30}$$

We therefore write

$$\int_A \hat{K} \rho \left(\mathbf{v} \cdot \mathbf{n} \right) dA = \int_A \frac{v^2}{2} \rho \left(\mathbf{v} \cdot \mathbf{n} \right) dA \qquad (3.1.2\text{-}31)$$

For a fluid where both ρ and the angle between the velocity vector and the outward normal are constant across A we may write

$$
\begin{aligned}
\int_A \hat{K} \rho \left(\mathbf{v} \cdot \mathbf{n} \right) dA &= \rho \int_A \frac{v^2}{2} \left(v \cos \alpha \right) dA \\
&= \frac{\rho \cos \alpha}{2} \int_A v^3 \, dA \\
&= \frac{\rho \cos \alpha}{2} \langle v^3 \rangle A
\end{aligned}
\qquad (3.1.2\text{-}32)
$$

(By definition the integral is the area-average of the cube of the velocity.[7])

We frequently assume that

$$\langle v^3 \rangle = \langle v \rangle^2 \langle v \rangle \qquad (3.1.2\text{-}33)$$

and write, for any given area A_i, also assuming that α and ρ are constant across A_i

$$\int_{A_i} \hat{K} \rho \left(\mathbf{v} \cdot \mathbf{n} \right) dA = \cos \alpha_i \frac{\langle v_i \rangle^2}{2} \rho_i \langle v_i \rangle A_i = \cos \alpha_i \frac{\langle v_i \rangle^2}{2} w_i$$

$$(3.1.2\text{-}34)$$

which requires the assumption that

$$\langle v^3 \rangle = \frac{\int_A v^3 \, dA}{\int_A dA} \Rightarrow \left[\frac{\int_A v \, dA}{\int_A dA} \right]^2 \left[\frac{\int_A v \, dA}{\int_A dA} \right] = \langle v \rangle^2 \langle v \rangle \qquad (3.1.2\text{-}35)$$

---

[7] Note that $\langle v^3 \rangle$ is not the same as $\langle v \rangle^3$ except for special cases

$$\langle v^3 \rangle = \frac{\int_A v^3 \, dA}{\int_A dA} \neq \left[\frac{\int_A v \, dA}{\int_A dA} \right]^3 = \langle v \rangle^3 = \langle v \rangle^2 \langle v \rangle$$

This is not, of course, true in general, but it is true for certain special models such as plug flow. We also use this as an approximation as discussed below.

The enthalpy term

This term requires that we evaluate differences in enthalpy, sometimes between streams that differ in temperature, phase, and/or composition. Enthalpy changes are evaluated using the concept of enthalpy as a state function, and choosing the appropriate path from initial to final state.

For changes in enthalpy which are expressed by changes in temperature or phase at constant composition, this term is easily represented using sensible or latent heats; that is, enthalpy changes tied to heat capacities per unit mass in the first case, and to unit masses in the second. For streams of differing compositions, heats of mixing are the usual model. For chemical reaction, heats of reaction serve the same purpose. We will not treat the enthalpy model itself in any detail here, but refer the reader to the many texts available on thermodynamics.

If we apply to this term the assumption that enthalpy and density are constant across any single inlet or outlet area (such as the cross-section of a pipe crossing the boundary), we may write, for each such single area

$$
\begin{aligned}
\int_{A_i} \hat{H} \rho \left(\mathbf{v} \cdot \mathbf{n} \right) dA &= \hat{H} \rho \int_{A_i} \left(\mathbf{v} \cdot \mathbf{n} \right) dA \\
&= \pm \hat{H}_i \rho_i \langle v \rangle_i A_i \\
&= \pm \hat{H}_i \rho_i Q_i \\
&= \pm \hat{H}_i w_i
\end{aligned}
\tag{3.1.2-36}
$$

To make it easier to relate temperature and enthalpy, we use the mixing-cup or bulk temperature, as defined in Chapter 1

$$
T_b \equiv \frac{\int_A T C_p \rho \left(\mathbf{v} \cdot \mathbf{n} \right) dA}{\int_A C_p \rho \left(\mathbf{v} \cdot \mathbf{n} \right) dA}
\tag{3.1.2-37}
$$

Averages and the macroscopic energy equations

The velocity distribution for turbulent flow in pipes is fairly well represented by the following empirical profile:

$$v = v_{max}\left(1 - \frac{r}{R}\right)^{\frac{1}{7}}$$

(3.1.2-38)

Energy balance approximation - turbulent flow

Let us consider the approximation in Equation (3.1.2-33) by taking the ratio of the average using the 1/7 power model for the turbulent velocity profile[8] to the approximation in Equation (3.1.2-33).

$$\frac{\langle v^3 \rangle}{\langle v \rangle^2 \langle v \rangle} = \frac{\langle v^3 \rangle}{\langle v \rangle^3} = \frac{\dfrac{\displaystyle\int_0^R \left[v_{max}\left[1 - \frac{r}{R}\right]^{\frac{1}{7}}\right]^3 2\,\pi\, r\, dr}{\displaystyle\int_0^R 2\,\pi\, r\, dr}}{\left\{\dfrac{\displaystyle\int_0^R \left[v_{max}\left[1 - \frac{r}{R}\right]^{\frac{1}{7}}\right] 2\,\pi\, r\, dr}{\displaystyle\int_0^R 2\,\pi\, r\, dr}\right\}^3}$$

(3.1.2-39)

Defining an auxiliary variable $t = 1 - r/R$, we have

$$dt = -\frac{dr}{R}$$
$$r = \left(1 - t\right) R$$

(3.1.2-40)

Substituting:

---

[8] The 1/7 power model itself is an approximation to the true velocity profile, but this is a different approximation than the one under discussion.

$$\frac{\langle v^3 \rangle}{\langle v \rangle^3} = \frac{\int_1^0 t^{3/7} \left[(1-t)R \right] \left[-R\, dt \right]}{\frac{R^2}{2}} \cdot \frac{\left(\frac{R^2}{2} \right)^3}{\left\{ \int_1^0 t^{1/7} \left[(1-t)R \right] \left[-R\, dt \right] \right\}^3}$$

$$= \frac{\left[-(7/10)\, t^{10/7} + (7/17)\, t^{17/7} \right]_1^0}{4 \left\{ \left[-(7/8)\, t^{8/7} + (7/15)\, t^{15/7} \right]_0^1 \right\}^3} = \frac{7/17 - 7/10}{4 \left(7/15 - 1/8 \right)^3} \approx 1.06$$

$$(3.1.2\text{-}41)$$

Therefore, the assumption creates an error of about

$$\% \text{ error} = 100 \, \frac{\text{approximate quantity} - \text{exact quantity}}{\text{exact quantity}}$$

$$= 100 \, \frac{1 - 1.06}{1.06} = -5.7\%$$

$$(3.1.2\text{-}42)$$

We could, of course, apply this correction, but it is seldom worthwhile because of a) the approximate nature of the velocity distribution used and b) other uncertainties that are usually present in real problems.

Energy balance approximation - laminar flow

As a matter of interest, let us see what the error would be if we made this approximation in the energy balance for *laminar* flow

$$\frac{\langle v^3 \rangle}{\langle v \rangle^2 \langle v \rangle} = \frac{\langle v^3 \rangle}{\langle v \rangle^3} = \frac{\dfrac{\int_0^R \left\{ v_{max} \left[1 - \left(\frac{r}{R} \right)^2 \right] \right\}^3 2\pi r\, dr}{\int_0^R 2\pi r\, dr}}{\left\{ \dfrac{\int_0^R \left\{ v_{max} \left[1 - \left(\frac{r}{R} \right)^2 \right] \right\} 2\pi r\, dr}{\int_0^R 2\pi r\, dr} \right\}^3}$$

$$(3.1.2\text{-}43)$$

This may be written as

$$\frac{\langle v^3 \rangle}{\langle v \rangle^3} = \frac{\frac{R^2}{2} \int_0^1 \left[1 - (r/R)^2 \right]^3 \frac{2r}{R} d\left(\frac{r}{R}\right)}{\frac{R^2}{2}}$$

$$\times \frac{\left(\frac{R^2}{2} \right)^3}{\frac{R^2}{2} \left\{ \int_1^0 \left[1 - (r/R)^2 \right] \frac{2r}{R} d\left(\frac{r}{R}\right) \right\}^3}$$

(3.1.2-44)

This time we define $t = (r/R)^2$ so that $dt = 2(r/R) \, d(r/R)$. Substituting

$$\frac{\langle v^3 \rangle}{\langle v \rangle^3} = \frac{\int_0^1 (1-t)^3 \, dt}{1} \frac{1}{\left[\int_0^1 (1-t) \, dt \right]^3}$$

$$= \frac{\left[(1-t)^4/4 \right]_1^0}{\left\{ \left[(1-t)^2/2 \right]_1^0 \right\}^3} = \frac{1/4}{1/8} = 2$$

(3.1.2-45)

Therefore, using the approximation would yield *half* the correct answer; therefore, this would be a poor model of the process.

Steady-state cases of the macroscopic total energy balance

A common special case of the macroscopic energy balance is that of a steady-state (no accumulation) system with a single inlet, a single outlet, $\cos \alpha = +1$ for the outlet and -1 for the inlet, and no variation of $\hat{H}$, $\hat{\Phi}$, or $\hat{K}$ across either inlet or outlet. For this case, dropping the accumulation term and recognizing the integrals as the mass flow rates, the macroscopic energy balance becomes

$$\left(\hat{H}_1 + \hat{\Phi}_1 + \hat{K}_1 \right) \int_{A_1} \rho \left(v \cdot n \right) dA$$

$$+ \left(\hat{H}_2 + \hat{\Phi}_2 + \hat{K}_2 \right) \int_{A_2} \rho \left(v \cdot n \right) dA = Q - W$$

$$\left(\hat{H}_2 + \hat{\Phi}_2 + \hat{K}_2 \right) w_2 - \left(\hat{H}_1 + \hat{\Phi}_1 + \hat{K}_1 \right) w_1 = Q - W \tag{3.1.2-46}$$

But a steady-state macroscopic mass balance gives

$$\int_{A_1} \rho \left(v \cdot n \right) dA + \int_{A_2} \rho \left(v \cdot n \right) dA = 0$$

$$w_2 - w_1 = 0 \tag{3.1.2-47}$$

so we can write $w_2 = w_1 = w$. Dividing the energy balance by w to put it on a unit mass basis

$$\left(\hat{H}_2 + \hat{\Phi}_2 + \hat{K}_2 \right) w - \left(\hat{H}_1 + \hat{\Phi}_1 + \hat{K}_1 \right) w = \hat{Q} - W$$

$$\left(\hat{H}_2 + \hat{\Phi}_2 + \hat{K}_2 \right) - \left(\hat{H}_1 + \hat{\Phi}_1 + \hat{K}_1 \right) = Q - W$$

$$\left(\hat{H}_2 - \hat{H}_1 \right) + \left(\hat{\Phi}_2 - \hat{\Phi}_1 \right) + \left(\hat{K}_2 - \hat{K}_1 \right) = Q - W$$

$$\Delta \hat{H} + \Delta \hat{\Phi} + \Delta \hat{K} = \hat{Q} - W \tag{3.1.2-48}$$

For turbulent flow we frequently assume that the velocity profile can be modeled as constant (plug flow) so as we noted above

$$\hat{K} = \left[\cos(\alpha) \right] \left[\frac{v^2}{2} \right] \tag{3.1.2-49}$$

The macroscopic energy balance then reduces to

$$\boxed{\Delta \hat{H} + g \Delta z + \Delta \left[\frac{v^2}{2} \right] = Q - W} \tag{3.1.2-50}$$

Conditions on this model are

- single inlet, single outlet
- steady state
- constant g, C_p, ρ
- no (i.e., negligible) variation of v, $\hat{H}$, $\hat{\Phi}$, or $\hat{K}$ across inlet or across outlet

Frequently the evaluation of the enthalpy change will involve a heat of combustion, reaction, phase change, etc. These are sometimes large terms which dominate the remaining enthalpy changes (from sensible heats, etc.) An example of phase change in a steady-state problem is illustrated below in Example 3.1.2-1. Table 3.1.2-1 below shows the relative magnitude of such terms for organic compounds. (Negative enthalpy changes imply that heat is evolved.) It is to be emphasized that this figure is for qualitative illustration only, not design.

Table 3.1.2-1 Qualitative comparison of ranges of enthalpy changes (kcal/mol) for processes involving organic compounds[9]

PROCESS	RANGE OF ENTHALPY CHANGE, kcal/mol	
Combustion	-2000	-500
Reaction	-100	100
Polymerization	-25	-5
Neutralization	-24	-13
Adsorption	-20	-2
Wetting	-6	-0.6
Mixing/Solution	-2	2
Dilution	-2	2
Fusion	0.3	5
Vaporization	5	20

Example 3.1.2-1   Relative magnitudes of mechanical and thermal energy terms with phase change

Water enters a boiler and leaves as steam as shown. Using 1 to designate the entering state and 2 for the exit, calculate the steady-state kinetic and potential energy and enthalpy changes per pound.

---

[9] Frurip, D. J., A. Chakrabarti, et al. (1995). *Determination of Chemical Heats by Experiment and Prediction*. International Symposium on Runaway Reactions and Pressure Relief Design, Boston, MA, American Institute of Chemical Engineers, p. 97.

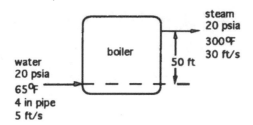

Figure 3.1.2-3 Mechanical energy and thermal energy terms compared for a boiler (I)

Solution

The kinetic energy term is

$$\hat{K}_1 = \frac{v_1^2}{2} = \frac{(5)^2 \left[\frac{ft}{s}\right]^2}{(2)} \frac{lbf\ s^2}{(32.2)\ lbm\ ft} \frac{Btu}{(778)\ ft\ lbf} = 0.0005 \frac{Btu}{lbm}$$

$$\hat{K}_2 = \frac{v_2^2}{2} = \frac{(30)^2 \left[\frac{ft}{s}\right]^2}{(2)} \frac{lbf\ s^2}{(32.2)\ lbm\ ft} \frac{Btu}{(778)\ ft\ lbf} = 0.0180 \frac{Btu}{lbm}$$

$$\hat{K}_2 - \hat{K}_1 = 0.0180 \frac{Btu}{lbm} - 0.0005 \frac{Btu}{lbm} = \boxed{0.0175 \frac{Btu}{lbm}} \qquad (3.1.2\text{-}51)$$

The potential energy term is, choosing $z = 0$ at the level of the entrance pipe

$$\Phi_1 = g\,z_1 = (32.2)\frac{ft}{s^2}(0)\ ft \frac{lbf\ s^2}{(32.2)\ lbm\ ft} \frac{Btu}{(778)\ ft\ lbf}$$

$$= (0) \frac{Btu}{lbm}$$

$$\Phi_2 = g\,z_2 = (32.2)\frac{ft}{s^2}(50)\ ft \frac{lbf\ s^2}{(32.2)\ lbm\ ft} \frac{Btu}{(778)\ ft\ lbf}$$

$$= (0.0643) \frac{Btu}{lbm}$$

$$\Phi_2 - \Phi_1 = \boxed{(0.0643) \frac{Btu}{lbm}} \qquad (3.1.2\text{-}52)$$

The enthalpy change is, using values from the steam table and ignoring the effect of pressure on density and/or enthalpy of liquids

$$\hat{H}_{\text{sat'd liquid}} = 33.05 \frac{\text{Btu}}{\text{lbm}} \text{ at } 65°F, 0.3056 \text{ psia}$$

$$\hat{H}_{\text{superheated vapor}} = 1191.6 \frac{\text{Btu}}{\text{lbm}} \text{ at } 300°F, 20 \text{ psia}$$

$$\hat{H}_2 - \hat{H}_1 = 1191.6 - 33.1 = \boxed{1158.5 \frac{\text{Btu}}{\text{lbm}}} \qquad (3.1.2\text{-}53)$$

Notice that the mechanical energy terms are insignificant compared to the thermal energy term. This is typical for many engineering problems.

Example 3.1.2-2   Steam production in a boiler

Liquid water enters a boiler at 5 psig and 65°F through a 4 in pipe at an area-average velocity of 5 ft/s. At steady state, steam is produced at 600°F and 400 psig. What is the rate of heat addition required?

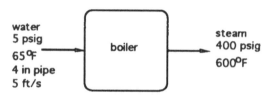

Figure 3.1.2-4 Mechanical and thermal energy terms compared for a boiler (II)

Solution

Potential and kinetic energy changes are negligible compared to the amount of energy required to accomplish the phase change. The time derivative is zero at steady state. The work term is zero.

The macroscopic energy balance

$$\int_A \left(\hat{H} + \hat{\Phi} + \hat{K}\right) \rho \left(v \cdot n\right) dA + \frac{d}{dt}\int_V \left(\hat{U} + \hat{\Phi} + \hat{K}\right) \rho \, dV = Q - W$$

$$(3.1.2\text{-}54)$$

then becomes

$$\int_A \left(\hat{H}\right) \rho \left(\mathbf{v} \cdot \mathbf{n}\right) dA = Q \qquad (3.1.2\text{-}55)$$

$$\hat{H}_1 \int_{A_1} \rho \left(\mathbf{v} \cdot \mathbf{n}\right) dA + \hat{H}_2 \int_{A_2} \rho \left(\mathbf{v} \cdot \mathbf{n}\right) dA = Q \qquad (3.1.2\text{-}56)$$

$$\hat{H}_2 \, w_2 - \hat{H}_1 \, w_1 = Q \qquad (3.1.2\text{-}57)$$

The macroscopic total mass balance yields

$$\int_A \rho \left(\mathbf{v} \cdot \mathbf{n}\right) dA + \frac{d}{dt} \int_V \rho \, dV = 0 \qquad (3.1.2\text{-}58)$$

$$\int_{A_1} \rho \left(\mathbf{v} \cdot \mathbf{n}\right) dA + \int_{A_2} \rho \left(\mathbf{v} \cdot \mathbf{n}\right) dA = 0 \qquad (3.1.2\text{-}59)$$

$$\rho_2 \langle v \rangle_2 A_2 - \rho_1 \langle v \rangle_1 A_1 = 0 \qquad (3.1.2\text{-}60)$$

$$w_2 - w_1 = 0$$

$$\Rightarrow \quad w_2 = w_1 \equiv w \qquad (3.1.2\text{-}61)$$

which reduces the energy balance to

$$\hat{H}_2 \, w - \hat{H}_1 \, w = Q$$

$$w \left(\hat{H}_2 - \hat{H}_1\right) = Q \qquad (3.1.2\text{-}62)$$

We can obtain property values from the steam tables.[10] We neglect the effect of pressure on the density and enthalpy of liquids.

$$\hat{H}_{\text{sat'd liquid}} = 33.05 \, \frac{\text{Btu}}{\text{lbm}} \text{ at } 65°\text{F}, 0.3056 \, \text{psia} \qquad (3.1.2\text{-}63)$$

$$\hat{V} = 0.01605 \, \frac{\text{ft}^3}{\text{lbm}} \text{ at } 65°\text{F}, 0.3056 \, \text{psia} \qquad (3.1.2\text{-}64)$$

$$\hat{H}_{\text{superheated vapor}} = 1305.7 \, \frac{\text{Btu}}{\text{lbm}} \text{ at } 600°\text{F}, 415 \, \text{psia} \qquad (3.1.2\text{-}65)$$

We find from pipe tables that the cross-sectional area for flow for 4 in nominal diameter Schedule 40 steel pipe is $0.0884 \, \text{ft}^2$. The mass flow rate, w, is then

---

[10] Keenan, J. H. and F. G. Keyes (1936). *Thermodynamic Properties of Steam*. New York, NY, John Wiley and Sons, Inc.

$$w = \rho_1 \langle v \rangle_1 A_1 = \frac{1}{\hat{V}} \langle v \rangle_1 A_1 \qquad (3.1.2\text{-}66)$$

$$= \frac{(1)}{(0.01605) \frac{ft^3}{lbm}} (5) \frac{ft}{s} (0.0884) \, ft^2 \qquad (3.1.2\text{-}67)$$

$$= 27.5 \frac{lbm}{s} \qquad (3.1.2\text{-}68)$$

Substituting the values obtained in the energy balance gives

$$Q = w \left(\hat{H}_2 - \hat{H}_1 \right)$$

$$Q = 27.5 \frac{lbm}{s} \left(1305.7 \frac{Btu}{lbm} - 33.05 \frac{Btu}{lbm} \right)$$

$$Q = 35{,}000 \frac{Btu}{s} \qquad (3.1.2\text{-}69)$$

Example 3.1.2-3   Temperature rise from conversion of mechanical to thermal energy

Water is pumped at 120 gpm through 300 ft of insulated 2 in Schedule 40 steel pipe, from a tank at ground level in a tank farm to a tank on the ground floor of a production building. The pump used supplies 5 HP to the fluid, and the pressure at the inlet and outlet of the pipe are identical. What is the temperature rise of the water?

Figure 3.1.2-5 Water supply system

Solution

Choose the system to be the fluid internal to the pipe, and apply the macroscopic energy balance. Since the pipe is of uniform diameter, the kinetic energy change will be zero. There is no potential energy change. The accumulation term is zero because of steady-state conditions. Therefore, the work supplied by the pump goes entirely into increasing the enthalpy of the fluid.

$$\int_A \left(\hat{H} + \Phi + \mathcal{K} \right) \rho \left(v \cdot n \right) dA + \frac{d}{dt} \int_V \left(\hat{U} + \Phi + \mathcal{K} \right) \rho \, dV = Q - \dot{W}$$

(3.1.2-70)

$$\int_A \left(\hat{H} \right) \rho \left(v \cdot n \right) dA = - \dot{W}$$
(3.1.2-71)

$$\hat{H}_1 \int_{A_1} \rho \left(v \cdot n \right) dA + \hat{H}_2 \int_{A_2} \rho \left(v \cdot n \right) dA = - \dot{W}$$
(3.1.2-72)

$$\hat{H}_2 \, w_2 - \hat{H}_1 \, w_1 = - \dot{W}$$
(3.1.2-73)

The macroscopic total mass balance shows that $w_1 = w_2 = w$

$$w \left(\hat{H}_2 - \hat{H}_1 \right) = - \dot{W}$$
(3.1.2-74)

But we can write

$$\left(\hat{H}_2 - \hat{H}_1 \right) = C_p \left(T_2 - T_1 \right)$$
(3.1.2-75)

Substituting

$$w \, C_p \left(T_2 - T_1 \right) = - \dot{W}$$
(3.1.2-76)

$$\left(T_2 - T_1 \right) = \frac{- \dot{W}}{w \, C_p}$$
(3.1.2-77)

$$= \frac{- \left(-5 \right) HP}{\left(120 \right) \frac{gal}{min} \left(1 \right) \frac{Btu}{lbm \, °F}} \frac{gal}{\left(8.33 \right) lbm}$$

$$\times \frac{Btu}{\left(778 \right) ft \, lbf} \frac{\left(33{,}000 \right) ft \, lbf}{HP \, min}$$
(3.1.2-78)

$$\left(T_2 - T_1 \right) = 0.21 \, °F$$
(3.1.2-79)

The very small temperature increase shows why one does not, for example, heat morning coffee by stirring the water with a spoon. A very small rise in temperature can consume a relatively enormous amount of mechanical energy.

Example 3.1.2-4   Heated tank, steady state in mass and unsteady state in energy

An insulated, perfectly mixed tank contains a heating coil. Water flows into and out of the tank at 10 ft³/min. The volume of the water in the tank is 100 ft³, the initial temperature is 70°F and the water flowing into the tank is at 150°F. If the heater adds 5000 BTU/min. and the horsepower added by the mixer is 5 hp, what is the tank temperature as a function of time?

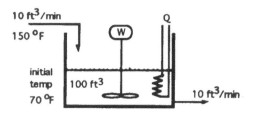

Figure 3.1.2-6 Heated tank

Solution

The tank is at steady state so far as the macroscopic total mass balance is concerned (accumulation is zero, $w_1 = w_2 = w$). It is not at steady state with respect to the macroscopic total energy balance, because the internal energy in the tank is increasing with time.

Neglecting kinetic and potential energy changes,[11] the macroscopic total energy balance becomes

$$\int_A \left(\hat{H}\right) \rho \left(v \cdot n\right) dA + \frac{d}{dt}\int_V \left(\hat{U}\right) \rho \, dV = Q - W \tag{3.1.2-80}$$

$$H_1 \int_{A_1} \rho \left(v \cdot n\right) dA + H_2 \int_{A_2} \rho \left(v \cdot n\right) dA + \frac{d}{dt}\int_V \left(\hat{U}\right) \rho \, dV$$
$$= Q - W \tag{3.1.2-81}$$

$$H_1 \left(-w_1\right) + H_2 \left(w_2\right) + \frac{d}{dt}\int_V \left(\hat{U}\right) \rho \, dV = Q - W \tag{3.1.2-82}$$

$$\left(\hat{H}_2 - H_1\right)\left(w\right) + \frac{d}{dt}\int_V \left(\hat{U}\right) \rho \, dV = Q - W \tag{3.1.2-83}$$

---

[11] Note that there also would be a volume change with temperature, which we ignore in this model, and which would probably be negligible for most applications.

We know from thermodynamics that

$$\hat{H} = C_p \left(T - T_{ref}\right)$$

$$\hat{U} = C_v \left(T - T_{ref}\right) \tag{3.1.2-84}$$

and that for an incompressible fluid

$$\hat{C}_p = \hat{C}_v \tag{3.1.2-85}$$

For a perfectly mixed tank, the internal energy is not a function of location within the tank, even though it is a function of time. Using this fact and the equality of the heat capacities, and assuming that the heat capacities are constant

$$\left[C_p\left(T_2 - T_{ref}\right) - C_p\left(T_1 - T_{ref}\right)\right]w + \frac{d}{dt}\int_v \left[C_p\left(T - T_{ref}\right)\right]\rho \, dV$$
$$= Q - W$$

$$\left[C_p\left(T_2 - T_1\right)\right]w + C_p\rho\frac{d}{dt}\left[\left(T - T_{ref}\right)\int_v dV\right] = Q - W$$

$$\left[C_p\left(T_2 - T_1\right)\right]w + C_p\rho\frac{d}{dt}\left[\left(T - T_{ref}\right)V\right] = Q - W \tag{3.1.2-86}$$

Expanding the derivative of the product, noting that V is constant as is T_{ref}

$$\left[C_p\left(T_2 - T_1\right)\right]w + C_p\rho\left[\left(T - T_{ref}\right)\frac{d}{dt}[V] + V\frac{d}{dt}\left(T - T_{ref}\right)\right] = Q - W$$

$$\left[C_p\left(T_2 - T_1\right)\right]w + C_p\rho\left[V\frac{dT}{dt}\right] = Q - W \tag{3.1.2-87}$$

Using the fact that for a perfectly mixed tank the outlet conditions are the same as those in the tank; specifically, that $T = T_2$

$$C_p\left(T - T_1\right)w + C_p\rho V\frac{dT}{dt} = Q - W \tag{3.1.2-88}$$

Rearranging

$$\frac{w\,C_p\,(T-T_1)}{(Q-W)} + \frac{\rho\,V\,C_p}{(Q-W)}\frac{dT}{dt} = 1$$

$$\frac{\rho\,V\,C_p}{(Q-W)}\frac{dT}{dt} = 1 - \frac{w\,C_p\,(T-T_1)}{(Q-W)} \tag{3.1.2-89}$$

Integrating and substituting known values

$$\frac{(62.4)\frac{lbm}{ft^3}(100)\,ft^3\,(1)\frac{Btu}{lbm\,°F}}{\left[(5000)\frac{Btu}{min}-(-5)\,HP\frac{(42.44)\,Btu}{HP\,min}\right]}\left(\frac{dT}{dt}\right)\frac{°F}{min}$$

$$= 1 - \frac{10\frac{ft^3}{min}(1)\frac{62.4\,lbm}{ft^3}\frac{Btu}{lbm\,°F}(T-150\,°F)}{\left[(5000)\frac{Btu}{min}-(-5)\,HP\frac{(42.44)\,Btu}{HP\,min}\right]}$$

$$\tag{3.1.2-90}$$

$$(1.197)\frac{dT}{dt} = 1 - 0.1197\,(T-150)$$

$$(1.197)\frac{dT}{dt} = (18.96 - 0.1197\,T) \tag{3.1.2-91}$$

$$(1.197)\int_{70}^{T}\frac{dT}{(18.96-0.1197\,T)} = \int_{0}^{t} dt \tag{3.1.2-92}$$

$$(-10)\ln\left[\frac{(18.96-0.1197\,T)}{10.6}\right] = t \tag{3.1.2-93}$$

Solving for T

$$0.1197\,T = 18.96 - 10.6\exp\left[-\frac{t}{(10)}\right]$$

$$T = \frac{18.96}{0.1197} - \frac{10.6}{0.1197}\exp\left[-\frac{t}{(10)}\right]$$

$$T = 158 - 88.6 \exp\left[-\frac{t}{(10)}\right]$$

(3.1.2-94)

As t approaches infinity, the tank temperature approaches 158°F.

3.1.3 The macroscopic mechanical energy balance

The work term, the kinetic energy term, the potential energy term **and the flow work part of the enthalpy term** all represent a special type of energy - **mechanical energy**. Mechanical energy is either work or a form of energy that may be directly converted into work. The other terms in the total energy balance (the internal energy and heat terms) do not admit of easy conversion into work. These we classify as **thermal energy**.

Mechanical energy need not go through a heat engine to be converted to work. A given change in potential energy can be very nearly completely converted into work (completely, if we could eliminate friction). For example, the potential energy in the weights of a grandfather clock are quite efficiently converted into the work to drive the clockworks.

Conversely, to obtain work from thermal energy we must go through some sort of heat engine, which is subject to efficiency limitations from the second law of thermodynamics, and which depends on the working temperatures of the heat engine. Because of the second law limitation, heat and internal energy represent a "lower quality" energy.

It therefore should come as no surprise that we might be interested in a balance on the "higher quality" energy represented by the mechanical energy terms. Even though we do not create or destroy energy, we are concerned about loss of the "high quality" energy through conversion to the "low quality" forms, much as we might be concerned with a conversion of crystal structure that changes our diamond to coke - even if we still have the same amount of carbon in the end.

The balance on macroscopic mechanical energy is yet another application of the entity balance, in this case incorporating a generation term which accounts for the destruction of mechanical energy - **mechanical** energy is not a conserved quantity. **Total** energy remains a conserved quantity, since any mechanical energy destroyed appears in the form of thermal energy, thus preserving the conservation of total energy.

When the brakes on a car are applied, the kinetic energy of the vehicle (the system) is converted to internal energy of the system at the point where the brake shoes and/or pads rub on the drums and/or disks and to heat, which crosses the boundary of the system where the tires rub the road. These converted forms of energy obviously cannot easily be recovered as work - say, to raise the elevation of the car. If the brakes had not been applied the kinetic energy could have been used to coast up a hill and thus preserve mechanical energy in the form of elevation (potential energy) increase.

The macroscopic total energy balance is

$$\int_A \left(\hat{H} + \Phi + K \right) \rho \left(v \cdot n \right) dA + \frac{d}{dt} \int_V \left(\hat{U} + \Phi + K \right) \rho \, dV$$
$$= Q - W \qquad (3.1.3-1)$$

Substituting the definition of enthalpy gives

$$\int_A \left(\boxed{\hat{U}} + p\hat{V} + \Phi + K \right) \rho \left(v \cdot n \right) dA + \frac{d}{dt} \int_V \left(\boxed{\hat{U}} + \Phi + K \right) \rho \, dV$$
$$= \boxed{Q} - W$$
$$(3.1.3-2)$$

where the terms classified as thermal energy are boxed. Since the conversion of mechanical energy into thermal energy represents a degradation of the possibility of recovering this energy as work, it has been conventional to define a term called **lost work** as

$$\dot{lw} \equiv \int_A \hat{U} \rho \left(v \cdot n \right) dA + \frac{d}{dt} \int_V \hat{U} \rho \, dV - Q$$
$$(3.1.3-3)$$

It would be more logical to call this term *lost mechanical energy,* but we will not assail tradition. **Lost work is not lost energy. Total** energy is still conserved - **mechanical** energy is not.

By the above definition we adopt the same sign convention for lw as that for W; that is, input is negative, output is positive. The lost work is a (negative) generation term in the entity balance. Substituting the definition of lost work and writing the relation in the form of the entity balance yields

$$\left| \begin{array}{c} \text{output} - \text{input} \\ \text{rate of} \\ \text{mechanical energy} \end{array} \right| + \left| \begin{array}{c} \text{accumulation} \\ \text{rate of} \\ \text{mechanical energy} \end{array} \right|$$

$$= \left| \begin{array}{c} \text{generation} \\ \text{rate of} \\ \text{mechanical energy} \end{array} \right| \qquad (3.1.3\text{-}4)$$

$$\left[\int_A \left(p\,\hat{V} + \hat{\Phi} + \hat{K} \right) \rho \left(v \cdot n \right) dA + W \right.$$

$$\left. + \left[\frac{d}{dt} \int_V \left(\hat{\Phi} + \hat{K} \right) \rho \, dV \right] \right] = \left[-\dot{lw} \right] \qquad (3.1.3\text{-}5)$$

$$\boxed{ \int_A \left(p\,\hat{V} + \hat{\Phi} + \hat{K} \right) \rho \left(v \cdot n \right) dA + \frac{d}{dt} \int_V \left(\hat{\Phi} + \hat{K} \right) \rho \, dV \\ = -\dot{lw} - W }$$
$$(3.1.3\text{-}6)$$

This is the *macroscopic mechanical energy balance.*

If we make approximately the same assumptions for the macroscopic mechanical energy balance that we did for the macroscopic total energy balance:

- single inlet, single outlet
- steady state
- constant g, $\hat{V}$
- no variation of v, p, $\hat{\Phi}$, or $\hat{K}$ across inlet or across outlet

and apply the assumptions as before, using the fact that $\hat{V} = 1/\rho$, we obtain for the simplified form of the macroscopic mechanical energy balance

$$\hat{V}\Delta p + \Delta\hat{\Phi} + \Delta\hat{K} = -\hat{lw} - \hat{W}$$

$$\boxed{ \frac{\Delta p}{\rho} + g\,\Delta z + \Delta\!\left(\frac{v^2}{2} \right) = -\hat{lw} - \hat{W} } \qquad (3.1.3\text{-}7)$$

In the absence of friction and shaft work this is called the *Bernoulli equation*

$$\boxed{\frac{\Delta p}{\rho} + g\,\Delta z + \Delta\left(\frac{v^2}{2}\right) = 0}$$

(3.1.3-8)

Example 3.1.3-1   Mechanical energy and pole vaulting

In pole vaulting, one of the "field" events from track and field, the idea is to speed down a runway holding one end of a pole, insert the opposite end of the pole into a slot-shaped socket fixed in the ground, and, through rotation of the pole about the end placed in the slot-shaped socket, to propel oneself over a bar resting on horizontal pegs attached to two uprights - without displacing the bar from its supports.

Regarded from the energy balance perspective, considering the body of the vaulter as the system, it is desired to convert as much internal energy of the body (at rest at the starting point on the runway) as possible into potential energy (height of center of gravity of the body above the ground) with just enough kinetic energy to move this center of gravity across the bar so the body can fall to the ground on the other side.

Initially, this is a problem for the total macroscopic energy balance, where the vaulter converts internal (chemically stored) energy of the system into the kinetic energy of the system (velocity of running down the runway). More internal energy is converted into vertical velocity as the vaulter jumps with legs and pulls upward with arms as the pole begins its rotation in the socket. At this point some part of the energy is converted into rotational energy as the vaulter's body rotates so as to pass over the bar in a more or less horizontal fashion.[12]

It is instructive to see how fast a vaulter would have to run in order to raise the center of gravity of his/her body 14 ft utilizing only kinetic energy. (Since a 6-ft vaulter might have a center of gravity 3 ft above the ground, this would correspond to a 17 ft vault.)

Solution

We assume that there is no dissipation of energy from

---

[12] Good vaulters actually curl over the bar so that at any one instant much of their mass remains below the level of the bar, much as an earthworm crawls over a soda straw.

- motion of arms and legs
- flexing and then straightening of the pole (we assume that energy absorbed by the pole as it flexes is totally returned to the system)
- wind resistance
- pole to ground friction as the pole slides into the socket and rotates, etc.

We also assume

- kinetic energy is converted totally to potential energy, which means that no linear velocity would remain to carry the vaulter across the bar

- an infinitesimal thickness to the vaulter's body or a curl over the bar so that a 17-foot elevation of the center of gravity would clear a 17-foot-high bar,

and we neglect energy in the form of angular velocity of the pole or vaulter's body.

Take the body of the vaulter as the system and take the initial state as the point where the vaulter is running and the pole just begins to slide into the socket, the final state as the vaulter at the height of the bar. Applying the mechanical energy balance, we see that there are no convective terms, no heat, no work, and no change in internal energy of the system (at this point the chemical energy of the body has been converted in to kinetic energy), leaving us with

$$\frac{d}{dt} \int_V \left(\Phi + \hat{K} \right) \rho \, dV = 0$$

$$\frac{d}{dt} \int_V \left(g \, z + \frac{v^2}{2} \right) \rho \, dV = 0 \tag{3.1.3-9}$$

Neglecting the variation of z over the height of the vaulter's body and assuming uniform density

$$\frac{d}{dt} \left[\left(g \, z + \frac{v^2}{2} \right) \rho \int_V dV \right] = 0$$

$$\frac{d}{dt} \left[\left(g \, z + \frac{v^2}{2} \right) \rho \, V \right] = 0$$

$$\frac{d}{dt}\left[\left(g\,z+\frac{v^2}{2}\right)m\right] = 0 \tag{3.1.3-10}$$

where m is the mass of the vaulter's body. Integrating with respect to time from the time the pole begins to slide into the socket to the time when the body of the vaulter is at maximum height

$$m\int_{\left(g\,z+\frac{v^2}{2}\right)=\frac{v^2}{2}}^{\left(g\,z+\frac{v^2}{2}\right)=g\,z} d\left(g\,z+\frac{v^2}{2}\right) = \int_{t=0}^{t=t_f} 0\,dt \tag{3.1.3-11}$$

$$m\left[\left(g\,z+\frac{v^2}{2}\right)\right]\Big|_{\left(g\,z+\frac{v^2}{2}\right)=\frac{v^2}{2}}^{\left(g\,z+\frac{v^2}{2}\right)=g\,z} = 0 \tag{3.1.3-12}$$

$$m\left(g\,z-\frac{v^2}{2}\right) = 0$$

$$g\,z = \frac{v^2}{2}$$

$$v^2 = 2\,g\,z$$

$$v = \sqrt{2\,g\,z} \tag{3.1.3-13}$$

Substituting known quantities

$$v = \sqrt{(2)\,(32.2)\,\frac{ft}{s^2}\,(14)\,ft}$$

$$= 30\,\frac{ft}{s} = 20.5\,mph \tag{3.1.3-14}$$

This is the equivalent of running 100 yards in 10 s - a pretty good pace, particularly when carrying an unwieldy pole - and this is without considering energy lost because of wind resistance, friction with the ground, etc., which also must be supplied by the vaulter. Obviously, legs, arms, shoulders, etc. (which contribute to the vertical component of velocity) are important.

Example 3.1.3-2   Calculation of lost work in pipe

Water is flowing at 230 gal/min. through 500 ft of fouled 3-inch Schedule 40 steel pipe. The pipe expands to a 4-inch pipe as shown. The gauge pressure at point 1 (in the 3-inch pipe) is 120 psig and at point 2 (in the 4 in pipe) is 15 psig. If the pipe rises in elevation 100 ft what is the lost work?

Solution

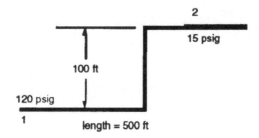

Figure 3.1.3-1 Pipe system

The system, if chosen to be the water in the pipe between points 1 and 2, is

- single inlet, single outlet
- steady state
- constant g, $\hat{V}$
- no variation of v, p, $\hat{\Phi}$, or $\hat{K}$ across inlet or across outlet

Applying the above assumptions and recognizing that there is no work term we can use the simplified form of the macroscopic mechanical energy balance

$$\frac{\Delta p}{\rho} + g\,\Delta z + \Delta\left(\frac{v^2}{2}\right) = -\hat{lw}$$

$$\frac{p_2 - p_1}{\rho} + g\left(z_2 - z_1\right) + \frac{\left[\left(\frac{Q}{A}\right)^2\right]}{2}\bigg|_2 - \frac{\left[\left(\frac{Q}{A}\right)^2\right]}{2}\bigg|_1 = -\hat{lw} \tag{3.1.3-15}$$

Evaluating term by term, choosing $z_1 = 0$ and substituting known information plus the values from the pipe table for cross-sectional flow area

$$\frac{\Delta p}{\rho} + g\,\Delta z + \Delta\left(\frac{v^2}{2}\right) = -\hat{lw} \tag{3.1.3-16}$$

Computing the velocities

$$v_1 = \frac{Q}{A_2} = \frac{(230)\frac{gal}{min}}{(0.0513)\,ft^2}\frac{ft^3}{(7.48)\,gal}\frac{min}{(60)\,s}$$
$$= 9.99\,\tfrac{ft}{s} \tag{3.1.3-17}$$

$$v_2 = \frac{Q}{A_2} = \frac{(230)\frac{gal}{min}}{(0.0884)\,ft^2}\frac{ft^3}{(7.48)\,gal}\frac{min}{(60)\,s}$$
$$= 5.80\,\tfrac{ft}{s} \tag{3.1.3-18}$$

Computing the individual terms of the equation

$$\frac{\Delta p}{\rho} = \frac{(15-120)\frac{lbf}{in^2}}{(62.4)\frac{lbm}{ft^3}}\frac{(144)\,in^2}{ft^2} = -242.3\,\tfrac{ft\,lbf}{lbm} \tag{3.1.3-19}$$

$$g\,\Delta z = (32.2)\tfrac{ft}{s^2}(100)\,ft\frac{lbf\,s^2}{(32.2)\,lbm\,ft} = 100\,\tfrac{ft\,lbf}{lbm} \tag{3.1.3-20}$$

$$\Delta\!\left(\frac{v^2}{2}\right) = \left\{\frac{\left[(5.80)\tfrac{ft}{s}\right]^2}{2} - \frac{\left[(9.99)\tfrac{ft}{s}\right]^2}{2}\right\}\left[\frac{lbf\,s^2}{(32.2)\,lbm\,ft}\right]$$
$$= -1.03\,\tfrac{ft\,lbf}{lbm} \tag{3.1.3-21}$$

$$-\widehat{lw} = -242.3\,\tfrac{ft\,lbf}{lbm} + 100\,\tfrac{ft\,lbf}{lbm} - 1.03\,\tfrac{ft\,lbf}{lbm}$$
$$= -143\,\tfrac{ft\,lbf}{lbm} \tag{3.1.3-22}$$

Notice that the kinetic energy change is negligible compared with the other terms. Also note that the lost work term is positive itself, which means that according to our sign convention that net mechanical energy is leaving the system, as we know it must for any real system.

3.1.4   The macroscopic thermal energy balance

We now take the macroscopic energy balance

$$\int_A \left(\hat{H}+\hat{\Phi}+\hat{K}\right)\rho\left(v\cdot n\right)dA+\frac{d}{dt}\int_V \left(\hat{U}+\hat{\Phi}+\hat{K}\right)\rho\,dV$$
$$= Q-W \qquad (3.1.4\text{-}1)$$

substitute the definition of enthalpy, and box the terms involving **mechanical energy**, regarding the remaining terms as thermal energy

$$\int_A \left(\hat{U}+\boxed{p\hat{V}+\hat{\Phi}+\hat{K}}\right)\rho\left(v\cdot n\right)dA+\frac{d}{dt}\int_V \left(\hat{U}+\boxed{\hat{\Phi}+\hat{K}}\right)\rho\,dV$$
$$= Q-\boxed{W}$$
$$(3.1.4\text{-}2)$$

The conversion of portions of these mechanical energy terms to thermal energy represents a **generation** of thermal energy (not of **total** energy - total energy is still conserved under processes of interest here). If we define

$$\gamma_\theta \equiv \text{rate of thermal energy generation}$$
$$\text{per unit volume} \qquad (3.1.4\text{-}3)$$

Probably the most common generation of thermal energy comes from the action of viscous forces to degrade mechanical energy into thermal energy (viscous dissipation), the thermal energy so generated appearing as internal energy of the fluid. However, it is also possible to generate thermal energy via forms of energy that we have classified as work by default - microwave energy, changing electrical fields, etc. - which are classified as work by our definition because they cross the system boundary not associated with mass but not flowing as a result of a temperature difference. Note that chemical reactions do not generate thermal energy in this sense because the heat of reaction is accounted for in the internal energy term.

By the above definition we adopt the same sign convention for γ_θ as that for Q; that is, input is positive, output is negative. Substituting the definition of thermal energy generation and writing the relation in the form of the entity balance yields

$$\left|\begin{array}{c}\text{output}-\text{input}\\\text{rate of}\\\text{thermal energy}\end{array}\right| + \left|\begin{array}{c}\text{accumulation}\\\text{rate of}\\\text{thermal energy}\end{array}\right| = \left|\begin{array}{c}\text{generation}\\\text{rate of}\\\text{thermal energy}\end{array}\right| \qquad (3.1.4\text{-}4)$$

$$\int_A \hat{U}\rho\left(\mathbf{v}\cdot\mathbf{n}\right)dA - Q + \frac{d}{dt}\int_V \hat{U}\rho\,dV = \int_V \gamma_\theta\,dV \qquad (3.1.4\text{-}5)$$

$$\boxed{\int_A \hat{U}\rho\left(\mathbf{v}\cdot\mathbf{n}\right)dA + \frac{d}{dt}\int_V \hat{U}\rho\,dV = Q + \int_V \gamma_\theta\,dV} \qquad (3.1.4\text{-}6)$$

This is the *macroscopic thermal energy balance*. It is common in practical applications for the heat transferred (Q) or the sensible and latent heat terms (the left-hand side of the equation) to be much larger than the thermal energy generation unless exceptionally large velocity gradients and/or unusually large viscosities are involved, so models using this equation will frequently neglect $\int_V \gamma_\theta\,dV$.

3.2 The Microscopic Energy Balances

3.2.1 The microscopic total energy balance

The **macroscopic** total energy balance was determined above to be

$$\int_A \left(\hat{H}+\hat{\Phi}+\hat{K}\right)\rho\left(\mathbf{v}\cdot\mathbf{n}\right)dA + \frac{d}{dt}\int_V \left(\hat{U}+\hat{\Phi}+\hat{K}\right)\rho\,dV$$
$$= Q - W \qquad (3.2.1\text{-}1)$$

We wish to develop a corresponding equation at the microscopic level.

$\dot{Q}$, the rate of heat transfer to the control volume, can be expressed in terms of the heat flux vector, **q**.

$$Q = -\int_A (q \cdot n)\, dA \tag{3.2.1-2}$$

where the negative sign is to ensure conformity with our convention that heat into the system is positive.

At the microscopic level W will not involve shafts or electrical leads crossing the boundary of the system (a point); instead, it will involve only the rate of work done per unit volume on the system at the surface via the total stress tensor **T**, which is related to the viscous stress tensor **τ** by

$$T = \tau + p\, \delta \tag{3.2.1-3}$$

Notice that the viscous stress tensor is simply the stress tensor without the thermodynamic pressure, a component of the normal force to the surface. This is desirable because the viscous forces go to zero in the absence of flow, while the thermodynamic pressure persists.

For the fluids which will be our primary concern, the viscous forces act only in shear and not normal to the surface, so the only normal force is the pressure; however, particularly in non-Newtonian fluids, the viscous forces often act normal to the surface as well as in shear.

The total stress tensor **T** incorporates the flow work from the enthalpy term. The dot product of the total stress tensor with the outward normal to a surface (a tensor operation which was defined in Table 1.6.9-1), yields the vector **T**^{n} (the surface traction) representing the total force acting on that surface (this resultant force need not - in fact, usually does not - act in the n direction).

$$T \cdot n = T^n \tag{3.2.1-4}$$

The dot product of this vector with the velocity gives the work passing into that surface, including the flow work.

$$W + \int_A p\, \hat{V}\left[\rho (v \cdot n)\right] dA = \int_A (T^n \cdot v)\, dA$$

$$W + \int_A p\, \tfrac{1}{\rho}\left[\rho (v \cdot n)\right] dA = \int_A (T^n \cdot v)\, dA$$

$$W + \int_A p\, (v \cdot n)\, dA = \int_A (T^n \cdot v)\, dA \tag{3.2.1-5}$$

Proceeding to use the above results in replacing first Q in the macroscopic energy balance

$$\int_A \left(\hat{U} + p\hat{V} + \hat{\Phi} + \hat{K}\right) \rho \left(v \cdot n\right) dA + \frac{d}{dt}\int_V \left(\hat{U} + \hat{\Phi} + \hat{K}\right)\rho \, dV$$
$$= -\int_A \left(q \cdot n\right) dA - \dot{W} \qquad (3.2.1\text{-}6)$$

and then W by moving the flow work contribution to the other side of the equation

$$\int_A \left(\hat{U} + \hat{\Phi} + \hat{K}\right) \rho \left(v \cdot n\right) dA + \frac{d}{dt}\int_V \left(\hat{U} + \hat{\Phi} + \hat{K}\right)\rho \, dV$$
$$= -\int_A \left(q \cdot n\right) dA - \dot{W} - \int_A p\, \hat{V} \rho \left(v \cdot n\right) dA \qquad (3.2.1\text{-}7)$$

Defining $\Omega = (U + \Phi + K)$ and using the fact that

$$\hat{V} = \frac{1}{\rho} \qquad (3.2.1\text{-}8)$$

gives

$$\int_A \hat{\Omega} \rho \left(v \cdot n\right) dA + \frac{d}{dt}\int_V \hat{\Omega} \rho \, dV$$
$$= -\int_A \left(q \cdot n\right) dA - \dot{W} - \int_A p \frac{1}{\rho} \rho \left(v \cdot n\right) dA$$
$$\int_A \hat{\Omega} \rho \left(v \cdot n\right) dA + \frac{d}{dt}\int_V \hat{\Omega} \rho \, dV$$
$$= -\int_A \left(q \cdot n\right) dA - \int_A \left(T^n \cdot v\right) dA \qquad (3.2.1\text{-}9)$$

Combining the area integrals

$$\int_A \left[\hat{\Omega} \rho \left(v \cdot n\right) + \left(q \cdot n\right) + \left(T^n \cdot v\right)\right] dA + \frac{d}{dt}\int_V \hat{\Omega} \rho \, dV = 0$$
$$(3.2.1\text{-}10)$$

and substituting for the surface traction in terms of the stress tensor[13]

$$\int_A \left[\Omega \rho \left(v \cdot n \right) + \left(q \cdot n \right) + \left([T \cdot n] \cdot v \right) \right] dA + \frac{d}{dt} \int_V \Omega \rho \, dV = 0$$

$$\int_A \left[n \cdot \left(\Omega \rho \, \tilde{v} \right) + \left(n \cdot q \right) + n \cdot \left(T \cdot v \right) \right] dA + \frac{d}{dt} \int_V \Omega \rho \, dV = 0$$

$$(3.2.1\text{-}11)$$

Applying the divergence theorem to the first term on the left-hand side to change the area integral to a volume integral, and applying the Leibnitz rule to the second term on the left-hand side to interchange differentiation and integration

$$\int_V \left\{ \nabla \cdot \left(\hat{\Omega} \rho \, v \right) + \left(\nabla \cdot q \right) + \left(\nabla \cdot [T \cdot v] \right) \right\} dV + \int_V \frac{\partial (\hat{\Omega} \rho)}{\partial t} \, dV = 0$$

$$\int_V \left\{ \nabla \cdot \left(\hat{\Omega} \rho \, v \right) + \left(\nabla \cdot q \right) + \left(\nabla \cdot [T \cdot v] \right) + \frac{\partial (\hat{\Omega} \rho)}{\partial t} \right\} dV = 0 \quad (3.2.1\text{-}12)$$

Since the limits of integration are arbitrary, the integrand must be identically zero.

Eulerian forms of the microscopic total energy balance

This gives the *microscopic total energy balance* in **Eulerian** form (which takes the point of view of an observer fixed in space)

$$\frac{\partial (\Omega \rho)}{\partial t} + \nabla \cdot \left(\Omega \rho \, v \right) + \left(\nabla \cdot q \right) + \left(\nabla \cdot [T \cdot v] \right) = 0$$

---

[13] Writing $[T \cdot n] \cdot v$ in rectangular Cartesian gives, since T_{ij} is symmetric: $T_{ij} n_j v_i = n_j T_{ij} v_i = n_j T_{ji} v_i$. But, in symbolic form, this is $n \cdot [T \cdot v]$.

$$\frac{\partial\left(\left[\hat{U}+\hat{\Phi}+\hat{K}\right]\rho\right)}{\partial t}+\nabla\cdot\left(\left[\hat{U}+\hat{\Phi}+\hat{K}\right]\rho\,\mathbf{v}\right)=-(\nabla\cdot\mathbf{q})-\left(\nabla\cdot[\mathbf{T}\cdot\mathbf{v}]\right)$$

$$(3.2.1\text{-}13)$$

This is a **scalar** equation. The first term represents accumulation of energy per unit volume at the point in question; the second, energy input per unit volume by convection; the third, energy input per unit volume by conduction; and the fourth, surface work (by both pressure and viscous forces) per unit volume done on the system.

This, however, is often not the most convenient form to use for many applications. To obtain a more convenient form, we first substitute for the kinetic energy term and then isolate the terms involving the potential energy

$$\frac{\partial(\hat{\Phi}\rho)}{\partial t}+\nabla\cdot\left(\hat{\Phi}\rho\,\mathbf{v}\right)+\frac{\partial\left(\left[\hat{U}+\frac{v^2}{2}\right]\rho\right)}{\partial t}+\nabla\cdot\left(\left[\hat{U}+\frac{v^2}{2}\right]\rho\,\mathbf{v}\right)$$
$$=-(\nabla\cdot\mathbf{q})-\left(\nabla\cdot[\mathbf{T}\cdot\mathbf{v}]\right)\qquad(3.2.1\text{-}14)$$

Examining only the potential energy terms

$$\frac{\partial(\hat{\Phi}\rho)}{\partial t}+\nabla\cdot\left(\hat{\Phi}\rho\,\mathbf{v}\right)$$
$$=\left[\rho\frac{\partial\hat{\Phi}}{\partial t}+\hat{\Phi}\frac{\partial\rho}{\partial t}\right]+\left[\rho\,\mathbf{v}\cdot\nabla\hat{\Phi}+\hat{\Phi}\,\nabla\cdot(\rho\,\mathbf{v})\right]\qquad(3.2.1\text{-}15)$$
$$=\rho\left[\frac{\partial\hat{\Phi}}{\partial t}+\mathbf{v}\cdot\nabla\hat{\Phi}\right]+\hat{\Phi}\left[\frac{\partial\rho}{\partial t}+\nabla\cdot(\rho\,\mathbf{v})\right]$$

But the last term in brackets is zero as a result of the continuity equation. Substituting

$$\rho\left[\frac{\partial\hat{\Phi}}{\partial t}+\mathbf{v}\cdot\nabla\hat{\Phi}\right]+\frac{\partial\left(\left[\hat{U}+\frac{v^2}{2}\right]\rho\right)}{\partial t}+\nabla\cdot\left(\left[\hat{U}+\frac{v^2}{2}\right]\rho\,\mathbf{v}\right)$$
$$=-(\nabla\cdot\mathbf{q})-\left(\nabla\cdot[\mathbf{T}\cdot\mathbf{v}]\right)\qquad(3.2.1\text{-}16)$$

Change in potential energy with time at a point in a stationary fluid ensues from changing the potential field; however, the only potential field of concern in our problems is gravitational, which is constant for problems of interest here. (We shall ignore all other field effects such as those engendered by electromagnetism, including any effects caused by fields that are unsteady. Our models therefore will not be valid for phenomena involving such field effects.)

We therefore set the partial derivative of the potential energy with time equal to zero. Using this and the fact that

$$\nabla \Phi = -g \qquad (3.2.1\text{-}17)$$

and moving the remaining term to the other side of the equation yields

$$\frac{\partial \left(\left[\hat{U} + \frac{v^2}{2} \right] \rho \right)}{\partial t} + \nabla \cdot \left(\left[\hat{U} + \frac{v^2}{2} \right] \rho \, v \right) = - \left(\nabla \cdot q \right) - \left(\nabla \cdot \left[T \cdot v \right] \right) + \rho \left(v \cdot g \right) \qquad (3.2.1\text{-}18)$$

Examining the last two terms

$$\left[- \left(\nabla \cdot \left[T \cdot v \right] \right) + \rho \left(v \cdot g \right) \right] \qquad (3.2.1\text{-}19)$$

We see that the first of these terms is work resulting from the action of surface forces (in cases of interest here, mainly the thermodynamic pressure acting opposite to the normal direction, and viscous forces acting in the shear directions - remembering, however, that, in general, it is possible to generate normal forces through the action of viscosity in addition to the thermodynamic pressure)

$$\nabla \cdot \left[T \cdot v \right] = \nabla \cdot \left[\left(\tau + p \, \delta \right) \cdot v \right] = \nabla \cdot \left[\tau \cdot v \right] + \nabla \cdot \left[p \, v \right] \qquad (3.2.1\text{-}20)$$

and the second

$$\rho \left(v \cdot g \right) \qquad (3.2.1\text{-}21)$$

is work resulting from the action of body forces (in cases of interest here, gravity).

We have chosen by this classification to regard the potential energy term as work rather than potential energy - the classification is, of course, arbitrary, but results from our regarding the potential energy term as a force acting through a distance rather than as a difference in two values of the potential function.

This gives the alternative form of the Eulerian equation as

$$\frac{\partial\left(\left[\hat{U}+\frac{v^2}{2}\right]\rho\right)}{\partial t} + \nabla \cdot \left(\left[\hat{U}+\frac{v^2}{2}\right]\rho\, v\right)$$
$$= -(\nabla \cdot q) - \nabla \cdot [\tau \cdot v] - \nabla \cdot [p\, v] + \rho\,(v \cdot g)$$

(3.2.1-22)

The terms on the left-hand side represent, respectively

   • the net rate of gain of internal and kinetic energy per unit volume at the point in question

   • the net rate of input of internal and kinetic energy per unit volume to the point by convection.

The terms on the right-hand side represent, respectively

   • the net rate of input of energy per unit volume to the point as heat (conduction)

   • the rate of work done on the fluid at the point per unit volume by viscous forces (including any normal forces from viscosity)

   • the rate of work done on the fluid at the point per unit volume by the thermodynamic pressure

   • the rate of work done on the fluid at the point per unit volume by gravitational forces.

Lagrangian forms of the microscopic total energy balance

To obtain the Lagrangian forms of the microscopic total energy balance, that is, from the viewpoint of an observer riding on a "particle" of fluid, we expand the derivatives on the left-hand side

$$\frac{\partial(\Omega \rho)}{\partial t} + \nabla \cdot (\Omega \rho \, \mathbf{v}) = -\nabla \cdot \mathbf{q} - \nabla \cdot [\mathbf{T} \cdot \mathbf{v}] \tag{3.2.1-23}$$

$$\Omega \frac{\partial \rho}{\partial t} + \rho \frac{\partial \Omega}{\partial t} + \Omega \nabla \cdot (\rho \, \mathbf{v}) + \rho \, \mathbf{v} \cdot \nabla \Omega = -\nabla \cdot \mathbf{q} - \nabla \cdot [\mathbf{T} \cdot \mathbf{v}] \tag{3.2.1-24}$$

$$\rho \left[\frac{\partial \Omega}{\partial t} + \mathbf{v} \cdot \nabla \Omega \right] + \Omega \left[\frac{\partial \rho}{\partial t} + \nabla \cdot (\rho \, \mathbf{v}) \right] = -\nabla \cdot \mathbf{q} - \nabla \cdot [\mathbf{T} \cdot \mathbf{v}] \tag{3.2.1-25}$$

But the factor in brackets in the first term is the substantial derivative by definition, and the factor in brackets in the second term is zero as a consequence of the continuity equation, so we have the **Lagrangian** form of the microscopic total energy balance as

$$\boxed{\rho \frac{D\Omega}{Dt} = -(\nabla \cdot \mathbf{q}) - \left(\nabla \cdot [\mathbf{T} \cdot \mathbf{v}] \right)} \tag{3.2.1-26}$$

Performing the same manipulations as with the Eulerian form we can obtain

$$\boxed{\rho \frac{D\left[\hat{U} + \frac{v^2}{2} \right]}{Dt} = -(\nabla \cdot \mathbf{q}) - \nabla \cdot [\tau \cdot \mathbf{v}] - \nabla \cdot [p \, \mathbf{v}] + \rho (\mathbf{v} \cdot \mathbf{g})} \tag{3.2.1-27}$$

3.2.2 The microscopic mechanical energy balance

By taking the dot product of the microscopic momentum balance with the velocity **v**, we obtain a microscopic balance that involves only mechanical energy - that is, energy directly convertible to work. This is the microscopic mechanical energy balance, below and in Equation (4.2-10), whose development is shown in Section 4.2.

$$\rho \frac{D\left[\frac{v^2}{2}\right]}{Dt} = -p\left(-\nabla \cdot v\right) - \nabla \cdot \left[p\,v\right] - \nabla \cdot \left[\tau \cdot v\right] - \left(-\tau : \nabla v\right) + \rho\left(v \cdot g\right)$$

(3.2.2-1)

3.2.3 The microscopic thermal energy balance

For many engineering applications, the energy equation is more useful written in terms of *thermal* energy, using the temperature and heat capacity of the fluid in question. The differential total energy balance can be converted to an equation expressing only thermal energy by subtracting from it the microscopic mechanical energy balance, which has the form[14]

$$\rho \frac{D\left[\frac{v^2}{2}\right]}{Dt} = p\left(\nabla \cdot v\right) - \nabla \cdot \left[\tau \cdot v\right] + \left(\tau : \nabla v\right) - \nabla \cdot \left[p\,v\right] + \rho\left(v \cdot g\right)$$

(3.2.3-1)

Subtracting gives

$$\rho \frac{D\hat{U}}{Dt} = -\left(\nabla \cdot q\right) - p\left(\nabla \cdot v\right) - \left(\tau : \nabla v\right)$$

(3.2.3-2)

where the left-hand side of the equation represents the net rate of gain of internal energy per unit volume of a "point" of the fluid, and the terms on the right-hand side represent, respectively,

> • the net rate of thermal energy input per unit volume by conduction

> • the reversible work input per unit volume by compression

> • the irreversible thermal energy input per unit volume from dissipation of kinetic energy into thermal energy via the action of viscosity (viscous dissipation)

From thermodynamics

---

[14] This equation ignores changing electrical fields, microwave energy, etc.

$$d\hat{U} = \left(\frac{\partial \hat{U}}{\partial \hat{V}}\right)_T d\hat{V} + \left(\frac{\partial \hat{U}}{\partial T}\right)_{\hat{V}} dT$$

$$= \left[-p + T\left(\frac{\partial p}{\partial T}\right)_{\hat{V}}\right] d\hat{V} + \hat{C}_{\hat{V}} dT \tag{3.2.3-3}$$

$$\rho \frac{D\hat{U}}{Dt} = \left[-p + T\left(\frac{\partial p}{\partial T}\right)_{\hat{V}}\right] \rho \frac{D\hat{V}}{Dt} + \rho \hat{C}_{\hat{V}} \frac{DT}{Dt} \tag{3.2.3-4}$$

Observing that

$$\rho \frac{D\hat{V}}{Dt} = \rho \left[\frac{D}{Dt}\left(\frac{1}{\rho}\right)\right] = \rho \left[-\frac{1}{\rho^2} \frac{D\rho}{Dt}\right] = -\frac{1}{\rho} \frac{D\rho}{Dt} \tag{3.2.3-5}$$

utilizing the continuity equation

$$\nabla \cdot (\rho \, v) + \frac{\partial \rho}{\partial t} = 0$$

and rewriting, illustrating using rectangular coordinates

$$\frac{\partial}{\partial x}(\rho \, v_x) + \frac{\partial}{\partial x}(\rho \, v_y) + \frac{\partial}{\partial x}(\rho \, v_z) + \frac{\partial \rho}{\partial t} = 0 \tag{3.2.3-6}$$

$$\left(\rho \frac{\partial v_x}{\partial x} + \rho \frac{\partial v_x}{\partial y} + \rho \frac{\partial v_z}{\partial z}\right)$$
$$+ \left(v_x \frac{\partial \rho}{\partial x} + v_y \frac{\partial \rho}{\partial y} + v_z \frac{\partial \rho}{\partial z} + \frac{\partial \rho}{\partial t}\right) = 0 \tag{3.2.3-7}$$

$$\rho \, (\nabla \cdot v) + \frac{D\rho}{Dt} = 0 \tag{3.2.3-8}$$

$$-\frac{1}{\rho} \frac{D\rho}{Dt} = (\nabla \cdot v) \tag{3.2.3-9}$$

Substituting Equation (3.2.3-9) into Equation (3.2.3-5), the result in Equation (3.2.3-4), and thence into Equation (3.2.3-2)

$$\rho \, C_\varphi \frac{DT}{Dt} = -\left(\nabla \cdot \mathbf{q}\right) - T\left(\frac{\partial p}{\partial T}\right)_\varphi \left(\nabla \cdot \mathbf{v}\right) - \left(\tau : \nabla \mathbf{v}\right) \tag{3.2.3-10}$$

As a first illustration, for **constant** thermal conductivity one obtains after substituting Fourier's law of heat conduction

$$\rho \, C_\varphi \frac{DT}{Dt} = -k \, \nabla^2 T - T\left(\frac{\partial p}{\partial T}\right)_\varphi \left(\nabla \cdot \mathbf{v}\right) + \gamma_\theta \tag{3.2.3-11}$$

where

$$\gamma_\theta \equiv \text{thermal energy generation}$$
$$= -\left[\left(\nabla \cdot \mathbf{q}\right) - k \, \nabla^2 T\right] - \left(\tau : \nabla \mathbf{v}\right) \tag{3.2.3-12}$$

The thermal energy generation term indicated contains both the effects of viscous dissipation and of externally imposed sources such as electromagnetic fields that represent non-conductive contributions to the heat flux.

As a second illustration, ignoring viscous dissipation and non-conductive contributions to the heat flux, rewriting Equation (3.2.3-11) in component form for a rectangular coordinate system yields

$$\rho \, C_\varphi \left(\frac{\partial T}{\partial t} + v_x \frac{\partial T}{\partial x} + v_y \frac{\partial T}{\partial y} + v_z \frac{\partial T}{\partial z}\right)$$
$$= -\left(\frac{\partial q_x}{\partial x} + \frac{\partial q_y}{\partial y} + \frac{\partial q_z}{\partial z}\right) - T\left(\frac{\partial p}{\partial T}\right)_\varphi \left(\frac{\partial v_x}{\partial x} + \frac{\partial v_y}{\partial y} + \frac{\partial v_z}{\partial z}\right) \tag{3.2.3-13}$$

After inserting Fourier's law

$$q_x = -k \frac{\partial T}{\partial x} \tag{3.2.3-14}$$

one obtains

$$\rho\, C_\varphi \left(\frac{\partial T}{\partial t} + v_x \frac{\partial T}{\partial x} + v_y \frac{\partial T}{\partial y} + v_z \frac{\partial T}{\partial z} \right)$$

$$= \left[\frac{\partial}{\partial x} \left(k \frac{\partial T}{\partial x} \right) + \frac{\partial}{\partial y} \left(k \frac{\partial T}{\partial y} \right) + \frac{\partial}{\partial z} \left(k \frac{\partial T}{\partial z} \right) \right]$$

$$- T \left(\frac{\partial p}{\partial T} \right)_\varphi \left(\frac{\partial v_x}{\partial x} + \frac{\partial v_y}{\partial y} + \frac{\partial v_z}{\partial z} \right)$$

(3.2.3-15)

where, since we have not assumed the thermal conductivity to be constant in space, it continues to reside inside the derivatives.

Third, in cylindrical coordinates, for unsteady-state heat transfer with flow in only in the z-direction, with transfer

> • only by conduction in the r-direction,
> • only by convection in the z-direction (neglecting conduction in the z-direction compared to convection in the z-direction)

for

> • an incompressible fluid (for which $\hat{C}_p = \hat{C}_\varphi$)

with

> • constant k,

the following equation is obtained

$$\rho\, C_p \frac{\partial T}{\partial t} + \rho\, C_p v_z \frac{\partial T}{\partial z} = k \left(\frac{\partial^2 T}{\partial r^2} + \frac{1}{r} \frac{\partial T}{\partial r} \right)$$

(3.2.3-16)

which is the differential equation describing unsteady-state temperature of an incompressible fluid flowing in a tube with no thermal energy generation in one dimensional axial flow at constant pressure, where k is the constant coefficient of thermal conductivity. For a solid rod fixed in space, as opposed to fluid in a tube, with no temperature gradient in the z-direction so that there still is no conduction in this direction, v_z is zero and the second term drops out of the equation. Solutions of this equation are discussed in Chapter 7.

Chapter 3   Problems

3.1 Initially, a mixing tank has fluid flowing in at a steady rate w1 and temperature T_1. The level remains constant and the exit temperature is $T_2 = T_1$. Then at time zero, Q Btu/sec flows through the tank walls and heats the fluid (Q-constant). Find T_2 as a function of time. Use $dH = C_p \, dT$ and $dU = C_v dT$ where $C_p = C_v$.

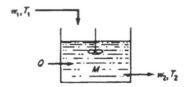

3.2 A perfectly mixed tank with dimension as in the sketch of liquid is at an initial temperature of $T_0 = 100°F$ at time t = 0 an electrical heating coil in the tank is turned on which applies 500,000 BTU/hr, and at the same time pumps are turned on which pump liquid into the tank at a rate of 100 gal/min. and a temperature of $T_1 = 80°F$ and out of the tank at a rate of 100 gal/min. Determine the temperature of the liquid in the tank as a function of time.

Properties of liquid (assume constant with temperature):

$\rho = 62.0$ lbm/ft³
$C_p = 1.1$ BTU/(lbm°F

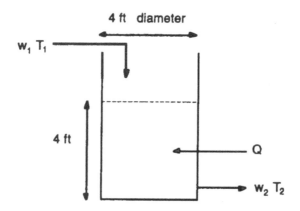

3.3 Two streams of medium oil are to be mixed and heated in a steady process. For the conditions shown in the sketch, calculate the outlet temperature T. The specific heat at constant pressure is 0.52 BTU/lbm°F and the base temperature for zero enthalpy is 32°F.

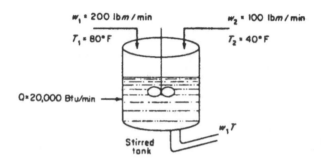

$w_1 = 200$ lbm/min $w_2 = 100$ lbm/min
$T_1 = 80°F$ $T_2 = 40°F$

Q=20,000 Btu/min

w, T

Stirred tank

3.4 Two fluids are mixed in a tank while a constant rate of heat of 350 kW is added to the tank. The specific heat of each stream 2000 J/(kg K). Determine the outlet temperature.

$w_1 = 200$ lbmass/min. $T_1 = 25°C$
$w_2 = 100$ lbmass/min. $T_2 = 10°C$

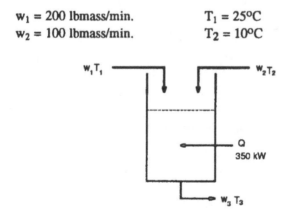

w, T_1 $w_2 T_2$

Q
350 kW

$w_3 T_3$

3.5 An organic liquid is being evaporated in a still using hot water flowing in a cooling coil. Water enters the coil at 130°F at the rate of 1,000 lbm/min., and exits at 140°F. The organic liquid enters the still at the same temperature as the exiting vapor. Calculate the steady-state vapor flow rate in lbm/s.

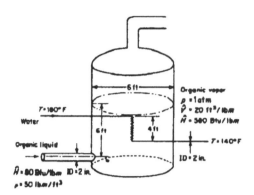

3.6 Steam at 200 psia, 600°F is flowing in a pipe. Connected to this pipe through a valve is an evacuated tank. The valve is suddenly opened and the tank fills with steam until the pressure is 200 psia and the valve is closed. If the whole process is insulated (no heat transfer) and K and f are negligible, determine the final internal energy of the steam.

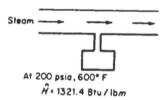

3.7 Oil (sp. gr. = 0.8) is draining from the tank shown. Determine

     a. efflux velocity as a function of height
     b. initial volumetric flow rate (gal/min.)
     c. mass flow rate (lbm/s)

How long will it take to empty the upper 5 ft of the tank?

3.8 Fluid with sp. gr. = 0.739 is draining from the tank in the sketch. Starting with the mechanical energy balance determine the volumetric flow rate and of the tank.

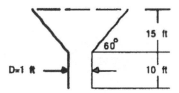

3.9 The system in the sketch shows an organic liquid ($\rho = 50$ lbm/ft^3) being siphoned from a tank. The velocity of the liquid in the siphon is 4 ft/s.

     a. Write down the macroscopic total energy balance for this problem and simplify it for this problem (state your assumptions).
     b. What is the magnitude of the lost work lw.
     c. If the lw is assumed linear with the length of the siphon, what is the pressure in psi at 1?

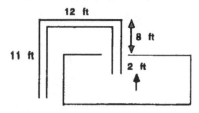

3.10 A siphon as shown in the sketch is used to drain gasoline (density = 50 lbm/ft^3) from a tank open to the atmosphere. The liquid is flowing through the tube at a rate of 5.67 ft/s.

a. What is the lost work lw?
b. If the lw term is assumed to be linear with tube length,
what is the pressure inside the bend at point A?

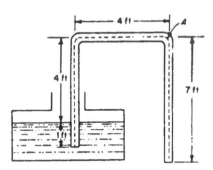

3.11 Cooling water for a heat exchanger is pumped from a lake through an insulated 3-in. diameter Schedule 40 pipe. The lake temperature is 50°F and the vertical distance from the lake to the exchanger is 200 ft. The rate of pumping is 150 gal/min. The electric power input to the pump is 10 hp.

a. Draw the flow system showing the control surfaces and control volume with reference to a zero energy reference plane.
b. Find the temperature at the heat exchanger inlet where the pressure is 1 psig.

3.12 It is proposed to generate hydroelectric power from a dammed river as shown below.

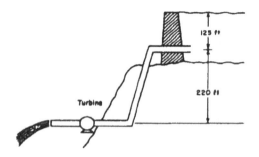

a. Draw a control volume for the system and label all control surfaces.
b. Find the power generated by a 90 percent efficient water turbine if lw = 90 ft. lbf/lbm.

3.13 River water at 70°F flowing at 10 mph is diverted into a 30' x 30' channel which is connected to a turbine operated 500 ft below the channel entrance. The water pressure at the channel inlet is 25 psia and the pressure 20 ft below the turbine is atmospheric. An estimate of the lost work in the channel is 90 ft lbf/lbm.

a. Draw a flow diagram showing all control volumes and control surfaces with reference to a zero energy plane.
b. Calculate the horsepower from a 90 percent efficient water turbine.

3.14 Water flows from the bottom of a large tank where the pressure is 90 psia to a turbine which produces 30 hp. The turbine is located 90 ft below the bottom of the tank. The pressure at the turbine exit is 30 psia and the water velocity is 30 ft/s. The flow rate in the system is 120 lbm/s. If the frictional loss in the system, not including the turbine, is 32 ft lbf/lbm

a. Draw a flow system showing the control volume and control surfaces with reference to a zero energy plane.
b. Find the efficiency of the turbine.

3.15 A pump operating at 80 percent efficiency delivers 30 gal/min. of water from a reservoir to a chemical plant 1 mile away. A Schedule 40 3-in. pipe is used to transport the water and the lost work due to pipe friction is 200 ft lbf/lbm. At the plant the fluid passes through a reactor to cool the reacting chemicals, and 800,000 BTU/hr are transferred before the water is discharged to the drain. The elevation of the plant is 873 ft above sea level; the elevation of the reservoir is 928 ft above sea level.

a. What is the minimum horsepower required for this pump?
b. The pump is removed from the system; will water flow?
c. Can you calculate the flow rate? Do so or state why not.

3.16   Nitrogen gas is flowing isothermally at 77°F through a horizontal 3-in. inside-diameter pipe at a rate of 0.28 lbm/sec. The absolute pressures at the inlet and outlet of the pipe are 2 atm and 1 atm, respectively. Determine the lost work lw in BTU/lbm.   The gas constant is 0.7302 ft³ atm/lbmole °F. Molecular weight of N_2 is 28.

3.17   Water is pumped from a lake to a standpipe. The input pipe to the pump extends 10 ft below the level of the lake; the pump is 15 ft above lake level and the water in the standpipe is almost constant at 310 ft above the pump inlet. The lost work due to friction is 140 ft lbf/lbm for pumping water through the 6,000 ft of 4-in. Schedule 40 pipe used. If the pump capacity is 6,000 gal/hr, what is the work required from the pump in ft lbf/lbm?

3.18   Water flows from the bottom of a large tank where the pressure is 100 psig through a pipe to a turbine which produces 5.82 hp. The pipe leading to the turbine is 60 ft below the bottom of the tank. In this pipe (at the turbine) the pressure is 50 psig, and the velocity is 70 ft/s. The flow rate is 100 lbm/s. If the friction loss in the system, not including the turbine, is 40 ft lbf/lbm, find the efficiency of the turbine.

3.19   Water flowing in a 1-in. ID pipe is suddenly expanded to a 9-in. ID pipe by means of an expanding conical section similar to a nozzle (here, a diffuser),. Heat is added at 3 BTU/lbm. Find the outlet temperature if the inlet temperature is 80°F, inlet pressure is 40 psig, outlet pressure is 30 psig, and the outlet velocity is 5 ft/s. Assume

$$\Delta H = 1\left(\frac{BTU}{lbmass\ °F}\right)\Delta T + \frac{\Delta p}{\rho}$$

with constant density $\rho = 62.4$ lbmass/ft³.

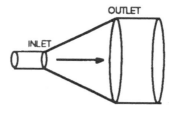

4

THE MOMENTUM BALANCES

4.1 The Macroscopic Momentum Balance

Momentum, in contrast to mass and energy, is not, in general, a **conserved** quantity. It can be created or destroyed by forces external to the system (either body or control volume) acting on the system. Moreover, momentum is a **vector** quantity, and so we will have to write vector equations.

Newton 's first law[1] states

> Every body persists in its state of rest or of uniform motion in a straight line unless it is compelled to change that state by forces impressed on it.

This law is often explained by assigning a property to bodies designated as *inertia*. Since this is a statement about acceleration (change in velocity), and, since acceleration depends on the reference frame with respect to which it is measured, this law is really a statement about reference frames. Reference frames in which this law is valid are called *inertial frames*. [2]

The momentum vector of a body with mass m_i is defined as

$$\mathbf{p}_i \equiv m_i \mathbf{v}_i \tag{4.1-1}$$

where $\mathbf{v}_i$ is the velocity of the body with respect to an inertial reference frame.

---

[1] Resnick, R. and D. Halliday (1977). *Physics: Part One*. New York, NY, John Wiley and Sons, p. 75.

[2] or Gallilean, or absolute frames: see, for example, Fanger, C. G. (1970). *Engineering Mechanics: Statics and Dynamics*. Columbus, OH, Charles E. Merrill Books, Inc., p. 19.

Newton's 2nd law for the ith body in an inertial frame can therefore be written as

$$\mathbf{F}_i = m_i \mathbf{a}_i = m_i \frac{d\mathbf{v}_i}{dt} = \frac{d(m_i \mathbf{v}_i)}{dt} = \frac{d\mathbf{p}_i}{dt} = \begin{pmatrix} \text{time rate of change} \\ \text{of momentum} \\ \text{of body} \end{pmatrix}_i$$

(4.1-2)

where $\mathbf{F}_i$ represents the net force acting on the ith body (sum of both body and surface forces).

Applying the entity balance for momentum

- in terms of rates

- in an inertial reference frame

- to a control volume either stationary or in uniform motion

- for a fluid flowing in and/or out of the control volume

- by adding the momentum contributions of the individual bodies (fluid particles) within or crossing the boundary of the control volume and taking the limit as the size of the fluid particle approaches zero (using the continuum assumption)

gives

$$\sum_{\substack{\text{body and} \\ \text{surface forces on} \\ \text{control volume}}} (\mathbf{F}) = \begin{pmatrix} \text{output} - \text{input} + \text{accumulation} \end{pmatrix} \text{rate of momentum}$$

(4.1-3)

which is a form of the entity balance written in terms of rates,[3] with the sum of forces representing the generation rate term. Note that the surface forces are those exerted **on the control volume,** not the forces exerted by the control volume on the surroundings.

---

[3] generation rate = output rate - input rate + accumulation rate

The discussion here will be concerned only with linear momentum, not with angular momentum.[4] Consider a control volume whose surfaces are stationary with respect to a fixed Cartesian axis system in a velocity field **v**. The sum of the external forces on the control volume is the rate of generation of momentum and so (output - input) rate of momentum transfer is constructed by multiplying the mass crossing the boundary by its velocity. The rate of mass crossing the boundary is its normal velocity relative to the boundary, $(\mathbf{v} \cdot \mathbf{n})$, multiplied by density and the area through which the mass passes. The momentum carried with the mass is the amount of mass multiplied by the velocity of the mass, which comes from the velocity field **v**.

$$\frac{\partial}{\partial t}\left[\left(\text{momentum output} - \text{momentum input}\right) \text{through} \, \Delta A\right]$$
$$= [\mathbf{v}]\left[\rho \left(\mathbf{v} \cdot \mathbf{n}\right) \Delta A\right] \tag{4.1-4}$$

Notice that this definition of relative velocity assigns the proper signs to input and output of momentum.

Summing over the whole surface of the system and letting ΔA approach zero

$$\frac{\partial}{\partial t}\left[\left(\text{momentum output} - \text{momentum input}\right) \text{total}\right]$$
$$= \lim_{\Delta A \to 0} \sum_A \left\{[\mathbf{v}]\left[\rho \left(\mathbf{v} \cdot \mathbf{n}\right) \Delta A\right]\right\} \tag{4.1-5}$$
$$= \int_A [\mathbf{v}] \rho \left(\mathbf{v} \cdot \mathbf{n}\right) dA \tag{4.1-6}$$

The rate of momentum accumulation is written by first considering a small volume ΔV and writing its mass times its velocity

---

[4] The analog of force for angular momentum is **torque**, τ
$$\tau = \mathbf{r} \times \mathbf{F}$$
where **r** is the displacement vector.
The analog of linear momentum is angular momentum,
$$\mathbf{P} = \mathbf{r} \times \mathbf{p}$$

$$\left[\text{rate of momentum accumulation in } \Delta V\right] = \frac{d}{dt}\left(v \, \rho \, \Delta V\right) \qquad (4.1\text{-}7)$$

The rate of momentum accumulation is found by summing over all the elemental volumes in V and taking the limit as ΔV goes to zero

$$\left[\text{total rate of momentum accumulation}\right] = \frac{d}{dt}\left\{\lim_{\Delta V \to 0}\left[\sum_V \left(v \, \rho \, \Delta V\right)\right]\right\}$$

$$= \frac{d}{dt}\int_V v \, \rho \, dV \qquad (4.1\text{-}8)$$

Substituting these terms in the entity balance gives the *macroscopic momentum balance*

$$\int_A v \, \rho \left(v \cdot n\right) dA + \frac{d}{dt}\int_V v \, \rho \, dV = \sum F \qquad (4.1\text{-}9)$$

This is a **vector** equation. It implies the equality of the resulting vector on the left to that on the right. Equality of vectors implies the equality of their components; therefore, the vector equation implies **one scalar equation** (which states the equality of one pair of components, one from the left and one from the right) **for each coordinate direction**.

Equation (4.1-9) can be written in component form for rectangular coordinates as

$$\int_A \left(\left[v_x\right]i + \left[v_y\right]j + \left[v_z\right]k\right)\rho\left(v \cos\alpha\right)dA + \frac{d}{dt}\int_V \left(v_x \, i + v_y \, j + v_z \, k\right)\rho \, dV$$

$$= \sum F_x \, i + \sum F_y \, j + \sum F_z \, k \qquad (4.1\text{-}10)$$

or, in index notation

$$\int_A v_i \, \rho \, v_j \, n_j \, dA + \frac{d}{dt}\int_V v_i \, \rho \, dV = F_i \qquad (4.1\text{-}11)$$

As is illustrated by the index form, equality of vectors implies equality of their corresponding components, so the vector equation yields three scalar equations in rectangular coordinates

$$\int_A v_x \rho \left(v \cos \alpha \right) dA + \frac{d}{dt} \int_V v_x \rho \, dV = \Sigma F_x \qquad (4.1\text{-}12)$$

$$\int_A v_y \rho \left(v \cos \alpha \right) dA + \frac{d}{dt} \int_V v_y \rho \, dV = \Sigma F_y \qquad (4.1\text{-}13)$$

$$\int_A v_z \rho \left(v \cos \alpha \right) dA + \frac{d}{dt} \int_V v_z \rho \, dV = \Sigma F_z \qquad (4.1\text{-}14)$$

In general, the convective terms as well as the sum of the forces may have a non-zero component in any direction, although, if possible, we usually try to choose a coordinate system which minimizes the number of non-zero components. Also notice that the forces exerted **by the surroundings on the control volume** (the forces in the momentum balance) are equal and opposite in sign to the forces exerted **by the control volume on the surroundings**.

Example 4.1-1   Momentum flux of fluid in laminar flow in circular pipe

The (axially symmetric) velocity profile for fully developed laminar flow in a smooth tube of constant circular cross-section is

$$v = v_{max} \left[1 - \left(\frac{r}{R} \right)^2 \right] \qquad (4.1\text{-}15)$$

Calculate the momentum flux.

Choose the flow direction to be the x-direction.

Solution

a) The momentum flux across any arbitrary pipe cross-section is

$$\text{Momentum flux} = v \int_A \left(v \cdot n \right) dA$$

$$= \int_0^R \left\{ v_{max} \left[1 - \left(\frac{r}{R} \right)^2 \right] (i) \right\} \left\{ v_{max} \left[1 - \left(\frac{r}{R} \right)^2 \right] (i \cdot n) \right\} 2 \pi r \, dr$$

$$(4.1\text{-}16)$$

Noting that for this choice of coordinates, the outward normal is in the x-direction

$$\left(\mathbf{i} \cdot \mathbf{n}\right) = 1 \tag{4.1-17}$$

Then

$$\text{Momentum flux} = 2\pi\, v_{max}^2\, \mathbf{i} \int_0^R \left\{\left[1 - \left(\tfrac{r}{R}\right)^2\right]\right\}\left\{\left[1 - \left(\tfrac{r}{R}\right)^2\right]\right\} r\, dr \tag{4.1-18}$$

$$= 2\pi\, v_{max}^2\, \mathbf{i} \int_0^R \left(r - \frac{2r^3}{R^2} + \frac{r^5}{R^4}\right) dr \tag{4.1-19}$$

$$= 2\pi\, v_{max}^2\, \mathbf{i} \left[\frac{r^2}{2} - \frac{r^4}{2R^2} + \frac{r^6}{6R^4}\right]\Bigg|_0^R \tag{4.1-20}$$

$$\text{Momentum flux} = \frac{\pi R^2 v_{max}^2}{3}\, \mathbf{i} \tag{4.1-21}$$

4.1.1 Types of forces

There are two types of forces with which we must contend:

• body forces

• surface forces.

Body forces arise from the same types of fields that we considered when discussing potential energy - gravitational, electromagnetic, etc. The change in energy from position changes in a potential field can be accomplished by a force acting on a body during its displacement. Such a force - for example, gravity - does not necessarily disappear when the body is at rest. All potential fields do not produce what we call body forces, but all body forces are produced by potential fields. Body forces are proportional to the amount of mass and act throughout the **interior** of the system or control volume.

The second type of force is the *surface force*. Surface forces are exerted on the system by the surroundings via components of the **traction** on the surface of the system. Traction (force) acting at a point on the surface is a vector since it has both magnitude and direction, and it can therefore be resolved into two components: one parallel to the surface (shear), and one component normal to the surface.

The normal component in many problems is constituted of only the negative of the pressure multiplied by the area on which it acts, although non-Newtonian fluids can also have viscous forces acting in the normal direction.

At a point in a fluid at rest, the normal force is the same in all directions and is called the **hydrostatic pressure**. Hydrostatic pressure is the same as the pressure used in a thermodynamic equation of state. There is no shear in static classical fluid systems because by definition such a fluid would deform continuously under shear. (This classic definition of a fluid becomes somewhat fuzzy when we later discuss non-Newtonian fluids, such a Bingham plastics, that exhibit characteristics of both fluids and solids.)

When a general fluid is flowing, the normal stresses are not always equal. If the fluid is incompressible, however, the pressure can be defined satisfactorily as the mean of the normal stresses in the three coordinate directions.

The pressure field in either a static or a moving fluid exhibits a gradient caused by gravity. This gradient can often be neglected except in special circumstances; for example, where buoyancy forces enter the problem.

4.1.2   Influence of uniform pressure over entire surface of irregular objects

To obtain the net force resulting from pressure on a control volume, we must integrate over the entire outside surface

$$\begin{bmatrix} \text{net force produced by pressure} \\ \text{on surface of control volume} \end{bmatrix} = \int_A \left(- p\,\mathbf{n} \right) dA \qquad (4.1.2\text{-}1)$$

For irregularly shaped volumes, this can lead to cumbersome integrals. We often can avoid integrating the force produced by pressure over the non-flow surface of irregular volumes by performing the integration in **gauge pressure**.

This artifice removes a constant pressure over the entire surface of the control volume from the integral. Adding or subtracting a constant normal stress over the entire outside surface of a control volume, no matter how irregularly shaped, does not give a net contribution to the external force term.

This can be seen by dividing an irregular control volume into elementary prisms as shown in Figure 4.1.2-1, adding the forces on the left- and right-hand faces, and taking the limit as the face areas of the prisms approach zero to obtain the area integral.

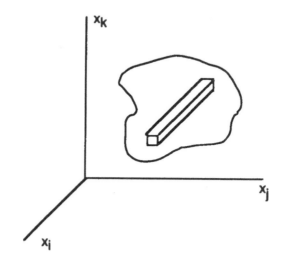

Figure 4.1.2-1 Approximation of solid by prisms

$$\int_A \left(\tau^n \cdot n \right) dA$$

$$= \lim_{dx_j, dx_k \to 0} \sum_A \left[\left(\text{normal stress} \times \text{area} \right)_{\text{left face}} + \left(\text{normal stress} \times \text{area} \right)_{\text{right face}} \right]$$

$$(4.1.2-2)$$

Consider the typical rectangular prism shown in Figure 4.1.2-2 of area dx_j by dx_k (the coordinate x_k is normal to the page) on its left face, **a**, and whose right face, **b**, is sliced at some angle β as indicated. The angle γ is a right angle, and the outward normal to **b** makes an angle α with the horizontal. The normal stress is assumed to arise only from the pressure.

Notice that if the magnitude of the normal stress

$$\left(\tau^n \cdot n \right)$$

$$(4.1-2-3)$$

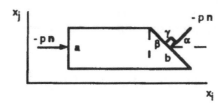

Figure 4.1.2-2 Detail of prism

is the same on each face, one obtains no net force in the x_i-direction, because the increase in area over which the normal force acts on the right face is always just compensated by the decrease in the component of the force acting in that direction

F_{x_i} on left face

$$= \left[\text{component of normal stress in } x_i\text{-direction} \right] \times \left[\text{area on which it acts} \right]$$

$$(4.1\text{-}2\text{-}4)$$

In this case the normal stress is just the pressure, which acts normal to the surface and in a direction opposed to that of the outward normal.

$$F_{x_i} \text{ on left face} = p \, dx_j \, dx_k \tag{4.1-2-5}$$

Considering the right face and noting that the angle α and the angle β are equal because their sides are perpendicular

F_{x_i} on right face

$$= \left[\text{component of normal stress in } x_i\text{-direction} \right] \times \left[\text{area on which it acts} \right]$$

$$= \left[-p \cos \alpha \right] \left[\frac{dx_j \, dx_k}{\cos \beta} \right] = \left[-p \cos \alpha \right] \left[\frac{dx_j \, dx_k}{\cos \alpha} \right]$$

$$= -p \, dx_j \, dx_k \tag{4.1-2-6}$$

The same conclusion would be reached if the left-hand face were sliced at some different angle.

Therefore, the imposition of a uniform pressure (normal force) over the outer surface of an irregularly shaped object contributes no net force the surface of the object.

4.1.3   Averages and the momentum equation

If we examine the first term in the macroscopic momentum balance in cylindrical coordinates as applied to constant-density steady axial flow in a uniform-diameter pipe between up-stream cross-section 1 and down-stream cross-section 2, with the positive z-coordinate in the direction of flow, we see that $u = 0$, that the magnitude of the z-component of v is the same as the magnitude of v itself, and that the cosine is -1 at cross-section 1 and +1 at cross-section 2, so the term reduces to

$$\int_A v_z \rho\, v \cos\alpha\, dA = \rho\left[\int_{A_2} v^2\, dA - \int_{A_1} v^2\, dA\right]$$
$$= \rho\left[\langle v_2^2\rangle - \langle v_1^2\rangle\right] A \qquad (4.1.3\text{-}1)$$

We often make the assumption in evaluating such a term that

$$\langle v^2\rangle \approx \langle v\rangle^2 \qquad\qquad (4.1.3\text{-}2)$$

In the following Sections we examine this assumption for a particular turbulent flow profile and for laminar flow.[5]

Momentum balance approximation - turbulent flow

The velocity distribution for turbulent flow in pipes is fairly well represented by the following empirical profile:

$$v = v_{max}\left(1 - \frac{r}{R}\right)^{\frac{1}{7}} \qquad\qquad (4.1.3\text{-}3)$$

In using the momentum balance for such flows, we often make the further approximation of replacing $<v^2>$ by $<v>^2$. Let us investigate the error

---

[5] Such an approximation in the momentum balance is similar to the assumption
$$\langle v^3\rangle \approx \langle v\rangle^3 \approx \langle v\rangle\langle v\rangle^2$$
that we made in the energy balance for some models.

introduced by this second approximation (this error is over and above the error introduced in using the 1/7 power approximate profile). For turbulent flow:

$$
\frac{\langle v^2 \rangle}{\langle v \rangle^2} = \frac{\dfrac{\displaystyle\int_0^R \left[v_{max} \left(1 - \frac{r}{R}\right)^{\frac{1}{7}} \right]^2 2\pi r\, dr}{\displaystyle\int_0^R 2\pi r\, dr}}{\left\{ \dfrac{\displaystyle\int_0^R \left[v_{max} \left(1 - \frac{r}{R}\right)^{\frac{1}{7}} \right] 2\pi r\, dr}{\displaystyle\int_0^R 2\pi r\, dr} \right\}^2}
\tag{4.1.3-4}
$$

$$
= \frac{\displaystyle\int_0^R \left[v_{max} \left(1 - \frac{r}{R}\right)^{\frac{1}{7}} \right]^2 2\pi r\, dr}{\left(\displaystyle\int_0^R \left[v_{max} \left(1 - \frac{r}{R}\right)^{\frac{1}{7}} \right] 2\pi r\, dr \right)^2} \int_0^R 2\pi r\, dr
\tag{4.1.3-5}
$$

Changing the variable of integration in the numerator and denominator of the quotient for convenience by defining

$$
t \equiv 1 - r/R
\tag{4.1.3-6}
$$

It then follows that

$$
r = R(1 - t)
$$

$$
dr = -R\, dt
\tag{4.1.3-7}
$$

$$
2\pi r\, dr = -2\pi R^2 (1 - t)\, dt
$$

and

$$r = 0 \quad \Rightarrow \quad t = 1$$
$$r = R \quad \Rightarrow \quad t = 0 \tag{4.1.3-8}$$

Substituting in terms of the new variable

$$\frac{\langle v^2 \rangle}{\langle v \rangle^2} = \frac{\int_1^0 \left[v_{max} \, t^{1/7} \right]^2 \left[-2 \, \pi \, R^2 \, (1-t) \right] dt}{\left(\int_1^0 \left[v_{max} \, t^{1/7} \right] \left[-2 \, \pi \, R^2 \, (1-t) \right] dt \right)^2} \int_0^R 2 \, \pi \, r \, dr \tag{4.1.3-9}$$

$$\frac{\langle v^2 \rangle}{\langle v \rangle^2} = -\frac{v_{max}}{2} \frac{\int_1^0 t^{2/7} \, (1-t) \, dt}{\left(\int_1^0 t^{1/7} \, (1-t) \, dt \right)^2} \tag{4.1.3-10}$$

$$\frac{\langle v^2 \rangle}{\langle v \rangle^2} = -\frac{v_{max}}{2} \frac{\left[\frac{7 \, t^{9/7}}{9} - \frac{7 \, t^{16/7}}{16} \right] \Big|_1^0}{\left[\left(\frac{7 \, t^{8/7}}{8} - \frac{7 \, t^{15/7}}{15} \right)^2 \right] \Big|_1^0} \tag{4.1.3-11}$$

$$= -\frac{v_{max}}{2} \frac{\left[\left(\frac{0-112}{144} \right) - \left(\frac{0-63}{144} \right) \right]}{\left[\left(\frac{0-105}{120} - \frac{0-56}{120} \right)^2 \right]} \tag{4.1.3-12}$$

$$\frac{\langle v^2 \rangle}{\langle v \rangle^2} = -\frac{v_{max}}{2} \frac{\left[-\frac{49}{144} \right]}{\left[\left(-\frac{49}{120} \right)^2 \right]}$$

$$= \frac{v_{max}}{2} \frac{120^2}{(49)(144)} = 1.02 \tag{4.1.3-13}$$

$$\% \text{ error} = 100 \frac{\left(\text{approximate} - \text{correct} \right)}{\left(\text{correct} \right)} \tag{4.1.3-14}$$

Momentum balance approximation - laminar flow

For laminar flow:

$$\frac{\langle v^2 \rangle}{\langle v \rangle^2} = \frac{\dfrac{\displaystyle\int_0^R \left\{ v_{max}\left[1 - \left(\frac{r}{R}\right)^2\right]\right\}^2 2\,\pi\,r\,dr}{\displaystyle\int_0^R 2\,\pi\,r\,dr}}{\left[\dfrac{\displaystyle\int_0^R \left\{ v_{max}\left[1 - \left(\frac{r}{R}\right)^2\right]\right\} 2\,\pi\,r\,dr}{\displaystyle\int_0^R 2\,\pi\,r\,dr}\right]^2}$$

$$= 100\,\frac{\left(1 - \dfrac{\text{correct}}{\text{approximate}}\right)}{\left(\dfrac{\text{correct}}{\text{approximate}}\right)} \tag{4.1.3-15}$$

Rearranging and defining $t = (r/R)^2$ so that $dt = (2r/R)d(r/R)$ as in the previous laminar flow case:

$$\frac{\langle v^2 \rangle}{\langle v \rangle^2} = \frac{\dfrac{\displaystyle\int_0^1 (1-t)^2\,dt}{1}}{\dfrac{1}{\left[\displaystyle\int_0^1 (1-t)\,dt\right]^2}}$$

$$= \frac{\left[(1-t)^3/3\right]_1^0}{\left\{\left[(1-t)^2/2\right]_1^0\right\}^2} = \frac{4}{3} \tag{4.1.3-16}$$

Therefore, if the approximation is used

$$\%\ \text{error} = 100\,\frac{3-4}{4} \approx -25\% \tag{4.1.3-17}$$

Example 4.1.3-1   Force on a nozzle

The sketch shows a horizontal nozzle which is attached to a fire hose (not shown) that delivers water at a steady rate of 500 gpm. Neglecting the friction loss in the nozzle, what is the force in the z-direction required in the threads to keep the hose attached to the nozzle?

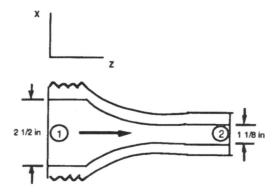

Solution

First, choose a control volume bounded by the outside surface of the nozzle and two planes, (1) and (2), normal to the flow direction. Apply the macroscopic momentum balance, noting that the velocity and density of the water within the nozzle does not change with time (steady-state)

$$\int_A \mathbf{v}\,\rho\left(\mathbf{v}\cdot\mathbf{n}\right)dA + \frac{d}{dt}\int_V \mathbf{v}\,\rho\,dV = \sum \mathbf{F}$$

$$\int_A \mathbf{v}\,\rho\left(\mathbf{v}\cdot\mathbf{n}\right)dA = \sum \mathbf{F} \qquad\qquad (4.1.3\text{-}18)$$

The left-hand side may be written as the sum of three integrals: the integral over flow area (1), the integral over flow area (2), and the integral over the remaining outside surface of the nozzle; however, the last integral is zero because the velocity normal to this surface is everywhere zero. Assuming the density of water to be constant

$$\rho \int_A \mathbf{v}\left(\mathbf{v}\cdot\mathbf{n}\right)dA = \sum \mathbf{F}$$

$$\rho \left[\int_{A_2} \mathbf{v}\, v_z\, dA - \int_{A_1} \mathbf{v}\, v_z\, dA \right] = \sum \mathbf{F} \tag{4.1.3-19}$$

This vector equation represents three algebraic equations for the components.

There is neither force acting nor momentum transferred in the y-direction, so this component equation yields merely

$$0 = 0 \tag{4.1.3-20}$$

There is no momentum transfer in the x-direction, and the only forces are the body force from gravity, any x-directed force in the threads, and any force the firefighter might apply in holding the nozzle, so this equation reduces to

$$0 = \left(F_x\right)_{\text{gravity}} + \left(F_x\right)_{\text{firefighter}} + \left(F_x\right)_{\text{threads}}$$
$$= m_{[\text{nozzle + water}]}\, g_x + \left(F_x\right)_{\text{firefighter}} + \left(F_x\right)_{\text{threads}} \tag{4.1.3-21}$$

which, since $g_x = -32.2$ ft/s^2, says that the firefighter must supply an upward force to counteract the weight of the nozzle and the water it contains, and since the x-component of the force in the threads is probably in the negative x-direction because of transmitted weight of the hose and the water it contains, the firefighter must supply this force as well.

We now turn to the z-component equation. If we assume the velocity profile to be uniform (that v_z is constant across flow area 2 and across flow area 1), we may write the z-component of the left-hand side of (a) as

$$\rho \left[\int_{A_2} \mathbf{v}\, v_z\, dA - \int_{A_1} \mathbf{v}\, v_z\, dA \right]_z = \rho \left[\int_{A_2} v_z\, v_z\, dA - \int_{A_1} v_z\, v_z\, dA \right]$$

$$= \rho \left([v_z^2]_2 \int_{A_2} dA - [v_z^2]_1 \int_{A_1} dA \right)$$

$$= [v_z]_2\, \rho \left([v_z]_2\, A_2 \right) - [v_z]_1\, \rho \left([v_z]_1\, A_1 \right) \tag{4.1.3-22}$$

$$= [v_z]_2\, \rho\, Q_2 - [v_z]_1\, \rho\, Q_1 \tag{4.1.3-23}$$

but a total mass balance at constant density and steady state shows that $Q_1 = Q_2$ = constant = Q. Using this fact and substituting for the velocities

$$\rho \left[\int_{A_2} v\, v_z\, dA - \int_{A_1} v\, v_z\, dA \right]_z = \rho\, Q \left([v_z]_2 - [v_z]_1 \right)$$

$$= \rho\, Q \left(\frac{Q}{A_2} - \frac{Q}{A_1} \right)$$

$$= \rho\, Q^2 \left(\frac{1}{A_2} - \frac{1}{A_1} \right) \qquad (4.1.3\text{-}24)$$

Turning now to the z-component of the right-hand side of (a), we observe that the significant external forces acting on the system in the z-direction are the force in the threads (which is mechanically transmitted by the coupling on the end of the hose) and the pressure forces.

We neglect any shear force on the outside surface of the nozzle from, for example, wind blowing past the nozzle giving frictional drag at the surface. Even if such a force existed, it would normally be small compared to the momentum change.

We also assume that the firefighter holding the nozzle is exerting a force only in the direction opposite to the force of gravity, and hence exerts no force in the z-direction.

$$[\Sigma F]_z = \Sigma F_z \qquad (4.1.3\text{-}25)$$

$$= [F_{threads}]_z + [F_{pressure}]_z \qquad (4.1.3\text{-}26)$$

$$= [F_{threads}]_z + [p_1\, A_1 - p_2\, A_2] \qquad (4.1.3\text{-}27)$$

But since we are working in gauge pressure, $p_2 = 0$. We do not know p_1 and therefore evaluate it by applying the mechanical energy balance to the water inside the nozzle, neglecting the lost work in the nozzle

$$\frac{\Delta p}{\rho} + g\, \Delta z + \Delta \left(\frac{v^2}{2} \right) = -l\tilde{w} - \hat{W}$$

$$\frac{p_2 - p_1}{\rho} + \frac{v_2^2 - v_1^2}{2} = 0 \qquad (4.1.3\text{-}28)$$

$$p_1 = p_2 + \rho\, \frac{\left(v_2^2 - v_1^2 \right)}{2}$$

$$p_1 = \rho \frac{\left(\left[\frac{Q}{A_2}\right]^2 - \left[\frac{Q}{A_1}\right]^2\right)}{2} \tag{4.1.3-29}$$

$$p_1 = \frac{\rho Q^2}{2}\left(\left[\frac{1}{A_2}\right]^2 - \left[\frac{1}{A_1}\right]^2\right) \tag{4.1.3-30}$$

Substituting

$$[\Sigma F]_z = [F_{threads}]_z + \left[\frac{\rho Q^2}{2}\left(\left[\frac{1}{A_2}\right]^2 - \left[\frac{1}{A_1}\right]^2\right)\right] A_1 \tag{4.1.3-31}$$

Substituting the left- and right-hand sides in the z-component macroscopic momentum balance

$$\rho Q^2\left(\frac{1}{A_2} - \frac{1}{A_1}\right) = [F_{threads}]_z + \left[\frac{\rho Q^2}{2}\left(\left[\frac{1}{A_2}\right]^2 - \left[\frac{1}{A_1}\right]^2\right)\right] A_1$$

$$[F_{threads}]_z = \rho Q^2\left(\frac{1}{A_2} - \frac{1}{A_1}\right) - \left[\frac{\rho Q^2}{2}\left(\left[\frac{1}{A_2}\right]^2 - \left[\frac{1}{A_1}\right]^2\right)\right] A_1$$

$$[F_{threads}]_z = \rho Q^2\left\{\left(\frac{1}{A_2} - \frac{1}{A_1}\right) - \left(\left[\frac{1}{A_2}\right]^2 - \left[\frac{1}{A_1}\right]^2\right)\frac{A_1}{2}\right\}$$

$$[F_{threads}]_z = \rho Q^2\left\{\left[\frac{1}{A_2}\right] - \frac{A_1}{2}\left[\frac{1}{A_2}\right]^2 - \frac{1}{2}\left[\frac{1}{A_1}\right]\right\} \tag{4.1.3-32}$$

Noting that

$$A_1 = \frac{\pi}{4}\left(\frac{2.5}{12}\right)^2 = 3.41 \times 10^{-2} \text{ ft}^2$$

$$A_2 = \frac{\pi}{4}\left(\frac{1.125}{12}\right)^2 = 6.90 \times 10^{-3} \text{ ft}^2 \tag{4.1.3-33}$$

$$\left\{ \left[\frac{1}{A_2} \right] - \frac{A_1}{2} \left[\frac{1}{A_2} \right]^2 - \frac{1}{2} \left[\frac{1}{A_1} \right] \right\}$$

$$= \frac{1}{\left(6.90 \times 10^{-3} \right)} - \frac{\left(3.41 \times 10^{-2} \right)}{2} \left[\frac{1}{6.90 \times 10^{-3}} \right]^2 - \frac{1}{2} \left[\frac{1}{3.41 \times 10^{-2}} \right]$$

$$= -\left(228 \right) \text{ft}^{-2}$$

$$(4.1.3\text{-}34)$$

Substituting in the z-component balance

$$\left[F_{\text{threads}} \right]_z = \rho Q^2 \left\{ \left[\frac{1}{A_2} \right] - \frac{A_1}{2} \left[\frac{1}{A_2} \right]^2 - \frac{1}{2} \left[\frac{1}{A_1} \right] \right\}$$

$$= \left(62.4 \right) \frac{\text{lbm}}{\text{ft}^3} \left(500 \right)^2 \frac{\text{gal}^2}{\text{min}^2} \left(\frac{\text{ft}^3}{7.48 \text{ gal}} \right)^2 \frac{\text{lbf s}^2}{\left(32.2 \right) \text{lbm ft}} \left[\frac{\text{min}}{\left(60 \right) \text{s}} \right]^2 \left\{ -\frac{228}{\text{ft}^2} \right\}$$

$$\left[F_{\text{threads}} \right]_z = -548 \text{ lbf}$$

$$(4.1.3\text{-}35)$$

Notice the force in the threads acts to the left on the nozzle - this is what we expect intuitively - if the force in the threads acted to the right, the threads would be unnecessary as the nozzle would be thrust onto the hose.

Example 4.1.3-2   Thrust of aircraft engine

A plane flies at 500 mph. Each engine uses 1000 lbm of air (including bypass air) per lbm of fuel, and each engine supplies 10,000 lbf of thrust. If each engine is burning 2 lbm/sec of fuel, what is the outlet velocity from an engine?

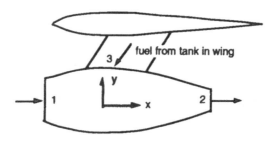

Solution

We choose coordinates attached to the plane as shown to make the control volume fixed with respect to the coordinate system. The velocity and density of the engine and contents are steady in time.

$$\int_A v \rho \left(v \cdot n\right) dA + \frac{d}{dt}\int_V v \rho \, dV = \sum F$$

$$\int_A v \rho \left(v \cdot n\right) dA = \sum F \qquad\qquad (4.1.3\text{-}36)$$

Using a control volume which covers the external surface of the engine, passes through the strut (which contains the fuel line) and cuts inlet area 1 and outlet area 2, we can see that mass enters and leaves only at 1, 2, and 3.

The fuel, which is already moving at the velocity of the airplane, enters at low velocity and flow rate relative to the control volume, so in our model we neglect the momentum entering the control volume at 3. We model the velocity as constant across 1 and 2 (although different at 1 and 2). We model the velocity as 600 mph across 1, even though the demand of the engine for air might differ from this in the real-world case.

We are not interested in the z- component equation because no momentum transfer or flow is occurring in the z- direction. The y-direction balance contains no momentum transfer terms, and so expresses the fact that the force in the strut must support the weight of the engine.

We model the pressure as atmospheric over the entire control surface, including areas 1 and 2; therefore, external forces act on the control volume in the x-direction only where the strut cuts the control surface, and where aerodynamic drag is exerted on the external surface of the control volume. Since the thrust (x-directed force in the strut) is used to overcome the drag over the entire surface of the aircraft, we neglect the drag on the engine surface compared to the thrust.

The velocities are parallel to the x-axis; therefore the magnitude of the velocity is the same as the magnitude of the x-component of the velocity. The x-component of our model then becomes

$$\int_{A_2} \rho \, v_x \, v \, dA - \int_{A_1} \rho \, v_x \, v \, dA = \sum F_x$$

$$v_2 \int_{A_2} \rho \, v \, dA - v_1 \int_{A_1} \rho \, v \, dA = F_{thrust}$$

$$v_2 \, w_2 - v_1 \, w_1 = F_{thrust} \qquad (4.1.3\text{-}37)$$

Note the differing signs from the cos a terms in the scalar product.

In the above equation we do not know the mass flow rates; however, we can obtain them from the fuel consumption rate, the fuel/air ratio, and a macroscopic total mass balance

$$1000 \, w_3 = w_1$$
$$w_2 - w_1 - w_3 = 0$$
$$w_2 - (1000)(2) - (2) = 0$$
$$w_2 = 2002 \, \tfrac{lbm}{s} \qquad (4.1.3\text{-}38)$$

Substituting in the model

$$v_2 \, w_2 - v_1 \, w_1 = F_{thrust}$$

$$(v_2)\left[\tfrac{ft}{s}\right](2002)\left[\tfrac{lbm}{s}\right] - (500)\left[\tfrac{mi}{hr}\right]\left[\frac{(5280)\,ft}{mi}\right]\left[\frac{hr}{(3600)\,s}\right](2000)\left[\tfrac{lbm}{s}\right]$$

$$= (10{,}000)\,[lbf]\left[\frac{(32.2)\,lbm\,ft}{lbf\,s^2}\right] \qquad (4.1.3\text{-}39)$$

$$v_2 = 893 \, \tfrac{ft}{s} \qquad (4.1.3\text{-}40)$$

Example 4.1.3-3   Piping support

The sketch shows a horizontal pipeline elbow carrying water at a rate of 7000 gal/min to supply cooling in an oil refinery.

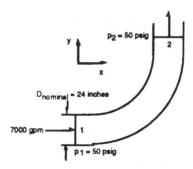

The line is 24 in nominal-diameter Schedule 40 steel pipe. The pressure at points 1 and 2 may both be modeled as 50 psig (negligible pressure drop due to viscous dissipation). The velocity profiles at 1 and 2 may be modeled as flat.

Calculate the magnitude and direction of the resultant force in the pipe and supports.

For 24-in. nominal-diameter pipe:

- ID = 22.626 in.

- flow cross-sectional area = 2.792 ft^2

Solution

Using a control volume which goes over the outside surface of the pipe and cuts across the flowing stream at 1 and 2, choosing coordinate axes as shown, and applying the macroscopic momentum balance at steady state gives

$$\int_A \mathbf{v}\,\rho\,(\mathbf{v}\cdot\mathbf{n})\,dA + \frac{d}{dt}\int_V \mathbf{v}\,\rho\,dV = \sum \mathbf{F}$$

$$\int_A \mathbf{v}\,\rho\,(\mathbf{v}\cdot\mathbf{n})\,dA = \sum \mathbf{F} \tag{4.1.3-41}$$

Since the only flow in or out of the control volume is at 1 and 2

$$\int_{A_2} \rho\,\mathbf{v}\,v\,dA - \int_{A_1} \rho\,\mathbf{v}\,v\,dA = \sum \mathbf{F} \tag{4.1.3-42}$$

The density can be modeled as constant.

$$\rho \int_{A_2} \left(v_y \, \mathbf{j} \right) v \, dA - \rho \int_{A_1} \left(v_x \, \mathbf{i} \right) v \, dA = F_x \, \mathbf{i} + F_y \, \mathbf{j} \qquad (4.1.3\text{-}43)$$

$$\rho \left(v_2 \, \mathbf{j} \right) v_2 \int_{A_2} dA - \rho \left(v_1 \, \mathbf{i} \right) v_1 \int_{A_1} dA = F_x \, \mathbf{i} + F_y \, \mathbf{j}$$

$$\left(\rho \, v_2 \, v_2 \, A_2 \right) \mathbf{j} - \left(\rho \, v_1 \, v_1 \, A_1 \right) \mathbf{i} = F_x \, \mathbf{i} + F_y \, \mathbf{j} \qquad (4.1.3\text{-}44)$$

But a macroscopic total mass balance shows

$$w_2 - w_1 = 0$$
$$w_2 = w_1 = w \qquad (4.1.3\text{-}45)$$

and the fact that w is constant implies that $v_1 = v_2 = v$ since ρ and A are constant.

$$w = \rho \, A \, v \qquad (4.1.3\text{-}46)$$

The fact that v is constant, combined with constant A, implies that Q is constant, since

$$Q = A \, v \qquad (4.1.3\text{-}47)$$

Using these facts

$$\left(-\rho \, v \, Q \right) \mathbf{i} + \left(\rho \, v \, Q \right) \mathbf{j} = F_x \, \mathbf{i} + F_y \, \mathbf{j} \qquad (4.1.3\text{-}48)$$

The external forces are pressure forces and the forces acting through the pipe walls and any supports. Noting that the y-direction pressure force is in the negative y-direction

$$\left(-\rho \, v \, Q \right) \mathbf{i} + \left(\rho \, v \, Q \right) \mathbf{j}$$
$$= \left([F_x]_{pipe} + [F_x]_{pressure} \right) \mathbf{i} + \left([F_y]_{pipe} + [F_y]_{pressure} \right) \mathbf{j} \qquad (4.1.3\text{-}49)$$

$$\left(-\rho \, \frac{Q}{A_1} \, Q \right) \mathbf{i} + \left(\rho \, \frac{Q}{A_2} \, Q \right) \mathbf{j}$$
$$= \left([F_x]_{pipe} + p_1 \, A_1 \right) \mathbf{i} + \left([F_y]_{pipe} - p_2 \, A_2 \right) \mathbf{j} \qquad (4.1.3\text{-}50)$$

$$\left(-\rho \frac{[Q]^2}{A}\right)i + \left(\rho \frac{[Q]^2}{A}\right)j = \left([F_x]_{pipe} + p\,A\right)i + \left([F_y]_{pipe} - p\,A\right)j$$

(4.1.3-51)

Equating x-components

$$-\rho \frac{[Q]^2}{A} = [F_x]_{pipe} + p\,A$$

$$[F_x]_{pipe} = -\left(\rho \frac{[Q]^2}{A}\right) - (p\,A)$$

(4.1.3-52)

$$= -(62.4)\left[\frac{lbm}{ft^3}\right]\frac{(7000)^2\left[\frac{gal}{min}\right]^2}{(2.792)\,ft^2}$$

$$\times \left[\frac{min}{(60)\,s}\right]^2\left[\frac{ft^3}{(7.48)\,gal}\right]^2\left[\frac{lbf\,s^2}{(32.2)\,lbm\,ft}\right]$$

$$-(50)\left[\frac{lbf}{in^2}\right](2.792)\left[ft^2\right]\left[\frac{(144)\,in^2}{ft^2}\right]$$

(4.1.3-53)

$$= -20{,}270\ lbf$$

(4.1.3-54)

A similar calculation equating the y-components gives

$$[F_y]_{pipe} = 20{,}270\ lbf$$

(4.1.3-55)

Therefore, the net force exerted on the control volume by the pipe supports and the pipe walls at 1 and 2 is

$$[F]_{pipe} = \left(-20{,}270\ lbf\right)i + \left(20{,}270\ lbf\right)j$$

(4.1.3-56)

We cannot separate the forces into the forces exerted by the pipe supports and by the pipe at 1 and 2 without knowing more about the system.

Example 4.1.3-4   Jet boat

A boat is propelled by a jet at water as in the sketch.

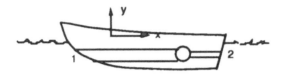

The inlet, at 1, is 6-in. inside diameter and the outlet, at 2, is 3-in. inside diameter. The motor and pump transfers water at a rate of 300 gal/min. If the drag on the hull can be expressed as

$$F = 0.25 \ v^2 \qquad\qquad (4.1.3\text{-}57)$$

where

> v = speed of boat relative to fixed water, ft/sec
> F = total drag force, lbf

and the constant 0.25 has dimensions of $[lbf \ sec^2]/ft^2$

Calculate the speed of the boat.

Solution

Using coordinates attached to the boat, and a control volume which runs over the exterior of the boat, cutting the inlet and outlet flow cross-sections, the overall momentum balance is

$$\int_A v \rho \left(v \cdot n \right) dA + \frac{d}{dt}\int_V v \rho \ dV = \sum F$$

$$\int_A v \rho \left(v \cdot n \right) dA = \sum F \qquad\qquad (4.1.3\text{-}58)$$

Since the only flow in or out of the control volume is at 1 and 2

$$\int_{A_2} \rho \, v \, v \, dA - \int_{A_1} \rho \, v \, v \, dA = \sum F \qquad\qquad (4.1.3\text{-}59)$$

Assuming plug flow at the inlet and outlet

$$\rho \int_{A_2} (v_y \mathbf{i}) v \, dA - \rho \int_{A_1} (v_x \mathbf{i}) v \, dA = F_x \mathbf{i} + F_y \mathbf{j} \tag{4.1.3-60}$$

$$\rho (v_2 \mathbf{i}) v_2 \int_{A_2} dA - \rho (v_1 \mathbf{i}) v_1 \int_{A_1} dA = F_x \mathbf{i} + F_y \mathbf{j}$$

$$\left[(v_2 \rho v_2 A_2) - (v_1 \rho v_1 A_1) \right] \mathbf{i} = F_x \mathbf{i} + F_y \mathbf{j}$$

$$\left[(v_2 \rho Q_2) - (v_1 \rho Q_1) \right] \mathbf{i} = F_x \mathbf{i} + F_y \mathbf{j} \tag{4.1.3-61}$$

But a macroscopic total mass balance shows that $w_1 = w_2 = w$, which, combined with the constant density implies that $Q = $ constant. The x-component of the force is the drag force, which acts to the right as the boat proceeds to the left. Equating the components in the x-direction yields an equation for the speed.

$$\rho Q (v_2 - v_1) = 0.025 \, v^2$$

$$\rho Q \left(\frac{Q}{A_2} - \frac{Q}{A_1} \right) = 0.025 \, v^2$$

$$\rho Q^2 \left(\frac{1}{A_2} - \frac{1}{A_1} \right) = 0.025 \, v^2 \tag{4.1.3-62}$$

Substituting known data

$$(62.4) \left[\frac{\text{lbm}}{\text{ft}^3} \right] (300)^2 \left[\frac{\text{gal}}{\text{min}} \right]^2 \left[\frac{\text{min}}{(60) \, \text{s}} \right]^2 \left[\frac{\text{ft}^3}{(7.48) \, \text{gal}} \right]^2$$

$$\times \left(\frac{1}{\left(\frac{\pi}{4} \right) \left(\frac{3}{12} \right)^2 \text{ft}^2} - \frac{1}{\left(\frac{\pi}{4} \right) \left(\frac{6}{12} \right)^2 \text{ft}^2} \right) \left[\frac{\text{lbf s}^2}{(32.2) \, \text{lbm ft}} \right] \tag{4.1.3-63}$$

$$= (0.025) \left[\frac{\text{lbf s}^2}{\text{ft}^2} \right] (v_b)^2 \left[\frac{\text{ft}^2}{\text{s}^2} \right]$$

$$v = 23 \tfrac{\text{ft}}{\text{s}} \tag{4.1.3-64}$$

(About 15 to 16 mph)

Example 4.1.3-5   Horizontal force on tank

Water is flowing at steady state into and out of a tank through pipes open to the atmosphere as shown

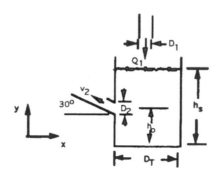

$D_T = 3\,m$
$D_1 = 10\,cm$
$D_2 = 4\,cm$
$\dot{Q}_1 = 0.05\,\dfrac{m^3}{min}$
$h_o = 0.5\,m$

The velocity profile through the exit orifice can be assumed to be flat and described by

$$v_2 = \sqrt{2\,g\,h}$$

(4.1.3-65)

where

$g = 9.8\,m/s^2$
$h = $ distance to surface, m

Find the horizontal component of the force of the floor supporting the tank in Newtons.

Solution

Writing the macroscopic momentum balance on a control volume that goes over the outside surface of the tank and exit pipe and cuts A_1 as shown

$$\int_A \mathbf{v} \rho \left(\mathbf{v} \cdot \mathbf{n}\right) dA + \frac{d}{dt} \int_V \mathbf{v} \rho \, dV = \sum \mathbf{F} \tag{4.1.3-66}$$

modeling the velocity of the contents of the tank as $\mathbf{v} = \mathbf{0}$ and using the fact that we are at steady state

$$\int_{A_1} \mathbf{v} \rho \left(\mathbf{v} \cdot \mathbf{n}\right) dA + \int_{A_2} \mathbf{v} \rho \left(\mathbf{v} \cdot \mathbf{n}\right) dA = \sum \mathbf{F}$$

$$-\rho \, v_1 \, \mathbf{v}_1 \int_{A_1} dA + \rho \, v_2 \, \mathbf{v}_2 \int_{A_2} dA = \sum \mathbf{F}$$

$$\rho \, v_2 \, \mathbf{v}_2 \, A_2 - \rho \, v_1 \, \mathbf{v}_1 \, A_1 = \sum \mathbf{F} \tag{4.1.3-67}$$

Writing the vectors in terms of their projections on the axes and equating the x-components

$$\rho \, v_2 \, A_2 \left[\left(v_x\right)_2 \mathbf{i} + \left(v_y\right)_2 \mathbf{j}\right] - \rho \, v_1 \, A_1 \left[\left(v_x\right)_1 \mathbf{i} + \left(v_y\right)_1 \mathbf{j}\right] = \sum \left[F_x \mathbf{i} + F_y \mathbf{j}\right]$$

$$\rho \, v_2 \, A_2 \left(v_x\right)_2 - \rho \, v_1 \, A_1 \left(v_x\right)_1 = \sum F_x \tag{4.1.3-68}$$

$$-\rho \, v_2 \, A_2 \, v_2 \cos 30° = F_x$$

$$F_x = -\rho \, v_2^2 \, A_2 \cos 30° \tag{4.1.3-69}$$

But a total mass balance at steady state gives

$$w_2 = w_1$$

$$\rho \, A_2 \, v_2 = \rho \, A_1 \, v_1 = \rho \, Q_1$$

$$v_2 = \frac{Q_1}{A_2} \tag{4.1.3-70}$$

so, substituting in the x-component equation

$$F_x = -\rho \, v_2^2 \, A_2 \cos 30°$$

$$F_x = -\rho \left(\frac{Q_1}{A_2}\right)^2 A_2 \left(0.866\right)$$

$$F_x = -\rho \frac{Q_1^2}{A_2} \left(0.866\right)$$

$$F_x = -\left(1000\right) \left[\frac{kg}{m^3}\right] \left(0.05\right)^2 \left[\frac{m^3}{min}\right]^2 \left(\frac{4}{\pi \, 4^2}\right) \left[\frac{1}{cm^2}\right] \left(0.866\right) \left[\frac{N \, s^2}{kg \, m}\right]$$

$$\times\left[\frac{(100)\ \text{cm}}{\text{m}}\right]^2\left[\frac{\text{min}}{(60)\ \text{s}}\right]^2 \tag{4.1.3-71}$$

$$F_x = -0.479\ \text{N} \tag{4.1.3-72}$$

4.2 The Microscopic Momentum Balance

The **macroscopic** momentum balance was determined in section 4.1 to be

$$\int_A v\,\rho\,(v\cdot n)\,dA + \frac{d}{dt}\int_V v\,\rho\,dV = \sum F \tag{4.2-1}$$

Normally we are concerned with the body force due to gravity only. The surface force may be expressed in terms of the stress vector T^n. For only these forces acting, Equation (4.2-1) becomes

$$\int_A v\,\rho\,(v\cdot n)\,dA + \frac{d}{dt}\int_V v\,\rho\,dV = \int_V \rho\,g\,dV + \int_A T^a\,dA \tag{4.2-2}$$

$$\int_A v\,\rho\,(v\cdot n)\,dA + \frac{d}{dt}\int_V v\,\rho\,dV = \int_V \rho\,g\,dV + \int_A (T\cdot n)\,dA \tag{4.2-3}$$

$$\int_A v\,\rho\,(v\cdot n)\,dA + \frac{d}{dt}\int_V v\,\rho\,dV = \int_V \rho\,g\,dV + \int_A \left(\left[\tau + p\,\delta\right]\cdot n\right)\,dA \tag{4.2-4}$$

Forming the scalar product with an arbitrary vector and applying the divergence theorem to replace the area integrals with volume integrals

$$\int_V \left[\frac{\partial(\rho\,v)}{\partial t} + \nabla\cdot(\rho\,v\,v) + \nabla p + (\nabla\cdot\tau) - \rho\,g\right]dV = 0 \tag{4.2-5}$$

Since the limits on this integration are arbitrary, the term in brackets must be zero and

$$\boxed{\frac{\partial(\rho \, v)}{\partial t} + \nabla \cdot (\rho \, v \, v) + \nabla p + (\nabla \cdot \tau) - \rho \, g = 0}$$ (4.2-6)

Equation (4.2-6) is the *microscopic momentum balance*. The terms represent the rate of increase per unit volume of momentum:

- the first, at a point;
- the second, convection of momentum into or from the point;
- the third, from pressure forces;
- the fourth, from the action of viscous forces; and
- the fifth, from gravitational forces.

The equation can also be written as

$$\rho \frac{Dv}{Dt} = -\nabla p - (\nabla \cdot \tau) + \rho \, g$$ (4.2-7)

If we take the dot product with the velocity, **v**

$$\rho \frac{D(v \cdot v)}{Dt} = -(v \cdot \nabla p) - (v \cdot [\nabla \cdot \tau]) + \rho \, (v \cdot g)$$ (4.2-8)

$$\rho \frac{D\left(\frac{v^2}{2}\right)}{Dt} = -(v \cdot \nabla p) - (v \cdot [\nabla \cdot \tau]) + \rho \, (v \cdot g)$$ (4.2-9)

$$\boxed{\begin{aligned} \rho \frac{D\left(\frac{v^2}{2}\right)}{Dt} &= -(\nabla \cdot p \, v) - p \, (-\nabla \cdot v) \\ &\quad - (\nabla \cdot [\tau \cdot v]) - (-[\tau : \nabla v]) \\ &\quad + \rho \, (v \cdot g) \end{aligned}}$$ (4.2-10)

This is the *microscopic mechanical energy balance*. It is in Lagrangian form, and therefore describes the rate of change per unit volume of kinetic energy as one follows the fluid motion. The terms on the right-hand side represent, respectively,

- rate of work done per unit volume by pressure forces

> • rate of reversible conversion to internal energy per unit
> volume
> • rate of work done per unit volume by viscous forces
> • rate of irreversible conversion to internal energy per unit
> volume
> • rate of work done per unit volume by gravitational forces

To illustrate a common application of Equation (4.2-7), for a Newtonian fluid, where x is the position vector

$$\tau = -\mu \frac{dv}{dx} \qquad\qquad (4.2\text{-}11)$$

Substituting this in Equation (4.2-7) and using the definitions

$$\frac{D}{Dt} \equiv \frac{\partial}{\partial t} + v_x \frac{\partial}{\partial x} + v_y \frac{\partial}{\partial y} + v_z \frac{\partial}{\partial z} \qquad\qquad (4.2\text{-}12)$$

$$\nabla^2 \equiv \frac{\partial^2}{\partial x^2} + \frac{\partial^2}{\partial y^2} + \frac{\partial^2}{\partial z^2} \qquad\qquad (4.2\text{-}13)$$

yields a differential equation called the **Navier-Stokes** equation.

$$\boxed{\rho \frac{Dv}{Dt} = -\nabla p + \mu \nabla^2 v + \rho g} \qquad\qquad (4.2\text{-}14)$$

The investigation of solutions of this equation for various system where momentum is transferred by viscous motion is a large part of the topic of momentum transport as a transport phenomenon.

For example, the axial flow of an incompressible Newtonian fluid in a horizontal pipe is described in cylindrical coordinates by

$$\rho \frac{\partial v_z}{\partial t} = -\frac{dp}{dx} + \mu \frac{1}{r} \frac{\partial}{\partial r} \left(r \frac{\partial v_z}{\partial r} \right) \qquad\qquad (4.2\text{-}15)$$

Chapter 6 is devoted entirely to the topic of momentum transfer in fluid flow and utilizes the momentum balance in both its integral and differential forms.

At this point it is useful to summarize both the macroscopic and microscopic equations of mass, energy, and momentum.

4.3 Summary of Balance Equations and Constitutive Relationships

The balance equations in Table 4.3-1 are not applicable to systems where there is interconversion of energy and mass.

Table 4.3-1 Tabulation of balance equations

	MACROSCOPIC	MICROSCOPIC
MOMENTUM	$\int_A \mathbf{v}\rho(\mathbf{v}\cdot\mathbf{n})\,dA$ $+\frac{d}{dt}\int_V \mathbf{v}\rho\,dV = \Sigma F$	$\rho\frac{D\mathbf{v}}{Dt} =$ $-\nabla p + \mu\nabla^2\mathbf{v} + \rho\mathbf{g}$
ENERGY - TOTAL	$\int_A \left(\hat{H}+\Phi+\hat{K}\right)\rho\left(\mathbf{v}\cdot\mathbf{n}\right)dA$ $+\frac{d}{dt}\int_V\left(\hat{U}+\Phi+\hat{K}\right)\rho\,dV$ $= Q - W$	$\rho\dfrac{D\left[\hat{U}+\frac{v^2}{2}\right]}{Dt} =$ $-(\nabla\cdot\mathbf{q}) - \nabla\cdot[\tau\cdot\mathbf{v}]$ $-\nabla\cdot[p\,\mathbf{v}] + \rho(\mathbf{v}\cdot\mathbf{g})$
ENERGY[6] - THERMAL	$\int_A \hat{U}\rho(\mathbf{v}\cdot\mathbf{n})\,dA$ $+\frac{d}{dt}\int_V \hat{U}\rho\,dV =$ $Q + \int_V \gamma_\theta\,dV$	$\frac{D\hat{U}}{Dt} = -(\nabla\cdot\mathbf{q})$ $-p(\nabla\cdot\mathbf{v}) - (\tau:\nabla\mathbf{v})$
ENERGY - MECHANICAL	$\int_A \left(p\hat{V}+\Phi+\hat{K}\right)\rho\left(\mathbf{v}\cdot\mathbf{n}\right)dA$ $+\frac{d}{dt}\int_V\left(\Phi+\hat{K}\right)\rho\,dV$ $= -lw - W$	$\rho\dfrac{D\left(\frac{v^2}{2}\right)}{Dt} =$ $-(\nabla\cdot p\mathbf{v}) - p(-\nabla\cdot\mathbf{v})$ $-(\nabla\cdot[\tau\cdot\mathbf{v}]) - (-[\tau:\nabla\mathbf{v}])$ $+\rho(\mathbf{v}\cdot\mathbf{g})$
MASS - TOTAL	$\int_A \rho(\mathbf{v}\cdot\mathbf{n})\,dA$ $+\frac{d}{dt}\int_V \rho\,dV = 0$	$\nabla\cdot(\rho\,\mathbf{v}) + \frac{\partial\rho}{\partial t} = 0$
MASS - SPECIES	$\int_A \rho_i(\mathbf{v}\cdot\mathbf{n})\,dA + \frac{d}{dt}\int_V \rho_i\,dV$ $= \int_V r_i\,dV$	$\nabla\cdot(\rho_i\mathbf{v}_i) + \frac{\partial\rho_i}{\partial t} = r_i$

---

[6] For one-dimensional flow of an incompressible fluid.

We summarize the common constitutive relationships - those equations that relate material properties to the fluxes - in Table 4.3-2.

Table 4.3-2 Tabulation of common constitutive relationships

	MACROSCOPIC	MICROSCOPIC[7]
MOMENTUM	$\dfrac{F}{A} = \frac{1}{2} C_D \rho \, v (v - 0)$ $\dfrac{F}{A} = \frac{1}{2} \rho \, f \langle v \rangle (\langle v \rangle - 0)$	$\tau_{yx} = -\mu \dfrac{dv_x}{dy}$
ENERGY	$\dfrac{Q}{A} = h (T - T_S)$	$q_x = -k \dfrac{dT}{dx}$
MASS	$N_A = k_x (x_A - x_{AS})$	$N_A = x_A (N_{Az} + N_{Bz})$ $- c \, \mathscr{D}_{AB} \dfrac{dx_A}{dz}$

4.4   The Momentum Equation in Non-Inertial Reference Frames

The momentum equation as we have written it applies to an **inertial** reference frame system. By an inertial frame we mean a frame which is not accelerated.[8] (We exclude rotation entirely and permit only uniform translation.) These frames are also referred to as Gallilean frames or Newtonian frames. The acceleration of frames attached to the earth's surface, however, is slight enough that our equations give good answers for most engineering problems - the error is of the order of one-third of 1 percent. For problems involving relatively large acceleration of the reference frame, however, Newtonian mechanics must be reformulated. An example of such a problem is the use of a reference frame attached to a rocket as it accelerates from zero velocity on the pad to some orbital velocity. For such a problem, observations from an inertial frame and from the accelerating frame can differ greatly.

For example, if an astronaut in zero gravity but in an accelerating rocket drops a toothbrush, he or she sees it apparently accelerate toward the "floor" (wherever the surface opposite to the direction of the acceleration vector of the spacecraft might be at the time). This observation is with respect to an accelerating frame (the spacecraft) within which the astronaut is fixed. To an observer in an inertial frame, however, the toothbrush simply continues at the

---

[7] One-dimensional.

[8] Fanger, C. G. (1970). *Engineering Mechanics: Statics and Dynamics.* Columbus, OH, Charles E. Merrill Books, Inc.

constant velocity it had when released, and the spacecraft accelerates past the toothbrush.

In other words, observed acceleration depends on the motion of the reference frame in which it is observed. If this is true, Newton's second law must change in form when written with respect to an accelerating reference frame, because the force experienced by a body should not depend on the motion of the observer. Here we will consider only a simple example of this - that of a system of constant mass.[9]

Return to the astronaut's toothbrush. There is no net external force on the toothbrush (we neglect any drag from the atmosphere within the spacecraft). Newton's law becomes

$$\Sigma\,F \;=\; m\,a_{\text{toothbrush, observed from inertial frame}}$$
$$0 \;=\; m\,a_{\text{toothbrush, observed from inertial frame}} \qquad\qquad (4.4\text{-}1)$$

which implies that the acceleration of the toothbrush with respect to an inertial coordinate frame is zero.

With respect to an accelerating frame (one attached to the spacecraft), however, there is an apparent acceleration of the toothbrush. This apparent acceleration is equal to the negative of the difference in acceleration between that of a frame attached to the spacecraft and that of the inertial frame. Since the acceleration of the inertial frame is zero, the relative acceleration is

$$a_{\text{relative}} \;=\; a_{\text{spacecraft}} - a_{\text{inertia}} \;=\; a_{\text{spacecraft}} - 0 \;=\; a_{\text{spacecraft}} \qquad (4.4\text{-}2)$$

Newton's law will still be operationally valid in the accelerating coordinate frame if we

> • interpret the acceleration on the right-hand side as the acceleration of the object as observed from the accelerating frame and also
> • add an appropriate force (the **inertial body force**) to the left-hand side.

The inertial body force is the negative of the product of the mass and the acceleration of the accelerating frame (the spacecraft)[10]

---

[9] We neglect the drag force from the atmosphere in this discussion.

[10] Hansen, A. G. (1967). *Fluid Mechanics*. New York, NY, Wiley.

$$\sum F + \text{(inertial body force)} = m\, a_{\text{relative to accelerating frame}} \qquad (4.4\text{-}3)$$

$$\sum F + (-m\, a_{\text{accelerating frame}}) = m\, a_{\text{relative to accelerating frame}} \qquad (4.4\text{-}4)$$

$$\sum F + (-m\, a_{\text{spacecraft}}) = m\left(-a_{\text{spacecraft}}\right)$$

$$\sum F = 0 \qquad (4.4\text{-}5)$$

Another example is a camera sitting on the ledge above the rear seat of your car. If you hit a cow which happens to be crossing the road, and therefore decelerate rapidly from a velocity of 60 mph, the camera flies forward with respect to the car as well as to the ground. The sum of forces on the camera is zero (again neglecting friction from the air). This time the coordinate frame is decelerating, not accelerating.

To the hitchhiker standing at the side of the road, the acceleration of the camera is zero - it continues at 60 mph until the intervention of a force (perhaps from the windshield - hopefully not the back of your head). In your accelerating (in this case, decelerating) reference frame, there is still no external force on the object, but there is an inertial body force of $\left(-m\, a_{\text{automobile}}\right)$, where $a_{\text{automobile}}$ is the relative acceleration between car and ground (inertial frame). This is balanced on the right-hand side of the equation by the same mass times the relative acceleration of the camera to the car, which is the same in magnitude as and opposite in direction to the relative acceleration of the car and the ground.

$$\sum F + \text{(inertial body force)} = m\, a_{\text{relative to automobile}}$$

$$\sum F + (m\, a_{\text{automobile}}) = m\, a_{\text{relative to automobile}} \qquad (4.4\text{-}6)$$

$$\sum F = 0$$

In the case of rotating frames things become more complex, and the reader is referred to texts such as that by Fanger,[11] which cover this subject in more detail.

---

[11] Fanger, C. G. (1970). *Engineering Mechanics: Statics and Dynamics*. Columbus, OH, Charles E. Merrill Books, Inc., p. 562.

Chapter 4 Problems

4.1 A fire hose has an inside diameter of 1 1/2 in. and a nozzle diameter of 0.5 in. The inlet pressure to the hose is 40 psig at a water flow of 8 ft³/min. If a flat velocity profile is assumed, what is the value of and the direction of the force exerted by the fireman holding the hose?

4.2 Your company is planning to manufacture a boat propelled by a jet of water. The boat must be capable of exerting a 200-lbf pull on a line which holds it motionless. The intake to the boat is inclined at an angle of 85° and is a pipe with a 6-in. inside diameter. The motor and pump are capable of delivering 150 gal/min through the horizontal outlet.

> a. Calculate the outlet pipe size which gives the required thrust.
> b. Calculate the minimum horsepower output required of the motor if its efficiency is 100 percent.

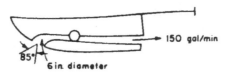

4.3 Water is flowing out of the frictionless nozzle shown in the following illustration, where $p_1 = 120$ psig and the nozzle discharges to the atmosphere.

> a. Show a control volume and all control surfaces with all forces labeled.
> b. Find the resultant force in the bolts at (1) which is necessary to hold the nozzle on the pipe.
> c. If mercury, sp. gr. = 13.6, were flowing through the nozzle instead of water at the same conditions, what would be the new resultant force on the bolts?

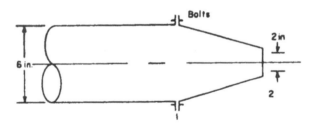

4.4 Consider the straight piece of pipe

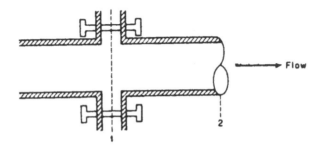

$p_1 = 82$ psig
$Q = 300$ gal/min
$\rho = 62.4$ lbm/ft^3; 0.1337 ft^3/gal
$D = 4$ in.

   a. Draw the control volume and label the control surfaces.
   b. Calculate the force on the bolts at point 1. Are the bolts in
   tension or compression?

4.5 Water is flowing out of a frictionless nozzle discharging to the atmosphere.
The inlet diameter and nozzle diameter are shown on the sketch. The pressure at
$p_2 = 80$ bar.

   a. Sketch a control volume and label it.
   b. What is the resultant force on the bolts holding the nozzle
   to the pipe?
   c. If a fluid of specific gravity = 10 flows through the nozzle
   what is the resulting force on the bolts?

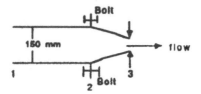

4.6 Consider the sketch below where
$p_1 = 300$ psig; $p_2 = 270$ psig; $Q = 400$ gpm
$D = 5$ in, $r = 62.4$ lbm/ft^3

    a. Draw a control volume and label it
    b. Calculate the force on the bolts
    c. Are the bolts in compression or tension?

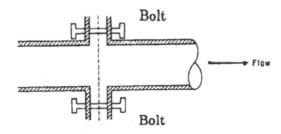

4.7 Water ($\rho = 62.4$ lbm/ft^3) is flowing through the 180∞ elbow represented by the sketch. The inlet gage pressure is 15 lbf/in^2. The diameter of the pipe at 1 is 2 inches. The diameter of the opening to the atmosphere at 2 is 1 inch $v_1 = 10$ ft/s.

    a. Write down the macroscopic total-momentum balance and simplify it for this problem (state your assumptions).
    b. What is the horizontal component of force required to hold the elbow in place?

4.8 Water is flowing through a 60° reducing elbow as in the sketch. The volume of the elbow is 28.3 in^3. Calculate the resultant force on the elbow.

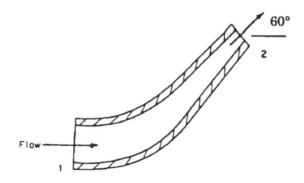

$p_1 = 30$ psig; $p_2 = 28$ psig
$A_1 = 12.56$ in^2; $A_2 = 3.14$ in^2
$w = 75$ lbm/sec

4.9 Consider steady, constant-temperature flow of water through the 45°
reducing elbow shown below. The volume of the elbow is 28.3 in^3.

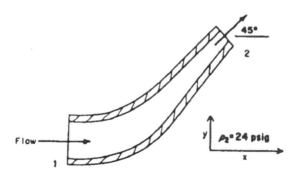

$p_1 = 26$ psig; $w = 85$ lbm/sec
$A_1 = 12.56$ in^2; $A_2 = 3.14$ in^2

Draw a control volume and label all control surfaces and forces on the
control volume. Calculate the total resultant force acting on the elbow.

4.10 Two streams of water join together at a reducing tee where the upstream
pressure of both is 40 psig. Stream 1 is flowing in a 1-in. pipe at 50 lbm/min,
stream 2 in a 2-in. pipe at 100 lbm/min, and stream 3 exits from the tee in a 3-

in. pipe. The straight run flow is from stream 2 to stream 3. Calculate the force of the pipe threads on the tee if the downstream pressure is 35 psig. Use positive coordinates in the direction of flow as shown in the following figure.

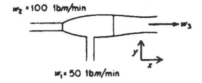

4.11 Consider the horizontal lawn sprinkler shown. Observe that if the sprinkler is split in half as shown, the two halves are identical and thus only one half need be considered. For one-half of the sprinkler:

> a. Use a mass balance to obtain an expression for v_0 in terms of v_1.
> b. Use the frictionless mechanical energy balance to find v_1.
> c. Find the value of F_R that will just prevent the sprinkler from rotating.

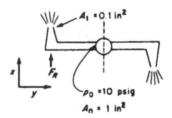

4.12 A stand is to be built to hold a rocket ship stationary while a lateral or side thrust engine is being fired. Twenty lbm/s of fuel is consumed and ejected only out of the side engine at a velocity of 4,000 ft/s. The direction of flow is as shown in the following diagram. Find the x and y components of the restraining force required of the stand.

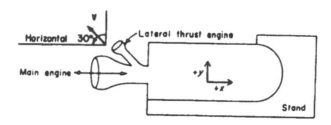

4.13 The tank shown below is secured to a concrete slab by bolts and receives an organic liquid (sp. gr. 0.72) from a filler line which enters the tank 1 ft above the tank floor. The liquid enters at a rate of 4,500 gal/min. What are the forces exerted on the restraining bolts? Assume that the filler line transmits no force to the tank (due to expansion joints).

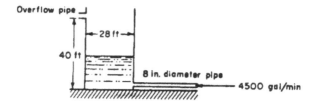

4.14 A cylindrical storage tank is connected to a nozzle by a short length of horizontal pipe as shown in the following diagram. At time equal to zero, the height of liquid in the tank is H ft above the axis of the discharge pipe. The liquid discharges through the nozzle into the atmosphere as the liquid level h(t) changes.

a. Using a mass balance, show that the liquid level h(t) is related to the nozzle discharge velocity v_2 by

$$\frac{dh}{dt} = \left(\frac{D_2}{D_T}\right)^2 v_2$$

b. Using a mechanical energy balance on the discharge pipe and the results of part a, show that the nozzle discharge velocity v_2 and the gage pressure at the pipe entrance $(p_1 - p_2)$ are given as functions of time by

$$v_2(t) = \left(\frac{2gH}{r(1-b^2)}\right)^{1/2} - \frac{g}{r(1-b^2)}\left(\frac{D_2}{D_T}\right)^2 t$$

where $b = (D_2/D_1)^2$

$$p_1 - p_2 = \left[H^{1/2} - \frac{1}{2}\left(\frac{D_2}{D_T}\right)^2\left(\frac{2g}{r(1-b^2)}\right)^{1/2}t\right]^2$$

c. Using a momentum balance on the nozzle and the results of part b, show that the force exerted on the threads of the nozzle F_T, is given as a function of time by

$$F_R = \left[rD_2^2(1-b^2)\left(\frac{2g}{r(1-b^2)}\right)^{1/2} - D_1^2\right]\frac{p}{4}rD_2^2\left[H^{1/2} - \frac{1}{2}\left(\frac{D_2}{D_T}\right)^2\left(\frac{2g}{1-b^2}\right)^{1/2}t\right]^2$$

Assume there is no pressure drop across the short section of pipe between the tank and the nozzle.

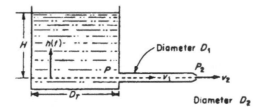

5

APPLICATION OF DIMENSIONAL ANALYSIS

5.1 Systems of Measurement

A **dimension** or **unit** is a way of assigning a numerical value to a property. For example, one could assign the number **66** to the property **height** as measured in the dimension or unit of **inches**. The appropriate number if the property were measured in cm would be 167.6; if the property were measured in feet, 5.5.

To have physical meaning, equations must be **dimensionally homogeneous**: in an equation of the form

$$f_1(x_1, x_2, x_3, \cdots x_n) + f_2(x_1, x_2, x_3, \cdots x_n) + f_3(x_1, x_2, x_3, \cdots x_n)$$
$$+ \cdots + f_m(x_1, x_2, x_3, \cdots x_n) = 0 \tag{5.1-1}$$

each of the terms f_1, f_2, f_3, ... f_m must have the same dimension for the equation to be meaningful.

For example, suppose that I have two bank accounts, one in the United States of America and one in England, and am asked, "How much money do you have?" Suppose further that my account in the United States has a balance of \$1000 and that my account in England has a balance of 500£. I do not get a meaningful application of the equation

$$\text{total money} = \text{money in US} + \text{money in England} \tag{5.1-2}$$

if I perform the calculation as

$$\text{total money} = \$1000 + 500£ = 1500 ? \tag{5.1-3}$$

The answer is meaningless, because I have added two numbers that result from measurements using two different scales.

The calculation is meaningful if (assuming the current exchange rate to be $1.50 = 1£) I convert the terms to the same units, either as

$$\text{total money} = \$1000 + 500£ \frac{\$1.5}{£} = \$1750 \tag{5.1-4}$$

or

$$\text{total money} = \$1000 \frac{£}{\$1.5} + 500£ = 1167£ \tag{5.1-5}$$

Notice that one can formally multiply and divide dimensions (multiplied by the appropriate scale factor) like algebraic quantities (even though they are not). In effect, this amounts to multiplication by unity - in the above example

$$\$1.50 = 1£ \Rightarrow \begin{Bmatrix} 1 = \frac{\$1.5}{1£} \\ 1 = \frac{1£}{\$1.5} \end{Bmatrix} \tag{5.1-6}$$

In this example we were concerned with measuring only one variable - money. In the usual engineering problem, however, we are concerned with many variables, and so we need many measures; in other words, a system of units, not just a single unit.

Useful **systems** of dimensions or units for engineering and scientific use must be consistent with fundamental physical laws - e.g., Newton's second law, which says that models of our physical systems must obey the law

$$\text{force} = \text{mass} \times \text{acceleration} \tag{5.1-7}$$

Because of this particular constraint, we can **independently** define only three of the four dimensions, mass, length, time, and force, because any combination of these units must satisfy Newton's second law.

For example, we could define the unit of mass as the pound (lbmass[1]) by measuring it in units of some mass we retain as a standard, the unit of length as the foot (ft.) in terms of the wavelength of certain prescribed radiation, and the unit of time as the second (s) in terms of the oscillations of a particular molecule under prescribed conditions. We then cannot define the unit of force independently, because Newton's law requires that

$$F \text{ [force units]} = m \text{ [lbmass units]} \times a \left[\frac{ft}{s^2} \text{ units} \right] \tag{5.1-8}$$

If we choose the unit force to be the force necessary to accelerate a unit mass 1 ft/s², Newton's law dictates that

$$\text{[force unit]} = 1 \text{ [lbmass]} \times 1 \left[\frac{ft}{s^2} \right] \tag{5.1-9}$$

This force unit is usually designated as the **poundal** to avoid writing [(lbmass ft)/s²] every time the force unit is required, so that

$$\text{[poundal]} = 1 \text{ [lbmass]} \times 1 \left[\frac{ft}{s^2} \right] \tag{5.1-10}$$

or

$$(1) = \left[\frac{1 \text{ lbmass ft}}{\text{poundal s}^2} \right] = g_c \tag{5.1-11}$$

For any system, g_c is common nomenclature for the conversion among force, length, mass and time units. Multiplying or dividing any term by g_c has no effect on the intrinsic value of the term, but only effects a change in units. In this system g_c has the convenient numerical value of 1.0.

The choice of g_c to designate this conversion factor for units is unfortunate, because it frequently leads to confusion with g, the acceleration of gravity. Some books write g_c as an integral part of equations; we shall not do this since all equations must be dimensionally homogeneous and, therefore, the need for g_c is self-evident when it is required.

---

[1] Most of the time we will write lbmass, kgmass, and gmmass when using these units; if used for brevity we will also restrict the notation lbm, kgm and gmm to mean mass; when we wish to refer to moles, we will write lbmol (not lbm) for pound moles, kgmol for kilogram moles, and gmmol for gram moles.

Similarly, if we use a system of units in which the fundamental units are kilograms for mass (kgmass), meters for length (m), and seconds for time (s), and choose the unit of force as that required to accelerate one kilogram one meter per (second)^2

$$[\text{force unit}] = 1 \,[\text{kgmass}] \times 1 \left[\frac{m}{s^2}\right] \qquad (5.1\text{-}12)$$

We designate this force unit as the Newton (N), and observe that

$$(1) = \left[\frac{1 \text{ kgmass m}}{N\,s^2}\right] = g_c \qquad (5.1\text{-}13)$$

so the numerical value of g_c equals 1.0 in this system of units as well.

If we choose instead to define the mass unit as pounds mass, the length unit as the foot, the time unit as the second, and then (as we often do) choose the force unit as the force exerted by a unit mass at sea level (lbf), the acceleration is that of gravity at sea level and Newton's law prescribes that

$$\text{force} = \text{mass} \times \text{acceleration} = m \times g \qquad (5.1\text{-}14)$$

where $g = 32.2$ ft/s^2, so the units must satisfy the equation

$$(1)\,[\text{lbf}] = (1)\,[\text{lbmass}] \times (32.2) \left[\frac{ft}{s^2}\right] \qquad (5.1\text{-}15)$$

In other words

$$(1) = \left[\frac{(32.2)\,\text{lbmass ft}}{\text{lbf s}^2}\right] = g_c \qquad (5.1\text{-}16)$$

and in this system g_c has the numerical value 32.2.

We encounter a similar problem if we work in a system which defines the unit of length as the meter, the unit of mass as the kilogram mass, the unit of time as the second, and then defines the kilogram force (kgf) as the force exerted by one kilogram of mass at sea level.

$$(1) \ [kgf] = (1) \ [kgmass] \times (9.80) \left[\frac{m}{s^2} \right] \tag{5.1-17}$$

giving

$$(1) = \left| \frac{(9.80) \ kgmass \ m}{kgf \ s^2} \right| = g_c \tag{5.1-18}$$

Example 5.1-1   Weight vs. mass; g vs. g_c

Weight refers to the force exerted by a given mass under the acceleration of local gravity. This means that the weight of a given mass will vary depending upon location - the weight of a certain mass is slightly less in an airplane flying at 30,000 feet than at the surface of the earth. The same mass will have a different weight on the surface of the moon. By virtue of the definition of pound force, one pound mass will weigh one pound force at the surface of the earth. On the moon, where g = 5.47 ft/s², [2] one pound mass would weigh

$$1 \ [lbmass] \times 5.47 \left[\frac{ft}{s^2} \right] \times \left[\frac{lbf \ s^2}{32.2 \ lbmass \ ft} \right] = 0.17 \ lbf \tag{5.1-19}$$

In other words, someone who weighs 150 lbf on earth would weigh 25.5 lbf on the moon.

On the surface of the sun, where g = 900.3 ft/s²,[3] one pound mass would weigh

$$1 \left[lbmass \right] \times 900.3 \left[\frac{ft}{s^2} \right] \times \left[\frac{lbf \ s^2}{32.2 \ lbmass \ ft} \right] = 27.96 \ lbf \tag{5.1-20}$$

and the same 150 lbf earthling would weigh 4194 lbf. **In each case the mass of the person would be the same, regardless of whether they are on the earth, the moon, or the sun.** This illustrates the difference between g, the acceleration of gravity, and g_c, the conversion among units, as well as the difference between mass and weight.

---

[2] Chemical Rubber Publishing Company, *CRC Handbook of Chemistry and Physics*. 50th ed. 1969, Cleveland, OH: Chemical Rubber Publishing Company, p. F-145.
[3] Chemical Rubber Publishing Company, *CRC Handbook of Chemistry and Physics*. 50th ed. 1969, Cleveland, OH: Chemical Rubber Publishing Company, p. F-145.

Units became essential when mankind needed to trade on a sight-unseen basis. One pile of goods can be traded for another if they are both physically present for inspection. One quantity of salt, for example, could be visually compared to another. For side-by-side in-sight trading to take place, however, the goods must be physically present before the trader. This is not an overwhelming restriction when applied to diamonds - and, in fact, some diamonds are still traded on a side-by-side basis. It is inconvenient when applied to, say, elephants, or impossible when applied to land, which cannot be moved (unless the two plots of land fortuitously adjoin each other).

One needs to assign a numerical quantity (value of a unit) to compare goods which are removed from each other in physical location. The assignment of values to units immediately requires standards. If we are to trade land on the basis of area determined by measuring dimensions with a stick, we had better both either use the same stick or sticks of equal length. Happily, there is some international cooperation in establishment of such standards.

In 1948, the 9th General Conference on Weights and Measures (CGPM[4]) by its Resolution 6 instructed the International Committee for Weights and Measures (CIPM[5]) "to study the establishment of a complete set of rules for units of measurement ..." The 11th CGPM (1960) by its Resolution 12 adopted the name **International System of Units** for the resulting system, with the international abbreviation SI (Système International d'Unités) and established rules for prefixes, derived and supplementary units, etc. The 10th CGPM (1954) by its Resolution 6, and the 14th CGPM (1971) by its Resolution 3 adopted the following seven base units.[6]

The current primary standards for the SI system follow[7]:

> • *Length: meter.* Originally defined by the distance between two marks on an international prototype of platinum-iridium, this original standard was replaced in 1960 by 1 650 763.73 wavelengths in vacuum of the orange-red line (between levels $2p_{10}$ and $5d_5$) of the spectrum of krypton-86; in 1983 the krypton standard was replaced by

---

[4] Conférence Générale des Poids et Mesures.

[5] Comité International des Poids et Mesures.

[6] Taylor, B. N., Ed. (1991). *The International System of Units (SI)*. NIST Special Publication 330. Washington, DC, National Institute of Standards and Technology.

[7] ASME (1975). *ASME Orientation and Guide for Use of SI (Metric) Units*. New York, NY, The American Society of Mechanical Engineers.

> The meter is the length of the path traveled by light in vacuum during a time interval of 1/299 792 458 of a second. [17th CGPM (1983) Resolution 1]

• *Mass: kilogram.* This is the only base unit still defined by an artifact.

> The kilogram is equal to the mass of a cylinder of platinum-iridium alloy. This bar is kept (under conditions specified by the first CGPM in 1889) by the International Bureau of Weights and Measures in Paris, France.

• *Time: second.* Originally defined as 1/86 400 of a mean solar day. The 11th CGPM adopted a definition by the International Astronomical Union based on the tropical year. This definition in turn has been replaced with

> The second is the duration of 9 192 631 770 periods of the radiation corresponding to the transition between the two hyperfine levels of the ground state of the cesium-133 atom. [13th CGPM (1967)]

• *Temperature: Kelvin.* The 10th CGPM (1954) by its Resolution 3 selected the triple point of water as the fundamental fixed point and assigned to it the temperature 273.16 K. The 13th CGPM (1967) by its Resolution 3

> a) adopted the name Kelvin and the symbol K (rather than **degree Kelvin** and symbol K), and

> b) decided that this unit and symbol should be used to express an interval or difference in temperature.

In its Resolution 4 it defined the unit of thermodynamic temperature as follows

> The Kelvin, unit of thermodynamic temperature, is the fraction 1/273.16 of the thermodynamic temperature of the triple point of water.

• *Amount of matter: mole (symbol mol).* Gram-atoms and gram-molecules are directly related to atomic and molecular weights, which are relative masses. Historically physicists used the atomic weight of 16 for a particular isotope of oxygen, while chemists used 16 for a mixture of isotopes. In 1959-60 an agreement between the International Union of Pure and Applied Physics (IUPAP) and the International Union of Pure and Applied Chemistry (IUPAC) assigned the value 12 to **isotope 12 of carbon**. The CIPM confirmed in 1969 the definition of the mole listed below, which was adopted by the 14th CGPM (1971) in its Resolution 3

> 1. The mole is the amount of substance of a system which contains as many elementary entities as there are atoms in 0.012 kg of carbon 12[8]
>
> 2. When the mole is used, the elementary entities must be specified and may be atoms, molecules, ions, electrons, other particles, or specified groups of such particles.

Note that **mass** and **amount of substance** are different quantities.

• *Luminous Intensity: candela.* Before 1948 there were in various countries a variety of standards based on flame or incandescent filaments. These were replaced by a standard based on the luminous intensity in the normal direction of a surface of 1/600 000 m^2 of a Planckian radiator (blackbody) at the temperature of freezing platinum under a pressure of 101 325 N/m^2. The 9th CGPM (1948) adopted the name **candela** (symbol cd) for this unit. The 16th CGPM (1979) in its Resolution 3 replaced this definition with the following definition

> The candela is the luminous intensity, in a given direction, of a source that emits monochromatic radiation of frequency 540×10^{12} hertz and that has a radiant intensity in that direction of 1/683 watt per steradian.

• *Electric current: ampere.* The "international" units for current and resistance, the ohm and the ampere, were introduced by the International Electrical Congress held in Chicago in 1893, and the definitions were confirmed by the International

---

[8] **Unbound** atoms of carbon 12 **at rest** and in their **ground state** are understood.

Conference of London in 1908. The definitions of these units were replaced [9th CGPM, (1948)] by

The ampere is that constant current which, if maintained in two straight parallel conductors of infinite length, of negligible circular cross-section, and placed one meter apart in a vacuum, would produce between these conductors a force equal to 2×10^{-7} Newton per meter of length.

[CIPM (1946) Resolution 2, approved by the 9th CGPM (1948)]

Table 5.1-1 shows the units of length, time, mass, force, work and weight for common systems of measurement. The SI system is the metric system being adopted as an international standard, and is the system that is recommended for engineering practice. Reality, however, dictates that engineers will still encounter other systems of units - particularly system (2), which uses the pound both as a force and as a mass unit.

Also shown are other typical systems - (3), the [kgmass m t kgf T] system, which is the metric analog of (2); (4), the British mass system which uses the poundals the mass unit; (5), the American engineering system which uses the slug as the mass unit; and (6), the cgs system which uses the dyne as the force unit and the erg as the energy unit.

We will use a variety of systems in this text because we believe that systems of units other than SI will be in use for many years, although we believe SI clearly to be a superior system.

Table 5.1-1a Systems of Units[9]

Unit/System	(1) SI	(2) FMLtT	(3) kgf kgmass M t T
F, force	Newton = kg m/s^2 [N]	*pound force = lbmass (32.2) ft/s^2 [lbf]	*kilogram force [kgf]
M, mass	*kilogram [kg]	*pound mass [lbmass]	*kilogram mass = kgf s^2/m [kgmass]
L, length	*meter [m]	*foot [ft]	*meter [m]
t, time	*second [s]	*second [s]	*second [s]
T, temperature	*Kelvin [K]	*Rankine [R]	*Kelvin [K]
gc	(1) kg m/(N s^2)	(32.2) lbmass ft/(lbf s^2)	(9.8) kgmass m/(kgf s^2)
W, work (energy)	Joule = N m [J]	Btu = (778) ft lbf	
P, power	Watt = J/s [W]	Horsepower = (550) ft lbf/s [HP]	
p, pressure	Pascal = N/m^2 [Pa]	psf = lbf/ft^2	
μ, viscosity[10]	Pascal second = kg/(m s) = N s/m^2 [Pa s]	Poise = 0.0672 lbmass/(ft s) = 242 lbmass/(ft hr) [P]	Poise = 0.1 kgmass/(m s) [P]

---

[9] Abbreviation shown in brackets. Asterisks denote a fundamental unit.

[10] A commonly used non-SI unit is the **centipoise [cP]**.

    1 cP = 0.01 poise [P]

In SI units

    1 P = 0.1 Pa s = 0.1 kg/(m s)

    1 cP = (0.001) Pa s = (0.001) kg/(m s) = (0.001) N s/m^2

In [kgf kgmass M t T] units (system 3 above)

    1 P = 0.1 kgm/(m s)

In FMLtT units

    1 cP = (6.72x10-4) lbmass/(ft s) = 2.42 lbmass/(ft hr).

Table 5.1-1b Systems of Units

Unit/System	(4) British (mass)	(5) American Engineering	(6) cgs
F, force	poundal = lb ft/s^2	*pound force = slug ft/s^2 [lbf]	dyne = gm cm/s^2
M, mass	*pound [lb]	slug = lbf s^2/ft	*gram [g]
L, length	*foot [ft]	*foot [ft]	*centimeter [cm]
t, time	*second [s]	*second [s]	*second [s]
gc	(1) lb ft/(poundal s^2)	(1) slug ft/(lbf s^2)	(1) g cm/(dyne s^2)
W, work (energy)			erg = dyne cm

SI prefixes are listed in Table 5.1-2. These prefixes permit easy designation of multiples of SI units. Those marked with plus signs are probably best shunned because of danger of typographical error or confusion with other symbols.

Table 5.1-2 SI Prefixes

Prefix	Factor	Symbol	Prefix	Factor	Symbol
exa	10^{18}	E	deci+	10^{-1}	d
peta	10^{15}	P	centi	10^{-2}	c
tera	10^{12}	T	milli	10^{-3}	m
giga	10^9	G	micro	10^{-6}	μ
mega	10^6	M	nano	10^{-9}	n
kilo	10^3	k	pico	10^{-12}	p
hecto+	10^2	h	femto	10^{-15}	f
deka+	10^1	da	atto	10^{-18}	a

5.2 Buckingham's Theorem

Dimensionally homogeneous functions are a special class of functions. The theory of **dimensional analysis** is the mathematical theory of this class of functions. Dimensional analysis is based on requiring the mathematical model for a given problem to be a dimensionally homogeneous equation.

A set of **dimensionless products** obtained from the set of variables incorporated in a dimensionally homogeneous equation is **complete** if (a) each product in the set is independent of all the other products and (b) every other dimensionless product of the variables is not independent but is a product of powers of dimensionless products in the set.

Buckingham's theorem states that if an equation is dimensionally homogeneous, it can be reduced to a relationship among a complete set of dimensionless products. One can show that if n variables Q_i [i = 1, 2, 3, ..., n] are related by

$$f(Q_1, Q_2, Q_3, \cdots Q_n) = 0 \qquad\qquad (5.2\text{-}1)$$

then there are (n - r) independent dimensionless products π_j [j = 1, 2, 3, ..., (n-r)] in a complete set of dimensionless products formed from these n variables where r is the rank of dimensional matrix.[11] These dimensionless products can in turn be related by another function

$$\phi(\pi^1, \pi^2, \pi^3, ..., \pi^{n\text{-}r}) = 0 \qquad\qquad (5.2\text{-}2)$$

where the π_i are the dimensionless products (a new set of variables). Usually r = k, where k is the number of fundamental dimensions; therefore, in most applications the number of dimensionless products is equal to the number of variables less the number of fundamental dimensions.

---

[11] The dimensional matrix is constructed by writing the variables as row headings and the fundamental dimensions as column headings, and filling in the matrix with the power of the dimension for each variable. The rank is the order of the highest order non-zero determinant that can be selected from the matrix.

Example 5.2-1   Dimensionless variables for pipe flow

The original set of variables chosen depends on what is important in the physical situation being modeled; however, for fluid flow in a pipe one possible set of variables is

- L, the length of the pipe
- D, the diameter of the pipe
- g, the acceleration of gravity
- <v>, the bulk velocity of the fluid
- ρ, the density of the fluid
- μ, the viscosity of the fluid
- σ, the surface tension of the fluid
- c, the speed of sound in the fluid
- Δp, the pressure difference across the length

What is a complete set of dimensionless variables (dimensionless products) for this set of primary variables?

Solution

First we must choose the system of units in which we propose to work. Assume that we choose an MLt system. We need not include temperature because temperature is neither a dimension for any of our variables nor a variable in itself; in other words, we are assuming either an isothermal system or one in which the effects of temperature changes are unimportant.

We have nine variables and three fundamental dimensions. The dimensions of the variables in terms of the fundamental dimensions are[12]

---

[12] The units of surface tension are energy/area or, equivalently, force/length. Since we are working with M, L, and t as our fundamental units, force must be expressed in these units using Newton's second law - $F = ML/t^2$. This makes the units on surface tension $(ML/t^2)/L = M/t^2$. Pressure is force per unit area: writing force in terms of our fundamental units by using Newton's law we have $F/L^2 = (ML/t^2)/L^2 = M/(L\,t^2)$.

	M	L	t
L		1	
D		1	
g		1	-2
<v>		1	-1
ρ	1	-3	
μ	1	-1	-1
σ	1		-2
c		1	-1
Δp	1	-1	-2

which gives the dimensional matrix as[13]

$$\begin{bmatrix} 0 & 1 & 0 \\ 0 & 1 & 0 \\ 0 & 1 & -2 \\ 0 & 1 & -1 \\ 1 & -3 & 0 \\ 1 & -1 & -1 \\ 1 & 0 & -2 \\ 0 & 1 & -1 \\ 1 & -1 & -2 \end{bmatrix} \qquad (5.2\text{-}3)$$

We look for the largest order non-zero determinant that can be found in this matrix. We note that determinants are square arrays, so a 3x3 determinant is clearly the largest order determinant that exists in the matrix, and we can easily find a non-zero 3x3 determinant by inspection[14] (there are more than one) as follows[15]

---

[13] Note that for our manipulations here it does not matter whether we start with this matrix or its transpose so long as we remember whether the rows or the columns correspond to the dimensions or the original variables, respectively.

[14] There are also systematic ways to find the rank of a matrix - for example, see Amundson, N. R. (1966). *Mathematical Methods in Chemical Engineering: Matrices and Their Application.* Englewood Cliffs, NJ, Prentice-Hall, Inc., Gerald, C. F. and P. O. Wheatley (1989). *Applied Numerical Analysis.* Reading, MA, Addison-Wesley, etc.

[15] The expansion of determinant $A = [a_{ij}]$ is

$$\sum_{j=1}^{n} \left[(-1)^{k+j} a_{kj} M_{kj} \right]$$

where M_{kj} is the determinant of the $(n-1) \times (n-1)$ matrix formed by deleting row i and column j from A (the minor of a_{ij} in A). O'Neil, P. V. (1991). *Advanced Engineering Mathematics.* Belmont, CA, Wadsworth, p. 689.

$$\begin{bmatrix} 0 & 1 & 0 \\ 0 & 1 & 0 \\ 0 & 1 & -2 \\ 0 & 1 & -1 \\ \hline 1 & -3 & 0 \\ 1 & -1 & -1 \\ 1 & 0 & -2 \\ \hline 0 & 1 & -1 \\ 1 & -1 & -2 \end{bmatrix} \Rightarrow$$

$$\begin{vmatrix} 1 & -3 & 0 \\ 1 & -1 & -1 \\ 1 & 0 & -2 \end{vmatrix} =$$

$$(1)[(-1)(-2)-(-1)(0)] - (-3)[(1)(-2) - (-1)(1)] + (0)[(1)(0) - (-1)(1)]$$

$$= 2 - 3 + 0 = -1 \tag{5.2-4}$$

Therefore, the rank of the dimensional matrix is three, so $r = k = 3$ and $n - k = 9 - 3 = 6$, and we should be able to find six independent dimensionless products (groups).

A systematic way of determining these groups is given below. For the moment, we simply list them with the name and abbreviation that they are customarily given[16]

$$\frac{D \langle v \rangle \rho}{\mu} = Re = \text{Reynolds number} = \frac{\text{inertial force}}{\text{viscous force}} \tag{5.2-5}$$

$$\frac{\Delta p}{\rho \langle v \rangle^2} = Eu = \text{Euler number} = \frac{\text{pressure force}}{\text{inertial force}} \tag{5.2-6}$$

$$\frac{\langle v \rangle^2}{L g} = Fr = \text{Froude number} = \frac{\text{inertial force}}{\text{gravity force}} \tag{5.2-7}$$

$$\frac{\langle v \rangle}{c} = Ma = \text{Mach number} = \frac{\text{fluid velocity}}{\text{sonic velocity}} \tag{5.2-8}$$

$$\frac{L \langle v \rangle^2 \rho}{\sigma} = We = \text{Weber number} = \frac{\text{inertial force}}{\text{interfacial force}} \tag{5.2-9}$$

$$\frac{L}{D} = \text{dimensionless length} \tag{5.2-10}$$

---

[16] The American Institute of Chemical Engineers officially suggests Buck, E. (1978). *Letter Symbols for Chemical Engineering. CEP*(October): 73-80, that the Reynolds, Euler, Froude, and Weber numbers be designated as N_{Re}, etc., but this is by no means universally followed. We do not do so here, because we feel that the notation we use is less confusing, since it does not repeat the same symbol (N) in every quantity.

It can easily be seen that these groups are independent, since comparing them pairwise shows that one member of each pair has (at least) one variable not present in the other of the pair. Further, any dimensionless combination of variables in the parent set can be expressed by multiplying and dividing these dimensionless products.

For example, if we construct the dimensionless product (group)

$$\frac{D \langle v \rangle^3 \sigma \rho}{L^2 g \mu \Delta p} \Rightarrow \frac{L \left(\frac{L}{t}\right)^3 \frac{M}{t^2} \frac{M}{L^3}}{L^2 \frac{L}{t^2} \frac{M}{L t} \frac{M}{L t^2}} \Rightarrow \text{dimensionless} \tag{5.2-11}$$

we can write it as the following combination of dimensionless products from our complete set

$$\frac{\text{Re Fr}}{\text{We Eu}} = \frac{\dfrac{D \langle v \rangle \rho}{\mu} \dfrac{\langle v \rangle^2}{L g}}{\dfrac{L \langle v \rangle^2 \rho}{\sigma} \dfrac{\Delta p}{\rho \langle v \rangle^2}} = \frac{D \langle v \rangle^3 \sigma \rho}{L^2 g \mu \Delta p} \tag{5.2-12}$$

Example 5.2-1 is a practical illustration of the application of Buckingham's theorem. As pointed out in the example, dimensionless groups have physical significance. We will see later that they are parameters in the dimensionless equations which are the mathematical models for the physical systems described.

We know that the n variables are related by some unknown but dimensionally homogeneous equation, for example, a) the differential equation describing the process and its boundary conditions, or, equivalently, b) the solution to this equation with its boundary conditions, but the form of this model equation is at this point unknown. Buckingham's theorem, however, tells us that the solution to this equation can be expressed as a functional relationship among a complete set of (n-r) dimensionless products equally as well as by the initial functional form relating the n original variables. This reduces the dimensionality of our original problem, an enormous advantage.

5.2.1 Friction factors and drag coefficients

In the course of making calculations regarding fluid flow, we are, in general, confronted with two situations

- flow **through** things (e.g., pipes, conduits, etc.)

- flow **around** things (e.g., drops, wings, hulls, etc.)

In each of these cases we need to relate the drag force to the flow situation.

In general, the drag force is some function (often very complicated in the case of complex shapes) of

- the properties of the fluid (density, viscosity)
- the properties of the object (usually characterized as some characteristic length)
- the flow field (usually described by a characteristic velocity)

Consider, then, the prediction of drag force as a function of these variables. We use v and D as generic variables at this point (for example, v refers to no particular velocity and D does not necessarily refer to diameter). The function

$$\varphi(F, \rho, \mu, D, v) = 0 \tag{5.2.1-1}$$

gives the dimensional matrix (for purposes of illustration, using a system with four fundamental dimensions; however, the outcome is the same if we use three fundamental dimensions - the only difference being that we must include g_c in our list of variables when using four fundamental dimensions)

$$
\begin{array}{c|cccc}
 & F & M & L & t \\
\hline
F & 1 & 0 & 0 & 0 \\
\rho & 0 & 1 & -3 & 0 \\
\mu & 0 & 1 & -1 & -1 \\
D & 0 & 0 & 1 & 0 \\
v & 0 & 0 & 1 & -1 \\
g_c & -1 & 1 & 1 & -2
\end{array}
\tag{5.2.1-2}
$$

It can be shown that this matrix is of rank 4, so in a complete set we should be able to obtain $6 - 4 = 2$ dimensionless products.

Carrying out the dimensional analysis yields

$$\Psi_1\left[\left(\frac{D\,v\,\rho}{\mu}\right),\left(\frac{F\,g_c}{\rho\,v^2\,D^2}\right)\right] = 0 \tag{5.2.1-3}$$

In principle, we can solve Ψ_1 explicitly for $\left(\dfrac{F\,g_c}{\rho\,v^2\,D^2}\right)$ to obtain a new function

$$\left(\frac{F\,g_c}{\rho\,v^2\,D^2}\right) = \Psi_2\left[\left(\frac{D\,v\,\rho}{\mu}\right)\right] \tag{5.2.1-4}$$

This relation may be rewritten as

$$F = \left(\frac{\rho\,v^2\,D^2}{g_c}\right)\Psi_2\left[\left(\frac{D\,v\,\rho}{\mu}\right)\right] \tag{5.2.1-5}$$

where $(\rho\,v^2)$ is proportional to the kinetic energy per unit volume and D^2 is proportional to a characteristic area.

For **internal** flows in **circular** conduits this may be expressed as

$$F = \left(\frac{\frac{1}{2}\rho\,\langle v\rangle^2\,2\pi R L}{\pi\,g_c}\right)\Psi_2\left[\left(\frac{D\,\langle v\rangle\,\rho}{\mu}\right)\right] \tag{5.2.1-6}$$

where we have used the bulk velocity as the characteristic velocity, the product $(R\,L)$ in place of the square of the characteristic length, and the conduit diameter as the characteristic length in the Reynolds number (for geometrically similar systems, defined below, these lengths remain in constant ratio to one another).

The **Fanning friction factor, f,** is then defined by the relation

$$F = \left(2\pi R L\right)\left(\tfrac{1}{2}\rho\,\langle v\rangle^2\right)f\left[\left(\frac{D\,\langle v\rangle\,\rho}{\mu}\right)\right] \tag{5.2.1-7}$$

where the first factor on the right-hand side is the lateral area of the conduit and the second factor on the right-hand side is the kinetic energy per unit volume. The function f is just the function Ψ_2 divided by (πg_c). Consistent with our practice throughout the text, we do not incorporate g_c in the formula, because the definition of f is the same for all systems of units, and when g_c is needed it is self-evident by dimensional homogeneity. The friction factor is a dimensionless function of Reynolds number.

A friction factor (unfortunately also designated by f) **four times the magnitude** of the Fanning friction factor is sometimes also defined in the literature. This friction factor is called the **Darcy** or **Blasius** friction factor. Two different friction factors, both designated by f, cause no real difficulty as long as one is consistent in which friction factor one uses and observes the factor of four. The biggest danger of mistake is usually in using a value from a chart or table which incorporates one friction factor in a formula using the other.

For **external** flows, Equation (5.2.1-5) may be written as

$$F = \left(\frac{\rho\, v^2\, D^2}{g_c}\right) \psi_3\left[\left(\frac{D\, v\, \rho}{\mu}\right)\right] \tag{5.2.1-8}$$

where we now have a different function because we have a different flow situation.

For flow **around** objects, we usually convert the square of the characteristic length to the cross-section of the object **normal to flow**. The reference velocity usually used in the kinetic energy term and in the Reynolds number is the **freestream velocity** (the velocity far enough away from the surface that there is no perturbation by the object). The characteristic length, if the cross-sectional area is circular, is the diameter of the circle - if not, a specific dimension must be specified. Then

$$F = \left(\tfrac{1}{2} \rho\, v_\infty^2\right)\left(A_{\mathrm{n}}\right) C_D\left[\left(\frac{D\, v_\infty\, \rho}{\mu}\right)\right] \tag{5.2.1-9}$$

where C_D, called the **drag coefficient**, is a dimensionless function of Reynolds number (C_D and ψ_3 are different functional forms because of the constants introduced). Again, we do not incorporate g_c because the definition is valid for any consistent system of units.

5.2.2   Shape factors

Dimensional analysis can seldom be used to predict ways in which phenomena are affected by different **shapes** of object. Therefore, when applying dimensional analysis to a problem it is usually convenient to eliminate the consideration of shape by considering bodies of similar shape, that is, bodies that are **geometrically similar**.

Example 5.2.2-1   Drag force on ship hull

The drag force exerted on the hull of a ship depends on the shape of the hull. Hulls of similar shape may be specified by a single characteristic dimension such as keel[17] length, L, and hulls of other lengths but similar shapes have their dimensions in the same ratio to this characteristic length - for example, the ratio of beam width (width across the hull) to length will be the same for all hulls of similar shape. If the hulls are considered to have the same relative water line (for example, with respect to freeboard, the distance of the top of the hull above the waterline), keel length will also be proportional to the draft (displacement).

The drag force will then depend on the keel length, L, the speed of the ship, v, the viscosity of the liquid, μ, the density of the liquid, ρ, and the acceleration of gravity (which affects the drag produced by waves). What are the dimensionless groups that describe this situation?

Solution

Our function is

$$\varphi_1(F, v, L, \mu, \rho, g) = 0 \qquad\qquad (5.2.2\text{-}1)$$

The rank of the dimensional matrix may be shown to be 3 for a system with three fundamental dimensions. We will therefore have $(6 - 3) = 3$ dimensionless groups.

If we carry out the dimensional analysis, we obtain

$$\varphi_2\left(\frac{F}{\rho\, v^2\, L^2}, \text{Re}, \text{Fr}\right) = 0 \qquad\qquad (5.2.2\text{-}2)$$

(the dimensionless group $F/[\rho v^2 L^2]$ is sometimes called the **pressure coefficient, P**).[18]

---

[17] The keel is the longitudinal member reaching from stem to stern to which the ribs are attached.

[18] Langhaar, H. L. (1951). *Dimensional Analysis and Theory of Models*. New York, NY, John Wiley and Sons, p. 21.

To run experiments, we would rearrange (5.2.2-1) as

$$F = \psi_1(v, L, \mu, \rho, g) \tag{5.2.2-3}$$

and would rearrange (5.2.2-2) as

$$\frac{F}{\rho\, v^2\, L^2} = P = \psi_2\!\left(Re, Fr\right) \tag{5.2.2-4}$$

In each case the experiments would consist of measuring the variable on the left-hand side of the equation as a function of the variables on the right-hand side of the equation, which would be varied from experiment to experiment.

Notice the numerous advantages of (5.2.2-4) over (5.2.2-3).

• Using (5.2.2-3), to run three levels of each variable would require $3^5 = 243$ experiments. Using (5.2.2-4), to run three levels of each variable would require only $3^2 = 9$ experiments.

• To use (5.2.2-4), we could easily vary both Reynolds and Froude numbers by varying velocity (which is easily accomplished in a towing tank) and hull length (which could be accomplished by using models of the same shape but different sizes).

To use (5.2.2-3), we would also have to find a way to vary μ (perhaps by varying temperature or fluid or adding a thickening agent to water), ρ (perhaps by varying fluid; or dissolving something in water to vary density; or varying temperature to vary density, even though this is not a very big effect), and g (perhaps by building two laboratories, one on a very high mountain and one in a mine, or building the experiment in a centrifuge). None of these alternatives is particularly appealing.

• To present the results of (5.2.2-3) in a compact form is difficult, as is extrapolating or interpolating the results. The results of (5.2.2-4), however, can be concisely presented on a two-dimensional graph, where the approximate results of interpolation or extrapolation can be easily visualized, as is illustrated in the following figure (the data is artificial and **must not be used for design purposes**). The figure shows the data from nine hypothetical experiments arranged in a factorial design (each of three levels of Re combined with each of three levels of Fr, and hypothetical curve fits which could be used for interpolation (or extrapolation, if the curves were to be extended beyond the range of the data - which would be dangerous).

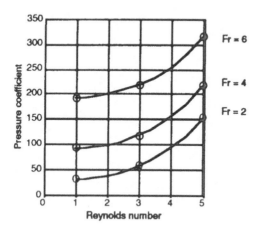

Example 5.2.2-2   Deceleration of compressible fluid

The important variables that determine the maximum pressure, p_{max}, resulting when flow of a compressible fluid is stopped instantly (for example, by shutting a valve rapidly) are the velocity, $\langle v \rangle$, of the fluid at shutoff, the density of the fluid, ρ, and the bulk modulus of the fluid, $\beta = \rho(\partial p/\partial \rho)_T$

$$\varphi(p_{max}, \langle v \rangle, \rho, \beta) = 0 \tag{5.2.2-5}$$

Find a complete set of dimensionless groups for this problem.

Solution

The dimensions of the bulk modulus are

$$\beta = \rho \left(\frac{\partial p}{\partial \rho} \right)_T = \frac{M}{L^3} \frac{ML}{t^2 L^2 \frac{M}{L^3}} = \frac{M}{L t^2} \tag{5.2.2-6}$$

The dimensional matrix in an MLt system is

$$\begin{array}{c} \\ p_{max} \\ \langle v \rangle \\ \rho \\ \beta \end{array} \begin{array}{ccc} M & L & t \\ \hline 1 & -1 & -2 \\ 0 & 1 & -1 \\ 1 & -3 & 0 \\ 1 & -1 & -2 \end{array} \qquad (5.2.2\text{-}7)$$

We have four rows that can be combined in various ways to make 3x3 determinants. From the properties of determinants, the only effect on a determinant of switching rows is to change the sign. Hence, only combinations of rows, not permutations of rows, are significant. We can therefore form[19]

$$\binom{4}{3} = \frac{P_3^4}{3!} = \frac{4!}{3!\,(4-3)!} = 4 \qquad (5.2.2\text{-}8)$$

different 3rd-order determinants from our dimensional matrix. We must see if any of these are not equal to zero.

From the properties of determinants we know that if any two rows of a determinant are identical, the determinant is zero, so it immediately follows that any determinant that contains both the first and fourth rows of our dimensional matrix will be zero. This eliminates two of our four combinations - the first and fourth rows combined with 1) the second row and 2) the third row.

This leaves us with two determinants to investigate further. One contains the first row in combination with the second and third rows, and the other contains the fourth row in combination with the second and third rows. However, since the first and fourth rows are identical, these last two determinants are identical, leaving only a single determinant to investigate

---

[19] The notation for combinations is read "n choose r," and the general formula in terms of permutations of r things from a set of n is

$$\binom{n}{r} = \frac{P_r^n}{r!} = \frac{n!}{r!\,(n-r)!}$$

Fraser, D. A. S. (1958). *Statistics: An Introduction.* New York, NY, John Wiley and Sons, Inc.

$$\begin{bmatrix} 1 & -1 & -2 \\ 0 & 1 & -1 \\ 1 & -3 & 0 \end{bmatrix} = -\begin{bmatrix} 1 & -3 & 0 \\ 0 & 1 & -1 \\ 1 & -1 & -2 \end{bmatrix} =$$

$$(1)\,[(1)\,(-2) - (-1)\,(-1)] - (-3)\,[(0)\,(-2) - (-1)\,(1)]$$
$$+ (0)[(0)\,(-1) - (1)\,(1)] = 0 \qquad (5.2.2\text{-}9)$$

and so we have a matrix of rank less than three since we have shown all possible third-order determinants to be zero.

We proceed to check second-order determinants. It is easy to find by inspection a 2x2 determinant that is not equal to zero, e.g.

$$\begin{bmatrix} 1 & -1 \\ 0 & 1 \end{bmatrix} = 1 \qquad (5.2.2\text{-}10)$$

so the rank of our dimensional matrix is two, and for this example, the rank of the dimensional matrix, 2, is not equal to the number of fundamental dimensions, 3. We will obtain, therefore, $(4 - 2) = 2$ dimensionless groups, not $(4 - 3) = 1$.

An acceptable set can be seen to be

$$\pi_1 = \frac{p_{max}}{\beta} \quad \text{and} \quad \pi_2 = \frac{\rho\,v^2}{\beta} \qquad (5.2.2\text{-}11)$$

5.3 Systematic Analysis of Variables

The proof of Buckingham's theorem is based on a set of algebraic theorems concerned with the class of functions known as **dimensionally homogeneous**.[20] A systematic approach to finding a set of independent dimensionless products can be based on this algebraic theory.

We now develop a systematic way to convert a functional relationship among n independent quantities Q_1, Q_2, ... Q_n to one involving $(n-r)$ dimensionless products π_1, π_2, π_3, ... π_{n-r}, where r is the rank of the dimensional matrix of the Q's

---

[20] Brand, L. (1957). The Pi Theorem of Dimensional Analysis. *Arch. Rat. Mech. Anal.* 1: 35.

$$f(Q_1, Q_2, Q_3, \ldots Q_n) = 0 \quad \Rightarrow \quad g(\pi_1, \pi_2, \pi_3, \ldots \pi_{n-r}) = 0 \qquad (5.3\text{-}1)$$

In general, the number of dimensionless products in a **complete set** is equal to the total number of **variables** minus the **maximum** number of these variables that will **not** form a dimensionless product (the rank of the dimensional matrix).

The algorithm to find the π's as follows:

1) Select the number of fundamental dimensions.

2) List the Q variables and their dimensions in terms of the fundamental dimensions.

3) Select a subset of the Q variables equal in number to the number of fundamental dimensions such that

- none of the selected quantities is dimensionless,
- the set includes all the fundamental dimensions and
- no two variables of the subset have the same dimensions.

4) The dimensionless products (p's) are found one at a time by forming the product of the subset variables each raised to an unknown power times one of the remaining Q variables to a known power. (This process is repeated n-k times, using all the remaining Q variables.)

5) Apply dimensional homogeneity to each of the products obtained to determine the unknown exponents.

Algebraic manipulation of the dimensionless products that does not change their number will not destroy the completeness of the set. Thus, a π may be replaced by any power of itself, by its product of any other π raised to any power, or by its product with a numerical constant.

Example 5.3-1 Drag force on a sphere

The drag force, F, on a smooth sphere of diameter D suspended in a stream of flowing fluid depends on the freestream velocity, v_∞, the fluid density, ρ, and the fluid viscosity, μ.

$$F = f(D, v_\infty, \rho, \mu) \qquad (5.3\text{-}2)$$

Using the FLt system,

     (a) Write the dimensional matrix.

     (b) Show that the rank of the dimensional matrix is 3.

     (c) Using (F, D, μ) as the fundamental set, determine the dimensionless group incorporating v_∞.

Solution

a) The dimensional matrix in the indicated system is

$$\begin{array}{c|ccccc} & F & D & v_\infty & \rho & \mu \\ \hline F & 1 & 0 & 0 & 1 & 1 \\ L & 0 & 1 & 1 & -4 & -2 \\ t & 0 & 0 & -1 & 2 & 1 \end{array} \qquad (5.3\text{-}3)$$

b) We can easily find a non-zero third-order determinant, e.g.,

$$\begin{vmatrix} 1 & 0 & 0 \\ 0 & 1 & 1 \\ 0 & 0 & -1 \end{vmatrix} = (1)\left[(1)(-1)-(1)(0)\right] = -1 \qquad (5.3\text{-}4)$$

c) Choosing $Q_1 = F$, $Q_2 = D$, and $Q_3 = \mu$ as the subset, $(n - k) = (6 - 3) = 3$

$$\pi = v_\infty F^a D^b \mu^c \quad \Rightarrow \quad \left[\frac{L}{t}\right][F]^a [L]^b \left[\frac{Ft}{L^2}\right]^c$$

F: $a + c = 0$ $\Rightarrow a = -1$

L: $1 + b - 2c = 0$ $\Rightarrow b = 1$

t: $-1 + c = 0$ $\Rightarrow c = 1$

$$\pi = \frac{v_\infty D \mu}{F} \qquad (5.3\text{-}5)$$

Example 5.3-2   Dimensionless groups for flow over a flat plate

For flow over a flat plate, the pertinent variables in predicting the velocity in the x-direction are

$$v_x = f(x, y, v_\infty, \rho, \mu) \tag{5.3-6}$$

Using the MLt system

> (a) Write the dimensional matrix.
> (b) Show that the rank of the dimensional matrix is 3.
> (c) Using (x, ρ, μ) as the fundamental set, determine the dimensionless group incorporating v_∞.

Solution

a)

$$\begin{array}{c|cccccc} & v_x & x & y & v_\infty & \rho & \mu \\ \hline M & 0 & 0 & 0 & 0 & 1 & 1 \\ L & 1 & 1 & 1 & 1 & -3 & -1 \\ t & -1 & 0 & 0 & -1 & 0 & -1 \end{array} \tag{5.3-7}$$

b)

$$\begin{vmatrix} 0 & 1 & 1 \\ 1 & -3 & -1 \\ -1 & 0 & -1 \end{vmatrix} = (0)\begin{vmatrix} -3 & -1 \\ 0 & -1 \end{vmatrix} - (1)\begin{vmatrix} 0 & 1 \\ -1 & -1 \end{vmatrix} + (1)\begin{vmatrix} 1 & -3 \\ -1 & 0 \end{vmatrix}$$

$$= 0 - 1 - 3 = -4 \neq 0 \quad \Rightarrow \quad \text{rank} = 3 \tag{5.3-8}$$

c) Choosing $Q_1 = x$, $Q_2 = \rho$, and $Q_3 = \mu$ as the subset

$$\pi = v_\infty x^a \rho^b \mu^c \quad \Rightarrow \quad \left[\frac{L}{t}\right]\left[L\right]^a \left[\frac{M}{L^3}\right]^b \left[\frac{M}{Lt}\right]^c$$

$$\begin{array}{llll} M: & b + c = 0 & \Rightarrow & a = 1 \\ L: & 1 + a - 3b - c = 0 & \Rightarrow & b = 1 \\ t: & -1 - c = 0 & \Rightarrow & c = -1 \end{array}$$

$$\pi = \frac{v_\infty x \rho}{\mu} \tag{5.3-9}$$

Example 5.3-3   Consistency of dimensionless groups across system of dimensions

The following table shows a list of possible variables for flow in a pipe. Apply the algorithm for obtaining dimensionless groups in each of the five fundamental systems of units MLt, FML, FLt, FMt, and FMLt, and show that the results obtained are equivalent. Details of the application of the algorithm for the FML, FLt, and FMt systems are omitted and left as an exercise in Problem 5.20. The rank of the dimensional matrix may be shown in each case to be equal to the number of fundamental dimensions. The algorithm presented will work for systems of four fundamental dimensions as well as three if the dimensional conversion factor g_c is included in the set of Q's purely as a formal procedure.

	MLt[21]	FML[22]	FLt[23]	FMt[24]	FMLt
$Q_1 = \Delta p$	$M/(Lt^2)$	F/L^2	F/L^2	$F/(Ft^2/M)^2$ $= M^2/(Ft^4)$	F/L^2
$Q_2 = L$	L	L	L	Ft^2/M	L
$Q_3 = D$	L	L	L	Ft^2/M	L
$Q_4 = v$	L/t	$L/(ML/F)^{0.5}$ $= (LF/M)^{0.5}$	L/t	$Ft^2/(Mt)$ $= Ft/M$	L/t
$Q_5 = \mu$	$M/(Lt)$	$M/[L(ML/F)^{0.5}]$ $= (FM)^{0.5}L^{-1.5}$	$(Ft^2/L)/(Lt)$ $= Ft/L^2$	$M/[(Ft^2/M)t]$ $= M^2/(Ft^3)$	$M/(Lt)$
$Q_6 = \rho$	M/L^3	M/L^3	$(Ft^2/L)/L^3$ $= Ft^2/L^4$	$M/(Ft^2/M)^3$ $= M^4/(F^3t^6)$	M/L^3
$Q_7 = k$	L	L	L	Ft^2/M	L
$Q_8 = g_c$					$ML/(Ft^2)$

Solution

For the MLt system

Choose subset

$$Q_1 = \Delta p \quad Q_3 = D \quad Q_4 = v$$

(5.3-10)

Form π_1

---

[21] From Newton's law, $F = ML/t^2$.
[22] From Newton's law, $t = (ML/F)^{0.5}$.
[23] From Newton's law, $M = Ft^2/L$.
[24] From Newton's law, $L = Ft^2/M$.

$$\pi_1 = L\left(\Delta p\right)^a D^b v^c \;\Rightarrow\; L\left(\frac{M}{Lt^2}\right)^a L^b \left(\frac{L}{t}\right)^c \qquad (5.3\text{-}11)$$

Apply dimensional homogeneity

M: $a = 0$ $a = 0$
L: $1 - a + b + c = 0 \;\Rightarrow\; b = -1$
t: $-2a - c = 0$ $c = 0$ $\qquad (5.3\text{-}12)$

Result

$$\boxed{\pi_1 = \frac{L}{D}} \qquad (5.3\text{-}13)$$

Form π_2

$$\pi_2 = \mu\left(\Delta p\right)^a D^b v^c \;\Rightarrow\; \left(\frac{M}{Lt}\right)\left(\frac{M}{Lt^2}\right)^a L^b \left(\frac{L}{t}\right)^c \qquad (5.3\text{-}14)$$

Apply dimensional homogeneity

M: $1 + a = 0$ $a = -1$
L: $-1 - a + b + c = 0 \;\Rightarrow\; b = -1$
t: $-1 - 2a - c = 0$ $c = 1$ $\qquad (5.3\text{-}15)$

Result

$$\boxed{\pi_2 = \frac{\mu v}{\Delta p\, D}} \qquad (5.3\text{-}16)$$

Form π_3

$$\pi_3 = \rho\left(\Delta p\right)^a D^b v^c \;\Rightarrow\; \left(\frac{M}{L^3}\right)\left(\frac{M}{Lt^2}\right)^a L^b \left(\frac{L}{t}\right)^c \qquad (5.3\text{-}17)$$

Apply dimensional homogeneity

M: $1 + a = 0$ $a = -1$
L: $-3 - a + b + c = 0 \;\Rightarrow\; b = 0$
t: $-2a - c = 0$ $c = 2$ $\qquad (5.3\text{-}18)$

Result

$$\boxed{\pi_3 = \frac{\rho v^2}{\Delta p}} \qquad (5.3\text{-}19)$$

Form π_4

$$\pi_4 = k\left(\Delta p\right)^a D^b v^c \;\Rightarrow\; L\left(\frac{M}{Lt^2}\right)^a L^b \left(\frac{L}{t}\right)^c \qquad (5.3\text{-}20)$$

Apply dimensional homogeneity

$$M: a = 0 \qquad\qquad a = 0$$
$$L: 1 - a + b + c = 0 \quad \Rightarrow \quad b = -1$$
$$t: -2a - c = 0 \qquad\qquad c = 0 \qquad\qquad (5.3\text{-}21)$$

Result

$$\boxed{\pi_4 = \frac{k}{D}} \qquad\qquad (5.3\text{-}22)$$

For the FML system[25]

Choose subset

$$Q_1 = \Delta p \quad Q_3 = D \quad Q_6 = \rho \qquad\qquad (5.3\text{-}23)$$

For the FLt system

Choose subset

$$Q_1 = \Delta p \quad Q_3 = D \quad Q_4 = v \qquad\qquad (5.3\text{-}24)$$

For the FMt system

Choose subset

$$Q_1 = \Delta p \quad Q_3 = D \quad Q_4 = v \qquad\qquad (5.3\text{-}25)$$

For the FMLt system

Choose subset

$$Q_1 = \Delta p \quad Q_3 = D \quad Q_4 = v \quad Q_6 = \rho \qquad\qquad (5.3\text{-}26)$$

Form π_1

---

[25] Since the algorithm is the same for each of the systems of three fundamental dimensions, we omit the development for the remaining three and present only the final results.

$$\pi_1 = L\left(\Delta p\right)^a D^b v^c \rho^d \;\Rightarrow\; L\left(\frac{F}{L^2}\right)^a L^b \left(\frac{L}{t}\right)^c \left(\frac{M}{L^3}\right)^d \tag{5.3-27}$$

Apply dimensional homogeneity

F: $a = 0$
M: $d = 0$ $\Rightarrow$
L: $1 - 2a + b + c - 3d = 0$
t: $c = 0$

$$\begin{aligned} a &= 0 \\ b &= -1 \\ c &= 0 \\ d &= 0 \end{aligned} \tag{5.3-28}$$

Result

$$\boxed{\pi_1 = \frac{L}{D}} \tag{5.3-29}$$

Form π_2

$$\pi_2 = \mu\left(\Delta p\right)^a D^b v^c \rho^d \;\Rightarrow\; \left(\frac{M}{Lt}\right)\left(\frac{F}{L^2}\right)^a L^b \left(\frac{L}{t}\right)^c \left(\frac{M}{L^3}\right)^d \tag{5.3-30}$$

Apply dimensional homogeneity

F: $a = 0$
M: $1 + d = 0$ $\Rightarrow$
L: $-1 - 2a + b + c - 3d = 0$
t: $-1 - c = 0$

$$\begin{aligned} a &= 0 \\ b &= -1 \\ c &= -1 \\ d &= -1 \end{aligned} \tag{5.3-31}$$

Result

$$\boxed{\pi_2 = \frac{\mu}{D v \rho}} \tag{5.3-32}$$

Note that

$$\left(\frac{\pi_3}{\pi_2}\right)_{MLt} = \left(\pi_2\right)_{FMLt} \tag{5.3-33}$$

Form π_3

$$\pi_2 = k\left(\Delta p\right)^a D^b v^c \rho^d \;\Rightarrow\; (L)\left(\frac{F}{L^2}\right)^a L^b \left(\frac{L}{t}\right)^c \left(\frac{M}{L^3}\right)^d \tag{5.3-34}$$

Apply dimensional homogeneity

F: $a = 0$
M: $d = 0$ $\Rightarrow$
L: $1 - 2a + b + c - 3d = 0$
t: $-1 - c = 0$

$$\begin{aligned} a &= 0 \\ b &= -1 \\ c &= 0 \\ d &= 0 \end{aligned} \tag{5.3-35}$$

Result

$$\boxed{\pi_3 = \frac{k}{D}}$$

(5.3-36)

Form π_4

$$\pi_4 = g_c \left(\Delta p\right)^a D^b v^c \rho^d \implies \left(\frac{ML}{Ft^2}\right)\left(\frac{F}{L^2}\right)^a L^b \left(\frac{L}{t}\right)^c \left(\frac{M}{L^3}\right)^d$$

(5.3-37)

Apply dimensional homogeneity

$$
\begin{array}{lll}
\text{F: } -1 + a = 0 & & a = 1 \\
\text{M: } 1 + d = 0 & & b = 0 \\
\text{L: } 1 - 2a + b + c - 3d = 0 & \implies & c = -2 \\
\text{t: } -2 - c = 0 & & d = 1
\end{array}
$$

(5.3-38)

Result

$$\boxed{\pi_4 = \frac{g_c \Delta p}{\rho v^2}}$$

(5.3-39)

The five sets of dimensionless groups are equivalent.

MLt	FML	FLt	FMt	FMLt
$\dfrac{L}{D}$	$\dfrac{L}{D}$	$\dfrac{L}{D}$	$\dfrac{L}{D}$	$\dfrac{L}{D}$
$\dfrac{\mu v}{\Delta p D}$	$\dfrac{\mu v}{\Delta p D} = \left(\left[\dfrac{\rho v^2}{\Delta p}\right]\left[\dfrac{\mu^2}{\Delta p D^2 \rho}\right]\right)^{\frac{1}{2}}$	$\dfrac{\mu v}{\Delta p D}$	$\dfrac{\mu v}{\Delta p D}$	$\dfrac{\mu v}{\Delta p D} = \dfrac{\left[\dfrac{\rho v^2}{\Delta p}\right]}{\left[\dfrac{\mu}{D v \rho}\right]}$
$\dfrac{\rho v^2}{\Delta p}$	$\dfrac{\rho v^2}{\Delta p} = \dfrac{\left[\dfrac{\mu v}{\Delta p D}\right]^2}{\left[\dfrac{\mu^2}{\Delta p D^2 \rho}\right]}$	$\dfrac{\rho v^2}{\Delta p}$	$\dfrac{\rho v^2}{\Delta p}$	$\dfrac{\rho v^2}{g_c \Delta p} = \dfrac{1}{\dfrac{g_c \Delta p}{\rho v^2}}$
$\dfrac{k}{D}$	$\dfrac{k}{D}$	$\dfrac{k}{D}$	$\dfrac{k}{D}$	$\dfrac{k}{D}$

Example 5.3-4 Capillary interface height via dimensional analysis

When a capillary (a tube of very small diameter) is dipped into a liquid, a concave or convex meniscus forms at the interface between the liquid and air, depending on the surface tension, and the interface inside the tube is raised or depressed some distance, h, relative to the interface of the surrounding liquid. Shown is a case where the meniscus is concave, and the liquid rises in the capillary.

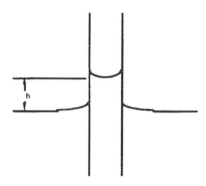

The displacement of the capillary interface relative to the surrounding liquid is a function of the diameter of the capillary tube, D; the density of the liquid, ρ; the acceleration of gravity, g; and the surface tension of the liquid, σ. Use the FMLt system of units to determine a complete set of dimensionless products for this problem.

Solution

Dimensions of the variables are

	FMLt
$Q_1 = h$	L
$Q_2 = D$	L
$Q_3 = \rho$	M/L^3
$Q_4 = g$	L/t^2
$Q_5 = \sigma$	F/L
$Q_6 = g_c$	$ML/(Ft^2)$

The dimensional matrix is

$$
\begin{array}{c|cccc}
 & F & M & L & t \\
\hline
Q_1 = h & 0 & 0 & 1 & 0 \\
Q_2 = D & 0 & 0 & 1 & 0 \\
Q_3 = \rho & 0 & 1 & -3 & 0 \\
Q_4 = g & 0 & 0 & 1 & -2 \\
Q_5 = \sigma & 1 & 0 & 1 & 0 \\
Q_6 = g_c & -1 & 1 & 1 & -2
\end{array}
\qquad (5.3\text{-}40)
$$

Looking for the largest non-zero determinant

$$
\begin{vmatrix}
0 & 0 & 1 & 0 \\
0 & 0 & 1 & 0 \\
0 & 1 & -3 & 0 \\
0 & 0 & 1 & -2 \\
1 & 0 & 1 & 0 \\
-1 & 1 & 1 & -2
\end{vmatrix} \Rightarrow
$$

$$
\begin{vmatrix}
0 & 1 & -3 & 0 \\
0 & 0 & 1 & -2 \\
1 & 0 & 1 & 0 \\
-1 & 1 & 1 & -2
\end{vmatrix} = -1 \cdot
\begin{vmatrix}
0 & 1 & -2 \\
1 & 1 & 0 \\
-1 & 1 & -2
\end{vmatrix} + (-3) \cdot
\begin{vmatrix}
0 & 0 & -2 \\
1 & 0 & 0 \\
-1 & 1 & -2
\end{vmatrix} =
$$

$$
-1\,[-1\,(-2)] + (-2)\,[(1) - (-1)] - 3\,[(-2)(1)] = (-1)\,(2 - 4) - 3\,(-2) =
$$

$$
2 + 6 = 8 \neq 0
$$

$$
(5.3\text{-}41)
$$

Therefore, the rank of the dimensional matrix is 4 and we expect 6 - 4 = 2 dimensionless groups.

Choose subset
$$
Q_2 = D \quad Q_3 = \rho \quad Q_4 = g \quad Q_6 = g_c \qquad (5.3\text{-}42)
$$

Form π_1

$$
\pi_1 = h\,D^a \rho^b g^c g_c^d \Rightarrow L\,L^a \left(\frac{M}{L^3}\right)^b \left(\frac{L}{t^2}\right)^c \left(\frac{ML}{Ft^2}\right)^d \qquad (5.3\text{-}43)
$$

Apply dimensional homogeneity

$$
\begin{array}{ll}
F: -d = 0 & a = -1 \\
M: b + d = 0 & b = 0 \\
L: 1 + a - 3b + c + d = 0 \quad \Rightarrow & c = 0 \\
t: -2c - 2d = 0 & d = 0
\end{array}
\qquad (5.3\text{-}44)
$$

Result

$$\pi_1 = \frac{h}{D} \qquad (5.3\text{-}45)$$

Form π_2

$$\pi_1 = \sigma D^a \rho^b g^c g_c^d \implies \left(\frac{F}{L}\right)\left(L^a\right)\left(\frac{M}{L^3}\right)^b\left(\frac{L}{t^2}\right)^c\left(\frac{ML}{Ft^2}\right)^d \qquad (5.3\text{-}46)$$

Apply dimensional homogeneity

$$
\begin{array}{ll}
\text{F: } 1 - d = 0 & a = -2 \\
\text{M: } b + d = 0 & b = -1 \\
\text{L: } -1 + a - 3b + c + d = 0 \quad\implies\quad & c = -1 \\
\text{t: } -2c - 2d = 0 & d = 1
\end{array}
\qquad (5.3\text{-}47)
$$

Result

$$\pi_2 = \frac{\sigma g_c}{D^2 \rho g} \qquad (5.3\text{-}48)$$

5.4 Dimensionless groups and differential models

We have been able to find complete sets of dimensionless products using systematic analysis of a set of variables without any knowledge of the underlying functional forms that describe the process. In principle, the functional form can then be elucidated by doing experiments in the laboratory. Laboratory experiments, however, are nothing more than a way of solving the differential equations that describe a process. If we have the correct differential equation model, by making it dimensionless we can find the dimensionless variables associated with the problem.

For example, if we consider the case of draining various tanks of differing sizes using a pipe that enters at the bottom of the tank, our laboratory experiment might be to observe the depth of water, h, in the tank as a function of time for various initial depths, h_o, (at $t_o = 0$). For convenience we assume t to start at 0; we could equally well vary the initial value of time.

Both the tank surface and the drain outlet are assumed to be at atmospheric pressure. As the depth of liquid in the tank decreases, the pressure drop across the drain pipe decreases, and so the flowrate through the drain pipe decreases. We will assume that the drain pipe remains in the laminar flow regime so that the instantaneous volumetric flowrate, Q, will be linear in the depth of liquid in the tank: $Q = kh$.

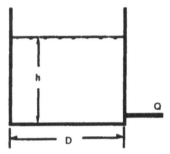

The functional relationship we would like to determine is

$$\varphi_1(D, h, h_0, t, k) = 0 \qquad\qquad (5.4\text{-}1)$$

and the dimensional matrix in an (L, t) system would be

$$
\begin{array}{c|cc}
 & L & t \\
\hline
D & 1 & 0 \\
h & 1 & 0 \\
h_0 & 1 & 0 \\
t & 0 & 1 \\
k & 2 & -1
\end{array}
\qquad\qquad (5.4\text{-}2)
$$

The rank of this matrix is two, and so we expect $(5 - 2) = 3$ dimensionless groups in a complete set. Such a set is (either from systematic analysis or by inspection, noting that $Qt/D^3 = kh_0t/D^3$ is dimensionless)

$$\pi_1 = \left[\frac{h}{D}, \frac{h_0}{D}, \frac{k\,h_0\,t}{D^3}\right] = \left[h^{*}, h_0^{*}, t^{*}\right] \qquad\qquad (5.4\text{-}3)$$

where we have designated the dimensionless groups as a dimensionless depth, h^{*}, a dimensionless initial depth, h_0^{*}, and a dimensionless time, t^{*}.

Since water is very nearly incompressible, a total mass balance on the tank contents reduces to a volume balance containing only output and accumulation terms. Hence, the basic differential equation that models the process is

$$Q = kh = -\frac{dV}{dt} = \frac{-d\left(\frac{\pi D^2 h}{4}\right)}{dt} = \frac{-\pi D^2}{4}\frac{dh}{dt} \quad \Rightarrow \quad \frac{dh}{dt} = \frac{-4k}{\pi D^2}h$$

$$(5.4\text{-}4)$$

with the initial condition $h = h_0$ at $t = 0$.[26] Notice that the constant k is dimensional, with the dimensions of L^2/t. We can de-dimensionalize this differential equation and its boundary condition using the dimensionless variables (indicated by an asterisk superscript) previously defined.

Using the chain rules of differentiation from calculus we can write the derivative term of Equation (5.4-4) in dimensionless form

$$\frac{dh}{dt} = \left(\frac{dh}{dh^*}\right)\left(\frac{dh^*}{dt^*}\right)\left(\frac{dt^*}{dt}\right) = D\left(\frac{dh^*}{dt^*}\right)\frac{k\,h_0}{D^3} \qquad (5.4\text{-}5)$$

Substituting in Equation (5.4-4), using the definition of $h_0{}^*$

$$D\left(\frac{dh^*}{dt^*}\right)\frac{k\,h_0}{D^3} = \frac{-4k}{\pi D^2}D\,h^* \quad \Rightarrow \quad \left(\frac{dh^*}{h^*}\right) = \frac{-4}{\pi h_0{}^*}dt^* \qquad (5.4\text{-}6)$$

The boundary condition de-dimensionalizes as

---

[26] Although the value of k is not important to this example, it could be determined from (where the subscript 0 refers to the surface of the liquid in the tank, 1 refers to the drain pipe inlet, and 2 refers to the drain pipe outlet)

$$p_2 - p_1 = -\frac{2fL\rho\langle v\rangle^2}{D_{pipe}} \quad \text{(mechanical energy balance on drain pipe)}$$

$$p_1 - p_0 = p_1 - 0 = \rho g(h - 0) \quad \text{(mechanical energy balance on tank)}$$

$$f = \frac{16}{Re} = \frac{16\mu}{D_{pipe}\langle v\rangle\rho} \quad \text{(Fanning friction factor, laminar regime)}$$

$$p_2 - \rho g h = 0 - \rho g h = -\frac{32\mu L\langle v\rangle}{D_{pipe}^2}$$

$$Q = \frac{\pi D_{pipe}^2}{4}\langle v\rangle = \frac{\frac{\pi D_{pipe}^2}{4}D_{pipe}^2\,\rho g h}{32\mu L} = \left(\frac{\pi D_{pipe}^4\,\rho g}{128\mu L}\right)h = kh$$

$$
\begin{bmatrix} h = h_0 \\ \text{at} \\ t = 0 \end{bmatrix}
\Rightarrow
\begin{bmatrix} D\,h^{\bullet} = h_0 \\ \text{at} \\ \dfrac{D^3 t^{\bullet}}{k\,h_0} = 0 \end{bmatrix}
\Rightarrow
\begin{bmatrix} h^{\bullet} = \dfrac{h_0}{D} \\ \text{at} \\ t^{\bullet} = 0 \end{bmatrix}
\qquad (5.4\text{-}7)
$$

Our dimensionless groups in Equation (5.4-3) appeared as dimensionless variables both in the differential equation and in the boundary condition, and as a dimensionless coefficient in the differential equation.

Integrating Equation (5.4-6) using Equation (5.4-7) gives

$$
\int \frac{dh^{\bullet}}{h^{\bullet}} = \int \left(\frac{-4}{\pi\,h_0^{\bullet}} \right) dt^{\bullet}
$$

$$
\ln\left[h^{\bullet} \right] = \left(\frac{-4}{\pi\,h_0^{\bullet}} \right) t^{\bullet} + C
\qquad (5.4\text{-}8)
$$

Applying the initial condition gives

$$
C = \ln\left[\frac{h_0}{D} \right] = \ln\left[h_0^{\bullet} \right]
\qquad (5.4\text{-}9)
$$

which yields

$$
\ln\left[\frac{h^{\bullet}}{h_0^{\bullet}} \right] = \left(\frac{-4}{\pi\,h_0^{\bullet}} \right) t^{\bullet}
\qquad (5.4\text{-}10)
$$

so it can be seen that all three of the dimensionless groups appear in the integrated relation.

This illustrates two things which we take without proof

• The dimensionless products we determine via the systematic method have significance in the fundamental differential equation(s) describing the process and its (their) boundary conditions.

• Another way to find sets of dimensionless products is via de-dimensionalizing the differential equation and boundary conditions (of course, this presupposes our ability to write the equations and boundary conditions in their dimensional form).

Example 5.4-1   Pipe flow of incompressible fluid with constant viscosity

The differential equation describing laminar flow of an incompressible Newtonian fluid of constant viscosity in the axial direction in a pipe, neglecting pressure gradients other than in the longitudinal direction, is

$$\rho \frac{\partial v_x}{\partial t} = -\frac{dp}{dx} + \mu \frac{1}{r} \frac{\partial}{\partial r}\left(r \frac{\partial v_x}{\partial r}\right) \tag{5.4-11}$$

Make this equation dimensionless by using the following dimensionless variables

$$v_x^* = \frac{v_x}{\langle v \rangle} \quad p^* = \frac{p}{\rho \langle v \rangle^2} \quad t^* = \frac{t \langle v \rangle}{D} \quad x^* = \frac{x}{D} \quad r^* = \frac{r}{D} \tag{5.4-12}$$

Solutions

Noting that

$$\frac{\partial v_x}{\partial v_x^*} = \frac{\partial\left(\langle v \rangle v_x^*\right)}{\partial v_x^*} = \langle v \rangle \tag{5.4-13}$$

$$\frac{\partial t^*}{\partial t} = \frac{\partial\left(\frac{t \langle v \rangle}{D}\right)}{\partial t} = \frac{\langle v \rangle}{D} \tag{5.4-14}$$

$$\frac{dp}{dp^*} = \frac{d\left(\rho \langle v \rangle^2 p^*\right)}{dp^*} = \rho \langle v \rangle^2 \tag{5.4-15}$$

$$\frac{dx^*}{dx} = \frac{d\left(\frac{x}{D}\right)}{dx} = \frac{1}{D} \tag{5.4-16}$$

$$\frac{\partial r^*}{\partial r} = \frac{\partial\left(\frac{r}{D}\right)}{\partial r} = \frac{1}{D} \tag{5.4-17}$$

Applying the chain rule and substituting in terms of dimensionless variables

$$\rho \frac{\partial v_x}{\partial v_x^*}\left(\frac{\partial v_x^*}{\partial t^*}\right)\frac{\partial t^*}{\partial t} = -\rho \frac{dp}{dp^*}\left(\frac{dp^*}{dx^*}\right)\frac{dx^*}{dx}$$

$$+\mu \frac{1}{(D\,r^*)}\frac{\partial r^*}{\partial r}\left((D\,r^*)\frac{\partial v_x}{\partial v_x^*}\left(\frac{\partial v_x^*}{\partial r^*}\right)\frac{\partial r^*}{\partial r}\right) \tag{5.4-18}$$

$$\rho \langle v\rangle \left(\frac{\partial v_x^*}{\partial t^*}\right)\frac{\langle v\rangle}{D} = -\rho \langle v\rangle^2 \left(\frac{dp^*}{dx^*}\right)\frac{1}{D}$$

$$+\mu \frac{1}{(D\,r^*)}\frac{1}{D}\frac{\partial}{\partial r^*}\left((D\,r^*)\langle v\rangle \left(\frac{\partial v_x^*}{\partial r^*}\right)\frac{1}{D}\right) \tag{5.4-19}$$

$$\frac{\rho \langle v\rangle^2}{D}\frac{\partial v_x^*}{\partial t^*} = -\frac{\rho \langle v\rangle^2}{D}\frac{dp^*}{dx^*} + \frac{\mu \langle v\rangle}{D^2 r^*}\frac{\partial}{\partial r^*}\left(r^* \frac{\partial v_x^*}{\partial r^*}\right) \tag{5.4-20}$$

$$\frac{\partial v_x^*}{\partial t^*} = -\frac{dp^*}{dx^*} + \frac{\mu}{D\langle v\rangle \rho}\frac{1}{r^*}\frac{\partial}{\partial r^*}\left(r^* \frac{\partial v_x^*}{\partial r^*}\right) \tag{5.4-21}$$

$$\frac{\partial v_x^*}{\partial t^*} = -\frac{dp^*}{dx^*} + \frac{1}{Re}\frac{1}{r^*}\frac{\partial}{\partial r^*}\left(r^* \frac{\partial v_x^*}{\partial r^*}\right) \tag{5.4-22}$$

For the same dimensionless boundary conditions, all systems described by the above differential equation will have the same solution

$$v_x^* = v_x^*(x^*, t^*) \tag{5.4-23}$$

Coincidence of the dimensionless velocity profiles indicates (by definition) **dynamic similarity** between two systems.

It does not matter whether a liquid or a gas flows in the pipe, whether the pipe is large or small, etc. The solution will be the same for the same values of the dimensionless groups. This means, for example, that we can generate (d) in the laboratory using convenient fluids such as air and water, yet obtain a solution applicable to corrosive, toxic, or highly viscous fluids, etc. (as long as they are Newtonian, which is assumed in the original equation).

Example 5.4-2   One-dimensional energy transport

The differential equation describing energy transport for a fluid at constant pressure in one dimension is

$$\rho \, C_p \frac{\partial T}{\partial t} + \rho \, C_p \, v_x \frac{\partial T}{\partial x} = k \frac{\partial^2 T}{\partial x^2} \qquad (5.4\text{-}24)$$

Make this equation dimensionless using the following dimensionless variables

$$v_x^{\bullet} = \frac{v_x}{\langle v \rangle} \qquad t^{\bullet} = \frac{t \langle v \rangle}{D} \qquad x^{\bullet} = \frac{x}{D} \qquad T^{\bullet} = \frac{T - T_0}{T_1 - T_0} \qquad (5.4\text{-}25)$$

Solution

Using the chain rule

$$\rho \, C_p \frac{\partial T}{\partial T^{\bullet}} \frac{\partial T^{\bullet}}{\partial t^{\bullet}} \frac{\partial t^{\bullet}}{\partial t} + \rho \, C_p \, v_x \frac{\partial T}{\partial T^{\bullet}} \frac{\partial T^{\bullet}}{\partial x^{\bullet}} \frac{\partial x^{\bullet}}{\partial x} = k \, \frac{\partial x^{\bullet}}{\partial x} \frac{\partial}{\partial x^{\bullet}} \left(\frac{\partial T}{\partial T^{\bullet}} \frac{\partial T^{\bullet}}{\partial x^{\bullet}} \frac{\partial x^{\bullet}}{\partial x} \right)$$

$$(5.4\text{-}26)$$

Using the definitions of the dimensionless variables

$$\rho \, C_p \, (T_1 - T_0) \frac{\partial T^{\bullet}}{\partial t^{\bullet}} \frac{\langle v \rangle}{D} + \rho \, C_p \, \langle v \rangle \, v_x^{\bullet} \, (T_1 - T_0) \frac{\partial T^{\bullet}}{\partial x^{\bullet}} \frac{1}{D}$$

$$= k \, \frac{1}{D} \frac{\partial}{\partial x^{\bullet}} \left(((T_1 - T_0) \frac{\partial T^{\bullet}}{\partial x^{\bullet}} \frac{1}{D} \right) \qquad (5.4\text{-}27)$$

which yields

$$\frac{\partial T^{\bullet}}{\partial t^{\bullet}} + v_x^{\bullet} \frac{\partial T^{\bullet}}{\partial x^{\bullet}} = \left[\frac{k}{D \, \langle v \rangle \, \rho \, C_p} \right] \frac{\partial^2 T^{\bullet}}{\partial x^{\bullet 2}} \qquad (5.4\text{-}28)$$

By multiplying and dividing by μ this can be written as

$$\frac{\partial T^{\bullet}}{\partial t^{\bullet}} + v_x^{\bullet} \frac{\partial T^{\bullet}}{\partial x^{\bullet}} = \left[\frac{k}{\mu \, C_p} \right] \left[\frac{\mu}{D \, \langle v \rangle \, \rho} \right] \frac{\partial^2 T^{\bullet}}{\partial x^{\bullet 2}} \qquad (5.4\text{-}29)$$

where the first bracketed term is the reciprocal of the Prandtl number and the second bracketed term is the reciprocal of the Reynolds number.

Two systems described by this equation, each system with the same Reynolds and Prandtl number, each subject to the same dimensionless boundary conditions, will each be described by the same solution function

$$T^\bullet = T^\bullet(x^\bullet, t^\bullet)$$

(5.4-30)

Such systems are designated as **thermally similar**.

Example 5.4-3   Mass transport in a binary mixture

The differential equation describing mass transport in one dimension of component, A, in a binary mixture of A and B is

$$\frac{\partial x_A}{\partial t} + v_z \frac{\partial x_A}{\partial z} = \mathcal{D}_{AB} \frac{\partial^2 x_A}{\partial z^2}$$

(5.4-31)

For this example we have switched to the z-direction rather than the x-direction for the differential equation to avoid possible confusion between the concentration x_A and the coordinate x.

Make this equation dimensionless using the dimensionless variables

$$v_z^\bullet = \frac{v_z}{\langle v \rangle} \qquad t^\bullet = \frac{t \langle v \rangle}{D} \qquad z^\bullet = \frac{z}{D} \qquad x_A^\bullet = \frac{x_A - x_{A0}}{x_{A1} - x_{A0}}$$

(5.4-32)

Solution

Using the chain rule

$$\frac{\partial x_A}{\partial x_A^\bullet} \frac{\partial x_A^\bullet}{\partial t^\bullet} \frac{\partial t^\bullet}{\partial t} + v_z \frac{\partial x_A}{\partial x_A^\bullet} \frac{\partial x_A^\bullet}{\partial z^\bullet} \frac{\partial z^\bullet}{\partial z} = \mathcal{D}_{AB} \frac{\partial z^\bullet}{\partial z} \frac{\partial}{\partial z^\bullet} \left(\frac{\partial x_A}{\partial x_A^\bullet} \frac{\partial x_A^\bullet}{\partial z^\bullet} \frac{\partial z^\bullet}{\partial z} \right)$$

(5.4-33)

Using the definitions of the dimensionless variables

$$\left(x_{A1} - x_{A0}\right) \frac{\partial x_A^{\bullet}}{\partial t^{\bullet}} \frac{\langle v \rangle}{D} + \langle v \rangle v_z^{\bullet} \left(x_{A1} - x_{A0}\right) \frac{\partial x_A^{\bullet}}{\partial z^{\bullet}} \frac{1}{D}$$

$$= \mathcal{D}_{AB} \frac{1}{D} \frac{\partial}{\partial z^{\bullet}} \left(\left(x_{A1} - x_{A0}\right) \frac{\partial x_A^{\bullet}}{\partial z^{\bullet}} \frac{1}{D} \right)$$

(5.4-34)

which reduces to

$$\frac{\partial x_A^{\bullet}}{\partial t^{\bullet}} + v_z^{\bullet} \frac{\partial x_A^{\bullet}}{\partial z^{\bullet}} = \frac{\mathcal{D}_{AB}}{\langle v \rangle D} \frac{\partial^2 x_A^{\bullet}}{\partial z^{\bullet 2}}$$

(5.4-35)

After multiplying and dividing by $(\rho\mu)$ we can write

$$\frac{\partial x_A^{\bullet}}{\partial t^{\bullet}} + v_z^{\bullet} \frac{\partial x_A^{\bullet}}{\partial z^{\bullet}} = \left[\frac{\mathcal{D}_{AB}\rho}{\mu} \right] \left[\frac{\mu}{D v \rho} \right] \frac{\partial^2 x_A^{\bullet}}{\partial z^{\bullet 2}}$$

(5.4-36)

where the first bracketed term is the reciprocal of the Schmidt number and the second bracketed term is the reciprocal of the Reynolds number.

Two systems described by this equation, each system with the same Reynolds and Schmidt number, each subject to the same dimensionless boundary conditions, will each be described by the same solution function

$$x_A^{\bullet} = x_A^{\bullet}(z^{\bullet}, t^{\bullet})$$

(5.4-37)

Example 5.4-4   Extrapolating model results from one category of momentum, heat, or mass transport to another

The concept of model can be extended even to permit taking data in one transport situation and applying it to another. Consider the following three cases, which are treated in detail later in each of the appropriate chapters.

CASE ONE

Consider a flat plate adjacent to a quiescent fluid of infinite extent in the positive x-direction and the z-direction. Let the plate be instantaneously set in motion at a constant velocity.

The governing differential equation

$$\frac{\partial v_y}{\partial t} = v \frac{\partial^2 v_y}{\partial x^2}$$

(5.4-38)

has boundary conditions

$t = 0$: $v_y = 0$, all x (initial condition)
$x = 0$: $v_y = V$, all $t > 0$ (boundary condition)
$x = \infty$: $v_y = 0$, all $t > 0$ (boundary condition)

If we de-dimensionalize the dependent variable in the equation using

$$v_y^* = \frac{v_y - v_i}{V - v_i} = \frac{v_y - 0}{V - 0} = \frac{v_y}{V}$$

(5.4-39)

and combine the independent variables in a dimensionless **similarity transformation**

$$\eta = \frac{x}{\sqrt{4 v t}}$$

(5.4-40)

we reduce the problem to a dimensionless ordinary differential equation where now, instead of velocity being a function of two independent variables,

$$v_y^* = v_y^* (x, t)$$

(5.4-41)

the dimensionless velocity is now a new function of only one independent variable, η. We therefore use a new symbol for the function, namely, ϕ

$$v_y^* = \varphi(\eta)$$

(5.4-42)

We have reduced the problem (omitting the intermediate steps which are covered later) to an ordinary differential equation

$$\varphi^{\cdot\cdot} + 2 \eta \, \varphi^{\cdot} = 0$$

(5.4-43)

with boundary conditions

$$\eta = \infty: \quad \varphi(\eta) = 0$$
$$\eta = 0: \quad \varphi(\eta) = 1 \tag{5.4-44}$$

whose solution is

$$\varphi = 1 - \text{erf}[\eta]$$
$$= \text{erfc}[\eta] \tag{5.4-45}$$

The solution of our problem in terms of the original variables is then

$$v_y^{\bullet} = 1 - \text{erf}\left[\frac{x}{\sqrt{4 v t}}\right]$$
$$= \text{erfc}\left[\frac{x}{\sqrt{4 v t}}\right] \tag{5.4-46}$$

CASE TWO

Consider a heat conduction in the x-direction in a semi-infinite (bounded only by one face) slab initially at a uniform temperature, T_i, whose face suddenly at time equal to zero is raised to and maintained at T_S.

The governing differential equation

$$\frac{\partial T}{\partial t} = \alpha \frac{\partial^2 T}{\partial x^2} \tag{5.4-47}$$

has boundary conditions

$$t = 0: \quad T = T_i, \text{ all } x \qquad \text{(initial condition)}$$
$$x = 0: \quad T = T_s, \text{ all } t > 0 \text{ (boundary condition)}$$
$$x = \infty: \quad T = T_i, \text{ all } t > 0 \text{ (boundary condition)}$$

If we de-dimensionalize the dependent variable in the equation using

$$T^{\bullet} = \frac{T - T_i}{T_s - T_i} \tag{5.4-48}$$

and combine the independent variables in a dimensionless **similarity transformation**

$$\eta = \frac{x}{\sqrt{4 \alpha t}} \tag{5.4-49}$$

we reduce the problem to a dimensionless ordinary differential equation where now, instead of temperature being a function of two independent variables,

$$T^{\bullet} = T^{\bullet}(x, t) \tag{5.4-50}$$

the dimensionless temperature is now a new function of only one independent variable, η. We therefore use a new symbol for the function, namely, ϕ

$$T^{\bullet} = \varphi(\eta) \tag{5.4-51}$$

We have reduced the problem (omitting the intermediate steps which are covered later) to an ordinary differential equation

$$\varphi^{\bullet} + 2 \eta \varphi^{'} = 0 \tag{5.4-52}$$

with boundary conditions

$$\begin{aligned} \eta = \infty: \quad \varphi(\eta) &= 0 \\ \eta = 0: \quad \varphi(\eta) &= 1 \end{aligned} \tag{5.4-53}$$

whose solution is

$$\begin{aligned} \varphi &= 1 - \mathrm{erf}[\eta] \\ &= \mathrm{erfc}[\eta] \end{aligned} \tag{5.4-54}$$

The solution of our problem in terms of the original variables is then

$$\begin{aligned} T^{\bullet} &= 1 - \mathrm{erf}\left[\frac{x}{\sqrt{4 \alpha t}}\right] \\ &= \mathrm{erfc}\left[\frac{x}{\sqrt{4 \alpha t}}\right] \end{aligned} \tag{5.4-55}$$

CASE THREE

Consider diffusion of component A in the x-direction in a semi-infinite (bounded only by one face) slab initially at a uniform concentration, c_{Ai}, whose face suddenly at time equal to zero is raised to and maintained at c_{AS}.

The governing differential equation

$$\frac{\partial c_A}{\partial t} = \mathcal{D}_{AB} \frac{\partial^2 c_A}{\partial x^2}$$ (5.4-56)

has boundary conditions

$t = 0$: $c_A = c_{Ai}$, all x   (initial condition)
$x = 0$: $c_A = c_{As}$, all $t > 0$ (boundary condition)
$x = \infty$: $c_A = c_{Ai}$, all $t > 0$ (boundary condition)

If we de-dimensionalize the dependent variable in the equation using

$$c_A^* = \frac{c_A - c_{Ai}}{c_{As} - c_{Ai}}$$ (5.4-57)

and combine the independent variables in a dimensionless **similarity transformation**

$$\eta = \frac{x}{\sqrt{4 \mathcal{D}_{AB} t}}$$ (5.4-58)

we reduce the problem to a dimensionless ordinary differential equation where now, instead of concentration being a function of two independent variables,

$$c_A^* = c_A^* (x, t)$$ (5.4-59)

the dimensionless concentration is now a new function of only one independent variable, η. We therefore use a new symbol for the function, namely, ϕ

$$c_A^* = \varphi(\eta)$$ (5.4-60)

We have reduced the problem (omitting the intermediate steps which are covered later) to an ordinary differential equation

$$\varphi'' + 2\eta\,\varphi' = 0 \tag{5.4-61}$$

with boundary conditions

$$\eta = \infty: \quad \varphi(\eta) = 0$$
$$\eta = 0: \quad \varphi(\eta) = 1 \tag{5.4-62}$$

whose solution is

$$\varphi = 1 - \mathrm{erf}\,[\eta]$$
$$= \mathrm{erfc}\,[\eta] \tag{5.4-63}$$

The solution of our problem in terms of the original variables is then

$$c_A^* = 1 - \mathrm{erf}\left[\frac{x}{\sqrt{4\,\mathcal{D}_{AB}\,t}}\right]$$

$$= \mathrm{erfc}\left[\frac{x}{\sqrt{4\,\mathcal{D}_{AB}\,t}}\right] \tag{5.4-64}$$

Notice that all three of these cases are described by the same dimensionless differential equation and dimensionless boundary conditions. The dimensionless solutions are likewise, therefore, the same. Below is a plot of the dimensionless solution.

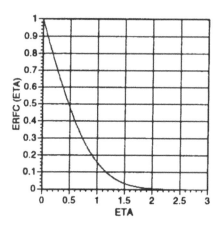

The plot can be taken as representing any of the three solutions

$$v_y^* = \text{erfc}\left[\frac{x}{\sqrt{4\,\nu\,t}}\right]$$

(5.4-65)

$$T^* = \text{erfc}\left[\frac{x}{\sqrt{4\,\alpha\,t}}\right]$$

(5.4-66)

$$c_A^* = \text{erfc}\left[\frac{x}{\sqrt{4\,\mathcal{D}_{AB}\,t}}\right]$$

(5.4-67)

This is not so significant in this case, because we have the form of the solution. **Frequently, however, we are confronted with physical situations for which we either cannot write or cannot solve the differential equation.**

If we are reasonably sure that the dimensionless differential equation and the dimensionless boundary conditions are the same between, for example, heat transfer and mass transfer, we know that the dimensionless solutions are the same.[27] In cases where we cannot solve the governing equation mathematically, we obtain the solution by experiment.

Since we would obtain the same solution whether we performed a heat transfer or a mass transfer experiment, we can choose the easier. In this example, it would usually be the heat transfer experiment, because, in general, temperatures are more easily measured than concentrations (because of such factors as the sensing element disturbing the flow, size of sensing element, sample size required, method of analysis, etc.).

We could, therefore, **determine data for mass transfer by taking heat transfer measurements,** converting them to the dimensionless solution, and then mapping the dimensionless solution into concentration variables.

Lists of common dimensionless variables, dimensionless equations, and dimensionless groups are listed in Tables 5.4-1, 5.4-2, and 5.4-3.

---

[27] This implies that the transport mechanisms are the same - so, for example, we would not expect similarity between momentum transport and either mass or heat transport in situations involving form drag, which has no counterpart in heat or mass transfer.

Table 5.4-1 Dimensionless variables

Time	$t^* = \dfrac{t\, v_0}{D}$
	$t^{**} = \dfrac{t\, \mu}{\rho\, D^2}$
Distance	$x_i^* = \dfrac{x_i}{D}$
Velocity	$v_i^* = \dfrac{v_i}{v_0}$
	$v_i^{**} = \dfrac{v_i\, D\, \rho}{\mu}$
Pressure	$p^* = \dfrac{p - p_0}{\rho\, v_0^2}$
Temperature	$T^* = \dfrac{T - T_0}{T_1 - T_0}$
Concentration	$x_A^* = \dfrac{\left(x_A\right) - \left(x_A\right)_0}{\left(x_A\right)_1 - \left(x_A\right)_0}$
Differential Operations	$\nabla^* = D\, \nabla$
	$\nabla^{*2} = D^2\, \nabla^2$
	$\dfrac{D}{Dt^*} = \dfrac{D}{v_0}\dfrac{D}{Dt}$

Table 5.4-2 Dedimensionalized balance equations

FORCED CONVECTION	
Total mass balance (Continuity)	$\nabla^{\bullet} \cdot \mathbf{v}^{*} = 0$
Species mass balance	$\dfrac{D x_A^{\bullet}}{D t^{\bullet}} = \dfrac{1}{Re\,Sc} \nabla^{\bullet 2} x_A^{\bullet}$
Energy balance	$\dfrac{D T^{\bullet}}{D t^{\bullet}} = \dfrac{1}{Re\,Pr} \nabla^{\bullet 2} T^{\bullet} + \dfrac{Br}{Re\,Pr} \Phi_v^{\bullet}$
Momentum balance	$\dfrac{D \mathbf{v}^{\bullet}}{D t^{\bullet}} = -\nabla^{\bullet} p^{\bullet} + \dfrac{1}{Re} \nabla^{\bullet 2} \mathbf{v}^{\bullet} + \dfrac{1}{Fr} \dfrac{\overline{g}}{g}$
NATURAL CONVECTION	
Total mass balance (Continuity)	$\nabla^{\bullet} \cdot \mathbf{v}^{**} = 0$
Species mass balance	$\dfrac{D x_A^{\bullet}}{D t^{\bullet\bullet}} = \dfrac{1}{Sc} \nabla^{\bullet 2} x_A^{\bullet}$
Temperature-difference induced flow	
Energy balance	$\dfrac{D T^{\bullet}}{D t^{\bullet\bullet}} = \dfrac{1}{Pr} \nabla^{\bullet 2} T^{\bullet}$
Momentum balance	$\dfrac{D \mathbf{v}^{\bullet\bullet}}{D t^{\bullet\bullet}} = \nabla^{\bullet 2} \overline{\mathbf{v}}^{\bullet\bullet} - T^{\bullet} Gr_h \dfrac{\overline{g}}{g}$
Concentration-difference induced flow	$\dfrac{D \mathbf{v}^{\bullet\bullet}}{D t^{\bullet\bullet}} = \nabla^{\bullet 2} \mathbf{v}^{\bullet\bullet} - x_A^{\bullet} Gr_m \dfrac{\overline{g}}{g}$

Table 5.4-3   Dimensionless numbers

Reynolds Number

$$Re = \frac{Lv\rho}{\mu} = \frac{inertial\ forces}{viscous\ forces}$$

Froude Number

$$Fr = \frac{v}{\sqrt{gL}} \implies Fr^2 = \frac{v^2}{gL} = \frac{\rho v^2 L^2}{\rho gL^3} = \frac{inertial\ forces}{gravitational\ forces}$$

Drag Coefficient

$$C_D = \frac{\tau_s}{\left[\frac{\rho\ v^2}{2}\right]} = dimensionless\ shear\ stress\ at\ surface$$

Friction Factor $\left(Fanning\ equal\ to\ \dfrac{Darcy\ [or\ Blasius]\ friction\ factor}{4} \right)$

$$f = \frac{\Delta p}{\left[\frac{L}{D}\right]\left[\frac{\rho\langle v\rangle^2}{2}\right]} = dimensionless\ pressure\ drop\ (internal\ flow)$$

Prandtl Number

$$Pr = \frac{\mu\hat{C}_p}{k} = \frac{v}{\alpha} = \frac{momentum\ diffusivity}{thermal\ energy\ diffusivity}$$

Graetz Number

(note close relationship to Peclet Number for heat transfer)

$$Gz = \frac{\langle v\rangle A\rho\hat{C}_p}{kL} = Re_D\ Pr\ \frac{D}{L} = \frac{bulk\ thermal\ energy\ transport}{rate\ of\ thermal\ energy\ conduction}$$

Schmidt Number

$$Sc = \frac{v}{\mathcal{D}_{AB}} = \frac{momentum\ diffusivity}{mass\ diffusivity}$$

Lewis Number

$$Le = \frac{Sc}{Pr} = \frac{thermal\ energy\ diffusivity}{mass\ diffusivity}$$

Table 5.4-3   Continued

Nusselt Number

$$Nu = \frac{hL}{k} = \text{dimensionless temperature gradient @ surface}$$

Sherwood Number

$$Sh = \frac{k_p L}{\mathcal{D}_{AB}} = \frac{k_x L}{c \, \mathcal{D}_{AB}} = \text{dimensionless concentration gradient @ surface}$$

Grashof Number for heat transfer

$$Gr_h = \frac{g\beta\rho^2 L^3 \Delta T}{\mu^2} = \frac{\text{temperature induced buoyancy forces}}{\text{viscous forces}}$$

$$\beta = -\frac{1}{\rho}\left(\frac{\partial \rho}{\partial T}\right)_p = \text{thermal coefficient of volume expansion}$$

Grashof Number for mass transfer

$$Gr_m = \frac{g\zeta\rho^2 L^3 \Delta x_A}{\mu^2} = \frac{\text{concentration induced buoyancy forces}}{\text{viscous forces}}$$

$$\zeta = -\frac{1}{\rho}\left(\frac{\partial \rho}{\partial x_A}\right)_T = \text{concentration coefficient of volume expansion}$$

Biot Number for heat transfer

$$Bi_h = \frac{hL}{k} = \frac{\text{internal resistance to conductive thermal energy transfer}}{\text{boundary layer resistance to convective thermal energy transfer}}$$

Biot Number for mass transfer

$$Bi_m = \frac{k_z L}{\mathcal{D}_{AB}} = \frac{\text{internal resistance to diffusive mass transfer}}{\text{boundary layer resistance to convective mass transfer}}$$

Fourier Number for heat transfer

$$Fo_h = \frac{\alpha t}{L^2} = \frac{\text{rate of thermal energy conduction}}{\text{rate of thermal energy accumulation}}$$

Fourier Number for mass transfer

$$Fo_m = \frac{\mathcal{D}_{AB} t}{L^2} = \frac{\text{rate of species diffusion}}{\text{rate of species accumulation}}$$

Table 5.4-3   Continued

Peclet Number for heat transfer **(note close relationship to Graetz number)**
$Pe_h = \dfrac{vL}{\alpha} = \left[\dfrac{vL\rho}{\mu}\right]\left[\dfrac{\mu C_p}{k}\right] = Re_L\,Pr$
$\qquad = \dfrac{vL\rho C_p}{k} = \dfrac{\text{bulk thermal energy transport}}{\text{rate of thermal energy conduction}}$
Peclet Number for mass transfer
$Pe_m = \dfrac{vL}{\mathcal{D}_{AB}} = \left[\dfrac{vL\rho}{\mu}\right]\left[\dfrac{\mu}{\rho\,\mathcal{D}_{AB}}\right] = \left[\dfrac{vL\rho}{\mu}\right]\left[\dfrac{v}{\mathcal{D}_{AB}}\right] = Re_L\,Sc$
$\qquad = \dfrac{vL\rho C_p}{k} = \dfrac{\text{bulk mass transport}}{\text{rate of mass diffusion}}$

Brinkman Number
$Br = \dfrac{\mu v^2}{k\,\Delta T} = \dfrac{\text{heating from viscous dissipation}}{\text{heating from temperature difference}}$
Eckert Number
$Ec = \dfrac{v^2}{C_p\left(T_s - T_\infty\right)} = \dfrac{Br}{Pr}$

5.5   Similarity, Models and Scaling

The use of scaling laws in designing physical models is based on the concept of **similarity**. In plane geometry, two polygons are **geometrically similar** when corresponding angles are equal and lengths of corresponding sides are in a constant ratio. Such figures have the same **shape** but may differ in size and/or **position**. Geometrical similarity is extended to bodies of arbitrary shape and size by requiring that corresponding lengths (whether of sides or distances between arbitrary points) be in constant ratio.

Physical similarity is more than geometrical similarity. A physical body has properties other than shape - mass, velocity, etc. We say two **physical** systems are similar if their relevant physical properties are in a constant ratio at geometrically corresponding points. Physical similarity presumes geometric similarity, but requires more than just geometrical similarity.

For example, **dynamic similarity** exists if two geometrically corresponding points experience **forces** in a constant ratio (which may or may not have the same value as the ratio for lengths). Two bodies are **thermally**

similar if at geometrically corresponding points the ratio of the **temperatures** is constant.

Points in the model and prototype that correspond are called **homologous points**. Phenomena occurring in the model and prototype[28] at homologous points are said to occur at **homologous times**, even though these events may not occur at the same real time. An equivalent definition of similarity is that two systems are similar if they behave in the same manner at homologous points and times.

The classical principle similarity from Buckingham π-theorem is

$$\pi_1 = f\left(\pi_2, \pi_3, \cdots, \pi_{n-r}\right) \tag{5.5-1}$$

Two systems in which all relevant physical quantities are in a constant ratio will satisfy Equation (5.5-1). The π's - dimensionless groups - are the criteria for geometric, dynamic, thermal, etc., similarity.

If we arbitrarily restrict the f of Equation (5.5-1) to a power function (a product of powers) then

$$\pi_1 = C\left(\pi_2\right)^{x_2}\left(\pi_3\right)^{x_3}\cdots\left(\pi_{n-r}\right)^{x_{n-r}} \tag{5.5-2}$$

where C is called the shape factor since it has been found to depend primarily on shape, and x_1, x_2 ... y_{n-r} are empirical exponents. In this relationship, which is called the extended principle (really, it is a **restricted** principle) of similarity, the shape factor depends on geometry; therefore, this equation is limited to systems that have the same shape.

The shape factor can be eliminated if two systems of similar geometric form are compared

$$\frac{\pi_1'}{\pi_1} = \left[\frac{\pi_2'}{\pi_2}\right]^{x_2}\left[\frac{\pi_3'}{\pi_3}\right]^{x_3}\cdots\left[\frac{\pi_{n-r}'}{\pi_{n-r}}\right]^{x_{n-r}} \tag{5.5-3}$$

---

[28] One dictionary definition of *prototype* is "the original on which a thing is modeled." This is a little ambiguous for common engineering usage, because we often use data from our *model* to design an object, in which case the *object is modeled* on the *model as the original*. We usually mean the laboratory or mathematical representation when we say "model," and the real world object under investigation or to be created when we say "prototype."

where the prime distinguishes between the systems. Use of the analysis in this manner is called **extrapolation** since the values of the exponents x_2, x_3, etc., are extrapolated from one system to the other.

It is possible to eliminate the effect of the exponents by comparing geometrically similar systems which have the same values of the corresponding dimensionless groups on the right-hand side. This makes all ratios on the right-hand side equal to unity, and therefore the right-hand side is equal to unity independent of the values of the exponents, thus

$$\pi_1{}' = \pi_1$$
$$\text{if: } \pi_2{}' = \pi_2, \pi_3{}' = \pi_3, \cdots, \pi_{n-r}{}' = \pi_{n-r} \qquad (5.5\text{-}4)$$

In general, when two physical systems are similar, knowledge about one system provides knowledge about the other. Equation (5.5-4) is called the **principle of corresponding states**.

Often it is not possible to match the values for every pair of corresponding groups as required above. In such cases we must run our experimental models so as to minimize the unmatched effects. If it is not possible to do this, we get a **scale effect**. For this reason, when running models, one has to be careful that values of the groups are appropriate to both the model and the prototype For example, if in testing a model of a breakwater in the laboratory, one uses a very small scale (shallow depth), in the data obtained the surface tension of the water may have a large effect on the behavior; however, at the scale of an actual breakwater the surface tension of the water would usually have no perceptible effect.

We can define **scale factors** as the ratio of a parameter in the model, X_{model}, to that in the prototype, X_{proto}.

$$K_X = \frac{X_{model}}{X_{proto}} \qquad (5.5\text{-}5)$$

Example 5.5-1   Drag on immersed body

Consider scaling the drag on a body immersed in the stream of incompressible fluid. The following equation describes the flow, where $\varphi\left(..\right)$ is functional notation.

$$\frac{F}{\rho\,v^2\,D^2} = \varphi\left(\frac{D\,v\,\rho}{\mu}\right) = \varphi\,(Re) \qquad (5.5\text{-}6)$$

If we wish to measure the same force in the model as in the prototype, and wish to use the same fluid in each, determine the parameters within which we must construct the model.

Solution

For φ_{model} to be the same as φ_{proto}, the Reynolds numbers of the model and prototype must be equal.

$$\left(\frac{D\,v\,\rho}{\mu}\right)_{model} = \left(\frac{D\,v\,\rho}{\mu}\right)_{proto} \qquad (5.5\text{-}7)$$

In terms of the scale factors

$$\frac{\left(\dfrac{D\,v\,\rho}{\mu}\right)_{model}}{\left(\dfrac{D\,v\,\rho}{\mu}\right)_{proto}} = 1$$

$$\frac{\left(\dfrac{D_{model}}{D_{proto}}\right)\left(\dfrac{v_{model}}{v_{proto}}\right)\left(\dfrac{\rho_{model}}{\rho_{proto}}\right)}{\left(\dfrac{\mu_{model}}{\mu_{proto}}\right)} = \frac{K_D\,K_v\,K_\rho}{K_\mu} = 1 \qquad (5.5\text{-}8)$$

If the same fluid is used in model and prototype

$$K_\rho = 1$$
$$K_\mu = 1 \qquad (5.5\text{-}9)$$

Then

$$\frac{K_D\,K_v\,K_\rho}{K_\mu} = \frac{K_D\,K_v\,(1)}{(1)} = 1$$
$$K_D\,K_v = 1 \qquad (5.5\text{-}10)$$

so the product of D and v needs to correspond in model and prototype. Then

$$\left(\frac{F}{\rho\,v^2\,D^2}\right)_{model} = \varphi_{model} = \varphi_{proto} = \left(\frac{F}{\rho\,v^2\,D^2}\right)_{proto} \tag{5.5-11}$$

$$\frac{\left(\dfrac{F}{\rho\,v^2\,D^2}\right)_{model}}{\left(\dfrac{F}{\rho\,v^2\,D^2}\right)_{proto}} = \frac{\varphi_{model}}{\varphi_{proto}} = 1 \tag{5.5-12}$$

$$\frac{\left(\dfrac{F_{model}}{F_{proto}}\right)}{\left(\dfrac{\rho_{mode}}{\rho_{proto}}\right)\left(\dfrac{v_{mode}}{v_{proto}}\right)^2\left(\dfrac{D_{mode}}{D_{proto}}\right)^2} = \frac{K_F}{K_\rho\,K_v^2\,K_D^2} = 1 \tag{5.5-13}$$

$$\frac{K_F}{K_\rho\,K_v^2\,K_D^2} = \frac{K_F}{(1)\,K_v^2\,K_D^2} = 1$$

$$\frac{K_F}{K_v^2\,K_D^2} = 1 \tag{5.5-14}$$

which implies

$$\frac{K_F}{K_v^2\,K_D^2} = \frac{K_F}{\left(K_v\,K_D\right)^2} = \frac{K_F}{(1)^2} = 1$$

$$K_F = 1 \tag{5.5-15}$$

Therefore, so long as the same fluid is used, K_F will remain equal to one for the same values of D and v in the model and prototype; that is, the force in the model and that in the prototype will be the same.

Example 5.5-2   Scale effects

We wish to estimate the pressure drop per unit length for laminar flow of ambient atmospheric air at 0.1 ft/s in a long duct. Because of space limitations, the duct must be shaped like a right triangle with sides of 3, 4, and 5 feet. We do not have a blower with sufficient capacity to test a short section of the full-size duct, so it has been proposed that we build a small-scale version of the duct, and further, that we use water as the fluid medium in order to achieve larger and therefore more accurately measurable pressure drops in the model.[29]

Discuss how to design and build the laboratory model.

Solution

Since there is no reason to create more scaling problems than we already have, we will make the laboratory model geometrically similar to the full-size duct. We will therefore be able to characterize it by a single length dimension, which we will designate as D.

In the laminar region, we can expect the friction factor, f, which is a vehicle (to be discussed in detail in Chapter 6) for predicting pressure drops in internal flows, to be a function of only the Reynolds number, the same dimensionless group as used in the previous example.

$$\text{friction factor} = f = f\left(\text{Re}\right) = f\left(\frac{D\,v\,\rho}{\mu}\right) \tag{5.5-16}$$

In the laminar region the friction factor is not a function of relative roughness of the surface; therefore, we are free to choose the most easily fabricated material to build the laboratory model.

The pressure drop is related to the friction factor for this case as

$$\frac{\Delta p}{L} = \frac{C\,f\,\rho\,v^2}{D} \tag{5.5-17}$$

where C is a constant.

If we choose to make the friction factors of the model and prototype the same, the scaling law follows as in the immediately preceding example

$$\frac{\left(\frac{D\,v\,\rho}{\mu}\right)_{\text{model}}}{\left(\frac{D\,v\,\rho}{\mu}\right)_{\text{proto}}} = 1$$

---

[29] This problem can be solved quite nicely by numerical methods if the necessary computing equipment is available. We ignore this alternative for pedagogical reasons in order to obtain a simple illustration.

$$\frac{\left(\dfrac{D_{model}}{D_{proto}}\right)\left(\dfrac{v_{model}}{v_{proto}}\right)\left(\dfrac{\rho_{model}}{\rho_{proto}}\right)}{\left(\dfrac{\mu_{model}}{\mu_{proto}}\right)} = \frac{K_D K_v K_\rho}{K_\mu} = 1 \tag{5.5-18}$$

This time, however, we have prescribed the scale factors for the fluids used. Approximate values at room temperature are

$$\left(\frac{\rho_{water}}{\rho_{air}}\right) = \frac{62.4 \dfrac{lbmass}{ft^3}}{0.0808 \dfrac{lbmass}{ft^3}} = 772 = K_\rho \tag{5.5-19}$$

$$\left(\frac{\mu_{water}}{\mu_{air}}\right) = \frac{1 \, cP}{1.8 \times 10^{-4} \, cP} = 5.56 \times 10^3 = K_\mu \tag{5.5-20}$$

Therefore, to keep the same friction factor will require

$$\frac{K_D K_v K_\rho}{K_\mu} = \frac{K_D K_v (772)}{\left(5.56 \times 10^3\right)} = 1$$

$$K_D K_v = 7.2 \tag{5.5-21}$$

Examining the scaling for pressure drop

$$\frac{\left(\dfrac{\Delta p}{L}\right)_{model}}{\left(\dfrac{\Delta p}{L}\right)_{proto}} = \frac{\left(\dfrac{C f \rho v^2}{D}\right)_{model}}{\left(\dfrac{C f \rho v^2}{D}\right)_{proto}} \tag{5.5-22}$$

$$K_{\left(\frac{\Delta p}{L}\right)} = \frac{K_f K_\rho K_v^2}{K_D} \tag{5.5-23}$$

However, we intend to make $K_f = 1.0$, so

$$K_{\left(\frac{\Delta p}{L}\right)} = \frac{(1)(772) K_v^2}{K_D} \tag{5.5-24}$$

$$K_{\left(\frac{\Delta p}{L}\right)} = (772) \frac{K_v^2}{K_D} \tag{5.5-25}$$

We now have considerable flexibility in choosing the size of our model and the velocity at which we pump the water.

For example, suppose that we decide it will be easy to fabricate a model duct with hypotenuse of 2 inches. This will set the scale factor for D

$$K_D = \frac{(2)\text{ in}}{(5)\text{ ft}}\frac{\text{ft}}{(12)\text{ in}} = 0.0333 \tag{5.5-26}$$

and we now have constrained our model to

$$K\left(\tfrac{\Delta p}{L}\right) = (772)\frac{K_v^2}{0.0333} = 2.32 \times 10^4\, K_v^2 \tag{5.5-27}$$

If we decide arbitrarily that it would be easy to measure pressure drops per unit length that are 10,000 times larger than we expect in the prototype, we have

$$K\left(\tfrac{\Delta p}{L}\right) = 10,000 \tag{5.5-28}$$

and we must have a velocity in the laboratory model of

$$K\left(\tfrac{\Delta p}{L}\right) = (772)\frac{K_v^2}{0.0333} = 10,000$$
$$K_v = 0.657 \tag{5.5-29}$$

Therefore, we need a water velocity in the laboratory model of

$$K_v = 0.657 = \frac{v_{model}}{v_{proto}} = \frac{v_{model}}{0.1\frac{\text{ft}}{\text{s}}}$$
$$v_{model} = 0.0657\frac{\text{ft}}{\text{s}} \tag{5.5-30}$$

This would require a laboratory pump of about

$$Q = vA = v\tfrac{1}{2}bh$$

$$= 0.0657 \frac{\text{ft}}{\text{s}} \frac{1}{2} \left(2\frac{3}{5}\right) \text{in} \left(2\frac{4}{5}\right) \text{in}$$

$$\times \left(\frac{1 \text{ ft}^2}{144 \text{ in}^2}\right) \left(\frac{7.48 \text{ gal}}{\text{ft}^3}\right) \left(\frac{60 \text{ s}}{\text{min}}\right)$$

$$= 0.197 \text{ gpm}$$

(5.5-31)

which is small, but well within the range of availability.

We must make the laboratory model long enough to make entrance and exit effects negligible. At such low velocities, we must also be careful of natural convection from heat transfer influencing the flow field in either model or prototype. We might find that the availability of pumps and/or pressure measuring devices restricts us to operate at a different velocity and/or pressure drop. There is obviously further flexibility available for the model design, as well as constraints we have not considered in detail.

Chapter 5   Problems

5.1 Write the fundamental units of the following quantities using the system of units indicated in parentheses:

    a. Diameter (FMt),
    b. density (FLt),
    c. acceleration (FML),
    d. viscosity (MLt),
    e. coefficient of thermal expansion (increase in volume per
    unit volume per degree change in temperature) (FMLtT)

5.2 The discharge through a horizontal capillary tube depends upon the pressure drop per unit length, the diameter, and the viscosity.

Quantity	Symbol	Dimensions
Discharge	Q	L^3/t
Pressure drop/length	$\Delta p/L$	$M/\left(L^2 t^2\right)$
Diameter	D	L
Viscosity	μ	$M/\left(L\,t\right)$

      a.   How many dimensionless groups are there?

      b.   Find the functional form of the equation.

5.3 Using dimensionless analysis, arrange the following variables into several dimensionless groups:

      pipe diameter D, ft
      fluid velocity v, ft/s
      pipe length L, ft
      fluid viscosity m, lbm/s ft
      fluid density r, lbm/ft^3

Show by dimensional analysis how these groups are related to pressure drop in lbf/in^2.

5.4 Using D (diameter), ρ (density), and g (acceleration of gravity) as Q's, form a dimensionless group using as the additional variable μ (viscosity). Use the MLt system and solve the equations (do not do by inspection).

5.5 The calibration curve for a particular flowmeter has the form

$$Q = f(D, \rho, \mu, \Delta p) = \text{volumetric flow rate}$$

      Using the dimensional analysis, find the required dimensionless variables p_1 and p_2 where $F(p_1, p_2) = 0$. Let $Q_1 = D$, $Q_2 = \rho$, $Q_3 = \Delta p$.

5.6 The pressure drop in a viscous incompressible fluid flowing in a length, L, may be represented functionally as

$$\Delta p = f(\mu, \rho, <v>, D, L, k)$$

where

      ρ = density
      (v) = bulk velocity
      D = diameter
      L = length
      k = roughness

Using dimensional analysis find the correct representation for the pressure drop in terms of dimensionless groups.

5.7 Flow in a long straight pipe is turbulent. The average velocity, v, at a distance, y, from the pipe wall depends on roughness height of the pipe, k, the pipe diameter, D, the kinematic viscosity, v ($v = \mu/\rho$), the mass density of the flowing fluid, ρ, and the shear stress at the pipe wall, τ_0.

    a. Write down the functional relation for these variables.
    b. Show the dimensional matrix for these variables in an M, L, t system
    c. If the functional relation in a is reduced to a complete set of dimensionless products, how many products are there?
    d. Determine the dimensionless products using the systematic algorithm (use v, y, ρ as the subset of Q's).

5.8 For uniform flow in open channel, Bernoulli's equation for energy loss is

$$s = \frac{f}{4R}\frac{v^2}{2g}$$

where s is the head loss (in ft per unit length measured along the bottom slope of the channel), f is the friction factor, v is the mean velocity, and R is the hydraulic radius (Area/wetted perimeter). The mean velocity, v, in an open channel depends on the hydraulic radius, R, the kinematic viscosity, v, the roughness height, k, and g, the acceleration of gravity. Show by dimensional analysis (apply the systematic analysis), Use v and R as the Q's.

$$f\left(\frac{v}{\sqrt{gsR}}, \frac{R}{n}\sqrt{gsR}, \frac{k}{R}\right) = 0$$

5.9 In heat transfer problems with flowing fluids, frequently the six variables, D, ρ, h, v, μ, k, may be combined into dimensionless groups which will characterize experimental data. Find these dimensionless groups by inspection with the restriction that h and v only appear in separate terms. D is the diameter, ρ is the density, v is the velocity, μ is the viscosity, h is the heat transfer coefficient and has dimensions Btu/hr ft^2°F, and k is the thermal conductivity and has dimensions Btu/hr ft °F.

5.10 In a forced convection heat transfer process the quantities listed below are pertinent to the problem:

> Tube diameter, D
> Thermal conductivity of the fluid, k
> Velocity of the fluid, v
> Density of the fluid, ρ
> Viscosity of the fluid, μ
> Specific heat of the fluid, C_p
> Heat transfer coefficient, h

Using dimensional analysis, find the functional relationship among these parameters.

5.11 Two similar centrifugal pumps are specific by the diameter of the impeller D[L]. The pumps are specified by their flow rate $Q[L^3/t]$ and the pump head H [L]. The size of a pump for a given head and flow rate is determined by an equation of the form

$$H = f\left(Q, N, D, \rho\right)$$

where N[1/t] is the angular speed and ρ [M/L^3]. Dimensional analysis shows

$$\frac{H}{\rho N^3 D^2} = f\left(\frac{Q}{N D^3}\right)$$

If a centrifugal pump with $D_1 = 1$ ft. has a $Q_1 = 400$ gal/min of water when tuning at $N_1 = 1800$ rpm Answer the following concerning a geometrically similar pump with $D_2 = 2$ ft. and $N_2 = 1200$ rpm pumping an organic liquid with sp. gr. = 0.82.

> a. What is Q_2?
> b. What is H_1/H_2?

5.12 Dimensional analysis is useful in the design of centrifugal pumps. The pressure rise across a pump (proportional to the head developed by the pump) is affected by the fluid density, ρ, the angular velocity, ω, the impeller diameter, D, the volumetric flow rate, Q, and the fluid viscosity, μ. Find the pertinent dimensionless groups, choosing them so the Δp, Q, and μ each appear in one group only. Find similar expressions replacing the pressure rise first by the power input to the pump, and then by the efficiency of the pump.

5.13 The scaling ratios for centrifugal pumps are Q/ND^3 = constant, $H/(ND)^2$ = constant, where Q is the volumetric flow rate (ft^3/sec), H is the pump head (ft), and N is rotational speed (rpm). Determine the head and discharge of a 1:4 model of a centrifugal pump that produces 20 ft^3/s at a head of 96 ft when turning 240 rpm. The model operates at 1,200 rpm.

5.14 The pressure, p, on the discharge side of a centrifugal compressor is given by an equation that has the functional form

$$p = f(p_0, \rho_0, m, n, D)$$

The dimensions of the variables in the MLt system are:

discharge pressure, p: $\left[\dfrac{M}{Lt^2}\right]$

inlet pressure, p_0: $\left[\dfrac{M}{Lt^2}\right]$

inlet air density, ρ_0: $\left[\dfrac{M}{L^3}\right]$

mass of air/s, m: $\left[\dfrac{M}{t}\right]$

revolutions/s, n: $\left[\dfrac{1}{t}\right]$

diameter, D: [L]

a. Use the MLt system of fundamental dimensions and show by dimensional analysis (use the algorithm) that:

$$p = p_o f\left(\frac{m}{nD^3 r_o}, \frac{p_o}{n^2 D^2 r_o}\right)$$

b. Two geometrically similar air compressors operate in the same room. One compressor is twice as large as the other. The larger operates at 1/2 the speed of the smaller.

(1) How do the mass rates of discharge of the two compressors compare?
(2) How do the discharge pressures compare?

5.15 You are trying to simulate dynamically the cooling system of a proposed sodium-cooled nuclear reactor by constructing a 1:4 model of the cooling system using water. These water experiments are to be carried out at 70°F and will simulate flowing sodium at 1000°F. Data on sodium reactor:

$$\text{Hydraulic diameter} = \frac{4\left(\dfrac{\text{flow cross-sectional area}}{\text{wetted perimeter}}\right)}{} = 1.768 \times 10^{-2} \text{ ft}$$

(This replaces the ordinary tube diameter for flows in noncircular channels.) Bulk sodium velocity = 20 ft/s. Pressure drop across the core = 30 psig. For the water model to be dynamically similar,

a. What must the water velocity be?
b. What is the expected pressure drop in the water model?

5.16 The following variables describe mixing in a tank f (P, N, D, ρ, μ) = 0

P is power, work/time - $[ML^2/t^3]$
N is rotational speed of impeller - $[1/t]$
D is diameter of the tank - $[L]$
ρ is density of the fluid - $[M/L^3]$
μ is viscosity of the fluid - $[M/Lt]$

a. Show the functional relationship among the variables is

$$F\left(\frac{P}{rN^3D^5}, \frac{D^2Nr}{m}\right) = 0$$

b. To scale a liquid mixing tank the power input per unit volume is kept constant. If it is desired to construct a prototype of an existing mixer by a 2:1 ratio, by what ratio must the tank diameter, D, and impeller speed, N, be changed?

5.17 Suppose a hiker falls into a crack in a glacier while mountain climbing. Assume the ice moves as a Newtonian fluid with viscosity = 250 x 10⁶ cP and

sp. gr. = 0.97 and assume the glacier is 15-m deep and is on a slope that falls
1.5 m every 1850 m. In order to determine how long before the professor
appears at the bottom of the glacier a model is to be constructed.

a. The variables that describe this problem are

$$V\left[\frac{L}{t}\right], r\left[\frac{m}{L^3}\right], g\left[\frac{L}{t^2}\right], m\left[\frac{M}{Lt}\right], D[L], H[L], L[L].$$

Use the MLt system of units, ρ, g, D, as repeating variables
and determine the dimensionless groups that describe this
problem (you may assume r = k and that two of the groups are
H/D, L/D)

b. If a model is constructed using a fluid with a viscosity of
250 cP and a sp. gr. = 1.26 and the model professor reappears
after 9.6 hours, when will the professor appear at the bottom
of the glacier?

5.18 The energy equation for free convection heat transfer from a flat plate of
length H at temperature T_0 suspended in a large body of fluid at temperature T_1
is

$$\rho C_p\left[v_y\frac{\partial(T-T_1)}{\partial y}+v_z\frac{\partial(T-T_1)}{\partial z}\right]=k\frac{\partial^2(T-T_1)}{\partial y^2}$$

This equation is nonlinear and is difficult to solve. Usually dimensional
analysis is applied to such equations. Given the following definitions

$$\Theta=\frac{T-T_1}{T_0-T_1} \quad \xi=\frac{z}{H} \quad \eta=\left(\frac{B}{\mu\alpha H}\right)^{1/4}y$$

$$\varphi_z=\left(\frac{\mu}{B\alpha H}\right)^{1/2}v_z \quad \varphi_y=\left(\frac{\mu H}{\alpha^3 B}\right)^{1/4}v_y$$

where $\alpha = k/\rho C_p$ and $B = \rho g \beta (T_0 - T_1)$, write the above equation in dimensionless form.

5.19 In a system of conductors and dielectrics Maxwell's equations are used to find the potential field, E, and the magnetic field, H, where

$$\text{curl } E + \mu \frac{\partial H}{\partial t} = 0$$
$$\text{curl } H - \varepsilon \frac{\partial E}{\partial t} = \kappa E$$

For two geometrically similar systems that have similar electromagnetic properties ε, μ, and κ and similar initial and boundary conditions, determine the following scale factors, where the prime refers to the model system

K_κ	where	$K_\kappa = \kappa'/\kappa$
K_t	where	$K_t = t'/t$
K_H/K_E	where	$K_H = H'/\mu$, $K_E = E'/E$

(This shows that if model and prototype have different sizes the two systems must be made of different materials.)

5.20 Apply the algorithm for obtaining dimensionless groups to the FML, FLt, and FMt systems for which the solution details were omitted in Example 5.3-3.

5.21 Consider the variables involved in flow (without heat transfer) of fluid through a porous medium such as a sand bed, limestone, filter cake, array of glass beads, etc.

$f(L, t, v, \mu, p, \mathcal{D}, k, g, v_\infty, D, \rho) = 0$
where
$\mathcal{D} = $ Dispersion coefficient
$k = $ Permeability

Find a complete set of dimensionless groups for this system using the MLt system of units.

5.22 The equation of continuity for flow through a porous medium is

$$-\varphi \frac{\partial c}{\partial t} - \frac{\partial}{\partial x_i} c\, v_i + \mathcal{D}\frac{\partial^2 c}{\partial x_i^2} = 0$$

where the repeated index i indicates summation over i = 1,2,3.

The equation of motion for flow through a porous medium is given by Darcy's law

$$v_i = -\frac{k}{\mu}\frac{\partial p}{\partial x_i} - \delta_3 \frac{k \rho g}{\mu}$$

where δ_3 is the magnitude of the unit vector in the z-direction.

Make these equations dimensionless using the following dimensionless variables and then indicate the functional form for flow of a fluid through a porous medium.

$$t^* = \frac{t\, v_\infty}{D} \qquad x_i^* = \frac{x_i}{D} \qquad v_i^* = \frac{v_i}{v_\infty}$$

6

MOMENTUM TRANSFER IN FLUIDS

6.1 Fluid Statics

Consider a static fluid - one that is not moving, i.e., where the velocity vector is everywhere zero. We further assume that density and viscosity are not functions of time (for example, as might be caused by an changing temperature field - perhaps from imposed microwave radiation). Starting with the equation of motion (the Navier-Stokes equation)

$$\rho \frac{Dv}{Dt} = -\nabla p + \mu \nabla^2 v + \rho g \tag{6.1-1}$$

and eliminating all the terms that depend on changes in velocity, we are left with a relationship between the pressure gradient, the density, and the acceleration of gravity.

$$0 = -\nabla p + \rho g \tag{6.1-2}$$

Equation (6.1-2) in component form gives the three equations

$$-\frac{\partial p}{\partial x} + \rho g_x = 0 \tag{6.1-3}$$

$$-\frac{\partial p}{\partial y} + \rho g_y = 0 \tag{6.1-4}$$

$$-\frac{\partial p}{\partial z} + \rho g_z = 0 \tag{6.1-5}$$

If the z-axis is chosen parallel to but in the opposite direction to the gravity vector, these three equations become

$$\frac{\partial p}{\partial x} = 0 \tag{6.1-6}$$

$$\frac{\partial p}{\partial y} = 0 \tag{6.1-7}$$

$$\frac{\partial p}{\partial z} = \rho g_z = -\rho g \tag{6.1-8}$$

where the minus sign appears because the gravity vector is in the opposite direction of that of the positive z-direction.[1]

Equations (6.1-6) and (6.1-7) show that pressure is not a function of x or y (because the gravity vector has no component acting in either of these two directions). Further, since we assume steady state for viscosity and density,

---

[1] c.f. Bird, R. B., W. E. Stewart, et al. (1960). *Transport Phenomena*. New York, Wiley, p. 45. One can combine the effects of static pressure and gravitational forces in a single quantity, P, defined as

$$\mathcal{P} = p + \rho g h$$

where h is defined as the component along a coordinate parallel to but in the **opposite** direction to the gravitational vector (**up** as opposed to **down**). This lets us write

$$\nabla \mathcal{P} = \nabla(p + \rho g h) = \nabla p + \nabla(\rho g h) = \nabla p - \rho g$$

where

$$\rho g \ \Rightarrow \ \left[\frac{M}{L^3}\right]\left[\frac{L}{t^2}\right] = \frac{\left[\frac{ML}{t^2}\right]}{\left[L^3\right]} = \frac{[F]}{\left[L^3\right]}$$

represents the force per unit volume of the vertical gravitational force.

The substitution used above

$$\nabla(\rho g h) = -\rho g$$

can be shown to be valid by considering h to be measured along the x_1 axis of a set of orthogonal Cartesian axes (x_1, x_2, x_3). The scalar field of **energy** per unit volume in this coordinate system is then a function of x_1 only, not of x_2 or x_3, and can be described by

$$-\rho g h = -\rho g x_1 \ \Rightarrow \ \left[\frac{M}{L^3}\right]\left[\frac{L}{t^2}\right][L] = \left[\frac{ML}{t^2}\right]\left[\frac{1}{L^3}\right][L] = \frac{[FL]}{\left[L^3\right]}$$

The gradient of this scalar field is then

$$\nabla(-\rho g x_1) = \partial_i(-\rho g x_1)\delta_i = (-\rho g)\delta_i = -\rho(g\,\delta_i) = \rho g$$

since the gravitational vector and the x_1 axis are collinear but in opposite directions. Therefore

$$\nabla(\rho g h) = -\rho g$$

which is the substitution used above.

pressure is not a function of t. This means that pressure is a function of z only, and therefore we can write the partial derivative in Equation (6.1-8) as an ordinary derivative. Equation (6.1-8) then becomes

$$\frac{dp}{dz} = -\rho g \tag{6.1-9}$$

For an incompressible fluid, that is, one where ρ is not a function of pressure, integrating Equation (6.1-9) from some datum z_0, where the pressure is denoted by p_0, to the height, z, where the corresponding pressure is p, yields

$$\int_{p_0}^{p} dp = -\rho g \int_{z_0}^{z} dz \tag{6.1-10}$$

Completing the integration yields the relationship for the pressure difference in terms of the difference in z.

$$(p - p_0) = -\rho g (z - z_0) = \rho g (z_0 - z) \tag{6.1-11}$$

We can apply Equation (6.1-11) at two arbitrary points, a and b, within the same phase[2] and take the difference of the results

$$\begin{aligned}(p_a - p_0) - (p_b - p_0) &= (p_a - p_b) \\ &= \left[-\rho g (z_a - z_0) \right] - \left[-\rho g (z_a - z_0) \right] \\ &= -\rho g (z_a - z_b) \end{aligned} \tag{6.1-12}$$

Notice the choice of location for the datum is immaterial, so long as a consistent datum is used.

If h is defined as (z_0 - z), Equation (6.1-11) reduces to

$$p = p_0 + \rho g h \tag{6.1-13}$$

This is the form of the equation commonly found in elementary physics texts.

---

[2] We cannot apply our equations across the boundaries between phases because the equations were derived on the assumption of a continuum, and at phase boundaries there is a discontinuity in properties.

6.1.1   Manometers

A practical application of static fluids occurs in a device called a **manometer**. Manometers are U-tubes (tubes bent into the shape of the letter U) partially filled with a liquid immiscible with and differing in density from the fluid in which the pressure is to be determined. The pressure difference between the points at the inlet of each of the legs (the vertical parts of the U) is inferred by measuring the difference in heights of the liquid interfaces in the legs and applying the equations for a static fluid.

Example 6.1.1-1   Pressure difference using a manometer

A manometer is used to determine the pressure difference between two vessels as shown in Figure 6.1.1-1.

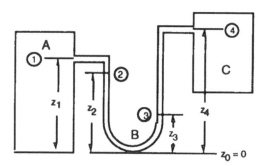

Figure 6.1.1-1 Measurement of pressure difference with manometer

The vessel on the left and the left leg of the manometer above point 2 are filled with fluid A; the manometer contains fluid B between points 2 and 3; the right leg of the manometer above point 3 and the tank on the right contain fluid C.

Find the difference in pressure between points 1 and 4.

Solution

We cannot apply our equations for a static fluid directly between points 1 and 4 because these equations were derived for a **continuum**, and the path between 1 and 4, if taken through the fluids, has abrupt discontinuities in properties at the two fluid interfaces. (If not taken through the fluids - that is, through the walls of the manometer - the discontinuities would require the introduction of solid mechanical properties.) We can, however, by writing the pressure difference as the sum of differences each in a single fluid, apply the equations to these single fluids. By adding and subtracting p_2 and p_3 we have

$$(p_1 - p_4) = (p_1 - p_2) + (p_2 - p_3) + (p_3 - p_4)$$ (6.1.1-1)

Applying Equation (6.1-14) to the various pressure drops in Figure (6.1.1-1)

$$(p_1 - p_4) = \left[-\rho_A g (z_1 - z_2)\right] + \left[-\rho_B g (z_2 - z_3)\right] + \left[-\rho_C g (z_3 - z_4)\right]$$ (6.1.1-2)

Example 6.1.1-2   Pressure difference between tanks

The two tanks in Figure 6.1.1-2 both contain water (fluids A and C). The manometer fluid B has a specific gravity of 1.1. Find the pressure difference (p_1 - p_4).

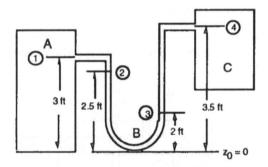

Figure 6.1.1-2 Pressure difference between tanks

Solution

Substituting in Equation (6.1.1-1) from the previous example

$$
\begin{aligned}
(p_1 - p_4) = \Bigg\{ &\left[-(62.4) \, \frac{lbm}{ft^3} \, (32.2) \, \frac{ft}{s^2} \, (3.0 \, ft - 2.5 \, ft) \right] + \\
&\left[-(1.1) \, (62.4) \, \frac{lbm}{ft^3} \, (32.2) \, \frac{ft}{s^2} \, (2.5 \, ft - 2.0 \, ft) \right] + \\
&\left[-(62.4) \, \frac{lbm}{ft^3} \, (32.2) \, \frac{ft}{s^2} \, (2.0 \, ft - 3.5 \, ft) \right] \Bigg\} \left\{ \frac{lbf \, s^2}{(32.2) \, lbm \, ft} \right\} \\
= \; &28 \left(\frac{lbf}{ft^2} \right)
\end{aligned}
$$

$$(6.1.1-3)$$

Example 6.1.1-3   Differential manometer

A differential manometer employs two fluids which differ very little in density. It normally also has reservoirs of a large cross-section to permit large changes in position of the interface level between the manometer liquids without appreciable change in interface levels between the manometer liquids and the fluid whose pressure is to be determined. A typical differential manometer is shown in Figure 6.1.1-3.

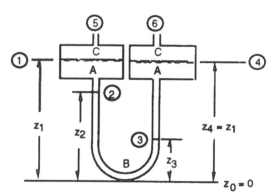

Figure 6.1.1-3 Differential manometer

Two fluids, A and B, are employed to measure the difference in pressure of a gas C at points 5 and 6 as shown. Develop an equation for the difference in pressure between point 5 and point 6.

Solution

The pressure difference is obtained in terms of the difference between 2 and 3 by writing

$$(p_5 - p_6) = (p_5 - p_1) + (p_1 - p_2) + (p_2 - p_3) + (p_3 - p_4) + (p_4 - p_6)$$

(6.1.1-4)

Since the density of a the gas is so low compared to the densities of the liquids, we neglect the pressure difference (the static head) of gas from 5 to 1 and from 6 to 4.

$$(p_5 - p_6) = (p_1 - p_2) + (p_2 - p_3) + (p_2 - p_3) + (p_3 - p_4)$$

(6.1.1-5)

Applying Equation (6.1.1-2)

$$(p_5 - p_6) = \left[-\rho_A g (z_1 - z_2) \right] + \left[-\rho_B g (z_2 - z_3) \right] + \left[-\rho_A g (z_3 - z_4) \right]$$

(6.1.1-6)

but since $z_4 = z_1$

$$(p_5 - p_6) = \left[-\rho_A g (z_3 - z_2) \right] + \left[-\rho_B g (z_2 - z_3) \right]$$

$$(p_5 - p_6) = g (z_2 - z_3)(\rho_A - \rho_B)$$

$$(z_2 - z_3) = \frac{(p_5 - p_6)}{g (\rho_A - \rho_B)}$$

(6.1.1-7)

which shows that, for a given pressure difference, as the densities of A and B approach each other, the difference in height between points 2 and 3 grows larger - that is, the manometer becomes more sensitive.[3]

---

[3] Another way to achieve more sensitivity in a manometer in an economical fashion is to incline one leg at an angle, viz.

so that a small vertical displacement will induce a proportionately larger displacement of the interface along the right leg of the tube. Such arrangements are

6.2 Description of Flow Fields

We have modeled fluids as continua even though they are made up of discrete molecules. It is, however, convenient to talk of fluid "particles," by which we mean aggregates of fluid that remain in proximity to one another.

Such a fluid "particle," if we could somehow paint it a different color to distinguish it from the surrounding fluid, would trace a **path line** if we took successive **multiple** exposures with a fixed camera of this particle as it passed through space.

If we think of instead continuously marking **each** fluid particle that passes through a given point in space - experimentally, for example, by supplying a tracer such as an injected dye or a stream of small bubbles - and taking an **single** photograph showing the instantaneous position of all particles so marked, the line established is called a **streakline**.

A **streamline** is yet another imaginary line in a fluid, this time established by moving in the direction of the instantaneous local fluid velocity to get from one point to the next on the line. A streamline is thereby everywhere in the direction of the velocity vector at a particular instant in time. By definition, then, there can be no flow across a streamline because there is no component of the velocity normal to the streamline.

If, to establish yet another imaginary line in the fluid, we mark a line of **adjacent fluid particles** at a given instant and observe the evolution of such a line with time, we obtain a **timeline** which may give us insights into the behavior of the fluid. For example, we could think of marking a line in an initially static fluid and then observing the behavior of this line as the fluid is set into motion.

In **steady** flow fluids move along streamlines, since the velocity at any point in the fluid is not changing with time; therefore, pathlines and streamlines are coincident. Since all fluid particles passing through a given point follow the same path in steady flow, streaklines also coincide with timelines and pathlines. In **unsteady** flow the streamline shifts as the direction of the velocity changes, and a fluid particle can shift from one streamline to another so the pathline is no longer necessarily a streamline. Similarly, successive particles of fluid passing through the same point no longer necessarily follow the same path, so

---

sometimes called **draft tubes,** probably because of their original application in measuring stack draft pressures for power plants.

streaklines do not necessarily coincide with streamlines and pathlines in unsteady flow.[4]

A **stream tube** is a imaginary tube whose surfaces are composed of streamlines. Since there is no flow across streamlines (because the component of the velocity normal to the streamline is zero), there is no flow through the walls of a stream tube.

The **stream function** Ψ is a quantity defined by

$$-\frac{\partial \Psi}{\partial y} \equiv v_x \quad \text{and} \quad \frac{\partial \Psi}{\partial x} \equiv v_y \tag{6.2-1}$$

Notice that defining the stream function in this way satisfies the continuity equation identically - for example, in two dimensions one has

$$\frac{\partial v_x}{\partial x} + \frac{\partial v_y}{\partial y} = \frac{\partial}{\partial x}\left(-\frac{\partial \Psi}{\partial y}\right) + \frac{\partial}{\partial y}\left(\frac{\partial \Psi}{\partial x}\right) = \frac{\partial^2 \Psi}{\partial x\, \partial y} - \frac{\partial^2 \Psi}{\partial y\, \partial x} = 0 \tag{6.2-2}$$

because the order of differentiation is immaterial if the partial derivatives are continuous, as they are in cases of interest here.

Consider an element of differential length, dr, lying along a streamline; i.e., tangent to the streamline. The condition that the velocity vector be collinear to the tangent to the streamline is expressed by

$$\nabla \times d\mathbf{r} = 0 \tag{6.2-3}$$

or, for example, for flow only in two dimensions

$$\left(v_x\, \mathbf{i} + v_y\, \mathbf{j}\right) \times \left(dx\, \mathbf{i} + dy\, \mathbf{j}\right) = \left(v_x\, dy - v_y\, dx\right)\mathbf{k} = 0 \tag{6.2-4}$$

which implies

$$\left(v_x\, dy - v_y\, dx\right) = 0 \tag{6.2-5}$$

---

[4] There is a very nice demonstration of the differences among these lines using flow past an oscillating flat plate in the NCFMF film *Flow Visualization*, Steven J. Kline, principal.

or, in terms of the stream function

$$\left(-\frac{\partial \Psi}{\partial y}\,dy - \frac{\partial \Psi}{\partial x}\,dx\right) = 0$$

$$\frac{\partial \Psi}{\partial y}\,dy + \frac{\partial \Psi}{\partial x}\,dx = 0 \tag{6.2-6}$$

However, for functions with continuous first partial derivatives near an arbitrary point in question (which, to reiterate, is true for functions of interest here), the function (here, the stream function) has a *total differential*. We consider a particular instant in time, say t_0, so the stream function is not a function of time and therefore we can write, again for two dimensions, the total differential as

$$d\Psi = \frac{\partial \Psi}{\partial x}\,dx + \frac{\partial \Psi}{\partial y}\,dy \tag{6.2-7}$$

We can see, then, that our collinearity condition requires that the total differential of the stream function be zero.

$$d\Psi = 0 \tag{6.2-8}$$

This is another way of saying that the **value of the stream function does not change along a given streamline**.

Since the stream function has an exact differential, we can integrate between two arbitrary points in the flow field to obtain

$$\int_{\Psi_1}^{\Psi_2} d\Psi = \Psi_2 - \Psi_1 = \text{constant} \tag{6.2-9}$$

that is, the integral is **independent of path**. A function whose integral between any two arbitrary points is independent of path is called a *point function*. Common examples of point functions are the functions referred to in thermodynamics as *state* functions, e.g., the enthalpy, internal energy, entropy, etc. - as opposed to heat and work, which are path-dependent.

To interpret the meaning of difference in the value of the stream function between streamlines, consider the two-dimensional (area) rate of flow across an arbitrary line, ℓ connecting two streamlines. By definition, the area rate of flow across such a line is

$$\int_\ell \left(\mathbf{v} \cdot \mathbf{n}\right) d\ell = Q_{2-\text{dimensional}} \qquad (6.2\text{-}10)$$

but this can be written as

$$\int_\ell \left[\left(v_x \,\mathbf{i} + v_y \,\mathbf{j}\right) \cdot \mathbf{n}\right] d\ell \qquad (6.2\text{-}11)$$

where ℓ is measured along the line connecting the two streamlines.

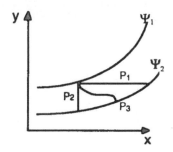

Figure 6.2-1 Paths between streamlines

Now consider this integral for a **particular** path between two streamlines, as illustrated in Figure 6.2-1, indicated by P_1 (at constant y), where the unit normal is directed in the positive y-direction

$$\int_\ell \left[\left(v_x \,\mathbf{i} + v_y \,\mathbf{j}\right) \cdot \mathbf{n}\right] d\ell = \int_x \left[\left(v_x \,\mathbf{i} + v_y \,\mathbf{j}\right) \cdot \mathbf{j}\right] dx \qquad (6.2\text{-}12)$$

$$\int_x \left(v_x \,\mathbf{i} \cdot \mathbf{j} + v_y \,\mathbf{j} \cdot \mathbf{j}\right) dx = \int_x v_y \,dx = \int_x \frac{\partial \Psi}{\partial x} \,dx \qquad (6.2\text{-}13)$$

$$= \int_{\Psi_1}^{\Psi_2} d\Psi = \Psi_2 - \Psi_1 \qquad (6.2\text{-}14)$$

In other words, the difference in the values of the stream functions on the two streamlines corresponds to the volumetric flow rate between these streamlines. Because of independence of path, we know that this is true for **any**

other path we might choose between the streamlines, whether it be at constant x, as is true for P_2 (which we could have chosen for the illustration rather than P_1 with equal ease), or some arbitrary convoluted path such as P_3.

6.2.1   Irrotational flow

The relative velocity between two fluid particles P and P′ whose coordinates are x_k and $x_k + dx_k$, respectively, in a velocity field instantaneously described by the vector field $v_i(x_k)$, is expressed by the total derivative

$$dv_k = dx_j \, \partial_j v_k \tag{6.2.1-1}$$

where the partial derivatives are evaluated at xk. We can write the vector gradient as the sum of its antisymmetric and symmetric parts

$$\partial_j v_k = \partial_{[j} v_{k]} + \partial_{(j} v_{k)} \tag{6.2.1-2}$$

and define

$$\begin{aligned} dx_j \partial_{[j} v_{k]} &\equiv dv_k^* \\ dx_j \partial_{(j} v_{k)} &\equiv dv_k^{**} \end{aligned} \tag{6.2.1-3}$$

to obtain

$$dv_k = dv_k^* + dv_k^{**} \tag{6.2.1-4}$$

We now proceed to show that dv_k^* corresponds to the relative velocities of a instantaneous rigid body rotation of the neighborhood of the particle P about an axis through P.[5]

We first define the dual vector of the antisymmetric part of the vector gradient as

$$2\,\omega_i \equiv \varepsilon_{ijk} \, \partial_{[j} v_{k]} \tag{6.2.1-5}$$

_____

[5] We can also show that dv_k^{**} corresponds to the instantaneous pure deformation of this neighborhood, for example, see Prager, W. (1961). *Introduction to Mechanics of Continua*. Boston, MA, Ginn and Company, p. 62 ff.; however, this is not essential to our argument at the moment.

which implies

$$\omega_i = \frac{1}{2}\varepsilon_{ijk}\,\partial_{[j}v_{k]} \tag{6.2.1-6}$$

The vector w_i is called the *vorticity* of the velocity field, and lines of constant vorticity are called *vortex lines*, terminology similar to that used for streamlines. We can rewrite this expression as

$$\omega_i = \frac{1}{2}\varepsilon_{ijk}\,\partial_{[j}v_{k]} \tag{6.2.1-7}$$

$$= \frac{1}{2}\varepsilon_{ijk}\left[\frac{1}{2}\left(\partial_j v_k - \partial_k v_j\right)\right] \tag{6.2.1-8}$$

$$= \frac{1}{2}\left(\frac{1}{2}\varepsilon_{ijk}\,\partial_j v_k + \frac{1}{2}\varepsilon_{ikj}\,\partial_k v_j\right) \tag{6.2.1-9}$$

We can obtain the dual tensor of the vorticity **vector** and use it to define the vorticity **tensor**

$$\frac{1}{2}\varepsilon_{ijk}\,\omega_i = \frac{1}{2}\varepsilon_{ijk}\left(\frac{1}{2}\varepsilon_{imn}\,\partial_{[m}v_{n]}\right) \tag{6.2.1-10}$$

$$= \frac{1}{2}\frac{1}{2}\varepsilon_{ijk}\,\varepsilon_{imn}\,\partial_{[m}v_{n]} \tag{6.2.1-11}$$

$$= \frac{1}{2}\frac{1}{2}\left(\delta_{jm}\delta_{kn} - \delta_{jn}\delta_{km}\right)\partial_{[m}v_{n]} \tag{6.2.1-12}$$

$$= \frac{1}{2}\frac{1}{2}\left(\partial_{[j}v_{k]} - \partial_{[k}v_{j]}\right) \tag{6.2.1-13}$$

$$= \frac{1}{2}\frac{1}{2}\left(\partial_{[j}v_{k]} + \partial_{[j}v_{k]}\right) \tag{6.2.1-14}$$

$$= \frac{1}{2}\frac{1}{2}\left(2\,\partial_{[j}v_{k]}\right) \tag{6.2.1-15}$$

$$= \frac{1}{2}\partial_{[j}v_{k]} \tag{6.2.1-16}$$

$$\varepsilon_{ijk}\,\omega_i = \partial_{[j}v_{k]} \equiv \omega_{jk} = \text{the vorticity tensor} \tag{6.2.1-17}$$

This permits us to write the antisymmetric part of the relative velocity, dv_k^*, as

$$dv_k^* = \varepsilon_{ijk}\,\omega_i\,dx_j \tag{6.2.1-18}$$

From kinematics we can recognize that the velocity of points on a solid body rotating about an axis through a point $x^{(0)}$ which can be seen from the sketch

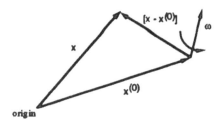

is described by

$$v_i = \varepsilon_{ijk}\,\omega_j\left(x_k - x_k^{(0)}\right) \quad \Rightarrow \quad v = \omega \times \left(x - x^{(0)}\right) \tag{6.2.1-19}$$

where w is the angular velocity, the angular velocity vector is normal to the plane of x_i and x_i^0 and points in the direction of a right-hand screw turning in the same direction as the rigid body. Comparing with the equation for dv_k^* we can see that dv_k^* represents a rigid body rotation about an axis through P.

If the vorticity vanishes identically, the motion of a continuum is called *irrotational*. For such a flow

$$\omega_i = \tfrac{1}{2}\varepsilon_{ijk}\,\partial_j v_k = 0 \quad \Rightarrow \quad \omega = \tfrac{1}{2}\left(\text{curl } v\right) = \tfrac{1}{2}\left(\nabla \times v\right) = 0$$

$$\varepsilon_{ijk}\,\partial_j v_k = 0 \quad \Rightarrow \quad \left(\text{curl } v\right) = \left(\nabla \times v\right) = 0 \tag{6.2.1-20}$$

However, because of the identity

$$\varepsilon_{ijk}\,\partial_j\,\partial_k\Phi = 0 \quad \Rightarrow \quad \text{curl grad } \Phi = \nabla \times \nabla\Phi = 0 \tag{6.2.1-21}$$

which is valid for any scalar function F, a velocity field with a vanishing curl is a gradient field, i.e.,

$$v_i = \partial_i\Phi \quad \Rightarrow \quad v = \nabla\Phi \tag{6.2.1-22}$$

Equating components

$$v_x = -\frac{\partial\Phi}{\partial x} \tag{6.2.1-23}$$

$$v_y = -\frac{\partial\Phi}{\partial y} \tag{6.2.1-24}$$

6.3 Potential Flow

In Section 6.1 we considered static fluids. In such fluids, viscosity played no part because there were no velocity gradients, the fluid being everywhere stationary. There is another important class of fluid behavior in which viscosity effects play no part, even though this time the fluid is in motion - behavior in which the viscosity of the fluid can be assumed to be zero. Such an assumption is useful in two situations:

- where the viscosity of the fluid is so small, or

- where the velocity gradients are so small

that viscous forces play a relatively unimportant role in the problem to be solved.

Many real-world flow problems can be successfully modeled as being

- **steady state**

- **irrotational** flow of an

- **incompressible** and

- **inviscid** (viscosity of zero) fluid.

Such fluids are called **ideal fluids**, and the corresponding flow model is called **potential flow**. We now assemble the basic equations constituting the model for such flows.

For **constant density** the continuity equation reduces to

$$\left(\nabla \cdot \rho\, v\right) + \frac{\partial \rho}{\partial t} = 0$$
$$\rho\left(\nabla \cdot v\right) + \frac{\partial \rho}{\partial t} = 0$$
$$\nabla \cdot v = 0 \tag{6.3-1}$$

and the **irrotationality** condition implies that

$$\nabla \times v = 0 \tag{6.3-2}$$

The equation of motion for a constant density Newtonian fluid - the Navier Stokes equation

$$\rho \frac{D\mathbf{v}}{Dt} = -\nabla p + \mu \nabla^2 \mathbf{v} + \rho \mathbf{g} \tag{6.3-3}$$

when applied to flow of an inviscid fluid reduces to Euler's equation

$$\rho \frac{D\mathbf{v}}{Dt} = -\nabla p + \rho \mathbf{g}$$

or

$$\rho \left[\frac{\partial \mathbf{v}}{\partial t} + (\mathbf{v} \cdot \nabla) \mathbf{v} \right] = -\nabla p + \rho \mathbf{g} \tag{6.3-4}$$

which at steady state becomes

$$\rho (\mathbf{v} \cdot \nabla) \mathbf{v} = -\nabla p + \rho \mathbf{g} \tag{6.3-5}$$

The left-hand side of this relation can be rewritten by defining[6]

$$(\mathbf{v} \cdot \nabla) \mathbf{v} \equiv \tfrac{1}{2} \nabla (\mathbf{v} \cdot \mathbf{v}) - \left[\mathbf{v} \times (\nabla \times \mathbf{v}) \right] \tag{6.3-6}$$

---

[6] c.f. Bird, R. B., W. E. Stewart, et al. (1960). *Transport Phenomena*. New York, Wiley, pp. 726, 731. This definition is applicable in both rectangular and curvilinear coordinates; however, for simplicity illustrating the equivalence using index notation (rectangular coordinates)

$$(\mathbf{v} \cdot \nabla) \mathbf{v} \equiv \tfrac{1}{2} \nabla (\mathbf{v} \cdot \mathbf{v}) - \left[\mathbf{v} \times (\nabla \times \mathbf{v}) \right] \;\Rightarrow$$

$$v_j \partial_j v_i = \partial_i \left(\tfrac{1}{2} v_j v_j \right) - \varepsilon_{ijk} v_j \varepsilon_{klm} \partial_l v_m$$

$$v_j \partial_j v_i = \partial_i \left(\tfrac{1}{2} v_j v_j \right) - \varepsilon_{ijk} \varepsilon_{klm} v_j \partial_l v_m$$

$$v_j \partial_j v_i = \partial_i \left(\tfrac{1}{2} v_j v_j \right) - \varepsilon_{kij} \varepsilon_{klm} v_j \partial_l v_m$$

$$v_j \partial_j v_i = \partial_i \left(\tfrac{1}{2} v_j v_j \right) - \left(\delta_{il} \delta_{jm} - \delta_{im} \delta_{jl} \right) v_j \partial_l v_m$$

$$v_j \partial_j v_i = \partial_i \left(\tfrac{1}{2} v_j v_j \right) - v_j \partial_i v_j + v_j \partial_j v_i$$

$$v_j \partial_j v_i = \tfrac{1}{2} v_j \partial_i v_j + \tfrac{1}{2} v_j \partial_i v_j - v_j \partial_i v_j + v_j \partial_j v_i$$

$$v_j \partial_j v_i = v_j \partial_i v_j - v_j \partial_i v_j + v_j \partial_j v_i$$

$$v_j \partial_j v_i = v_j \partial_j v_i \quad \text{QED}$$

which gives

$$\rho \left\{ \tfrac{1}{2} \nabla(\mathbf{v} \cdot \mathbf{v}) - \left[\mathbf{v} \times (\nabla \times \mathbf{v}) \right] \right\}$$
$$= -\nabla p + \rho\, \mathbf{g} = -\nabla(p + \rho\, g\, h) = -\nabla \mathcal{P} \tag{6.3-7}$$

Introducing the irrotationality condition reduces the relation to

$$\tfrac{1}{2} \rho\, \nabla(\mathbf{v} \cdot \mathbf{v}) = -\nabla \mathcal{P} \tag{6.3-8}$$

$$\nabla\!\left(\tfrac{1}{2} \rho\, v^2 \right) = -\nabla \mathcal{P} \tag{6.3-9}$$

$$\nabla\!\left(\tfrac{1}{2} \rho\, v^2 \right) + \nabla \mathcal{P} = 0 \tag{6.3-10}$$

$$\nabla\!\left(\tfrac{1}{2} \rho\, v^2 + \mathcal{P} \right) = 0 \tag{6.3-11}$$

which upon integration gives the conclusion that

$$\tfrac{1}{2} \rho\, v^2 + \mathcal{P} = \text{constant} \tag{6.3-12}$$

In two dimensions

$$\tfrac{1}{2} \rho\left(v_x^2 + v_y^2 \right) + \mathcal{P} = \text{constant} \tag{6.3-13}$$

Since we have shown in Section 6.2 that

$$-\frac{\partial \Psi}{\partial y} \equiv v_x = -\frac{\partial \Phi}{\partial x}$$
$$\frac{\partial \Psi}{\partial x} \equiv v_y = -\frac{\partial \Phi}{\partial y} \tag{6.3-14}$$

we have the pair of equations

$$\frac{\partial \Psi}{\partial y} = \frac{\partial \Phi}{\partial x}$$
$$\frac{\partial \Psi}{\partial x} = -\frac{\partial \Phi}{\partial y} \tag{6.3-15}$$

These correspond to the **Cauchy-Riemann** conditions on the real functions that form a function, $\Omega(z)$, of a complex variable, $z = (x + i\,y)$[7]

$$\Omega(z) = \Phi(x, y) + i\,\Psi(x, y) \qquad\qquad (6.3\text{-}16)$$

The Cauchy-Riemann conditions are both necessary and sufficient to ensure that the function has a derivative in a neighborhood about a point.[8] Functions that satisfy the Cauchy-Riemann conditions are known as *analytic*, or *holomorphic*, or *regular* functions.

If we differentiate the first of the equations with respect to x and the second with respect to y

$$-\frac{\partial^2 \Psi}{\partial x\,\partial y} = -\frac{\partial^2 \Phi}{\partial x^2}$$

$$\frac{\partial^2 \Psi}{\partial y\,\partial x} = -\frac{\partial^2 \Phi}{\partial y^2} \qquad\qquad (6.3\text{-}17)$$

and add, noting that the order of differentiation is immaterial, we obtain

$$\frac{\partial^2 \Phi}{\partial x^2} + \frac{\partial^2 \Phi}{\partial y^2} = 0 \qquad\qquad (6.3\text{-}18)$$

If we instead differentiate the first of the equations with respect to y and the second with respect to x

$$-\frac{\partial^2 \Psi}{\partial y^2} = -\frac{\partial^2 \Phi}{\partial y\,\partial x}$$

$$\frac{\partial^2 \Psi}{\partial x^2} = -\frac{\partial^2 \Phi}{\partial x\,\partial y} \qquad\qquad (6.3\text{-}19)$$

and add, again noting that the order of differentiation is immaterial, we obtain

---

[7] In most mathematics texts, the function $\Omega(z) = \Phi(x, y) + i\,\Psi(x, y)$ is written as $w(z) = u(x, y) + i\,v(x, y)$.

[8] Churchill, R. V. (1948). *Introduction to Complex Variables and Applications*. New York, NY, McGraw-Hill, p. 30.

$$\frac{\partial^2 \Psi}{\partial x^2} + \frac{\partial^2 \Psi}{\partial y^2} = 0 \tag{6.3-20}$$

Equations (6.3-18) and (6.3-20) are forms of Laplace's equation.

Each such pair of Ψ and Φ are orthogonal - that is, streamlines and potential functions intersect at 90 degrees. This can be seen by considering the fact that the values of the stream function and the velocity potential are constant along a streamline or a line of constant potential, respectively,

$$d\Psi = 0$$
$$d\Phi = 0 \tag{6.3-21}$$

However, because the Cauchy-Riemann conditions are satisfied, we know that we can write

$$d\Psi = \frac{\partial \Psi}{\partial x} dx + \frac{\partial \Psi}{\partial y} dy = 0$$
$$d\Phi = \frac{\partial \Phi}{\partial x} dx + \frac{\partial \Phi}{\partial y} dy = 0 \tag{6.3-22}$$

from which we can extract the respective slopes

$$\left(\frac{dy}{dx}\right)_{\Psi = \text{constant}} = \frac{\left(\frac{\partial \Psi}{\partial x}\right)}{\left(\frac{\partial \Psi}{\partial y}\right)}$$

$$\left(\frac{dy}{dx}\right)_{\Phi = \text{constant}} = \frac{\left(\frac{\partial \Phi}{\partial x}\right)}{\left(\frac{\partial \Phi}{\partial y}\right)} \tag{6.3-23}$$

But from the Cauchy-Riemann conditions we see that

$$\left(\frac{dy}{dx}\right)_{\Psi = \text{constant}} = \frac{\left(\frac{\partial \Psi}{\partial x}\right)}{\left(\frac{\partial \Psi}{\partial y}\right)} = \frac{\left(-\frac{\partial \Phi}{\partial y}\right)}{\left(\frac{\partial \Phi}{\partial x}\right)} = -\frac{1}{\left(\frac{dy}{dx}\right)_{\Phi = \text{constant}}} \tag{6.3-24}$$

Since the slopes are negative reciprocals of each other, the intersections are orthogonal.

Any pair of functions Ψ and Φ that satisfy the Cauchy-Riemann conditions furnishes streamlines and potential lines of some particular two-dimensional, incompressible, irrotational flow field. **Either function** can be interpreted as the stream function and the other of the pair as the velocity potential. Since there is no flow across streamlines, any streamline in such a flow field can be interpreted as a solid boundary.

This means that each such pair constitutes the real part and the imaginary part of a particular analytic function of the complex variable z, and each member of the pair represents a solution of Laplace's equation. This has two implications

a) Every analytic function of a complex variable will furnish, upon being put in the form

$$\Omega(z) = \Phi(x, y) + i\, \Psi(x, y)$$

(6.3-25)

a pair of functions Φ and Ψ that represent a solution to some set of problems in ideal fluid flow. Such a manipulation is relatively straightforward **given the analytic function**.

The set of problems for which the pair represent a solution can be defined by interpreting various streamlines as solid boundaries. Whether the boundaries represented by constant values of Ψ are of any real-world problem of interest is another matter.

We will see below that often only a particular streamline may be of interest to interpret as a boundary, as in the case of two-dimensional flow past a cylinder.

The various values of Φ and Ψ have been worked out for many flow problems. See Table 6.3-1 for elementary plane flows and Table 6.3-2 for superposition flows.

b) **If we do not have the analytic function**, we can construct it by solving the Laplace equation with appropriate boundary conditions to determine Φ and Ψ and then assembling them in the form

$$\Omega(z) = \Phi(x, y) + i\, \Psi(x, y)$$

(6.3-26)

Usually this is a much more difficult problem than (a). We frequently do not bother to construct the function Ω once we determine Φ and Ψ because we can determine the velocity field directly from Φ and Ψ.

Thus, Equations (6.3-26) and (6.3-26) model ideal flow in two dimensions.

Tables 6.3-1 and 6.3-2 give examples of pairs of functions that satisfy the Laplace equation.

Table 6.3-1   Elementary plane flows[9]

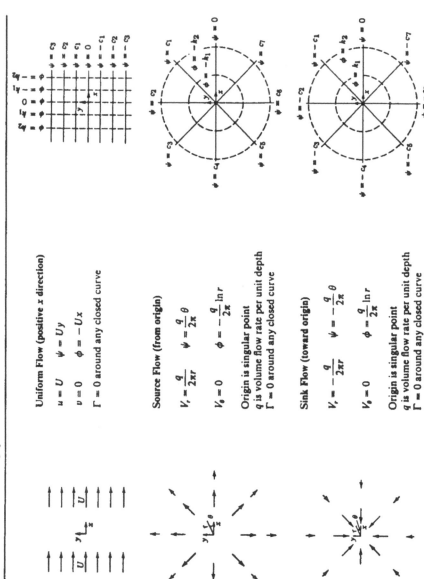

Uniform Flow (positive x direction)

$u = U \qquad \psi = Uy$

$v = 0 \qquad \phi = -Ux$

$\Gamma = 0$ around any closed curve

Source Flow (from origin)

$V_r = \dfrac{q}{2\pi r} \qquad \psi = \dfrac{q}{2\pi}\theta$

$V_\theta = 0 \qquad \phi = -\dfrac{q}{2\pi}\ln r$

Origin is singular point

q is volume flow rate per unit depth

$\Gamma = 0$ around any closed curve

Sink Flow (toward origin)

$V_r = -\dfrac{q}{2\pi r} \qquad \psi = -\dfrac{q}{2\pi}\theta$

$V_\theta = 0 \qquad \phi = \dfrac{q}{2\pi}\ln r$

Origin is singular point

q is volume flow rate per unit depth

$\Gamma = 0$ around any closed curve

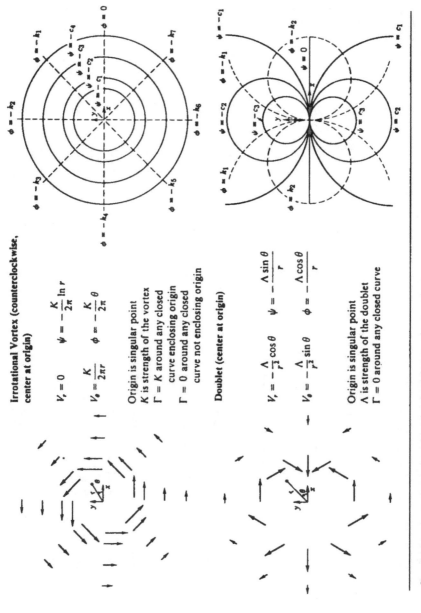

Irrotational Vortex (counterclockwise, center at origin)

$$V_r = 0 \qquad \psi = -\frac{K}{2\pi}\ln r$$

$$V_\theta = \frac{K}{2\pi r} \qquad \phi = -\frac{K}{2\pi}\theta$$

Origin is singular point
K is strength of the vortex
$\Gamma = K$ around any closed
    curve enclosing origin
$\Gamma = 0$ around any closed
    curve not enclosing origin

Doublet (center at origin)

$$V_r = -\frac{\Lambda}{r^2}\cos\theta \qquad \psi = -\frac{\Lambda\sin\theta}{r}$$

$$V_\theta = -\frac{\Lambda}{r^2}\sin\theta \qquad \phi = -\frac{\Lambda\cos\theta}{r}$$

Origin is singular point
Λ is strength of the doublet
$\Gamma = 0$ around any closed curve

[9] Fox, R. W. and A. T. McDonald (1985). *Introduction to Fluid Mechanics*, New York, John Wiley and Sons, Inc., pp. 276-277.

Table 6.3-2 Superposition of elementary plane flows[10]

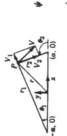
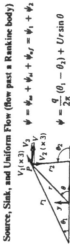

Source and Uniform Flow (flow past a half-body)

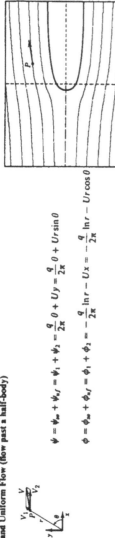

$$\psi = \psi_{\infty} + \psi_{sf} = \psi_1 + \psi_2 = \frac{q}{2\pi}\theta + Uy = \frac{q}{2\pi}\theta + Ur\sin\theta$$

$$\phi = \phi_{\infty} + \phi_{sf} = \phi_1 + \phi_2 = -\frac{q}{2\pi}\ln r - Ux = -\frac{q}{2\pi}\ln r - Ur\cos\theta$$

Source and Sink (equal strength, separation distance on x axis = 2a)

$$\psi = \psi_{\infty} + \psi_{ss} = \psi_1 + \psi_2 = \frac{q}{2\pi}\theta_1 - \frac{q}{2\pi}\theta_2 = \frac{q}{2\pi}(\theta_1 - \theta_2)$$

$$\phi = \phi_{\infty} + \phi_{ss} = \phi_1 + \phi_2 = -\frac{q}{2\pi}\ln r_1 + \frac{q}{2\pi}\ln r_2 = \frac{q}{2\pi}\ln\frac{r_2}{r_1}$$

Source, Sink, and Uniform Flow (flow past a Rankine body)

$$\psi = \psi_{\infty} + \psi_{ss} + \psi_{sf} = \psi_1 + \psi_2 + \psi_3 = \frac{q}{2\pi}\theta_1 - \frac{q}{2\pi}\theta_2 + Uy$$

$$\psi = \frac{q}{2\pi}(\theta_1 - \theta_2) + Ur\sin\theta$$

$$\phi = \phi_{\infty} + \phi_{ss} + \phi_{sf} = \phi_1 + \phi_2 + \phi_3 = -\frac{q}{2\pi}\ln r_1 + \frac{q}{2\pi}\ln r_2 - Ux$$

$$\phi = \frac{q}{2\pi}\ln\frac{r_2}{r_1} - Ur\cos\theta$$

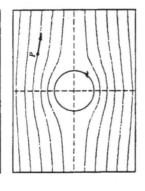
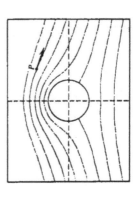

Vortex (clockwise) and Uniform Flow

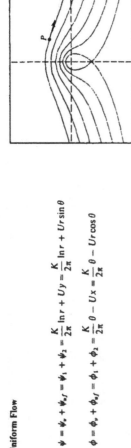

$$\psi = \psi_v + \psi_{uf} = \psi_1 + \psi_2 = \frac{K}{2\pi}\ln r + Uy = \frac{K}{2\pi}\ln r + Ur\sin\theta$$

$$\phi = \phi_v + \phi_{uf} = \phi_1 + \phi_2 = \frac{K}{2\pi}\theta - Ux = \frac{K}{2\pi}\theta - Ur\cos\theta$$

Doublet and Uniform Flow (flow past a cylinder)

$$\psi = \psi_d + \psi_{uf} = \psi_1 + \psi_2 = -\frac{A\sin\theta}{r} + Uy = -\frac{A\sin\theta}{r} + Ur\sin\theta$$

$$\psi = U\left(r - \frac{A}{Ur}\right)\sin\theta = Ur\left(1 - \frac{a^2}{r^2}\right)\sin\theta \qquad a = \sqrt{\frac{A}{U}}$$

$$\phi = \phi_d + \phi_{uf} = \phi_1 + \phi_2 = -\frac{A\cos\theta}{r} - Ux = -\frac{A\cos\theta}{r} - Ur\cos\theta$$

$$\phi = -U\left(r + \frac{A}{Ur}\right)\cos\theta = -Ur\left(1 + \frac{a^2}{r^2}\right)\cos\theta$$

Doublet, Vortex (clockwise), and Uniform Flow (flow past a cylinder with circulation)

$$\psi = \psi_d + \psi_v + \psi_{uf} = \psi_1 + \psi_2 + \psi_3 = -\frac{A\sin\theta}{r} + \frac{K}{2\pi}\ln r + Uy$$

$$\psi = -\frac{A\sin\theta}{r} + \frac{K}{2\pi}\ln r + Ur\sin\theta = Ur\left(1 - \frac{a^2}{r^2}\right)\sin\theta + \frac{K}{2\pi}\ln r$$

$$\phi = \phi_d + \phi_v + \phi_{uf} = \phi_1 + \phi_2 + \phi_3 = -\frac{A\cos\theta}{r} + \frac{K}{2\pi}\theta - Ux$$

$$\phi = -\frac{A\cos\theta}{r} + \frac{K}{2\pi}\theta - Ur\cos\theta = -Ur\left(1 + \frac{a^2}{r^3}\right)\cos\theta + \frac{K}{2\pi}\theta$$

$$a = \sqrt{\frac{A}{U}}; \quad K < 4\pi a U$$

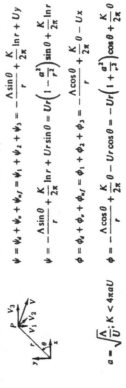

Table 6.3-2 Continued

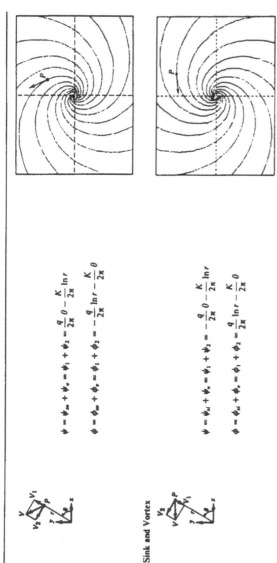
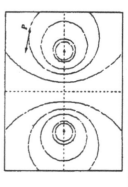

$$\psi = \psi_\infty + \psi_v = \psi_1 + \psi_2 = \frac{q}{2\pi}\theta - \frac{K}{2\pi}\ln r$$

$$\phi = \phi_\infty + \phi_v = \phi_1 + \phi_2 = -\frac{q}{2\pi}\ln r - \frac{K}{2\pi}\theta$$

Sink and Vortex

$$\psi = \psi_{si} + \psi_v = \psi_1 + \psi_2 = -\frac{q}{2\pi}\theta - \frac{K}{2\pi}\ln r$$

$$\phi = \phi_{si} + \phi_v = \phi_1 + \phi_2 = \frac{q}{2\pi}\ln r - \frac{K}{2\pi}\theta$$

Vortex Pair (equal strength, opposite rotation, separation distance on x axis = 2a)

$$\psi = \psi_{v_1} + \psi_{v_2} = \psi_1 + \psi_2 = -\frac{K}{2\pi}\ln r_1 + \frac{K}{2\pi}\ln r_2 = \frac{K}{2\pi}\ln\frac{r_2}{r_1}$$

$$\phi = \phi_{v_1} + \phi_{v_2} = \phi_1 + \phi_2 = -\frac{K}{2\pi}\theta_1 + \frac{K}{2\pi}\theta_2 = \frac{K}{2\pi}(\theta_2 - \theta_1)$$

[10] Fox, R. W. and A. T. McDonald (1985). *Introduction to Fluid Mechanics*, New York, John Wiley and Sons, Inc., pp. 279-281.

Example 6.3-1 Flow around a circular cylinder

a) Show that the following complex potential which is the sum of uniform flow and a doublet describes flow of an ideal fluid with approach velocity v_∞ around a circular cylinder of radius R.

$$\Omega(z) = v_\infty \left(z + \frac{R^2}{z} \right) \tag{6.3-27}$$

b) Plot typical streamlines and lines of constant velocity potential.

c) Find the stagnation points for the flow.

Solution

a) Substituting in the complex potential using the definition of z

$$\Omega = v_\infty \left(x + iy + \frac{R^2}{x + iy} \right) \tag{6.3-28}$$

Multiplying and dividing the fraction by the complex conjugate, $\bar{z}$

$$\Omega = v_\infty \left(x + iy + \frac{R^2 [x - iy]}{x^2 + y^2} \right)$$

$$= v_\infty x \left(1 + \frac{R^2}{x^2 + y^2} \right) + i v_\infty y \left(1 - \frac{R^2}{x^2 + y^2} \right) \tag{6.3-29}$$

which using Equation (6.3-25) permits us to identify

$$\Phi = v_\infty x \left(1 + \frac{R^2}{x^2 + y^2} \right)$$

$$\Psi = v_\infty y \left(1 - \frac{R^2}{x^2 + y^2} \right) \tag{6.3-30}$$

Since there is no flow across a streamline, any streamline can be interpreted as the boundary of a solid object. If we examine the streamline $\Psi = 0$, we see that it is satisfied by

$$y = 0$$

$$\left(1 - \frac{R^2}{x^2 + y^2}\right) = 0 \quad \Rightarrow \quad x^2 + y^2 = R^2 \tag{6.3-31}$$

The first equation is simply the x-axis. The second is the locus of a circle of radius R with center at the origin, so we can use this streamline as a solid boundary to describe flow around a right circular cylinder normal to the x-y plane.

b) Substituting arbitrary values into Φ and Ψ, choosing arbitrary values for y and solving for x or vice versa yields streamlines and potential lines

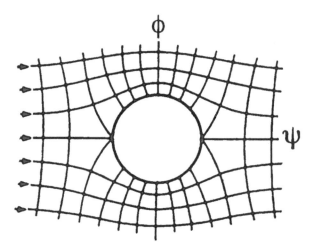

Figure 6.3-1 Flow around circular cylinder

c) From the velocity potential

$$\Phi = v_\infty x \left(1 + \frac{R^2}{x^2 + y^2}\right) \tag{6.3-32}$$

$$v_x = -\frac{\partial \Phi}{\partial x}$$

$$v_y = -\frac{\partial \Phi}{\partial y}$$

(6.3-33)

It is easiest to use polar coordinates

$$\Phi = v_\infty r \cos \theta \left(1 + \frac{R^2}{r^2} \right)$$

(6.3-34)

$$v_r = -\frac{\partial \Phi}{\partial r} = v_\infty \cos(\theta) \left(1 - \frac{R^2}{r^2} \right)$$

$$v_\theta = -\frac{1}{r}\frac{\partial \Phi}{\partial \theta} = -v_\infty r \sin(\theta) \left(1 + \frac{R^2}{r^2} \right)$$

(6.3-35)

At the stagnation points the velocity is zero, which implies that both v_r and v_θ are zero

$$v_\infty \cos(\theta) \left(1 - \frac{R^2}{r^2} \right) = 0$$

$$-v_\infty r \sin(\theta) \left(1 + \frac{R^2}{r^2} \right) = 0$$

(6.3-36)

The first equation can be seen to be zero for $r = \pm R$; the second for $\theta = 0$. Therefore, the stagnation points lie on the surface of the cylinder at its two points of intersection with the x-axis.

Example 6.3-2   Flow of an ideal fluid through a corner

Show that the complex potential $\Omega = -v_\infty z^2$ can describe the flow through a corner. What are the velocity components $v_x(x, y)$ and $v_y(x, y)$? What is the physical significance of v_∞?

Solution

$$\Omega = -v_\infty z^2 = -v_\infty (x + i y)^2$$

(6.3-37)

$$\Omega = \left[-v_{\infty}\left(x^2 - y^2\right)\right] + i\left[-v_{\infty}\left(2\,x\,y\right)\right] = [\Phi] + i\,[\Psi] \qquad (6.3\text{-}38)$$

Note that the streamline $\Psi = 0$ consists of the x and y axes - these form the corner.

Therefore,

$$v_x = -\frac{\partial \Phi}{\partial x} = -\frac{\partial \Psi}{\partial y} = 2\,x\,v_{\infty}$$

$$v_y = -\frac{\partial \Phi}{\partial y} = \frac{\partial \Psi}{\partial x} = -2\,y\,v_{\infty}$$

$$v^2 = v_x^2 + v_y^2 = 4\,v_{\infty}^2\left(x^2 + y^2\right) = 4\,v_{\infty}^2\,r^2$$

$$v_{\infty} = \tfrac{1}{2}\tfrac{v}{r} = \tfrac{1}{2}\left[\text{angular velocity}\right] \qquad (6.3\text{-}39)$$

The solution is shown below. This solution can be generalized to non-90° corners.

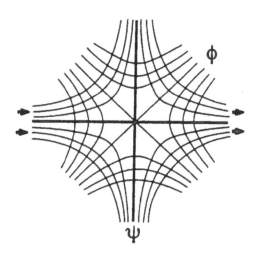

Figure 6.3-2 Flow through a corner

Example 6.3-3   Flow around a rotating cylinder

For potential flow, addition of uniform flow and a doublet yields flow around a circular cylinder of radius R. If we now add a **vortex** to this flow, we have

$$\varphi = -v_\infty r\left(1 + \frac{R^2}{r^2}\right)\cos(\theta) - \frac{\Gamma}{2\pi}\theta$$

$$\psi = v_\infty r\left(1 - \frac{R^2}{r^2}\right)\sin(\theta) - \frac{\Gamma}{2\pi}\ln r \qquad (6.3\text{-}40)$$

which describes flow around a **rotating** circular cylinder of radius R, where Γ is a constant related to the speed of rotation called the **circulation**, and v_∞ and p_∞, respectively, are the velocity and pressure far from the surface of the cylinder.

Remembering that in cylindrical coordinates

$$v_r = -\frac{\partial\varphi}{\partial r} = \frac{1}{r}\frac{\partial\psi}{\partial\theta}$$

$$v_\theta = -\frac{1}{r}\frac{\partial\varphi}{\partial\theta} = -\frac{\partial\psi}{\partial r} \qquad (6.3\text{-}41)$$

a) show that $v_r = 0$ along the line $(r, \theta) = (r, \pi/2)$, $r \neq 0$

b) for $v_\infty = 1$ m/s, $\Gamma = -20$ m²/s and $R = 3$ m, find the stagnation points.

Solution

a) One can use either

$$v_r = -\frac{\partial\varphi}{\partial r} = -\frac{\partial}{\partial r}\left[-v_\infty r\left(1 + \frac{R^2}{r^2}\right)\cos(\theta) - \frac{\Gamma}{2\pi}\theta\right]$$

$$= -\frac{\partial}{\partial r}\left[-v_\infty r\left(1 + \frac{R^2}{r^2}\right)\cos(\theta)\right] = v_\infty\cos(\theta)\frac{\partial}{\partial r}\left[\left(r + \frac{R^2}{r}\right)\right]$$

$$= v_\infty\cos(\theta)\left(1 - \frac{R^2}{r^2}\right) \qquad (6.3\text{-}42)$$

or

$$v_r = \frac{1}{r}\frac{\partial \psi}{\partial \theta} = \frac{1}{r}\frac{\partial}{\partial \theta}\left[v_\infty r\left(1 - \frac{R^2}{r^2}\right)\sin(\theta) - \frac{\Gamma}{2\pi}\ln r\right]$$

$$= \frac{1}{r}\frac{\partial}{\partial \theta}\left[v_\infty r\left(1 - \frac{R^2}{r^2}\right)\sin(\theta)\right] = \frac{1}{r}\left[v_\infty r\left(1 - \frac{R^2}{r^2}\right)\cos\theta\right]$$

$$= v_\infty \cos(\theta)\left(1 - \frac{R^2}{r^2}\right) \tag{6.3-43}$$

each of which gives the same result. It follows that $v_r = 0$ for $\cos(\theta) = 0$, $r > 0$; a special case of which is $\theta = \pi/2$, $r > 0$.

b) At the stagnation points, $v_r = 0$ and $v_\theta = 0$. We can use the expression for v_r derived in part (a) and obtain v_θ from either

$$v_\theta = -\frac{1}{r}\frac{\partial \varphi}{\partial \theta} = -\frac{1}{r}\frac{\partial}{\partial \theta}\left[-v_\infty r\left(1 + \frac{R^2}{r^2}\right)\cos(\theta) - \frac{\Gamma}{2\pi}\theta\right]$$

$$= -\frac{1}{r}\left[v_\infty r\left(1 + \frac{R^2}{r^2}\right)\sin\theta - \frac{\Gamma}{2\pi}\right]$$

$$= -v_\infty\left(1 + \frac{R^2}{r^2}\right)\sin(\theta) + \frac{\Gamma}{2\pi r} \tag{6.3-44}$$

or

$$v_\theta = -\frac{\partial \psi}{\partial r}$$

$$v_\theta = -\frac{\partial}{\partial r}\left[v_\infty r\left(1 - \frac{R^2}{r^2}\right)\sin(\theta) - \frac{\Gamma}{2\pi}\ln r\right]$$

$$= -v_\infty\left(1 + \frac{R^2}{r^2}\right)\sin(\theta) + \frac{\Gamma}{2\pi r} \tag{6.3-45}$$

so we require that

$$v_\theta = -v_\infty \left(1 + \frac{R^2}{r^2}\right) \sin(\theta) + \frac{\Gamma}{2\pi r} = 0$$

$$v_r = v_\infty \cos(\theta)\left(1 - \frac{R^2}{r^2}\right) = 0 \qquad\qquad (6.3-46)$$

This is true for the second equation at r = R, which means that from the first equation

$$\frac{\Gamma}{2\pi R v_\infty} = \left(1 + \frac{R^2}{R^2}\right)\sin(\theta)$$

$$\sin(\theta) = \frac{\Gamma}{4\pi R v_\infty}$$

$$\theta = \arcsin\left(\frac{\Gamma}{4\pi R v_\infty}\right) = \arcsin\left|\frac{-(20)\left[\frac{m^2}{s}\right]}{(4)(\pi)(3)[m](1)\left[\frac{m}{s}\right]}\right|$$

$$= \arcsin(-0.531) = \left\{\begin{array}{l} -0.559 \text{ (radians)} = -32 \text{ (degrees)} \\ 3.7 \text{ (radians)} = 212 \text{ (degrees} \end{array}\right\}$$

$$\qquad\qquad (6.3-47)$$

Therefore, the stagnation points lie at

$$(r, \theta) = \left(3\text{ m}, 212\text{ degrees}\right) \text{ and } \left(3\text{ m}, -32\text{ degrees}\right) \qquad (6.3-48)$$

Many potential flow solutions can be obtained from the theory of complex variables via use of **conformal mapping**. A conformal mapping or transformation is one that preserves angles and orientations[11] - that is, an angle in the first space is mapped into the same angle in the second space, and a direction of rotation in the first space is mapped into the same direction of rotation in the second space.

---

[11] O'Neil, P. V. (1991). *Advanced Engineering Mathematics*. Belmont, CA, Wadsworth, p. 1405.

Although it can be proved[12] that either of two bounded simply connected regions in the (x, y) and (Φ, Ψ)[13] planes can be mapped conformally onto the other, finding the transformation to do so is not, in general, a simple task. Such transformations, however, often permit problems with difficult boundary configurations to be mapped into a simpler geometry, in which the problem can be more easily solved, and the solution subsequently mapped back to the original plane containing the problem boundary configuration.

A conformal mapping of particular utility is the *Schwarz-Christoffel* transformation, which maps the upper half-plane into the interior of any domain bounded by a polygon.

6.4 Laminar Flow

The equations used to describe the movement of fluids are formulated assuming a continuum. For flow which we call **laminar**, layers of fluid exchange momentum as they slide past one another smoothly - much as when one pushes on the top of a deck of cards, the top card slides over the next card, which transfers momentum to the succeeding card, etc., and so on down through the deck. At a solid surface, the velocity of the fluid is that of the surface, which can be demonstrated by experiment.

In laminar flow, the fluid particles move in smooth paths. It is important to recognize that fluid "particle" does not refer to a molecule of fluid. Rather, by fluid "particle" we mean some identifiable region in the fluid that is moving as a unit at the moment.

If the relative velocity of the fluid layers becomes high enough, the layers lose identity and the fluid "tumbles." Momentum is transferred not only by the mechanism seen in laminar flow where molecules jump from regions of one velocity to regions of a different velocity, but also by sizable clumps of fluid possessing the velocity of a given region moving from that region to a region of different velocity. This type of flow is called **turbulent**. Turbulent flow is not easy to describe in detail, and we therefore usually must resort to experiment and statistical theories to describe such flow.

The result of momentum exchange within the fluid and between the fluid and solid surfaces is the development of a **velocity profile**.

---

[12] Wylie, C. R. and L. C. Barrett (1995). *Advanced Engineering Mathematics*. New York, McGraw-Hill, p. 1262.

[13] (u, v) in the usual mathematical notation.

6.4.1 Laminar flow between infinite parallel plates

Suppose that we have two parallel infinite plates with a Newtonian fluid in steady flow in the gap between them, as in Figure 6.4.1-1.

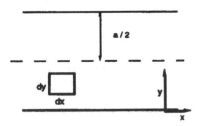

Figure 6.4.1-1 Steady flow between infinite stationary parallel plates

We can write the x-direction momentum balance on a control volume of dimensions Δx by Δy by a unit distance in the z-direction for this situation. Since the flow is steady, there is no net transfer of momentum in or out of the control volume, and there are no body forces acting in the x-direction, this balance reduces to

$$\sum F_{xx} = 0 \tag{6.4.1-1}$$

We now proceed to write and sum the surface forces in the x-direction. The force on the left face of the control volume is

$$(p)\Big|_{x-\Delta x} \Delta y\, \Delta z \tag{6.4.1-2}$$

The force on the right face of the control volume is

$$-(p)\Big|_{x+\Delta x} \Delta y\, \Delta z \tag{6.4.1-3}$$

The shear force on the top of the control volume is

$$-(\tau_{yx})\Big|_{y+\Delta y} \Delta x\, \Delta z \tag{6.4.1-4}$$

The shear force on the bottom of the control volume is

$$\left(\tau_{yx}\right)\Big|_{y-\Delta y} \Delta x\, \Delta z \tag{6.4.1-5}$$

Summing the surface forces, setting the result equal to zero

$$(p)\Big|_{x-\Delta x} \Delta y\, \Delta z - (p)\Big|_{x+\Delta x} \Delta y\, \Delta z$$
$$+\left(\tau_{yx}\right)\Big|_{y-\Delta y} \Delta x\, \Delta z -\left(\tau_{yx}\right)\Big|_{y+\Delta y} \Delta x\, \Delta z = 0 \tag{6.4.1-6}$$

rearranging, dividing by $(\Delta x\, \Delta y\, \Delta z)$, and taking the limit

$$\lim_{\substack{\Delta x \to 0 \\ \Delta y \to 0}} \left[\frac{(p)\Big|_{x+\Delta x} - (p)\Big|_{x-\Delta x}}{\Delta x}\right] = -\lim_{\substack{\Delta x \to 0 \\ \Delta y \to 0}} \left[\frac{\left(\tau_{yx}\right)\Big|_{y+\Delta y} - \left(\tau_{yx}\right)\Big|_{y-\Delta y}}{\Delta y}\right]$$

$$\tag{6.4.1-7}$$

$$\frac{dp}{dx} = -\frac{d\tau_{yx}}{dy} \;\Rightarrow\; f(x) = g(y) \tag{6.4.1-8}$$

Observing that the only way a function of x can be equal to a function of y for arbitrary values is for each of the functions to be equal to the same constant

$$\frac{dp}{dx} = \frac{d\tau_{yx}}{dy} = \text{constant} \tag{6.4.1-9}$$

Integrating with respect to y

$$\tau_{yx} = -\frac{dp}{dx}\, y - C_1 \tag{6.4.1-10}$$

Substituting for the shear stress in terms of the Newtonian viscosity

$$-\mu\frac{dv_x}{dy} = -\frac{dp}{dx}\, y - C_1 \tag{6.4.1-11}$$

Integrating once again with respect to y

$$v_x = \frac{1}{2\mu}\frac{dp}{dx}y^2 + \frac{C_1}{\mu}y + C_2 \qquad (6.4.1\text{-}12)$$

Applying the boundary conditions

$$\begin{aligned}
\text{at } y = 0 \quad v_x = 0\\
\text{at } y = a \quad v_x = 0
\end{aligned} \qquad (6.4.1\text{-}13)$$

gives find the constants C_1 and C_2, which upon substitution give

$$v_x = \frac{a^2}{2\mu}\frac{dp}{dx}\left[\left(\frac{y}{a}\right)^2 - \frac{y}{a}\right] \qquad (6.4.1\text{-}14)$$

Notice that the velocity profile is parabolic.

The shear stress distribution is given by

$$\tau_{yx} = \frac{dp}{dx}y + C_1 = a\frac{dp}{dx}\left[\frac{y}{a} - \frac{1}{2}\right] \qquad (6.4.1\text{-}15)$$

The volumetric flow rate per unit width in the z-direction (normal to the page) is given by

$$Q = \int_A \left(v\cdot n\right)dA = \int_0^a v_x\,dy = -\frac{1}{12\mu}\frac{dp}{dx}a^3 \qquad (6.4.1\text{-}16)$$

The average velocity is given by

$$\langle v\rangle = \frac{Q}{A} = -\frac{1}{12\mu}\frac{dp}{dx}\frac{a^3}{a} = -\frac{a^2}{12\mu}\frac{dp}{dx} \qquad (6.4.1\text{-}17)$$

The maximum velocity is obtained by setting dv_x / dy equal to zero

$$\frac{dv_x}{dy} = 0 = \frac{a^2}{2\mu}\frac{dp}{dx}\left[\frac{2y}{a^2} - \frac{1}{a}\right] \qquad (6.4.1\text{-}18)$$

which is satisfied at $y = a / 2$, giving the maximum velocity as

$$v_{x\,max} = -\frac{1}{8\mu}\frac{dp}{dx} = \frac{3}{2}\langle v\rangle \qquad (6.4.1\text{-}19)$$

Flow is laminar in this instance as long as the Reynolds number $Re = \rho \langle v \rangle a/\mu < 1400$.

Example 6.4.1-1   Steady flow between infinite parallel plates

An oil is flowing through a slit 0.01-mm thick. The length of the slit is 20 mm and its width, W, is 10 mm. The fluid has a specific gravity of 0.9 and a viscosity of 0.02 kg/(m s) at the temperature of the system.

   a) Determine the flow rate through the slit if the pressure drop causing flow is 15 mPa.

   b) What if the top plate were moving at a constant velocity with respect to the bottom plate?

Solution

$$\langle v \rangle = \frac{Q}{A} = \frac{Q}{Wa} = -\frac{a^2}{12\mu}\frac{dp}{dx} \tag{6.4.1-20}$$

$$Q = -\frac{Wa^3}{12\mu}\frac{dp}{L} \tag{6.4.1-21}$$

$$Q = \frac{(10)\,[\text{mm}]\,(.01)^3\,[\text{mm}]^3}{(12)\,(0.02)\left[\frac{\text{kg}}{\text{m s}}\right]}\left(15\times10^6\right)\left[\frac{N}{m^2}\right]$$
$$\times \frac{1}{(20)\,[\text{mm}]}\left[\frac{\text{kg m}}{N s^2}\right] \tag{6.4.1-22}$$

$$Q = 31.2\,\frac{\text{mm}^3}{s} \tag{6.4.1-23}$$

check Re:

$$\langle v \rangle = \frac{Q}{A} \tag{6.4.1-24}$$

$$= (31.2)\left[\frac{\text{mm}^3}{s}\right]\frac{1}{(10)\,[\text{mm}]\,(0.01)\,[\text{mm}]}\left(\frac{1}{10^3}\right)\left[\frac{m}{\text{mm}}\right]$$

$$= 0.312\,\frac{m}{s} \tag{6.4.1-25}$$

$$Re = \frac{\rho\langle v\rangle a}{\mu} \tag{6.4.1-26}$$

$$= (0.9)\,(999)\left[\frac{\text{kg}}{m^3}\right](0.312)\left[\frac{m}{s}\right](0.01)\,[\text{mm}]\left(\frac{1}{0.02}\right)\left[\frac{\text{m s}}{\text{kg}}\right]\left(\frac{1}{10^3}\right)\left[\frac{m}{\text{mm}}\right]$$

$$= 0.14 \quad \Rightarrow \quad \text{laminar} \tag{6.4.1-27}$$

We now consider the case where one of the plates is moving at a uniform velocity. Often this model is used in viscometry or for flow in a journal bearing. For situations where the gap is very small and the radius large, this model can be used as an approximation in cylindrical systems.

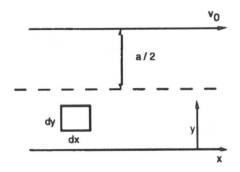

Figure 6.4.1-2 Flow between infinite parallel plates, top plate moving at v_0

Figure 6.4.1-2 is similar to Figure 6.4.1-1, except the upper plate is now considered to be moving with a constant velocity, v_0. The momentum balance before introducing boundary conditions is the same as in the previous case

$$v_x = \frac{1}{2\mu} \frac{dp}{dx} y^2 + \frac{C_1}{\mu} y + C_2 \tag{6.4.1-28}$$

In this case the boundary conditions are

$$\begin{aligned} &\text{at } y = 0 \quad v_x = 0 \\ &\text{at } y = a \quad v_x = v_0 \end{aligned} \tag{6.4.1-29}$$

Applying these boundary conditions to Equation (6.4.1-28) yields the following

$$v_x = \frac{v_0 y}{a} + \frac{a^2}{2\mu} \frac{dp}{dx} \left[\left(\frac{y}{a}\right)^2 - \left(\frac{y}{a}\right) \right] \tag{6.4.1-30}$$

Note that this profile is linear in y for the case of zero pressure gradient.

The shear stress is found by taking the derivative of velocity and substituting Newton's law of viscosity

$$\frac{dv_x}{dy} = \frac{v_0}{a} + \frac{a^2}{2\mu}\frac{dp}{dx}\left[\left(\frac{2\,y}{a^2}\right)-\left(\frac{1}{a}\right)\right]$$

$$-\tau_{yx} = \mu\frac{dv_x}{dy} = \mu\frac{v_0}{a} + \frac{a}{dx}\frac{dp}{dx}\left[\left(\frac{y}{a}\right)-\left(\frac{1}{2}\right)\right] \qquad (6.4.1\text{-}31)$$

The volumetric flow rate per unit width in the z-direction is

$$Q = \int_A \left(v\cdot n\right)dA = \int_0^a v_x\,dy = \frac{v_0\,a}{2} - \frac{1}{12\mu}\frac{dp}{dx}\,a^3 \qquad (6.4.1\text{-}32)$$

The average velocity is

$$\langle v\rangle = \frac{Q}{A} = \frac{v_0}{2} - \frac{1}{12\mu}\frac{dp}{dx}\,a^2 \qquad (6.4.1\text{-}33)$$

The maximum velocity is at the point where

$$\frac{dv_x}{dy} = 0 = \frac{v_0}{a} + \frac{a^2}{2\mu}\frac{dp}{dx}\left[\left(\frac{2\,y}{a^2}\right)-\left(\frac{1}{a}\right)\right] \qquad (6.4.1\text{-}34)$$

which gives

$$y = \left(\frac{a}{2}\right) - \frac{\left(\frac{v_0}{a}\right)}{\left(\frac{1}{\mu}\right)\left(\frac{dp}{dx}\right)} \qquad (6.4.1\text{-}35)$$

The velocity profile as a function of pressure drop is shown in Figure 6.4.1-3.

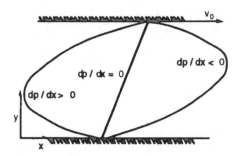

Figure 6.4.1-3 Velocity profiles for laminar flow of Newtonian fluid between parallel plates with imposed pressure drop, top plate moving at steady velocity

---

Example 6.4.1-2   Flow between infinite rotating concentric cylinders

A Stormer viscometer consists of concentric inner and outer cylinders. The inner cylinder is rotated with fluid in between them. The viscosity of the fluid can be determined from the torque required to rotate the inner cylinder. Assume a fluid of viscosity 0.002 lbf s/ft^2 is placed in a viscometer of length 4 inches and spacing .01 inch between the cylinders. If the inside 7.5-inch diameter cylinder is rotated at 360 rpm, determine the torque required. Assume the viscometer can be modeled as flow between two parallel plates.

Determine the torque.

Solution

$$\tau_{yx} = \mu \frac{dv_x}{dy} = \mu \frac{U}{a} a \left(\frac{dp}{dx}\right)\left[\frac{\mu}{a} - \frac{1}{2}\right]$$
$$= \mu \frac{U}{a} + a \frac{dp}{dx}\left[\left(\frac{y}{a}\right) - \left(\frac{1}{2}\right)\right] \tag{6.4.1-36}$$

Since

$$\frac{dp}{dx} = 0$$

$$\tau_{yx} = \mu \frac{U}{a} = \mu \frac{\omega R}{a} = \mu \frac{\omega D}{2 a} \tag{6.4.1-37}$$

$$\tau_{yx} = (0.0002)\left[\frac{lbf\,s}{ft^2}\right](360)\left[\frac{rev}{min}\right](2\,\pi)\left[\frac{radians}{rev}\right]$$
$$\times\left(\frac{1}{60}\right)\left[\frac{min}{s}\right]\left(\frac{2.5}{2}\right)[in]\frac{1}{(0.01)[in]} \tag{6.4.1-38}$$

$$\tau_{yx} = 0.942\ lbf\,/\,ft^2 \tag{6.4.1-39}$$

$$T = FR = \tau_{yx}\,\pi\,D\,L\,R = \frac{\pi}{2}\tau_{yx}\,D^2\,L$$

$$T = \left(\frac{\pi}{2}\right)\left(0.942\,\frac{lbf}{ft^2}\right)(2.5\ in)^2\left(\frac{ft^2}{144\ in}\right)(4\ in) = 0.1\ in\ lbf \tag{6.4.1-40}$$

The reader should check the Reynolds number.

6.4.2   Laminar flow in a circular pipe

One of the most useful laminar velocity profiles is for flow in a circular conduit such as a pipe. Assume a fluid is in steady laminar flow in a pipe of radius, R, and length, L, under the influence of a pressure drop, Δp. We have already indicated that one can obtain the appropriate differential equation to model a particular physical situation by eliminating extraneous or negligible terms from the general balance (here, that of momentum).

One can also develop the appropriate differential equation by writing it from first principles - but one must be careful not to leave out any significant terms. Here we choose to illustrate the latter approach, although the former would be equally valid.

Velocity profiles for steady laminar flow can be determined by writing a force balance on a right cylindrical control volume, as indicated in Figure 6.4.2-1.

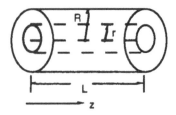

Figure 6.4.2-1 Control volume for force balance on fluid in pipe

The overall momentum balance for this situation, since we have steady flow and neglect body forces, reduces to the requirement that the sum of the components in the z-direction of the surface forces must equal zero. Therefore, the net pressure force in the z-direction on the ends of this circular column must be balanced by the viscous force on the lateral surface. The net pressure force is

$$F_{pz} = \left(p \Big|_{z=0} - p \Big|_{z=L} \right) \pi r^2 \tag{6.4.2-1}$$

The viscous force opposing the pressure force is the shear stress (or momentum flux), τ_{rz}, multiplied by the area on which this shear stress acts (or through which the momentum flux flows), $2\pi rL$

$$F_{vz} = \tau_{rz} \left(2 \pi r L \right) \tag{6.4.2-2}$$

At steady state $F_p + F_v = 0$ and

$$\tau_{rz} = \frac{\left(p \Big|_{z=L} - p \Big|_{z=0} \right) r}{2L} \tag{6.4.2-3}$$

Substituting Newton's law of viscosity

$$-\mu \frac{dv_z}{dr} = \frac{\left(p \Big|_{z=L} - p \Big|_{z=0} \right) r}{2L} \tag{6.4.2-4}$$

Integrating and subsequently applying the conditions that a) at $r = R$, $v_z = 0$ and b) at $r = 0$, $dv_z / dr = 0$ (since there is no source or sink present on the centerline and hence by symmetry there can be no momentum transfer across the centerline) yields

$$v_z = -\frac{r^2}{4\mu} \frac{\Delta p}{L} + \frac{C_1}{\mu} \ln [r] + C_2 \tag{6.4.2-5}$$

$$v_z = -\frac{\Delta p R^2}{4 \mu L} \left[1 - \left(\frac{r}{R} \right)^2 \right] \tag{6.4.2-6}$$

Note that Δp is defined as

$$\Delta p = \left(p \Big|_{z=L} - p \Big|_{z=0} \right) \tag{6.4.2-7}$$

Since Equation (6.4.2-6) gives the velocity as a function of the radius, we can differentiate and set the derivative equal to zero to determine the maximum velocity

$$\frac{dv_z}{dr} = 0 = -\frac{\Delta p \, r}{2 \mu L} \tag{6.4.2-8}$$

It is seen that the maximum velocity occurs at $r = 0$. Substituting $r = 0$ into the velocity profile gives the maximum velocity as

$$v_{z\,max} = -\frac{\Delta p \, R^2}{4 \mu L} \tag{6.4.2-9}$$

Figure 6.4.2-2 shows a graph of the predicted velocity profile.

The drag on the fluid from the pipe wall (the negative of the drag on the pipe wall by the fluid) resulting from the shear force transmitted from the fluid to the wall is the product of the wetted area of the wall and the shear stress at the wall. The shear stress at the wall can be calculated as

$$\tau_{rz} \Big|_{r=R} = \frac{\left(p \Big|_{z=L} - p \Big|_{z=0} \right) R}{2 L} = 2 \pi R L \left(-\mu \frac{dv_z}{dr} \Big|_{r=R} \right) \tag{6.4.2-10}$$

Note that the shear stress varies linearly from a maximum at the wall to zero at the center of the pipe, as shown in Figure 6.4.2-3.

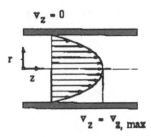

Figure 6.4.2-2 Velocity profile for laminar flow of a Newtonian fluid in a pipe or duct of circular cross-section[14]

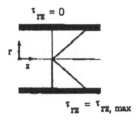

Figure 6.4.2-3 Shear stress profile for laminar flow of a Newtonian fluid in a pipe or duct of circular cross-section

An average velocity for a cross-section is found by integrating

$$\langle v_z \rangle = \frac{\int_0^R v_z \, 2\pi r \, dr}{\int_0^R 2\pi r \, dr} = -\frac{\Delta p \, R^2}{8\mu L} = \frac{1}{2} v_{zmax} \qquad (6.4.2\text{-}11)$$

The ratio of the local velocity, v_z, to the maximum velocity is

$$\frac{v_z}{v_{zmax}} = 1 - \left(\frac{r}{R}\right)^2 \qquad (6.4.2\text{-}12)$$

---

[14] Adapted from Bird, R. B., W. E. Stewart, et al. (1960). *Transport Phenomena*. New York, Wiley, p. 45.

The volumetric flow rate can be obtained from

$$Q = \langle v_z \rangle \pi R^2 = -\frac{\pi R^4 \Delta p}{8 \mu L} \qquad (6.4.2\text{-}13)$$

Equation (6.4.2-13) is the **Hagen-Poiseuille** equation.

Example 6.4.2-1   Flow in a capillary viscometer

A capillary viscometer 3 m long and 0.5 mm in diameter passes a fluid of density 999 kg/m³ at a volumetric flow of 800 mm³/s under a pressure drop of 3 mPa. What is the viscosity of the fluid?

Solution

$$Q = -\frac{\pi \Delta p \, D^4}{128 \mu L} \qquad (6.4.2\text{-}14)$$

$$\mu = \frac{\pi \Delta p \, D^4}{128 L Q} = \frac{(\pi)\left(3 \times 10^6 \, \frac{N}{m^2}\right)(0.5)^4 \left(mm^4\right)}{(128)\left(800 \, \frac{mm^3}{s}\right)(3 \, m)\left(10^3 \, \frac{mm}{m}\right)} \qquad (6.4.2\text{-}15)$$

$$\mu = 1.83 \times 10^{-3} \, \frac{N \, s}{m^2} \qquad (6.4.2\text{-}16)$$

Checking to ensure that using a model valid only for laminar flow was correct

$$\langle v \rangle = \frac{Q}{A} = \frac{4Q}{\pi D^2} = \frac{(4)\left(800 \, \frac{mm^3}{s}\right)}{\pi (0.5)^2 \, mm^2} \, \frac{m}{1000 \, mm}$$

$$= 4.074 \, \frac{m}{s} \qquad (6.4.2\text{-}17)$$

$$Re = \frac{\rho \langle v \rangle D}{\mu} = \frac{\left(\frac{999 \, kg}{m^3}\right)\left(4.074 \, \frac{m}{s}\right)(0.5 \, mm)\left(\frac{N \, s^2}{kg \, m}\right)}{\left(1.83 \times 10^{-3} \, \frac{N \, s}{m^2}\right)\left(10^3 \, \frac{mm}{m}\right)} \qquad (6.4.2\text{-}18)$$

$$Re = 1112 \quad \Rightarrow \quad \text{laminar flow} \qquad (6.4.2\text{-}19)$$

Example 6.4.2-2   Flow between two concentric cylinders

An incompressible Newtonian fluid is in upward steady axial flow in an annular region between two concentric vertical circular cylinders of radii R and aR, as shown in Figure 6.4.2-4. Determine the expressions for shear stress, velocity profile, maximum velocity, average velocity and volumetric flow rate.

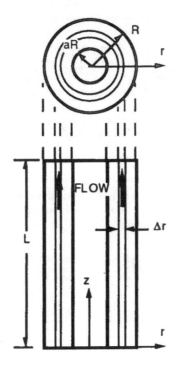

Figure 6.4.2-4 Viscometric flow between cylinders

Solution

rate of momentum across cylindrical surface	in at r = r $\left(2\pi r L \tau_{rz}\right)\Big|_{r=r}$	out at r = r + Δr $\left(2\pi r L \tau_{rz}\right)\Big|_{r=r+\Delta r}$
rate of momentum across annular cross-sectional area	in at z=0 $\left(2\pi r \Delta r\, v_z\right)\left(\rho\, v_z\right)\Big|_{z=0}$	out at z = L $\left(2\pi r \Delta r\, v_z\right)\left(\rho\, v_z\right)\Big|_{z=L}$
gravity force on cylindrical shell	$\left(2\pi r \Delta r L\right)\rho\, g$	
pressure force on annular cross-sectional area	at z = 0 $\left(2\pi r \Delta r\right)p\Big|_{z=0}$	at z = L $-\left(2\pi r \Delta r\right)p\Big|_{z=L}$

Substitute in momentum balance

$$\left(2\pi r L \tau_{rz}\right)\Big|_{r} - \left(2\pi r L \tau_{rz}\right)\Big|_{r+\Delta r} + \left(2\pi r \Delta r\, \rho\, v_z^2\right)\Big|_{z=0} - \left(2\pi r \Delta r\, \rho\, v_z^2\right)\Big|_{z=L}$$
$$+\, 2\pi r \Delta r L \rho\, g + 2\pi r \Delta r \left(p_0 - p_L\right) = 0 \qquad\qquad \text{(6.4.2-20)}$$

For an incompressible fluid in axial flow, the continuity equation shows that the velocity is the same at z = 0 and z = L, so the two terms containing velocity cancel each other. Rearranging

$$\lim_{\Delta r \to 0}\left[\frac{\left(r\,\tau_{rz}\right)\Big|_{r+\Delta r} - \left(r\,\tau_{rz}\right)\Big|_{r}}{\Delta r}\right] = \left(\frac{p_0 - p_L}{L} + \rho\, g\right)r \qquad\qquad \text{(6.4.2-21)}$$

$$\frac{d}{dr}\left(r\,\tau_{rz}\right) = \left(\frac{\mathcal{P}_0 - \mathcal{P}_L}{L}\right)r \qquad\qquad \text{(6.4.2-22)}$$

where[15]

---

[15] This is the same approach we took in hydrostatics:

$$\left(\frac{p_0 - p_L}{L} + \rho\, g\right) = \left(\frac{p_0 - p_L + \rho\, g L}{L}\right) =$$

$$\left(\frac{\left(p + \rho\, g L\right)\Big|_{L=0} - \left(p + \rho\, g L\right)\Big|_{L=L}}{L}\right) = \left(\frac{\mathcal{P}_0 - \mathcal{P}_L}{L}\right)$$

$$\mathcal{P} = p + \rho\, g\, z \tag{6.4.2-23}$$

Upon integrating with respect to r

$$\tau_{rz} = \left(\frac{\mathcal{P}_0 - \mathcal{P}_L}{2L}\right) r + \frac{C_1}{r} \tag{6.4.2-24}$$

(Note: for $\alpha \to 0$, $c_1 \to 0$, and we approach the same solution as for a single hollow cylinder - a tube.)

Let the radius at which the maximum velocity occurs (we know there will be a maximum velocity because the velocity goes to zero at each of the walls and is finite in between) be designated by $r = \lambda r$. At this point the momentum flux will be zero because of Newton's law of viscosity.

$$\tau_{rz} = -\mu \frac{dv_z}{dr} \tag{6.4.2-25}$$

Then it follows that

$$C_1 = -\frac{\left(\mathcal{P}_0 - \mathcal{P}_L\right)\left(\lambda R\right)^2}{2L} \tag{6.4.2-26}$$

$$\therefore\ \tau_{rz} = \frac{\left(\mathcal{P}_0 - \mathcal{P}_L\right) R}{2L}\left[\frac{r}{R} - \lambda^2\left(\frac{R}{r}\right)\right] \tag{6.4.2-27}$$

Substituting using Newton's law of viscosity and integrating

$$v_z = -\frac{\left(\mathcal{P}_0 - \mathcal{P}_L\right) R^2}{4\mu L}\left[\left(\frac{r}{R}\right)^2 - 2\lambda^2 \ln\!\left(\frac{r}{R}\right) + C_2\right] \tag{6.4.2-28}$$

$$\begin{aligned} &\text{at } r = \alpha R, \quad v_z = 0 \\ &\text{at } r = R, \quad\ \ v_z = 0 \end{aligned} \tag{6.4.2-29}$$

$$0 = -\frac{\left(\mathcal{P}_0 - \mathcal{P}_L\right) R^2}{4\mu L}\left(\alpha^2 - 2\lambda^2 \ln[\alpha] + C_2\right)$$

$$0 = -\frac{\left(\mathcal{P}_0 - \mathcal{P}_L\right) R^2}{4\mu L}\left(1 + C_2\right)$$

Note the above is valid independent of the choice of datum for the z axis.

$$\therefore \quad C_2 = -1, \qquad 2\lambda^2 = \frac{1-\alpha^2}{\ln\left[\frac{1}{\alpha}\right]} \qquad\qquad (6.4.2\text{-}30)$$

Substituting in Equation (6.4.2-27)

$$\tau_\alpha = \frac{\left(\mathcal{P}_0 - \mathcal{P}_L\right)R}{2L}\left[\left(\frac{r}{R}\right) - \left(\frac{1-\alpha^2}{2\ln\left[1/\alpha\right]}\right)\left(\frac{R}{r}\right)\right] \qquad (6.4.2\text{-}31)$$

$$v_z = \frac{\left(\mathcal{P}_0 - \mathcal{P}_L\right)R^2}{4\mu L}\left[1 - \left(\frac{r}{R}\right)^2 + \left(\frac{1-\alpha^2}{\ln\left[1/\alpha\right]}\right)\ln\left[\frac{r}{R}\right]\right] \qquad (6.4.2\text{-}32)$$

$$v_{z,max} = v_z\big|_{r=\lambda r} = \frac{\left(\mathcal{P}_0 - \mathcal{P}_L\right)R^2}{4\mu L}\left\{1 - \left(\frac{1-\alpha^2}{2\ln\left[\frac{1}{\alpha}\right]}\right)\left[1 - \ln\left[\frac{1-\alpha^2}{2\ln\left[\frac{1}{\alpha}\right]}\right]\right]\right\}$$

$$(6.4.2\text{-}33)$$

$$\langle v_z\rangle = \frac{\displaystyle\int_0^{2\pi}\int_{\alpha R}^R v_z\, r\, dr\, d\theta}{\displaystyle\int_0^{2\pi}\int_{\alpha R}^R r\, dr\, d\theta} = \frac{\left(\mathcal{P}_0 - \mathcal{P}_L\right)R^4}{8\mu L}\left(\frac{1-\alpha^4}{1-\alpha^2} - \frac{1-\alpha^2}{\ln\left[1/\alpha\right]}\right)$$

$$(6.4.2\text{-}34)$$

$$Q = \pi R^2\left(1 - \alpha^2\right)v_z = \frac{\pi\left(\mathcal{P}_0 - \mathcal{P}_L\right)R^4}{8\mu L}\left(1 - \alpha^4 - \frac{\left(1-\alpha^2\right)^2}{\ln\left[1/\alpha\right]}\right)$$

$$(6.4.2\text{-}35)$$

Example 6.4.2-3   Film flow down a wall

Find the velocity profile, average velocity and free surface velocity for a Newtonian fluid flowing down a vertical wall of length, L, and width, W.

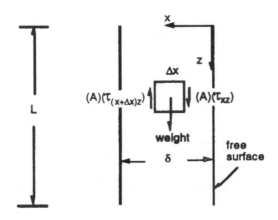

Figure 6.4.2-5 Film flow down wall

Solution

$$\Sigma F_z = \rho \left(W \, \Delta x \, L \right) g + \tau_{xz} \Big|_x \left(W L \right) - \tau_{xz} \Big|_{x+\Delta x} \left(W L \right) = 0 \qquad (6.4.2\text{-}36)$$

$$\lim_{\Delta x \to 0} \left[\frac{\tau_{xz} \Big|_{x+\Delta x} - \tau_{xz} \Big|_x}{\Delta x} \right] = \rho \, g \qquad (6.4.2\text{-}37)$$

$$\frac{d\tau_{xz}}{dx} = \rho \, g \qquad (6.4.2\text{-}38)$$

$$\tau_{xz} = -\mu \frac{dv_z}{dx} \qquad (6.4.2\text{-}39)$$

$$\mu \frac{d^2 v_z}{dx^2} = -\rho \, g \qquad (6.4.2\text{-}40)$$

at $x = 0,\quad \dfrac{dv_z}{dx} = 0$

at $x = \delta,\quad v_z = 0$

$$v_z = \frac{\rho \, g \, \delta^2}{2\mu} \left(1 - \frac{x^2}{\delta^2} \right) \qquad (6.4.2\text{-}41)$$

$$\langle v_z \rangle = \frac{w \displaystyle\int_0^\delta v_z \, dx}{w \displaystyle\int_0^\delta dx} = \frac{\displaystyle\int_0^\delta \frac{\rho \, g \, \delta^2}{2\mu} \left(1 - \frac{x^2}{\delta^2} \right) dx}{\delta} \qquad (6.4.2\text{-}42)$$

$$\langle v_z \rangle = \frac{\rho g \delta^2}{3 \mu}$$

(6.4.2-43)

At the free surface, $x = 0$, and

$$v_{\text{free surface}} = v_{\text{max}} = v_z \Big|_{x=0} = \frac{\rho g \delta^2}{2 \mu}$$

(6.4.2-44)

Example 6.4.2-4   Flow adjacent to a flat plate instantaneously set in motion

Consider a flat plate adjacent to a quiescent fluid of infinite extent in the positive x-direction (normal to the plate; $x = 0$ at the plate surface), y-direction (along the plate surface) and the z-direction (along the plate surface). Let the plate be instantaneously set in motion in the y-direction at a constant velocity, V. The velocity profile induced in the fluid by drag from the plate will "penetrate" into the fluid with time in the general fashion shown in Figure 6.4.2-6.

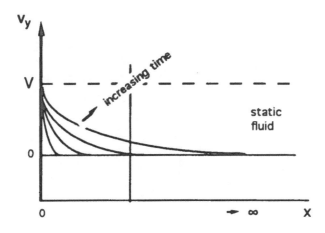

Figure 6.4.2-6 Flow adjacent to flat plat instantaneously set in motion

The partial differential equation describing the momentum transfer, which is the microscopic momentum balance, is

$$\frac{\partial v_y}{\partial t} = v \frac{\partial^2 v_y}{\partial x^2}$$

(6.4.2-45)

has initial and boundary conditions

$t = 0$: $v_y = 0$, all x   (initial condition)
$x = 0$: $v_y = V$, all $t > 0$ (boundary condition)
$x = \infty$: $v_y = 0$, all $t > 0$ (boundary condition)

(6.4.2-46)

De-dimensionalize the dependent variable in the equation using

$$v_y^* = \frac{v_y - v_i}{V - v_i} = \frac{v_y - 0}{V - 0} = \frac{v_y}{V}$$

(6.4.2-47)

Treating individual factors

$$\frac{\partial v_y}{\partial t} = \frac{\partial v_y}{\partial v_y^*} \frac{\partial v_y^*}{\partial t} = V \frac{\partial v_y^*}{\partial t}$$

(6.4.2-48)

$$\frac{\partial^2 v_y}{\partial x^2} = \frac{\partial}{\partial x}\left[\frac{\partial v_y}{\partial x}\right] = \frac{\partial}{\partial x}\left[\frac{\partial v_y}{\partial v_y^*} \frac{\partial v_y^*}{\partial x}\right] = V \frac{\partial^2 v_y^*}{\partial x^2}$$

(6.4.2-49)

Substituting in the differential equation

$$V \frac{\partial v_y^*}{\partial t} = V v \frac{\partial^2 v_y^*}{\partial x^2}$$

(6.4.2-50)

$$\frac{\partial v_y^*}{\partial t} = v \frac{\partial^2 v_y^*}{\partial x^2}$$

(6.4.2-51)

with transformed initial and boundary conditions

$t = 0$: $v_y^* = 0$, all x   (initial condition)
$x = 0$: $v_y^* = 1$, all $t > 0$ (boundary condition)
$x = \infty$: $v_y^* = 0$, all $t > 0$ (boundary condition)

(6.4.2-52)

This is the analogous problem to that of temperature profiles in a semi-infinite slab initially at uniform temperature whose face is suddenly increased to a different temperature, and to that of concentration profiles resulting from certain binary diffusion processes in a semi-infinite medium. The dimensionless velocity, v_y^*, is analogous to the dimensionless temperature defined in Chapter 7

$$T^* = \frac{T - T_i}{T_s - T_i}$$

(6.4.2-53)

or the dimensionless concentration defined in Chapter 8

$$c_A^* = \frac{c_A - c_{Ai}}{c_{As} - c_{Ai}}$$

(6.4.2-54)

and the kinematic viscosity (momentum diffusivity) is analogous to the thermal diffusivity, α, or the binary diffusivity, $\mathcal{D}_{AB}$.

The solution to the differential equation is

$$v_y^* = f_1(x, t, v)$$

(6.4.2-55)

and can be written as

$$f_2(v_y^*, x, t, v) = v_y^* - f_1(x, t, v) = 0$$

(6.4.2-56)

where we observe that $f_2(v_y^*, x, t, v)$ must be a dimensionally homogeneous function.

Applying the principles of dimensional analysis in a system of two fundamental dimensions (L, t) indicates that we should obtain two dimensionless groups. One of our variables, v_y^*, is already dimensionless and therefore serves as the first member of the group of two. A systematic analysis such as we developed in Chapter 5 will give the other dimensionless group as the similarity variable

$$\eta = \frac{x}{\sqrt{4 v t}}$$

(6.4.2-57)

which implies that

$$f_3(v_y^*, \eta) = 0$$

(6.4.2-58)

or that

$$v_y^* = \varphi(\eta) \tag{6.4.2-59}$$

Transforming the terms in the differential equation we have

$$
\begin{aligned}
\frac{\partial v_y^*}{\partial t} &= \frac{dv_y^*}{d\eta}\frac{\partial \eta}{\partial t} = \frac{d[\varphi(\eta)]}{d\eta}\frac{\partial\left[\frac{x}{\sqrt{4\,v\,t}}\right]}{\partial t} \\
&= \varphi'\left(\frac{x}{\sqrt{4\,v}}\right)\left(-\tfrac{1}{2}t^{-\frac{3}{2}}\right) = \varphi'\left(\frac{x}{\sqrt{4\,v\,t}}\right)\left(-\tfrac{1}{2t}\right) \\
&= -\tfrac{1}{2}\frac{\eta}{t}\varphi'
\end{aligned}
\tag{6.4.2-60}
$$

$$
\begin{aligned}
\frac{\partial^2 v_y^*}{\partial x^2} &= \frac{\partial}{\partial x}\frac{\partial v_y^*}{\partial x} = \frac{\partial}{\partial x}\frac{\partial[\varphi(\eta)]}{dx} = \frac{\partial \eta}{\partial x}\frac{d}{d\eta}\left\{\frac{d[\varphi(\eta)]}{d\eta}\frac{\partial \eta}{\partial x}\right\} \\
&= \frac{\partial\left[\frac{x}{\sqrt{4\,v\,t}}\right]}{\partial x}\frac{d}{d\eta}\left\{\frac{d[\varphi(\eta)]}{d\eta}\frac{\partial\left[\frac{x}{\sqrt{4\,v\,t}}\right]}{\partial x}\right\} \\
&= \frac{1}{\sqrt{4\,v\,t}}\frac{d}{d\eta}\left\{\frac{d[\varphi(\eta)]}{d\eta}\frac{1}{\sqrt{4\,v\,t}}\right\} \\
&= \frac{1}{\sqrt{4\,v\,t}}\frac{1}{\sqrt{4\,v\,t}}\frac{d}{d\eta}\left\{\frac{d[\varphi(\eta)]}{d\eta}\right\} = \frac{1}{4\,v\,t}\varphi''
\end{aligned}
\tag{6.4.2-61}
$$

Substituting in the differential equation

$$-\tfrac{1}{2}\frac{\eta}{t}\varphi' = v\left(\frac{1}{4\,v\,t}\varphi''\right) \tag{6.4.2-62}$$

Rearranging

$$\varphi'' + 2\,\eta\,\varphi' = 0 \tag{6.4.2-63}$$

which leaves us with an ordinary differential equation, not a partial differential equation as we originally had, because of the combining of two independent variables into one via the similarity transformation.

The initial and boundary conditions transform as

$$t = 0 \Rightarrow \eta = \infty: \quad v_y^* = 0 \Rightarrow \varphi(\eta) = 0$$
$$x = 0 \Rightarrow \eta = 0: \quad v_y^* = 1 \Rightarrow \varphi(\eta) = 1 \qquad (6.4.2\text{-}64)$$
$$x = \infty \Rightarrow \eta = \infty: \quad v_y^* = 0 \Rightarrow \varphi(\eta) = 0$$

which gives the transformed boundary conditions as

$$\eta = \infty: \quad \varphi(\eta) = 0$$
$$\eta = 0: \quad \varphi(\eta) = 1 \qquad (6.4.2\text{-}65)$$

This equation is easily solved because it can be seen to be **separable** via the simple expedient of substituting ψ for φ'.

$$\psi' + 2\eta\psi = 0 \qquad (6.4.2\text{-}66)$$

$$\frac{d\psi}{d\eta} + 2\eta\psi = 0 \qquad (6.4.2\text{-}67)$$

$$\int \frac{d\psi}{\psi} = -\int 2\eta\,d\eta \qquad (6.4.2\text{-}68)$$

$$\ln \psi = -\eta^2 + C_1 \qquad (6.4.2\text{-}69)$$

$$\psi = C_1 e^{-\eta^2} \qquad (6.4.2\text{-}70)$$

Back substituting for ψ

$$\varphi' = C_1 e^{-\eta^2}$$
$$\int \varphi'\,d\eta = C_1 \int e^{-\eta^2}\,d\eta$$
$$\varphi = C_1 \int_0^\eta e^{-\zeta^2}\,d\zeta + C_2 \qquad (6.4.2\text{-}71)$$

The integral here is one that we cannot evaluate in closed form. We choose to base our tables of numerical evaluation of the form on a lower limit of zero for the integral. This arbitrary choice is permissible because changing the basis point for the integration from one point to another will affect only the value of the constant, C_2.

Introducing the boundary conditions and substituting the appropriate value for the definite integrals[16]

$$1 = C_1 \int_0^0 e^{-\zeta^2} d\zeta + C_2 \quad \Rightarrow \quad C_2 = 1 - C_1(0) = 1$$

$$0 = C_1 \int_0^\infty e^{-\zeta^2} d\zeta + 1 \quad \Rightarrow \quad C_2 = -\frac{1}{\int_0^\infty e^{-\zeta^2} d\zeta} = -\frac{2}{\sqrt{\pi}} \qquad (6.4.2\text{-}72)$$

Substituting the constants in the equation and re-ordering the terms gives

$$\varphi = 1 - \frac{2}{\sqrt{\pi}} \int_0^\eta e^{-\zeta^2} d\zeta \qquad (6.4.2\text{-}73)$$

We define the *error function* and the *complementary error function*

$$\boxed{\begin{aligned} \text{error function} &= \text{erf}\,(\eta) \equiv \frac{2}{\sqrt{\pi}} \int_0^\eta e^{-\zeta^2} d\zeta \\ \text{complementary error function} &= \text{erfc}(\eta) \equiv 1 - \text{erf}\,(\eta) \end{aligned}} \qquad (6.4.2\text{-}74)$$

which allows us to write[17]

$$\begin{aligned} \varphi &= 1 - \text{erf}\left[\eta\right] \\ &= \text{erfc}\left[\eta\right] \end{aligned} \qquad (6.4.2\text{-}75)$$

so the solution of our problem is

$$\begin{aligned} v_y^* &= 1 - \text{erf}\left[\frac{x}{\sqrt{4\nu t}}\right] \\ &= \text{erfc}\left[\frac{x}{\sqrt{4\nu t}}\right] \end{aligned} \qquad (6.4.2\text{-}76)$$

---

[16] The error function plays a key role in statistics. We do not have space here to show the details of the integration of the function to obtain the tables listed; for further details the reader is referred to any of the many texts in mathematical statistics.

[17] Engineers have a certain regrettable proclivity to use the same symbol for the dummy variable of integration that appears in the integrand and for the variable in the limits. This often does not cause confusion; here, however, it is clearer to make the distinction.

A table of values of the error function is listed in Appendix B.

6.5   Turbulent Flow

Both laminar and turbulent flow are characterized by a pressure loss caused by momentum transfer from the fluid to containing walls or to solid objects in the flow stream. There are two principal mechanisms by which momentum is transferred from one location to another within a flowing fluid.

The dominant force in laminar flow is the viscous force operating at the molecular level. In turbulent flow, the mechanism of momentum transfer is mainly the migration of relatively large aggregates of fluid which preserve their identity for some small but finite time as they move from one point to another (eddies). The dominant forces in turbulent flow are inertial forces. These inertial forces arise from the rapid acceleration and deceleration of the fluid due to the velocity fluctuations as the eddies move from regions of one velocity to regions of a different velocity and mix with the new region.

The Reynolds number is the dimensionless product that represents the ratio of inertial to viscous forces. As the velocity gradients in a fluid increase because the fluid moves faster relative to solid boundaries, the inertial forces become larger relative to the viscous forces, and eventually flow shifts from laminar (viscous-dominated) to turbulent (inertial-dominated) flow.

The transition occurs at about Re = 2100 for flow in a pipe. For a flat plate the transition occurs at about $Re_x = 3 \times 10^6$. Note that the characteristic length in Re_x is the distance from the nose (front) of the plate and the characteristic velocity is the freestream velocity. It is, therefore, not surprising that transition occurs at a different numerical value than in a pipe - in addition, one is external flow where the velocity profile continues to develop and the other is internal flow which reaches a fully developed state under steady-flow conditions.

In turbulent flow, the velocity at a point varies in time stochastically both in magnitude and in direction. If the velocity in a particular direction is measured at a point in a fluid in turbulent flow, a result similar to that shown in Figure 6.5-1 is obtained.

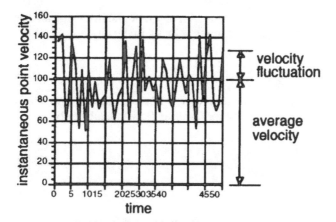

Figure 6.5-1 Local velocity in turbulent flow as a function of time

Although we can measure the (virtually) instantaneous velocity in turbulent flow,[18] we cannot, in general, solve the basic differential equation models describing such flows for these instantaneous velocities. We resort instead to models of **time averaged** velocities, which we can treat in some cases.

The average velocity used with turbulent flow is a time-averaged velocity defined as

$$\bar{v} = \frac{\int_0^{t_0} v\, dt}{\int_0^{t_0} dt} = \frac{\int_0^t v\, dt}{t_0} \tag{6.5-1}$$

---

[18] Such measurements are sometimes made in gases with hot-wire anemometers - very fine short wires through which electrical current is passed and which are cooled by the passing gas stream. The cooling effect of the gas changes the temperature and therefore the electrical resistance of the wire, which can be translated into fluid velocity. Such wires can be run at constant voltage and the fluctuation in current measured, or vice versa. Liquids, because of their larger drag, would destroy a wire, so conductive films deposited on a non-conducting substrate are used in a similar fashion.

Wires oriented at an angle to each other can be used to measure more than one fluctuating component at the same time.

In Doppler anemometry, the Doppler effect is used to determine velocity by examining change in frequency of light signals reflected from particles in the fluid.

where v is the instantaneous velocity. At any time the instantaneous velocity is the sum of the average velocity and the velocity fluctuation.

$$v = \bar{v} + v'$$

(6.5-2)

In Equations (6.4.2-11) and (6.4.2-12) we above saw the velocity distribution and average velocity for **laminar** flow in a tube to be

$$v_z = v_{zmax}\left[1 - \left(\frac{r}{R}\right)^2\right]$$

(6.5-3)

$$\langle v_z \rangle = \frac{1}{2} v_{zmax}$$

(6.5-4)

and the pressure drop to be directly proportional to the volume rate of flow.

For **turbulent** flow in pipes, experiments show that the distribution of the time-averaged velocity is approximately

$$\bar{v}_z \cong \bar{v}_{zmax}\left[1 - \frac{r}{R}\right]^{\frac{1}{7}}$$

(6.5-5)

and the average velocity (as shown above in Example 1.5.2-3) to be approximately

$$\bar{v}_z \cong \frac{4}{5} \bar{v}_{zmax}$$

(6.5-6)

Figure 6.5-2 shows a qualitative comparison of the laminar and turbulent velocity profiles in a pipe.

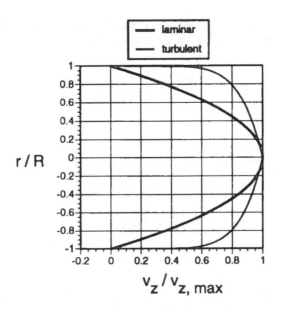

Figure 6.5-2 Laminar and time-smoothed turbulent (1/7 power model) velocity profiles in steady pipe flow

6.5.1 Time averaging the equations of change

Let us now examine the result of time averaging some typical equations that are frequently encountered. First, starting with the continuity equation in rectangular coordinates and restricting it to **an incompressible fluid**

$$\frac{\partial \rho}{\partial t} + \frac{\partial(\rho v_x)}{\partial x} + \frac{\partial(\rho v_y)}{\partial y} + \frac{\partial(\rho v_z)}{\partial z} = 0$$

$$0 + \rho\left(\frac{\partial v_x}{\partial x} + \frac{\partial v_y}{\partial y} + \frac{\partial v_z}{\partial z}\right) = 0 \tag{6.5.1-1}$$

$$\frac{\partial v_x}{\partial x} + \frac{\partial v_y}{\partial y} + \frac{\partial v_z}{\partial z} = 0 \tag{6.5.1-2}$$

Substituting using Equation (6.5-2) gives

$$\frac{\partial(\overline{v}_x + v'_x)}{\partial x} + \frac{\partial(\overline{v}_y + v'_y)}{\partial y} + \frac{\partial(\overline{v}_z + v'_z)}{\partial z} = 0 \tag{6.5.1-3}$$

Taking the time average each side of the equation, and writing the integral on the left-hand side as the sum of the three integrals as shown

$$\frac{1}{t_0}\int_0^{t_0}\frac{\partial\left(\overline{v}_x + v_x'\right)}{\partial x}\,dt + \frac{1}{t_0}\int_0^{t_0}\frac{\partial\left(\overline{v}_y + v_y'\right)}{\partial y}\,dt + \frac{1}{t_0}\int_0^{t_0}\frac{\partial\left(\overline{v}_z + v_z'\right)}{\partial z}\,dt = \frac{1}{t_0}\int_0^{t_0}0\,dt$$

(6.5.1-4)

Writing the result for a typical term on the left-hand side and applying the Leibnitz rule

$$\frac{1}{t_0}\int_0^{t_0}\frac{\partial\left(\overline{v}_y + v_y'\right)}{\partial y}\,dt = \frac{1}{t_0}\frac{\partial}{\partial y}\int_0^{t_0}\left(\overline{v}_y + v_y'\right)dt$$

(6.5.1-5)

$$\frac{1}{t_0}\frac{\partial}{\partial y}\int_0^{t_0}\overline{v}_y\,dt + \frac{1}{t_0}\frac{\partial}{\partial y}\int_0^{t_0}v_y'\,dt$$

$$= \frac{1}{t_0}\frac{\partial\overline{v}_y}{\partial y} + \frac{1}{t_0}\frac{\partial\overline{v}_y'}{\partial y} = \frac{1}{t_0}\frac{\partial\overline{v}_y}{\partial y} + 0 = \frac{1}{t_0}\frac{\partial\overline{v}_y}{\partial y}$$

(6.5.1-6)

Applying the result to the whole equation

$$\frac{1}{t_0}\frac{\partial\overline{v}_x}{\partial x} + \frac{1}{t_0}\frac{\partial\overline{v}_y}{\partial y} + \frac{1}{t_0}\frac{\partial\overline{v}_z}{\partial z} = 0$$

(6.5.1-7)

$$\frac{\partial\overline{v}_x}{\partial x} + \frac{\partial\overline{v}_y}{\partial y} + \frac{\partial\overline{v}_z}{\partial z} = 0$$

(6.5.1-8)

so we find that for an incompressible fluid, the time-averaged equation is identical in form to the instantaneous equation, but with velocities replaced by their respective time averages. This is not true in general.

Next, as a typical example, examine in the same fashion the z-component of the equation of motion in rectangular coordinates for a Newtonian fluid with constant density and viscosity, observing that the pressure at any time can be written as the time average plus the fluctuation (just as with the velocity)

$$p = \overline{p} + p'$$

(6.5.1-9)

The z-component of the equation of motion is

$$\rho \left(\frac{\partial v_z}{\partial t} + v_x \frac{\partial v_z}{\partial x} + v_y \frac{\partial v_z}{\partial y} + v_z \frac{\partial v_z}{\partial z} \right)$$

$$= -\frac{\partial p}{\partial z} + \mu \left(\frac{\partial^2 v_z}{\partial x^2} + \frac{\partial^2 v_z}{\partial y^2} + \frac{\partial^2 v_z}{\partial z^2} \right) + \rho\, g_z \tag{6.5.1-10}$$

If we multiply the continuity equation for an incompressible fluid, Equation (6.5.1-2), by ρ and v_z, and add it to Equation (6.5.1-10), the right-hand side will remain unchanged and the left-hand side will become

$$\rho \left(\frac{\partial v_z}{\partial t} + v_x \frac{\partial v_z}{\partial x} + v_y \frac{\partial v_z}{\partial y} + v_z \frac{\partial v_z}{\partial z} \right) + \rho\, v_z \left(\frac{\partial v_x}{\partial x} + \frac{\partial v_y}{\partial y} + \frac{\partial v_z}{\partial z} \right)$$

$$= \rho \left[\frac{\partial v_z}{\partial t} + \left(v_x \frac{\partial v_z}{\partial x} + v_z \frac{\partial v_x}{\partial x} \right) + \left(v_y \frac{\partial v_z}{\partial y} + v_z \frac{\partial v_y}{\partial y} \right) + \left(v_z \frac{\partial v_z}{\partial z} + v_z \frac{\partial v_z}{\partial z} \right) \right]$$

$$\tag{6.5.1-11}$$

$$= \rho \left[\frac{\partial v_z}{\partial t} + \frac{\partial (v_x v_z)}{\partial x} + \frac{\partial (v_y v_z)}{\partial y} + \frac{\partial (v_z v_z)}{\partial z} \right] \tag{6.5.1-12}$$

which, upon substitution in Equation (6.5.1-10), gives

$$\rho \left[\frac{\partial v_z}{\partial t} + \frac{\partial (v_x v_z)}{\partial x} + \frac{\partial (v_y v_z)}{\partial y} + \frac{\partial (v_z v_z)}{\partial z} \right]$$

$$= -\frac{\partial p}{\partial z} + \mu \left(\frac{\partial^2 v_z}{\partial x^2} + \frac{\partial^2 v_z}{\partial y^2} + \frac{\partial^2 v_z}{\partial z^2} \right) + \rho\, g_z \tag{6.5.1-13}$$

Time averaging a typical term on the left-hand side of Equation (6.5.1-13) as an example, remembering that time-averaged velocities by definition are constant with respect to integration over time, and that likewise the time average of velocity fluctuations is zero yields

$$\frac{1}{t_0} \int_0^{t_0} \rho \frac{\partial (v_x v_z)}{\partial x}\, dt = \frac{\rho}{t_0} \int_0^{t_0} \frac{\partial \left[\left(\bar{v}_x + v_x' \right) \left(\bar{v}_z + v_z' \right) \right]}{\partial x}\, dt \tag{6.5.1-14}$$

$$= \frac{\rho}{t_0} \int_0^{t_0} \frac{\partial\left[\left(\overline{v}_x + v_x'\right)\left(\overline{v}_z + v_z'\right)\right]}{\partial x} dt = \rho \frac{\partial}{\partial x} \frac{1}{t_0} \int_0^{t_0} \left[\left(\overline{v}_x + v_x'\right)\left(\overline{v}_z + v_z'\right)\right] dt$$

$$(6.5.1\text{-}15)$$

$$= \rho \frac{\partial}{\partial x} \frac{1}{t_0} \int_0^{t_0} \left(\overline{v}_x \overline{v}_z + \overline{v}_x v_z' + v_x' \overline{v}_z + v_x' v_z'\right) dt \qquad (6.5.1\text{-}16)$$

$$= \rho \frac{\partial}{\partial x} \frac{1}{t_0} \int_0^{t_0} \left(\overline{v}_x \overline{v}_z\right) dt + \rho \frac{\partial}{\partial x} \frac{1}{t_0} \int_0^{t_0} \left(\overline{v}_x v_z'\right) dt$$

$$+ \rho \frac{\partial}{\partial x} \frac{1}{t_0} \int_0^{t_0} \left(v_x' \overline{v}_z\right) dt + \rho \frac{\partial}{\partial x} \frac{1}{t_0} \int_0^{t_0} \left(v_x' v_z'\right) dt$$

$$(6.5.1\text{-}17)$$

$$= \rho \frac{\partial}{\partial x} \left[\overline{v}_x \overline{v}_z \left(\frac{1}{t_0} \int_0^{t_0} dt\right)\right] + \rho \frac{\partial}{\partial x} \left[\overline{v}_x \left(\frac{1}{t_0} \int_0^{t_0} v_z' dt\right)\right]$$

$$+ \rho \frac{\partial}{\partial x} \left[\overline{v}_z \left(\frac{1}{t_0} \int_0^{t_0} v_x' dt\right) + \rho \frac{\partial}{\partial x} \left(\frac{1}{t_0} \int_0^{t_0} v_x' v_z' dt\right)\right]$$

$$(6.5.1\text{-}18)$$

$$= \rho \frac{\partial}{\partial x}\left(\overline{v}_x \overline{v}_z\right) + \rho \frac{\partial}{\partial x}\left(\overline{v}_x \overline{v_z'}\right) + \frac{\partial}{\partial x}\left(\overline{v}_z \overline{v_x'}\right) + \rho \frac{\partial}{\partial x}\left(\overline{v_x' v_z'}\right) \qquad (6.5.1\text{-}19)$$

$$= \rho \frac{\partial}{\partial x}\left(\overline{v}_x \overline{v}_z\right) + \rho \frac{\partial}{\partial x}\left(\overline{v_x' v_z'}\right) \qquad (6.5.1\text{-}20)$$

In this instance, as opposed to the case of the continuity equation, we see that we obtain an "extra" term (enclosed in the box) from the time averaging process - in other words, we cannot obtain the correct result by formally replacing time average of the velocity product by the product of the time-averaged velocities.

$$\frac{\rho}{t_0} \frac{\partial}{\partial x} \int_0^{t_0} \left(v_x v_z\right) dt = \rho \frac{\partial}{\partial x}\left(\overline{v}_x \overline{v}_z\right) + \boxed{\rho \frac{\partial}{\partial x}\left(\overline{v_x' v_z'}\right)} \neq \rho \frac{\partial}{\partial x}\left(\overline{v}_x \overline{v}_z\right)$$

$$(6.5.1\text{-}21)$$

Notice that the units on this sort of term can be interpreted as either momentum flux or shear stress

$$\left[\frac{M}{L^3}\right]\left[\frac{L}{t}\right]\left[\frac{L}{t}\right] = \frac{M}{Lt} = \frac{\left[M\frac{L}{t}\right]}{L^2 t} = \frac{\left[M\frac{L}{t^2}\right]}{L^2}$$

momentum   shear
flux   stress

(6.5.1-22)

The same sort of result is obtained from the second, third, and fourth terms on the left-hand side of Equation (6.5.1-13).

Time averaging the first term on the left-hand side of Equation (6.5.1-13) gives

$$\frac{1}{t_0}\int_0^{t_0} \rho \frac{\partial v_z}{\partial t} dt = \rho \frac{1}{t_0} \frac{\partial}{\partial t} \int_0^{t_0} v_z dt = \rho \frac{\partial \bar{v}_z}{\partial t} \tag{6.5.1-23}$$

For this term the formal substitution gives the correct result.

Time averaging the first term on the right-hand side of Equation (6.5.1-13) gives

$$-\frac{1}{t_0}\int_0^{t_0} \frac{\partial p}{\partial z} dt = -\frac{1}{t_0}\int_0^{t_0} \frac{\partial(\bar{p}+p')}{\partial z} dt = -\frac{\partial}{\partial z}\left[\frac{1}{t_0}\int_0^{t_0} (\bar{p}+p') dt\right]$$

$$-\frac{\partial}{\partial z}\left[\frac{1}{t_0}\int_0^{t_0} \bar{p}\, dt\right] - \frac{\partial}{\partial z}\left[\frac{1}{t_0}\int_0^{t_0} p'\, dt\right] = -\frac{\partial(\bar{p}+\bar{p'})}{\partial z} = -\frac{\partial(\bar{p}+0)}{\partial z} = -\frac{\partial \bar{p}}{\partial z}$$

(6.5.1-24)

Again, formal substitution of the average for the instantaneous value gives the same result as time averaging.

Time averaging a typical term from the second, third, and fourth terms on the right-hand side of Equation (6.5.1-13) and applying the Leibnitz rule twice gives

$$\frac{1}{t_0} \int_?^? \mu \frac{\partial^2 v_z}{\partial y^2} \, dt = \mu \frac{1}{t_0} \int_?^? \left[\frac{\partial}{\partial y} \left(\frac{\partial}{\partial y} v_z \right) \right] dt$$

$$= \mu \frac{\partial}{\partial y} \left[\frac{\partial}{\partial y} \left(\frac{1}{t_0} \int_?^? \mu \frac{\partial^2 v_z}{\partial y^2} \, dt \right) \right] = \mu \frac{\partial^2 \bar{v}_z}{\partial y^2}$$

(6.5.1-25)

Again, the formal substitution gives the same result as the time-averaging process.

Time averaging of the last term on the right-hand side returns the last term unchanged, because we have assumed density to be constant and we do not consider time-changing gravitational fields; therefore, this term is constant in time.

$$\frac{1}{t_0} \int_0^{t_0} \rho \, g_z \, dt = \rho \, g_z \frac{1}{t_0} \int_0^{t_0} dt = \rho \, g_z$$

(6.5.1-26)

Using the above results and Equation (6.5.1-13) gives for the time-averaged equation for the z-component (similar equations exist for the other directions)

$$\rho \left[\frac{\partial \bar{v}_z}{\partial t} + \frac{\partial}{\partial x} (\bar{v}_x \bar{v}_z) + \frac{\partial}{\partial y} (\bar{v}_y \bar{v}_z) + \frac{\partial}{\partial z} (\bar{v}_z \bar{v}_z) \right]$$

$$+ \boxed{\rho \left[\frac{\partial}{\partial x} (\overline{v'_x v'_z}) + \frac{\partial}{\partial y} (\overline{v'_y v'_z}) + \frac{\partial}{\partial z} (\overline{v'_z v'_z}) \right]}$$

$$= -\frac{\partial \bar{p}}{\partial z} + \mu \left[\frac{\partial^2 \bar{v}_z}{\partial x^2} + \frac{\partial^2 \bar{v}_z}{\partial y^2} + \frac{\partial^2 \bar{v}_z}{\partial z^2} \right] + \rho \, g_z$$

(6.5.1-27)

where again the terms in the box are "extra" terms when compared with Equation (6.5.1-13), and these terms have the equivalent units of either momentum flux or shear stress. Interpreting the units as those of shear stress has led to these terms being commonly called the "Reynolds stresses" (in honor of Osborne Reynolds of the Reynolds number), even though they are made up of velocity fluctuations and therefore intuitively share more of the character of momentum flux.

Example 6.5-2   Time averaging of velocity product

Time average the term

$$v_z v_y'$$ (6.5.1-28)

Solution

$$\overline{v_z v_y'} = \frac{1}{t_0} \int_0^{t_0} \left(v_z v_y' \right) dt$$ (6.5.1-29)

$$= \frac{1}{t_0} \int_0^{t_0} \left(\overline{v}_z + v_z' \right) \left(v_y' \right) dt$$ (6.5.1-30)

$$= \frac{1}{t_0} \int_0^{t_0} \overline{v}_z v_y' dt + \frac{1}{t_0} \int_0^{t_0} v_z' v_y' dt$$ (6.5.1-31)

$$= \frac{\overline{v}_z}{t_0} \int_0^{t_0} v_y' dt + \frac{1}{t_0} \int_0^{t_0} v_z' v_y' dt$$ (6.5.1-32)

$$= \frac{\overline{v}_z}{t_0} \left(0 \right) + \overline{v_z' v_y'}$$ (6.5.1-33)

$$= \overline{v_z' v_y'}$$ (6.5.1-34)

6.5.2   The mixing length model

A model equation for velocity distribution in turbulent flow in a tube can be developed by assuming the shear stress to be proportional to the fluctuating velocities v_y' and v_z'. This model was originally formulated by Prandtl[19] who assumed a "particle" of fluid to be displaced a distance, ℓ (called the **mixing length**), before its momentum was changed by the new environment, as shown in Figure 6.5.2-1.

Prandtl related the velocity fluctuations to the time-averaged velocity gradient as

$$v_z' = \ell \frac{d\overline{v}_z}{dy}$$ (6.5.2-1)

---

[19] Prandtl, L. (1925). *ZAMM* **5**: 136.

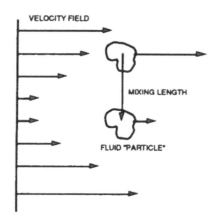

Figure 6.5.2-1 Mixing length model

assuming v_y' is approximately equal to v_z'. If the shear stress in the fluid is assumed to retain the form of Newton's law of viscosity, then

$$\tau_{yz} = -(\mu + \eta)\frac{dv_z}{dy} = \tau_{yz}^l + \tau_{yz}^t \tag{6.5.2-2}$$

where μ is the fluid viscosity and η is called the eddy viscosity and

$$\tau_{yz}^t = \overline{\rho\, v_y'\, v_z'} \tag{6.5.2-3}$$

This quantity is called the Reynolds **stress** even though it is the product of two velocities and a density, because it has the dimensions of stress. Examining the units, one sees it has units of stress which are the same as momentum flux. Therefore,

$$\tau_{yz}^t = \rho\, \ell^2 \left|\frac{dv_z}{dy}\right| \frac{dv_z}{dy} \tag{6.5.2-4}$$

where ℓ is the mixing length and

$$\eta = \rho\, \ell^2 \left|\frac{dv_z}{dy}\right| \tag{6.5.2-5}$$

Note that η is not a fluid property - it depends on the flow field, and is defined in such a way that it is related to the shear stress. As turbulence decreases, Equation (6.5.2-2) approaches Newton's law of viscosity, because η approaches zero. For turbulent flow past a boundary, η is equal to zero at the surface, while μ is negligible at a distance from the surface that is large when compared to δ, the thickness of the viscous sublayer.

(In earlier literature one may find references to the **laminar** sublayer, but this usage was found not to be correct, because turbulent eddies can be photographed penetrating this layer and impacting the wall. In this layer, however, viscous forces do predominate.)

The concept of mixing length, together with the time-averaged velocity, may be used to obtain an expression for **average** velocity during steady-state turbulent flow in a pipe. Close to the wall, where η is negligible, the shear stress is assumed to be given by Equation (6.5.2-2), where the thickness for which μ dominates is δ, the thickness of the viscous sublayer. The shear stress at the wall is

$$\tau_0 = -\mu \frac{d\bar{v}_z}{dy}\bigg|_{y=0} \qquad (6.5.2\text{-}6)$$

or

$$\frac{\tau_0}{\rho} = -\nu \frac{d\bar{v}_z}{dy}\bigg|_{y=0}; \quad y \leq \delta \qquad (6.5.2\text{-}7)$$

In the latter equation, $\nu = \mu/\rho$ (the kinematic viscosity). We assume that τ is equal to τ_0 throughout the viscous sublayer and that δ is much less than R, the radius of the tube; therefore, we can use a rectangular coordinate.

Defining a shear velocity (the **friction velocity**)

$$v_* \equiv \sqrt{\frac{\tau_0}{\rho}} = \sqrt{\left|\overline{u'\,v'}\right|} \qquad (6.5.2\text{-}8)$$

which measures the intensity of eddies in two-dimensional flow and the consequent momentum transfer, Equation (6.5.2-8) may be written in the following form:

$$\frac{\bar{v}_z}{v_*} = \frac{v_* y}{\nu}; \quad y \leq \delta \qquad (6.5.2\text{-}9)$$

For y > δ, μ is assumed to be much less than η and

$$\tau_{yz} = \rho \, \ell^2 \left| \frac{d\vartheta_z}{dy} \right| \frac{d\vartheta_z}{dy} \qquad (6.5.2\text{-}10)$$

Since ℓ has dimensions of length, we may assume as a simple model that ℓ = Ky and substitute into Equation (6.5.2-10)

$$\frac{d\vartheta_z}{\vartheta_*} = \frac{1}{K} \frac{dy}{y} \qquad (6.5.2\text{-}11)$$

Assuming the shear stress to be constant at τ_0 throughout the turbulent core, and integrating Equation (6.5.2-11)

$$\frac{\vartheta_z}{\vartheta_*} = \frac{1}{K} \ln \left[y \right] + \text{constant} \qquad (6.5.2\text{-}12)$$

or

$$\frac{\vartheta_z}{\vartheta_*} = \frac{1}{K} \ln \left[\frac{y}{y_0} \right] = \frac{1}{K} \left(\ln \left[\frac{y \, \vartheta_*}{\nu} \right] - \ln \left[\frac{y_0 \, \vartheta_*}{\nu} \right] \right) \qquad (6.5.2\text{-}13)$$

The constant of integration may be determined by assuming that v_z is 0 at some value of y, say y_0, (but not at y = 0 since log[0] = -∞).

$$\frac{\vartheta_z}{\vartheta_*} = \frac{1}{K} \ln \left[\frac{y \, \vartheta_*}{\nu} \right] + C \qquad (6.5.2\text{-}14)$$

where C is 0 in the viscous sublayer, and is determined from experiment for the buffer zone and the turbulent core. The logarithmic model (as opposed to the 1/7 power law model) expressed by Equation (6.5.2-14) is called the **universal velocity distribution**. Data on turbulent flow in pipes with 4000 < Re < 3.2 x 10⁶ may be fit by the following forms.

$$\frac{\vartheta_z}{\vartheta_*} = \frac{y \, \vartheta_*}{\nu}; \quad 0 < \frac{y \, \vartheta_*}{\nu} < 5 \quad \text{(viscous sublayer)} \qquad (6.5.2\text{-}15)$$

$$\frac{\vartheta_z}{\vartheta_*} = 5 \ln \left[\frac{y \, \vartheta_*}{\nu} \right] - 3.05; \quad 5 < \frac{y \, \vartheta_*}{\nu} < 30 \quad \text{(buffer zone)} \qquad (6.5.2\text{-}16)$$

$$\frac{\vartheta_z}{\vartheta_*} = 2.5 \ln \left[\frac{y \, \vartheta_*}{\nu} \right] + 5.5; \quad 30 < \frac{y \, \vartheta_*}{\nu} \quad \text{(turbulent core)} \qquad (6.5.2\text{-}17)$$

Figure 6.5.2-2 is a plot of this model, where v^+ and y^+ are, respectively, a dimensionless velocity and a dimensionless distance from the wall[20] defined using the friction velocity in the following fashion

$$v^+ \equiv \frac{v}{v_*}$$

(6.5.2-18)

and

$$y^+ \equiv \frac{y\, v_*}{v}$$

(6.5.2-19)

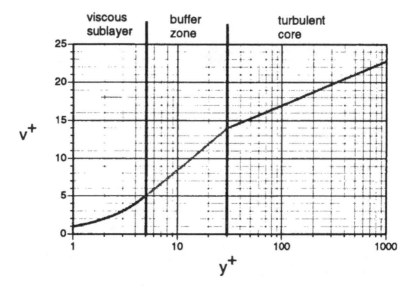

Figure 6.5.2-2 Universal velocity distribution

Example 6.5.2-1   Size of sublayer and buffer zone in turbulent flow

Water is flowing in a 1-in. Schedule 40 steel pipe at a bulk velocity of 5 ft/s. If the pressure drop over 100 ft of pipe is 30 psig, what is the thickness of the viscous sublayer and the buffer zone?

---

[20] Note that y^+ is a Reynolds number based on distance from the wall.

Solution

$$Re = \frac{D\langle v\rangle \rho}{\mu} = \frac{\left(\frac{1.049}{12}\right)[ft](5)\left[\frac{ft}{s}\right](62.4)\left[\frac{lbm}{ft^3}\right]}{(1.13)[cP]\left(6.72\times 10^{-4}\right)\left[\frac{lbm}{ft\,s\,cP}\right]}$$

$$= 35,900 \tag{6.5.2-20}$$

Sublayer

$$\frac{v_x}{v^*} = \frac{y\,v^*}{v} = 5$$

$$y = \frac{5\,v}{v^*} = \frac{5\,\mu}{\rho}\sqrt{\frac{\rho}{\tau_o}} = \frac{5\,\mu}{\sqrt{\rho\,\tau_o}} \tag{6.5.2-21}$$

Force balance on fluid

$$\tau_o\,\pi\,D\,L = -\Delta p\,\frac{\pi\,D^2}{4} \tag{6.5.2-22}$$

$$y = 5\,\mu\sqrt{\frac{4L}{\rho\,D\left(-\Delta p\right)}} = 10\,\mu\sqrt{\frac{L}{\rho\,D\left(-\Delta p\right)}} \tag{6.5.2-23}$$

$$y = (10)(1.13)[cP]\left(6.72\times 10^{-4}\right)\left[\frac{lbm}{cP\,ft\,sec}\right]$$

$$\times \left\{\frac{(100)[ft]}{(62.4)\left[\frac{lbm}{ft^3}\right]\left(\frac{1.049}{12}\right)[ft](32.2)\left[\frac{lbm\,ft}{lbf\,sec^2}\right](30)(144)\left[\frac{lbf}{ft^2}\right]}\right\}^{1/2}$$

$$y = 0.000268\,ft \;\Rightarrow\; 0.00322\,in \tag{6.5.2-24}$$

Edge of buffer zone

$$y = \frac{30\,v}{v^*} = (6)\left(0.000268\,ft\right) = 0.00161\,ft \;\;or\;\; 0.0193\,in \tag{6.5.2-25}$$

Buffer zone

$$0.00161 - 0.000268 = 0.00134\,ft \;\;or\;\; 0.0161\,in. \tag{6.5.2-26}$$

6.6 The Boundary Layer Model

Consider flow over a flat plate from a uniform velocity profile as shown in Figure 6.6-1. (Only the upper flow field is shown - there is a mirror image field on the lower surface of the plate.)

turbulent

laminar

Figure 6.6-1 Boundary layer development on flat plate

Flow will be stopped on the surface of the plate, and this effect will gradually transmit itself into the main body of fluid via viscosity. The fluid which is retarded near the plate transfers momentum from the fluid which is further from the plate, and so on; thus, drag is transmitted into the main stream of fluid.

The transition from laminar to turbulent flow in the boundary layer occurs at approximately

$$Re_x = \frac{\rho\, v_\infty\, x}{\mu} \cong 5 \times 10^5 \qquad (6.6\text{-}1)$$

The point where the velocity reaches 99% of the freestream velocity marks the edge of what is defined as the **boundary layer**. The boundary layer thickness, δ, is the distance from the surface to this point. The value of 99% is arbitrary.

The purpose for creating this model is to divide the flow field into two parts, the boundary layer and the freestream, with almost all the viscous momentum transfer inside the boundary layer. In the region next to the solid boundary, viscous effects are predominant, and in the second region, "far" from the boundary, flow may be modeled as inviscid (an ideal fluid, if pressure effects are small enough that density can also be modeled as constant).

The mass flow rate in the boundary layer is less than the mass flow rate outside the boundary layer. The decrease in mass flow rate due to the retarding effect of the viscous forces in the boundary layer can be described using the **displacement thickness, δ^***, which is defined by the relation

$$\rho \, v_\infty \, \delta^* \equiv \int_0^\infty \rho \left(v_\infty - v_x \right) dy \qquad\qquad (6.6\text{-}2)$$

δ^* is the distance the solid boundary would have to be displaced to give the same mass flow rate if the flow field remained uniform (no drag at the surface - inviscid flow). For incompressible flow

$$\delta^* = \int_0^\infty \left(1 - \frac{v_x}{v_\infty} \right) dy \cong \int_0^\delta \left(1 - \frac{v_x}{v_\infty} \right) dy \qquad\qquad (6.6\text{-}3)$$

The flow slowing next to the boundary also results in the reduction in momentum flux in the x-direction (the flux in the y-direction correspondingly increases) compared to that for inviscid flow. The **momentum thickness, θ,** is defined as the thickness of a layer of fluid of uniform velocity, v_∞, for which momentum flux would be equal to the decrease in the x-momentum caused by the boundary layer.

$$\rho \, v_\infty^2 \, \theta \equiv \int_0^\infty \rho \, v_x \left(v_\infty - v_x \right) dy \qquad\qquad (6.6\text{-}4)$$

For incompressible flow

$$\theta = \int_0^\infty \frac{v_x}{v_\infty} \left(1 - \frac{v_x}{v_\infty} \right) dy \cong \int_0^\delta \frac{v_x}{v_\infty} \left(1 - \frac{v_x}{v_\infty} \right) dy \qquad\qquad (6.6\text{-}5)$$

Example 6.6-1   Displacement thickness

A wind tunnel has a 250-mm square test section. Boundary layer velocity profiles are measured at two cross-sections. Displacement thicknesses are calculated from the measured profiles. At 1 where the free-stream speed is $v_1 = 30 \, \text{m/s}$, the displacement thickness is $\delta_1^* = 1.5 \, \text{mm}$. At 2 $\delta_2^* = 2.0 \, \text{mm}$. What is the change in static pressure between 1 and 2?

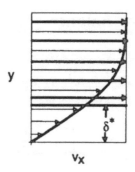

Solution

$$0 = \frac{\partial}{\partial t} \int_V^0 \rho \, dV + \int_A \rho \, (\mathbf{v} \cdot \mathbf{n}) \, dA \qquad (6.6\text{-}6)$$

$$\frac{p_1}{\rho} + \frac{v_1^2}{2} + g z_1 = \frac{p_2}{\rho} + \frac{v_2^2}{2} + g z_2 \qquad (6.6\text{-}7)$$

$$g z_1 = g z_2 \qquad (6.6\text{-}8)$$

$$p_1 - p_2 = \frac{1}{2} \rho \left(v_2^2 - v_1^2 \right) = \frac{1}{2} \rho \, v_1^2 \left[\left(\frac{v_2}{v_1} \right)^2 - 1 \right] \qquad (6.6\text{-}9)$$

$$\frac{p_1 - p_2}{\frac{1}{2} \rho \, v_1^2} = \left(\frac{v_2}{v_1} \right)^2 - 1 \qquad (6.6\text{-}10)$$

Continuity:

$$v_1 A_1 = v_2 A_2 \qquad (6.6\text{-}11)$$

$$\frac{v_2}{v_1} = \frac{A_1}{A_2} \quad \text{where} \quad A = \left(L - 2 \, \delta^* \right)^2 \qquad (6.6\text{-}12)$$

$$\frac{p_1 - p_2}{\frac{1}{2} \rho \, v_1^2} = \left(\frac{A_1}{A_2} \right)^2 - 1 = \left[\frac{\left(L - 2 \, \delta_1^* \right)^2}{\left(L - 2 \, \delta_2^* \right)^2} \right]^2 - 1 \qquad (6.6\text{-}13)$$

$$\frac{p_1 - p_2}{\frac{1}{2} \rho \, v_1^2} = \left[\frac{250 - 2 \, (1.5)}{250 - 2 \, (2.0)} \right]^4 - 1 = 0.004 \quad \text{or} \quad 0.4\% \qquad (6.6\text{-}14)$$

6.6.1   Momentum balance - integral equations

Figure 6.6.1-1 shows a boundary layer and identifies a small differential element, Δx, of this boundary layer. An x-direction momentum balance can be applied to this element in the boundary layer in steady flow. The net force in the x-direction is equal to the net rate of momentum transferred to the element.

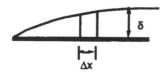

Figure 6.6.1-1 Element in boundary layer

The force on the element is made up of the pressure force on the two faces normal to flow and the drag force at the plate surface. (The drag force on the top of the element is neglected because of the small velocity gradient at this point. The pressure force is neglected on the top face because the area normal to the x-direction is small and the external pressure gradient is assumed to be small also. There are no body forces acting in the x-direction.)

$$\text{external force on element} = -\tau_0 \, \Delta x - \frac{dp}{dx} \, \Delta x \, \delta \tag{6.6.1-1}$$

From a steady-state total mass balance

mass flow rate in top of element

$$= \text{mass flow rate out right side} - \text{mass flow rate in left side}$$

$$= \int_0^\delta \rho \, v_x \, dy \bigg|_{x \,=\, x_0 + \Delta x} - \int_0^\delta \rho \, v_x \, dy \bigg|_{x \,=\, x_0} \tag{6.6.1-2}$$

From a steady-state x-momentum balance

output flow rate of momentum – input flow rate of momentum
   = flow rate of momentum out right side
   – flow rate of momentum in left side
   – flow rate of momentum in top

$$= \int_0^\delta \rho\, v_x\, v_x\, dy \Bigg|_{x=x_0+\Delta x} - \int_0^\delta \rho\, v_x\, v_x\, dy \Bigg|_{x=x_0}$$

$$- \left[\int_0^\delta \rho\, v_x\, v_\infty\, dy \Bigg|_{x=x_0} - \int_0^\delta \rho\, v_x\, v_\infty\, dy \Bigg|_{x=x_0+\Delta x} \right] \qquad (6.6.1\text{-}3)$$

Substituting these terms in the x-momentum balance, dividing by Δx, and taking the limit as Δx approaches zero yields

$$-\tau_0 - \frac{dp}{dx}\,\delta = \frac{d}{dx}\left(\int_0^\delta \rho\, v_x^2\, dy - \int_0^\delta \rho\, v_x\, v_\infty\, dy \right) \qquad (6.6.1\text{-}4)$$

For a flat plate with $dp/dx = 0$, $v_\infty = $ constant

$$\tau_0 = \frac{d}{dx}\int_0^\delta \rho\left(v_\infty - v_x\right) v_x\, dy \qquad (6.6.1\text{-}5)$$

Assuming a velocity distribution of the form

$$\frac{v_x}{v_\infty} = f\left(\frac{y}{\delta}\right) \qquad (6.6.1\text{-}6)$$

where δ is the boundary layer thickness, expressions for the thickness of the boundary layer and the shear stress on the flat plate can be determined.

For laminar flow, a possible form of Equation (6.6.1-6) is

$$\frac{v_x}{v_\infty} = \frac{3}{2}\frac{y}{\delta} - \frac{1}{2}\left(\frac{y}{\delta}\right)^2; \quad 0 \le y \le \delta$$

$$\frac{v_x}{v_\infty} = 1; \quad y > \delta \qquad (6.6.1\text{-}7)$$

$$\tau_0 = -\mu\frac{dv_x}{dy}\Bigg|_{y=0} = \frac{d}{dx}\int_0^\delta \rho\left(v_\infty - v_x\right) v_x\, dy \qquad (6.6.1\text{-}8)$$

Substituting Equation (6.6.1-7) into Equation (6.6.1-8) yields

$$\frac{\delta}{x} = \frac{4.64}{\sqrt{Re_x}}$$
$$\frac{\delta}{x} = 4.64 \sqrt{\frac{\nu}{v_\infty x}}$$
$$\delta = 4.64 \sqrt{\frac{\nu x}{v_\infty}}$$

(6.6.1-9)

and

$$\tau_0 = 0.332 \sqrt{\frac{\nu \rho v_\infty^3}{x}}$$

(6.6.1-10)

Turbulent flow may be modeled by the logarithmic universal velocity distribution or an arbitrary law such as the 1/7th power law for pipe flow, depending on the particular situation. In the latter case

$$\frac{v_x}{v_\infty} = \left(\frac{y}{\delta}\right)^{\frac{1}{7}}$$

(6.6.1-11)

From Equation (6.6.1-11) the boundary layer thickness and the shear stress are

$$\frac{\delta}{x} = \frac{0.38}{Re_x^{1/5}}$$

(6.6.1-12)

$$\tau_0 = 0.023 \, \rho \, v_\infty^2 \left(\frac{1}{Re_\delta}\right)^{1/4}$$

(6.6.1-13)

The drag coefficient, which can be defined as the shear stress divided by the kinetic energy, may be expressed as

$$C_f = \frac{\tau_0}{\frac{1}{2} \rho v_\infty^2} = \frac{0.058}{Re_x^{1/5}}$$

(6.6.1-14)

The boundary layer concept can be used to model entrance developing flow entering a pipe. Figure 6.6.1-2 shows the entrance to a pipe for (1) laminar flow progressing to fully developed flow, and (2) transition to turbulent flow before the laminar region becomes fully developed.

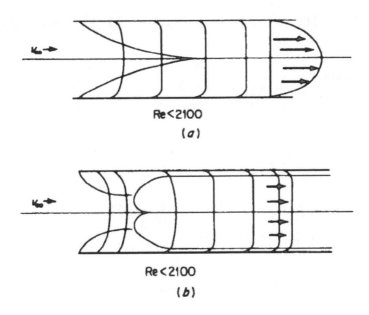

Figure 6.6.1-2 Velocity profile development in the entrance region to a pipe

(a) laminar flow
(b) turbulent flow

For the first situation, the laminar boundary layer grows until it fills the pipe. Downstream from that point the flow is laminar, the velocity profile is parabolic, and the Reynolds number is less than about 2100.

In the second situation, after the fluid enters the pipe, the boundary layer becomes turbulent before filling the pipe. Downstream of the point where the boundary layer fills the pipe, the flow is turbulent, the velocity profile is blunt, and the Reynolds number is greater than 2100. We usually assume the viscous sublayer to persist the entire length of the pipe.

6.6.2 De-dimensionalization of the boundary layer equations

The transfers of heat and of mass are also calculated using a boundary layer model, and discussion of this model will be augmented in Chapters 7 and 8. The general boundary layer model is based on the following approximations

$$
\left.
\begin{array}{l}
v_x \gg v_y \\[4pt]
\dfrac{\partial v_x}{\partial y} \gg \dfrac{\partial v_x}{\partial x}, \ \dfrac{\partial v_y}{\partial x}, \ \dfrac{\partial v_y}{\partial y}
\end{array}
\right\} \quad \text{velocity boundary layer}
$$

$$
\left.
\dfrac{\partial T}{\partial y} \gg \dfrac{\partial T}{\partial x}
\right\} \qquad\qquad \text{thermal boundary layer}
$$

$$
\left.
\dfrac{\partial c_A}{\partial y} \gg \dfrac{\partial c_A}{\partial x}
\right\} \qquad\qquad \text{concentration boundary layer} \qquad (6.6.2\text{-}1)
$$

With these assumptions, the microscopic differential equations, of change - that is, the balance equations describing continuity (total mass), momentum, energy and mass of individual species - simplify to the boundary layer equations in two dimensions.

continuity:
$$
\frac{\partial v_x}{\partial x} + \frac{\partial v_y}{\partial y} = 0
\qquad (6.6.2\text{-}2)
$$

momentum:
$$
v_x \frac{\partial v_x}{\partial x} + v_y \frac{\partial v_x}{\partial y} = -\frac{1}{\rho}\frac{\partial p}{\partial x} + v \frac{\partial^2 v_x}{\partial y^2}
\qquad (6.6.2\text{-}3)
$$

energy:
$$
v_x \frac{\partial T}{\partial x} + v_y \frac{\partial T}{\partial y} = \alpha \frac{\partial^2 T}{\partial y^2}
\qquad (6.6.2\text{-}4)
$$

species mass:
$$
v_x \frac{\partial c_A}{\partial x} + v_y \frac{\partial c_A}{\partial y} = \mathcal{D}_{AB} \frac{\partial^2 c_A}{\partial y^2}
\qquad (6.6.2\text{-}5)
$$

These equations may be made dimensionless using

$$
x^{*} = \frac{x}{L}; \quad y^{*} = \frac{y}{L}
$$
$$
v_x^{*} = \frac{v_x}{V_\infty}; \quad v_y^{*} = \frac{v_y}{V_\infty}
$$
$$
T^{*} = \frac{T - T_s}{T_\infty - T_s}; \quad c_A^{*} = \frac{c_A - c_{As}}{c_{A\infty} - c_{As}}; \quad p^{*} = \frac{p}{\rho\, v_\infty^2}
\qquad (6.6.2\text{-}6)
$$

and defining

$$Re_L = \frac{v_\infty L}{\nu}$$
$$Pr = \frac{C_p \mu}{k} = \frac{\nu}{\alpha}$$
$$Sc = \frac{\nu}{\mathcal{D}_{AB}}$$

(6.6.2-7)

which will give

continuity: $$\frac{\partial v_x^\bullet}{\partial x} + \frac{\partial v_y^\bullet}{\partial y} = 0$$

(6.6.2-8)

momentum: $$v_x^\bullet \frac{\partial v_x^\bullet}{\partial x^\bullet} + v_y^\bullet \frac{\partial v_x^\bullet}{\partial y^\bullet} = -\frac{\partial p^\bullet}{\partial x^\bullet} + \frac{1}{Re_L} \frac{\partial^2 v_x^\bullet}{\partial y^{\bullet 2}}$$

(6.6.2-9)

energy: $$v_x^\bullet \frac{\partial T^\bullet}{\partial x} + v_y^\bullet \frac{\partial T^\bullet}{\partial y^\bullet} = \frac{1}{Re_L Pr} \frac{\partial^2 T^\bullet}{\partial y^{\bullet 2}}$$

(6.6.2-10)

species mass: $$v_x^\bullet \frac{\partial c_A^\bullet}{\partial x^\bullet} + v_y^\bullet \frac{\partial c_A^\bullet}{\partial y^\bullet} = \frac{1}{Re_L Sc} \frac{\partial^2 c_A^\bullet}{\partial y^{\bullet 2}}$$

(6.6.2-11)

for momentum transfer

$$v_x^\bullet = f_1\left[x^\bullet, y^\bullet, Re_L, \frac{\partial p^\bullet}{\partial x^\bullet}\right]$$

(6.6.2-12)

$$C_f = \frac{\tau_0}{\frac{1}{2} \rho v_\infty^2} = \frac{2}{Re_L} \frac{\partial v_x^\bullet}{\partial y^\bullet}\bigg|_{y^\bullet = 0}^? = f_2\left[x^\bullet, Re_L\right]$$

(6.6.2-13)

where

$$\tau_0 = -\mu \frac{\partial v_x^\bullet}{\partial y^\bullet}\bigg|_{y^\bullet = 0}$$

(6.6.2-14)

6.6.3   Exact solution of the momentum boundary layer equations via similarity variables

The solution of the momentum boundary layer equations may be obtained by using similarity variables. For this case it is assumed that the pressure gradient is zero. We look for solutions of the form

$$\frac{v_x}{v_\infty} = g[\eta]$$

(6.6.3-1)

where $\eta = y / \delta$. Assume

$$\delta \approx \sqrt{\frac{v x}{v_x}}$$

(6.6.3-2)

therefore,

$$\eta = y\sqrt{\frac{v}{v x}}$$

(6.6.3-3)

In Equation (6.6.3-3) it is called a **similarity variable**.

Example 6.6.3-1   Similarity variable developed from dimensional analysis

For flow over a flat plate, the pertinent variables in predicting the velocity in the x-direction are

$$v_x = f(x, y, v_\infty, \rho, \mu)$$

(6.6.3-4)

so

$$\varphi(v_x, x, y, v_\infty, \rho, \mu) = 0$$

(6.6.3-5)

Using the MLt system, the dimensional matrix is

$$\begin{array}{c@{\ }c@{\ }c@{\ }c} & \text{M} & \text{L} & \text{t} \\ v_x & 0 & 1 & -1 \\ x & 0 & 1 & 0 \\ y & 0 & 1 & 0 \\ v_\infty & 0 & 1 & -1 \\ \rho & 1 & -3 & 0 \\ \mu & 1 & -1 & -1 \end{array}$$

(6.6.3-6)

from which we can show the rank is three from the non-zero third-order determinant

$$\begin{vmatrix} 0 & 1 & 0 \\ 1 & -3 & 0 \\ 1 & -1 & -1 \end{vmatrix} = (-1)\left[(1)(-1) - (0)(1)\right] = 1$$

(6.6.3-7)

Choosing as the fundamental set (x, ρ, μ), and anticipating $(6 - 3) = 3$ dimensionless groups in a complete set

Form π_1

$$\pi_1 = v_x (x)^a \rho^b \mu^c \ \Rightarrow \ \left(\frac{L}{t}\right)(L)^a \left(\frac{M}{L^3}\right)^b \left(\frac{M}{Lt}\right)^c$$

(6.6.3-8)

Apply dimensional homogeneity

$$\begin{array}{lll} \text{M: } b + c = 0 & & a = 1 \\ \text{L: } 1 + a - 3b - c = 0 & \Rightarrow & b = 1 \\ \text{t: } -1 - c = 0 & & c = -1 \end{array}$$

(6.6.3-9)

Result

$$\pi_1 = \frac{v_x x \rho}{\mu}$$

(6.6.3-10)

Note that this is a local Reynolds number since it is based on the local velocity and local distance from the front of the plate

$$Re_{x, v_x} = \frac{v_x x \rho}{\mu}$$

(6.6.3-11)

___

Form π_2

$$\pi_2 = y (x)^a \rho^b \mu^c \ \Rightarrow \ (L)(L)^a \left(\frac{M}{L^3}\right)^b \left(\frac{M}{Lt}\right)^c$$

(6.6.3-12)

Apply dimensional homogeneity

M: $b + c = 0$ $a = -1$
L: $1 + a - 3b - c = 0$ $\Rightarrow$ $b = 0$
t: $c = 0$ $c = 0$

$$(6.6.3\text{-}13)$$

Result

$$\pi_2 = \frac{y}{x} \tag{6.6.3-14}$$

Form π_3

$$\pi_3 = v_\infty \left(x\right)^a \rho^b \mu^c \;\Rightarrow\; \left(\frac{L}{t}\right)\left(L\right)^a \left(\frac{M}{L^3}\right)^b \left(\frac{M}{Lt}\right)^c \tag{6.6.3-15}$$

Apply dimensional homogeneity

M: $b + c = 0$ $a = -1$
L: $1 + a - 3b - c = 0$ $\Rightarrow$ $b = 0$
t: $c = 0$ $c = 0$

$$(6.6.3\text{-}16)$$

Result

$$\pi_3 = \frac{v_\infty x \rho}{\mu} \tag{6.6.3-17}$$

This is a Reynolds number that is intermediate between the Reynolds number defined by π_1 and a Reynolds number based on the total length of the plate and the freestream velocity, since it contains the local distance but the freestream velocity.

$$\text{Re}_{x, v_\infty} = \frac{v_\infty x \rho}{\mu} \tag{6.6.3-18}$$

We can obtain our similarity variable, η, by

$$\eta = \pi_2 \pi_3^{0.5} = \frac{y}{x}\sqrt{\frac{v_\infty x \rho}{\mu}} = y\sqrt{\frac{v_\infty}{\nu x}} \tag{6.6.3-19}$$

Introducing the stream function (which satisfies the continuity equation identically)

$$v_x = \frac{\partial \Psi}{\partial y}$$

$$v_y = -\frac{\partial \Psi}{\partial x}$$

(6.6.3-20)

The dimensionless stream function is defined as

$$f(\eta) \equiv \frac{\Psi}{\sqrt{\nu\, x\, v_\infty}}$$

(6.6.3-21)

Velocity components are given by

$$v_x = \frac{\partial \Psi}{\partial y} = \frac{\partial \Psi}{\partial \eta}\frac{\partial \eta}{\partial y} = \sqrt{\nu\, x\, v_\infty}\,\frac{df}{d\eta}\sqrt{\frac{v_\infty}{\nu x}} = v_\infty \frac{df}{d\eta}$$

$$v_y = -\frac{\partial \Psi}{\partial x} = \frac{1}{2}\sqrt{\frac{\nu\, v_\infty}{x}}\left(\eta\frac{df}{d\eta} - f\right)$$

(6.6.3-22)

Differentiating

$$\frac{\partial v_x}{\partial x} = -\frac{v_\infty}{2x}\eta\frac{d^2 f}{d\eta^2}$$

$$\frac{\partial v_x}{\partial y} = v_\infty\sqrt{\frac{v_\infty}{\nu x}}\frac{d^2 f}{d\eta^2}$$

$$\frac{\partial^2 v_x}{\partial y^2} = \frac{v_\infty^2}{\nu x}\frac{d^3 f}{d\eta^3}$$

(6.6.3-23)

Substituting these expressions into the momentum balance Equation (6.6.2-9) yields the **Blasius** boundary layer equation

$$2\frac{d^3 f}{d\eta^3} + f\frac{d^2 f}{d\eta^2} = 0$$

Boundary conditions:

$$\eta = 0; \quad f = 0$$

$$\eta = 0; \quad \frac{df}{d\eta} = 0$$

$$\eta = \infty; \quad \frac{df}{d\eta} = 1$$

(6.6.3-24)

Applying the boundary conditions one can solve either by a series expansion or by various numerical methods, one of which is detailed in the example below. The solution is shown in Figure 6.6.3-1.

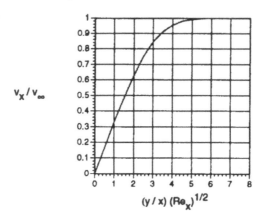

Figure 6.6.3-1 Solution to Blasius boundary layer equation

Example 6.6.3-2   Runge-Kutta solution of Blasius problem

Solve the Blasius equation for two-dimensional boundary layer flow past a flat plate

$$2 f''' + f f'' = 0$$

(6.6.3-25)

with boundary conditions

$$
\begin{aligned}
&\text{for } \eta = 0, \quad f = f' = 0 \\
&\text{for } \eta = 1, \quad f' = 1
\end{aligned}
$$

(6.6.3-26)

by reducing the equation to a system of first-order ordinary differential equations and using a fourth-order Runge-Kutta[21] method.

---

[21] Runge-Kutta methods use weighted averages of derivatives across intervals to estimate the value of the dependent variable at the end of the interval. For a simpler version of the same approach, see Example 9.2-1, Design of a cooling tower, which uses Heun's method.

Use a step size of 0.05 for η, and carry the computation to $\eta = 8.0$, which can be regarded as ∞. The initial value for f'' should be between 0.30 and 0.35.

Plot the results.

Solution

The nonlinear ordinary differential equation and associated initial and boundary conditions, using ϕ in place of f, can be written as

$$2\varphi''' + \varphi\varphi'' = 0 \implies \varphi''' = -\frac{\varphi\varphi''}{2} \tag{6.6.3-27}$$

with initial conditions

$$\eta = 0: \quad \varphi = 0, \varphi' = 0 \tag{6.6.3-28}$$

and boundary condition

$$\eta = \infty: \quad \varphi' = 1 \tag{6.6.3-29}$$

Define

$$\varphi'' = \frac{d\psi}{d\eta} = \xi \tag{6.6.3-30}$$

$$\varphi' = \frac{d\varphi}{d\eta} = \psi \tag{6.6.3-31}$$

giving a set of ordinary first-order differential equations

$$\varphi''' = \frac{d\xi}{d\eta} = -\frac{\varphi\varphi''}{2} \tag{6.6.3-32}$$

$$\varphi'' = \frac{d\psi}{d\eta} = \xi \tag{6.6.3-33}$$

$$\varphi' = \frac{d\varphi}{d\eta} = \psi \tag{6.6.3-34}$$

with initial conditions

$$\eta = 0: \quad \varphi = 0, \varphi' = 0 \tag{6.6.3-35}$$

and boundary condition

$$\eta = \infty: \quad \varphi' = 1 \tag{6.6.3-36}$$

The three equations are of the form

$$\frac{d\xi}{d\eta} = f_a(\eta, \varphi, \psi, \xi) = -\frac{\varphi\xi}{2} \tag{6.6.3-37}$$

$$\frac{d\psi}{d\eta} = f_b(\eta, \varphi, \psi, \xi) = \xi \tag{6.6.3-38}$$

$$\frac{d\varphi}{d\eta} = f_c(\eta, \varphi, \psi, \xi) = \psi \tag{6.6.3-39}$$

with initial conditions

$$\eta = 0: \quad \varphi = 0, \psi = 0 \tag{6.6.3-40}$$

and boundary condition

$$\eta = \infty: \quad \psi = 1 \tag{6.6.3-41}$$

For a **single** equation

$$y' = \frac{dy}{dx} = f(x, y) \tag{6.6.3-42}$$

with initial condition

$$x = 0: \quad f = f_0 \tag{6.6.3-43}$$

the fourth-order Runge-Kutta algorithm is

$$y_{n+1} = y_n + \frac{1}{6}\left(k_1 + 2k_2 + 2k_3 + k_4\right)$$

$$k_1 = h\,f(x_n, y_n)$$

$$k_2 = h\,f\left[\left(x_n + \frac{1}{2}h\right), \left(y_n + \frac{1}{2}k_1\right)\right]$$

$$k_3 = h\,f\left[\left(x_n + \frac{1}{2}h\right), \left(y_n + \frac{1}{2}k_2\right)\right] \tag{6.6.3-44}$$

$$k_4 = h\,f\left[\left(x_n + h\right), \left(y_n + k_3\right)\right]$$

We apply the fourth-order Runge-Kutta method to Equations (6.6.3-32), (6.6.3-33), and (6.6.3-34).

The difficulty is that we have initial conditions on only two of the variables (ϕ and ψ), while the remaining condition is a boundary condition on ψ. This means that we must estimate an initial condition for ξ, carry out the computation, and see if we have satisfied the boundary condition on ψ (a so-called "shooting" method). Since the solution is relatively well behaved, this presents no great difficulty.

If we apply the algorithm using a spreadsheet, it is easy to insert various guesses for ξ_0 and observe the final value for ψ. In general, for such a system of equations we could apply a more sophisticated numerical technique such as some variation on Newton's method if we so desired, but for this problem, convergence using this simple trial and error method is rapid. Since the problem at hand does not demand high sophistication, we will simply use the trial-and-error calculation.

Noting that each equation will have its own values of k_1, k_2, k_3, and k_4, first calculate k_1 for each equation,

$$k_{1\xi} = h\,f_a(\eta_n, \varphi_n, \psi_n, \xi_n) = -h\,\frac{\varphi_n \xi_n}{2} \tag{6.6.3-45}$$

$$k_{1\psi} = h\,f_b(\eta_n, \varphi_n, \psi_n, \xi_n) = h\,\xi_n \tag{6.6.3-46}$$

$$k_{1\varphi} = h\,f_c(\eta_n, \varphi_n, \psi_n, \xi_n) = h\,\psi_n \tag{6.6.3-47}$$

then k_2

$$k_{2\xi} = h f_a\left[\left(\eta_n + \frac{1}{2}h\right), \left(\phi_n + \frac{1}{2}k_{1\phi}\right), \left(\psi_n + \frac{1}{2}k_{1\psi}\right), \left(\xi_n + \frac{1}{2}k_{1\xi}\right)\right]$$

$$= -h\left[\frac{\left(\phi_n + \frac{1}{2}k_{1\phi}\right)\left(\xi_n + \frac{1}{2}k_{1\xi}\right)}{2}\right] \tag{6.6.3-48}$$

$$k_{2\psi} = h f_b\left[\left(\eta_n + \frac{1}{2}h\right), \left(\phi_n + \frac{1}{2}k_{1\phi}\right), \left(\psi_n + \frac{1}{2}k_{1\psi}\right), \left(\xi_n + \frac{1}{2}k_{1\xi}\right)\right]$$

$$= -h\left(\xi_n + \frac{1}{2}k_{1\xi}\right) \tag{6.6.3-49}$$

$$k_{2\phi} = h f_c\left[\left(\eta_n + \frac{1}{2}h\right), \left(\phi_n + \frac{1}{2}k_{1\phi}\right), \left(\psi_n + \frac{1}{2}k_{1\psi}\right), \left(\xi_n + \frac{1}{2}k_{1\xi}\right)\right]$$

$$= -h\left(\psi_n + \frac{1}{2}k_{1\psi}\right) \tag{6.6.3-50}$$

then k_3

$$k_{3\xi} = h f_a\left[\left(\eta_n + \frac{1}{2}h\right), \left(\phi_n + \frac{1}{2}k_{2\phi}\right), \left(\psi_n + \frac{1}{2}k_{2\psi}\right), \left(\xi_n + \frac{1}{2}k_{2\xi}\right)\right]$$

$$= -h\left[\frac{\left(\phi_n + \frac{1}{2}k_{2\phi}\right)\left(\xi_n + \frac{1}{2}k_{2\xi}\right)}{2}\right] \tag{6.6.3-51}$$

$$k_{3\psi} = h f_b\left[\left(\eta_n + \frac{1}{2}h\right), \left(\phi_n + \frac{1}{2}k_{2\phi}\right), \left(\psi_n + \frac{1}{2}k_{2\psi}\right), \left(\xi_n + \frac{1}{2}k_{2\xi}\right)\right]$$

$$= -h\left(\xi_n + \frac{1}{2}k_{2\xi}\right) \tag{6.6.3-52}$$

$$k_{3\phi} = h f_c\left[\left(\eta_n + \frac{1}{2}h\right), \left(\phi_n + \frac{1}{2}k_{2\phi}\right), \left(\psi_n + \frac{1}{2}k_{2\psi}\right), \left(\xi_n + \frac{1}{2}k_{2\xi}\right)\right]$$

$$= -h\left(\psi_n + \frac{1}{2}k_{2\psi}\right) \tag{6.6.3-53}$$

and then k_4

$$k_{4\xi} = h \, f_a\left[\left(\eta_n + h\right), \left(\varphi_n + k_{3\varphi}\right), \left(\psi_n + k_{3\psi}\right), \left(\xi_n + k_{3\xi}\right)\right]$$

$$= -h \left[\frac{\left(\varphi_n + k_{3\varphi}\right)\left(\xi_n + k_{3\xi}\right)}{2}\right] \qquad (6.6.3\text{-}54)$$

$$k_{4\psi} = h \, f_b\left[\left(\eta_n + h\right), \left(\varphi_n + k_{3\varphi}\right), \left(\psi_n + k_{3\psi}\right), \left(\xi_n + k_{3\xi}\right)\right]$$

$$= -h \left(\xi_n + k_{3\xi}\right) \qquad (6.6.3\text{-}55)$$

$$= -h \left(\psi_n + k_{3\psi}\right) \qquad (6.6.3\text{-}56)$$

The functions can next be updated

$$\xi_{n+1} = \xi_n + \frac{1}{6}\left(k_{1\xi} + 2\,k_{2\xi} + 2\,k_{3\xi} + k_{4\xi}\right) \qquad (6.6.3\text{-}57)$$

$$\psi_{n+1} = \psi_n + \frac{1}{6}\left(k_{1\psi} + 2\,k_{2\psi} + 2\,k_{3\psi} + k_{4\psi}\right) \qquad (6.6.3\text{-}58)$$

$$\varphi_{n+1} = \varphi_n + \frac{1}{6}\left(k_{1\varphi} + 2\,k_{2\varphi} + 2\,k_{3\varphi} + k_{4\varphi}\right) \qquad (6.6.3\text{-}59)$$

This algorithm is easily adapted to spreadsheet use.

6.7 Drag Coefficients

Drag coefficients are used to relate flow rate to momentum transfer for both external (over the outside surfaces of a solid such as over a flat plate, a cylinder, a sphere, etc.) and internal flows (inside a conduit such as a pipe). The basic concept is the same in both cases, but with external flows the velocity used is the velocity "far" from the surface, i.e., the velocity of approach or the freestream velocity, and for internal flows the velocity used is usually the area-average velocity.

Drag is usually divided into two classes, skin friction and form drag. **Skin friction** is the term used for drag force caused by momentum transfer from the fluid to a solid surface. Form drag is the term used for force resulting from a pressure gradient from the front to the back of a solid object.

For example, if one holds a circular disk of cardboard (perhaps the base from a now-consumed pizza) out the window of a car as it travels down the highway, the force in the direction of travel of the car necessary to hold the cardboard in place will vary depending on the orientation of the disk. If the axis of the disk is normal to the direction of travel of the car, the main force experienced is a result of the flow of air over the top and bottom of the disk - skin friction. If, however, the axis of the disk is held parallel to the direction of travel of the car, a considerably larger force will be experienced, which is caused by the pressure difference that develops between the forward and rearward surfaces of the disk - form drag. Form drag does exist in the first instance, but is much less that the skin friction because the area normal to flow on which the pressure drop acts is very small. Similarly, skin friction exists in the second case, but the skin friction over the forward and rearward surfaces of the disk has no component in the direction of travel, and the skin friction on the edge of the disk acts on such a small area as to be relatively insignificant.

Consider the case of flow over a curved surface as shown in Figure 6.7-1 (such as an airplane wing). At very low speeds, flow is essentially laminar and the air follows the contour of the wing as is shown in Figure 6.7-1a. The majority of the drag in this case results from skin friction, which is the transfer of momentum from the wing, which is moving, to the surrounding air, which is stationary. Under certain circumstances, however, as flow is increased, the boundary layer can separate from the wing, as shown in Figure 6.7-1b.

To understand why the boundary layer separates, imagine a small particle of air which is going to pass over the wing. The mechanical energy balance developed in Chapter 4 can be applied to this particle of air as a system. To simplify things, let us assume that the particle does not change position significantly in a gravitational field, and therefore the change in potential energy

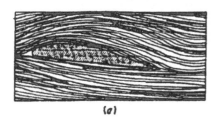 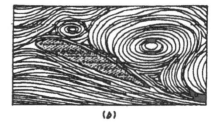

(a)   (b)

Figure 6.7-1 Flow around an airfoil (a) without and (b) with separation. (From L. Prandtl and O.G. Tietjens. Fundamentals of Hydro- and Aerodynamics, New York, Dover, reprinted by permission of the publishers.)

is negligible in passing across the wing. The air, therefore, possesses two types of mechanical energy: kinetic energy and pressure energy.

The mechanical energy does not stay constant because the air undergoes momentum interchange with surrounding air as it goes across the wing; therefore, the important terms in the mechanical energy balance for the air particle are the kinetic energy, the pressure energy, and the lost work due to friction. (There is no shaft work.)

The lift on the airfoil results from the greater increase in kinetic energy across the upper surface of the airfoil. This form of mechanical energy can come only from the pressure energy, so the pressure decreases, giving lift. As the air comes off the back edge of the airfoil, it once again slows down and its kinetic energy is reconverted to pressure energy. There is, however, a net loss of mechanical energy because of the frictional processes as it traverses the airfoil. This means that the remaining kinetic energy, when transformed to pressure energy, is insufficient to recoup the original level of pressure. The pressure **gradient**, in fact, becomes negative. Since flow tends to go down the pressure gradient, it may actually reverse next to the airfoil, the boundary layer separate, and the wing **stalls**. This condition is of great concern to a pilot, because stalling means that he is progressively losing the lift on the latter part of the wing.

The region downstream from the separation point is called the **wake**. The effect of separation and the subsequent wake is to decrease the amount of kinetic energy transfer back to pressure energy and to increase transfer to internal energy (in other words, loss of mechanical energy).

Bodies are often streamlined to cause the separation point to be as far downstream as possible. If there is no separation, the primary loss is caused by the shear in the boundary layer, that is, skin friction. Remember that even though we call this term skin friction, it is not really due to the layer of air molecules adjacent to the surface of the body rubbing on the surface. We know that these molecules are almost always attached to the surface. Skin friction is simply a term we use to describe the momentum transfer from boundary layer formation.

The general drag coefficient is defined by the equation

$$F_{drag} \equiv \frac{1}{2} C_D \rho v^2 A \qquad\qquad (6.7\text{-}1)$$

where A is a characteristic area - usually either the cross-sectional area normal to flow for bodies with significant form drag, or the wetted area for bodies with primarily skin friction (e.g., flat plates) - and

$$C_D = C_D' + C_D''$$

(6.7-2)

C_D' is the drag coefficient defined on the basis of drag force from skin friction and C_D'' is the corresponding coefficient for form drag. Their sum is the coefficient of total drag.

The form drag coefficient, C_D'', is usually determined from experiments using some sort of scaled (or sometimes full-size) physical model because there is seldom an adequate model to calculate it from first principles.

Skin friction is a function of the shear stress at the surface

$$F_{drag, skin friction} = \frac{1}{2} C_D' \rho v_\infty^2 A = \int_A \tau_0 \, dA$$

(6.7-3)

The shear stress at the surface is, in turn, a function of Reynolds number and is often determinable by calculation, as we shall see below.

To use a given drag coefficient appropriately, it is necessary to know the velocity and the area that was used in its definition (as well as whether it is a coefficient of total drag, form drag, or skin friction). We distinguish two cases.

For **internal flows**, the velocity in the kinetic energy term is normally the **bulk** velocity. This velocity could be **any** velocity one wishes to choose, but, conventionally, the bulk velocity is used. The area that is conventionally used for internal flows is the inside surface area of the conduit, i.e., the circumference multiplied by the length. Internal flows usually do not involve form drag, but only skin friction.

For **external flows**, completely different definitions are used. External flows usually involve both skin friction and form drag (except for such special cases as an infinitesimally thick plate positioned parallel to flow, where the cross-sectional area normal to flow is zero, obviating form drag). The characteristic velocity is usually taken to be the freestream velocity. Notice that there is equivalence from the standpoint of the drag force generated between fluid flowing **past** an object and the object moving **through** the fluid. Therefore, when we say "freestream velocity" for the case of an object moving through a quiescent fluid, we mean the relative velocity of the fluid as seen from the object (the velocity of the object). Since form drag often dominates the case of external flow, the characteristic area usually used is the cross-sectional area of the object normal to flow (because this is the effective area upon which the pressure difference acts to produce form drag). We could use any other well-defined area (for example, the external surface area of the object) because, for objects of geometrically similar shape, the ratios of the various areas remain constant as the size of the object is changed; however, one conventionally uses the cross-sectional area normal to flow. For external flow, the drag coefficient is normally defined in terms of total drag.

6.7.1 Drag on immersed bodies (external flow)

Figure 6.7.1-1 shows empirical curves fit to the data for drag coefficient for a smooth flat plate aligned parallel to the flow stream.

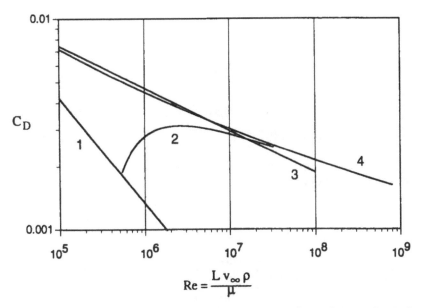

$$Re = \frac{L \, v_\infty \, \rho}{\mu}$$

Figure 6.7.1-1 Drag coefficient for smooth flat plate oriented parallel to flow stream[22]

---

[22] Curves shown are empirical fits to experimental data: (continued next page)

The local shear stress at the surface of a flat plate was found in Section 6.6.1 using the integral momentum balance to be

$$\tau_0 = 0.332 \sqrt{\frac{\nu \rho v_\infty^3}{x}} \tag{6.7.1-1}$$

Another way to determine the same result is to integrate the Blasius solution. The laminar curve in Figure 6.7.1-1 comes from integrating the local shear stress found using the Blasius solution as developed in Section 6.6.3. The drag force per unit width for one side of the plate is

$$F_D = \int_0^L \tau_0 \, dx \tag{6.7.1-2}$$

where

$$\tau_0 = \mu \left[\frac{dv_x}{dy} \right]_{y=0} \tag{6.7.1-3}$$

From the Blasius solution

$$v_x = \frac{\partial \Psi}{\partial y} = \frac{\partial \Psi}{\partial \eta} \frac{\partial \eta}{\partial y} = \sqrt{\nu x v_\infty} \, \frac{df}{d\eta} \sqrt{\frac{v_\infty}{\nu x}} = v_\infty f' \tag{6.7.1-4}$$

$$\frac{dv_x}{dy} = v_\infty \sqrt{\frac{v_\infty}{\nu x}} \, f'' \tag{6.7.1-5}$$

$$\left[\frac{dv_x}{dy} \right]_{y=0} = v_\infty \sqrt{\frac{v_\infty}{\nu x}} \, f''(0) \tag{6.7.1-6}$$

From Table 6.6.3-1 one can find

Curve 1: $c_D = \dfrac{1.328}{\sqrt{Re_L}}$ Laminar, Blasius

Curve 2: $c_D = \dfrac{0.455}{\left(\log Re_L \right)^{2.58}} - \dfrac{1700}{Re_L}$ Transitional

Curve 3: $c_D = \dfrac{0.074}{\left(Re_L \right)^{0.2}}$ Turbulent, Prandtl

Curve 4: $c_D = \dfrac{0.455}{\left(\log Re_L \right)^{2.58}}$ Turbulent, Prandtl-Schlichting

For details see Schlichting, H. (1960). *Boundary Layer Theory*, New York, McGraw-Hill, p. 538.

$$f''(0) = \xi_1(0) = 0.332 \tag{6.7.1-7}$$

Substituting, performing the integration, doubling the result (so as to account for both the top and the bottom surfaces of the plate) and applying the relationship between the drag force and the drag coefficient, where W is the width of the plate,[23] one obtains

$$C_D = \frac{1.328}{\sqrt{Re_L}} \tag{6.7.1-8}$$

where the Reynolds number is based on the freestream velocity and the length of the plate.

The Prandtl solution is based on the 1/7 power velocity distribution model and the wall shear stress model for turbulent pipe flow[24]

$$C_D = \frac{0.074}{Re_L^{1/5}} \quad \Rightarrow \quad 5 \times 10^5 < Re_L < 10^7 \tag{6.7.1-9}$$

This model is applicable to plates where the boundary layer is turbulent starting at the leading edge. Usually, however, there is a laminar section preceding the turbulent section, leading to lower drag. The transition is normally somewhere between Reynolds numbers of 3×10^5 and 3×10^6.

The Prandtl-Schlichting curve comes from applying the logarithmic universal velocity profile for pipe flow, assuming that transition takes place at a critical Reynolds number of 3×10^5, empirically modifying constants in the model to account for experimental data, and allowing for the laminar initial section, resulting in the equation

---

[23] Note that the relationship of drag force to drag coefficient gives

$$C_D = C_D' = \frac{\tau_0}{\frac{1}{2}\rho v_\infty^2} = \frac{\frac{F_{drag}}{x\,W}}{\frac{1}{2}\rho v_\infty^2} = \frac{0.664}{\sqrt{\frac{x\,v_\infty\,\rho}{\mu}}}$$

$$F_{drag} = \frac{1}{2}\rho v_\infty^2\, x\, W \frac{0.664}{\sqrt{\frac{x\,v_\infty\,\rho}{\mu}}}$$

$$F_{drag} = \frac{1}{2} C_D' \rho v_\infty^2\, x\, W$$

[24] For details of development of this model and the models immediately below, see Schlichting, H. (1960). *Boundary Layer Theory*, New York, McGraw-Hill, p. 536.

$$C_D = \frac{0.455}{\left(\log \text{Re}_L\right)^{2.58}} - \frac{1700}{\text{Re}_L} \quad \Rightarrow \quad 5 \times 10^5 < \text{Re}_L < 10^9 \qquad (6.7.1\text{-}10)$$

Schultz-Grunow corrected for the upward deviation of the outward boundary layer velocity of a flat plate compared to a pipe, and developed the relation

$$C_D = \frac{0.427}{\left(\log \text{Re}_L - 0.407\right)^{2.64}} \qquad (6.7.1\text{-}11)$$

Example 6.7.1-1   Drag on a flat plate

Assume a water ski can be modeled as a flat plate. The ski is 2-m long, 0.25-m wide, and is immersed to a depth of 0.05 m. The boat dragging the ski is moving at 25 knots. What is the skin friction force?

Solution

$$v_\infty = (25)\,\text{knots}\left[\frac{\text{nautical mi}}{\text{knot hr}}\right]\left[\frac{6076\,\text{ft}}{\text{nautical mi}}\right]\left[\frac{0.3048\,\text{m}}{\text{ft}}\right]\left[\frac{\text{hr}}{3600\,\text{sec}}\right]$$

$$= 12.86\,\tfrac{\text{m}}{\text{s}} \qquad\qquad (6.7.1\text{-}12)$$

Use as the viscosity of water $v = 1.4 \times 10^{-6}$ m²/s.

$$\text{Re}_L = \frac{v_\infty L}{v} = \left(12.86\,\tfrac{\text{m}}{\text{s}}\right)(2\,\text{m})\left(\frac{\text{sec}}{1.4 \times 10^{-6}\,\text{m}^2}\right)$$

$$= 1.837 \times 10^7 \qquad\qquad (6.7.1\text{-}13)$$

From Figure 6.7.1-1 $C_D = 0.0026$

$$F_D = C_D\,A\,\tfrac{1}{2}\,\rho\,v_\infty^2 \qquad\qquad (6.7.1\text{-}14)$$

$$A = L\left(w + 2\,d\right) \qquad\qquad (6.7.1\text{-}15)$$

$$F_D = \tfrac{1}{2}(0.0026)(2\,\text{m})(0.25\,\text{m} + 0.1\,\text{m})\left(1020\,\tfrac{\text{kg}}{\text{m}^3}\right)\left(12.86\,\tfrac{\text{m}}{\text{s}}\right)^2\left(\frac{\text{N sec}^2}{\text{kg m}}\right)$$

$$\qquad\qquad (6.7.1\text{-}16)$$

$$F_D = 153\,\text{N} \quad \Rightarrow \quad \left(34.4\,\text{lbf}\right) \qquad\qquad (6.7.1\text{-}17)$$

For flow past a circular cylinder, Figure 6.7.1-2 shows (a) ideal flow and (b) turbulent flow with boundary layer separation and wake formation. For the ideal flow situation, there is no separation of the boundary layer because the presence of the body does not result in momentum exchange.

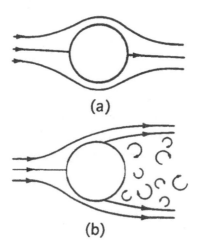

Figure 6.7.1-2 Flow past circular cylinder

Figure 6.7.1-3 shows the drag coefficient for a circular cylinder as a function of Reynolds number.

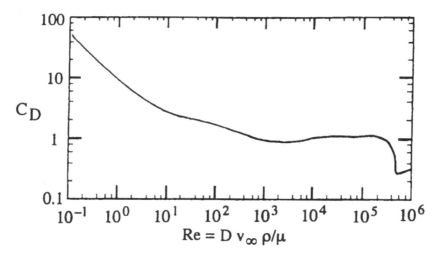

Figure 6.7.1-3 Drag coefficient for circular cylinder[25]

Example 6.7.1-2   Wind force on a distillation column

Calculate the drag force on a distillation column 50 ft high by 8-ft diameter in a wind of 40 mph. For air $\nu = 16.78 \times 10^{-5}$ ft²/s, $\rho = 0.0735$ lbm/ft³.

Solution

$$v_{\infty} = \left(\frac{40\,mi}{hr}\right)\left(\frac{hr}{3600\,s}\right)\left(\frac{5280\,ft}{mi}\right) = 58.7\,\tfrac{ft}{s} \qquad (6.7.1\text{-}18)$$

$$Re = \frac{D\,v_{\infty}\,\rho}{\mu} = \frac{(8\,ft)\left(58.7\,\tfrac{ft}{s}\right)}{\left(16.88 \times 10^{-5}\,\tfrac{ft^2}{s}\right)} = 2.78 \times 10^6 \qquad (6.7.1\text{-}19)$$

From Figure 6.7.1-3

$$C_D = 0.4$$
$$F_{drag} = \tfrac{1}{2}\,C_D\,\rho\,v_{\infty}^2\,(D\,L)$$

---

[25] Adapted from data in Schlichting, H. (1960). *Boundary Layer Theory*, New York, McGraw-Hill, p. 16.

$$F_{drag} = \frac{(0.4)\left(0.0735\,\frac{lbm}{ft^3}\right)\left(58.7\,\frac{ft}{s}\right)^2(8\,ft)(50\,ft)}{(2)\left(32.2\,\frac{lbm\,ft}{lbf\,sec^2}\right)} \tag{6.7.1-20}$$

$$F_{drag} = 630\,lbf \tag{6.7.1-21}$$

Note: The drag coefficient for a cylinder is relatively constant at these Reynolds numbers. For a hurricane wind at 200 mph the drag force increases ~25 times.

The drag coefficient for flow past a sphere is given in Figure 6.7.1-4.

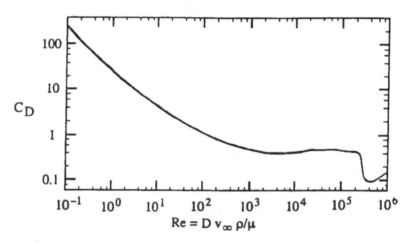

Figure 6.7.1-4 Drag coefficient for sphere[26]

The leftward part of the figure where the Reynolds number is less than 0.1 is called the **creeping flow** regime. Various parts of the curve can be fitted algebraically as follow

$$C_D = \frac{24}{Re}; \qquad Re < 0.1 \tag{6.7.1-22}$$

$$C_D = \frac{18.5}{Re^{3/5}}; \quad 2 < Re < 500 \tag{6.7.1-23}$$

---

[26] Adapted from data in Schlichting, H. (1960). *Boundary Layer Theory*, New York, McGraw-Hill, p. 16.

$$C_D = 0.44; \quad 500 < Re < 10^5$$

(6.7.1-24)

Example 6.7.1-3   Terminal velocity of a polymer sphere in water

Calculate the terminal velocity in room temperature water of a 3-mm diameter sphere made of a polymer with density 1.15 gm/cm^3.

Solution

We make a balance at terminal velocity on the vertical components of the force vectors (assuming that the positive direction, chosen as the z-direction, is upward, which gives the vertical component of the acceleration of gravity a negative sign).

A balance on the z-components of the force vectors in the vertical direction is the scalar equation.

$$\left[F_{buoyancy}\right]_z + \left[F_{drag}\right]_z = 0$$

(6.7.1-25)

$$F_{buoyancy} = V_{sphere}\left(\rho_{water} - \rho_{sphere}\right) g$$

(6.7.1-26)

The net buoyancy force is negative. The velocity is downward, so the drag force will be in the positive direction. The magnitude of the drag force is

$$F_{drag} = C_D \frac{1}{2} \rho_{water} A_{xs} v_{terminal}^2$$

(6.7.1-27)

Substituting

$$V_{sphere}\left(\rho_{water} - \rho_{sphere}\right) g + C_D \frac{1}{2} \rho_{water} A_{xs} v_{terminal}^2 = 0$$

(6.7.1-28)

This is one nonlinear algebraic equation in two unknowns, C_D and $v_{terminal}$

The other relationship we have is the graph of C_D vs. Re.

$$F_{buoyancy} = \frac{\pi D^3}{6}\left(\rho_{water} - \rho_{sphere}\right)g$$

$$= \frac{(\pi)(3)^3 [mm]^3}{(6)}\left(1.00\frac{gmm}{cm^3} - 1.15\frac{gmm}{cm^3}\right)$$

$$\times (9.8)\left[\frac{m}{s^2}\right]\left[\frac{N s^2}{kgm\ m}\right]\left[\frac{kgm}{(1000)\ gmm}\right]\left[\frac{cm^3}{(10)^3\ mm^3}\right]$$

$$= -6.93 \times 10^{-6}\,N \tag{6.7.1-29}$$

$$F_{buoyancy} + F_{drag} = 0$$

$$-6.93 \times 10^{-6}\,N + F_{drag} = 0$$

$$F_{drag} = 6.93 \times 10^{-6}\,N \tag{6.7.1-30}$$

The drag force is given by

$$F_{drag} = C_D \frac{1}{2}\rho_{water} A_{xs} v_{terminal}^2 = 6.93 \times 10^{-6}\,N \tag{6.7.1-31}$$

$$C_D = \frac{6.93 \times 10^{-6}\,N}{\rho_{water}\frac{\pi D^2}{4}v_{terminal}^2}$$

$$= \frac{\left(6.93 \times 10^{-6}\right)N}{(1.00)\left[\frac{gmm}{cm^3}\right]\frac{(\pi)(3)^2\ mm^2}{(4)}\left(v_{terminal}^2\right)\left[\frac{cm}{s}\right]^2}$$

$$\times \frac{(10)\ mm}{cm}\frac{(1000)\ mm}{m}\left[\frac{kg\ m}{N s^2}\right]$$

$$C_D = \frac{9.80 \times 10^{-3}}{v_{terminal}^2} \tag{6.7.1-32}$$

To obtain an initial guess for $v_{terminal}$ assume $C_D = 1.0$

$$1.0 = \frac{9.80 \times 10^{-3}}{v_{terminal}^2}$$

$$v_{terminal}^2 = 9.80 \times 10^{-3}\left[\frac{cm}{s}\right]^2$$

$$v_{terminal} = 9.90 \times 10^{-2}\left[\frac{cm}{s}\right] \tag{6.7.1-33}$$

Calculating the corresponding Re

$$Re = \frac{D\,v_{terminal}\,\rho}{\mu}$$

$$= \frac{3\,[mm]\,\left(9.9 \times 10^{-2}\right)\left[\frac{cm}{s}\right]\,(1.00)\left[\frac{gmm}{cm^3}\right]}{(1)\,cP\left[\frac{(0.01)\,gmm}{cm\;s\;cP}\right]}\left[\frac{cm}{(10)\,mm}\right]$$

$$Re = 2.97 \tag{6.7.1-34}$$

Reading the drag coefficient from the graph

$$C_D = 10 \tag{6.7.1-35}$$

To get a second guess for $v_{terminal}$, use a successive substitution technique

$$C_D = 10 = \frac{9.80 \times 10^{-3}}{v_{terminal}^2}$$

$$v_{terminal} = \sqrt{\frac{9.80 \times 10^{-3}}{10}} = 0.031\,\frac{cm}{s} \tag{6.7.1-36}$$

$$Re = \frac{D\,v_{terminal}\,\rho}{\mu} = \frac{3\,[mm]\,(0.031)\left[\frac{cm}{s}\right]\,(1.00)\left[\frac{gmm}{cm^3}\right]}{(1)\,cP\,\frac{(0.01)\,gmm}{cm\;s\;cP}}\frac{cm}{(10)\,mm}$$

$$Re = 0.94 \tag{6.7.1-37}$$

Reading the drag coefficient from the graph

$$C_D = 30 \tag{6.7.1-38}$$

$$C_D = 30 = \frac{9.80 \times 10^{-3}}{v_{terminal}^2}$$

$$v_{terminal} = \sqrt{\frac{9.80 \times 10^{-3}}{30}} = 0.018\,\frac{cm}{s} \tag{6.7.1-39}$$

$$Re = \frac{D\,v_{terminal}\,\rho}{\mu} = \frac{3\,[mm]\,(0.018)\left[\frac{cm}{s}\right]\,(1.00)\left[\frac{gmm}{cm^3}\right]}{(1)\,cP\left[\frac{(0.01)\,gmm}{cm\;s\;cP}\right]}\left[\frac{cm}{(10)\,mm}\right]$$

$$Re = 0.542 \tag{6.7.1-40}$$

Reading the drag coefficient from the graph

$$C_D \approx 35 \qquad\qquad (6.7.1\text{-}41)$$

Therefore, the terminal velocity is somewhat less than 0.018 cm/s.

6.7.2 Drag in conduits - pipes (internal flow)

Common properties of pipe are shown in Table 6.7.2-1.[27]

The friction factor chart for pipe flows can be derived based on dimensional analysis principles via the dimensionless boundary layer equations. In the usual flow problem, we are not concerned with all the possible variables. Normally, we will be most concerned (for incompressible fluid flowing in a horizontal uniform pipe) with the pressure drop, Δp; the length, L; the diameter, D; the roughness height of the pipe surface, k; and, for the fluid, the average velocity, $<v>$; the viscosity, μ; and the density, ρ. (The interpretation of roughness height is a statistical problem that has not yet been resolved satisfactorily - we will not consider the statistics here but will use an effective roughness height.) The relationship among the variables is

$$g_1 \left(\Delta p, L, D, \langle v \rangle, \mu, \rho, k \right) = 0 \qquad\qquad (6.7.2\text{-}1)$$

By Buckingham's theorem, a complete set of dimensionless products for these variables is

$$G_1 \left(Eu, Re, \frac{L}{D}, \frac{k}{D} \right) = 0 \qquad\qquad (6.7.2\text{-}2)$$

The pressure drop in a uniform horizontal pipe is related to the shear stress at the wall, which in fully developed steady flow is constant down the pipe; therefore, the pressure drop is proportional to the length of the pipe and

$$\Delta p = \frac{1}{2} \rho \langle v \rangle^2 \frac{L}{D} G_2 \left(Re, \frac{k}{D} \right) \qquad\qquad (6.7.2\text{-}3)$$

---

[27] Reprinted from ASME A17.4 by permission of The American Society of Mechanical Engineers

Table 6.7.2-1 Properties of pipe

Nominal pipe size, in.	Outside diameter, in.	Schedule	Wall thickness, in.	Inside diameter, in.	Cross-sectional area Metal, in.²	Cross-sectional area Flow, ft²	Circumference, ft, or surface, ft²/ft of length Outside	Circumference, ft, or surface, ft²/ft of length Inside	Capacity at 1 ft/sec velocity U.S. gall/min	Capacity at 1 ft/sec velocity lb/hr water	Weight of plain-end pipe, lb/ft
⅛	0.405	10S	0.049	0.307	0.055	0.00051	0.106	0.0804	0.231	115.5	0.19
		40ST, 40S	0.068	0.269	0.072	0.00040	0.106	0.0705	0.179	89.5	0.24
		80XS, 80S	0.095	0.215	0.093	0.00025	0.106	0.0563	0.113	56.5	0.31
¼	0.540	10S	0.065	0.410	0.097	0.00092	0.141	0.107	0.412	206.5	0.33
		40ST, 40S	0.088	0.364	0.125	0.00072	0.141	0.095	0.323	161.5	0.42
		80XS, 80S	0.119	0.302	0.157	0.00050	0.141	0.079	0.224	112.0	0.54
⅜	0.675	10S	0.065	0.545	0.125	0.00162	0.177	0.143	0.727	363.5	0.42
		40ST, 40S	0.091	0.493	0.167	0.00133	0.177	0.129	0.596	298.0	0.57
		80XS, 80S	0.126	0.423	0.217	0.00098	0.177	0.111	0.440	220.0	0.74
½	0.840	5S	0.065	0.710	0.158	0.00275	0.220	0.186	1.234	617.0	0.54
		10S	0.083	0.674	0.197	0.00248	0.220	0.176	1.112	556.0	0.67
		40ST, 40S	0.109	0.622	0.250	0.00211	0.220	0.163	0.945	472.0	0.85
		80XS, 80S	0.147	0.546	0.320	0.00163	0.220	0.143	0.730	365.0	1.09
		160	0.188	0.464	0.385	0.00117	0.220	0.122	0.527	263.5	1.31
		XX	0.294	0.252	0.504	0.00035	0.220	0.066	0.155	77.5	1.71
¾	1.050	5S	0.065	0.920	0.201	0.00461	0.275	0.241	2.072	1036.0	0.69
		10S	0.083	0.884	0.252	0.00426	0.275	0.231	1.903	951.5	0.86
		40ST, 40S	0.113	0.824	0.333	0.00371	0.275	0.216	1.665	832.5	1.13
		80XS, 80S	0.154	0.742	0.433	0.00300	0.275	0.194	1.345	672.5	1.47
		160	0.219	0.612	0.572	0.00204	0.275	0.160	0.917	458.5	1.94
		XX	0.308	0.434	0.718	0.00103	0.275	0.114	0.461	230.5	2.44

Nom.	OD	Schedule								
1	1.315	5S	0.065	1.185	0.255	0.00768	0.344	0.310	3.449	1725	0.87
		10S	0.109	1.097	0.413	0.00656	0.344	0.287	2.946	1473	1.40
		40ST, 40S	0.133	1.049	0.494	0.00600	0.344	0.275	2.690	1345	1.68
		80XS, 80S	0.179	0.957	0.639	0.00499	0.344	0.250	2.240	1120	2.17
		160	0.250	0.815	0.836	0.00362	0.344	0.213	1.625	812.5	2.84
		XX	0.358	0.599	1.076	0.00196	0.344	0.157	0.878	439.0	3.66
1¼	1.660	5S	0.065	1.530	0.326	0.01277	0.435	0.401	5.73	2865	1.11
		10S	0.109	1.442	0.531	0.01134	0.435	0.378	5.09	2545	1.81
		40ST, 40S	0.140	1.380	0.668	0.01040	0.435	0.361	4.57	2285	2.27
		80XS, 80S	0.191	1.278	0.881	0.00891	0.435	0.335	3.99	1995	3.00
		160	0.250	1.160	1.107	0.00734	0.435	0.304	3.29	1645	3.76
		XX	0.382	0.896	1.534	0.00438	0.435	0.235	1.97	985	5.21
1½	1.900	5S	0.065	1.770	0.375	0.01709	0.497	0.463	7.67	3835	1.28
		10S	0.109	1.682	0.614	0.01543	0.497	0.440	6.94	3465	2.09
		40ST, 40S	0.145	1.610	0.800	0.01414	0.497	0.421	6.34	3170	2.72
		80XS, 80S	0.200	1.500	1.069	0.01225	0.497	0.393	5.49	2745	3.63
		160	0.281	1.338	1.429	0.00976	0.497	0.350	4.38	2190	4.86
		XX	0.400	1.100	1.885	0.00660	0.497	0.288	2.96	1480	6.41
2	2.375	5S	0.065	2.245	0.472	0.02749	0.622	0.588	12.34	6170	1.61
		10S	0.109	2.157	0.776	0.02538	0.622	0.565	11.39	5695	2.64
		40ST, 40S	0.154	2.067	1.075	0.02330	0.622	0.541	10.45	5225	3.65
		80ST, 80S	0.218	1.939	1.477	0.02050	0.622	0.508	9.20	4600	5.02
		160	0.344	1.687	2.195	0.01552	0.622	0.436	6.97	3485	7.46
		XX	0.436	1.503	2.656	0.01232	0.622	0.393	5.53	2765	9.03
2½	2.875	5S	0.083	2.709	0.728	0.04003	0.753	0.709	17.97	8985	2.48
		10S	0.120	2.635	1.039	0.03787	0.753	0.690	17.00	8500	3.53
		40ST, 40S	0.203	2.469	1.704	0.03322	0.753	0.647	14.92	7460	5.79
		80XS, 80S	0.276	2.323	2.254	0.02942	0.753	0.608	13.20	6600	7.66
		160	0.375	2.125	2.945	0.2463	0.753	0.556	11.07	5535	10.01
		XX	0.552	1.771	4.028	0.01711	0.753	0.464	7.68	3840	13.70

Table 6.2.7-1   Continued

Nominal pipe size, in.	Outside diameter, in.	Schedule	Wall thickness, in.	Inside diameter, in.	Cross-sectional area Metal, in.²	Flow, ft²	Circumference, ft, or surface, ft²/ft of length Outside	Inside	Capacity at 1 ft/sec velocity U.S. gall min	lb/hr water	Weight of plain-end pipe, lb/ft
3	3.500	5S	0.083	3.334	0.891	0.06063	0.916	0.873	27.21	13,605	3.03
		10S	0.120	3.260	1.274	0.05796	0.916	0.853	26.02	13,010	4.33
		40ST, 40S	0.216	3.068	2.228	0.05130	0.916	0.803	23.00	11,500	7.58
		80XS, 80S	0.300	2.900	3.016	0.04587	0.916	0.759	20.55	10,275	10.25
		160	0.438	2.624	4.213	0.03755	0.916	0.687	16.86	8430	14.31
		XX	0.600	2.300	5.466	0.02885	0.916	0.602	12.95	6475	18.58
3½	4.0	5S	0.083	3.834	1.021	0.08017	1.047	1.004	35.98	17,990	3.48
		10S	0.120	3.760	1.463	0.07711	1.047	0.984	34.61	17,305	4.97
		40ST, 40S	0.226	3.548	2.680	0.06870	1.047	0.929	30.80	15,400	9.11
		80XS, 80S	0.318	3.364	3.678	0.06170	1.047	0.881	27.70	13,850	12.51
4	4.5	5S	0.083	4.334	1.152	0.10245	1.178	1.135	46.0	23,000	3.92
		10S	0.120	4.260	1.651	0.09898	1.178	1.115	44.4	22,200	5.61
		40ST, 40S	0.237	4.026	3.17	0.08840	1.178	1.054	39.6	19,800	10.79
		80XS, 80S	0.337	3.826	4.41	0.07986	1.178	1.002	35.8	17,900	14.98
		120	0.438	3.624	5.58	0.07170	1.178	0.949	32.2	16,100	19.01
		160	0.531	3.438	6.62	0.06647	1.178	0.900	28.9	14,450	22.52
		XX	0.674	3.152	8.10	0.05419	1.178	0.825	24.3	12,150	27.54
5	5.563	5S	0.109	5.345	1.87	0.1558	1.456	1.399	69.9	34,950	6.36
		10S	0.134	5.295	2.29	0.1529	1.456	1.386	68.6	34,300	7.77
		40ST, 40S	0.258	5.047	4.30	0.1390	1.456	1.321	62.3	31,150	14.62
		80XS, 80S	0.375	4.813	6.11	0.1263	1.456	1.260	57.7	28,850	20.78
		120	0.500	4.563	7.95	0.1136	1.456	1.195	51.0	25,500	27.04
		160	0.625	4.313	9.70	0.1015	1.456	1.129	45.5	22,750	32.96
		XX	0.750	4.063	11.34	0.0900	1.456	1.064	40.4	20,200	38.55

Nominal	OD	Schedule									
6	6.25	5S	0.109	6.407	2.23	0.2239	1.734	1.667	100.5	50,250	7.60
		10S	0.134	6.357	2.73	0.2204	1.734	1.664	98.9	49,450	9.29
		40ST, 40S	0.280	6.065	5.58	0.2006	1.734	1.588	90.0	45,000	18.97
		80XS, 80S	0.432	5.761	8.40	0.1810	1.734	1.508	81.1	40,550	28.57
		120	0.562	5.501	10.70	0.1650	1.734	1.440	73.9	36,950	36.39
		160	0.719	5.187	13.34	0.1467	1.734	1.358	65.9	32,950	45.35
		XX	0.864	4.897	15.64	0.1308	1.734	1.282	58.7	29,350	53.16
8	8.625	5S	0.109	8.407	2.915	0.3855	2.258	2.201	173.0	86,500	9.93
		10S	148	8.329	3.941	0.3784	2.258	2.180	169.8	84,900	13.40
		20	0.250	8.125	6.578	0.3601	2.258	2.127	161.5	80,750	22.36
		30	0.277	8.071	7.260	0.3553	2.258	2.113	159.4	79,700	24.70
		40ST, 40S	0.322	7.981	8.396	0.3474	2.258	2.089	155.7	77,850	28.55
		60	0.406	7.813	10.48	0.3329	2.258	2.045	149.4	74,700	35.64
		80XS, 80S	0.500	7.625	12.76	0.3171	2.258	1.996	142.3	71,150	43.39
		100	0.594	7.437	14.99	0.3017	2.258	1.947	135.4	67,700	50.95
		120	0.719	7.187	17.86	0.2817	2.258	1.882	126.4	63,200	60.71
		140	0.812	7.001	19.93	0.2673	2.258	1.833	120.0	60,000	67.76
		XX	0.875	6.875	21.30	0.2578	2.258	1.800	115.7	57,850	72.42
		160	0.906	6.813	21.97	0.2532	2.258	1.784	113.5	56,750	74.69
10	10.75	5S	0.134	10.842	4.47	0.5993	2.814	2.744	269.0	134,500	15.23
		10S	0.165	10.420	5.49	0.5922	2.814	2.728	265.8	132,900	18.70
		20	0.250	10.250	8.25	0.5731	2.814	2.685	257.0	128,500	28.04
		30	0.307	10.136	10.07	0.5603	2.814	2.655	252.0	126,000	34.24
		40ST, 40S	0.365	10.020	11.91	0.5475	2.814	2.620	246.0	123,000	40.48
		80S, 60XS	0.500	9.750	16.10	0.5185	2.814	2.550	233.0	116,500	54.74
		80	0.594	9.562	18.95	0.4987	2.814	2.503	223.4	111,700	64.43
		100	0.719	9.312	22.66	0.4729	2.814	2.438	212.3	106,150	77.03
		120	0.844	9.062	26.27	0.4479	2.814	2.372	201.0	100,500	89.29
		140, XX	1.000	8.750	30.63	0.4176	2.814	2.291	188.0	94,000	104.13
		160	1.125	8.500	34.02	0.3941	2.814	2.225	177.0	88,500	115.64

Table 6.2.7-1 Continued

Nominal pipe size, in.	Outside diameter, in.	Schedule	Wall thickness, in.	Inside diameter, in.	Cross-sectional area Metal, in.²	Cross-sectional area Flow, ft²	Circumference, ft, or surface, ft²/ft of length Outside	Circumference, ft, or surface, ft²/ft of length Inside	Capacity at 1 ft/sec velocity U.S. gall/min	Capacity at 1 ft/sec velocity lb/hr water	Weight of plain-end pipe, lb/ft
12	12.75	5S	0.156	12.438	6.17	0.8438	3.338	3.26	378.7	189,350	22.22
		10S	0.180	12.390	7.11	0.8373	3.338	3.24	375.8	187,900	24.20
		20	0.250	12.250	9.82	0.8185	3.338	3.21	367.0	183,500	33.38
		30	0.330	12.090	12.88	0.7972	3.338	3.17	358.0	179,000	43.77
		ST, 40S	0.375	12.000	14.58	0.7854	3.338	3.14	352.5	176,250	49.56
		40	0.406	11.938	15.74	0.7773	3.338	3.13	349.0	174,500	53.52
		XS, 80S	0.500	11.750	19.24	0.7530	3.338	3.08	338.0	169,000	65.42
		60	0.562	11.626	21.52	0.7372	3.338	3.04	331.0	165,500	73.15
		80	0.688	11.374	26.07	0.7056	3.338	2.98	316.7	158,350	88.63
		100	0.844	11.062	31.57	0.6674	3.338	2.90	299.6	149,800	107.32
		120, XX	1.000	10.750	36.91	0.6303	3.338	2.81	283.0	141,500	125.49
		140	1.125	10.500	41.09	0.6013	3.338	2.75	270.0	135,000	139.67
		160	1.312	10.126	47.14	0.5592	3.338	2.65	251.0	125,500	160.27
14	14	5S	0.156	13.688	6.78	1.0219	3.665	3.58	459	229,500	22.76
		10S	0.188	13.624	8.16	1.0125	3.665	3.57	454	227,000	27.70
		10	0.250	13.500	10.80	0.9940	3.665	3.53	446	223,000	36.71
		20	0.312	13.376	13.42	0.9750	3.665	3.50	438	219,000	45.61
		30, ST	0.375	13.250	16.05	0.9575	3.665	3.47	430	215,000	54.57
		40	0.438	13.124	18.66	0.9397	3.665	3.44	422	211,000	63.44
		XS	0.500	13.000	21.21	0.9218	3.665	3.40	414	207,000	72.09
		60	0.594	12.812	25.02	0.8957	3.665	3.35	402	201,000	85.05
		80	0.750	12.500	31.22	0.8522	3.665	3.27	382	191,000	106.13
		100	0.938	12.124	38.49	0.8017	3.665	3.17	360	180,000	130.85
		120	1.094	11.812	44.36	0.7610	3.665	3.09	342	171,000	150.79
		140	1.250	11.500	50.07	0.7213	3.665	3.01	324	162,000	170.21
		160	1.406	11.188	55.63	0.6827	3.665	2.93	306	153,000	181.11

16

Schedule									
5S	0.165	15.670	8.18	1.3393	4.189	4.10	601	300,500	27.87
10S	0.188	15.624	9.34	1.3314	4.189	4.09	598	299,000	31.62
10	0.250	15.500	12.37	1.3104	4.189	4.06	587	293,500	42.05
20	0.312	15.376	15.38	1.2985	4.189	4.03	578	289,000	52.27
30, ST	0.375	15.250	18.41	1.2680	4.189	3.99	568	284,000	62.58
40, XS	0.500	15.000	24.35	1.2272	4.189	3.93	550	275,000	82.77
60	0.656	14.688	31.62	1.1766	4.189	3.85	528	264,000	107.50
80	0.844	14.312	40.19	1.1171	4.189	3.75	501	250,500	136.51
100	1.031	13.938	48.48	1.0596	4.189	3.65	474	237,000	164.82
120	1.219	13.562	56.61	1.0032	4.189	3.55	450	225,000	192.43
140	1.438	13.124	65.79	0.9394	4.189	3.44	422	211,000	223.57
160	1.594	12.812	72.14	0.8953	4.189	3.35	402	201,000	245.22

18

Schedule									
5S	0.165	17.670	9.25	1.7029	4.712	4.63	764	382,000	31.32
10S	0.188	17.624	10.52	1.6941	4.712	4.61	760	379,400	35.48
10	0.250	17.500	13.94	1.6703	4.712	4.58	750	375,000	47.39
20	0.312	17.376	17.34	1.6468	4.712	4.55	739	369,500	58.94
ST	0.375	17.250	20.76	1.6230	4.712	4.52	728	364,000	70.59
30	0.438	17.124	24.16	1.5993	4.712	4.48	718	359,000	82.15
XS	0.500	17.000	27.49	1.5763	4.712	4.45	707	353,500	93.45
40	0.562	16.876	30.79	1.5533	4.712	4.42	697	348,500	104.67
60	0.750	16.500	40.64	1.4849	4.712	4.32	666	333,000	138.17
80	0.938	16.124	50.28	1.4180	4.712	4.22	636	318,000	170.92
100	1.156	15.688	61.17	1.3423	4.712	4.11	602	301,000	207.96
120	1.375	15.250	71.82	1.2684	4.712	3.99	569	284,500	244.14
140	1.562	14.876	80.66	1.2070	4.712	3.89	540	270,000	274.22
160	1.781	14.438	90.75	1.1370	4.712	3.78	510	255,000	308.50

Table 6.2.7-1   Continued

Nominal pipe size, in.	Outside diameter, in.	Schedule	Wall thickness, in.	Inside diameter, in.	Cross-sectional area Metal, in.²	Cross-sectional area Flow, ft²	Circumference, ft, or surface, ft²/ft of length Outside	Circumference, ft, or surface, ft²/ft of length Inside	Capacity at 1 ft/sec velocity U.S. gall/min	Capacity at 1 ft/sec velocity lb/hr water	Weight of plain-end pipe, lb/ft
20	20	5S	0.188	19.624	11.70	2.1004	5.236	5.14	943	471,500	39.76
		10S	0.218	19.564	13.55	2.0878	5.236	5.12	937	467,500	45.98
		10	0.250	19.500	15.51	2.0740	5.236	5.11	930	465,000	52.73
		20, ST	0.375	19.250	23.12	2.0211	5.236	5.04	902	451,000	78.60
		30, XS	0.500	19.000	30.63	1.9689	5.236	4.97	883	441,500	104.13
		40	0.594	18.812	36.21	1.9302	5.236	4.92	866	433,000	123.11
		60	0.812	18.376	48.95	1.8417	5.236	4.81	826	413,000	166.40
		80	1.031	17.938	61.44	1.7550	5.236	4.70	787	393,500	208.87
		100	1.281	17.438	75.33	1.6585	5.236	4.57	744	372,000	256.10
		120	1.500	17.000	87.18	1.5763	5.236	4.45	707	353,500	296.37
		140	1.750	16.500	100.3	1.4849	5.236	4.32	665	332,500	341.09
		160	1.969	16.062	111.5	1.4071	5.236	4.21	632	316,000	379.17
24	24	5S	0.218	23.564	16.29	3.0285	6.283	6.17	1359	679,500	55.08
		10,10S	0.250	23.500	18.65	3.012	6.283	6.15	1350	675,000	63.41
		20,ST	0.375	23.250	27.83	2.948	6.283	6.09	1325	662,500	94.62
		XS	0.500	23.000	36.90	2.885	6.283	6.02	1295	642,500	125.49
		30	0.562	22.876	41.39	2.854	6.283	5.99	1281	640,500	140.68
		40	0.688	22.624	50.39	2.792	6.283	5.92	1253	626,500	171.29
		60	0.969	22.062	70.11	2.655	6.283	5.78	1192	596,000	238.35
		80	1.219	21.562	87.24	2.536	6.283	5.64	1138	569,000	296.58
		100	1.531	20.938	108.1	2.391	6.283	5.48	1073	536,500	367.39
		120	1.812	20.376	126.3	2.264	6.283	5.33	1016	508,000	429.39
		140	2.062	19.876	142.1	2.155	6.283	5.20	965	482,500	483.12
		160	2.344	19.312	159.5	2.034	6.283	5.06	913	456,500	542.13

30	30									
	5S	0.250	29.500	23.37	4.746	7.854	7.72	2130	1,065.000	79.43
	10.10S	0.312	29.376	29.10	4.707	7.854	7.69	2110	1,055.000	98.93
	ST	0.375	29.250	34.90	4.666	7.854	7.66	2094	1,084.000	118.65
	20.XS	0.500	29.000	46.34	4.587	7.854	7.59	2055	1,027.500	157.53
	30	0.625	28.750	57.68	4.508	7.854	7.53	2020	1,010.000	196.08

Extracted from American National Standard Wrought Steel and Wrought Iron Pipe, ANSI B36.10-1970, and Stainless Steel Pipe, ANSI B36.19-1965, with permission of the publishers, the American Society of Mechanical Engineers, New York, NY.

ST = standard wall; XS = extra strong wall; XX = double extra strong wall. Wrought iron pipe has slightly thicker walls, approximately 3 percent, but the same weight per foot as wrought steel pipe, because of lower density. Schedules 10, 20, 30, 40, 60, 80, 100, 120, 140, and 160 apply to steel pipe only. Decimal thicknesses for respective pipe sizes represent their nominal or average wall dimensions. Mill tolerances as high as 12 ½ percent are permitted.

Plain-end pipe is produced by a square cut. Pipe is also shipped from the mills threaded, with a threaded coupling on one end, or with the ends beveled for welding, or grooved or sized for patented couplings. Weights per foot for threaded and coupled pipe are slightly greater because of the weight of the coupling, but it is not available larger than 12 in., or lighter than schedule 30 sizes 8 through 12 in., or schedule 40 6 in. and smaller.

This equation is referred to as the Darcy equation. The function G(Re, k / D) is called the pipe *friction factor, f*. A chart that graphs the friction factor is known as a Stanton or Moody diagram. Figure 6.7.2-3 is such a diagram.

The point of transition from laminar flow to turbulent flow is at a critical velocity, $<v>_{critical}$. By dimensional analysis we can show

$$g_2 \left(\left\langle v \right\rangle_{critical}, D, \rho, \mu \right) = 0 \tag{6.7.2-4}$$

$$G_2 \left(Re_{critical} \right) = 0 \tag{6.7.2-5}$$

From experiment, Re_{cr} is approximately 2100 for circular pipes in ordinary applications. The transition is not well defined; in fact, one can obtain laminar flow up to about a Reynolds number of 10^4 in systems carefully isolated from disturbances such as vibration, etc. In ordinary pipes, however, there are sufficient external disturbances to trigger turbulence far below this point. It is not possible in going to laminar **from** turbulent flow to maintain **turbulent** flow much below a Reynolds number of about 2100.

The friction factor is commonly defined as some multiple of the proportionality factor in a model relating the drag force to the product of a characteristic area and a characteristic kinetic energy per unit volume.

$$F_{drag} = \left[(A)(KE) \right] f = \left[(2 \pi R L) \left(\tfrac{1}{2} \rho \left\langle v \right\rangle^2 \right) \right] f \tag{6.7.2-6}$$

Unfortunately, there are two friction factors commonly used to describe flow in pipes. The friction factor defined by Equation (6.7.2-6) is known as the **Fanning friction factor**. The other friction factor that is used is called the **Darcy (or Blasius) friction factor**. The relationship between the Fanning friction factor and the Darcy or Blasius friction factor is

$$f_{Darcy} = f_{Blasius} = 4 f_{Fanning} \tag{6.7.2-7}$$

In this text we shall assume an **unsubscripted** f to be understood as referring to the **Fanning** friction factor. (Conversion of formulas and/or graphical relationships from the Fanning to the Blasius or Darcy friction factor is, of course, trivial, but it is essential to keep in mind which friction factor is being used; otherwise an error of a factor of four is likely.)

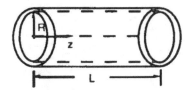

Figure 6.7.2-1 Momentum balance on cylindrical fluid element in horizontal pipe

The energy dissipated by friction is termed lost work or head loss. If we make a momentum balance on the system in Figure 6.7.2-1, assuming steady flow in a horizontal pipe and using the friction factor to calculate pressure drop, we see that in the x-direction the drag force at the wall balances the difference in the pressure forces on either end of the cylinder.

$$F_{pressure} + F_{drag} = 0 \tag{6.7.2-8}$$

$$p_1 \left(\pi R^2\right) - p_2 \left(\pi R^2\right) - \left(2 \pi R L\right)\left(\tfrac{1}{2} \rho \langle v \rangle^2\right) f = 0 \tag{6.7.2-9}$$

Solving

$$p_2 - p_1 = -\frac{2 f L \rho \langle v \rangle^2}{D} \tag{6.7.2-10}$$

The mechanical energy balance for this case reduces to

$$\frac{p_2 - p_1}{\rho} = -l\hat{w} \tag{6.7.2-11}$$

Substituting for ΔP in terms of the friction factor

$$l\hat{w} = \frac{2 f L \langle v \rangle^2}{D} \tag{6.7.2-12}$$

so the friction factor gives us a way to evaluate both pressure drop and lost work in horizontal pipes. We can extend the approach to non-horizontal pipes, as we now see.

If the pipe is not horizontal we have an additional force involved - the body (gravity) force on the fluid in the system.

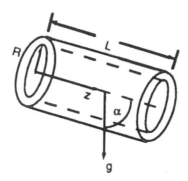

Figure 6.7.2-2 Momentum balance on cylindrical fluid element in non-horizontal pipe

Choosing the z-direction as the direction of flow, and letting α be the angle between the z-direction and the gravity vector as shown in Figure 6.7.2-2, the momentum balance yields

$$\sum F_z = 0 = F_{z,\text{pressure}} + F_{z,\text{drag}} + F_{z,\text{gravity}} \qquad (6.7.2\text{-}13)$$

The friction factor gives the drag force or lost work in **horizontal** pipe. We usually assume (an excellent assumption) that the lost work is unaltered by pipe orientation.

$$\left(p_1 - p_2\right)\left(\pi R^2\right) - \left(2\pi R L\right)\left(\tfrac{1}{2}\rho \langle v \rangle^2\right) f + g\rho\left(\pi R^2 L\right)\cos(\alpha) = 0$$

$$(6.7.2\text{-}14)$$

Solving

$$\left(p_2 - p_1\right) = -\frac{2fL\rho \langle v \rangle^2}{D} + g\rho L \cos\alpha \qquad (6.7.2\text{-}15)$$

If we rearrange the above equation, noting that $(L \cos \alpha) = -\Delta h$

$$\frac{\left(p_2 - p_1\right)}{\rho} + g\left(h_2 - h_1\right) = -\frac{2fL\langle v \rangle^2}{D} \qquad (6.7.2\text{-}16)$$

and compare with the mechanical energy balance for a uniform-size pipe with no shaft work

$$\frac{\left(p_2 - p_1\right)}{\rho} + g\left(h_2 - h_1\right) = -\hat{lw}$$

(6.7.2-17)

we see that

$$\hat{lw} = \frac{2fL\langle v\rangle^2}{D}$$

(6.7.2-18)

Thus, the friction factor gives us a way to evaluate lost work in non-horizontal pipes as well.

We now must have a way to determine f. If we make actual experiments (usually in horizontal, uniform-diameter pipe) we find

$$f = f\left(Re, \frac{k}{D}\right)$$

(6.7.2-19)

The group k/D is the dimensionless pipe roughness. In theory, it is supposed to represent the ratio of some sort of average roughness height to the pipe diameter. In fact, data were obtained by gluing particles of known uniform size to the wall of a smooth pipe, plotting the resulting pressure drop data (for which k/D is simply the ratio of particle diameter to pipe diameter), then taking pressure drop data for various types of commercial pipe and assigning values of k/D to the pipes, depending upon which data for artificially roughened pipe to which they correspond. Figure 6.7.2-3 is a plot obtained in this manner.

Figure 6.7.2-3 is called a Moody friction factor chart or Stanton diagram. It is based on data of Moody. The data can be presented in such a way as to give either a Fanning friction factor or a Darcy (or Blasius) friction factor. The commonly used chart shown here has as its ordinate the Darcy (or Blasius) friction factor, which is four times the Fanning friction factor.

The values of relative roughness for many commercial types of pipe are given in Figure 6.7.2-4.

We can obtain an analytic expression for the friction factor in laminar flow. Making a force balance on an element of fluid previously gave us

$$p_2 - p_1 = -\frac{2fL\rho\langle v\rangle^2}{D}$$

(6.7.2-20)

But the Hagen-Poiseuille equation gives us a relation between Δp and the other variables in laminar flow

$$p_2 - p_1 = -\pi R^2 \langle v \rangle \frac{8 \mu L}{\pi R^4} = \frac{32 \langle v \rangle \mu L}{D^2} \qquad (6.7.2\text{-}21)$$

Notice that in laminar flow, fluid density does not affect pressure drop. Substituting

$$\frac{32 \langle v \rangle \mu L}{D^2} = \frac{2 f L \rho \langle v \rangle^2}{D} \qquad (6.7.2\text{-}22)$$

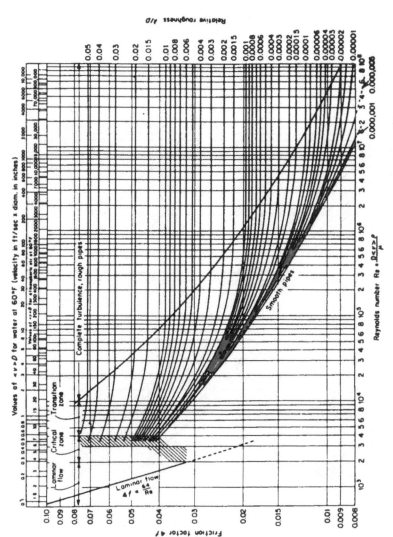

Figure 6.7.2-3 Moody friction factor chart[28]

$f_{Fanning} = f = f_{Darcy} / 4 = f_{Blasius} / 4$ - Note that the ordinate is in terms of 4f.

[28] Reprinted from the Pipe Friction Manual, 3rd ed., copyright 1961 by the Hydraulic Institute, 122 East 42nd St., New York, NY 10017. See Moody, L. F. Friction Factors for Pipe Flow. *Trans. ASME* 66(8): 671-684.

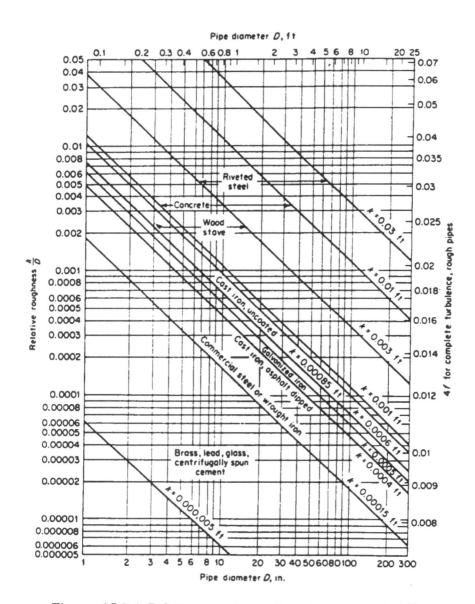

Figure 6.7.2-4 Relative roughness for clean new pipes[29]

---

[29] Moody, L. F. Friction Factors for Pipe Flow. Trans. *ASME* **66**(8): 671-684.

or

$$f = \frac{16}{Re} \tag{6.7.2-23}$$

Note that for laminar flow, f is not a function of pipe roughness. This line is labeled on Figure 6.7.2-3.

For the turbulent region, f gradually approaches a fairly constant value - in other words, f no longer depends on Re but is primarily a function of pipe roughness.

A qualitative explanation of the behavior of f is that at low Reynolds numbers, the flow is streamlined, and roughness protuberances contribute only skin friction, not form drag. At higher Reynolds numbers, the roughness protrudes beyond the viscous sublayer and form drag becomes important, overshadowing the skin friction contribution.

The universal velocity distribution can be used to fit the turbulent region of the friction factor chart. By substituting the universal velocity profile into the definition of the friction factor, and going through a great deal of algebraic manipulation, one can show that, in essence, the data of the friction factor chart for hydraulically smooth pipes is fitted by

$$\frac{1}{\sqrt{f}} = 4.06 \log\left(Re \sqrt{f}\right) - 0.6 \tag{6.7.2-24}$$

A portion of the turbulent region can also be represented by an empirical curve fit using the Blasius relation:

$$f = \frac{0.0791}{Re^{1/4}}; \qquad 2.1 \times 10^3 < Re < 10^5 \tag{6.7.2-25}$$

For rough pipes the universal velocity distribution yields the equation

$$\frac{1}{\sqrt{f}} = 4.06 \log\left(\frac{R}{k}\right) + 3.36 \tag{6.7.2-26}$$

In many practical situations it is more convenient to use the friction factor chart than the above algebraic forms. However, with computer calculations the various equations are often convenient to use.

In a general flow system, the pipe may not be of a uniform size and/or there may be various kinds of fittings, valves, and changes in pipe diameter which will cause losses. The lost work term in the mechanical energy balance must include these losses.

The losses from changes in pipe diameter can be estimated from the kinetic energy term in the energy balance. Losses caused by fittings and valves are most often treated by a model that replaces the fitting or valve with an equivalent length of straight pipe - that is, a pipe length that would give the same pressure loss as the fitting or valve at the same flow rate.

Equivalent lengths for standard-size fittings and valves are given in the nomograph of Figure 6.7.2-5. This nomograph will also give expansion and contraction losses for most cases except for sudden expansions into large regions, e.g., tanks. The latter type of situation is treated in Example 6.7-3.

Example 6.7.2-1   Expansion losses

The nomograph in Figure 6.7.2-5 does not give losses for sudden enlargements with diameter ratio greater than 4/1. Develop an appropriate formula for calculation of enlargement losses and use it to calculate the lost work involved in pumping water at 25 gal/min from a 1-in. Schedule 40 steel pipe into a pipe that is 4 times larger in diameter.

Solution

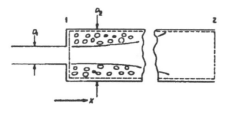

$$\frac{\Delta p}{\rho} + \frac{\Delta \langle v \rangle^2}{2} = -\hat{lw} \qquad\qquad (6.7.2\text{-}27)$$

Assume: Uniform pressure across the piping at 1 and 2, wall stress on the short pipe negligible, plug flow (turbulent)

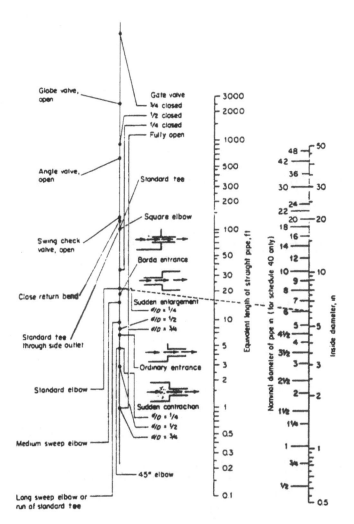

Figure 6.7.2-5 Equivalent lengths for losses in pipes[30]

[30] Crane Company, *Technical Paper No. 410: Flow of Fluids Through Valves, Fittings, and Pipe*, Twenty-fifth printing, 1991, Engineering Department: Joliet, IL. This report has its genesis in a booklet originally published by Crane in 1935, *Flow of Fluids and Heat Transmission*, which was published in a revised edition under the present title in 1942. A new edition was published as Technical Paper No. 410 in 1957, and has been continually updated. The present printing contains a more elaborate calculation method than in the nomograph above and should be consulted for more accurate estimates; however, the nomograph above is more compact and suffices for preliminary design calculations.

$$\Sigma F_x = p_1 A_2 - p_2 A_2 = \rho A_2 \langle v_2 \rangle \left[\langle v_2 \rangle - \langle v_1 \rangle \right] \qquad (6.7.2\text{-}28)$$

$$p_1 - p_2 = -\Delta p \qquad (6.7.2\text{-}29)$$

$$-\Delta p = \rho \frac{A_2}{A_1} \langle v_2 \rangle \left(\langle v_2 \rangle - \langle v_1 \rangle \right) \qquad (6.7.2\text{-}30)$$

$$-\hat{lw} = -\langle v_2 \rangle \left(\langle v_2 \rangle - \langle v_1 \rangle \right) + \frac{\langle v_2 \rangle^2 - \langle v_1 \rangle^2}{2} \qquad (6.7.2\text{-}31)$$

$$\langle v_1 \rangle A_1 \rho = \langle v_2 \rangle A_2 \rho \qquad (6.7.2\text{-}32)$$

$$\langle v_2 \rangle = \langle v_1 \rangle \frac{A_1}{A_2} \qquad (6.7.2\text{-}33)$$

$$-\hat{lw} = -\langle v_1 \rangle \frac{A_1}{A_2} \left(\langle v_1 \rangle \frac{A_1}{A_2} - \langle v_1 \rangle \right) + \left(\frac{\langle v_1 \rangle^2}{2} \frac{A_1^2}{A_2^2} - \frac{\langle v_1 \rangle^2}{2} \right) \qquad (6.7.2\text{-}34)$$

$$-\hat{lw} = -\langle v_1 \rangle^2 \left[\left(\frac{A_1}{A_2} \right)^2 - \frac{A_1}{A_2} - \frac{1}{2} \frac{A_1^2}{A_2^2} + \frac{1}{2} \right] \qquad (6.7.2\text{-}35)$$

$$\hat{lw} = \frac{\langle v_1 \rangle^2}{2} \left(1 - \frac{A_1}{A_2} \right)^2 = \frac{\langle v_1 \rangle^2}{2} \left[1 - \left(\frac{D_1}{D_2} \right)^2 \right]^2 \qquad (6.7.2\text{-}36)$$

For our situation we use the model that

$$\frac{D_1}{D_2} \cong 0 \qquad (6.7.2\text{-}37)$$

$$\hat{lw} = \frac{1}{2}\langle v_1 \rangle^2 = \frac{1}{2} \left[\left(\frac{25 \text{ gal}}{\text{min}} \right) \left(\frac{\text{ft}^3}{7.48 \text{ gal}} \right) \left(\frac{\text{min}}{60 \text{ sec}} \right) \left(\frac{1}{0.006 \text{ ft}^3} \right) \right]^2$$
$$\times \left(\frac{\text{lbf s}^2}{32.2 \text{ lbm ft}} \right)$$

$$\hat{lw} = 1.34 \frac{\text{ft lbf}}{\text{lbm}} \qquad (6.7.2\text{-}38)$$

<u>*Example 6.7.2-2*   *Direction of flow between tanks at differing*</u>
<u>*pressures and heights*</u>

Two tanks containing water and pressurized with air are connected as shown.

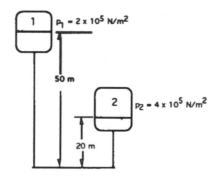

Prove whether the water flows from tank 1 to tank 2 or vice versa.

Solution

Writing the mechanical energy balance between surfaces 1 and 2

$$\frac{\Delta p}{\rho} + g \Delta z + \Delta\left(\frac{v^2}{2}\right) = -\hat{lw} - \hat{W}$$

$$\frac{\Delta p}{\rho} + g \Delta z = -\hat{lw} \qquad\qquad (6.7.2\text{-}39)$$

We know that the numerical value of the lost work term, $\left[-\hat{lw}\right]$, must be negative because the term represents generation in the entity balance on mechanical energy, and since it represents loss (destruction, negative generation) of mechanical energy its net sign must be negative. However, the lost work itself is made up of only positive terms

$$\hat{lw} = \frac{2fL\langle v\rangle^2}{D} \qquad\qquad (6.7.2\text{-}40)$$

Therefore, $\left[-\hat{lw}\right]$ must be negative.

Assume flow is from 1 to 2

$$\frac{(p_2 - p_1)}{\rho} + g(z_2 - z_1) = -\hat{lw} \tag{6.7.2-41}$$

$$\hat{lw} = -\left[\frac{(p_2 - p_1)}{\rho} + g(z_2 - z_1)\right] \tag{6.7.2-42}$$

$$\hat{lw} = -\left[\frac{\left(4 \times 10^5 - 2 \times 10^5\right)\left[\frac{N}{m^2}\right]}{(1000)\left[\frac{kg}{m^3}\right]} + (9.8)\left[\frac{m}{s^2}\right](20 - 50)[m]\left[\frac{N s^2}{kg\, m}\right]\right]$$

$$\hat{lw} = 94\,\frac{N\,m}{kg} \tag{6.7.2-43}$$

$$-\hat{lw} = -94\,\frac{N\,m}{kg} \tag{6.7.2-44}$$

Therefore, the flow is from 1 to 2.

Example 6.7.2-3   Friction loss in a piping system

A distilled oil is transferred from a tank at 1 atm absolute pressure to a pressure vessel at 50 psig by means of the piping arrangement shown. The oil flows at a rate of 23,100 lbm/hr through a 3-in. Schedule 40 steel pipe. The total length of straight pipe in the system is 450 ft. Calculate the minimum horsepower input to a pump having an efficiency of 60 percent. The oil has a density of 52 lbm/ft^3 and viscosity of 3.4 cP.

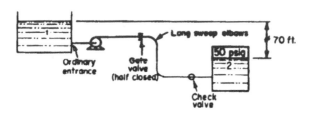

Solution

We first evaluate the terms in the mechanical energy balance using a system which is from liquid surface 1 to liquid surface 2:

$$\frac{\Delta p}{\rho} = \left[(50-0)\frac{lb_f}{in^2}\right]\left(144\frac{in^2}{ft^2}\right)\left(\frac{ft^3}{52\,lbm}\right) = 138\,\frac{ft\,lbf}{lbm} \qquad (6.7.2\text{-}45)$$

$$g\,\Delta z = \left(\frac{32.2\,ft}{s^2}\right)\left[\frac{(0-70)\,ft\,lbf\,s^2}{32.2\,lbm\,ft}\right] = -70\,\frac{ft\,lbf}{lbm} \qquad (6.7.2\text{-}46)$$

$$\frac{\Delta\langle v\rangle^2}{2} = 0 \qquad (6.7.2\text{-}47)$$

The equivalent length of pipe in the system is

$$L_{eq} = 450 + 4.5 + 50 + (2)\,(5) + 20 = 535\,ft \qquad (6.7.2\text{-}48)$$

$$\langle v\rangle = \frac{Q}{A} = \left(\frac{23{,}100\,lbm}{hr}\right)\left(\frac{ft^3}{52\,lbm}\right)\left(\frac{hr}{3{,}600\,s}\right)\left(\frac{1}{0.0513\,ft^2}\right)$$

$$= 2.41\,\frac{ft}{s} \qquad (6.7.2\text{-}49)$$

$$Re = \left(\frac{3.068\,ft}{12}\right)\left(\frac{2.41\,ft}{s}\right)\left(\frac{52\,lbm}{ft^3}\right)\left[\frac{cP\,ft\,s}{(3.4\,cP)\left(6.72\times10^{-4}\,lbm\right)}\right]$$

$$= 14{,}000 \qquad (6.7.2\text{-}50)$$

Using the friction factor plot and pipe roughness chart:

$$\frac{k}{D} = 0.0006$$

$$f = 0.007 \qquad (6.7.2\text{-}51)$$

The total lost work is the sum of the loss calculated using the friction factor chart plus the loss from the sudden expansion into tank 2 (which we cannot get from the friction factor approach, and therefore calculate as in Example 6.7.2-1).

$$\hat{lw} = \frac{2fL\,\langle v\rangle^2}{D} + \frac{\langle v\rangle^2}{2}\left[1-\left(\frac{D_1}{D_2}\right)^2\right]^2$$

$$= \left[\frac{(2)\,(0.007)\,(535)\,ft\,(2.41)^2}{3.068\,in}\right]\left(\frac{ft^2}{s^2}\right)\left(\frac{12\,in.}{ft}\right)\left(\frac{1\,lbf\,s^2}{32.2\,lbm\,ft}\right)$$

$$+ \left[\frac{(2.41)^2}{2}\right]\left(\frac{ft^2}{sec^2}\right)\left(\frac{lbf\,s^2}{32.2\,lbm\,ft}\right)$$

$$= 5.30 + 0.0903 = 5.31 \frac{\text{ft lbf}}{\text{lbm}}$$

(6.7.2-52)

Notice that the sudden expansion loss is negligible for this problem; it was included simply to illustrate the method of its inclusion. Substituting in the mechanical energy balance,

$$138 - 70 + 0 = -5.3 - \text{W}$$

$$\text{W} = -73.3 \frac{\text{ft lbf}}{\text{lbm}}$$

(6.7.2-53)

The sign is negative, indicating that the work is done on the fluid and not by the fluid. Since W is the actual work done on the fluid, the pump must have an input of

$$0.6 \, \text{W}_{\text{to pump}} = -73.3 \frac{\text{ft lbf}}{\text{lbm}}$$

$$\text{W}_{\text{to pump}} = -122 \frac{\text{ft lbf}}{\text{lbm}}$$

(6.7.2-54)

or

Power to pump

(6.7.2-55)

$$= -\left(122 \frac{\text{ft lbf}}{\text{lbm}}\right)\left(\frac{\text{hp min}}{33,000 \text{ ft lbf}}\right)\left(23,100\frac{\text{lbm}}{\text{hr}}\right)\left(\frac{\text{hr}}{60 \text{ min}}\right)$$

Power to pump $= -1.42$ hp

(6.7.2-56)

Friction factor calculations - serial paths

For serial path piping systems, the total lost work consists of losses due to friction in the straight pipe sections plus losses in fittings, etc.

$$\Delta p = f\left(L, Q, D, k, \Delta h, \rho, \mu, \text{configuration}\right)$$

(6.7.2-57)

If we assume for the moment that fluid properties are constant (incompressible isothermal Newtonian flow), and that roughness, elevation change and system configuration depend on choice of pipe and layout, then

$$\Delta p = f\left(L, Q, D\right) \qquad\qquad (6.7.2\text{-}58)$$

and there are four general cases which we will address by means of examples in the order

- Δp unknown

- D unknown

- L unknown

- $\dot{Q}$ (i.e., velocity) unknown

Case 1:   Pressure drop unknown

Example 6.7.2-4   Pressure loss for flow between tanks

Water is pumped at a rate of 150 gpm from tank A to tank B as shown. Each tank is 20 ft in diameter. The pressure in tank A is 10 psig and in tank B is 35 psig. All pipe is 2-inch Schedule 40 steel and the only fittings are the elbows shown, which are all standard.

Calculate the pressure at the pump outlet in psig.

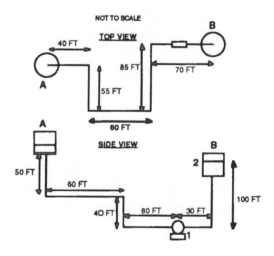

Solution

All that is necessary is to find the pressure drop from point 1 to point 2 and add it to the pressure in tank B. The lost work results from 130 feet of straight pipe plus one elbow.

From the pipe tables

   inside diameter = 2.067 inches
   flow cross sectional area = 0.02330 ft^2 (6.7.2-59)

and from the pipe nomographs

   relative roughness = 0.00085
   $L_{equivalent, standard elbow}$ = 5.5 ft   (6.7.2-60)

Then

$$\langle v \rangle = \frac{Q}{A_{xs}} = \frac{(150) \frac{gal}{min}}{(0.02330) \, ft^2} \frac{ft^3}{(7.48) \, gal} = 861 \, \frac{ft}{min} \qquad (6.7.2\text{-}61)$$

$$\text{Re} = \frac{D \langle v \rangle \rho}{\mu}$$

$$= \frac{\left(\frac{2.067}{12}\right) ft \, (861) \frac{ft}{min} \, (62.4) \frac{lbm}{ft^3} \, (60) \frac{min}{hr}}{(1) \, cP \, \frac{(2.42) \, lbm}{ft \, hr \, cP}}$$

$$= 2.29 \times 10^5 \qquad (6.7.2\text{-}62)$$

and from the friction factor plot

   4f = 0.018
   f = 0.0045   (6.7.2-63)

Applying the mechanical energy balance between 1 and 2

$$g \, \Delta z + \frac{\Delta p}{\rho} = - \frac{2 \, f \, L \, \langle v \rangle^2}{D} \qquad (6.7.2\text{-}64)$$

$$-\Delta p = \rho \left[\frac{2 f L \langle v \rangle^2}{D} + g \Delta z \right] \tag{6.7.2-65}$$

$$= 62.4 \frac{lbm}{ft^3} \left[\frac{(2)(0.0045)(135.5) \, ft \, (861)^2 \left[\frac{ft}{min}\right]^2}{\left(\frac{2.067}{12}\right) ft} \frac{min^2}{(3600) \, s^2} \right.$$

$$\left. + (32.2) \frac{ft}{s^2} (100) \, ft \right] \frac{lbf \, s^2}{32.2 \, lbm \, ft}$$

$$-\Delta p = 3415 \frac{mbf}{ft^2} \;\Rightarrow\; 23.7 \frac{lbf}{in^2} \tag{6.7.2-66}$$

Therefore, the pressure at the pump outlet is 35 + 23.7 = 58.7 psig.

Case 2: Diameter unknown

Example 6.7.2-5   Transfer line from tank to column

Benzene (density = 55 lbm/ft^3, viscosity = 0.65 cP) is stored in a vented tank and is to be pumped to the top of a vented absorption column at a controlled rate of 120 lbm/min. You have a pump with a 1/2-hp motor, which operates with an 80 percent total efficiency, and which has an outlet for 1-in. steel pipe.

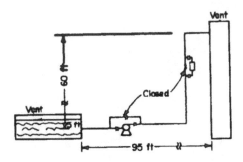

1. Can 1-in. Schedule 40 steel pipe be used throughout?

2. Could a smaller standard pipe size (still in Schedule 40 steel) be used throughout?

The system is shown in the sketch. The rotameter has flow resistance equivalent to 20 ft of pipe; standard elbows, standard tees, and gate valves are to be used in the system. (Neglect kinetic energy changes).

Solution

From the mechanical energy balance and Figures 6.2-4 to 6.2-7

	Equivalent length, ft
1 Ordinary entrance	1.5
2 Runs of tee	3.2
4 Standard elbows	11.2
2 Tees (side outlet)	10.0
4 Open gate valves	2.4
1 Rotameter	20.0
Straight pipe (neglecting the expansion loss)	160.0
TOTAL	**208.3**

$$\langle v \rangle = \left(\frac{w}{\rho A}\right) = \left(120 \, \frac{lbm}{min}\right)\left(\frac{min}{60 \, s}\right)\left(\frac{ft^3}{55 \, lbm}\right)\left(\frac{1}{0.006 \, ft^2}\right)$$

$$= 6.06 \, \frac{ft}{s} \qquad\qquad (6.7.2\text{-}67)$$

$$Re = \left(\frac{1.049 \, ft}{12}\right)\left(6.06 \, \frac{ft}{s}\right)\left(\frac{55 \, lbm}{ft^3}\right)\left(\frac{1}{0.65 \, cP}\right)\left(\frac{cP \, ft \, s}{6.72 \times 10^{-4} \, lbm}\right)$$

$$= 6.68 \times 10^4 \qquad\qquad (6.7.2\text{-}68)$$

for 1-in. pipe

$$\frac{k}{D} = 0.0018. \qquad\qquad (6.7.2\text{-}69)$$

Then

$$f = 0.0063 \qquad\qquad (6.7.2\text{-}70)$$

Applying the mechanical energy balance:

$$- W = l\hat{w} + g \, \Delta z$$

$$-W = \frac{2fL\langle v \rangle^2}{D} + g\,\Delta z = \left[(2)(0.0063)(208)\,\text{ft}\right]\left[(6.06)^2\,\frac{\text{ft}^2}{\text{s}^2}\right]$$

$$(6.7.2\text{-}71)$$

$$(1.049\text{ in})\left(\frac{12\text{ in.}}{\text{ft}}\right)\left(\frac{\text{lbf s}^2}{32.2\text{ lbm ft}}\right) + \left(\frac{32.2,\ \text{ft}}{\text{s}^2}\right)(60\text{ ft})\left(\frac{\text{lbf s}^2}{32.2\text{ lbm ft}}\right)$$

$$(6.7.2\text{-}72)$$

Therefore, the energy required is

$$W = -\left(33.4\,\frac{\text{ft lbf}}{\text{lbm}}\right) - \left(60\,\frac{\text{ft lbf}}{\text{lbm}}\right) = 93.4\,\frac{\text{ft lbf}}{\text{lbm}} \qquad (6.7.2\text{-}73)$$

The energy available to the fluid is

$$\hat{W}_{\text{avail}} = -\left(\frac{0.8}{2}\text{hp}\right)\left(\frac{33,000\text{ ft lbf}}{\text{hp min}}\right)\left(\frac{\text{min}}{120\text{ lbm}}\right) = -110\,\frac{\text{ft lbf}}{\text{lbm}} \quad (6.7.2\text{-}74)$$

Thus, we can use 1-in. pipe throughout.

If we go to the next smaller size of standard pipe, 3/4 in., the equivalent length in feet will change but the equivalent length in diameters will be about the same, and so

$$L_{eq} = 160 + \left(\frac{3/4}{1}\right)(48) = 186\text{ ft} \qquad (6.7.2\text{-}75)$$

(We use nominal diameter with sufficient accuracy.) Also

$$\langle v \rangle_2 = 6.06\,\frac{A_1}{A_2} = 6.06\,\frac{0.006\text{ ft}^2}{0.00371\text{ ft}^2} = 9.8\,\frac{\text{ft}}{\text{s}} \qquad (6.7.2\text{-}76)$$

The Reynolds number is

$$Re = \left(6.68\times 10^4\right)\left(\frac{0.92\text{ in}}{1.049\text{ in}}\right)\left(\frac{9.8\,\frac{\text{ft}}{\text{s}}}{6.06\,\frac{\text{ft}}{\text{s}}}\right) = 9.46\times 10^4 \qquad (6.7.2\text{-}77)$$

$$\frac{k}{D} = \frac{0.00015}{3/4}(12) = 0.0024 \qquad (6.7.2\text{-}78)$$

f = 0.0061   (6.7.2-79)

and so

$$- \dot{W} = \frac{2 f L \langle v \rangle^2}{D} + g \Delta z$$

$$= \frac{(2)(0.0061)(186)(9.8)^2(12)}{(0.92)(32.2)} + \frac{(32.2)(60)}{(32.2)}$$

$$- \dot{W} = \left(89 \frac{ft \, lbf}{lbm}\right) + \left(60 \frac{ft \, lbf}{lbm}\right) = 149 \frac{ft \, lbf}{lbm} \qquad (6.7.2-80)$$

Since only $110 \frac{ft \, lbf}{lbm}$ is available, 3/4-in. pipe is too small.

Example 6.7.2-6   Minimum pipe diameter

Water is to be pumped through a 1000 ft of pipe at 1500 gal/min. The discharge pressure from the pump must not exceed approximately 140 psig. The outlet pressure must be at least 30 psig. What is the smallest pipe diameter that can be used?

Solution

$$\frac{\Delta p}{\rho} + g \Delta z = - \hat{lw} \qquad (6.7.2-81)$$

$$p_1 - p_2 = \frac{2 f L \rho \langle v \rangle^2}{D} \qquad (6.7.2-82)$$

$$\langle v \rangle = \frac{Q}{A} = \frac{4 Q}{\pi D^2} \qquad (6.7.2-83)$$

$$p_1 - p_2 = \frac{32 f L \rho Q^2}{\pi^2 D^5} \qquad (6.7.2-84)$$

$$Re = \frac{D \langle v \rangle \rho}{\mu} = \frac{4 Q}{\pi v D} \qquad (6.7.2-85)$$

Assume D = 4 in

$$Q = \left(1500\frac{\text{gal}}{\text{min}}\right)\left(\frac{\text{min}}{60\,\text{s}}\right)\left(\frac{\text{ft}^3}{7.48\,\text{gal}}\right) = 3.34\frac{\text{ft}^3}{\text{s}} \tag{6.7.2-86}$$

$$\text{Re} = \left(\frac{4}{\pi}\right)\left(3.34\frac{\text{ft}^3}{\text{s}}\right)\left(\frac{\text{s}}{1.2\times10^{-5}\,\text{ft}^2}\right)\left(\frac{1}{4.026\,\text{in}}\right)\left(\frac{12\,\text{in}}{\text{ft}}\right)$$
$$= 1.06\times10^6 \tag{6.7.2-87}$$

$$\frac{k}{D} = 0.000016 \;\Rightarrow\; 4\,f = 0.012 \tag{6.7.2-88}$$

$$p_1 - p_2 = \frac{8\,(4\,f)\,L\,\rho\,Q^2}{\pi^2\,D^5}$$
$$= \left(\frac{8}{\pi^2}\right)(0.012)(1000\,\text{ft})\left(\frac{62.4\,\text{lbm}}{\text{ft}^3}\right)\left(3.34\frac{\text{ft}^3}{\text{s}}\right)^2\left(\frac{1}{4.026\,\text{in}}\right)^5$$
$$\times\left(\frac{1728\,\text{in}^3}{\text{ft}^3}\right)\left(\frac{\text{lbf s}^2}{32.2\,\text{lbm ft}}\right)$$
$$= 343\frac{\text{lbf}}{\text{in}^2} > 110\frac{\text{lbf}}{\text{in}^2} = \Delta p_{\text{max}} \tag{6.7.2-89}$$

Assume D = 6 in

$$\text{Re} = \left(\frac{4.056}{6.065}\right)\left(1.06\times10^6\right) = 7.01\times10^5 \tag{6.7.2-90}$$

$$\frac{k}{D} = 0.000010 \;\Rightarrow\; 4\,f = 0.013 \tag{6.7.2-91}$$

$$p_1 - p_2 = (344)\left(\frac{0.013}{0.12}\right)\left(\frac{4.026}{6.065}\right)^5$$
$$= 48\frac{\text{lbf}}{\text{in}^2} < 110\frac{\text{lbf}}{\text{in}^2} = \Delta p_{\text{max}} \tag{6.7.2-92}$$

Assume D = 5 in

$$\text{Re} = \left(\frac{4.026}{5.047}\right)\left(1.06\times10^6\right) = 8.43\times10^5 \tag{6.7.2-93}$$

$$\frac{k}{D} = 0.000012 \;\Rightarrow\; 4\,f = 0.012 \tag{6.7.2-94}$$

$$p_1 - p_2 = (344)\left(\frac{.012}{.012}\right)\left(\frac{4.026}{5.0017}\right)^5$$
$$= 111\frac{\text{lbf}}{\text{in}^2} \cong 110\frac{\text{lbf}}{\text{in}^2} = \Delta p_{\text{max}} \tag{6.7.2-95}$$

Therefore, 5-inch nominal diameter pipe would seem to be appropriate.

Case 3:   Length unknown

Example 6.7.2-7   Air supply through hose

Air is to be supplied to a tool through a hose at 0.2 kg/sec and maximum pressure drop of 50 kPa/m^2 (which is low enough that we can consider the air as an incompressible fluid). The hose is to be 30 mm in diameter. The density and viscosity of the air at the conditions in the hose are 8.81 kg/m^3 and 1.8 x 10^{-5} kg/(m sec), respectively. What is the maximum length of the hose?

Solution

$$\left(\frac{p_2}{\rho} + \frac{\langle v_2 \rangle^2}{2} + g\, z_2 \right) - \left(\frac{p_1}{\rho} + \frac{\langle v_1 \rangle^2}{2} + g\, z_1 \right) = -\hat{lw} - \hat{W} \qquad (6.7.2\text{-}96)$$

$$\frac{p_1 - p_2}{\rho} = \hat{lw} = \frac{2\,f\,L \langle v \rangle^2}{D} \qquad (6.7.2\text{-}97)$$

$$\langle v \rangle = \frac{\omega}{\rho\,A} = \frac{4\,\omega}{\pi\,D^2\,\rho} = \left(\frac{4}{\pi}\right)\left(\frac{0.2\,\text{kg}}{\text{s}}\right)\left(\frac{1000}{30\,\text{m}}\right)^2 \left(\frac{\text{m}^3}{8.81\,\text{kg}}\right)$$

$$= 32.1 \frac{\text{m}}{\text{s}} \qquad (6.7.2\text{-}98)$$

$$\text{Re} = \frac{\rho \langle v \rangle D}{\mu} = \frac{\left(8.81 \frac{\text{kg}}{\text{m}^3}\right)\left(\frac{32.1\,\text{m}}{\text{sec}}\right)\left(\frac{30}{1000}\right)\text{m}}{\left(1.8 \times 10^{-5} \frac{\text{kg}}{\text{m s}}\right)}$$

$$= 4.71 \times 10^5 \qquad (6.7.2\text{-}99)$$

From the friction factor chart assuming hydraulically smooth hose

$$(4\,f) = 0.0135 \qquad (6.7.2\text{-}100)$$

Then

$$L = \left(\frac{p_1 - p_2}{\rho}\right)\left(\frac{2D}{f\langle v \rangle^2}\right)$$

$$= \left(\frac{0.5 \times 10^5 \, N}{m^2}\right)(2)\left(\frac{30}{1000} \, m\right)\left(\frac{m^3}{8.81 \, kg}\right)\left(\frac{4}{0.0135}\right)\left(\frac{s}{32.1 \, m}\right)^2\left(\frac{kg \, m}{N \, sec^2}\right)$$

$$L = 97.9 \, m$$

$$\text{(6.7.2-101)}$$

Case 4: Flow rate unknown

One method of supplying a relatively constant pressure to water lines used for fire fighting, drinking water, sanitary use, etc., is to maintain a rather large reservoir of water at sizable elevation in a standpipe, more commonly known as a water tower.

The idea behind using a standpipe is to use a supply pump running at a more or less steady volumetric rate to fill the standpipe, even though the demand for water withdrawal will at different times be greater or less than the rate of this supply pump. The storage capacity of the standpipe supplying the surge capacity is then utilized either to supply any extra demand above the pump capacity or to store the excess capacity when demand falls below pump rate. This permits the supply pump to be sized to supply only the average demand, rather than the peak demand.

The standpipe furnishes a pressurized water supply (for example, for fire fighting) that is independent of electrical power. In addition, it furnishes a water supply at constant pressure, independent of pulsations from the supply pump.

The kind of pump one would probably utilize for such an application is a centrifugal pump, which is typical of the class of pumps most widely used in manufacturing plants. Applications include booster pumps on the heat exchangers, product transfer pumps, etc.

A centrifugal pump consists of an impeller with vanes mounted on a shaft. As the impeller rotates, the centrifugal forces developed in the water contained in the case (volute) hurl the water from its entrance at the center of the impeller to the periphery, where it exits into the pipe it supplies.

One of the advantages of a centrifugal pump is that one can throttle the outlet. As a consequence of throttling, the fluid tends to recirculate within the casing of the pump and increase in temperature from the lost work. However, fairly extensive degrees of throttling are possible before this becomes a serious problem. The result of throttling a centrifugal pump is expressed in the form of a "characteristic curve" of the pump, which gives the volumetric output in terms of **head** (pressure at outlet divided by the density of the fluid being pumped) versus the volumetric throughput of the liquid.

A centrifugal pump does not supply a constant volumetric flow of water; rather, the volumetric flow supplied depends upon the hydraulic resistance of the system to which the water is supplied. An example follows of how one uses the characteristic curve for a centrifugal pump to determine the flow rate in the pipe network which the pump supplies.

Example 6.7.2-8   Flow rate unknown

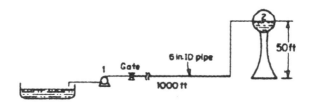

Water is supplied to a standpipe thorough 1000 equivalent feet of straight pipe, as shown above, by a pump with characteristic curve shown below. Assume that $\mu = 1$ cP, that $\rho = 62.4$ lbm/ft^3, and that the pipe is smooth.

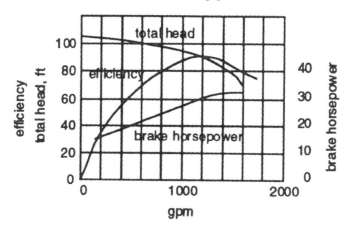

What will the flow rate of water be if the gate valve is fully open?

Solution

A mechanical energy balance from 1 to 2 (note this excludes the pump) yields

$$\Delta \frac{\langle v \rangle^2}{2} + g \, \Delta z + \frac{\Delta p}{\rho} = - l\bar{w} - W \tag{6.7.2-102}$$

$$\left(50 \text{ ft}\right)\left(32.2 \frac{\text{ft}}{\text{s}^2}\right)\left(\frac{\text{lbf s}^2}{32.2 \text{ lbm ft}}\right) + \left(\frac{0 - p_1}{62.4}\right)\left(\frac{\text{lbf ft}^3}{\text{in}^2 \text{ lbm}}\right)\left(144 \frac{\text{in}^2}{\text{ft}^2}\right)$$

$$= -\left(2 \text{ f}\right)\left(1{,}000 \text{ ft}\right) v^2 \left(\frac{\text{ft}^2}{\text{s}^2}\right)\left(\frac{\text{lbf s}^2}{32.2 \text{ lbm ft}}\right)\left(\frac{1}{0.5 \text{ ft}}\right)$$

$$50 - 2.31 \, p_1 = -124.4 \, \text{f} \, v^2 \tag{6.7.2-103}$$

p_1 is given by the pump characteristic curve
as a function of $\dot{Q}$, i.e., $\langle v \rangle$ A $\hspace{3cm}$ (6.7.2-104)

f is given by the friction factor chart
as a function of <v>, i.e., Re $\hspace{3cm}$ (6.7.2-105)

Therefore we have three equations, Equation (6.7.2-103), Equation (6.7.2-104), and Equation (6.7.2-105), in three unknowns, <v>, f, and p_1. Equation (6.7.2-104) and Equation (6.7.2-105) are in graphical form.

This solution converges relatively easily by trial, so we will not resort to a more sophisticated numerical method. Proceed as follows:

(1) Assume <v>.

(2) Calculate Re and read f from the friction factor chart.

(3) Determine p_1 from the pump characteristic curve.

(4) Substitute in Equation (6.7.2-103) to see if an identity is obtained.

Thus:

Assume <v> = 10 ft/s

$$\text{Re} = \frac{(0.5)(10)(62.4)}{(1)(6.72 \times 10^{-4})} = 4.64 \times 10^5 \quad \Rightarrow \quad f = 0.0033 \quad \text{(6.7.2-106)}$$

Then

$$Q = \langle v \rangle A = \left(10 \tfrac{\text{ft}}{\text{s}}\right)\left(\tfrac{\pi}{4}\right)\left[(0.5)^2 \text{ft}^2\right]\left(7.48 \tfrac{\text{gal}}{\text{ft}^3}\right)\left(60 \tfrac{\text{s}}{\text{min}}\right)$$

$$= 800 \text{ gpm} \qquad\qquad\qquad\qquad\qquad\qquad \text{(6.7.2-107)}$$

$$\tfrac{\Delta p}{\rho} = 95 \tfrac{\text{ft lbf}}{\text{lbm}} \qquad\qquad\qquad\qquad\qquad\qquad \text{(6.7.2-108)}$$

From the pump characteristic

$$\Delta p = \left(95 \tfrac{\text{ft lbf}}{\text{lbm}}\right)\left(62.4 \tfrac{\text{lbm}}{\text{ft}^3}\right)\left(\tfrac{\text{ft}^2}{144 \text{ in}^2}\right) = 41.2 \tfrac{\text{lbf}}{\text{in}^2} \qquad \text{(6.7.2-109)}$$

$$50 - (2.31)(41.2) = -(124.4)(0.0033)(100)$$
$$= -45.2 \neq -41.2 \qquad\qquad\qquad \text{(6.7.2-110)}$$

Making a new assumption:

Assume $\langle v \rangle$ = 5 ft/s

$$\text{Re} = 2.3 \times 10^5 \quad \Rightarrow \quad f = 0.0037 \qquad\qquad\qquad \text{(6.7.2-111)}$$

$$Q = 800 \left(\tfrac{5}{10}\right) = 400 \text{ gpm} \qquad\qquad\qquad\qquad \text{(6.7.2-112)}$$

$$\tfrac{\Delta p}{\rho} = 102 \tfrac{\text{ft lbf}}{\text{lbm}} \qquad\qquad\qquad\qquad\qquad\qquad \text{(6.7.2-113)}$$

$$\Delta p = 41.2 \left(\tfrac{102}{95}\right) = 44.2 \tfrac{\text{lbf}}{\text{in}^2} \qquad\qquad\qquad \text{(6.7.2-114)}$$

$$50 - (2.31)(44.2) = -(124.4)(0.0037)(100)$$
$$= -52 \neq -44.2 \qquad\qquad\qquad \text{(6.7.2-115)}$$

Making a new assumption:

Assume $\langle v \rangle$ = 4 ft/s

$$\text{Re} = 1.86 \times 10^5 \quad \Rightarrow \quad f = 0.0042 \qquad\qquad\qquad \text{(6.7.2-116)}$$

$$Q = 800 \left(\frac{4}{10}\right) = 320 \text{ gpm} \tag{6.7.2-117}$$

$$\frac{\Delta p}{\rho} = 103 \frac{\text{ft lbf}}{\text{lbm}} \tag{6.7.2-118}$$

$$\Delta p = 41.2 \left(\frac{103}{95}\right) = 44.6 \frac{\text{lbf}}{\text{in}^2} \tag{6.7.2-119}$$

$$50 - (2.31)(44.6) = -(124.4)(0.0042)(100)$$
$$= -53 \cong -52.7 \tag{6.7.2-120}$$

$$\therefore \ Q = 320 \text{ gpm} \tag{6.7.2-121}$$

In problems where the velocity is unknown, but instead the pressure drop is known, it is necessary to do a iterative calculation if the ordinary friction factor chart is used, as in Example 6.7.2-8, since the velocity is necessary to find f.

Problems of this kind are simplified by plotting the friction factor, f, as a function of the **Karman number**.

$$\text{Re} \sqrt{f} = \frac{D\rho}{\mu} \sqrt{\frac{\tilde{\text{lw}} D}{2L}} \equiv \text{Karman number} = \lambda \tag{6.7.2-122}$$

The motivation to construct such a plot comes from the fact that, in evaluating lost work when velocity is unknown, neither the ordinate nor the abscissa of the ordinary friction factor plot is known. When flow rate (i.e., velocity) is given, we normally first calculate from the chart the friction factor, which is basically

$$f = \varphi_1 \left[\langle v \rangle \right] \tag{6.7.2-123}$$

and then substitute f and <v> in the relation for lost work

$$\tilde{\text{lw}} = \frac{2fL \langle v \rangle^2}{D} = \varphi_2 \left[f, \langle v \rangle \right] \tag{6.7.2-124}$$

When pressure drop (i.e., lost work) is given, however, we cannot solve the above set of equations by successive substitution because two unknowns (<v> and f) appear in the relationship given in graphical form. We therefore rearrange the second equation to read

$$\frac{D \, \hat{lw}}{2 \, L \, \langle v \rangle^2} = f \tag{6.7.2-125}$$

and multiply both sides by $(D \langle V \rangle \, \rho/\mu)^2$ to obtain

$$\frac{D^2 \rho^2}{\mu^2} \frac{\hat{lw} \, D}{2 \, L} = Re^2 \, f \tag{6.7.2-126}$$

or

$$\frac{D \rho}{\mu} \sqrt{\frac{\hat{lw} \, D}{2 \, L}} = Re \sqrt{f} \tag{6.7.2-127}$$

This equation involves only one unknown, f, since we can calculate $\hat{lw}$ from the mechanical energy balance if we are given Δp (assuming no large kinetic energy changes). We can easily plot the relation between f and $Re \sqrt{f}$ from the friction factor plot by picking values of Re, reading f, calculating $Re \sqrt{f}$ and plotting. Such a plot is given in Figure 6.7.2-6.

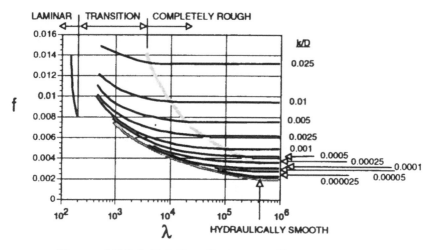

Figure 6.7.2-6 Friction factor vs. Karman number

$$\lambda = \frac{D \langle v \rangle \rho}{\mu} \sqrt{f} = \frac{D \rho}{\mu} \sqrt{\frac{\hat{lw} \, D}{2 \, L}}$$

If the pipe has the same diameter throughout the system, the velocity is not needed to evaluate the Karman number and f may be found directly. If, on the other hand, there is a velocity change large enough to be a significant part of $\hat{lw}$, even Figure 6.7.2-7 will not avoid the iterative solution.

Knowing f and $\hat{lw}$, we can easily get <v> from

$$\hat{lw} = \frac{2fL\langle v\rangle^2}{D} \tag{6.7.2-128}$$

Example 6.7.2-9   Calculation of flow rate via Karman number when pressure drop is known

Water at 110°F flows through a horizontal 2-in Schedule 40 pipe with equivalent length of 100 ft. A pressure gauge at the upstream end reads 30 psi, while one at the downstream end reads 15 psi. What is the flow rate in gallons per minute?

Solution

Applying the mechanical energy balance:

$$\frac{\Delta p}{\rho} + \frac{\Delta\langle v\rangle^2}{2} + g\,\Delta z = -\hat{lw} \tag{6.7.2-129}$$

$$-\hat{lw} = \left[(15-30)\frac{lbf}{in^2}\right]\left(\frac{ft^3}{61.8\ lbm}\right)$$

$$\hat{lw} = 35\,\frac{ft\ lbf}{lbm} \tag{6.7.2-130}$$

Calculating the Karman number:

$$\lambda = Re\sqrt{f} = \frac{D\rho}{\mu}\sqrt{\frac{\hat{lw}\,D}{2\,L}} \tag{6.7.2-131}$$

$$\lambda = \frac{\left[\frac{2.067}{12}\ ft\right]\left(61.8\,\frac{lbm}{ft^3}\right)}{(0.62\ cP)\left(6.72\times10^{-4}\frac{lbm}{ft\ s\ cP}\right)}$$

$$\times\sqrt{\frac{(35)\frac{ft\ lbf}{lbm}\left[\frac{2.067}{12}\ ft\right]\left(32.2\,\frac{lbm\ ft}{lbf\ s^2}\right)}{(2)(100)\ ft}} \tag{6.7.2-132}$$

$$\lambda = 2.51 \times 10^4$$

$$(6.7.2\text{-}133)$$

From the Karman plot, with $k/D = 0.0009$,

$$f = 0.005$$

$$(6.7.2\text{-}134)$$

Then

$$\hat{lw} = \frac{2\,f\,L\,\langle v \rangle^2}{D}$$

$$(6.7.2\text{-}135)$$

$$35\,\frac{\text{ft lbf}}{\text{lbm}} = \left\{ \frac{\left[(2)\,(0.005)\,(100)\,\text{ft}\right]\langle v \rangle^2}{\frac{2.067}{12}\,\text{ft}} \right\} \left(\frac{\text{lbf s}^2}{32.2\ \text{lbm ft}} \right)$$

$$(6.7.2\text{-}136)$$

$$\langle v \rangle = 13.9\,\tfrac{\text{ft}}{\text{s}}$$

$$(6.7.2\text{-}137)$$

Non-circular conduits

For non-circular conduits with turbulent flow, the hydraulic radius concept is often used. The hydraulic radius is a useful model which assumes that the resistance of a circular pipe with some equivalent diameter is the same as the non-circular conduit in question.

The **hydraulic radius** is defined as the flow cross-sectional area divided by the wetted perimeter of the non-circular conduit in question . The **equivalent diameter** is defined as **four** times the hydraulic radius (as opposed to the ordinary diameter, which is twice the radius). One then uses the equivalent diameter of the pipe in the Reynolds number for use in determining friction factors as usual.

$$R_H = \frac{S}{Z}$$

$$(6.7.2\text{-}138)$$

$$D_{eq} = 4\,R_H \quad \Rightarrow \quad Re = \frac{D_{eq}\,\langle v \rangle\,\rho}{\mu}$$

$$(6.7.2\text{-}139)$$

Example 6.7.2-10   Flow in a smooth annulus

Water is pumped at a rate of 0.8 cm³/s through the annular space formed by 10-cm-long horizontal glass tubes, the inner tube with outer diameter 2 mm and the outer tube with inner diameter 4 mm. Calculate the pressure drop in Pa.

Solution

$$R_H = \frac{S}{Z} = \frac{(\pi)\frac{(4)^2 \text{ mm}^2}{(4)} - (\pi)\frac{(2)^2 \text{ mm}^2}{(4)}}{(\pi)(2)\text{ mm} + (\pi)(4)\text{ mm}} = 0.5 \text{ mm} \qquad (6.7.2\text{-}140)$$

$$D_{eq} = 4R_H = (4)(0.5)\text{ mm} = 2 \text{ mm} \qquad (6.7.2\text{-}141)$$

$$\langle v \rangle = \frac{Q}{A} = \frac{(0.8)\frac{\text{cm}^3}{\text{s}}}{\left[(\pi)\frac{(4)^2 \text{ mm}^2}{(4)} - (\pi)\frac{(2)^2 \text{ mm}^2}{(4)}\right]} \frac{(100)\text{ mm}^2}{\text{cm}^2}$$

$$= 8.49 \frac{\text{cm}}{\text{s}} \qquad (6.7.2\text{-}142)$$

$$Re = \frac{D_{eq}\langle v \rangle \rho}{\mu}$$

$$= \frac{(2)\text{ mm}(8.49)\frac{\text{cm}}{\text{s}}(1)\frac{\text{gmm}}{\text{cm}^3}}{(1)\text{ cP}\frac{(0.01)\text{ gmm}}{\text{cm s cP}}} \frac{\text{cm}}{(10)\text{ mm}}$$

$$= 170 \qquad (6.7.2\text{-}143)$$

Since this is in the laminar flow regime, roughness is not a factor, and

$$4f = \frac{64}{Re}$$

$$f = \frac{16}{Re} = \frac{16}{170} = 0.094 \qquad (6.7.2\text{-}144)$$

$$-\Delta p = \frac{2\rho f L \langle v \rangle^2}{D_{eq}}$$

$$= \frac{(2)(1)\frac{gmm}{cm^3}(0.094)(10)\,cm\,(8.49)^2\left[\frac{cm}{s}\right]^2(10)\frac{mm}{cm}}{(2)\,mm}$$

$$\times\frac{Pa\,m^2}{N}\frac{N\,s^2}{kgm\,m}\frac{kgm}{(1000)\,gmm}\frac{(100)\,cm}{m}$$

$$-\Delta p = 6.78\,Pa$$

$$(6.7.2\text{-}145)$$

Example 6.7.2-11   Pressure drop in a pipe annulus

Water at 70°F is flowing at 10 ft/sec in the annulus between a 1-in. OD tube and 2-in. ID tube of a concentric tube heat exchanger. If the exchanger is 100-ft long, what is the pressure drop in the annulus? Neglect end effects.

Solution

Note that if we are going to manufacture an actual exchanger for a plant application, we certainly would not make one which was 100-ft long, but rather would make it in several sections and place the sections side by side.

$$R_H = \frac{\frac{\pi D_2^2}{4}-\frac{\pi D_1^2}{4}}{\pi D_2 + \pi D_1} = \frac{D_2^2 - D_1^2}{4(D_2 + D_1)} = \frac{(D_2 - D_1)(D_2 + D_1)}{4(D_2 + D_1)}$$

$$= \frac{D_2 - D_1}{4}$$

$$(6.7.2\text{-}146)$$

$$Re = \frac{D_{eq}\langle v\rangle\rho}{\mu} = \frac{4R_H\langle v\rangle\rho}{\mu}$$

$$= \frac{\left[\left(\frac{4}{4}\right)\left(\frac{2}{12}-\frac{1}{12}\right)ft\right]\left(10\,\frac{ft}{s}\right)\left(62.4\,\frac{lbm}{ft^3}\right)}{(1\,cP)\left(6.72\times10^{-4}\frac{lbm}{cP\,ft\,s}\right)}$$

$$= 7.75\times10^4$$

$$(6.7.2\text{-}147)$$

$$\frac{k}{D} = 0.0018 \quad \left(\text{based on } D_2 - D_1\right)$$

$$(6.7.2\text{-}148)$$

$$f = 0.0062$$

$$(6.7.2\text{-}149)$$

$$\Delta p = -\frac{2fL\rho\langle v\rangle^2}{D_{eq}}$$

$$= \left[(2) \, (0.0062) \, (100) \text{ ft} \right] \left(\frac{62.4 \text{ lbm}}{\text{ft}^3} \right)$$

$$\times \left\{ (10^2) \frac{\text{ft}^2}{\text{s}^2 \left[\left(\frac{4}{4} \right) \left(\frac{2}{12} - \frac{1}{12} \right) \text{ft} \right]} \right\} \left(\frac{\text{lbf s}^2}{32.2 \text{ lbm ft}} \right)$$

$$\Delta p = -2{,}880 \frac{\text{lbf}}{\text{ft}^2}$$

$$= -20 \text{ psi} \qquad\qquad (6.7.2\text{-}150)$$

Note that what we have done above is consistent with our equations for pipe, because the limit as D_1 approaches zero gives $D_2/4$ as the hydraulic radius. Four times the hydraulic radius is the pipe diameter D_2.

Friction factor calculations - parallel paths

The calculation of flow in piping systems containing branches (parallel paths) - pipe **networks** - is more complex, requiring numerical solutions most often done using a computer. In principle, the calculations are the same as for serial paths; however, systems of simultaneous nonlinear equations are generated, whereas for the serial path there was but a single equation.

Example 6.7.2-12   Pipe network with imposed pressure drop

The simple pipe network shown is constructed of Schedule 40 steel pipe with the dimensions indicated. The network has pressures imposed at the outlets as labeled.

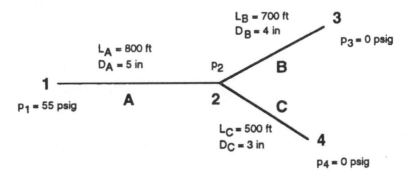

Calculate p_2 and the velocities in the pipes.

Solution

A mechanical energy balance on the ith pipe is

$$-\frac{(\Delta p)_i}{\rho} = \frac{2 f_i L_i \langle v_i \rangle^2}{D_i} \qquad\qquad (6.7.2\text{-}151)$$

Assume that the flow is in the completely rough regime so the friction factors will be constant - this assumption must be checked for accuracy at the end of the calculation.

Mechanical energy balance on pipe A:

From the pipe nomograph, for 5-in. diameter pipe in the completely turbulent region one can read directly from the right-hand ordinate

$$4f_A = 0.0153$$
$$f_A = 0.00383 \qquad\qquad (6.7.2\text{-}152)$$

Substituting in the mechanical energy balance for pipe A and simplifying

$$\frac{(55-p_2)\frac{lbf}{in^2}}{(62.4)\frac{lbm}{ft^3}} = \left[\frac{(2)(.0038)(800)\,ft\,\langle v_A\rangle^2\,\frac{ft^2}{s^2}}{\left(\frac{5}{12}\right)ft}\right]\left[\frac{lbf\,s^2}{(32.2)\,lbm\,ft}\right]\left[\frac{ft^2}{(144)\,in^2}\right]$$

$$(55-p_2) = 0.196\,\langle v_A\rangle^2 \qquad\qquad (6.7.2\text{-}153)$$

Mechanical energy balance on pipe B:

From the nomograph, for 4-in. pipe

$$4f_B = 0.0162$$
$$f_B = 0.00405 \qquad\qquad (6.7.2\text{-}154)$$

Substituting in the mechanical energy balance for pipe B and simplifying

$$\frac{(p_2 - 0)\frac{lbf}{in^2}}{(62.4)\frac{lbm}{ft^3}} = \left[\frac{(2)(.00405)(700) \, ft \, \langle v_B \rangle^2 \frac{ft^2}{s^2}}{\left(\frac{4}{12}\right) ft}\right]\left[\frac{lbf \, s^2}{(32.2) \, lbm \, ft}\right]\left[\frac{ft^2}{(144) \, in^2}\right]$$

$$\tag{6.7.2-155}$$

$$p_2 = 0.229 \langle v_B \rangle^2 \tag{6.7.2-156}$$

Mechanical energy balance on pipe C:

From the nomograph, for 3-in. pipe

$$4f_C = 0.0173$$
$$f_C = 0.00433 \tag{6.7.2-157}$$

Substituting in the mechanical energy balance for pipe C and simplifying

$$\frac{(p_2 - 0)\frac{lbf}{in^2}}{(62.4)\frac{lbm}{ft^3}} = \left[\frac{(2)(.00433)(500) \, ft \, \langle v_C \rangle^2 \frac{ft^2}{s^2}}{\left(\frac{3}{12}\right) ft}\right]\left[\frac{lbf \, s^2}{(32.2) \, lbm \, ft}\right]\left[\frac{ft^2}{(144) \, in^2}\right]$$

$$\tag{6.7.2-158}$$

$$p_2 = 0.233 \langle v_C \rangle^2 \tag{6.7.2-159}$$

A total mass balance on a control volume whose surface surrounds the pipes and cuts the entrance plane and each of the exit planes yields

$$(5)^2 \langle v_A \rangle = (4)^2 \langle v_B \rangle + (3)^2 \langle v_C \rangle \tag{6.7.2-160}$$

The system of equations (c), (e), (h), and (i) can be regarded as

$$f_1(x_1, x_2, x_3, x_4) = 0 = p_2 + 0.196 \langle v_A \rangle^2 - 55 \tag{6.7.2-161}$$
$$f_2(x_1, x_2, x_3, x_4) = 0 = p_2 - 0.229 \langle v_B \rangle^2 \tag{6.7.2-162}$$
$$f_3(x_1, x_2, x_3, x_4) = 0 = p_2 - 0.233 \langle v_C \rangle^2 \tag{6.7.2-163}$$
$$f_4(x_1, x_2, x_3, x_4) = 0 = (5)^2 \langle v_A \rangle - (4)^2 \langle v_B \rangle - (3)^2 \langle v_C \rangle \tag{6.7.2-164}$$

This is a system of non-linear algebraic equations which we can solve by Newton's method. We need the first partial derivatives

$$\frac{\partial f_1}{\partial p_2} = 1 \qquad \frac{\partial f_1}{\partial \langle v_A \rangle} = 0.392 \langle v_A \rangle \qquad \frac{\partial f_1}{\partial \langle v_B \rangle} = 0 \qquad \frac{\partial f_1}{\partial \langle v_C \rangle} = 0$$

$$\frac{\partial f_2}{\partial p_2} = 1 \qquad \frac{\partial f_2}{\partial \langle v_A \rangle} = 0 \qquad \frac{\partial f_2}{\partial \langle v_B \rangle} = -0.458 \langle v_B \rangle \qquad \frac{\partial f_2}{\partial \langle v_C \rangle} = 0$$

$$\frac{\partial f_3}{\partial p_2} = 1 \qquad \frac{\partial f_3}{\partial \langle v_A \rangle} = 0 \qquad \frac{\partial f_3}{\partial \langle v_B \rangle} = 0 \qquad \frac{\partial f_3}{\partial \langle v_C \rangle} = -0.466 \langle v_C \rangle$$

$$\frac{\partial f_4}{\partial p_2} = 0 \qquad \frac{\partial f_4}{\partial \langle v_A \rangle} = 25 \qquad \frac{\partial f_4}{\partial \langle v_B \rangle} = -16 \qquad \frac{\partial f_4}{\partial \langle v_C \rangle} = -9$$

Newton's method becomes a solution of

$$\begin{bmatrix} 1 & 0.392 \langle v_A \rangle_n & 0 & 0 \\ 1 & 0 & -0.458 \langle v_B \rangle_n & 0 \\ 1 & 0 & 0 & -0.466 \langle v_C \rangle_n \\ 0 & 25 & -16 & -9 \end{bmatrix} \begin{bmatrix} [\Delta p_2]_{n+1} \\ [\Delta \langle v_A \rangle]_{n+1} \\ [\Delta \langle v_B \rangle]_{n+1} \\ [\Delta \langle v_C \rangle]_{n+1} \end{bmatrix}$$

$$= - \begin{bmatrix} [p_2]_n + 0.196 \langle v_A \rangle_n^2 - 55 \\ [p_2]_n - 0.229 \langle v_B \rangle_n^2 \\ [p_2]_n - 0.233 \langle v_C \rangle_n^2 \\ (5)^2 \langle v_A \rangle - (4)^2 \langle v_B \rangle - (3)^2 \langle v_C \rangle \end{bmatrix} \qquad (6.7.2\text{-}165)$$

for the incremental values for the (n+1)st step [the values to be added to the nth step to get the values for the (n+1)st step] using the values for the nth step, followed by the addition of the incremental values to the values for the nth step to obtain the values for the (n+1)st step.

$$
\begin{bmatrix}
(p_2)_{n+1} \\
\langle v_A \rangle_{n+1} \\
\langle v_B \rangle_{n+1} \\
\langle v_C \rangle_{n+1}
\end{bmatrix}
=
\begin{bmatrix}
(p_2)_n \\
\langle v_A \rangle_n \\
\langle v_B \rangle_n \\
\langle v_C \rangle_n
\end{bmatrix}
+
\begin{bmatrix}
\Delta p_2 \\
\Delta \langle v_A \rangle \\
\Delta \langle v_B \rangle \\
\Delta \langle v_C \rangle
\end{bmatrix}
\qquad (6.7.2\text{-}166)
$$

For such a small system, the system of linear equations (n) can be solved easily using the matrix inversion program accompanying spreadsheet software. Results converge very rapidly, as shown in Table 6.7.2-2. The initial estimate of values was made on the basis that p_2 would be somewhere between the inlet and outlet pressures,[31] and the velocities were estimated using rule of thumb for economical pipe velocities, which should put the flow in the turbulent regime.

Table 6.7.2-2   Convergence of Newton's Method

n	0	1	2	3
p_2	30	29.72	29.72	29.72
v_A	10	11.45	11.36	11.36
v_B	10	11.49	11.39	11.39
v_C	10	11.38	11.29	11.29

We indicate the calculations for the first step, $n = 0$ to $n = 1$:

The coefficient matrix for the first step is

$$
\begin{bmatrix}
1 & 3.92 & 0 & 0 \\
1 & 0 & -4.58 & 0 \\
1 & 0 & 0 & -4.66 \\
0 & 25 & -16 & -9
\end{bmatrix}
\qquad (6.7.2\text{-}167)
$$

and its inverse can be computed to be

$$
\begin{bmatrix}
0.54036367 & 0.29599659 & 0.16363974 & -0.084729 \\
0.11725417 & -0.0755093 & -0.0417448 & 0.02161455 \\
0.11798333 & -0.1537125 & 0.0357292 & -0.0184998 \\
0.11595787 & 0.06351858 & -0.1794764 & -0.0181822
\end{bmatrix}
\qquad (6.7.2\text{-}168)
$$

---

[31] The pipeline is horizontal, so nowhere should pressure increase because of a potential energy change; likewise, there is no point in the network where there is conversion of kinetic to pressure energy - therefore, pressure should decrease monotonically.

The vector of constants on the right-hand sides for the first time step is

$$
\begin{array}{r}
5.4 \\
-7.1 \\
-6.7 \\
0
\end{array}
$$
(6.7.2-169)

Combining these results with the initial estimate of values gives for the incremental values

$$
\begin{array}{r}
-0.2799982 \\
1.44897915 \\
1.48908335 \\
1.37768278
\end{array}
$$
(6.7.2-170)

which, when added to the initial estimates, yields the values at n = 1 shown in Table 6.7.2-2.

Convergence was assumed to be satisfactory at n = 3, giving the final result that

p_2	29.72
v_A	11.36
v_B	11.39
v_C	11.29

The number of decimal places shown is far in excess of the accuracy of the calculation, and is shown only for the sake of illustration of the numerical method.

Checking the Reynolds numbers, using values for density and viscosity as indicated

$$
Re = \frac{D[in]\left[\dfrac{ft}{12\,in}\right]\langle v\rangle\left[\dfrac{ft}{s}\right]\rho\left[\dfrac{lb}{ft^3}\right]}{\mu[cP]\left[\dfrac{\left(6.72\times10^{-4}\right)lb}{ft\,s\,cP}\right]}
$$

$$= \frac{\frac{D}{(12)} \langle v \rangle (62.4)}{(1.0)(6.72 \times 10^{-4})} = (7.74 \times 10^3) D \langle v \rangle \qquad (6.7.2\text{-}171)$$

	A	B	C
D, in	5	4.00	3.00
v, ft/s	11.36	11.36	11.29
Re	4.40E+05	3.52E+05	2.62E+05

As can be seen, the assumption of turbulent flow was justified for all three pipes.

Example 6.7.2-13   Flow in a parallel piping system

A pipeline 20 miles long delivers 5,000 barrels of petroleum per day. The pressure drop over the line is 500 psig. Find the new capacity of the total system if a parallel and identical line is laid along the last 12 miles. The pressure drop remains at 500 psig and flow is laminar (see sketch).

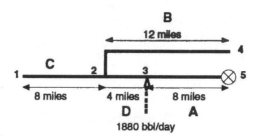

Solution

A mass balance yields

$$Q_T = Q_A + Q_B \qquad (6.7.2\text{-}172)$$

Neglecting the effect of the short connecting length

$$Q_A = Q_B \tag{6.7.2-173}$$

$$\therefore \; Q_B = \frac{Q_T}{2} \tag{6.7.2-174}$$

The overall pressure drop we denote as

$$\Delta p_T = 500 \, \text{psig} \tag{6.7.2-175}$$

$$\left(\frac{\Delta p}{\rho}\right)_A = \left(\frac{\Delta p}{\rho}\right)_B \tag{6.7.2-176}$$

$$\therefore \; \Delta p_T = \Delta p_C + \Delta p_B = \Delta p_C + \Delta p_A \tag{6.7.2-177}$$

But if we assume laminar flow

$$-\Delta p = \frac{32 \mu L \langle v \rangle}{D^2} = \frac{32 \mu L Q}{D^2 A} = \alpha L Q \tag{6.7.2-178}$$

We do not have physical properties, and so we evaluate α using the information given for the original line.

$$\left(144 \frac{\text{in}^2}{\text{ft}^2}\right)\left(500 \frac{\text{lbf}}{\text{in}^2}\right) = \alpha \, (20 \, \text{mi})\left(5{,}280 \frac{\text{ft}}{\text{mi}}\right)$$

$$\times \left(5{,}000 \frac{\text{barrels}}{\text{day}}\right)\left(\frac{\text{day}}{24 \, \text{hr}}\right)\left(\frac{\text{hr}}{3{,}600 \, \text{s}}\right)\left(\frac{42 \, \text{gal}}{\text{barrel}}\right)\left(\frac{\text{ft}^3}{7.48 \, \text{gal}}\right)$$

$$\tag{6.7.2-179}$$

$$\alpha = 2.1 \, \text{lbf} \, \text{sec} / \text{ft}^6 \tag{6.7.2-180}$$

Substituting in Equation (6.7.2-177) using Equation (6.7.2-178)

$$(500)\,(144) = (2.1)\,(8)\,(5{,}280)\,\dot{Q} + (2.1)\,(12)\,(5.280)\frac{\dot{Q}}{2} \tag{6.7.2-181}$$

$$Q = \frac{(500)\,(144)}{(2.1)\,(5{,}280)\,(14)}\left(\frac{\text{ft}^3}{\text{s}}\right)\left(7.48 \frac{\text{gal}}{\text{ft}^3}\right)\left(\frac{\text{barrel}}{42 \, \text{gal}}\right)\left(\frac{3{,}600 \, \text{s}}{\text{hr}}\right)\left(\frac{24 \, \text{hr}}{\text{day}}\right)$$

$$= 7{,}200 \, \text{barrels} / \text{day} \tag{6.7.2-182}$$

Example 6.7.2-14   Input of additional fluid to an existing pipe network

It is desired to add an additional plant to the 24-in. pipeline network designed in Example 6.7.2-13. The plant will produce 1,880 barrels/day and will be located at point 3 (see sketch). The pipeline network is to supply equal quantities of product to the plants at points 4 and 5.

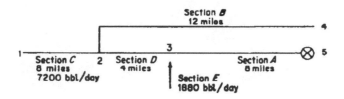

What pressure must a throttling valve at 5 maintain in order that $\dot{Q}_A = \dot{Q}_B$ (p_4 remains at 0 psig and the pressure at 1 is changed in such a way that the flow in Section C is maintained at 7,200 barrels/day).

Solution

The Hagen-Poiseuille equation yields

$$\Delta p = \frac{-32\,\mu\,L\,\langle v \rangle}{D^2} = \frac{-32\,\mu\,L\,Q}{D^2\,A} \tag{6.7.2-183}$$

$$\Delta p \,\frac{lbf}{in^2} = -\alpha\,L\,Q \tag{6.7.2-184}$$

or, converting the units of α so that we can use Δp in psi, L in miles, and $\dot{Q}$ in barrels/day,

$$\Delta p \,\frac{lbf}{in^2} = \left(-5 \times 10^{-3}\right)\left(\frac{psi\ day}{mi\ barrel}\right) L\ Q\left(\frac{mi\ barrel}{day}\right) \tag{6.7.2-185}$$

We now write down the equations describing our system. First, we apply the Hagen-Poiseuille law to each section of pipe:

Section C:

$$p_2 - p_1 = \left[-5 \times 10^{-3} (8)(7{,}200) \right] \qquad (6.7.2\text{-}186)$$

Section D:

$$p_3 - p_2 = \left(-5 \times 10^{-3} \right)(4) Q_D \qquad (6.7.2\text{-}187)$$

Section B:

$$0 - p_2 = \left(-5 \times 10^{-3} \right)(12) Q_B \qquad (6.7.2\text{-}188)$$

Section A:

$$p_5 - p_3 = \left(-5 \times 10^{-3} \right)(8) Q_A \qquad (6.7.2\text{-}189)$$

Total mass balances around points 2 and 3 yield

Point 2:

$$7{,}200 - Q_D - Q_B = 0 \qquad (6.7.2\text{-}190)$$

Point 3:

$$1{,}880 + Q_D - Q_A = 0 \qquad (6.7.2\text{-}191)$$

We also have the imposed restriction

$$Q_A - Q_B = 0 \qquad (6.7.2\text{-}192)$$

Note: We assume a direction of flow for stream D in Equations (6.7.2-190) and (6.7.2-191). If the direction assumed is wrong, the only effect is that the flow rate calculated for stream D will be negative; however, we must assume a consistent direction for any given stream in all our equations.

Equations (6.7.2-186) through (6.7.2-192) constitute a set of seven linearly independent equations in the seven unknowns p_1, p_2, p_3, p_5, $\dot{Q}_A$, $\dot{Q}_B$, and $\dot{Q}_D$, and can be written in matrix notation as

$$\begin{bmatrix} -1 & 0 & 0 & 0 & 0 & 0 & 0 \\ 0 & -1 & 1 & 0 & 0 & 0 & 0.02 \\ 0 & -1 & 0 & 0 & 0 & 0.06 & 0 \\ 0 & 0 & -1 & 1 & -0.04 & 0 & 0 \\ 0 & 0 & 0 & 0 & 1 & 1 & 0 \\ 0 & 0 & 0 & 1 & 0 & -1 & 0 \\ 0 & 0 & 0 & 1 & -1 & 0 & 0 \end{bmatrix} \begin{bmatrix} p_1 \\ p_2 \\ p_3 \\ p_5 \\ \dot{Q}_A \\ \dot{Q}_B \\ \dot{Q}_D \end{bmatrix} = \begin{bmatrix} -288 \\ 0 \\ 0 \\ 0 \\ 7200 \\ 1880 \\ 0 \end{bmatrix}$$

The set may be solved easily by elementary matrix methods, or alternatively solved as follows:

Use Equation (6.7.2-192) to substitute for $\dot{Q}_A$ in Equation (6.7.2-191), then add the result to Equation (6.7.2-190) to yield

$$-2\dot{Q}_B + 9{,}080 = 0 \qquad (6.7.2\text{-}193)$$
$$\dot{Q}_B = 4{,}540 \qquad (6.7.2\text{-}194)$$

Therefore, from Equation (6.7.2-192)

$$\dot{Q}_A = 4{,}540 \qquad (6.7.2\text{-}195)$$

and Equation (6.7.2-190) will give

$$\dot{Q}_D = 2{,}660 \qquad (6.7.2\text{-}196)$$

Equation (6.7.2-187) will then give

$$p_2 = 272 \qquad (6.7.2\text{-}197)$$

Equation (6.7.2-186) will give

$$p_3 = 219 \qquad (6.7.2\text{-}198)$$

following which Equation (6.7.2-185) gives

$$p_1 = 560 \tag{6.7.2-199}$$

and Equation (6.7.2-188) gives

$$p_5 = 37 \, \text{psig} \tag{6.7.2-200}$$

Notice that the pump at point 1 must now supply 560 psig, and that the fluid added at point 3 must come from a pump which will supply 219 psig. Both of these requirements would probably be feasible in a real-world situation.

This problem illustrates the general attack on pipe network problems: Write the independent equations describing the system; if these are fewer than the number of unknowns, the system is underdetermined and values must be assigned to some of the unknowns; if there are fewer unknowns than independent equations, the system is overdetermined and some variables that have been fixed will have to be considered unknowns.

6.8   Non-Newtonian Flow

6.8.1   Bingham plastics

As we discussed in Chapter 1, a Bingham plastic is modeled by

$$\tau_{yx} = \tau_0 - \mu_0 \frac{dv_z}{dy} \tag{6.8.1-1}$$

Consider the flow of Bingham plastic in a circular tube as shown in Figure 6.8.1-1.

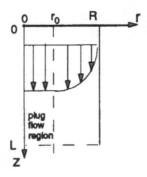

Figure 6.8.1-1 Tube flow of Bingham plastic

In cylindrical coordinates for a Bingham fluid with $r \geq r_0$, where $\tau = \tau_0$ at $r = r_0$ and $\mu = \mu_0$ where $\tau > \tau_0$

$$\tau_0 - \mu_0 \frac{dv_z}{dr} = \left(\frac{\mathcal{P}_0 - \mathcal{P}_L}{2L}\right) r \qquad (6.8.1-2)$$

Integrating

$$v_z = -\left(\frac{\mathcal{P}_0 - \mathcal{P}_L}{4\mu_0 L}\right) r^2 + \frac{\tau_0}{\mu_0} r + C_2 \qquad (6.8.1-3)$$

At $r = R$, $v_z = 0$, giving for $r \geq r_0$

$$0 = -\left(\frac{\mathcal{P}_0 - \mathcal{P}_L}{4\mu_0 L}\right) R^2 + \frac{\tau_0}{\mu_0} R + C_2 \qquad (6.8.1-4)$$

$$C_2 = \left(\frac{\mathcal{P}_0 - \mathcal{P}_L}{4\mu_0 L}\right) R^2 - \frac{\tau_0}{\mu_0} R \qquad (6.8.1-5)$$

So for $r \geq r_0$

$$\boxed{v_z = \left(\frac{\mathcal{P}_0 - \mathcal{P}_L}{4\mu_0 L}\right) R^2 \left[1 - \left(\frac{r}{R}\right)^2\right] - \frac{\tau_0 R}{\mu_0} \left[1 - \frac{r}{R}\right]}$$

$$(6.8.1-6)$$

Then for $r \leq r_0$, $v_z = v_z(r_0)$

$$v_z = \left(\frac{\mathcal{P}_0 - \mathcal{P}_L}{4\mu_0 L}\right) R^2 \left[1 - \left(\frac{r_0}{R}\right)^2\right] - \frac{\tau_0 R}{\mu_0}\left[1 - \frac{r_0}{R}\right] \tag{6.8.1-7}$$

But a force balance on the cylinder that comprises the plug flow zone shows

$$2\pi r_0 L \tau_0 = \left(\mathcal{P}_0 - \mathcal{P}_L\right)\pi r_0^2 \tag{6.8.1-8}$$

$$\tau_0 = \frac{\left(\mathcal{P}_0 - \mathcal{P}_L\right)}{2L} r_0 \tag{6.8.1-9}$$

Substituting in Equation (6.8.1-7)

$$v_z = \left(\frac{\mathcal{P}_0 - \mathcal{P}_L}{4\mu_0 L}\right) R^2 \left[1 - \left(\frac{r_0}{R}\right)^2\right] - \left[\frac{\left(\mathcal{P}_0 - \mathcal{P}_L\right)}{2L} r_0\right]\frac{R}{\mu_0}\left[1 - \frac{r_0}{R}\right] \tag{6.8.1-10}$$

$$v_z = \left(\frac{\mathcal{P}_0 - \mathcal{P}_L}{4\mu_0 L}\right) R^2 \left[1 - \left(\frac{r_0}{R}\right)^2\right] - \left[\frac{\left(\mathcal{P}_0 - \mathcal{P}_L\right)}{4\mu_0 L}\right] 2R^2 \frac{r_0}{R}\left[1 - \frac{r_0}{R}\right] \tag{6.8.1-11}$$

$$v_z = \left(\frac{\mathcal{P}_0 - \mathcal{P}_L}{4\mu_0 L}\right) R^2 \left[1 - 2\frac{r_0}{R} + \left(\frac{r_0}{R}\right)^2\right] \tag{6.8.1-12}$$

Which yields, for $r \le r_0$

$$v_z = \left(\frac{\mathcal{P}_0 - \mathcal{P}_L}{4\mu_0 L}\right) R^2 \left[1 - \left(\frac{r_0}{R}\right)\right]^2 \tag{6.8.1-13}$$

6.8.2 Power-law fluids

The power-law fluid analog of the Hagen-Poiseuille equation for a Newtonian fluid can be derived using the power law model. From the constitutive equation

$$\tau_{rz} = -m\left|\frac{dv_z}{dr}\right|^{n-1}\frac{dv_z}{dr}$$

$$= -m\left|-\frac{dv_z}{dr}\right|^{n-1}\left(-\frac{dv_z}{dr}\right)$$

$$= m \left| -\frac{dv_z}{dr} \right|^n \tag{6.8.2-1}$$

For a tube

$$\tau_{rz} = \left(\frac{\mathscr{P}_0 - \mathscr{P}_L}{2L} \right) r \tag{6.8.2-2}$$

Equating

$$m \left| -\frac{dv_z}{dr} \right|^n = \left(\frac{\mathscr{P}_0 - \mathscr{P}_L}{2L} \right) r \tag{6.8.2-3}$$

Letting $s = 1/n$

$$-\frac{dv_z}{dr} = \left(\frac{\mathscr{P}_0 - \mathscr{P}_L}{2\,mL} \right)^s r^s \tag{6.8.2-4}$$

Integrating

$$-v_z = \left(\frac{\mathscr{P}_0 - \mathscr{P}_L}{2\,mL} \right)^s \left[\frac{r^{(s+1)}}{s+1} + C \right] \tag{6.8.2-5}$$

At $r = R$, $v_z = 0$, giving

$$C = -\left(\frac{\mathscr{P}_0 - \mathscr{P}_L}{2\,mL} \right)^s \left[\frac{R^{(s+1)}}{s+1} \right] \tag{6.8.2-6}$$

Substituting

$$v_z = \left(\frac{\mathscr{P}_0 - \mathscr{P}_L}{2\,mL} \right)^s \left[\frac{R^{s+1}}{s+1} \right] \left[1 - \left(\frac{r}{R} \right)^{(s+1)} \right] \tag{6.8.2-7}$$

Determining the flow rate

$$Q = \int_0^R v_z\, 2\,\pi\, r\, dr = 2\,\pi\, R^2 \int_0^1 v_z \frac{r}{R}\, d\frac{r}{R} \tag{6.8.2-8}$$

$$Q = 2\pi \left(\frac{\mathcal{P}_0 - \mathcal{P}_L}{2\,m\,L}\right)^s \frac{R^{s+3}}{s+1} \int_0^1 \left[1 - \left(\frac{r}{R}\right)^{(s+1)}\right]\left(\frac{r}{R}\right) d\left(\frac{r}{R}\right) \qquad (6.8.2\text{-}9)$$

$$Q = 2\pi \left(\frac{\mathcal{P}_0 - \mathcal{P}_L}{2\,m\,L}\right)^s \frac{R^{s+3}}{s+1} \left[\frac{1}{2}\left(\frac{r}{R}\right)^2 - \frac{1}{s+3}\left(\frac{r}{R}\right)^{s+3}\right]_{\frac{r}{R}=0}^{\frac{r}{R}=1} \qquad (6.8.2\text{-}10)$$

$$= 2\pi \left(\frac{\mathcal{P}_0 - \mathcal{P}_L}{2\,m\,L}\right)^s \frac{R^{s+3}}{s+1}\left[\frac{1}{2} - \frac{1}{s+3}\right] \qquad (6.8.2\text{-}11)$$

$$= 2\pi \left(\frac{\mathcal{P}_0 - \mathcal{P}_L}{2\,m\,L}\right)^s \frac{R^{s+3}}{s+1}\left[\frac{s+3-2}{2(s+3)}\right] \qquad (6.8.2\text{-}12)$$

$$= \pi \left(\frac{\mathcal{P}_0 - \mathcal{P}_L}{2\,m\,L}\right)^s \frac{R^{s+3}}{(s+3)} \qquad (6.8.2\text{-}13)$$

For $n = 1/s = 1$ and $m = \mu$, Equation (6.8.2-13) reduces to the Hagen-Poiseuille equation.

For a power law fluid, a friction factor approach can be used with a modified Reynolds number written as

$$Re_m = \frac{D^n \langle v \rangle^{(2-n)} \rho}{\gamma} \qquad (6.8.2\text{-}14)$$

where n and γ are properties of the particular non-Newtonian fluid. (For a Newtonian fluid $n = 1$ and $\gamma = \mu$.) This definition permits the usual chart for the friction factor to be used for the laminar flow regime.

Another modification uses the apparent viscosity. The parameters n and γ are characteristic of the particular non-Newtonian fluid and can be determined by doing flow measurements in the capillary tube. The diameter and velocity can be incorporated in the apparent viscosity by defining

$$\mu_{app} = \gamma \left(\frac{D}{\langle v \rangle}\right)^{(1-n)} \qquad (6.8.2\text{-}15)$$

$$Re_m = \frac{D \langle v \rangle \rho}{\mu_{app}} \qquad (6.8.2\text{-}16)$$

Note that the apparent viscosity is not a simple property of the fluid but depends on flow rate and geometry.

Example 6.8-1   Flow of polymer melt

Molten polyethylene is to be pumped through a 1-inch Schedule 40 steel pipe at the rate of .25 gal/min. Calculate the pressure drop per unit length.

Properties:

$$n = 0.49$$
$$\rho = 58 \text{ lbmass} / \text{ft}^3$$
$$\mu_{ap} = 22,400 \text{ Poise}$$

Solution

The Reynolds number is

$$Re_m = \frac{D \langle v \rangle \rho}{\mu_{ap}} = D \frac{\dot{Q}}{A} \frac{\rho}{\mu_{ap}}$$

$$= \left(\frac{1.049}{12} \text{ft} \right) \left[\left(\frac{0.25 \text{ gal}}{\text{min}} \right) \left(\frac{\text{ft}^3}{7.48 \text{ gal}} \right) \left(\frac{1}{.00211 \text{ ft}} \right) \right]$$

$$\times \left[\frac{\left(58 \frac{\text{lbmass}}{\text{ft}^3} \right) \left(\frac{\text{min}}{60 \text{ s}} \right)}{\left(2.2 \times 10^4 \text{ Poise} \right) \left(6.72 \times 10^{-2} \frac{\text{lbmass}}{\text{ft s Poise}} \right)} \right]$$

$$= 9.05 \times 10^{-4} \tag{6.8-30}$$

$$f = \frac{16}{Re} = \frac{16}{9.05 \times 10^{-4}} = 1.77 \times 10^4 \tag{6.8-31}$$

$$\Delta p = \frac{2 f L \rho \langle v \rangle^2}{D} \tag{6.8-32}$$

$$\frac{\Delta p}{L} = \frac{2 f \rho \langle v \rangle^2}{D}$$

$$= \left[(2) \left(1.77 \times 10^4 \right) (54) \frac{\text{lbmass}}{\text{ft}^3} \right] \left(\frac{12}{1.049 \, \text{ft}} \right)$$

$$\times \left(\frac{\text{lbforce s}^2}{32.2 \, \text{lbmass ft}} \right) \left[\left(\frac{0.25 \, \text{gal}}{\text{min}} \right) \left(\frac{\text{ft}^3}{7.48 \, \text{gal}} \right) \left(\frac{1}{0.00211 \, \text{ft}^2} \right) \left(\frac{\text{min}}{60 \, \text{s}} \right) \right]^2$$

$$= 4.59 \times 10^4 \, \frac{\text{lbforce} / \text{ft}^2}{\text{ft}}$$

$$= 318 \, \text{psi} / \text{ft} \qquad\qquad\qquad\qquad\qquad (6.8\text{-}33)$$

The apparent viscosity and therefore the pressure drop decreases with increasing flow rate.

6.9   Flow in Porous Media

Many process fluid flow operations involve porous media - flow in filters, packed bed catalytic reactors, absorption, desorption, and distillation units, to name just a few. Other examples of porous media flows involve environmental flows: flow of petroleum, natural gas, and water in underground reservoirs; flow of agricultural fertilizers, herbicides, and pesticides in soils; transport, absorption and adsorption in natural soils, specially prepared beds of rocks, sand, etc., or underground formations used for waste disposal of naturally occurring chemicals or water-borne contaminants such as sewage, microorganisms, etc. The bodies of mammals and fish offer multiple illustrations of flow in porous media: the general microcirculation in the capillary system, membrane flows, the flow in the alveolar system of the lungs, flow in the kidney, liver, and other internal organs, etc. Plants also use flow in porous media for the transport of nutrients and disposal of waste products.

A porous medium can be anything from a (consolidated) solid with holes in it - like a sponge - to an (unconsolidated) assembly of solid particles in contact at various points but not bonded together - like a bag of marbles. In the first case models tend to emphasize flow through passages (capillary models and models using the hydraulic radius approach); in the second case, models tend to be derived from flow around things (drag models).

The **matrix** is the material in which the holes are imbedded or past which the fluid flows. **Porosity** (void fraction) is a measure of the void space (usually the **effective porosity,** that is, the void space accessible to the face of the media via connected passages as opposed to voids that are not accessible from the surface) and is often modeled either as uniform or distributed according to some random scheme.

$$\text{porosity} \equiv \frac{\text{accessible void volume}}{\text{total volume (media volume plus all void volume)}} \equiv \varepsilon$$

One of the problems in characterizing real porous media is that the porosity is frequently a function of the applied pressure - that is, the media (solids) are **compressible,** which introduces complications in the same sort of way as going from an incompressible fluid to a compressible one.

The **accessible** void volume includes both **flow-through** pores, which have at least one inlet and at least one outlet that connects to the surface either directly or via other pores, and **dead-end** pores, which are cul-de-sacs, having diffusive inlets only (dead-end pores are important primarily in multicomponent systems). The inaccessible void volume consists of pores whose inlet and outlet (if either or both even exist) do not intersect a path which connects to the surface - isolated pores. In the following sections we will drop the adjective *accessible,* and assume that reference to "pore" is to be understood as "accessible pore," and references to "porosity" to be to "effective porosity" unless specifically otherwise stated.

The following sections illustrate some simple but typical models of the flow behavior of porous media.

6.9.1   Darcy's law

Flow at low superficial velocities through materials that exhibit relatively low porosity is often modeled by Darcy's "law," an empirical equation of motion describing flow rate as a function of pressure drop in a porous medium.[32] Darcy

---

[32] H. Darcy, *Les Fontaines publiques de la ville de Dijon* (Paris: Victor Dalmont, 1856) cited in Crichlow, H. B. (1977). *Modern Reservoir Engineering - A Simulation Approach.* Englewood Cliff, NJ, Prentice-Hall, Inc., p. 39. Darcy's law is not a law in the same sense as Newton's laws of motion; we will, however, drop the quotation marks as is common practice.

arrived at his law in the course of investigations of the hydrology of the fountains of Dijon, France.[33]

In words, **Darcy's law** states that[34]

> **The rate of flow of a fluid through a homogeneous medium is directly proportional to the pressure or hydraulic gradient and to the cross-sectional area normal to the direction of flow and inversely proportional to the viscosity of the fluid.**

In one-dimensional (x-direction) rectangular Cartesian coordinates the Darcy equation[35] takes the macroscopic form

$$[v_x]_\infty = -\frac{k}{\mu}\left[\frac{\Delta p}{\Delta x} + \rho\, g\,\frac{\Delta h}{\Delta x}\right] \tag{6.9.1-1}$$

and the microscopic form

$$[v_x]_\infty = -\frac{k}{\mu}\left[\frac{dp}{dx} + \rho\, g\,\frac{dh}{dx}\right] \tag{6.9.1-2}$$

The subscript infinity on the velocity means that it refers to the **approach** (or **superficial**, or **Darcy**, or **filter**) **velocity,** which is the velocity obtained by dividing the volumetric flow rate by the **total** cross-sectional area normal to flow (including both the area of the pores and the area of the media)

---

[33] Nield, D. A. and A. Bejan (1992). *Convection in Porous Media*. New York, NY, Springer-Verlag, p. 5.

[34] Crichlow, H. B. (1977). *Modern Reservoir Engineering - A Simulation Approach*. Englewood Cliff, NJ, Prentice-Hall, Inc., p. 22.

[35] The original form proposed for the Darcy equation was

$$Q = K\, A\,\frac{\Delta H}{L}$$

where the proportionality constant K (with dimension L/t) is called the hydraulic conductivity and H is the head of fluid (with dimension L)

$$\Delta H = \Delta z + \frac{\Delta p}{\rho\, g}$$

The hydraulic conductivity contains properties of both the fluid and the medium, and so it is more useful to split these effects by introducing the intrinsic permeability

$$K = [k]\left[\frac{\rho}{\mu}\right] g$$

$$v_{\infty} \equiv \frac{Q}{A_{total}}$$

(6.9.1-3)

The constant of proportionality, **k**, is called the **specific permeability** or the **intrinsic permeability**. It depends in a relatively complicated way on the geometry of the medium, and, in general, is a second-order tensor, not a simple scalar. We here see only one component of the second-order tensor. It would be convenient if the permeability bore some simple relationship to the porosity, but since pores generally vary widely in geometry, which is not described by porosity, such a relationship is possible only among media with similar (scalable) geometry.

For example, as shown in Figure 6.9.1-1, a reasonable correlation between permeability and sieve opening can be constructed over a wide range of particle diameters for uniform diameter spherical beads. If the beads are packed in a consistent manner (e.g., close packed, dumped, etc.) the void fraction will be independent of bead size (if the volume considered is large compared to the bead diameter) and

$$\text{screen size} \simeq \text{bead diameter} \simeq \varepsilon$$

so with a little algebra we could replace the ordinate with porosity.

If we examine the dimensions of permeability in the FMLt system using the units conversion represented by g_c, we see that the permeability must have units of L^2 (the same conclusion would, of course, be reached in any other fundamental set of units).

$$\left[\frac{L}{t}\right] \Leftrightarrow \frac{[?]}{\left[\frac{M}{Lt}\right]} \left\{ \frac{\left[\frac{F}{L^2}\right]}{[L]} \left[\frac{ML}{Ft^2}\right] + \left[\frac{M}{L^3}\right]\left[\frac{L}{t^2}\right]\left[\frac{L}{L}\right] \right\}$$

$$\left[\frac{L}{t}\right] \Leftrightarrow \frac{[?]}{\left[\frac{M}{Lt}\right]} \left[\frac{M}{L^2 t^2}\right]$$

$$[?] \Leftrightarrow \left[\frac{L}{t}\right]\left[\frac{M}{Lt}\right]\frac{1}{\left[\frac{M}{L^2 t^2}\right]} \Rightarrow [L^2]$$

(6.9.1-4)

The common unit of permeability for geophysical problems (such as are involved in underground reservoirs) is the **Darcy**. The Darcy is defined such that for one-dimensional horizontal flow of a fluid of one cP viscosity through a porous medium infinite in the y- and z-directions and 1-cm thick in the x-direction, subjected to a pressure drop of one atm between the faces, a volumetric

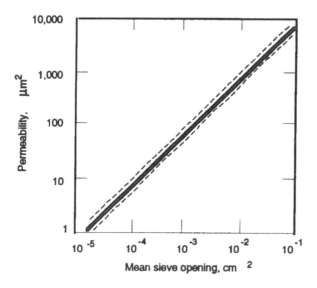

Figure 6.9.1-1 Permeability as a function of porosity for a bed of spheres (adapted from Stegemeier, G.L., Laumbach, D.D., and Volek, C.W. Representing Steam Processes with Vacuum Models. SPE 6787, 52nd Annual Fall Technical Conference and Exhibition, Society of Petroleum Engineers, Denver, CO (1977)

flux of one $(cm^3/s)/cm^2$ will be obtained if the medium has a permeability of one Darcy. (Note that $\Delta p/\Delta x$ is negative for flow in the positive x-direction.)

$$\frac{Q}{A_{total}} = -\frac{k}{\mu}\left[\frac{\Delta p}{\Delta x}\right]$$

$$\left|\frac{1\,\frac{cm^3}{s}}{1\,cm^2}\right| \Leftrightarrow \left[\frac{1\,Darcy}{1\,cP}\right]\left[\frac{1\,atm}{1\,cm}\right]\left[\frac{1.013\times10^5\,Pa}{atm}\right]$$

$$\left[\frac{\left(\frac{N}{m^2}\right)}{Pa}\right]\left[\frac{cP\,cm\,s}{0.01\,gmm}\right]\left[\frac{kgm\,m}{N\,s^2}\right]\left[\frac{100\,cm}{m}\right]\left[\frac{1000\,gmm}{kgm}\right]$$

1 Darcy $\Leftrightarrow$ 0.987 x 10^{-12} m^2

$$(6.9.1-5)$$

As is the case with the poise for viscosity, the Darcy is a relatively large permeability unit for many applications, so the milliDarcy is often more convenient to use.

We can see how the superficial velocity relates to our earlier definition of bulk velocity (which for porous media defines what is called the **pore velocity** or **interstitial velocity**) by writing the mass flow integral (from the macroscopic total mass balance) over the flow cross-section and recognizing that the normal component of the velocity is zero over the area occupied by media

$$Q = v_{\infty} A_{total} = \int_{A_{total}} \rho \left(v \cdot n \right) dA$$

$$= \int_{A_{pores}} \rho \left(v \cdot n \right) dA + \int_{A_{media}} \rho \left(v \cdot n \right) dA$$

$$= \int_{A_{pores}} \rho \left(v \cdot n \right) dA + 0$$

$$= \langle v \rangle A_{pores} \tag{6.9.1-6}$$

$$\frac{\text{pore velocity}}{\text{filter velocity}} = \frac{\langle v \rangle}{v_{\infty}} = \frac{A_{total}}{A_{pores}} \tag{6.9.1-7}$$

If we make the assumption that

$$\frac{A_{total}}{A_{pores}} = \frac{V_{total}}{V_{pores}} \tag{6.9.1-8}$$

we conclude that

$$\frac{\text{pore velocity}}{\text{filter velocity}} = \frac{\langle v \rangle}{v_{\infty}} = \frac{V_{total}}{V_{pores}} = \frac{1}{\varepsilon} \tag{6.9.1-9}$$

This is usually referred to as the Dupuit-Forcheimer relationship.

Darcy's original model was one dimensional and further assumed

1. A homogeneous and single-phase fluid

2. No chemical reactions between the media and the fluid

3. Permeability to be independent of fluid, temperature, pressure, and position

4. Laminar flow

5. No electrokinetic effects exist (potential differences that can be produced by flow - the **streaming** potential which is measured as the **zeta** potential)

6. Pore sizes to be sufficiently larger than the mean free path of the fluid molecules that no slippage at the pore walls occurs (slippage is called **Knudsen diffusion** or the **Klinkenberg effect**)

Darcy's law is valid for creeping flow, as defined by

$$Re_p = \frac{D_p \, v_\infty \, \rho}{\mu (1 - \varepsilon)} \le 1.0 \qquad (6.9.1\text{-}10)$$

where D_p is the effective particle diameter of the porous material and ε is the porosity.

An analog of a slight modification of Darcy's law is Ohm's law

$$E = I R$$

$$I = \frac{E}{R}$$

$$\left[v_x \right]_\infty A_{total} = Q = -\frac{k \, A_{total}}{\mu} \left[\frac{\Delta p}{\Delta x} + \rho \, g \, \frac{\Delta h}{\Delta x} \right]$$

$$I \Rightarrow Q$$

$$E \Rightarrow - \left[\frac{\Delta p}{\Delta x} + \rho \, g \, \frac{\Delta h}{\Delta x} \right]$$

$$R \quad \Rightarrow \frac{1}{\left(\frac{k \, A_{total}}{\mu} \right)} = \frac{\mu}{k \, A_{total}} \qquad (6.9.1\text{-}11)$$

Examine the units on the term analogous to the voltage (the potential term). Using the FMLt system and using the unit conversion represented by g_c

$$\left\{ \frac{\left[\frac{F}{L^2}\right]}{[L]} \left[\frac{ML}{Ft^2}\right] + \left[\frac{M}{L^3}\right]\left[\frac{L}{t^2}\right]\left[\frac{L}{L}\right] \right\} \Rightarrow \left[\frac{M}{L^2 t^2}\right]$$

$$\left[\frac{M}{L^2 t^2}\right]\left[\frac{Ft^2}{ML}\right] \Rightarrow \frac{[FL]}{[L][L^3]} = \frac{(\text{mechanical energy})}{(\text{length})(\text{volume})} \qquad (6.9.1\text{-}12)$$

The first and second terms incorporate two of the forms of energy from the mechanical energy balance: pressure energy and gravitational potential energy.

A logical question might be "why doesn't Darcy's law contain a term corresponding to the kinetic energy term of the mechanical energy balance?" The answer is that it should, but applications of the law are normally to situations where changes in kinetic energy are negligible at the scale over which the law is applied. Obviously, however, if we wish to model flows on the scale of individual pores which vary widely in effective diameter, such effects must be incorporated into any models we apply to predict pressure drop and flow rate.

Darcy's law is an empirical, macroscopic momentum equation. A large body of literature exists[36] which attempts by a variety of approaches (e.g., volume averaging) to relate it to the microscopic description of flow (the Navier-Stokes equation).

Some typical values of porosity and permeability for various porous materials are given in Table 6.9.1-1 and Table 6.9.1-2. Note that the permeability units are different in the two sections of the table.

---

[36] Kaviany, M. (1991). *Principles of Heat Transfer in Porous Media*. New York, NY, Springer-Verlag.

Table 6.9.1-1 Porosities (void fractions) for dumped packings[37]

$D_{particle}/D_{bed}$	$\varepsilon_{spheres}$	$\varepsilon_{cylinders}$
0.0	0.34	0.34
0.1	0.38	0.35
0.2	0.42	0.39
0.3	0.46	0.45
0.4	0.50	0.53
0.5	0.55	0.60

Extending the one-dimensional Darcy model to three dimensions in Cartesian coordinates

$$\left[v_i\right]_- = -\frac{k_{ij}}{\mu}\left[\frac{\partial p}{\partial x_j} + \rho\, g\, \frac{dh}{dx_j}\right]$$

$$\left[v_i\right]_- = -\frac{k_{ij}}{\mu}\left[\frac{\partial p}{\partial x_j} + \rho\, g_j\right] \tag{6.9.1-13}$$

In symbolic notation, valid for curvilinear coordinates

$$v_- = -\frac{k}{\mu}\left[\nabla p + \rho\, g\right] \tag{6.9.1-14}$$

---

[37] Leva, M. and M. Grummer (1947). *Chem Eng Prog* **43**: 713.

Table 6.9.1-2 Porosity and permeability for typical materials[38]

MATERIAL[39]	POROSITY	k (Darcies)
Berl Saddles	0.68 - 0.83	130,000. - 390,000.
brick	0.12 - 0.34	0.0048 - 0.22
catalyst granules	0.45	
cigarette filters	0.17 - 0.49	
coal	0.02 - 0.12	
concrete	0.02 - 0.07	
crushed rock, granular	0.44 - 0.45	
Fiberglass	0.88 - 0.93	
leather	0.56 - 0.59	
limestone	0.02 - 0.20	0.000001 - 2.
limestone (dolomite)	0.04 - 0.10	
Raschig rings	0.56 - 0.65	
sand	0.37 - 0.50	
sand	0.31 - 0.50	2. - 180.
sandstone	0.08 - 0.38	
sandstone	0.08 - 0.40	0.00000011 - 11.
silica powder	0.013 - 0.051	0.37 - 0.49
soil	0.43 - 0.54	0.29 - 14.
spherical packings, shaken	0.36 - 0.43	
wire rings	0.68 - 0.76	3,800. - 10,000.

SOILS	$k \ (m^2)$
clean gravel	$10^{-7} - 10^{-9}$
clean sand	$10^{-9} - 10^{-12}$
peat	$10^{-11} - 10^{-13}$
stratified clay	$10^{-13} - 10^{-16}$
unweathered clay	$10^{-16} - 10^{-20}$

Notice that the permeability has now become a second-order tensor rather than a scalar. This is because for three-dimensional flow the velocity does not

---

[38] Scheidegger, A. E. (1974). *The Physics of Flow Through Porous Media*. Toronto, Canada, University of Toronto Press.

[39] Values in part from Nield, D. A. and A. Bejan (1992). *Convection in Porous Media*. New York, NY, Springer-Verlag, p. 6.

necessarily point in the same direction as the pressure gradient or the gravitational field (it does for **isotropic media**, it does not for **anisotropic** media). The permeability, as a second-order tensor, relates these two directions - it associates a direction for the velocity with every direction for the pressure gradient or gravitational field and vice versa. The resistance to flow experienced by the fluid depends upon the direction in which it is moving.

Single-phase flow is described by the equation of continuity, using the relationship between the local and superficial velocities is

$$\int_A \rho \left(\epsilon \, v_- \cdot n\right) dA + \frac{d}{dt}\int_V \rho \, dV = 0 \qquad\qquad (6.9.1\text{-}15)$$

and Darcy's law, which is the equation of motion.

Example 6.9.1-1   Flow of water in sandstone

What is the volumetric flow rate of water through a horizontal sandstone core 10 cm long and 5 cm in diameter if k = 2 Darcies and $\Delta p = 1$ atm?

Solution

$$
\begin{aligned}
\left[v_x\right]_- &= -\frac{k}{\mu}\left[\frac{\Delta p}{\Delta x}\right] \\[4pt]
&= -\frac{(2)\,\text{Darcy}}{(1)\,\text{cP}}\frac{(1)\,\text{atm}}{(10)\,\text{cm}}\frac{\left(1.013\times 10^5\right)\text{Pa}}{\text{atm}}\frac{N}{\text{Pa}\,m^2}\frac{\text{kgm m}}{N\,s^2} \\[4pt]
&\quad \times \frac{\left(0.987\times 10^{-12}\right)m^2}{\text{Darcy}}\frac{(100)\,\text{cP cm s}}{\text{gmm}}\frac{(100)\,\text{gmm}}{\text{kgm}} \\[4pt]
&= 2\times 10^{-4}\,\frac{m}{s}
\end{aligned}
\qquad\qquad (6.9.1\text{-}16)
$$

Using the definition of superficial velocity

$$Q = \left[v_x\right]_- A = \left(2\times 10^{-4}\right)\frac{m}{s}\frac{(\pi)\,(5)^2\,cm^2}{(4)}\frac{(100)\,cm}{m} = 0.393\,\frac{cm^3}{s}$$

$$(6.9.1\text{-}17)$$

6.9.2   Packed beds[40]

Packed beds are just specific instances of flow through unconsolidated porous media (where we now refer to what we previously called "media" as "packing"). However, flow in what we consider a "packed bed" as opposed to "porous media" tends, in general, to approach more closely flow around objects than flow through channels, and interstitial velocities also tend to be higher. Hence, the appropriate mathematical models often have somewhat differing features.

Packed bed models are often based on friction factors in a similar fashion to the way we modeled flow in conduits. A friction factor for flow in a packed bed can be defined by using the same form that we used for flow in conduits of non-circular cross-section

$$\hat{lw} = \frac{2 f L v_{int}^2}{D_{eq}} \tag{6.9.2-1}$$

where v_{int} is the interstitial velocity, L is the thickness of the bed (length in the flow direction) and D_{eq} is the equivalent diameter of the **pores**.

Since we are usually interested in the pressure drop per unit length for systems in which the potential energy and kinetic energy changes are negligible, using the mechanical energy balance to substitute for the lost work term and rearranging, we write

$$-\frac{\Delta p}{L} = \frac{2 f \rho v_{int}^2}{D_{eq}} \tag{6.9.2-2}$$

We now proceed to replace the equivalent diameter of the pores, about which for a packed bed we usually know little, with a corresponding diameter of the packing, of which we usually know more. If we apply our definition of void fraction to find

---

[40] The important topic of momentum transfer in countercurrent gas-liquid flow in packed beds is deferred until design of mass transfer columns, because they constitute its primary application.

$$\frac{\text{void volume}}{\text{packing volume}} = \frac{\left(\frac{\text{void volume}}{\text{total volume}}\right)}{\left(\frac{\text{packing volume}}{\text{total volume}}\right)} \tag{6.9.2-3}$$

$$= \frac{\left(\frac{\text{void volume}}{\text{total volume}}\right)}{\left(\frac{[\text{total volume} - \text{void volume}]}{\text{total volume}}\right)} \tag{6.9.2-4}$$

$$= \frac{\left(\frac{\text{void volume}}{\text{total volume}}\right)}{\left(\frac{\text{total volume}}{\text{total volume}}\right) - \left(\frac{\text{void volume}}{\text{total volume}}\right)} \tag{6.9.2-5}$$

$$\frac{\text{void volume}}{\text{packing volume}} = \frac{\varepsilon}{1 - \varepsilon} \tag{6.9.2-6}$$

and further define for the packing (where surface area means area in contact with the fluid, and volume includes isolated pore spaces)

$$S \equiv \text{specific surface area of packing} \equiv \frac{\text{surface area packing}}{\text{packing volume}} = \frac{S_p}{V_p}$$

$$\tag{6.9.2-7}$$

the hydraulic radius of a packed bed may then be defined by

$$R_H = \frac{\text{void volume}}{\text{surface area packing}}$$

$$= \frac{\text{void volume}}{\text{packing volume}} \frac{\text{packing volume}}{\text{surface area packing}}$$

$$R_H = \left(\frac{\varepsilon}{1 - \varepsilon}\right)\left(\frac{1}{S}\right) \tag{6.9.2-8}$$

As before, we define $D_{eq} = 4\,R_H$. We also continue to make the assumption analogous to the Dupuit-Forcheimer relation

$$\frac{A_{total}}{A_{pores}} = \frac{V_{total}}{V_{pores}} = \frac{1}{\varepsilon} \tag{6.9.2-9}$$

The appropriate Reynolds number is then

$$Re = \frac{D_{eq} \, v_{int} \, \rho}{\mu} = \frac{(4 \, R_H) \left(\frac{v_\infty}{\epsilon} \right) \rho}{\mu}$$

$$= \frac{\left[4 \left(\frac{\epsilon}{1-\epsilon} \right) \left(\frac{1}{S} \right) \right] \left(\frac{v_\infty}{\epsilon} \right) \rho}{\mu} = \frac{4 \, v_\infty \, \rho}{\mu (1-\epsilon) \, S} \qquad (6.9.2\text{-}10)$$

where we are now using the superficial velocity.

We now define an equivalent **particle** diameter in terms of a sphere diameter, that is, a sphere with the same surface area and volume as the packing in question (this is **not** precisely the D_{eq} we have used previously for non-circular conduits, although it is similar). As can be seen from the above

$$D_{eq} = 4 \, R_H = 4 \left(\frac{\epsilon}{1-\epsilon} \right) \left(\frac{1}{S} \right)$$

$$D_{eq} = 4 \left(\frac{\epsilon}{1-\epsilon} \right) \left(\frac{D_p}{6} \right)$$

$$D_{eq} = \frac{2}{3} \left(\frac{\epsilon}{1-\epsilon} \right) D_p \qquad (6.9.2\text{-}11)$$

For a sphere

$$S = \frac{\left(\pi D_p^{\,2} \right)}{\left(\pi \frac{D_p^{\,3}}{6} \right)} = \frac{6}{D_p} \qquad (6.9.2\text{-}12)$$

so for a non-spherical particle we **define**

$$D_p \equiv \frac{6}{S} \qquad (6.9.2\text{-}13)$$

which gives as the Reynolds number

$$Re = \frac{4 \, v_\infty \, \rho}{\mu (1-\epsilon) \left(\frac{6}{D_p} \right)}$$

$$Re = \frac{2}{3} \frac{D_p \, v_\infty \, \rho}{\mu (1-\epsilon)} \qquad (6.9.2\text{-}14)$$

and as the friction factor relationship

$$
-\frac{\Delta p}{L} = \frac{2 f \rho v_{int}^2}{D_{eq}} = \frac{2 f \rho \left(\frac{v_-}{\varepsilon}\right)^2}{\left(\frac{2}{3}\left(\frac{\varepsilon}{1-\varepsilon}\right)D_p\right)}
$$

$$
= 3\frac{f \rho v_-^2}{D_p}
$$

$$(6.9.2\text{-}15)$$

Since the constant terms in each of the above do nothing to help us, we drop them and replace f by f_p, leaving us with the working relationships

$$
Re_p = \frac{D_p v_- \rho}{\mu\left(1-\varepsilon\right)}
$$

$$(6.9.2\text{-}16)$$

$$
-\frac{\Delta p}{L} = \left(\frac{\rho v_-^2}{D_p}\right)\left(\frac{1-\varepsilon}{\varepsilon^3}\right)\left(f_p\right)
$$

$$(6.9.2\text{-}17)$$

Experimentally, the relationship between friction factor and Reynolds number has been found to be reasonably modeled by

Range of Re_p	f_p	Appellation
< 1.0	$150/Re_p$	Blake-Kozeny
> 10^4	1.75	Burke-Plummer
$1.0 < Re_p < 10^4$	$150/Re_p + 1.75$	Ergun

Example 6.9.2-1   Pressure drop for air flowing though bed of spheres

Calculate the pressure drop for air at 100°F and 1 atm flowing axially at 2.1 lbm/sec through a 4-ft diameter x 8-ft high bed of 1-inch spheres. Porosity of the bed is 38%.

Solution

Using the fact that

$$(w) \frac{lbm}{s} = (Q) \frac{ft^3}{s} (\rho) \frac{lbm}{ft^3} = (v_\infty) \frac{ft}{s} (A) ft^2 (\rho) \frac{lbm}{ft^3} \quad (6.9.2\text{-}18)$$

we can determine the Reynolds number as

$$Re = \frac{D_p (v_\infty \rho)}{\mu (1-\varepsilon)} = \frac{D_p \left(\frac{w}{A}\right)}{\mu (1-\varepsilon)}$$

$$= \frac{\left(\frac{1}{12}\right) ft \left[\dfrac{(2.1) \frac{lbm}{s}}{(\pi) \dfrac{(4)^2 ft^2}{(4)}} \right]}{(0.0182) \, cP \left(6.72 \times 10^{-4}\right) \dfrac{lbm}{cP \, ft \, s} (1-0.38)} = 1837 \quad (6.9.2\text{-}19)$$

For this Reynolds number we use the Ergun equation

$$f_p = \frac{150}{Re_p} + 1.75 = \frac{150}{1837} + 1.75 = 1.83 \quad (6.9.2\text{-}20)$$

which then gives for the pressure drop[41]

$$-\Delta p = L \left(\frac{\rho \, v_\infty^2}{D_p}\right) \left(\frac{1-\varepsilon}{\varepsilon^3}\right) (f_p) \quad (6.9.2\text{-}21)$$

$$= (L) \left(\frac{\rho \left[\frac{w}{A\rho}\right]^2}{D_p}\right) \left(\frac{1-\varepsilon}{\varepsilon^3}\right) (f_p) = (L) \left(\frac{\frac{1}{\rho}\left[\frac{w}{A}\right]^2}{D_p}\right) \left(\frac{1-\varepsilon}{\varepsilon^3}\right) (f_p) \quad (6.9.2\text{-}22)$$

---

[41] Note that we have neglected the effect of change in pressure on the density. To get a slightly better model we might use the calculated pressure drop to obtain an average pressure from which we could calculate a more nearly correct Reynolds number, friction factor and pressure drop. To convert the problem to a differential equation with continuously changing density and hence Reynolds number, friction factor, and pressure (now) gradient is probably not justified considering the accuracy of the model and empirical parameters involved.

$$= (8)\, \text{ft} \left[\frac{\dfrac{1}{(0.0808)\,\dfrac{\text{lbm}}{\text{ft}^3}\,\dfrac{(492)\,\text{R}}{(560)\,\text{R}}} \left[\dfrac{(2.1)\,\dfrac{\text{lbm}}{\text{s}}}{\left(\dfrac{\pi\,4^2}{4}\right)\text{ft}^2} \right]^2}{\left(\dfrac{1}{12}\right)\text{ft}} \right]$$

$$\times \left[\frac{\text{lbf s}^2}{(32.2)\,\text{lbm ft}} \right] \left(\frac{1-0.38}{0.38^3} \right) (1.83) \qquad (6.9.2\text{-}23)$$

$$= 305\, \frac{\text{lbf}}{\text{ft}^2} \Rightarrow 2.1\, \text{psi} \qquad\qquad\qquad (6.9.2\text{-}24)$$

Example 6.9.2-2   ***Pressure drop for water flowing though bed of cylinders***

A cylindrical column 8-mm in inside diameter and 40-mm in height is filled with dumped cylinders 3-mm in diameter and 2-mm long. Water flows through the bed at a rate of 5 cm^3/s.

The void fraction of the bed is estimated to be 0.47.

Calculate the pressure drop due to **friction** in Pa between the bottom and the top of the bed (ignoring the hydrostatic head).

Use the following physical properties.

	Density (gm/cm^3)	Viscosity (cP)
Water	1.0	1.0

Solution

$$S = \frac{A}{V} = \frac{\text{lateral area} + \text{end areas}}{\text{volume}}$$

$$= \frac{\pi\,D\,L_p + 2\left(\dfrac{\pi\,D^2}{4}\right)}{\left(\dfrac{\pi\,D^2}{4}\right)L_p}$$

$$= \frac{(\pi)(3)\,mm\,(2)\,mm + (2)\dfrac{(\pi)\left[(3)\,mm\right]^2}{(4)}}{\dfrac{(\pi)\left[(3)\,mm\right]^2}{(4)}(2)\,mm}$$

$$= 2.33\ mm^{-1} \tag{6.9.2-25}$$

$$D_p = \frac{6}{S} = \frac{6}{2.33\ mm^{-1}} = 2.58\ mm \tag{6.9.2-26}$$

$$\frac{D_p}{D_t} = \frac{2.58\ mm}{8\ mm} = 0.32 \tag{6.9.2-27}$$

$$\varepsilon = 0.47 \tag{6.9.2-28}$$

$$v_\infty = \frac{\dfrac{(5)\ cm^3}{s}}{\dfrac{(\pi)(8)^2\ mm^2}{(4)}}\frac{100\ mm^2}{cm^2} = 9.95\ \frac{cm}{s} \tag{6.9.2-29}$$

$$Re_p = \frac{D_p\, v_\infty\, \rho}{\mu\,(1-\varepsilon)}$$

$$Re_p = \frac{(2.58)\,mm\,(9.95)\frac{cm}{s}\,(1)\frac{gmm}{cm^3}}{(1)\,cP\dfrac{(0.01)\,gmm}{cm\,s\,cP}(1-0.47)}\frac{cm}{(10)\,mm} = 484 \tag{6.9.2-30}$$

$$f_p = \frac{150}{Re_p} + 1.75 = \frac{150}{484} + 1.75 = 2.06 \tag{6.9.2-31}$$

$$-\Delta p = L\,\frac{\rho\,v_\infty^2}{D_p}\,\frac{1-\varepsilon}{\varepsilon}\,f_p$$

$$-\Delta p = (40)\,mm\,\frac{(1)\frac{gmm}{cm^3}(9.95)^2\left[\frac{cm}{s}\right]^2}{(2.58)\,mm}$$

$$\times\frac{Pa\,m^2}{N}\,\frac{N\,s^2}{kgm\,m}\,\frac{kgm}{(1000)\,gmm}\,\frac{(100)\,cm}{m}\frac{(1-0.47)}{(0.47)}(2.06)$$

$$-\Delta p = 357\ Pa \tag{6.9.2-32}$$

6.9.3   Filters

An important application of flow in porous media is the operation and design of filters. Filtration is basically the removal of either a solid phase or a phase dispersed as liquid drops from a gas phase or a liquid phase through interaction with a porous medium.

If the amount of solid or liquid to be removed is small, the porous medium is usually some material other than the material to be removed; for example, a surgical mask which is designed to remove dispersions of liquid droplets or solid particles in air by collecting them on fibers of a woven material.

On the other hand, if the material to be collected is large in amount, sometimes the bed of collected material itself is used as the filtration medium. In order to support the process at the outset - to collect the initial **filter cake** - it is necessary to have some sort of support, frequently in the form of a cloth or wire screen. In addition, if the material to be collected is dispersed in a sufficiently small size, a **pre-coat** of some material which yields a cake with smaller pore size than the cloth or screen support is employed. A classical pre-coat material is diatomaceous earth.[42]

The driving force for filtration can be as simple as gravity - as is the case in filtration in the chemistry lab with a funnel and filter paper or sand bed filters in water treatment plants. Since gravity is a relatively mild driving force, most production processes utilize either applied pressure on the upstream side of the filter or vacuum on the downstream side.

It is also possible to use a different body force from gravity by creating **centrifugal filters**, which use centrifugal force to drive the fluid through a porous medium as opposed to keeping the fluid medium stationary and simply letting the particles pile up at the face of a solid (an enhanced **settling** process), as in centrifugation in test tubes or production-scale solid-bowl filters.

Filters, like most pieces of equipment, can be run in either batch or continuous modes. By the very nature of the increasing collection of particles in a cake or on a support, filtration tends to have an unsteady-state nature.

No matter what the specific process, the fundamentals of modeling filtration remain the same. The filtering medium and/or the filter cake and/or the support and/or the pre-coat are porous media.

---

[42] Diatoms are the skeletal remains of prehistoric marine invertebrates. Such material is commercially available.

The nature of many filtration processes lends itself to a Darcy's law type of model. For one-dimensional flow, the pressure drop across a filter cake which is incompressible[43] and for which gravitational driving forces are negligible can be calculated using Darcy's law.

$$v_\infty = -\frac{k}{\mu}\left[\frac{\Delta p}{L}\right] \qquad\qquad (6.9.3\text{-}1)$$

or

$$Q = v_\infty A = -\frac{k}{\mu}A\left[\frac{\Delta p}{L}\right] \qquad\qquad (6.9.3\text{-}2)$$

Solving this equation for the pressure drop

$$-\Delta p = Q\left(\frac{\mu L}{k A}\right) \qquad\qquad (6.9.3\text{-}3)$$

Drawing an analogy to Ohm's law, $E = I R$, the term in parentheses represents the resistance to flow. As opposed to the Ohm's law case, the resistance changes with time in this model because of the increase in L.

The support (e.g., filter cloth) and precoat (if any) have constant resistance if they are not compressible. The increasing resistance represented by the cake thickness, L, is proportional to the volume of filtrate collected, V_f, again assuming incompressibility.[44] Therefore we can write

$$-\Delta p = Q\left(K_1 V_f + K_2\right) \qquad\qquad (6.9.3\text{-}4)$$

But by definition

$$Q = \frac{dV_f}{dt} \qquad\qquad (6.9.3\text{-}5)$$

---

[43] We assume throughout this section that permeability is independent of pressure - i.e., that the cake is incompressible, so pressure does not change its structure and hence its resistance to flow.

[44] For a more extensive discussion of filtration, including, for example, continuous filters and compressible cakes, the reader is referred to texts such as McCabe, W. L. and J. C. Smith (1976). *Unit Operations of Chemical Engineering*, McGraw-Hill. or Geankoplis, C. J. (1993). *Transport Processes and Unit Operations*. Englewood Cliffs, NJ, Prentice Hall.

giving

$$-\Delta p = \left(K_1 V_f + K_2\right) \frac{dV_f}{dt} \qquad (6.9.3-6)$$

Two models described by this equation are of particular interest. It is relatively easy to operate filters at constant pressure drop or at constant rate of filtrate flow. In other cases models combining these two extremes offer good approximation to actual operation.

Since it is difficult to predict porous media properties *a priori*, we usually must collect laboratory data on the actual system and from this data design and operate production units. For constant pressure and constant flow rate modes of operation, therefore, we wish to put our model in such a form that we can use laboratory data to infer K_1 and K_2.

First, for constant pressure drop denoted by Δp

$$-\Delta p_0 = \left(K_1 V_f + K_2\right) \frac{d V_f}{dt} \qquad (6.9.3-7)$$

Integrating

$$\int_0^t \left(-\Delta p_0\right) dt = \int_0^{V_f} \left(K_1 V_f + K_2\right) d V_f$$

$$\left(-\Delta p_0\right) t = K_1 \frac{V_f^2}{2} + K_2 V_f \qquad (6.9.3-8)$$

This result may be rearranged to the convenient form

$$\frac{t}{V_f} = \left[\frac{K_1}{2\left(-\Delta p_0\right)}\right] V_f + \left[\frac{K_2}{\left(-\Delta p_0\right)}\right] \qquad (6.9.3-9)$$

This is a linear relationship between t/V_f and V_f.

We apply the relationship by taking laboratory data at constant pressure drop, recording t and V_f. Next we compute t/V_f, and fit the above equation either by linear regression techniques or simply by plotting and fitting a line by eye using graph paper (if we plot V_f versus t_f/V_f we will get a straight line with slope $K_1/2\Delta p_0$ and intercept $K_2/\Delta p_0$).

Second, for a constant flow rate of filtrate we have

$$\frac{dV_f}{dt} = \dot{Q}_0 = \text{constant} \tag{6.9.3-10}$$

and

$$-\Delta p = \left(K_1 V_f + K_2 \right) Q_0 \tag{6.9.3-11}$$

Integrating

$$\int_0^{V_f} dV_f = Q_0 \int_0^t dt$$

$$V_f = Q_0 t \tag{6.9.3-12}$$

$$-\Delta p = \left(K_1 Q_0 t + K_2 \right) Q_0 \tag{6.9.3-13}$$

and

$$-\Delta p = K_1 Q_0^2 t + K_2 Q_0 \tag{6.9.3-14}$$

Here Δp is linear in t for a given filtrate rate $\dot{Q}_0$. This time the data collected consist of pressure drop vs. time, and again the data can be fitted with a regression calculation or simply plotted on graph paper to determine K_1 and K_2.

Many filter operations can be modeled as constant filtrate rate until the pressure builds up to some maximum determined by pump characteristics, and then by constant pressure until the filtrate rate drops below an economically determined limit. Of course, the actual performance depends on pump characteristics and valving, and the transition from one regime to another is not sharp, but, within the accuracy usually required, this two-stage model is often useful.

Example 6.9.3-1   Production scale filter performance prediction from pilot plant data

A 20-gallon sample of 0.5% solids in water is filtered on a 3-ft^2 pilot plant filter using the same cloth and pre-coat thickness as to be used in the production plant. The filter cake may be assumed to be incompressible. The pilot unit was run at a filtrate rate of 5 gpm per ft^2, the same as to be used in the production unit, and the following data were obtained

Pressure Drop (psi)	Time (min)
0.2	0
0.6	40
1.4	120

The production filter is to have an area of 400 ft^2. How long will it take the production unit to reach a pressure drop of 2.5 psi?

Solution

$$- \Delta p = K_1 Q_0^2 t + K_2 Q_0 \tag{6.9.3-15}$$

Using the point at t = 0

$$(0.2)\, psi = K_1 \left(5 \frac{gal}{min\ ft^2}\right)^2 (0) + K_2 \left(5 \frac{gal}{min\ ft^2}\right)$$

$$K_2 = 0.04 \frac{psi\ min\ ft^2}{gal} \tag{6.9.3-16}$$

One can then use either of the points remaining to get K_1. Using the second point

$$(0.6) \frac{lbf}{in^2} = K_1 \left[(5)\frac{gal}{min\ ft^2}\right]^2 (40)\, min + (0.04) \frac{\frac{lbf}{in^2}\ min\ ft^2}{gal} \left[(5)\frac{gal}{min\ ft^2}\right]$$

$$K_1 = 4.0 \times 10^{-4} \frac{\frac{lbf}{in^2}\ min\ ft^2}{gal} \tag{6.9.3-17}$$

The time to reach 25 psi is then

$$- \Delta p = K_1 Q_0^2 t + K_2 Q_0$$

$$(2.5)\frac{lbf}{in^2} = \left(4.0 \times 10^{-4}\right)\frac{\frac{lbf}{in^2} \, min \, ft^4}{gal^2}\left[(5)\frac{gal}{min \, ft^2}\right]^2 (t) \, min$$
$$+ (0.04)\frac{\frac{lbf}{in^2} \, min \, ft^2}{gal}(5)\frac{gal}{min \, ft^2}$$
$$t = 230 \, min$$

$$(6.9.3-18)$$

Example 6.9.3-2   Filter performance from data

A filter is run at constant rate for 20 min. The amount of filtrate collected is 30 gal. The initial pressure drop is 5 psig. The pressure drop at the end of 20 min is 50 psig. If the filter is operated for an additional 20 min at a constant pressure of 50 psig, how much total filtrate will have been collected?

Solution

For the constant rate period

$$Q_0 = \frac{V_f}{t} = \frac{30 \, gal}{20 \, min} = 1.5 \frac{gal}{min}$$
$$-\Delta p = K_1 Q_0^2 t + K_2 Q_0 \qquad\qquad (6.9.3-19)$$

At constant rate we have initially

$$(5) \, psig = K_1 (1.5)^2 \left[\frac{gal}{min}\right]^2 (0) \, min + K_2 (1.5)\frac{gal}{min}$$
$$K_2 = 3.33 \frac{psig \, min}{gal} \qquad\qquad (6.9.3-20)$$

At the end of the constant rate period

$$(50) \, psig = K_1 (1.5)^2 \frac{gal^2}{min^2} (20) \, min + (3.33)\frac{psig \, min}{gal} (1.5)\frac{gal}{min}$$
$$K_1 = 1.00 \frac{psig \, min}{gal^2} \qquad\qquad (6.9.3-21)$$

For the constant pressure drop period

$$\left(-\Delta p_0\right) t = K_1^* \frac{V_f^2}{2} + K_2^* V_f$$

$$t = \frac{K_1^*}{2\left(-\Delta p_0\right)} V_f^2 + \frac{K_2^*}{\left(-\Delta p_0\right)} V_f \qquad (6.9.3\text{-}22)$$

Remember that we obtained the equation for the constant pressure drop period by integrating, assuming that at $t = 0$, $V_f = 0$. This means that the constant K_2^* *will incorporate the resistance of cake deposited during the constant rate operation in addition to that of the medium and pre-coat.* The constant K_1^* will be the same as the constant K_1 because the first term remains proportional to V_f in the same manner as before.

Recognizing that the end of the constant rate cycle, where $Q = 1.5$ and $(-\Delta p) = 50$, is also the beginning of the constant pressure cycle, where $V_f = 0$

$$-\Delta p = \left(K_1^* V_f + K_2^*\right) Q_0$$

$$\left(50\right) \text{psig} = \left[\left(1.00\right) \frac{\text{psig min}}{\text{gal}^2} \left(0\right) \text{gal} + K_2^*\right] \left(1.5\right) \frac{\text{gal}}{\text{min}}$$

$$K_2^* = 33.3 \frac{\text{psig min}}{\text{gal}} \qquad (6.9.3\text{-}23)$$

Substituting in the constant pressure expression

$$\frac{t}{V_f} = \left[\frac{K_1^*}{2\left(-\Delta p_0\right)}\right] V_f + \left[\frac{K_2^*}{\left(-\Delta p_0\right)}\right]$$

$$\frac{\left(20\right) \text{min}}{V_f} = \left[\frac{\left(1.00\right) \frac{\text{psig min}}{\text{gal}^2}}{\left(2\right)\left(50\right) \text{psig}}\right] V_f + \left[\frac{\left(33.3\right) \frac{\text{psig min}}{\text{gal}}}{\left(50\right) \text{psig}}\right]$$

$$V_f^2 + 66.6 \, V_f - 2000 = 0$$

$$V_f = 22.4 \text{ gal} \quad \text{(the other root is negative)} \qquad (6.9.3\text{-}24)$$

The total filtrate collected is then

$$30 \text{ gal} + 22.4 \text{ gal} = 52.4 \text{ gal}$$

(6.9.3-25)

Example 6.9.3-3   Adapting existing filter to new product

Our present production plant contains a 100-ft^2 batch filter that consists of a supporting screen and cloth, on which is deposited 0.25 in. of a diatomaceous earth pre-coat. This filter is presently used at a constant pressure drop of 50 psi to filter substance A from a 5% by weight water solution. The plant is now to be converted to produce a new product, B, which is somewhat similar to product A. This product will also be in a water solution, but 10% by weight. It has been proposed that the present filter, modified to operate at 75 psi, be used in the process to produce B.

What production rate of B in lbm for 24 hrs of continuous filtration can we expect using our present filter with the present pre-coat if we convert it to a pressure capability of 75 psi?

Information at our disposal:

A 50-gallon sample of the new material (10% B in water) was shipped to a filter manufacturer to be tested on a 1-ft^2 laboratory-scale model of the filter being used in the plant. Unfortunately, the manufacturer ran the test at a pressure drop of 25 psi, and used a pre-coat of only 0.125 inch, but returned the following data.

Volume of Filtrate collected, Gal.	Time
0	0
10	30 min
30	3.5 hrs

Current production shift records show for a typical run

Volume of Filtrate collected, Gal.	Time
0	9:00 am
1500	10:00 am
6000	2:30 PM

Solution

For constant pressure operation we may write

$$t_f = \frac{K_1}{2\Delta p} V_f^2 + \frac{K_2}{\Delta p} V_f$$

$$\frac{t_f}{V_f} = \frac{K_1}{2\Delta p} V_f + \frac{K_2}{\Delta p} \qquad (6.9.3\text{-}26)$$

Taking the derivative, recognizing the result as linear, and substituting the lab results

$$\frac{\Delta\left(\frac{t_f}{V_f}\right)_{B,\,lab}}{\Delta(V_f)_{B,\,lab}} = \frac{(K_1)_{B,\,lab}}{(2)(25)\frac{lbf}{in^2}} = \frac{\left(\frac{3.5}{30} - \frac{0.5}{10}\right)\frac{hr}{gal}}{(30-10)\,gal}$$

$$(K_1)_{B,\,lab} = 0.17\frac{hr\,lbf}{in^2\,gal} \qquad (6.9.3\text{-}27)$$

We can then calculate K_2 for the laboratory setup

$$\frac{(3.5)\,hrs}{(30)\,gal} = \frac{(0.17)\frac{hr\,lbf}{in^2\,gal^2}}{(2)(25)\frac{lbf}{in^2}}(30)\,gal + \frac{(K_2)_{media,\,lab}}{(25)\frac{lbf}{in^2}}$$

$$(K_2)_{media,\,lab} = 0.37\frac{lbf\,hrs}{in^2\,gal} \qquad (6.9.3\text{-}28)$$

K_1 (the cake resistance) should be the same for the laboratory and the production units in processing product B, but K_2 for the lab will differ from that for the production unit.

Performing the same calculation for the production data, but on a 1-ft^2 basis

$$\frac{\Delta\left(\frac{t_f}{V_f}\right)_{A,\,prodn}}{\Delta\left(V_f\right)_{prodn}} = \frac{\left(K_1\right)_{A,\,prodn}}{(2)(50)\frac{lbf}{in^2}} = \frac{\left(\frac{1}{15} - \frac{5.5}{16}\right)\frac{hr}{gal}}{(60-15)\,gal}$$

$$\left(K_1\right)_{A,\,prodn} = -0.62\,\frac{hr\,lbf}{in^2\,gal^2} \qquad\qquad (6.9.3\text{-}29)$$

$$\frac{(5.5)\,hrs}{(60)\,gal} = \frac{(-0.62)\frac{hr\,lbf}{in^2\,gal^2}}{(2)(50)\frac{lbf}{in^2}}(60)\,gal + \frac{\left(K_2\right)_{media,\,prodn}}{(50)\frac{lbf}{in^2}}$$

$$\left(K_2\right)_{media,\,prodn} = 23.2\,\frac{lbf\,hrs}{in^2\,gal} \qquad\qquad (6.9.3\text{-}30)$$

In using the production filter for product **B**, we will have a constant media resistance represented by K_2 from the **production** data, but a varying cake resistance represented by K_1 from the **lab** data.

$$t_f = \frac{0.17\,\frac{hr\,lbf}{in^2\,gal^2}}{2\,\Delta p}\,V_f^2 + \frac{23.2\,\frac{lbf\,hrs}{in^2\,gal}}{\Delta p}\,V_f \qquad\qquad (6.9.3\text{-}31)$$

At 75-psi in a 24 hour period, we can therefore filter per square foot of the production filter

$$(24)\,hr = \frac{(0.17)\frac{hr\,lbf}{in^2\,gal^2}}{(2)(75)\frac{lbf}{in^2}}\,V_f^2 + \frac{(23.2)\frac{lbf\,hrs}{in^2\,gal}}{(75)\frac{lbf}{in^2}}\,V_f$$

$$0 = V_f^2 + \frac{(2)(23.2)}{(0.17)}\,V_f - (24)\frac{(2)(75)}{(0.17)} = V_f^2 + 273\,V_f - 21176$$

$$V_f = \frac{-273 \pm \sqrt{273^2 + (4)(1)(21176)}}{2} = 63\,gal\,(\text{reject negative root})$$

$$(6.9.3\text{-}32)$$

giving us a total filtration volume of 6300 gal. Since the concentration of B is 10% by weight, we can therefore produce

$$6300 \text{ gal water} \frac{8.33 \text{ lb water}}{\text{gal water}} \frac{10 \text{ lbm B}}{90 \text{ lbm water}} = 5831 \text{ lbm of B}$$

(6.9.3-33)

6.10   Flow Measurement

Instruments used for the measurement of flow of gases and liquids furnish convenient illustrations of application of the principles of momentum, heat, and mass transfer. The following sections consider some of these devices which rely on momentum transfer for their function.

6.10.1   Pitot tube

Figure 6.10.1-1 shows a schematic diagram of a device called a **pitot tube**. This particular pitot tube uses a manometer to measure the pressure drop between points 1 and 4. Although a manometer is shown as the pressure measuring device, any appropriate pressure measuring device could be used.

The physical configurations of actual pitot tubes are much more streamlined and sophisticated than would be suspected from this sketch, but their principle of operation is exactly the same. Such devices, more frequently used to measure gas velocities than liquid velocities, are inserted into the flow stream with the opening (point 1) perpendicular to the flow stream (facing upstream).

The idea behind the device is that the flow supposedly decelerates to a standstill at point 1, thus converting most of the energy from the kinetic energy term in the mechanical energy balance to the pressure energy term (some also goes into the lost work term). Since the kinetic energy of the flow is a measure of the local velocity, the difference between the dynamic pressure obtained at point 1 and the static pressure obtained by openings parallel to the flow at point 4 is a measure of the velocity of the fluid.

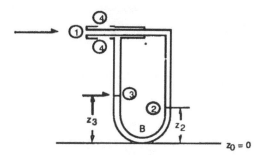

Figure 6.10.1-1 Pitot tube schematic

As the opening at point 1 of the pitot tube is made progressively smaller, velocities closer and closer to the velocity at a point are obtained. The pitot tube, however, by its very nature is cumbersome enough that it is seldom used to approximate point measurements. For measurements over very small regions and of rapidly fluctuating velocities, other techniques are usually used; e.g., hot film or hot wire anemometers or laser Doppler techniques.

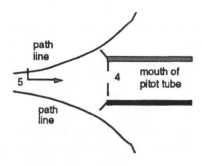

Figure 6.10.1-2 Flow at mouth of pitot tube

We may describe the rise in pressure associated with the deceleration of the fluid by applying the mechanical energy balance through a system swept out by the fluid which decelerates to nearly zero velocity at point 4 at the mouth of pitot tube, as shown in Figure 6.10.1-2. The boundaries of the system are pathlines. Writing the mechanical energy balance for this system

$$\frac{\langle v_5 \rangle^2 - \langle v_4 \rangle^2}{2} + g\,\Delta z + \frac{p_5 - p_4}{\rho} = -\hat{lw} - \hat{W}$$

(6.10.1-1)

If we arbitrarily represent the lost work term as some fraction C_a of the kinetic energy change

$$C_a \left[\frac{\langle v_5 \rangle^2 - \langle v_4 \rangle^2}{2} \right] = -\hat{lw}$$

(6.10.1-2)

we may write, assuming that $\langle v_4 \rangle^2$ is much less than $\langle v_5 \rangle^2$

$$\left(1 - C_a\right)\langle v_5 \rangle^2 = 2\frac{p_5 - p_4}{\rho}$$

(6.10.1-3)

or

$$\langle v_5 \rangle = \sqrt{\frac{1}{1 - C_a}}\sqrt{\frac{2\,\Delta p}{\rho}}$$

(6.10.1-4)

Since $\sqrt{\frac{1}{1 - C_a}}$ is simply another coefficient: let us rename it C_{pitot}, the pitot tube coefficient, so

$$\langle v_5 \rangle = C_{pitot}\sqrt{\frac{2\,\Delta p}{\rho}}$$

(6.10.1-5)

The coefficient C_p accounts for the mechanical energy lost while decelerating and accelerating the fluid. This quantity cannot generally be calculated, and C_p is usually determined experimentally for any given instrument. $\langle v_5 \rangle$ is the average velocity over the inlet area to the system, not for the total pipe cross-section. As the area of the pitot tube is made smaller and smaller, $\langle v_5 \rangle$ approaches the local velocity. The density used in Equation (6.10.1-4) is that of the **flowing** fluid, **not the manometer** fluid.

Example 6.10.1-1   Pitot tube traverse

A cylindrical duct 2 ft in diameter carries air at atmospheric pressure. A pitot tube traverse is made in the duct using a water-filled manometer giving

Radial distance from wall (ft.)	Height difference in manometer fluid levels (in.)
0.1	0.77
0.3	1.99
0.5	4.00
0.7	5.14
0.9	7.76

Calculate the approximate velocity profile and volumetric flow rate.

Further data:

$$\rho_{air} = 0.0808 \frac{lbm}{ft^3}$$

$$\rho_{HOH} = 62.4 \frac{lbm}{ft^3}$$

$$C_{pitot} = 0.99$$

Solution

The height differences may be converted to pressures and the pressures to velocities as follows:

Observe that the density of air is much less than that of water. If the pitot tube locations are numbered as shown in Figure 6.10.1-1, we may write

$$p_1 - p_4 = (p_1 - p_2) + (p_2 - p_3) + (p_3 - p_4)$$
$$= -\rho_{air} g (z_1 - z_2) - \rho_{HOH} g (z_2 - z_3) - \rho_{air} g (z_3 - z_4) \quad (6.10.1-6)$$

which can be written, noting that $z_4 = z_1$

$$p_1 - p_4 = -\rho_{air} g \left[(z_4 - z_3) + (z_3 - z_2)\right] - \rho_{HOH} g (z_2 - z_3) - \rho_{air} g (z_3 - z_4)$$
$$(6.10.1-7)$$

giving

$$p_1 - p_4 = g (z_2 - z_3) (\rho_{HOH} - \rho_{air}) \quad (6.10.1-8)$$

But, in addition, $\rho_{HOH} \gg \rho_{air}$, so

$$p_1 - p_2 = \rho_{HOH}\, g\left(z_3 - z_2\right) \tag{6.10.1-9}$$

Substituting in the expression relating the pressure drop to velocity

$$v = C_{pilot}\sqrt{\frac{2\,\Delta p}{\rho}} = C_{pilot}\sqrt{\frac{2\,\rho_{HOH}\,g\left(z_3 - z_2\right)}{\rho_{air}}}$$

$$(v)\frac{ft}{s} = 0.99\sqrt{\frac{(2)\left(62.4\frac{lbm}{ft^3}\right)\left(32.2\frac{ft}{s^2}\right)\left(z_3 - z_2\right)in\left(\frac{ft}{12\ in}\right)}{0.0808\frac{lbm}{ft^3}}}$$

$$(v)\frac{ft}{s} = 11.2\frac{ft}{in^{1/2}\,s}\sqrt{\left(z_3 - z_2\right)in} \tag{6.10.1-10}$$

From this relation we can calculate, for each height difference, a velocity:

Radial distance from wall, d (ft)	<v> (ft/s)
0.1	9.8
0.3	15.8
0.5	22.4
0.7	25.4
1	31.2

This profile is plotted in the following figure. Note that, as usual, our experimental data scatter somewhat.

We would have to decide in a real-world situation whether we had good reason for drawing a smooth curve such as shown approximating the data and assuming a symmetrical velocity profile or if, in fact, the curve should go through the individual data points. (Upstream disturbances can severely distort velocity profiles.)

We fit a second-order polynomial to the data, forcing it to pass through (0,0), because we know the fluid sticks to the wall. The polynomial fit is

$$\langle v \rangle = -26.5\,d^2 + 57.1\,d \tag{6.10.1-11}$$

This simple curve fit does not yield a zero derivative at the center of the duct as the symmetrical velocity profile model we will assume when integrating would demand, but is reasonably close.

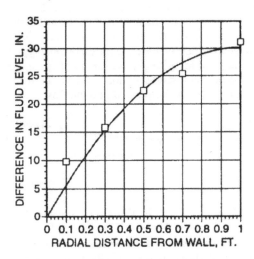

We now obtain the volumetric flow rate from the model by

$$Q = \int_A v \, dA \tag{6.10.1-12}$$

This relation may be written as

$$Q = \int_0^R v\,(2\pi r)\, dr \tag{6.10.1-13}$$

if we assume no variation in the θ direction. Noting that $d = (1 - r)$

$$Q = \int_0^1 \left(-26.5\,[1-r]^2 + 57.1\,[1-r]\right)(2\pi r)\, dr \tag{6.10.1-14}$$

$$Q = 45.9\,\frac{\text{ft}^3}{\text{s}} \tag{6.10.1-15}$$

6.10.2   Venturi meter

The flow of a fluid through the constriction of a venturi meter as shown in Figure 6.10.2-1, causes a pressure drop which may be related to the average flow velocity with a mechanical energy balance and the mass balance. As can be seen from Figure 6.10.2-2, a properly designed venturi profile permits almost frictionless flow in some situations; i.e., the pressure energy that is converted to kinetic energy is almost all recovered downstream of the meter, so a very small permanent pressure drop (represented by lost work in the form of its conversion to internal energy of the fluid) is introduced into the flow.

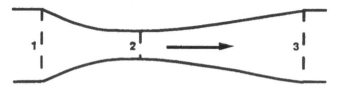

Figure 6.10.2-1 Venturi schematic

From a mass balance, the volumetric flow rate at points 1 (the entering pipe inside diameter) and 2 (the throat diameter) is the same if $\rho_1 = \rho_2$.

$$\rho_1 A_1 \langle v_1 \rangle = \rho_2 A_2 \langle v_2 \rangle \qquad\qquad (6.10.2\text{-}1)$$
$$Q_1 = Q_2 \qquad\qquad (6.10.2\text{-}2)$$

A mechanical energy balance yields

$$\frac{\langle v_2 \rangle^2 - \langle v_1 \rangle^2}{2} + g\,\Delta z + \frac{p_2 - p_1}{\rho} = -\hat{l_w} - \hat{W} \qquad\qquad (6.10.2\text{-}3)$$

As with the pitot tube, the lost work term is represented as some fraction of the kinetic energy change.

$$C_b \left[\frac{\langle v_2 \rangle^2 - \langle v_1 \rangle^2}{2} \right] = -\hat{l_w} \qquad\qquad (6.10.2\text{-}4)$$

then

$$\left(1 - C_b\right)\left[\frac{\langle v_2\rangle^2 - \langle v_1\rangle^2}{2}\right] = -\frac{p_2 - p_1}{\rho} \tag{6.10.2-5}$$

But a mass balance allows replacing $\langle v_1\rangle$ in terms of $\langle v_2\rangle$.

$$\langle v_1\rangle = \frac{A_2}{A_1}\langle v_2\rangle \tag{6.10.2-6}$$

Substituting and rearranging

$$\langle v_2\rangle^2\left[1 - \left(\frac{A_2}{A_1}\right)^2\right] = -\frac{2}{1 - C_b}\frac{p_2 - p_1}{\rho} \tag{6.10.2-7}$$

Defining the venturi coefficient as C_v

$$C_v = \sqrt{\frac{1}{1 - C_b}} \tag{6.10.2-8}$$

then

$$\langle v_2\rangle = C_v\sqrt{\frac{-2\left(p_2 - p_1\right)}{\rho\left[1 - \left(\frac{A_2}{A_1}\right)^2\right]}} \tag{6.10.2-9}$$

It is customary to define

$$\frac{D_2}{D_1} \equiv \beta \tag{6.10.2-10}$$

and thus

$$\langle v_2\rangle = C_v\sqrt{\frac{-2\left(p_2 - p_1\right)}{\rho\left[1 - \beta^4\right]}} \tag{6.10.2-11}$$

The venturi coefficient in the above equations can be determined approximately using Figure 6.10.2-2, or by calibration of the actual unit.

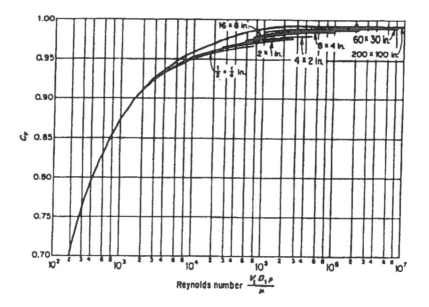

Figure 6.10.2-2 Venturi meter coefficient (Reynolds number based on entrance diameter)[45]

Example 6.10.2-1   Flow measurement with venturi meter

A venturi meter with a throat diameter of 1 1/2 in. is installed in a 4-in. Schedule 40 pipe line to measure the flow of water in the line. Using a manometer containing mercury as the manometric fluid, the pressure drop observed is 50 in Hg from entrance to vena contracta.

Data:

$$\left(\text{specific gravity}\right)_{Hg} = \left(13.6\right)$$

$$\rho_{HOH} = 62.4 \frac{lbm}{ft^3}$$

$$C_{venturi} = 0.975$$

What is the flow rate of water in the pipe?

---

[45] *Fluid Meters: Their Theory and Application, 4th ed., ASME (1937).*

Solution

Calculating the pressure drop from the manometer data, letting points 1 and 2 be the low and high levels of the mercury, respectively, and remembering that the legs of the manometer above the mercury are filled with water

$$(p_1 - p_2) = \left[-\rho_{Hg}\, g\,(z_1 - z_2)\right] + \left[-\rho_{HOH}\, g\,(z_2 - z_1)\right]$$

$$(p_1 - p_2) = \left[-g\,(z_1 - z_2)\right]\left[\rho_{Hg} - \rho_{HOH}\right] \qquad (6.10.2\text{-}12)$$

$$(p_1 - p_2) = -(62.4)\left[\frac{lbm}{ft^3}\right](13.6 - 1.0)(32.2)\left[\frac{ft}{s^2}\right]$$

$$\times (-50\text{ in})\left[\frac{ft}{(12)\text{ in}}\right]\left[\frac{lbf\ s^2}{(32.2)\ lbm\ ft}\right]$$

$$= 3276\left(\frac{lbf}{ft^2}\right) \qquad (6.10.2\text{-}13)$$

$$Q = \langle v_2 \rangle A_2 = C_v A_2 \sqrt{\frac{-2\,(p_2 - p_1)}{\rho\left[1 - \beta^4\right]}} \qquad (6.10.2\text{-}14)$$

$$Q = (0.975)\frac{\pi}{4}\left(\frac{1.5}{12}\right)^2 ft^2 \sqrt{\frac{-(2)(-3276)\left[\frac{lbf}{ft^2}\right]}{\left(62.4\frac{lbm}{ft^3}\right)\left[1 - \left(\frac{1.5}{4.026}\right)^4\right]}(32.2)\left[\frac{lbm\ ft}{lbf\ s^2}\right]}$$

$$(6.10.2\text{-}15)$$

$$Q = 0.70\frac{ft^3}{s} \ \Rightarrow\ (0.70)\left[\frac{ft^3}{s}\right]\left[\frac{(7.48)\ gal}{ft^3}\right]\left[\frac{(60)\ s}{min}\right] = 315\text{ gpm}$$

$$(6.10.2\text{-}16)$$

(This is in the turbulent flow regime, so the coefficient used is reasonable.)

6.10.3 Orifice meter and flow nozzle

An orifice or a flow nozzle is essentially a restriction placed normal to the flow in a pipe to cause a pressure drop. The pressure is related to the average velocity of fluid via the mechanical energy balance. Figure 6.10.3-1 shows both an orifice and a nozzle. The main difference between the orifice and the nozzle is that the orifice is more or less a simple hole in a plate, while the flow nozzle has a shape intermediate between an orifice and a venturi, in an attempt to decrease the amount of lost work experienced with the orifice without the high cost of fabrication of the venturi.

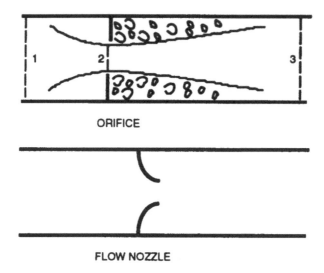

ORIFICE

FLOW NOZZLE

Figure 6.10.3-1 Orifice meter, flow nozzle

We apply the mechanical energy balance, replacing lost work with a constant times kinetic energy as in the case of the venturi; however, in the case of the orifice and flow nozzle, the position of the *vena contracta* (minimum effective flow area) changes with velocity. Since location of the pressure taps remains fixed we use a coefficient defined as

$$C_o = \sqrt{\frac{1}{1-C}} \sqrt{\frac{1-\left(\frac{A_o}{A_1}\right)^2}{1-\left(\frac{A_2}{A_1}\right)^2}} \qquad (6.10.3\text{-}1)$$

Solving as before for $<v_o>$ in terms of β

$$\langle v_o \rangle = C_o \sqrt{\frac{-2\left(p_2 - p_1\right)}{\rho\left[1 - \beta^4\right]}} \qquad (6.10.3\text{-}2)$$

The value of the orifice coefficient can be shown to be a function of the Reynolds number and β as in Figure 6.10.3-2, which shows a typical orifice coefficient.

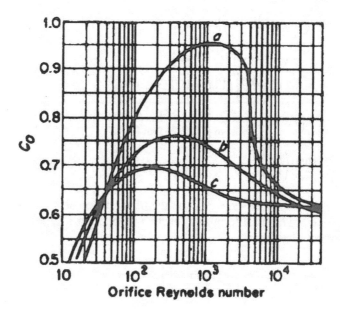

Figure 6.10.3-2 Orifice coefficient[46]: (a) $D_o/D_1 = 0.80$, (b) $D_o/D_1 = 0.60$, (c) $D_o/D_1 = 0.20$

Example 6.10.3-1   Metering of crude oil with orifice

A 2-in. diameter orifice with flange taps is used to measure the flow rate of 100°F crude oil flowing in a 4-in. Schedule 40 steel pipe. The manometer reads 30 in Hg with a 1.1 specific gravity fluid filling the manometer above the mercury to isolate it from the crude.

If the viscosity of the oil is 5 cP at 100°F and the specific gravity of the oil is 0.9, what is the flow rate?

---

[46] Tuve and Sprenkle (1933). *Instruments* 6: 201.

Solution

Calculate Δp

$$\left(p_1 - p_2\right) = \left[-\rho_{Hg}\, g\left(z_1 - z_2\right)\right] + \left[-\rho_{fluid}\, g\left(z_2 - z_1\right)\right]$$

$$\left(p_1 - p_2\right) = \left[-g\left(z_1 - z_2\right)\right]\left[\rho_{Hg} - \rho_{fluid}\right] \qquad (6.10.3\text{-}3)$$

$$\left(p_1 - p_2\right) = -(62.4)\left[\frac{lbm}{ft^3}\right](13.6 - 1.1)(32.2)\left[\frac{ft}{s^2}\right]$$

$$\times\left(-30\ in\right)\left[\frac{ft}{(12)\ in}\right]\left[\frac{lbf\ s^2}{(32.2)\ lbm\ ft}\right]$$

$$= 1950\left(\frac{lbf}{ft^2}\right) \qquad\qquad (6.10.3\text{-}4)$$

We do not know either $<v_o>$ or C_o in the flow relation below; however, C_o is also available in graphical form as a function of $<v_o>$ (as it appears in the Re on the orifice coefficient plot).

$$Q = \langle v_o\rangle\, A_o = C_o\, A_o \sqrt{\frac{-2\left(p_2 - p_1\right)}{\rho\left[1 - \beta^4\right]}} \qquad (6.10.3\text{-}5)$$

We have, therefore, two equations in two unknowns, but one is not in analytical form. We could fit the orifice coefficient chart numerically to some analytic function and solve simultaneously using either an explicit analytical or a numerical technique, but this would be overkill for such a simple problem. We proceed instead by trial.

Assume $C_o = 0.61$

$$Q = v_o\, A_o$$

$$= (0.61)\,\pi\left[\frac{2}{(4)\,(12)}\right]^2 ft^2$$

$$\times \left[\frac{-(2)\,(-1950)\left[\frac{lbf}{ft^2}\right]}{(0.9)\,(62.4)\left[\frac{lbm}{ft^3}\right]\left[1 - \left(\frac{2}{4.026}\right)^4\right]}(32.2)\left[\frac{lbm\ ft}{lbf\ s^2}\right]\right]^{1/2} \qquad (6.10.3\text{-}6)$$

$$Q = 0.16 \frac{ft^3}{s} \implies 73 \text{ gpm} \qquad (6.10.3-7)$$

Chapter 6   Problems

6.1 An orifice is being used to measure the flow of water through a line. A mercury-water manometer is connected across the orifice as shown below, the lines being filled with water up to the mercury-water interface. If the manometer reading is 5 in., what is the pressure difference between the two locations in the pipe? Suppose the pipe had been vertical with water flowing upward, and the distance between the two manometer taps was 2 ft. What would the pressure difference between the two taps have been when the manometer reading was 5 in.?

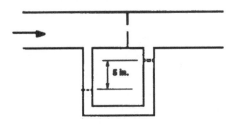

In the vertical case, what would the pressure difference have been if ethyl acetate were substituted for the water?

6.2 For the following situation

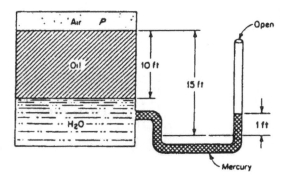

What is the pressure p? The specific gravity of oil is 0.8. The specific gravity of mercury is 13.6.

6.3 For the following situation

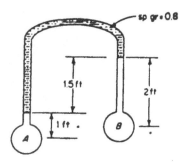

The liquid at A and B is water.

a. Find $p_A - p_B$.

b. If the atmospheric pressure is 14.7 psia, find the gauge pressure at A.

6.4 For the following situation

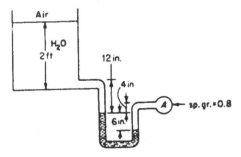

Find the pressure in the drum A at position A. The manometer fluid is Hg.

6.5 For the following situation find the pressure at P. The specific gravity of Hg is 13.6

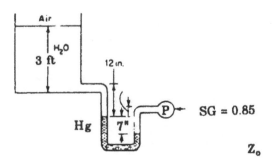

6.6 An inclined tube reservoir manometer is shown in the sketch.

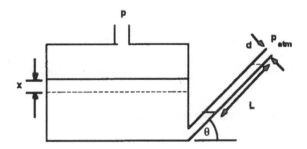

a. Show the expression for the liquid deflection, L, is given in terms of the pressure difference Δp by the following expression, where x is the change in level of the reservoir in inches

$$x = L \frac{A_{tube}}{A_{res}}$$

$$L = \frac{\Delta p}{(SG) \rho_{H_2O} \, g \left[\sin \theta + \left(\frac{d}{D} \right)^2 \right]}$$

b. If the height of an equivalent column of water is $\Delta p = \rho_{H_2O} g \Delta h_e$ what is the expression for the manometer sensitivity $\dfrac{L}{\Delta h_e}$?

c. If $D = 96$ mm, $d = 6$ mm, what angle θ is necessary to provide a 5:1 increase in liquid deflection, L, compared to the total deflection in a U-tube manometer (i.e. $\Delta p = \rho g h$)?

6.7 Consider the ideal flow field determined by

$$\psi = ax^2 - ay^2 \left(a \left[\frac{1}{sec} \right] \right)$$

a.   Show this flow field is irrotational.
b.   What is the velocity potential for this flow?

6.8 The complex potential $w = -v_0 z^2$ describes ideal flow near a plane stagnation point.

a. Determine the expression for the potential and stream functions.

b. What are the velocity components v_x, v_y?

c. What is v_0?

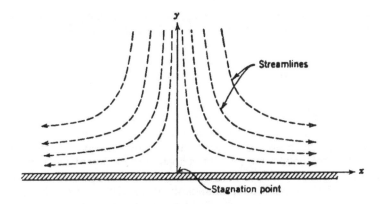

6.9 Consider the two-dimensional ideal flow described by the potential function

$$\phi = -\frac{\Gamma}{2\pi}\theta$$

and stream function

$$\psi = -\frac{\Gamma}{2\pi}\ln r$$

where r and θ are the usual polar coordinates. Sketch the flow field. This flow field represents a vortex - similar to the vortex seen in a draining bathtub. What does the ideal solution yield as r approaches zero? The ideal solution does not describe the real-world situation for small r because of large viscous forces in this region.

6.10 Sketch the flow field for the potential function given by

$$\phi = \frac{K}{2\pi}\ln r$$

and stream function

$$\psi = \frac{K}{2\pi}\theta$$

where K is a constant. Note that these are the interchanged functions for the vortex above. Depending on the sign of K, the flow represents a source or a sink. The point at $r = 0$ is again a singularity. This type of mathematical description is sometimes used for flow of oil into a well bore from an underground formation. The solution is not good close to the well.

6.11 Two horizontal square parallel plates measuring 10 by 10 in. are separated by a film of water at 80°F. The bottom plate is bolted down, and the top plate, in contact with the water, is free to move. A spring gauge is attached to the top plate and a horizontal force of 1 lbf is applied to the top plate. If the distance between the plates were 0.1 in., how fast would the top plate move at steady state? How fast would the top plate move if the distance were 1/8 in.? If the horizontal force were increased to 1.6 lbf and the distance between the plates were 1/6 in., how fast would the top plate move?

6.12 A Newtonian fluid is flowing between two parallel plates of length L. The two plates are separated by a distance h. The flow is steady and only in the y direction. Sufficient forces are applied to both plates so the upper plate has a velocity v_B while the bottom plate moves at v_0.

    a. Draw a control volume and label it.

    b. What is the pressure drop per unit length?

    c. Derive the appropriate 2nd-order differential equation for
       velocity and put it in dimensionless form.

    d. Calculate the dimensionless average velocity.

    e. If the fluid were pseudoplastic (non-Newtonian) how would
       you change a, b, c, d? (Don't do it - explain.)

6.13 Find the velocity profile, average velocity, and free surface velocity for a Newtonian fluid flowing down a vertical wall of length L. Use a control volume (CV) of thickness Δx inside the fluid and measure x from the free surface. Remember that because the liquid surface is "free" (unbounded physically) at $x = 0$, the pressure at all points, vertical as well as horizontal, is atmospheric.

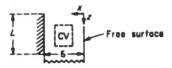

6.14 Derive the equations for the velocity distribution and for the pressure drop in laminar, Newtonian flow through a slit of height y_0 and infinite width.

6.15 A fluid flows down the wall of a wetted wall absorber. A gaseous material is adsorbed into the liquid causing a viscosity variation according to

$$\frac{\mu}{\mu_o} = e^{\alpha x}$$

where μ_0 is the value of viscosity at the film surface. To determine the expression for velocity:

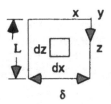

a. Draw the control volume (dx)(dz) and show the forces acting on it (assume width of film is w). Write the momentum balance and simplify it for this problem, indicating your assumptions for the simplification.

b. Derive the 2nd-order differential equation describing the problem to be solved for velocity as a function of film thickness.

c. What are the boundary conditions for this problem?

6.16 A Newtonian fluid is moving in laminar flow down an inclined wall under the influence of gravity. Assume a width of w and determine the:

a. Velocity profile

b. Maximum velocity

c. Average velocity

d. Volumetric flow rate

e. Film thickness, δ

f. Drag force

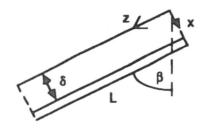

6.17 Compare the plots of τ_{yz} versus (dv_z/dy) for a Newtonian fluid in laminar flow in a pipe and for a pseudoplastic fluid with $n = 0.8$. Assume the viscosity of the Newtonian fluid is $\mu = 2$ cP and the apparent viscosity of the non-Newtonian fluid $\mu = 2$ cP. (Use the velocity gradient from the Newtonian fluid in the power-law expression.)

6.18 Glycerine at 26.5°C is flowing through a horizontal tube 1 ft in length and 0.1-in. ID. For a pressure drop of 40 psi the flow rate is 0.00398 ft³/min. The density of glycerine at 26.5°C is 1.261 gm/cm³. What is the viscosity of the glycerine? Glycerine may be regarded as a Newtonian fluid.

6.19 An incompressible fluid is in steady-state laminar flow in the annular space between two horizontal, concentric pipes of length L. To find the expressions for the velocity distribution and pressure drop, the momentum balance may be applied to a control volume which is an annular shell of thickness dr. Show the differential equation obtained

$$\frac{d}{dr} r\tau_n = \frac{p_o - p_L}{L} r$$

where r is the radius and τ_{rz} is shear stress in the z direction on the face of the shell perpendicular to the r direction.

6.20 A fluid with a density of 1.261 gm/cm^3 is flowing through a tube 3 feet long with a diameter of 0.1 inch. If the pressure drop is 120 psi and the flow rate is 0.004 ft^3/min, what is the viscosity of the fluid?

6.21 The sketch shows a rod moving through a fluid at velocity V. The rod and the cylinder are coaxial. Use the 1-D equation of motion for a Newtonian fluid.

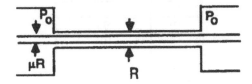

a. Show by appropriate assumptions that the equation of motion simplifies to

$$\frac{d}{dr}\left(r\frac{dv_z}{dr} \right) = 0$$

b. What are the boundary conditions that apply to this equation?

c. Integrate the equation and apply the boundary condition to show

$$\frac{v_z}{V} = \frac{\ln\left(\frac{r}{R} \right)}{\ln \kappa}$$

d. Show the volumetric flow rate is

$$Q = \frac{\pi R^2 V}{z}\left[\frac{1-\kappa^2}{\ln \frac{1}{\kappa}} - 2\kappa^2 \right]$$

6.22 Shown are data taken at short equal time intervals from a hot wire anemometer mounted in a turbulent air stream. Plot the data, and determine the mean velocity and the mean square fluctuation.

Time	Velocity, ft/s
1	76
2	103
3	118
4	102
5	112
6	100
7	97
8	87
9	96
10	105
11	85
12	80
13	92
14	78
15	100
16	101
17	92
18	115
19	112
20	81

6.23 The velocity data below were measured for a fluid in turbulent flow. Plot the data, determine the velocity and the mean square fluctuation.

Time	Velocity (ft/s)	Time	Velocity (ft/s)
1	103	11	76
2	118	12	118
3	112	13	112
4	87	14	97
5	105	15	96
6	80	16	85
7	78	17	92
8	101	18	100
9	115	19	92
10	81	20	112

6.24 A smooth pipe carries water at a bulk velocity of 15 ft/s. The inside diameter of the pipe is 1 in. Use the universal velocity profile to calculate the shear stress at the wall.

6.25 Water at 60°F flows through a straight section of 100 ft of 1-in. Schedule 40 pipe. At a bulk velocity of 5 ft/s the total pressure drop over the 100 ft is 3 psi. Calculate the thickness of the viscous sublayer inside the pipe. What is the equivalent roughness height for this pipe? Would the pipe be considered completely rough or hydraulically smooth under the given conditions?

6.26 For steady flow: Briefly discuss (a) how ideal flow differs from laminar flow, and (b) how laminar flow differs from turbulent flow. (c) On a sketch such as below, draw in a laminar boundary layer forming at the tip of the flat plate (upper side only). At some point downstream, show on the sketch how transition to a turbulent boundary layer occurs. Label where the laminar sublayer, the buffer layer, and the turbulent core should be on the diagram. (d) Describe what would happen if we increased the approach velocity.

6.27 Assume the boundary layer which forms at the entrance to a circular pipe behaves like that on a flat plate. Estimate the length in diameters to reach fully developed flow (for the edge of the boundary layer to reach the center line of the pipe) for water at 68°F entering a pipe of D inches in diameter with a uniform velocity of 0.05 ft/s. (At this velocity the flow is laminar.)

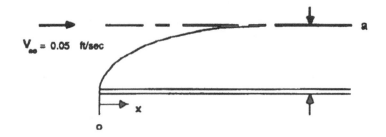

6.28 The boundary layer which forms at the entrance to a circular pipe may be considered to behave like the boundary layer on a flat plate. Estimate the length needed to reach fully developed flow (i.e., for the edge of the boundary layer to reach the center line) for water at 68°F entering a pipe of 2-in. ID at a uniform velocity of (a) 50 ft/s, (b) 5 ft/s, (c) 0.05 ft/s.

6.29 Assume that the velocity distribution for laminar flow over a flat plate is given by $v_x/v_\bullet = 3/2\ (y/\delta)^3$. Obtain the expression for the boundary layer thickness relative to distance along the plate, and the expression for shear stress at the plate.

6.30 Section 6.6.2 lists the assumptions for the boundary layer model. Show how to derive Eq. 6.6.2-2 and 6.6.2-3 starting with these assumptions and the microscopic mass and momentum balance.

6.31 Obtain a computer solution to the Blasius equation for two-dimensional boundary layer flow past a flat plate

$$2\frac{d^3f}{d\eta^3} + f\frac{d^2f}{d\eta^2} = 0$$

$$\eta = 0 \qquad f = 0$$

$$\eta = 0 \qquad \frac{df}{d\eta} = 0$$

$$\eta = \infty \qquad \frac{df}{d\eta} = 1$$

Using a fourth-order Runge-Kutta method, use a step size at .01 for h, obtain a solution for h = 2 to h = 5.

6.32 Calculate the total drag force on a bridge cable which is 8 in. in diameter and extends a distance of 1,000 ft between support towers when a wind is blowing across the line with a velocity of 40 mph. Use properties of air at 60°F.

6.33 Calculate the force acting on a circular bridge pier 5 ft in diameter which is immersed 60 ft in a river which is traveling at 3 mph. River temperature is at 70°F.

6.34 In calculating the drag force on a wing, we might consider it as two flat plates for the upper and lower surfaces plus a half-cylinder on the leading edge. Estimate the drag force for a wing approximately 5 inches thick, 30 ft long and 6 feet wide traveling at 500 ft/s in air at 32°F. Data for air at 32°F, 1 atm: $\rho = 0.081$ lbm/ft³; $\mu = 0.023$ cP.

6.35 In order to estimate the drag force on an airplane wing, model the wing as a half-cylinder for the leading edge and as two flat plates for the upper and lower surfaces. Estimate the drag force for a wing approximately 15 cm thick, 10 m long and 2 m wide on an airplane traveling at 100 m/s in air at 0°C. The density of the air is 1.3 kg/m³ and viscosity is .023 cP at this temperature.

6.36 For the situation sketched, find the required power input to the fluid from the pump.

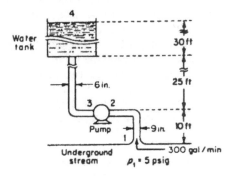

6.37 A 25-hp pump is employed to deliver an organic liquid (ρ=50 lbm/ft³) through a piping system, as shown. Under the existing conditions, the liquid flow rate is 4,488 gal/min.

    a. Calculate the frictional losses ("lost work") in the piping system between points 1 and 2 in ft lbf/s.

    b. Assuming that the frictional losses vary as the square of the flow rate, calculate the required pump horsepower if the delivery rate is to be doubled.

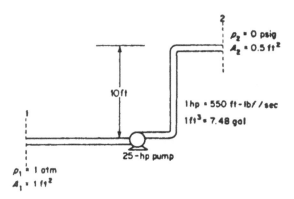

6.38 Water at 60°F (viscosity 1.05 cP) is being pumped at a rate of 185 gal/min for a total distance of 200 ft from a lake to the bottom of a tank in which the water surface is 50 ft above the level of the lake. The flow from the lake first passes through 100 ft of 3-1/2-in. Schedule 40 pipe (ID = 3.548 in). This section contains the pump, one open globe valve and three standard elbows. The remaining distance to the tank is through two parallel 100-ft lengths of 2-1/2 in. Schedule 40 pipe (ID = 2.469), each including four standard elbows. A reducing tee is used for the junction between the 3-1/2-in. line and the two 2-1/2-in. lines. Calculate the lost work in the piping system.

6.39 Water at 20°C is pumped from a tank to another tank at a volumetric flow rate of 1 x 10⁻³ m³/s. All the pipe is Schedule 40 and has an inside diameter of 0.1 m. For water at this temperature ρ = 998 kg/m³, μ = 1 x 10⁻³ Pa s. If the pump has an efficiency of 60%, what is the power required of the pump in kW?

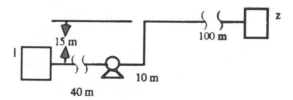

6.40 Water is pumped from an open tank to a pressure vessel at 45 psig as shown. The difference in liquid levels between the two tanks is 50 ft. The flow rate is 20,000 lbm/hr in a 3-inch Schedule 40 steel pipe. The total length of pipe is 300 ft. What is the horsepower required if the pump has an efficiency of 50%?

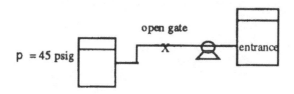

6.41 An 80°F organic liquid ($\rho = 52$ lbm/ft^3, $\mu = 3.4$ cP) is pumped from an open vessel to a pressurized storage tank as shown in the sketch. The difference in liquid levels between the two tanks is 60 ft The pipe is 3-inch Schedule 40 steel and the sum of all straight lengths of pipe is 400 ft. The liquid is pumped at a rate of 20,000 lbm/hr.

    a. Show the equivalent length of the piping system is 485 feet. (You may assume entrance to pressure tank does not contribute.)

    b. Show how to find the friction factor.

    c. If the pump has an efficiency of 65%, what is the minimum horsepower required for the pump?

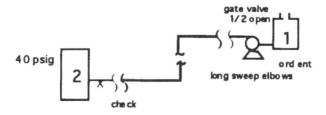

6.42 Water, at 60°F, is to be pumped at 200 gal/min for a total distance of 200 ft from a lake to the bottom of a tank in which the water surface is 50 ft above the level of the lake. The flow from the lake first passes through 100 ft of 3-1/2-in. Schedule 40 commercial steel pipe (ID = 3.548 in.). This section contains the pump, one open globe valve, and three standard elbows. The remaining distance to the tank is through two parallel 100 ft lengths of 2-1/2-in. Schedule 40 cast iron pipe (ID = 2.469 in.), each including four standard elbows. A reducing tee is used for the junction between the 3-1/2- and the two 2-1/2-in. lines. Calculate the lost work in the system.

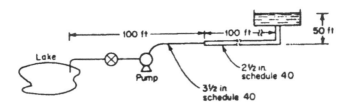

6.43 Calculate the pump horsepower required for the system below if the pump operates at an efficiency of 70 percent.

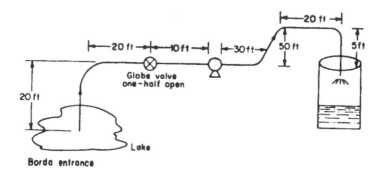

Data: The fluid is water; $\rho = 62.4$ lbm/ft^3, $\mu = 1$ cP. The pipe is galvanized iron and is 3 in. ID. $Q = 300$ gal/min. All elbows are standard.

6.44 An airplane owner wants to install a fuel filter into his fuel line. In the event the engine-driven fuel pump (inside the engine) fails, the electric fuel pump will have to deliver the fuel to the engine at a minimum of 5 psig. On takeoff, the plane will be 12 degrees nose up and the engine will require 38 gallons per hour.

    a. Can the filter be safely installed?

    b. If not, can the size of the electric fuel pump be increased to
      allow a safe installation?

    c. How big a pump is necessary?

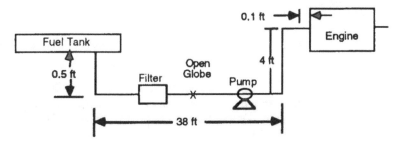

Pipe 3/8-inch Schedule 40 Steel $\rho = 6$ lbm/gal
Elbows Standard $\mu = 3.1$ cP
Pump power = 1 W, Efficiency = 55%

6.45 A tank contains water at 180°F. The discharge rate at the end of the pipe is to be 100 gal/min. All pipe is Schedule-40 commercial steel.

    For H$_2$O @ 180°F

    $\rho = 60.5$ lbm/ft^3

    $\mu = .35$ cP

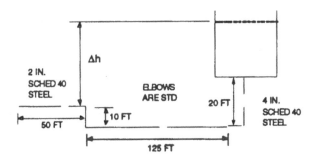

The value of Δh is the elevation of the surface of the water in the tank from the discharge point.

    a. Show the equivalent length of the system 4-in. pipe (include 4-in. elbow plus contraction to 2-in. line is 40 ft and the 2-in. line is 197 ft.

    b. Determine the lost work in each pipe.

    c. Calculate Δh in ft.

6.46 Water is flowing from a pressurized tank as in the sketch. The pipe is an 8-in. Schedule 40 steel pipe. Determine the gauge pressure at 1. The flow rate is 5 ft^3/s. Use 1 cP as the viscosity of water.

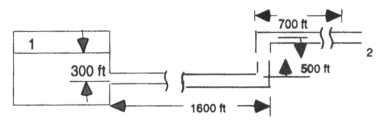

6.47 An experiment is performed to check the equivalent length of a standard elbow in a galvanized iron pipe system. The pressure drop p_1-p_2 between two taps 10 ft on either side of the elbow is found to be 11 psig when the mass flow rate of water is 26 lbm/s. Calculate the equivalent length of the elbow.

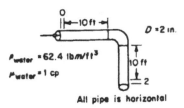

All pipe is horizontal

6.48 Water at 68°F flowing at an average velocity of 9.7 ft/s through an old rusty pipe having an ID of 4.026 in. gave a net loss of head of 12.4 ft for a 51.5 ft test length. How does the experimental friction factor compare with that calculated for new commercial pipe?

6.49 Shown is an open waste-surge tank with diameter of 30 ft. This tank is fed by a pump supplying a steady 100 gal/min. When the pump is first turned on, the level in the tank is 10.0 ft. The tank drains to the ocean through 3-in. cast iron pipe as shown. Will the tank empty, overflow, or reach a constant level? Support your answer with calculations.

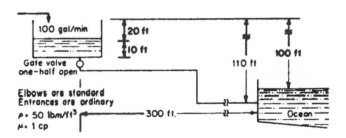

6.50 For the following system a head of 20 ft. is required to overcome friction loss. Calculate the pipe diameter.

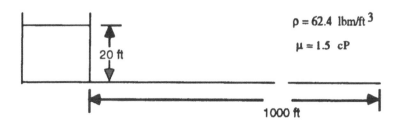

$$\rho = 62.4 \ \text{lbm/ft}^3$$

$$\mu = 1.5 \ \text{cP}$$

20 ft

1000 ft

6.51 Water is to be pumped through 1000 ft. of a horizontal Schedule 40 steel pipe at a flow rate of 1000 gpm. The maximum discharge pressure of the pump is 75 psig. The water pressure necessary at the end of the pipe must exceed 45 psig. Determine the smallest pipe size that may be used.

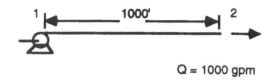

1   1000'   2

Q = 1000 gpm

$p_1 \leq 75$ psig

$p_2 \geq 45$ psig

$v = 1.2 \times 10^{-5} \ \text{ft}^2/\text{s}$

$\rho = 1.94 \ \text{slug/ft}^3$

6.52 An organic liquid flows from a reactor through a pipe to a storage tank in an oil refinery at a rate of 3 m³/min. The pipe is a commercial steel pipe with an inside diameter of 0.2 m. The gauge pressure inside the reactor vessel is 110 kPa. What is the total length of the pipe?

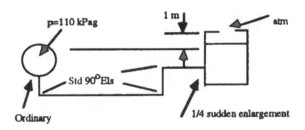

6.53 A fluid with viscosity of 100 cP and density 77.5 lbm/ft³ is flowing through a 100 ft length of vertical pipe (Schedule 40 steel). If the pressure at the bottom of the pipe is 20 psi greater than at the top, what is the mass flow rate?

6.54 Water at 60°F is flowing from a very large constant head reservoir to an irrigation ditch through a 1-1/2-in. Schedule 40 cast iron pipe as shown below. All elbows are 90° standard elbows. Find the water flow rate in gallons per minute.

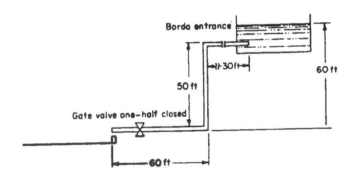

6.55 A hypodermic syringe is used to inject acetone at 70°F into a laboratory equilibrium still operating at 5.0 psig. The syringe has a 1.5-in. barrel ID and is connected to the still by 5.09 ft of special hypodermic tubing having an ID of 0.084 in. What injection rate in gallons per minute would you expect for a force of 15 lbf applied to the syringe plunger? State your assumptions.

Barometer = 750 mm Hg

Viscosity = 0.35 cP

Specific gravity = 0.7

6.56 An industrial furnace uses natural gas (assume 100% CH_4) as a fuel. The natural gas is burned with 20% excess air, and is converted completely to CO_2 and water. The combustion products leave the furnace at 500°F and pass into the stack, which is 150 ft high and 3.0 ft ID. The temperature of the air outside of the stack is 65°F and barometric pressure is 745 mm Hg.

       a. If the gases in the stack were standing still, what would be the stack draft (difference in pressure inside and outside the bottom of the stack)?

       b. If the gases were moving upward in the stack at a rate of 10 ft/s, what would be the stack draft, neglecting friction?

       c. Including the effect of friction, estimate the expected flow rate. (The 10 ft/s given previously no longer applies). Make whatever simplifying assumptions you may feel necessary. μ = 0.03 cP.

6.57 A liquid with ρ = 54 lbm/ft^3 and μ = 1.5 cP flows from a tank maintained at constant level through 70 ft of 1-1/2-inch Schedule 40 steel pipe to another constant level tank 60 ft. below the other tank. There are two standard elbows, 2 ordinary entrances and an open gate valve in the piping system. What is the flow rate in the 1-1/2-inch pipe?

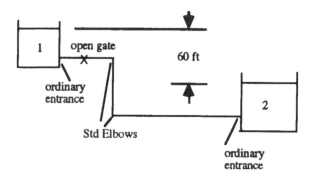

6.58 Water (μ = 1 cP) flows from a constant level tank through a 3-in. inside diameter drawn tubing with an equivalent length of 200 ft. The exit of the pipe is 30 ft. below the water level. Calculate the velocity of flow at 2 in ft/s. (Use k/D = 0.0005).

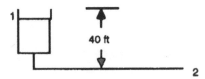

6.59 Water is flowing in the pipe network shown below. The water flows through a process in branch C, which for the purpose of calculating pressure drop can be considered as 300 ft of straight pipe. A bypass section is provided with a throttling valve so that the water flow to the process can be adjusted.

For the system shown, find p_2, p_3, $<v_A>$, $<v_B>$, and $<v_c>$.

All pipe is horizontal 3-in. Schedule 40 steel. Pressures at points 1 and 4, respectively, are 30 psig and 2 psig.

$$\rho = 62.4$$
$$\mu = 1 \text{ cP}$$

Branch	Length, ft
A	400
B	400
C	300
D	200

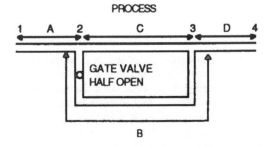

6.60 A large constant head water reservoir feeds two irrigation lines as shown below. If we neglect any kinetic energy change due to an area change of the piping and any losses which are associated with the tee joint at point A, find the bulk velocities in each line.

D_I = 1-1/2-in Schedule 40 commercial pipe

D_{II} = 2-in Schedule 40 commercial pipe

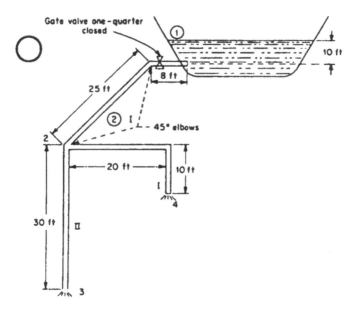

6.61 A horizontal pipeline system, composed of 6-in. ID pipe, carries cooling water at about 60°F for a lean oil cooler and a product cooler. The product cooler needs much less cooling water than does the lean oil cooler. (Lengths given are equivalent lengths; that is, they include bends, fittings, etc.) Outlets exist at points 3 (lean oil cooler) and 5 (product cooler) and the outlet pressures may both be taken to be 0 psig. The inlet is supplied by a pump capable of maintaining a 30-psig pressure. It has become necessary to increase the flow supplied at point 3, with less flow required at point 5. If a 6-in. line is installed from point 4 to point 3, as indicated, what will be the flow rate in this branch? Assume f - 0.015 over the entire system.

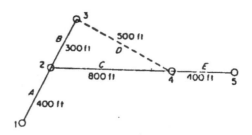

6.62 Calculate the equivalent diameter of the annular space between a 1-1/2- and a 2-in. Schedule 40 pipe.

6.63 A rectangular enclosed channel of asphalted cast iron is used to transport water at 30 ft/s. Find the pressure drop for each foot of channel length.

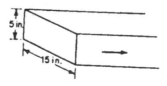

6.64 A conduit whose cross-section is an equilateral triangle (with side 5 in.) has a pressure drop of 1 psi/ft. How much air at standard conditions will the conduit deliver?

6.65 In many nuclear reactors, the nuclear fuel is assembled in solid form in the form of a lattice through which liquid sodium coolant is pumped. This assembly is usually built up of fuel plates which are thin rectangular prisms. In one such assembly, the openings for coolant are 2.620 in. wide by 0.118 in. The plates are 25 in. long. Manifolding at the ends is such that an infinite diameter may be assumed. Calculate the pressure drops in psi between the inlet and outlet manifolds of the fuel assembly for both water at an average temperature of 100°F and sodium at 400°F, assuming that the surface roughness of the plates is equivalent to drawn tubing and that fluid velocities of 20 ft/s are used.

$\mu_{Na} = 0.428$ cP

ρNa $= 0.901$ g/cm^3

6.66 A polymer with n $= 0.5$ g has a numerical value of 1500 and $\rho = 58$ lbm/ft^3 is flowing at a velocity of 0.25 ft/s through a square conduit 1/4 inch on a side.

   a. What are the units of g?

   b. What is the pressure drop per foot (psi/ft)?

6.67 For flow of a fluid through a porous medium the equation of continuity is

$$\varphi \frac{\partial \rho}{\partial t} + \left(\nabla \cdot \rho\, q \right) = 0$$

where f is porosity and q is the superficial (Darcy) velocity.

The equation of motion is Darcy's law

$$q = -\frac{k}{\mu}\left(\nabla p + \rho\, g \right)$$

where k is permeability. Define

$$\Phi = p + \rho\, g\, z$$

(gravity in the z-direction).

   a. Show by combining the above that the equation of condition for single fluid flow in a porous medium is

$$\varphi \mu \frac{\partial \rho}{\partial t} = \left(\nabla \cdot \rho\, k\, \nabla \Phi \right)$$

   b. Show for the flow of an incompressible flow

$$\nabla \cdot k \nabla \Phi = 0 \quad \text{or} \quad \nabla^2 \Phi = 0$$

   if k is constant

c. For a compressible liquid

$$c_1 = \frac{1}{\rho} \frac{\partial \rho}{\partial p}$$

If c_e is constant and assume gravity is negligible and show

$$\phi \mu c_r \frac{\partial p}{\partial t} = \nabla \cdot k \, \nabla p$$

d. For the flow of an ideal gas

$$\rho = \frac{M}{RT} p$$

Assume gravity is negligible and show

$$\phi \mu \frac{\partial p}{\partial t} = \nabla \cdot k \, \nabla p^2$$

6.68 A filter is to be operated as follows:

a. At a constant rate of 2 gal/min until the pressure drop reaches 5 psi, and then

b. At a constant pressure drop of 5 psi to the end of the run.

How long will it take to collect 500 gal of filtrate operating as above? You may make appropriate assumptions. The following data have been taken at a constant rate of 1 gal/min.

t	Δp
100 min	1 psi
200 min	3 psi

6.69 Calculate the pressure drop for water at 100°F and 1 atm flowing at 500 lb/hr through a bed of 1/2-in. spheres. The bed is 4 in. in diameter and 8 in. high. The porosity (void fraction) of the bed is 0.38. Assume gravity is negligible.

6.70 A 12-in. circular duct in which a gas with $\rho = 0.001$ lbm ft^3 and $\mu = 0.1$ cP is flowing is shown below. A pitot tube attached to a differential manometer is inserted in the duct as shown. Flow is laminar, and the pitot tube coefficient may be taken as 1.0.

  a. Calculate the velocity (if any) in ft/s using the pitot tube equation. If no flow is present, explain.

  b. Calculate the mass flow rate in lbm/s if possible - if not, explain why in detail.

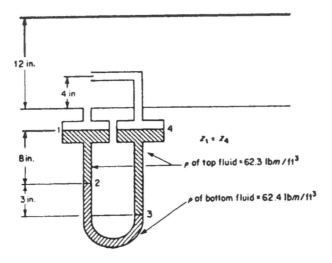

6.71 A pitot tube traverse was made in a 3-in. Schedule 40 steel pipe in which water at 58°F was flowing. By integrating

$$\int_A v\, dA = \int_0^{R^2} v\, 2\pi r\, dr = \pi \int_0^{R^2} y\, dr^2$$

$$= \pi \int_0^{R^2} C \sqrt{\frac{2 g \left(\rho_m - \rho \right)}{\rho}}\, \Delta h\, dr^2$$

graphically, we get a flow rate of 0.1336 ft³/s (C is the coefficient). At the same time as the traverse the discharge into a weigh tank was 1,400 lbm of water in 194.4 s. Find the pitot tube coefficient. Explain the integration above.

6.72 You are calibrating a pitot tube in a 4-in. Schedule 40 pipe in which water is flowing at 60°F. The following data are recorded:

y/R	h(inches of manometer fluid, sp. gr. = 1.2)
0.1	13.60
0.2	16.59
0.3	18.49
0.5	21.55
0.7	23.70
0.9	25.55
1.0	26.30

where y is the distance measured from the wall. At the same time, a sample of the flow is taken by discharging the entire stream into a weighing tank. The discharge rate was found to be 1,410 lbm in 60 s. Assuming that the velocity profile is symmetrical, find the pitot tube coefficient.

6.73 Fuel oil is flowing into an open-hearth-furnace complex and is entering through a venturi meter as shown below.

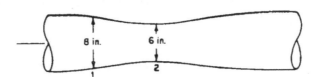

The lost work between points 1 and 2 is estimated at 5 percent of $-\Delta p/r$ between 1 and 2. If the specific gravity of the oil is 1.4 and the measured pressure drop is $p_2 - p_1 = -8$ psig,

    a. Show a control volume and label all control surfaces and forces.

b. Calculate the oil flow rate in gallons per minute.

6.74 A venturi meter is used to determine rate of flow in a pipe. The diameter at the entrance (1) is 6 in. and at the vena contracta (2) is 4 in. Find the discharge through the pipe when $p_1 - p_2 = 3$ psi, for oil with specific gravity = 0.6.

6.75 A venturi meter is placed in a 4-in. ID pipe on an angle of 30° with the horizontal. Water at room temperature flows through the venturi, which has a throat of 2 in. If the pressure drop in the meter is measured by a mercury manometer, find the velocity of water in the 4-in. pipe.

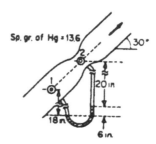

6.76 Water at 70°F is flowing in a 3-in. ID pipeline and is being metered by an orifice meter where b = 0.5. Calculate the flow rate in gallons per minute if the Δp across the orifice is 28 in. of manometer fluid where the manometer fluid has a specific gravity of 1.7. Justify your assumption on C_0, the orifice discharge coefficient.

6.77 Water at 70°F is flowing in a 4-in. ID pipeline and is being metered by an orifice meter where b = 0.6. Calculate the flow rate in gallons per minute if the Δp across the orifice is 26 in. of manometer fluid where the manometer fluid has a specific gravity of 1.6. Justify your assumption on C_0, the discharge coefficient.

6.78 Water is being metered by an orifice as shown below. Calculate the water flow rate in gallons per minute.

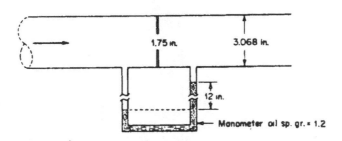

6.79 Water is metered by an orifice (see sketch). Calculate the flow rate in gallons per minute.

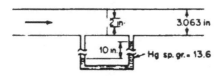

7

HEAT TRANSFER MODELS

7.1 The Nature of Heat

One term in our overall energy balance is "heat" and was defined as energy transferred across the boundary of the system as a result of temperature difference. The overall energy balance tells us only that the sum of all the terms in the equation must remain constant - that a *balance* be maintained - that is, that energy be a *conserved* quantity. We can, therefore, calculate rate of transfer of any of the various forms of energy directly from the overall energy balance only if we know the total of the remaining terms. The overall energy balance by itself (or even when combined with the overall momentum and mass balances) is therefore seldom sufficient to determine the state of our system.

In construction of the overall energy balance, however, we did not make any statement about how the rate of transfer of heat depended on the distribution in space of temperature or the thermal resistance of the medium. We will now consider *rate* equations (as opposed to *balance* equations) which will give us additional information for modeling.

Elementary physics teaches us that heat transfers by three mechanisms:

- conduction
- convection
- radiation

Conduction is the transfer of energy on a molecular scale independent of net bulk flow of the material. For example, in a solid the molecules are held in a more or less rigid structure, so that there is no large amount of molecular translation from point to point. As long as the molecules have significant amounts of thermal energy, however, they do vibrate, with the vibrations increasing in amplitude with the thermal energy level.

These vibrations can be transmitted from one layer of molecules to an adjacent layer of molecules at a lower thermal energy level, one of the ways in which heat conduction takes place. Another mechanism of heat conduction on a molecular level involves the so-called free electrons which are more abundant in metals than other substances. These "free" electrons are electrons which are very loosely bound to atoms and are therefore relatively free to move about in the solid. Sometimes these electrons are referred to as the "electron gas." These electrons are excellent carriers of both thermal energy and electrical energy and, therefore, substances which have large numbers of free electrons are good conductors of heat as well as good conductors of electricity - for example, silver and aluminum.

In liquids, molecules also vibrate as a result of thermal energy. In liquids, however, they are not held in a rigid lattice structure and the molecules can undergo a process of molecular diffusion, which is the interchange of position of molecules on a molecular scale. In this process, energy is not transferred from one molecule to the next, but rather by the interchange of a high-energy molecule with a low-energy molecule.

If one looks at the problem microscopically, that is to say, on a molecular level, some of these mechanisms of conduction look much like what we define as convection. The question is really one of scale. In convection we are concerned with groups of molecules which contain many millions of molecules, so we simply ignore the molecular level.

A classical example of energy transfer by conduction is that of transfer of thermal energy down a poker thrust into a fire to warm the hand holding it; a more literary example is a cat on a hot tin roof, with conduction from roof to feline paw.

Convection is the transfer of energy from place to place by a bulk flow of matter from a hotter region to a colder, or vice versa. The thermal energy is carried by the matter rather than being transmitted by it. Convection is usually subdivided into two categories: free convection and forced convection.

By *forced convection*, we mean convection which is induced by some prime mover such as a pump, fan, blower, etc. By *free (or natural) convection*, we mean convection which results from density differences within the system.[1] This type of convection can, for example, result from thermal gradients or concentration gradients. Either of these mechanisms of convection can give rise

---

[1] These terms are also applicable to mass transfer; for example, density differences can result from either temperature or concentration gradients, or both.

to either laminar or turbulent flow, although turbulent flow is perhaps more common in forced convection problems than in free convection problems.

In flowing systems, the classification of heat transfer as convection rather than conduction is largely a question of scale, with larger scale associated with convection. In fact, if we wish to be rigorous, what we refer to as heat transfer by convection is really a sequential process. For example, in transfer of heat from a warmer to a cooler region by what we call convection, a mass of material moves by bulk flow from a hotter region to a colder region, at which point it mixes with the local fluid and ultimately gives up the energy by conduction on a molecular scale to the colder surrounding fluid.

Typical examples illustrating convective heat transfer are forced air furnaces in homes and transfer of thermal energy on the earth by large-scale movements of the atmosphere.

Radiation is the transfer of energy by the emission and subsequent absorption of electromagnetic radiation. A heat lamp transfers heat primarily by radiation, as does a campfire. The energy from the sun arrives via radiation. The mechanism of heat transfer by radiation does not fall into the convenient parallelism with mass and momentum transfer that conductive and convective heat transfer do. There is no mass or momentum transfer process analogous to that of radiation for cases of interest here.

We know, of course, from physics that radiation is associated with an equivalent amount of mass and, therefore, when radiation is emitted mass is transferred. The *amount* of mass, however, is insignificant in any present-day engineering process.

Convective and conductive heat transfer are intimately related, because heat transferred by convection ultimately involves conduction. Perhaps a crude analogy will make this clearer. Suppose we have a group of people assembled, and someone in the back of the group wishes to communicate with someone in the front. One approach to the problem is for the person to write a note and hand it to his neighbor toward the front, and have the neighbor pass it to his neighbor to the front, etc. This is analogous to conduction. Another alternative is for the person in the back row to walk through the group (convection) and hand the note to the individual in front (conduction). Here, although the mechanism is convection, the ultimate transfer is still by conduction. The radiation analog is for the person standing at the rear of the group to shout the message, which utilizes a different medium (air) to carry the message - a medium in which the message moves as sound waves, of a very different nature from electromagnetic waves.

7.1.1   Forced convection heat transfer

For forced convection, the microscopic thermalenergy equation, incorporating Fourier's law and assuming a Newtonian fluid of constant density and thermal conductivity[2] with viscous dissipation the only generator of thermal energy, as shown in Section 3.2.3 is

$$\rho \, \hat{C}_p \frac{DT}{Dt} = k \, \nabla^2 T + \mu \, \Phi, \tag{7.1.1-1}$$

In rectangular coordinates, this equation takes the form

$$
\rho \, \hat{C}_p \left(\frac{\partial T}{\partial t} + v_1 \frac{\partial T}{\partial x_1} + v_2 \frac{\partial T}{\partial x_2} + v_3 \frac{\partial T}{\partial x_3} \right) = k \left(\frac{\partial^2 T}{\partial x_1^{\,2}} + \frac{\partial^2 T}{\partial x_2^{\,2}} + \frac{\partial^2 T}{\partial x_3^{\,2}} \right)
$$

$$
+ 2\mu \left[\left(\frac{\partial v_1}{\partial x_1} \right)^2 + \left(\frac{\partial v_2}{\partial x_2} \right)^2 + \left(\frac{\partial v_3}{\partial x_3} \right)^2 \right]
$$

$$
+ \mu \left[\left(\frac{\partial v_1}{\partial x_2} + \frac{\partial v_2}{\partial x_1} \right)^2 + \left(\frac{\partial v_1}{\partial x_3} + \frac{\partial v_3}{\partial x_1} \right)^2 + \left(\frac{\partial v_2}{\partial x_3} + \frac{\partial v_3}{\partial x_2} \right)^2 \right] \tag{7.1.1-2}
$$

The last two terms on the right-hand side account for viscous dissipation (the conversion of mechanical energy into thermal energy) and are frequently negligible, even for forced convection models, except for systems where either velocity gradients or viscosity are unusually large. Since energy is conserved in systems we consider in this text, the lost mechanical energy appears in the thermal energy balance as part[3] of the generation of thermal energy, which we designate below as γ_θ (for generation$_{thermal}$).

Let us use the following dimensionless variables to de-dimensionalize this equation for forced convection models. The single superscript * is used for forced convection, the double superscript ** for free convection.

---

[2] This equation omits certain forms of energy input; i.e., microwave radiation.
[3] One can generate thermal energy in other ways; e.g., through effects produced by fluctuating electromagnetic fields.

$$\boxed{v_i^{\bullet} = \frac{v_i}{V}} \qquad \boxed{t^{\bullet} = \frac{t\,V}{D}} \qquad \boxed{x_i^{\bullet} = \frac{x_i}{D}}$$

$$\boxed{T^{\bullet} = \frac{(T - T_0)}{(T_1 - T_0)}} \qquad \boxed{p^{\bullet} = \frac{(p - p_0)}{\rho\,V^2}} \qquad (7.1.1\text{-}3)$$

Calculating some derivatives to be used later

$$\frac{\partial v_i^{\bullet}}{\partial v_i} = \frac{\partial}{\partial v_i}\left(\frac{v_i}{V}\right) = \frac{1}{V} \qquad \frac{\partial t^{\bullet}}{\partial t} = \frac{\partial}{\partial t}\left(\frac{t\,V}{D}\right) = \frac{V}{D}$$

$$\frac{\partial x_i^{\bullet}}{\partial x_i} = \frac{\partial}{\partial x_i}\left(\frac{x_i}{D}\right) = \frac{1}{D} \qquad \frac{\partial T^{\bullet}}{\partial T} = \frac{\partial}{\partial T}\left(\frac{T - T_0}{T_1 - T_0}\right) = \frac{1}{(T_1 - T_0)}$$

$$(7.1.1\text{-}4)$$

Using the chain rule for the time derivative

$$\frac{\partial T}{\partial t} = \left[\frac{\partial t^{\bullet}}{\partial t}\right]\left[\frac{\partial T}{\partial T^{\bullet}}\right]\left[\frac{\partial T^{\bullet}}{\partial t^{\bullet}}\right] = \left[\frac{V}{D}\right]\left[T_1 - T_0\right]\left[\frac{\partial T^{\bullet}}{\partial t^{\bullet}}\right] \qquad (7.1.1\text{-}5)$$

and for a typical space coordinate

$$\frac{\partial T}{\partial x_i} = \left[\frac{\partial x_i^{\bullet}}{\partial x_i}\right]\left[\frac{\partial T}{\partial T^{\bullet}}\right]\left[\frac{\partial T^{\bullet}}{\partial x_i^{\bullet}}\right] = \left[\frac{1}{D}\right]\left[T_1 - T_0\right]\left[\frac{\partial T^{\bullet}}{\partial x_i^{\bullet}}\right] \qquad (7.1.1\text{-}6)$$

$$\frac{\partial^2 T}{\partial x_i^2} = \frac{\partial}{\partial x_i}\left[\frac{\partial T}{\partial x_i}\right] = \left[\frac{\partial x_i^{\bullet}}{\partial x_i}\right]\frac{\partial}{\partial x_i^{\bullet}}\left[\frac{\partial T}{\partial x_i}\right]$$

$$= \left[\frac{1}{D}\right]\frac{\partial}{\partial x_i^{\bullet}}\left[\frac{1}{D}\right]\left[T_1 - T_0\right]\left[\frac{\partial T^{\bullet}}{\partial x_i^{\bullet}}\right]$$

$$= \left[\frac{T_1 - T_0}{D^2}\right]\left[\frac{\partial^2 T^{\bullet}}{\partial x_i^{\bullet 2}}\right] \qquad (7.1.1\text{-}7)$$

$$\frac{\partial v_i}{\partial x_j} = \left[\frac{\partial x_j^{\bullet}}{\partial x_j}\right]\left[\frac{\partial v_i}{\partial v_i^{\bullet}}\right]\left[\frac{\partial v_i^{\bullet}}{\partial x_j^{\bullet}}\right] = \left[\frac{1}{D}\right]\left[\frac{\mu}{D\rho}\right]\left[\frac{\partial v_i^{\bullet}}{\partial x_j^{\bullet}}\right] \qquad (7.1.1\text{-}8)$$

Substituting in the energy equation

$$\rho \hat{C}_p \left\{ \left[\frac{V}{D} \right] \left[(T_1 - T_0) \right] \left[\frac{\partial T^*}{\partial t^*} \right] + [V \, v_1^*] \left[\frac{1}{D} \right] [T_1 - T_0] \left[\frac{\partial T^*}{\partial x_1^*} \right] \right.$$

$$\left. + [V \, v_2^*] \left[\frac{1}{D} \right] [T_1 - T_0] \left[\frac{\partial T^*}{\partial x_2^*} \right] + [V \, v_3^*] \left[\frac{1}{D} \right] [T_1 - T_0] \left[\frac{\partial T^*}{\partial x_3^*} \right] \right\}$$

$$= k \left\{ \left[\frac{T_1 - T_0}{D^2} \right] \left[\frac{\partial^2 T^*}{\partial x_1^{*2}} \right] + \left[\frac{T_1 - T_0}{D^2} \right] \left[\frac{\partial^2 T^*}{\partial x_2^{*2}} \right] + \left[\frac{T_1 - T_0}{D^2} \right] \left[\frac{\partial^2 T^*}{\partial x_3^{*2}} \right] \right\}$$

$$+ 2 \mu \left\{ \left(\left[\frac{1}{D} \right] \left[\frac{\mu}{D \rho} \right] \left[\frac{\partial v_1^*}{\partial x_1^*} \right] \right)^2 + \left(\left[\frac{1}{D} \right] \left[\frac{\mu}{D \rho} \right] \left[\frac{\partial v_2^*}{\partial x_2^*} \right] \right)^2 + \left(\left[\frac{1}{D} \right] \left[\frac{\mu}{D \rho} \right] \left[\frac{\partial v_3^*}{\partial x_3^*} \right] \right)^2 \right\}$$

$$+ \mu \left\{ \left(\left[\frac{1}{D} \right] \left[\frac{\mu}{D \rho} \right] \left[\frac{\partial v_1^*}{\partial x_2^*} \right] + \left[\frac{1}{D} \right] \left[\frac{\mu}{D \rho} \right] \left[\frac{\partial v_2^*}{\partial x_1^*} \right] \right)^2 \right.$$

$$+ \left(\left[\frac{1}{D} \right] \left[\frac{\mu}{D \rho} \right] \left[\frac{\partial v_1^*}{\partial x_3^*} \right] + \left[\frac{1}{D} \right] \left[\frac{\mu}{D \rho} \right] \left[\frac{\partial v_3^*}{\partial x_1^*} \right] \right)^2$$

$$+ \left. \left(\left[\frac{1}{D} \right] \left[\frac{\mu}{D \rho} \right] \left[\frac{\partial v_2^*}{\partial x_3^*} \right] + \left[\frac{1}{D} \right] \left[\frac{\mu}{D \rho} \right] \left[\frac{\partial v_3^*}{\partial x_2^*} \right] \right)^2 \right\}$$

$$(7.1.1\text{-}9)$$

Factoring constants

$$\rho \hat{C}_p \left[\frac{V}{D}\right]\left[(T_1 - T_0)\right] \left\{\left[\frac{\partial T^\bullet}{\partial t}\right] + \left[v_1^\bullet\right]\left[\frac{\partial T^\bullet}{\partial x_1^\bullet}\right] + \left[v_2^\bullet\right]\left[\frac{\partial T^\bullet}{\partial x_2^\bullet}\right] + \left[v_3^\bullet\right]\left[\frac{\partial T^\bullet}{\partial x_3^\bullet}\right]\right\}$$

$$= k\left[\frac{T_1 - T_0}{D^2}\right]\left\{\left[\frac{\partial^2 T^\bullet}{\partial x_1^{\bullet 2}}\right] + \left[\frac{\partial^2 T^\bullet}{\partial x_2^{\bullet 2}}\right] + \left[\frac{\partial^2 T^\bullet}{\partial x_3^{\bullet 2}}\right]\right\}$$

$$+ \frac{2\mu^2}{D^2 \rho}\left\{\left[\frac{\partial v_1^\bullet}{\partial x_1^\bullet}\right]^2 + \left[\frac{\partial v_2^\bullet}{\partial x_2^\bullet}\right]^2 + \left[\frac{\partial v_3^\bullet}{\partial x_3^\bullet}\right]^2\right\}$$

$$+ \frac{\mu^2}{D^2 \rho}\left\{\left(\left[\frac{\partial v_1^\bullet}{\partial x_2^\bullet}\right] + \left[\frac{\partial v_2^\bullet}{\partial x_1^\bullet}\right]\right)^2 + \left(\left[\frac{\partial v_1^\bullet}{\partial x_3^\bullet}\right] + \left[\frac{\partial v_3^\bullet}{\partial x_1^\bullet}\right]\right)^2 + \left(\left[\frac{\partial v_2^\bullet}{\partial x_3^\bullet}\right] + \left[\frac{\partial v_3^\bullet}{\partial x_2^\bullet}\right]\right)^2\right\} \quad (7.1.1\text{-}10)$$

Rearranging

$$\left\{\left[\frac{\partial T^\bullet}{\partial t^\bullet}\right] + \left[v_1^\bullet\right]\left[\frac{\partial T^\bullet}{\partial x_1^\bullet}\right] + \left[v_2^\bullet\right]\left[\frac{\partial T^\bullet}{\partial x_2^\bullet}\right] + \left[v_3^\bullet\right]\left[\frac{\partial T^\bullet}{\partial x_3^\bullet}\right]\right\}$$

$$= \left(\frac{k}{\mu \hat{C}_p}\right)\left(\frac{\mu}{D V \rho}\right)\left\{\left[\frac{\partial^2 T^\bullet}{\partial x_1^{\bullet 2}}\right] + \left[\frac{\partial^2 T^\bullet}{\partial x_2^{\bullet 2}}\right] + \left[\frac{\partial^2 T^\bullet}{\partial x_3^{\bullet 2}}\right]\right\}$$

$$+ \left(\frac{k}{\mu \hat{C}_p}\right)\left(\frac{\mu}{D V \rho}\right)\left(\frac{\mu V^2}{k(T_1 - T_0)}\right)\left[\left\{\left[\frac{\partial v_1^\bullet}{\partial x_1^\bullet}\right]^2 + \left[\frac{\partial v_2^\bullet}{\partial x_2^\bullet}\right]^2 + \left[\frac{\partial v_3^\bullet}{\partial x_3^\bullet}\right]^2\right\}\right.$$

$$\left. + \left\{\left(\left[\frac{\partial v_1^\bullet}{\partial x_2^\bullet}\right] + \left[\frac{\partial v_2^\bullet}{\partial x_1^\bullet}\right]\right)^2 + \left(\left[\frac{\partial v_1^\bullet}{\partial x_3^\bullet}\right] + \left[\frac{\partial v_3^\bullet}{\partial x_1^\bullet}\right]\right)^2 + \left(\left[\frac{\partial v_2^\bullet}{\partial x_3^\bullet}\right] + \left[\frac{\partial v_3^\bullet}{\partial x_2^\bullet}\right]\right)^2\right\}\right] \quad (7.1.1\text{-}11)$$

Using the definitions of the Reynolds number, the Prandtl number, and the Brinkman number (all dimensionless)

$$Re = \frac{D V \rho}{\mu} \qquad Pr = \frac{\mu C_p}{k} \qquad Br = \frac{\mu V^2}{k\left(T_1 - T_0\right)} \qquad (7.1.1\text{-}12)$$

we can write

$$
\left\{ \left[\frac{\partial T^*}{\partial t^*} \right] + \left[v_1^* \right] \left[\frac{\partial T^*}{\partial x_1^*} \right] + \left[v_2^* \right] \left[\frac{\partial T^*}{\partial x_2^*} \right] + \left[v_3^* \right] \left[\frac{\partial T^*}{\partial x_3^*} \right] \right\}
$$

$$
= \left(\frac{1}{Re\,Pr} \right) \left\{ \left[\frac{\partial^2 T^*}{\partial x_1^{*2}} \right] + \left[\frac{\partial^2 T^*}{\partial x_2^{*2}} \right] + \left[\frac{\partial^2 T^*}{\partial x_3^{*2}} \right] \right\}
$$

$$
+ \left(\frac{Br}{Re\,Pr} \right) \left[\left\{ \left(\left[\frac{\partial v_1^*}{\partial x_1^*} \right] \right)^2 + \left(\left[\frac{\partial v_2^*}{\partial x_2^*} \right] \right)^2 + \left(\left[\frac{\partial v_3^*}{\partial x_3^*} \right] \right)^2 \right\} \right.
$$

$$
\left. + \left\{ \left(\left[\frac{\partial v_1^*}{\partial x_2^*} \right] + \left[\frac{\partial v_2^*}{\partial x_1^*} \right] \right)^2 + \left(\left[\frac{\partial v_1^*}{\partial x_3^*} \right] + \left[\frac{\partial v_3^*}{\partial x_1^*} \right] \right)^2 + \left(\left[\frac{\partial v_2^*}{\partial x_3^*} \right] + \left[\frac{\partial v_3^*}{\partial x_2^*} \right] \right)^2 \right\} \right] \qquad (7.1.1\text{-}13)
$$

Using Φ^* to denote the dissipation[4] function written in dimensionless variables, applying the definition of the substantial derivative, and defining the dimensionless operator in rectangular coordinates as

$$\boxed{\nabla^* \equiv \delta_1 \frac{\partial}{\partial x_1^*} + \delta_2 \frac{\partial}{\partial x_2^*} + \delta_3 \frac{\partial}{\partial x_3^*}} \qquad (7.1.1\text{-}14)$$

the dimensionless energy balance for forced convection becomes

$$\boxed{\frac{D T^*}{D t^*} = \left(\frac{1}{Re\,Pr} \right) \nabla^{*2} T^* + \left(\frac{Br}{Re\,Pr} \right) \Phi_v^*} \qquad (7.1.1\text{-}15)$$

where the definition of the dimensionless substantial derivative operator is

---

[4] The *dissipation function* is comprised of those terms in the balance representing conversion of mechanical to thermal energy via the action of viscosity.

$$\frac{D}{Dt^*} \equiv \left[\frac{\partial}{\partial t^*}\right] + [v_1^*]\left[\frac{\partial}{\partial x_1^*}\right] + [v_2^*]\left[\frac{\partial}{\partial x_2^*}\right] + [v_3^*]\left[\frac{\partial}{\partial x_3^*}\right] \qquad (7.1.1\text{-}16)$$

7.1.2 Free convection heat transfer

For free convection, we begin with the energy equation incorporating Fourier's law and assuming a Newtonian fluid of constant density and thermal conductivity. The terms which result from viscous dissipation are usually considered negligible for free convection models, and consequently are omitted.

$$\rho \hat{C}_p \frac{DT}{Dt} = k \nabla^2 T \qquad (7.1.2\text{-}1)$$

and its rectangular coordinate form

$$\rho \hat{C}_p \left(\frac{\partial T}{\partial t} + v_1 \frac{\partial T}{\partial x_1} + v_2 \frac{\partial T}{\partial x_2} + v_3 \frac{\partial T}{\partial x_3}\right) = k \left(\frac{\partial^2 T}{\partial x_1^2} + \frac{\partial^2 T}{\partial x_2^2} + \frac{\partial^2 T}{\partial x_3^2}\right)$$

$$(7.1.2\text{-}2)$$

Use the following dimensionless variables to de-dimensionalize this equation for free convection models. (The single superscript * is used for forced convection, the double superscript ** for free convection.) We need new definitions for dimensionless velocity and time because there is no "natural" time or velocity because motion is induced by density differences caused by temperature differences, not forced by an outside agent.

$$v_i^{**} = \frac{v_i D \rho}{\mu} \qquad t^{**} = \frac{t \mu}{\rho D^2} \qquad x_i^* = \frac{x_i}{D}$$

$$T^* = \frac{(T - T_0)}{(T_1 - T_0)} \qquad p^* = \frac{(p - p_0)}{\rho V^2} \qquad (7.1.2\text{-}3)$$

Calculating some derivatives we will need below

$$\frac{\partial v_i^{**}}{\partial v_i} = \frac{\partial}{\partial v_i}\left(\frac{v_i D \rho}{\mu}\right) = \frac{D \rho}{\mu} \qquad \frac{\partial t^{**}}{\partial t} = \frac{\partial}{\partial t}\left(\frac{t \mu}{\rho D^2}\right) = \frac{\mu}{\rho D^2}$$

$$\frac{\partial x_i^\bullet}{\partial x_i} = \frac{\partial}{\partial x_i}\left(\frac{x_i}{D}\right) = \frac{1}{D} \qquad\qquad \frac{\partial T^\bullet}{\partial T} = \frac{\partial}{\partial T}\left(\frac{T-T_0}{T_1-T_0}\right) = \frac{1}{(T_1-T_0)}$$

$$(7.1.2\text{-}4)$$

Using the chain rule for the time derivative gives a different result than in the previous example

$$\frac{\partial T}{\partial t} = \left[\frac{\partial t^{\bullet\bullet}}{\partial t}\right]\left[\frac{\partial T}{\partial T^\bullet}\right]\left[\frac{\partial T^\bullet}{\partial t^{\bullet\bullet}}\right] = \left[\frac{\mu}{\rho D^2}\right]\left[(T_1-T_0)\right]\left[\frac{\partial T^\bullet}{\partial t^{\bullet\bullet}}\right] \qquad (7.1.2\text{-}5)$$

as does application to the derivative of velocity with respect to a typical space coordinate

$$\frac{\partial v_i}{\partial x_j} = \left[\frac{\partial x_j^\bullet}{\partial x_j}\right]\left[\frac{\partial v_i}{\partial v_i^{\bullet\bullet}}\right]\left[\frac{\partial v_i^{\bullet\bullet}}{\partial x_j^\bullet}\right] = \left[\frac{1}{D}\right]\left[\frac{\mu}{D\rho}\right]\left[\frac{\partial v_i^{\bullet\bullet}}{\partial x_j^\bullet}\right] \qquad (7.1.2\text{-}6)$$

The space derivatives of the temperature remain the same as in the previous example

$$\frac{\partial T}{\partial x_i} = \left[\frac{1}{D}\right][T_1-T_0]\left[\frac{\partial T^\bullet}{\partial x_i^\bullet}\right] \qquad\qquad (7.1.2\text{-}7)$$

$$\frac{\partial^2 T}{\partial x_i^2} = \left[\frac{T_1-T_0}{D^2}\right]\left[\frac{\partial^2 T^\bullet}{\partial x_i^{\bullet 2}}\right] \qquad\qquad (7.1.2\text{-}8)$$

Substituting in the energy equation after deleting the viscous dissipation terms

$$\rho \hat{C}_p \left\{ \left[\frac{\mu}{\rho D^2} \right] \left[(T_1 - T_0) \right] \left[\frac{\partial T^{\bullet}}{\partial t^{**}} \right] \right.$$

$$+ \left[\frac{\mu}{\rho D} v_1^{\bullet\bullet} \right] \left[\frac{1}{D} \right] \left[T_1 - T_0 \right] \left[\frac{\partial T^{\bullet}}{\partial x_1^{\bullet}} \right] + \left[\frac{\mu}{\rho D} v_2^{\bullet\bullet} \right] \left[\frac{1}{D} \right] \left[T_1 - T_0 \right] \left[\frac{\partial T^{\bullet}}{\partial x_2^{\bullet}} \right]$$

$$\left. + \left[\frac{\mu}{\rho D} v_3^{\bullet\bullet} \right] \left[\frac{1}{D} \right] \left[T_1 - T_0 \right] \left[\frac{\partial T^{\bullet}}{\partial x_3^{\bullet}} \right] \right\}$$

$$= k \left\{ \left[\frac{T_1 - T_0}{D^2} \right] \left[\frac{\partial^2 T^{\bullet}}{\partial x_1^{\bullet 2}} \right] + \left[\frac{T_1 - T_0}{D^2} \right] \left[\frac{\partial^2 T^{\bullet}}{\partial x_2^{\bullet 2}} \right] + \left[\frac{T_1 - T_0}{D^2} \right] \left[\frac{\partial^2 T^{\bullet}}{\partial x_3^{\bullet 2}} \right] \right\}$$

$$(7.1.2-9)$$

Factoring constants and rearranging

$$\frac{\partial T^{\bullet}}{\partial t^{**}} + v_1^{\bullet\bullet} \frac{\partial T^{\bullet}}{\partial x_1^{\bullet}} + v_2^{\bullet\bullet} \frac{\partial T^{\bullet}}{\partial x_2^{\bullet}} + v_3^{\bullet\bullet} \frac{\partial T^{\bullet}}{\partial x_3^{\bullet}}$$

$$= \frac{k}{\mu \hat{C}_p} \left[\frac{\partial^2 T^{\bullet}}{\partial x_1^{\bullet 2}} + \frac{\partial^2 T^{\bullet}}{\partial x_2^{\bullet 2}} + \frac{\partial^2 T^{\bullet}}{\partial x_3^{\bullet 2}} \right]$$

$$(7.1.2-10)$$

Adjusting the definition of the dimensionless substantial derivative operator to use dimensionless variables from free convection

$$\boxed{ \frac{D}{Dt^{\bullet\bullet}} \equiv \left[\frac{\partial}{\partial t^{**}} \right] + \left[v_1^{\bullet\bullet} \right] \left[\frac{\partial}{\partial x_1^{\bullet}} \right] + \left[v_2^{\bullet\bullet} \right] \left[\frac{\partial}{\partial x_2^{\bullet}} \right] + \left[v_3^{\bullet\bullet} \right] \left[\frac{\partial}{\partial x_3^{\bullet}} \right] }$$
$$(7.1.2-11)$$

gives

$$\boxed{ \frac{DT^{\bullet}}{Dt^{**}} = \left(\frac{1}{Pr} \right) \nabla^{\bullet 2} T^{\bullet} }$$
$$(7.1.2-12)$$

Table 7.1.2-1 summarizes the equations for forced and free convection transfer in Newtonian fluids with constant density and thermal conductivity.

Table 7.1.2-1 Dimensionless Forms: Mass, Energy, and Momentum Equations for Natural and Forced Convection

	FORCED CONVECTION	FREE CONVECTION
TOTAL MASS BALANCE (CONTINUITY)	$\nabla^* \cdot v^* = 0$	$\nabla^* \cdot v^* = 0$
SPECIES MASS BALANCE	$\dfrac{Dx_A^*}{Dt^*} = \dfrac{1}{Re\,Sc}\nabla^{*2}x_A^*$	$\dfrac{Dx_A^*}{Dt^*} = \dfrac{1}{Re\,Sc}\nabla^{*2}x_A^*$
ENERGY	$\dfrac{DT^*}{Dt^*} = \dfrac{1}{Re\,Pr}\nabla^{*2}T^* + \gamma_\theta$	$\dfrac{DT^*}{Dt^{**}} = \dfrac{1}{Pr}\nabla^{*2}T^*$
MOMENTUM	$\dfrac{Dv^*}{Dt^*} = -\nabla^* p^* + \dfrac{1}{Re}\nabla^{*2}v^*$ $+ \dfrac{1}{Fr}\dfrac{g}{g}$	$\dfrac{Dv^{**}}{Dt^{**}} = \nabla^{*2}v^{**}$ $- x_A^*\,Gr_{AB}\dfrac{g}{g}$

7.2 Conduction Heat Transfer Models

Previously we accounted for the conversion of mechanical energy into either heat or internal energy by categorizing it as "lost work." Simply naming the category did not give a means of relating the category to the physical variables of the system so that **prediction** could be added to the model as well as **description**.

In carrying the model to the next level of detail, "lost work" was related to momentum transfer using Newton's law of viscosity:

$$\tau_{yz} = -\mu\frac{dv_z}{dy} \tag{7.2-1}$$

We now use an analogous equation for heat transfer, which is (in its one-dimensional form)

$$q_x = -k\frac{dT}{dx}$$

$$\left[= \frac{\dot{Q}}{A} = \frac{\text{rate of heat flow in x-direction}}{\text{area normal to x-direction}} \right]$$

(7.2-2)

where $\dot{Q}$ designates heat **flow** and q designates heat **flux**. This relationship describes heat transfer by *conduction*. The isotropic thermal conductivity k, as will be mentioned later, is a function of pressure and (primarily) of temperature.[5]

Equation (7.2-2), known as Fourier's law, is a phenomenological law; that is, it is true not because of our ability to derive it from first principles (although we are coming closer with models based on molecular dynamics), but rather because we define the thermal conductivity as the number which makes the equation true.

Conveniently, in doing this the thermal conductivity turns out to be relatively constant - otherwise, the attempt would be futile. In fact, some efforts[6] have been made to find higher-order dependence of heat transfer upon the temperature gradient, but such efforts to date have been unsuccessful.

7.2.1 Three-dimensional conduction in isotropic media

We wrote the Fourier equation in one-dimensional form in Cartesian coordinates to show that it parallels our previous equation (Newton's law of viscosity) for momentum transfer. The general form of the Fourier equation for isotropic media is

$$q = -k\nabla T$$

(7.2.1-1)

---

[5] For cases where the thermal conductivity varies with direction in the material (anisotropy) the thermal conductivity becomes a tensor; we will treat in this text primarily materials which may be considered isotropic. An example of an anisotropic material is a fibrous material such as wood, where the resistance to heat conduction varies depending on whether one is going along the grain or across the grain.

[6] Flumerfeld, R. V. and J. C. Slattery (1969). *AIChEJ* **15**: 291.

where

> q = heat transfer flux [F/(t L) = M/t^2], with units of energy [F
> L = M L^2/t^2] per unit time [t] and unit area [L^2] - typically,
> Btu/(hr ft^2), a **vector** quantity

> k = thermal conductivity in units of energy per unit time, unit
> area, and unit temperature gradient [F L/(t L^2 T/L) = F/(t T) =
> M L/(t^3 T)]; typically, Btu/(hr ft^2 °F/ft) or W/(m K), here
> written as a **scalar** quantity (this assumes an **isotropic**
> medium - in a more general medium the thermal conductivity
> is a second-order tensor - but for most of the problems of
> interest herein, we can assume isotropy)

> ∇T = gradient of the temperature field: a vector whose
> dimensions are temperature per unit length [T/L] (typically,
> °F/ft), a **vector** quantity

The negative sign is chosen for the equation because heat flows in the opposite
direction to the temperature gradient: heat runs downhill, as it were, just as a
skier progresses downhill. A negative temperature gradient (derivative) in the x-
direction gives a positive heat flux in the x-direction.

Remember also that the gradient operator takes different forms in various
coordinate systems as can be seen in the fluxes listed in Table 7.2.1-1.

**Table 7.2.1-1 Components of Fourier Equation in Various
Coordinate Systems**

Rectangular Cartesian	Cylindrical	Spherical
$q_x = -k \dfrac{\partial T}{\partial x}$	$q_r = -k \dfrac{\partial T}{\partial r}$	$q_r = -k \dfrac{\partial T}{\partial r}$
$q_y = -k \dfrac{\partial T}{\partial y}$	$q_\theta = -k \dfrac{1}{r} \dfrac{\partial T}{\partial \theta}$	$q_\theta = -k \dfrac{1}{r} \dfrac{\partial T}{\partial \theta}$
$q_z = -k \dfrac{\partial T}{\partial z}$	$q_z = -k \dfrac{\partial T}{\partial z}$	$q_\varphi = -k \dfrac{1}{r \sin\theta} \dfrac{\partial T}{\partial \varphi}$

When considering energy transfer in solids rather than liquids, there is no
viscous dissipation because there is no flow; however, thermal energy can be
generated by imposed fluctuating electromagnetic fields, so, in general, one can
have a thermal energy generation term γ_θ in the unsteady-state energy balance
even in the absence of dissipation in the mechanical energy balance

$$\rho \, C_p \frac{\partial T}{\partial t} = -(\nabla \cdot \mathbf{q}) + \gamma_\theta \tag{7.2.1-2}$$

For a solid

$$\mathbf{q} = -\bar{\mathbf{k}} \cdot \nabla T \tag{7.2.1-3}$$

Substituting Equation 7.2.1-3 into Equation 7.2.1-2

$$\rho \, C_p \frac{\partial T}{\partial t} = -\left(\nabla \cdot \left[-\bar{\mathbf{k}} \cdot \nabla T\right]\right) + \gamma_\theta \tag{7.2.1-4}$$

In rectangular coordinates

$$\rho \, C_p \frac{\partial T}{\partial t} = \frac{\partial}{\partial x}\left(k_x \frac{\partial T}{\partial x}\right) + \frac{\partial}{\partial y}\left(k_y \frac{\partial T}{\partial y}\right) + \frac{\partial}{\partial z}\left(k_z \frac{\partial T}{\partial z}\right) + \gamma_\theta \tag{7.2.1-5}$$

If the material is isotropic, k is independent of direction and may be factored out of the derivative, then both sides divided by k

$$\frac{\rho \, C_p}{k} \frac{\partial T}{\partial t} = \frac{\partial^2 T}{\partial x^2} + \frac{\partial^2 T}{\partial y^2} + \frac{\partial^2 T}{\partial z^2} + \frac{\gamma_\theta}{k} \tag{7.2.1-6}$$

The **thermal diffusivity, α,** so called because it both has the same units and serves an analogous function in heat transfer to the mass diffusivity (i.e., the diffusion coefficient, $\mathcal{D}$) in mass transfer, is defined in terms of the resulting group

$$\frac{\rho \, C_p}{k} = \frac{1}{\alpha} \tag{7.2.1-7}$$

$$\alpha \equiv \text{thermal diffusivity} \equiv \frac{k}{\rho \, C_p}$$

At steady state for an isotropic material and if there is no source for thermal energy (for example, from microwave heating, etc.)

$$\frac{\partial^2 T}{\partial x^2} + \frac{\partial^2 T}{\partial y^2} + \frac{\partial^2 T}{\partial z^2} = 0 \tag{7.2.1-8}$$

or, in symbolic notation

$$\nabla^2 T = 0 \tag{7.2.1-9}$$

This is the well-known Laplace equation, which appears in many contexts in engineering, and has been extensively studied for many years. As is typical for differential equations, there are an infinite number of solutions $T(x_i, t)$ from which the boundary conditions of the particular problem select the appropriate set for the physical situation(s) considered.

7.2.2   Boundary conditions at solid surfaces

For heat transfer problems considered here, there are three types of boundary conditions at solid surfaces that are pertinent

• Constant surface temperature (the Dirichlet condition), typically,

$$T = T_s \tag{7.2.2-1}$$

• Constant surface heat flux (the Neumann condition), typically,

$$q_s = -k \left. \frac{dT}{dx} \right|_{x=x_s} \tag{7.2.2-2}$$

where $q_s = 0$ for an insulated boundary (*adiabatic* condition)

• Convection coefficient prescribed at surface, typically,

$$-k \left. \frac{dT}{dx} \right|_{x=x_s} = h \left(T_s - T \right) \tag{7.2.2-3}$$

where h is the heat transfer coefficient and T is the ambient temperature

For unsteady-state problems an initial profile must also be specified: $T(x_i, 0)$.

7.2.3 Thermal conductivity

Typical values for thermal conductivities of some common materials are given in Table 7.2.3-1.

Table 7.2.3-1 Relative Values of Thermal Conductivity, Btu/[hr ft °F] (Qualitative use only - not for design calculations)

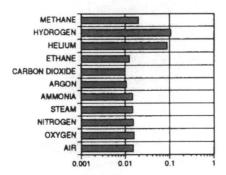

**Table 7.2.3-1 (continued) Relative Values of Thermal
Conductivity, Btu/[hr ft °F] (Qualitative use only - not for
design calculations)**

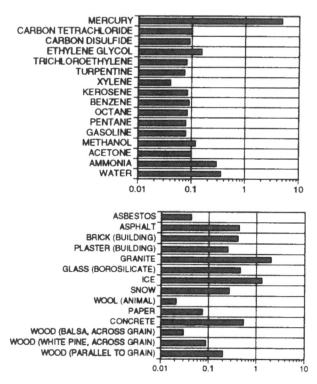

In general, the thermal conductivity is a fairly weak function of pressure and
a much stronger one of temperature. In our work we will not be greatly
concerned with pressure effects on thermal conductivity. We will, however, treat
cases in which temperature effects on thermal conductivity are important.

Data are available[7] or may be estimated by a variety of equations based on molecular principles.[8,9,10]

The high thermal conductivity of metals is largely due to the great number of free electrons - the so-called electron gas. Since these electrons are responsible for electrical conductivity as well as thermal conductivity, these two properties are highly correlated. Non-conducting (electrically) solids must rely mostly on lattice vibrations to conduct thermal energy, except for some cases where radiation is exchanged between molecules.

There are four mechanisms for conduction in solids:

- motion of free electrons
- vibrations of the lattice
- coupling of molecular magnetic dipoles
- electromagnetic radiation between atoms

Thermal conductivity of gases and non-metallic liquids is small.

For a stationary solid the velocity terms in the energy balance are zero, leaving

$$\rho\, \hat{C}_p \frac{\partial T}{\partial t} = -\left(\nabla \cdot \mathbf{q}\right) + \gamma_\theta \qquad (7.2.3\text{-}1)$$

Using as the constitutive relationship the Fourier equation

$$\mathbf{q} = -\bar{\mathbf{k}} \cdot \nabla T \qquad (7.2.3\text{-}2)$$

gives

$$\rho\, \hat{C}_p \frac{\partial T}{\partial t} = \nabla \cdot \left[\bar{\mathbf{k}} \cdot \nabla T\right] + \gamma_\theta \qquad (7.2.3\text{-}3)$$

---

[7] For example, the Thermophysical Properties Research Center (TPRC) in the Center for Information and Numerical Data Analysis and Synthesis (CINDAS) at Purdue University, West Lafayette, IN, maintains large data bases on thermal conductivity as well as other thermophysical properties.

[8] Bird, R. B., W. E. Stewart, et al. (1960). *Transport Phenomena*. New York, Wiley, pp. 247 ff.

[9] Hsu, S. T. (1963). *Engineering Heat Transfer*, D. Van Nostrand, pp. 9 ff., 569 ff.

[10] Perry, R. H., C. H. Chilton, et al., Eds. (1963). *Chemical Engineer's Handbook*. New York, McGraw-Hill, pp. 3-224.

or, in rectangular coordinates

$$\rho \, C_p \frac{\partial T}{\partial t} = \frac{\partial}{\partial x}\left(k_x \frac{\partial T}{\partial x}\right) + \frac{\partial}{\partial y}\left(k_y \frac{\partial T}{\partial y}\right) + \frac{\partial}{\partial z}\left(k_z \frac{\partial T}{\partial z}\right) + \gamma_\theta \qquad (7.2.3\text{-}4)$$

If the thermal conductivity is constant, $k_x = k_y = k_z = k$, giving

$$\rho \, C_p \frac{\partial T}{\partial t} = k\left[\frac{\partial^2 T}{\partial x^2} + \frac{\partial^2 T}{\partial y^2} + \frac{\partial^2 T}{\partial z^2}\right] + \gamma_\theta \qquad (7.2.3\text{-}5)$$

If, in addition, steady-state conditions prevail and there is no thermal energy generation

$$\frac{\partial^2 T}{\partial x^2} + \frac{\partial^2 T}{\partial y^2} + \frac{\partial^2 T}{\partial z^2} = 0 \qquad (7.2.3\text{-}6)$$

7.2.4   One-dimensional steady-state conduction in rectangular coordinates

One-dimensional conduction in single or multi-layer media in the absence of thermal energy generation is relatively simple to treat by analytical methods, so these problems will be the primary focus of the analytical treatment in this section.

We then use the case of conduction with thermal energy generation to illustrate the finite element numerical method. The finite element method is one of two numerical methods for the solution of elliptic and parabolic partial differential equations that we illustrate by simple examples. Our purpose is not to become involved in the details of the solution methods, but only to give the reader a basic understanding of upon what the particular method is based.

Both of these numerical methods are normally implemented at a scale where the numerical method used for machine solution of the large systems of equations is an important item. Both methods use the generation of a mesh to cover the domain of interest. The finite element method in particular relies heavily upon software for mesh generation. We confine ourselves to essentially trivial examples in order not to become involved with details of the numerical analysis/computer implementation.

Analytical solution

If we assume the x axis to be in the direction of heat flow, and further assume that no heat is being conducted in the y- or z-directions (or equivalently, that no temperature gradients exist in those directions), the Fourier equation reduces to[11]

$$q_x = -k \frac{dT}{dx} \tag{7.2.4-1}$$

If we consider the case of a single homogeneous solid as shown in Figure 7.2.4-1, where the area normal to heat flow is constant and the faces are maintained at T_1 and T_2, we may integrate the equation as follows:

$$q_x \int_{x_1}^{x_2} dx = -\int_{T_1}^{T_2} k \, dT \tag{7.2.4-2}$$

$$q_x (x_2 - x_1) = -\int_{T_1}^{T_2} k \, dT \tag{7.2.4-3}$$

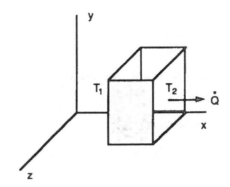

Figure 7.2.4-1 Homogeneous solid

A total energy balance shows us that Q is constant; since A is constant, q_x is constant and may be removed from the integrand above. If we know k as a function of T, we can integrate the right-hand side. For various functional forms this might be done analytically, numerically, or graphically.

---

[11] An energy balance shows that Q_x is constant. In the present geometry, the area is constant as we move in the x-direction, so q_x is constant. This is not true in general - for example, in cylindrical and spherical coordinates, the area usually changes as we go in the r-direction.

The simplest case is that of constant **k**, where Equation (7.2.4-3) integrates as

$$q_x (x_2 - x_1) = -k \int_{T_1}^{T_2} dT$$

$$q_x (x_2 - x_1) = -k (T_2 - T_1)$$

(7.2.4-4)

If the limits on the integration are taken rather as between x_1 (where we have T_1) and some arbitrary point x in the solid (where we have T) we obtain the temperature profile

$$q_x (x - x_1) = -k (T - T_1)$$

(7.2.4-5)

This is the equation of a straight line. Therefore, for conduction problems

- at steady state
- where the cross-sectional area normal to the flow of heat is constant
- k is constant

the temperature profile is linear, as shown in Figure 7.2.4-2.

By using the fact that $\dot{Q}_x / A = q_x$, we can rearrange this equation to the form[12]

$$\text{resistance} = \frac{\text{potential difference}}{\text{current}}$$

(7.2.4-6)

as

$$\frac{(x_2 - x)}{k} = \frac{-(T_2 - T)}{q_x} = \frac{-\Delta T}{q_x}$$

---

[12] The parallel is obvious to Ohm's law in the form

$$R = \frac{E}{I}$$

where E = voltage difference, I = current, and R = resistance. However, by convention E represents a voltage difference $E_1 - E_2 = \Delta E = E$. Our convention is that the Δ symbol represents the value at point 2 minus the value at point 1, so one must be careful of signs.

$$\boxed{\frac{\Delta x}{k\,A} = \frac{-\Delta T}{\dot{Q}_x}}$$ (7.2.4-7)

$$\text{resistance} = \frac{\Delta x}{k\,A}$$

$$\frac{\text{potential difference}}{\text{current}} = \frac{-\Delta T}{\dot{Q}_x}$$

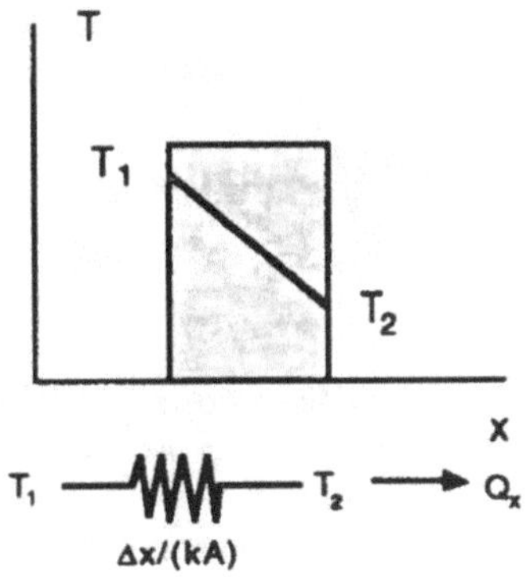

Figure 7.2.4-2 Temperature profile in 1-D heat transfer by conduction, rectangular coordinates

Equation (7.2.4-7) is satisfactory for a single solid; however, we frequently have a series of different solids, as in the case where we have a composite wall (for example, a layer of insulation plus the wall itself). Before we discuss this case we must briefly treat a very important point: the interface condition between solids.

Interface condition between solids - series conduction

In this text we will make the assumption that there is **no thermal resistance** between solids in direct contact for series conduction unless we specifically state otherwise. Another way to state this assumption is to say that the contact faces of two solids A and B have at the same surface temperature - that in Figure 7.2.4-3 both the right-hand face of solid A and the left-hand face of solid B are at T_3. For clean, smooth solids in intimate contact this is a good assumption.

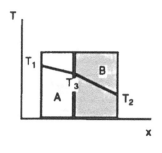

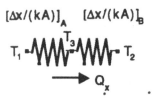

Figure 7.2.4-3 Conduction with two solids in contact assuming no temperature drop at interface

However, dirt at the interface or a poor contact which leaves void space along the contact line (for example, because of an irregular surface) can make it a poor assumption, so the assumption does not hold, as in Figure 7.2.4-4. The right-hand temperature of solid A is no longer the same as the left-hand temperature of solid B.

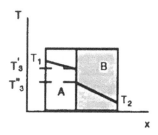

Figure 7.2.4-4 Temperature profile with interfacial resistance

When interfacial resistance is present, one way it can be modeled is by treating it as if it were another solid in the series of solids in the conduction path, as in Figure 7.2.4-5, by defining an **effective resistance** (see fouling coefficients in heat exchangers later in Section 7.6 for a similar treatment.) The interfacial resistance becomes just another in the series of resistances in the conduction path.

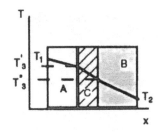

Figure 7.2.4-5 Contact resistance treated as an intermediate solid

We could calculate the heat flow through the wall in Figure 7.2.4-3 by applying Equation (7.2.4-2) to either solid A or solid B - both approaches would give the same answer because

$$\dot{Q}_x = q_x A = -\frac{k_A A (T_3 - T_1)}{(x_3 - x_1)}$$ (7.2.4-8)

and

$$\dot{Q}_x = q_x A = -\frac{k_B A (T_2 - T_3)}{(x_2 - x_3)}$$ (7.2.4-9)

However, to solve the problem by this method requires a knowledge of T_3, and this temperature is often hard to obtain experimentally. For example, suppose that solid B is merely a coat of paint instead of a fairly thick coat of insulation - how does one measure T_3?

The usual information we have at our disposal is the *overall* temperature drop, that is, $T_1 - T_2$. (In fact, we usually do not even have the surface temperatures T_1 and T_2, but rather the temperatures somewhere in the fluid phase adjoining the wall. This will be discussed further when we study convection; for the moment assume that we know T_1 and T_2.)

Equivalent thermal resistance - series conduction

Let's see if we can develop an equation for heat conduction through solids in series in terms of the *overall* temperature drop $(T_1 - T_2)$ and an **equivalent resistance** by incorporating the individual thicknesses, Δx_A and Δx_B, and the thermal conductivities, k_A and k_B.

In other words, we want an equation which gives

$$
\begin{aligned}
\dot{Q}_x &= q_x A \\
&= \frac{-(\Delta T)_{overall}}{[\text{equivalent resistance}]} \\
&= \frac{-(T_2 - T_1)}{[\text{equivalent resistance}]} \\
&= \frac{(T_1 - T_2)}{[\text{equivalent resistance}]}
\end{aligned}
\tag{7.2.4-10}
$$

for the situation in Figure 7.2.4-6.

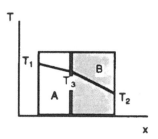

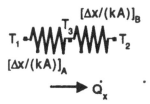

Figure 7.2.4-6 Development of equivalent conductance for series conduction

Thus far we know how to relate q_x only to the **individual** temperature drops. However, observe that

$$-(\Delta T)_{overall} = (T_1 - T_2) = [(T_1 - T_3) + (T_3 - T_2)] \qquad (7.2.4\text{-}11)$$

and that for each solid individually

$$-(\Delta T)_i = \left[\frac{q_x \, \Delta x}{k}\right]_i = \left[\frac{\dot{Q}_x \, \Delta x}{k \, A}\right]_i \qquad (7.2.4\text{-}12)$$

Substituting Equation (7.2.4-12) in Equation (7.2.4-11)

$$-(\Delta T)_{overall} = (T_1 - T_2) = \left(\left[\frac{\dot{Q}_x \, \Delta x}{k \, A}\right]_A + \left[\frac{\dot{Q}_x \, \Delta x}{k \, A}\right]_B\right) \qquad (7.2.4\text{-}13)$$

and the result in Equation (7.2.4-10), we have

$$\dot{Q}_x = q_x \, A$$
$$= \left[\frac{\dot{Q}_x \, (x_3 - x_1)}{k_A \, A} + \frac{\dot{Q}_x \, (x_2 - x_3)}{k_B \, A}\right] \frac{1}{\text{(equivalent resistance)}} \qquad (7.2.4\text{- }14)$$

This yields

$$\text{(equivalent resistance)} = \left[\frac{(x_3 - x_1)}{k_A \, A} + \frac{(x_2 - x_3)}{k_B \, A}\right] \qquad (7.2.4\text{-}15)$$

The first term involves only properties of solid A and the second only of solid B, so the equivalent resistance for two solids in series can be seen to be the sum of two resistances, one for solid A and one for solid B. Substituting in Equation (7.2.4-10)

$$\dot{Q}_x = q_x \, A = \frac{(T_1 - T_2)}{\left[\dfrac{(x_3 - x_1)}{k_A \, A} + \dfrac{(x_2 - x_3)}{k_B \, A}\right]} \qquad (7.2.4\text{-}16)$$

Note that the area is constant.

Observe that although this example was done for only two solids, the same procedure is valid for any number of layers of solids **in series**. We simply add the individual temperature differences to get the overall ΔT, substitute for the individual temperature differences using Equation (7.2.4-12), and obtain [the minus sign occurs because by definition $\Delta T = (T_2 - T_1)$ rather than $(T_1 - T_2)$.

$$\dot{Q}_x = q_x A = \frac{-1}{\sum\limits_{i=A,B,\ldots} \left(\dfrac{\Delta x}{k A}\right)_i} (\Delta T)_{overall} \qquad\qquad (7.2.4\text{-}17)$$

Equivalent thermal resistance - parallel conduction

Now let us consider conduction in **parallel** layers. The physical situation being considered is shown in Figure 7.2.4-7.

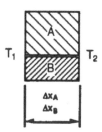

Figure 7.2.4-7 Development of equivalent conductance for parallel conduction

Again we are pursuing an equivalent resistance to use in the formula

$$
\begin{aligned}
\dot{Q}_x &= q_x A \\[6pt]
&= \frac{-(\Delta T)_{overall}}{[\text{equivalent resistance}]} \\[6pt]
&= \frac{-(T_2 - T_1)}{[\text{equivalent resistance}]} \\[6pt]
&= \frac{(T_1 - T_2)}{[\text{equivalent resistance}]} \qquad\qquad (7.2.4\text{-}18)
\end{aligned}
$$

However, in the parallel case the temperature drops are identical for the two layers, whereas in the series case the fluxes were the same.

If we assume that thermal energy flows in only the x-direction (a good assumption in this case, since the linear x-direction temperature profiles will coincide in the two materials for constant thermal conductivities, thus giving rise to no temperature gradients in the y-direction), the electrical analog would be as shown in Figure 7.2.4-8, where $R_i = [\Delta x/(kA)]_i$ for the appropriate solid.

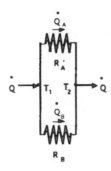

Figure 7.2.4-8 Equivalent circuit for parallel conduction

An energy balance around the point at either T_1 or T_2 gives us

$$\dot{Q} = \dot{Q}_A + \dot{Q}_B \qquad (7.2.4\text{-}19)$$

Substituting for $\dot{Q}_A$ and $\dot{Q}_B$ in terms of the driving force and corresponding resistances

$$\dot{Q} = \frac{\left(T_1 - T_2\right)}{\left[\dfrac{\Delta x}{k\,A}\right]_A} + \frac{\left(T_1 - T_2\right)}{\left[\dfrac{\Delta x}{k\,A}\right]_B} \qquad (7.2.4\text{-}20)$$

$$\dot{Q} = \left(T_1 - T_2\right)\left(\frac{1}{\left[\dfrac{\Delta x}{k\,A}\right]_A} + \frac{1}{\left[\dfrac{\Delta x}{k\,A}\right]_B}\right) \qquad (7.2.4\text{-}21)$$

$$\left(T_1 - T_2\right) = \dot{Q}\,\frac{1}{\left(\dfrac{1}{\left[\dfrac{\Delta x}{k\,A}\right]_B} + \dfrac{1}{\left[\dfrac{\Delta x}{k\,A}\right]_B}\right)} \qquad (7.2.4\text{-}22)$$

$$\left(T_1 - T_2\right) = \dot{Q}\frac{1}{\left(\left[\frac{kA}{\Delta x}\right]_A + \left[\frac{kA}{\Delta x}\right]_B\right)}$$

$$\left(T_1 - T_2\right) = \dot{Q}\left(\left[\frac{kA}{\Delta x}\right]_A + \left[\frac{kA}{\Delta x}\right]_B\right)^{-1} \qquad (7.2.4\text{-}23)$$

Extending this pattern to more branches in parallel would give

$$\boxed{\dot{Q}_x = q_x A = \frac{-1}{\left[\sum_{i=A,B,\dots}\left(\frac{kA}{\Delta x}\right)_i\right]^{-1}}(\Delta T)_{overall}} \qquad (7.2.4\text{-}24)$$

Example 7.2.4-4 shows how to evaluate the equivalent resistance for situations involving both series and parallel paths for the heat conducted.

Example 7.2.4-1   Series conduction through layers - constant temperature at external surfaces

A thermocouple attached to the combustion side of the inside wall of a furnace indicates 1500°F. The inside wall is firebrick, 3 inches thick, and is in intimate contact with a 0.25-inch steel wall. The steel wall is to be coated with a layer of magnesia insulation. For economic reasons, the heat loss through the composite wall must not exceed 50 Btu/(hr ft^2). The outside temperature of the insulation at steady state is not to exceed 150°F. How thick a layer of magnesia is required?

Solution

By applying in succession a total energy balance to each of the sections as shown, we see that the only terms involved are the input of heat to the left face and the output of heat at the right face of each. (There is no convection, we are at steady state, and there is no work.) It follows that the heat in and out of each of the faces must be equal, and, since the areas normal to the heat transfer are equal, the fluxes are also equal. Using subscripts f, s, and m to refer to firebrick, steel, and magnesia, respectively,

$$\dot{Q}_f = \dot{Q}_s = \dot{Q}_m = \dot{Q} = \text{constant}$$

$$\frac{\dot{Q}_f}{A} = \frac{\dot{Q}_s}{A} = \frac{\dot{Q}_m}{A} = q = \text{constant} \qquad (7.2.4\text{-}25)$$

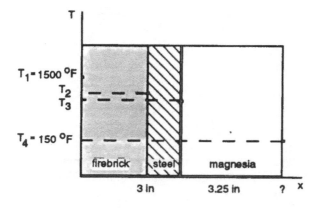

Figure 7.2.4-9 Series conduction through layers with constant temperature at external surfaces

But we can express the heat flux in each of the sections in terms of the driving force and the resistance of that section

$$q_x = -\left[\frac{k}{\Delta x}\right]_f (T_2 - T_1)$$

$$q_x = -\left[\frac{k}{\Delta x}\right]_s (T_3 - T_2)$$

$$q_x = -\left[\frac{k}{\Delta x}\right]_m (T_4 - T_3) \qquad (7.2.4\text{-}26)$$

Substituting known values

$$50\frac{Btu}{hr\,ft^2} = -\left[\frac{(0.6)\frac{Btu}{hr\,ft\,°F}}{(0.25)\,ft}\right]_f (T_2 - 1500)\ °F$$

$$50\frac{Btu}{hr\,ft^2} = -\left[\frac{(26)\frac{Btu}{hr\,ft\,°F}}{\frac{(1)}{(4)(12)}\,ft}\right]_s (T_3 - T_2)\,°F$$

$$50\frac{Btu}{hr\,ft^2} = -\left[\frac{0.03\frac{Btu}{hr\,ft\,°F}}{(\Delta x)\,ft}\right]_m (150 - T_3)\,°F \tag{7.2.4-27}$$

This is a set of three independent equations in three unknowns (Δx, T_2, and T_3), but the set does not require simultaneous solution. The first equation can immediately be solved for T_2

$$\left(T_2 - 1500\right) = -50\frac{0.25}{0.6} = -20.8$$
$$T_2 = -20.8 + 1500 = 1479.2 \tag{7.2.4-28}$$

Substituting this result in the second equation

$$\left(T_3 - 1479\right) = -\left(50\right)\frac{1}{(4)(12)(26)} = -0.004$$
$$T_3 = -0.004 + 1479.2 = 1479.2 \tag{7.2.4-29}$$

and substituting in the third equation (be careful to write 150, not 1500, for T_4)

$$\Delta x_m = -\frac{\left(0.03\right)\left(150 - 1479.2\right)}{\left(50\right)} = 0.80\,ft \tag{7.2.4-30}$$

Notice the order of magnitude of the temperature drops - the steel contributes virtually nothing, the firebrick only a very modest amount; the magnesia takes almost the total temperature drop. This is a function primarily of the large range in thermal conductivities, and also partly because of differing thicknesses, although the thickness effect is much smaller. We can see the reason by comparing thermal resistances (remembering the areas are all the same)

$$R_f = \left[\frac{\Delta x}{k\,A}\right]_f = \frac{\left(\frac{3}{12}\right)}{(0.6)\,(A)} = \frac{4.17}{A}$$

$$R_s = \left[\frac{\Delta x}{k\,A}\right]_s = \frac{\left[\frac{(1)}{(4)\,(12)}\right]}{(26)\,(A)} = \frac{0.0008}{A} \qquad (7.2.4\text{-}31)$$

$$R_m = \left[\frac{\Delta x}{k\,A}\right]_m = \frac{(0.80)}{(0.03)\,(A)} = \frac{26.7}{A}$$

It is frequently not worth including some resistance terms in models because of their relatively minor contribution to the overall temperature drop.

___

Example 7.2.4-2   Series conduction through layers - constant convective heat transfer coefficient at external surfaces

A flat composite furnace wall consists of 9 in. of firebrick [k = 0.96 Btu/(hr ft °F)], 4.5 in. of insulating brick [k = 0.183 Btu/(hr ft °F)], and 4.5 in. of building brick [k = 0.40 Btu/(hr ft °F)]. The individual convective heat transfer coefficients at the inside and outside walls are 10 Btu/(hr ft^2 °F) and 10 Btu/(hr ft^2 °F), respectively.

The model for the heat transfer coefficient is

$$\dot{Q}_x = h\,A\,\Delta T \qquad (7.2.4\text{-}32)$$

where ΔT is defined in terms of the surface temperature and the ambient temperature in such a way as to give the proper sign to Q_x. Consequently, the resistances represented by the internal and external ambient phases are

$$\text{resistance} = \frac{\text{potential difference}}{\text{current}} \qquad (7.2.4\text{-}33)$$

$$\dot{Q}_x = h_i\,A\left(T_i - T_1\right)$$
$$\frac{1}{h_i\,A} = \frac{-\Delta T}{\dot{Q}_x} \qquad (7.2.4\text{-}34)$$
$$\text{resistance} = \frac{1}{h_i\,A}$$

and

$$\dot{Q}_x = h_o A \left(T_4 - T_0 \right)$$

$$\frac{1}{h_o A} = \frac{-\Delta T}{\dot{Q}_x}$$ (7.2.4-35)

$$\text{resistance} = \frac{1}{h_o A}$$

The temperature of the gas within the furnace is 2400°F and that of the ambient air outside the furnace is 100°F.

    a) What is the heat transfer rate per square foot of wall area?
    b) What is the temperature at the surface of each layer of brick?

Solution

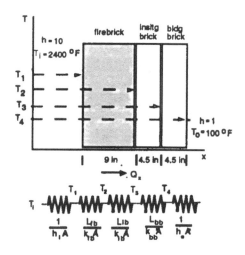

Figure 7.2.4-10 Series conduction through layers with constant convective heat transfer coefficient at external surfaces

(a)

$$A = 1 \text{ ft}^2$$

$$\frac{1}{h_i A} = \frac{(1)}{(10)(1)} = 0.1 \frac{\text{hr ft}^2 \, °F}{\text{Btu}}$$ (7.2.4-36)

$$\frac{L_{fb}}{k_{fb} A} = \frac{\left(\frac{9}{12}\right)}{(0.96)(1)} = 0.78 \frac{\text{hr ft}^2 \text{ °F}}{\text{Btu}} \tag{7.2.4-37}$$

$$\frac{L_{ib}}{k_{ib} A} = \frac{\left(\frac{4.5}{12}\right)}{(0.183)(1)} = 2.06 \frac{\text{hr ft}^2 \text{ °F}}{\text{Btu}} \tag{7.2.4-38}$$

$$\frac{L_{bb}}{k_{bb} A} = \frac{\left(\frac{4.5}{12}\right)}{(0.40)(1)} = 2.94 \frac{\text{hr ft}^2 \text{ °F}}{\text{Btu}} \tag{7.2.4-39}$$

$$\frac{1}{h_o A} = \frac{(1)}{(1)(1)} = 1 \frac{\text{hr ft}^2 \text{ °F}}{\text{Btu}} \tag{7.2.4-40}$$

$$\sum_{n=1}^{5} R_n = 4.88 \frac{\text{hr ft}^2 \text{ °F}}{\text{Btu}} \tag{7.2.4-41}$$

$$\dot{Q} = \frac{\Delta T}{\sum_{n=1}^{5} R_n} = \frac{(2400 - 100)^\circ \text{F}}{4.88 \frac{\text{hr ft}^2 \text{ °F}}{\text{Btu}}} = 471 \frac{\text{Btu}}{\text{hr}} \tag{7.2.4-42}$$

(b) Since $\dot{Q}$ is constant, we can write for any individual resistance step

$$\dot{Q} = \frac{(\Delta T)_i}{R_i} \tag{7.2.4-43}$$

So for the individual steps

$$\dot{Q} = \frac{(T_o - T_1)}{\frac{1}{h_i A}} \quad \Rightarrow \quad 471 \frac{\text{Btu}}{\text{hr}} = \frac{(2400 - T_1)^\circ \text{F}_i}{0.1 \frac{\text{hr ft}^2 \text{ °F}}{\text{Btu}}}$$
$$\Rightarrow \quad T_1 = 2353 \text{°F} \tag{7.2.4-44}$$

$$\dot{Q} = \frac{(T_1 - T_2)}{\frac{L_{fb}}{k_{fb} A}} \quad \Rightarrow \quad 471 \frac{\text{Btu}}{\text{hr}} = \frac{(2353 - T_2)^\circ \text{F}_i}{0.78 \frac{\text{hr ft}^2 \text{ °F}}{\text{Btu}}}$$
$$\Rightarrow \quad T_2 = 1985 \text{°F} \tag{7.2.4-45}$$

$$\dot{Q} = \frac{\left(T_2 - T_3\right)}{\frac{L_{ib}}{k_{ib}\,A}} \quad \Rightarrow \quad 471\,\frac{Btu}{hr} = \frac{\left(1985 - T_3\right)^{\circ}F_i}{2.06\,\frac{hr\,ft^2\,^{\circ}F}{Btu}}$$

$$\Rightarrow \quad T_3 = 1013^{\circ}F \qquad\qquad (7.2.4\text{-}46)$$

$$\dot{Q} = \frac{\left(T_3 - T_4\right)}{\frac{L_{bb}}{k_{bb}\,A}} \quad \Rightarrow \quad 471\,\frac{Btu}{hr} = \frac{\left(1013 - T_4\right)^{\circ}F_i}{2.94\,\frac{hr\,ft^2\,^{\circ}F}{Btu}}$$

$$\Rightarrow \quad T_4 = 570^{\circ}F \qquad\qquad (7.2.4\text{-}47)$$

Example 7.2.4-3   Conduction with variable thermal conductivity

A firebrick whose thermal conductivity variation with temperature can be described as

$$k = \left(0.1\right) + \left(5.0 \times 10^{-5}\right)T \qquad\qquad (7.2.4\text{-}48)$$

where

$k = Btu/(hr\,ft\,^{\circ}F)$
$T = ^{\circ}F$

is to be used in a 4-inch thick furnace liner. The outside temperature of the liner is to be about 100°F and the inside temperature 1500°F. Calculate the heat flux and the temperature profile.

Solution

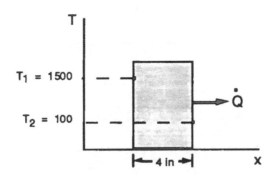

Figure 7.2.4-11 Conduction through firebrick with variable thermal conductivity

$$q_x = -k \frac{dT}{dx} \tag{7.2.4-49}$$

Here q is constant because of the constancy of A, so

$$q_x \int_{x_1}^{x_2} dx = -\int_{T_1}^{T_2} k \, dT$$

$$q_x \Delta x = -\int_{T_1}^{T_2} \left(0.1 + 5.0 \times 10^{-5} \, T\right) dT$$

$$q_x \left(\frac{4}{12}\right) ft = -\left[0.1 \, T + 5.0 \times 10^{-5} \frac{T^2}{2}\right]_{1500}^{100} \frac{Btu}{hr \, ft \, °F} \tag{7.2.4-50}$$

$$q_x = -\left(\frac{12}{4}\right) \left[(0.1)(100 - 1500) + \left(5.0 \times 10^{-5}\right)\left(\frac{100^2}{2} - \frac{1500^2}{2}\right)\right]$$

$$q_x = 588 \frac{Btu}{hr \, ft^2 \, °F} \tag{7.2.4-51}$$

To obtain the profile one simply chooses as a system the solid between one face (located at x = 0) and an unspecified point x where the temperature is T. Here, arbitrarily choosing the left face and proceeding as before

$$q_x \int_0^x dx = -\int_{T_1}^T k \, dT$$

$$q_x x = -\int_{T_1}^{T} \left(0.1 + 5.0 \times 10^{-5}\, T\right) dT$$

$$(588)\, x = -\left[0.1\, T + 5.0 \times 10^{-5}\, \frac{T^2}{2}\right]_{1500}^{T} \qquad (7.2.4\text{-}52)$$

$$x = -\left(\frac{1}{588}\right)\left[(0.1)\left(T - 1500\right) + \left(5.0 \times 10^{-5}\right)\left(\frac{T^2}{2} - \frac{1500^2}{2}\right)\right]$$

$$x = \left(0.351\right) - \left(1.7 \times 10^{-4}\right) T - \left(4.25 \times 10^{-8}\right) T^2 \qquad (7.2.4\text{-}53)$$

where x is in feet and T in degrees Fahrenheit.

Note that this is a nonlinear relationship which is in the form x = f(T). We could solve for T to obtain T = g(x); instead, we will simply plot the result. Preparing a table of values from the above equation we have

T °F	x ft
1500	0.000
1400	0.030
1300	0.058
1200	0.086
1100	0.113
1000	0.139
900	0.164
800	0.188
700	0.211
600	0.234
500	0.255
400	0.276
300	0.296
200	0.315
100	0.334

which yields the plot shown in Figure 7.2.4-12.

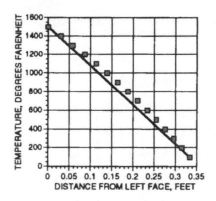

**Figure 7.2.4-12 Temperature profile
(straight line shown for purpose of comparison)**

Example 7.2.4-4   One-dimensional steady-state conduction with parallel path

For the composite wall shown in Figure 7.2.4-13, draw the electrical analog and write the conduction equation in terms of an equivalent resistance and the overall temperature drop. The appropriate areas in Equations (7.2.4-55) to (7.2.4-64) are normal to the x-direction.

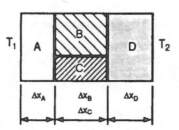

Figure 7.2.4-13 Conduction with parallel paths

Solution

If we assume that thermal energy flows only in the x-direction (as if there were an infinitesimally thin, perfectly insulating wall between B and C, and the temperature gradients normal to the x-direction in A and D were insignificant) the electrical analog would be as shown in Figure 7.2.4-14, where $R_i = [\Delta x/(kA)]_i$ for the appropriate solid.

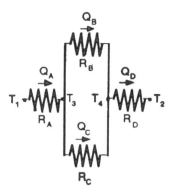

Figure 7.2.4-14 Analogous circuit

Energy balances around the points at T_3 and T_4 give us

$$\dot{Q}_A = \dot{Q}_B + \dot{Q}_C$$
$$\dot{Q}_B + \dot{Q}_C = \dot{Q}_D \qquad\qquad (7.2.4\text{-}54)$$

which gives $\dot{Q}_A = \dot{Q}_D$, which we designate simply as $\dot{Q}$.

Substituting for $\dot{Q}_A$ and $\dot{Q}_B$ in the first equation in terms of the conductance and driving force

$$\dot{Q} = \frac{\left(T_3 - T_4\right)}{\left[\dfrac{\Delta x}{k\,A}\right]_B} + \frac{\left(T_3 - T_4\right)}{\left[\dfrac{\Delta x}{k\,A}\right]_C}$$

$$\dot{Q} = \left(T_3 - T_4\right)\left(\frac{1}{\left[\dfrac{\Delta x}{k\,A}\right]_B} + \frac{1}{\left[\dfrac{\Delta x}{k\,A}\right]_C}\right) \qquad\qquad (7.2.4\text{-}55)$$

$$\left(T_3 - T_4\right) = \dot{Q} \dfrac{1}{\left(\dfrac{1}{\left[\dfrac{\Delta x}{kA}\right]_B} + \dfrac{1}{\left[\dfrac{\Delta x}{kA}\right]_C}\right)} \qquad (7.2.4\text{-}56)$$

(Use of the second equation in a similar fashion yields the same result.)

As for the case of purely series resistances, we wish to write

$$\dot{Q} = \dfrac{T_1 - T_2}{R_{equivalent}} \qquad (7.2.4\text{-}57)$$

or

$$T_1 - T_2 = \dot{Q}\,R_{equivalent} \qquad (7.2.4\text{-}58)$$

Writing the overall driving force in terms of the individual driving forces

$$\left(T_1 - T_2\right) = \left(T_1 - T_3\right) + \left(T_3 - T_4\right) + \left(T_4 - T_2\right) \qquad (7.2.4\text{-}59)$$

and substituting

$$\dot{Q}\,R_{equivalent} = \left(T_1 - T_3\right) + \left(T_3 - T_4\right) + \left(T_4 - T_2\right) \qquad (7.2.4\text{-}60)$$

$$\dot{Q}\,R_{equivalent} = \dot{Q}\left[\dfrac{\Delta x}{kA}\right]_A + \dot{Q}\dfrac{1}{\left(\dfrac{1}{\left[\dfrac{\Delta x}{kA}\right]_B} + \dfrac{1}{\left[\dfrac{\Delta x}{kA}\right]_C}\right)} + \dot{Q}\left[\dfrac{\Delta x}{kA}\right]_D$$

$$(7.2.4\text{-}61)$$

$$R_{equivalent} = \left[\dfrac{\Delta x}{kA}\right]_A + \dfrac{1}{\left(\dfrac{1}{\left[\dfrac{\Delta x}{kA}\right]_B} + \dfrac{1}{\left[\dfrac{\Delta x}{kA}\right]_C}\right)} + \left[\dfrac{\Delta x}{kA}\right]_D \qquad (7.2.4\text{-}62)$$

which gives

$$\dot{Q} = \frac{T_1 - T_2}{\left[\frac{\Delta x}{kA}\right]_A + \cfrac{1}{\left(\cfrac{1}{\left[\frac{\Delta x}{kA}\right]_B} + \cfrac{1}{\left[\frac{\Delta x}{kA}\right]_C}\right)} + \left[\frac{\Delta x}{kA}\right]_D} \qquad (7.2.4\text{-}63)$$

This may be written more compactly as

$$\dot{Q} = \frac{T_1 - T_2}{\left[\frac{\Delta x}{kA}\right]_A + \left(\left[\frac{kA}{\Delta x}\right]_B + \left[\frac{kA}{\Delta x}\right]_C\right)^{-1} + \left[\frac{\Delta x}{kA}\right]_D} \qquad (7.2.4\text{-}64)$$

where, by comparison with the results above, the middle term can be seen to give the equivalent resistance of two solids with parallel, one-dimensional conduction paths connected in series with two regular series resistances.

Numerical solution

We introduce at this point the finite element method, which, with the finite difference method, comprise the two main numerical techniques used for the solution of heat transfer conduction problems with awkward boundaries, irregular source terms, etc. It is emphasized that the isolation of this method to one-dimensional problems is purely for the purpose of pedagogy, and must not be construed to mean that the method is either valid or useful only for such simple problems.

The finite element method

One approach to the solution of partial differential equations

$$F\left(u \text{ and its derivatives}\right) = 0 \qquad (7.2.4\text{-}65)$$

is to attempt to find a **single** approximating function, u, that satisfies the equation **throughout** the domain of interest in some average sense, rather than satisfying the equation exactly, as is the objective with analytical methods (an interpolation problem).

For certain cases, these methods give satisfactory results - for example, use of

• the Raleigh-Ritz method, based on minimizing a member of the class of functions called **functionals** using the calculus of variations

• collocation methods, which make the residual between the true solution and a sum of trial functions (usually polynomials) equal to zero at a series of selected points

• Galerkin methods, which weights the residual with the trial functions of the collocation method

Obtaining a good fit to the partial differential equation over the whole domain is not easy; however, if the domain is subdivided into smaller regions, a number of fairly low-degree polynomials (often linear functions) can be used to make up the approximating function with good success. One then has, of course, the additional problem of joining the solutions within each element to the solution for adjoining elements at their common boundaries.

The process of subdivision leaves us with **finite elements**, hence the name of the method now to be considered.

We can characterize how well we have fit the differential equation with an approximation, $\tilde{u}$, by examining the **residual**, R, which will remain after using the approximating function, $\tilde{u}$, in the differential equation rather than the true solution, u

$$F\left(u\right) = 0 \tag{7.2.4-66}$$

$$F\left(\tilde{u}\right) \neq 0 \tag{7.2.4-67}$$

$$R \equiv F\left(\tilde{u}\right) \tag{7.2.4-68}$$

The overall process is to

• Subdivide the domain of solution into subdomains (the finite elements). For the one- and two-dimensional problems most of interest here, the elements are line segments for the one-dimensional problem, and usually triangles or quadrilaterals for the two-dimensional problems, examples of which are shown in Figure 7.2.4-14.

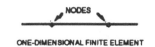

ONE-DIMENSIONAL FINITE ELEMENT

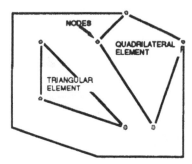

TWO-DIMENSIONAL FINITE ELEMENTS

Figure 7.2.4-14 Examples of 1-D and 2-D finite elements

• Select an approximation method/functions to be used within the elements.

• Match the approximation functions for the elements at their common boundaries by adjusting the coefficients in the functions

• Apply the boundary conditions (in the finite element method, this becomes an intrinsic part of the solution rather than being accomplished at the end of the process)

• Solve the resulting system of equations

We now consider a one-dimensional problem. We will use as approximation functions first-order polynomials

$$u\left(x\right) = a_0 + a_1\, x \tag{7.2.4-69}$$

where u is the approximation to the dependent variable, x is the independent variable, and a_0 and a_1 are constants.

If the distance variable z goes over the interval [0, L], we subdivide the interval into elements, and denote the left and right nodes of a typical element by

z_1 and z_2, respectively. (This is a **local** numbering, i.e., applicable to only nodes in the element in question. We shall also need a **global** numbering system, which will permit distinguishing among all the nodes taken as a group.)

For our illustration, we shall choose a first-order (linear) polynomial for the approximation function, which we will call u(z). At the ends of the element, the approximation function will assume the values u_1 and u_2, respectively. The approximation function must pass through the values of u(x) at the ends of the element

$$u\left(x_1\right) = u_1 = a_0 + a_1 x_1$$
$$u\left(x_2\right) = u_2 = a_0 + a_1 x_2 \qquad (7.2.4\text{-}70)$$

This system of equations can readily be solved for the values of a_0 and a_1

$$a_0 = \frac{u_1 x_2 - u_2 x_1}{x_2 - x_1}$$
$$a_0 = \frac{u_2 - u_1}{x_2 - x_1} \qquad (7.2.4\text{-}71)$$

Substituting in Equation (7.2.4-69)

$$\bar{u} = a_0 + a_1 x$$
$$\bar{u} = \left[\frac{u_1 x_2 - u_2 x_1}{x_2 - x_1}\right] + \left[\frac{u_2 - u_1}{x_2 - x_1}\right] x \qquad (7.2.4\text{-}72)$$

and rearranging

$$\bar{u} = \left[\frac{x_2 - x}{x_2 - x_1}\right] u_1 + \left[\frac{x - x_1}{x_2 - x_1}\right] u_2 \qquad (7.2.4\text{-}73)$$

Defining

$$N_1 \equiv \left[\frac{x_2 - x}{x_2 - x_1}\right]$$
$$N_2 \equiv \left[\frac{x - x_1}{x_2 - x_1}\right] \qquad (7.2.4\text{-}74)$$

gives

$$\mathbf{\hat{u}} = \mathbf{N}_1\,\mathbf{u}_1 + \mathbf{N}_2\,\mathbf{u}_2$$

$$(7.2.4\text{-}75)$$

For historical reasons, this is usually referred to as the **shape function**[13] and N_1 and N_2 are called the **interpolation functions**, and their sum is always unity, as illustrated in Figure 7.2.4-15.

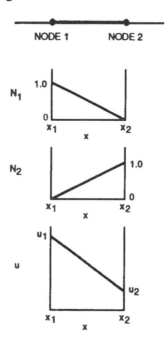

Figure 7.2.4-15 Interpolation functions (above), Shape function (below)

We now must choose a way to represent the fit of the function to the differential equation. Common methods are

---

[13] It is the Lagrange first-order interpolating polynomial. When we choose first degree functions, the shape functions are often referred to as *chapeau* functions, from the French word for hat, because of the resemblance to the cartoon witch's hat or dunce's cap when regarded for adjoining elements, viz.

• the direct approach

• the method of weighted residuals, which minimizes the weighted integral of the residual

$$\int_D R\, W_i\, dD = 0 \quad i = 1, 2, \dots, n \tag{7.2.4-76}$$

where D is the domain of the solution and W_i are linearly independent weighting functions[14]

• the variational approach

These methods specify relationships among the unknowns in Equation (7.2.4-75) so as to satisfy the differential equation in some optimal sense.

Example 7.2.4-5   Solution by finite elements of steady-state conduction with generation

Consider the case of steady-state conduction in one dimension with Dirichlet boundary conditions and a thermal energy source. The governing differential equation in this case is

$$0 = k\frac{d^2T}{dx^2} + \gamma_\theta \tag{7.2.4-77}$$

which is an ordinary, not a partial differential equation and so will offer simplicity for pedagogical purposes. We rewrite, defining

---

[14] Several alternatives exist for choosing the weighting functions. For example

• Collocation methods make the residual vanish at certain selected points by adjusting coefficients (basically, weighting with a Dirac delta function, which vanishes everywhere except at the collocation points

• Subdomain methods, which make the average value of the residual in chosen subdomains vanish by adjusting coefficients
• Least squares methods, which minimize the integral of the square of the residual

• Galerkin's method, which uses the interpolation functions as the weighting functions

$$\Gamma \equiv \frac{\gamma_\theta}{k} \tag{7.2.4-78}$$

so our problem consists of solving the differential equation and boundary conditions

$$\frac{d^2T}{dx^2} = -\Gamma \tag{7.2.4-79}$$

at x = 0   T = T₁

at x = L   T = T₂ $\tag{7.2.4-80}$

We shall assume L = 10, T₁ = 200, T₂ = 120, and Γ = 5.

Let us choose to use four equal-length elements as shown in Figure 7.2.4-16.

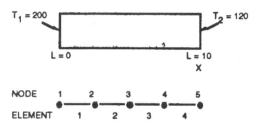

Figure 7.2.4-16 Bar with thermal energy source

We will use the approximation function

$$\tilde{T} = N_1 T_1 + N_2 T_2 \tag{7.2.4-81}$$

where N_1 and N_2 are the functions shown above in (7.2.4-27). We will illustrate a solution using the example of application of Galerkin's method.

The residual is

$$R = \frac{d^2\tilde{T}}{dx^2} + \Gamma \tag{7.2.4-82}$$

We wish to minimize this residual subject to

$$\int_D R W_i \, dD = 0 \quad i = 1, 2, \dots, n \qquad (7.2.4\text{-}83)$$

Application of Galerkin's method implies that the weighting functions be the interpolation functions N_1 and N_2.

$$\int_{x_1}^{x_2} \left[\frac{d^2T}{dx^2} + \Gamma \right] N_i \, dx = 0 \quad i = 1, 2 \qquad (7.2.4\text{-}84)$$

$$\int_{x_1}^{x_2} \frac{d^2T}{dx^2} N_i \, dx = - \int_{x_1}^{x_2} \Gamma N_i \, dx \quad i = 1, 2 \qquad (7.2.4\text{-}85)$$

It is typical of finite element methods to integrate the left-hand side by parts[15] at this juncture. This reduces the order of the highest order term in the integral and, it is to be hoped, replaces it with another integral that is more tractable than the original. Here we choose

$$N_i = u$$
$$\frac{d^2T}{dx^2} dx = dv \qquad (7.2.4\text{-}86)$$

giving

$$\int_{x_1}^{x_2} N_i \left[\frac{d^2T}{dx^2} \, dx \right] = \left[N_i \frac{dT}{dx} \right]_{x_1}^{x_2} - \int_{x_1}^{x_2} \frac{dT}{dx} \left[\frac{dN_i}{dx} \, dx \right] \quad i = 1, 2 \qquad (7.2.4\text{-}87)$$

Next we evaluate the first term on the right-hand side, beginning with $i = 1$

$$\left[N_1 \frac{dT}{dx} \right]_{x_1}^{x_2} = N_1(x_2) \left[\frac{dT}{dx} \right]_{x_2} - N_1(x_1) \left[\frac{dT}{dx} \right]_{x_1}$$

---

[15] Application of the rule for differentiation of a product followed by integration gives

$$\int_a^b u \, dv = \left[u \, v \right]_a^b - \int_a^b v \, du$$

$$= (0)\left[\frac{dT}{dx}\right]_{x_2} - (1)\left[\frac{dT}{dx}\right]_{x_1}$$

$$= -\left[\frac{dT}{dx}\right]_{x_1}$$

(7.2.4-88)

so, for i = 1

$$\int_{x_1}^{x_2} N_1\left[\frac{d^2T}{dx^2}\,dx\right] = -\left[\frac{dT}{dx}\right]_{x_1} - \int_{x_1}^{x_2}\frac{dT}{dx}\frac{dN_1}{dx}\,dx$$

(7.2.4-89)

and in the same manner, for i = 2[16]

$$\left[N_2\frac{dT}{dx}\right]_{x_1}^{x_2} = N_2(x_2)\left[\frac{dT}{dx}\right]_{x_2} - N_2(x_1)\left[\frac{dT}{dx}\right]_{x_1}$$

$$= (1)\left[\frac{dT}{dx}\right]_{x_2} - (0)\left[\frac{dT}{dx}\right]_{x_1}$$

$$= \left[\frac{dT}{dx}\right]_{x_2}$$

(7.2.4-90)

$$\int_{x_1}^{x_2} N_2\frac{d^2T}{dx^2}\,dx = \left[\frac{dT}{dx}\right]_{x_2} - \int_{x_1}^{x_2}\frac{dT}{dx}\frac{dN_2}{dx}\,dx$$

(7.2.4-91)

Substituting these results into (i)

$$-\left[\frac{dT}{dx}\right]_{x_1} - \int_{x_1}^{x_2}\frac{dT}{dx}\frac{dN_1}{dx}\,dx = -\int_{x_1}^{x_2}\Gamma N_1\,dx$$

$$\left[\frac{dT}{dx}\right]_{x_2} - \int_{x_1}^{x_2}\frac{dT}{dx}\frac{dN_2}{dx}\,dx = -\int_{x_1}^{x_2}\Gamma N_2\,dx$$

(7.2.4-92)

rearranging

---

[16] The first term on the right in (l) and the second term on the right in (n) represent the **natural** boundary conditions at the ends of the element, that is, the case of an insulated boundary where the flux is zero.

$$\int_{x_1}^{x_2} \frac{dT}{dx} \frac{dN_1}{dx}\, dx = -\left[\frac{dT}{dx}\right]_{x_1} + \int_{x_1}^{x_2} \Gamma N_1\, dx$$

$$\int_{x_1}^{x_2} \frac{dT}{dx} \frac{dN_2}{dx}\, dx = \left[\frac{dT}{dx}\right]_{x_2} + \int_{x_1}^{x_2} \Gamma N_2\, dx \qquad (7.2.4\text{-}93)$$

Thus, the integration by parts has not only lowered the highest order derivative from a second to a first derivative, but it has also incorporated the boundary conditions directly into the element equations. The first of these effects means that the approximation functions need to preserve only continuity at the nodes, not slope as well.

The left-hand side represents the element properties. The first term on the right represents one of boundary conditions imposed on the element, and the second (the thermal energy source) is a forcing function. Examining the left-hand sides

$$\int_{x_1}^{x_2} \frac{dT}{dx} \frac{dN_1}{dx}\, dx$$

$$\int_{x_1}^{x_2} \frac{dT}{dx} \frac{dN_2}{dx}\, dx \qquad (7.2.4\text{-}94)$$

the linearity of the shape functions makes evaluation simple. For $n = 1$ and $n = 2$, respectively

$$\frac{dN_1}{dx} = \frac{d}{dx}\left[\frac{x_2 - x}{x_2 - x_1}\right] = -\frac{1}{x_2 - x_1}$$

$$\frac{dN_2}{dx} = \frac{d}{dx}\left[\frac{x - x_1}{x_2 - x_1}\right] = \frac{1}{x_2 - x_1} \qquad (7.2.4\text{-}95)$$

Writing the derivative

$$\bar{T} = N_1 T_1 + N_2 T_2$$

$$\frac{d\tilde{T}}{dx} = \left[-\frac{1}{x_2 - x_1}\right] T_1 + \left[\frac{1}{x_2 - x_1}\right] T_2$$

$$\frac{d\tilde{T}}{dx} = \frac{T_2 - T_1}{x_2 - x_1} \qquad (7.2.4\text{-}96)$$

Substituting in the left-hand sides

$$\int_{x_1}^{x_2} \frac{dT}{dx} \frac{dN_1}{dx} \, dx = \int_{x_1}^{x_2} \left[\frac{T_2 - T_1}{x_2 - x_1} \right] \left[-\frac{1}{x_2 - x_1} \right] dx$$

$$= \int_{x_1}^{x_2} \frac{T_1 - T_2}{(x_2 - x_1)^2} \, dx = \frac{T_1 - T_2}{x_2 - x_1}$$

$$\int_{x_1}^{x_2} \frac{dT}{dx} \frac{dN_2}{dx} \, dx = \int_{x_1}^{x_2} \left[\frac{T_2 - T_1}{x_2 - x_1} \right] \left[\frac{1}{x_2 - x_1} \right] dx$$

$$= \int_{x_1}^{x_2} \frac{-T_1 + T_2}{(x_2 - x_1)^2} \, dx = \frac{-T_1 + T_2}{x_2 - x_1}$$

(7.2.4-97)

Substituting in (q) and writing the result in matrix form

$$\underbrace{\frac{1}{x_2 - x_1} \begin{bmatrix} 1 & -1 \\ -1 & 1 \end{bmatrix} \begin{bmatrix} T_1 \\ T_2 \end{bmatrix}}_{\text{Element stiffness matrix}} = \underbrace{\begin{bmatrix} -\left[\frac{dT}{dx}\right]_{x_1} \\ \left[\frac{dT}{dx}\right]_{x_2} \end{bmatrix}}_{\text{Boundary condition}} + \underbrace{\begin{bmatrix} \int_{x_1}^{x_2} \Gamma N_1 \, dx \\ \int_{x_1}^{x_2} \Gamma N_2 \, dx \end{bmatrix}}_{\text{External Effects}}$$

(7.2.4-98)

We now apply the results to our specific problem. Evaluating the source terms

$$\int_{x_1}^{x_2} \Gamma N_1 \, dx = \int_0^{2.5} (5) \left(\frac{2.5 - x}{2.5} \right) dx = 6.25$$

$$\int_{x_1}^{x_2} \Gamma N_2 \, dx = \int_0^{2.5} (5) \left(\frac{x - 0}{2.5} \right) dx = 6.25$$

(7.2.4-99)

Substituting

$$\frac{1}{2.5 - 0} \begin{bmatrix} 1 & -1 \\ -1 & 1 \end{bmatrix} \begin{bmatrix} T_1 \\ T_2 \end{bmatrix} = \begin{bmatrix} -\left[\frac{dT}{dx}\right]_{x_1} \\ \left[\frac{dT}{dx}\right]_{x_2} \end{bmatrix} + \begin{bmatrix} 6.25 \\ 6.25 \end{bmatrix}$$

(7.2.4-100)

gives the set of equations for an element

$$0.4\,T_1 - 0.4\,T_2 = -\left[\frac{dT}{dx}\right]_{x_1} + 6.25$$

$$-0.4\,T_1 + 0.4\,T_2 = \left[\frac{dT}{dx}\right]_{x_2} + 6.25 \qquad (7.2.4\text{-}101)$$

We next need to establish a global numbering scheme to permit assembling the element equations. Here the establishment of such a scheme is trivial; however, in two- or three-dimensional problems it is not so trivial. Table 7.2.4-1 shows the scheme we shall use.

Table 7.2.4-1 Global node numbering scheme

	Node number	
Element	Local	Global
1	1	1
1	2	2
2	1	2
2	2	3
3	1	3
3	2	4
4	1	4
4	2	5

Next we write the element equations using the global coordinates, and add them one by one to obtain the system matrix. During this process the internal boundary conditions cancel as indicated.

First element

$$\begin{bmatrix} 0.4 & -0.4 & 0 & 0 & 0 \\ -0.4 & 0.4 & 0 & 0 & 0 \\ 0 & 0 & 0 & 0 & 0 \\ 0 & 0 & 0 & 0 & 0 \\ 0 & 0 & 0 & 0 & 0 \end{bmatrix} \begin{bmatrix} T_1 \\ T_2 \\ 0 \\ 0 \\ 0 \end{bmatrix} = \begin{bmatrix} -\left[\frac{dT}{dx}\right]_{x_1} + 6.25 \\ \left[\frac{dT}{dx}\right]_{x_2} + 6.25 \\ 0 \\ 0 \\ 0 \end{bmatrix} \qquad (7.2.4\text{-}102)$$

Add second

$$
\begin{bmatrix}
0.4 & -0.4 & 0 & 0 & 0 \\
-0.4 & 0.4+0.4 & -0.4 & 0 & 0 \\
0 & -0.4 & +0.4 & 0 & 0 \\
0 & 0 & 0 & 0 & 0 \\
0 & 0 & 0 & 0 & 0
\end{bmatrix}
\begin{bmatrix}
T_1 \\ T_2 \\ T_3 \\ 0 \\ 0
\end{bmatrix}
=
\begin{bmatrix}
-\left[\dfrac{dT}{dx}\right]_{x_1}+6.25 \\
6.25+6.25 \\
\left[\dfrac{dT}{dx}\right]_{x_2}+6.25 \\
0 \\ 0
\end{bmatrix}
$$

(7.2.4-103)

Third

$$
\begin{bmatrix}
0.4 & -0.4 & 0 & 0 & 0 \\
-0.4 & 0.8 & -0.4 & 0 & 0 \\
0 & -0.4 & 0.4+0.4 & -0.4 & 0 \\
0 & 0 & -0.4 & +0.4 & 0 \\
0 & 0 & 0 & 0 & 0
\end{bmatrix}
\begin{bmatrix}
T_1 \\ T_2 \\ T_3 \\ T_4 \\ 0
\end{bmatrix}
=
\begin{bmatrix}
-\left[\dfrac{dT}{dx}\right]_{x_1}+6.25 \\
12.5 \\
6.25+6.25 \\
\left[\dfrac{dT}{dx}\right]_{x_2}+6.25 \\
0
\end{bmatrix}
$$

(7.2.4-104)

Fourth

$$
\begin{bmatrix}
0.4 & -0.4 & 0 & 0 & 0 \\
-0.4 & 0.8 & -0.4 & 0 & 0 \\
0 & -0.4 & 0.8 & -0.4 & 0 \\
0 & 0 & -0.4 & 0.4+0.4 & -0.4 \\
0 & 0 & 0 & -0.4 & +0.4
\end{bmatrix}
\begin{bmatrix}
T_1 \\ T_2 \\ T_3 \\ T_4 \\ T_5
\end{bmatrix}
=
\begin{bmatrix}
-\left[\dfrac{dT}{dx}\right]_{x_1}+6.25 \\
12.5 \\
12.5 \\
6.25+6.25 \\
\left[\dfrac{dT}{dx}\right]_{x_2}+6.25
\end{bmatrix}
$$

(7.2.4-105)

$$
\begin{bmatrix}
0.4 & -0.4 & 0 & 0 & 0 \\
-0.4 & 0.8 & -0.4 & 0 & 0 \\
0 & -0.4 & 0.8 & -0.4 & 0 \\
0 & 0 & -0.4 & 0.8 & -0.4 \\
0 & 0 & 0 & -0.4 & +0.4
\end{bmatrix}
\begin{bmatrix}
T_1 \\ T_2 \\ T_3 \\ T_4 \\ T_5
\end{bmatrix}
=
\begin{bmatrix}
-\left[\dfrac{dT}{dx}\right]_{x_1}+6.25 \\
12.5 \\
12.5 \\
12.5 \\
\left[\dfrac{dT}{dx}\right]_{x_2}+6.25
\end{bmatrix}
$$

(7.2.4-106)

Since only the temperatures at the ends of the bar are specified, the derivatives at the ends of the bar are unknowns. We therefore bring them to the other side of the equations and transfer the known temperatures T_1 and T_2 to the right-hand side (after multiplying by their respective coefficients from the left-hand side).

$$
\begin{bmatrix}
1.0 & -0.4 & 0 & 0 & 0 \\
0 & 0.8 & -0.4 & 0 & 0 \\
0 & -0.4 & 0.8 & -0.4 & 0 \\
0 & 0 & -0.4 & 0.8 & 0 \\
0 & 0 & 0 & -0.4 & -1.0
\end{bmatrix}
\begin{bmatrix}
\left[\dfrac{dT}{dx}\right]_{x_1} \\
T_2 \\
T_3 \\
T_4 \\
\left[\dfrac{dT}{dx}\right]_{x_2}
\end{bmatrix}
=
\begin{bmatrix}
-54.25 \\
60.5 \\
12.5 \\
92.5 \\
-73.75
\end{bmatrix}
\tag{7.2.4-107}
$$

The solution to the set of equations above is

$$
\begin{bmatrix}
\left[\dfrac{dT}{dx}\right]_{x_1} \\
T_2 \\
T_3 \\
T_4 \\
-\left[\dfrac{dT}{dx}\right]_{x_2}
\end{bmatrix}
=
\begin{bmatrix}
20.5 \\
186.9 \\
222.5 \\
226.9 \\
-17
\end{bmatrix}
\tag{7.2.4-108}
$$

We can solve this problem analytically for comparison

$$
\int \frac{d^2T}{dx^2}\,dx = -\int \Gamma\,dx
$$

$$
\frac{dT}{dx} = -\Gamma x + C_1
\tag{7.2.4-109}
$$

$$
\int dT = -\int \left(\Gamma x + C_1\right) dx
\tag{7.2.4-110}
$$

$$
T = -\left[\Gamma \frac{x^2}{2} + C_1 x + C_2\right]
\tag{7.2.4-111}
$$

At $x = 0$, $T = T_1$

$$
T_1 = -C_2
\tag{7.2.4-112}
$$

At $x = L$, $T = T_2$

$$T_2 = T_1 - C_1 L - \Gamma \frac{L^2}{2} \qquad\qquad (7.2.4\text{-}113)$$

$$C_1 = \frac{T_1 - T_2}{L} - \Gamma \frac{L}{2} \qquad\qquad (7.2.4\text{-}114)$$

Substituting for C_1 and C_2

$$T = -\left[\Gamma \frac{x^2}{2} + \left(\frac{T_1 - T_2}{L} - \Gamma \frac{L}{2} \right) x - T_1 \right] \qquad (7.2.4\text{-}115)$$

$$T = -\left[5 \frac{x^2}{2} + \left(\frac{120 - 200}{10} - 5 \frac{10}{2} \right) x - 120 \right] \qquad (7.2.4\text{-}116)$$

$$T = 120 + 33\, x - 2.5\, x^2 \qquad\qquad (7.2.4\text{-}117)$$

The results are plotted in Figure 7.2.4-17.

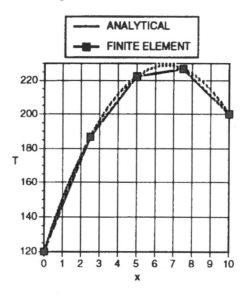

Figure 7.2.4-17 Comparison of finite element and analytic solution

The finite element solution coincides with the analytic solution at the nodes, but because of the linearity of the shape functions does not do as well interior to nodes where the analytic solution is highly curved.

Finite element method in higher dimensions

Extension of the one-dimensional finite element method to two dimensions is straightforward, although increasing in complexity. Therefore, we will not treat the higher dimensional applications other than for passing mention (nor will we treat other than the simple linear interpolation functions already discussed).

The range of element shapes expands considerably as one goes from one to two dimensions, with triangular and quadrilateral shapes both in wide use. Other shapes are also possible. Automatic mesh generation software becomes very useful with the expansion to higher dimensions.

We will not attempt to treat this very important subject in any more detail, but instead refer the reader to the many texts available specifically on the finite element technique. Two very lucid introductions to finite element methods as well as other numerical techniques useful to engineers are

Chapra, S. C. and R. P. Canale (1988). *Numerical Methods for Engineers*. New York, McGraw-Hill.

Gerald, C. F. and P. O. Wheatley (1994). *Applied Numerical Analysis*. Reading, MA, Addison-Wesley.

For more complete treatments of finite elements the reader is referred to, for example,

Baker, A. J. (1983). *Finite Element Computational Fluid Mechanics*. New York, Hemisphere.

Becker, E. B., G. F. Carey, et al. (1981). *Finite Elements: An Introduction*. Englewood Cliffs, NJ, Prentice-Hall.

Carey, G. F. (1986). *Finite Elements: Fluid Mechanics*. Englewood Cliffs, NJ, Prentice-Hall.

Finlayson, B. A. (1980). *Nonlinear Analysis in Chemical Engineering*. New York, NY, McGraw-Hill.

Finlayson, B. A. (1992). *Numerical Methods for Problems with Moving Fronts*. Seattle, WA, Ravenna Park Publishing, Inc.

Hughes, T. J. R. (1987). *The Finite Element Method: Linear Static and Dynamic Finite Element Analysis*. Englewood Cliffs, NJ, Prentice-Hall.

Segerlind, G. J. (1984). *Applied Finite Element Analysis*. New York, John Wiley and Sons.

Stasa, F. L. (1985). *Applied Finite Element Analysis for Engineers.* New York, Holt, Rinehart and Winston.

Taylor, C. and T. G. Hughes (1981). *Finite Element Programming of the Navier-Stokes Equations.* Swansea, U.K., Pineridge Press Limited.

7.2.5   One-dimensional steady-state conduction in cylindrical coordinates

We next consider a derivation similar to that of the preceding section but for cylindrical rather than rectangular coordinates. This time we consider conduction in **only the r-direction,** as shown in Figure 7.2.5-1.

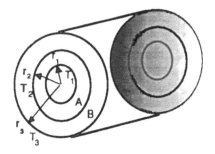

Figure 7.2.5-1 Conduction through the wall of a composite cylinder

Using Table 7.2.1-1, the Fourier equation reduces to

$$\left(\frac{Q}{A}\right)_r = q_r = -k\frac{dT}{dr} \qquad (7.2.5\text{-}1)$$

An energy balance on a system between any two surfaces of constant r tells us that $\dot{Q}$ is constant, since we have no thermal energy generation in the model under consideration.

To integrate we must account for the fact that A (and therefore q) is not **constant** as in the rectangular coordinate case considered previously, but is instead **a function of r.** The relationship is $A = 2\pi rL$.

Considering the case of **constant k** and integrating between any two values of r, r_1 and r_2 where the corresponding temperatures are T_1 and T_2

$$\frac{\overline{Q_r}}{2\pi r L}\, dr = -k\, dT$$

$$\frac{Q_r}{2\pi L} = \int_{R_1}^{R_2} \frac{dr}{r} = -k \int_{T_1}^{T_2} dT \qquad (7.2.5\text{-}2)$$

and so

$$\frac{Q_r}{2\pi L}\, \ln\frac{r_2}{r_1} = -k\,(T_2 - T_1) \qquad (7.2.5\text{-}3)$$

This parallels Equation (7.2.4-5) but is not the form of the equation used most by engineers. Many engineers prefer to say, "I intend to apply the rectangular coordinate *form* [Equation (7.2.4-5)] to the case where the area is changing." Very well, let us put Equation (7.2.5-3) in this form; that is,

$$\frac{\text{Heat transfer} \times \text{distance}}{\text{Area}} = -\,\text{thermal conductivity} \times \text{temperature drop}$$

$$(7.2.5\text{-}4)$$

Comparing with Equation (7.2.4-5), we see we lack

    (1) an appropriate area
    (2) a distance

To introduce the distance, we multiply and divide Equation (7.2.5-3) by $(r_2 - r_1)$, and rearrange slightly

$$\frac{Q_r\,(r_2 - r_1)}{\left(\dfrac{2\pi\,(r_2 - r_1)\,L}{\ln\dfrac{r_2}{r_1}}\right)} = -k\,(T_2 - T_1) \qquad (7.\,2.5\text{-}5)$$

in fact. we can rewrite as

$$\left(\frac{2\pi(r_2-r_1)L}{\ln\frac{r_2}{r_1}}\right) = \left(\frac{2\pi r_2 L - 2\pi r_1 L}{\ln\frac{2\pi r_2 L}{2\pi r_1 L}}\right) = \frac{A_2 - A_1}{\ln\left(\frac{A_2}{A_1}\right)} \qquad (7.2.5\text{-}6)$$

and **define** this quantity to be the **logarithmic mean area, A_{lm}**.

$$A_{lm} \equiv \frac{A_2 - A_1}{\ln\left(\frac{A_2}{A_1}\right)} \qquad\qquad (7.2.5\text{-}7)$$

We then may write Equation (7.2.5-2) as

$$\frac{\dot{Q}_r (r_2 - r_1)}{A_{lm}} = -k(T_2 - T_1) \qquad\qquad (7.2.5\text{-}8)$$

which corresponds directly to Equation (7.2.4-5).

Referring to Figure 7.2.5-1 and using the same argument for a series of resistances (now cylinders) in series as we used for the rectangular coordinate case, we write the sum of the individual temperature differences as

$$\Delta T_{overall} = -\dot{Q}_r \left(\sum_{i=A,B,C,\dots} \left[\frac{\Delta r}{k A_{lm}} \right]_i \right) \qquad\qquad (7.2.5\text{-}9)$$

Rearranging

$$\dot{Q}_r = \frac{-1}{\sum_{i=A,B,C,\dots} \left[\frac{\Delta r}{k A_{lm}} \right]_i} \Delta T_{overall} \qquad\qquad (7.2.5\text{-}10)$$

The denominator of the right-hand side is the thermal resistance term, which represents a series of resistances of the individual materials.

Example 7.2.5-1   Conduction in a fuel rod

A nuclear reactor contains a bundle of cylindrical fuel rods, with each rod having a radius of 25 mm. The rod may be modeled as a solid with average properties

$$\rho = 1000 \frac{kg}{m^3}$$

$$C_p = 800 \frac{J}{kg\,K}$$

$$k = 25 \frac{W}{m\,K}$$

At steady state the fission of the fuel produces a temperature distribution in the rod described by

$$T(r) = (750)[°C] - \left(4 \times 10^5\right)\left[\frac{°C}{m^2}\right](r^2)[m^2]$$

What is the rate of heat transfer at the surface for a unit length of the rod?

Solution

$$\left(\frac{Q}{A}\right)_r = -k\frac{dT}{dr}$$

$$\frac{Q_r}{2\pi r L} = -k\frac{dT}{dr} \tag{7.2.5-11}$$

$$\frac{Q_r}{L} = -2\pi r k\frac{dT}{dr}$$

Using the given temperature profile to evaluate the derivative at the surface

$$\frac{dT}{dr} = \frac{d\left[(750) - \left(4 \times 10^5\right)(r^2)\right]}{dr} \tag{7.2.5-12}$$

$$\left.\frac{dT}{dr}\right|_{r = 0.025\,m} = -(2)\left(4 \times 10^5\right)(0.025) = 2 \times 10^4 \left[\frac{°C}{m}\right]$$

$$\frac{dT}{dr} = -(2)\left(4 \times 10^5\right)(r) \tag{7.2.5-13}$$

Substituting to evaluate the heat transfer per unit length

$$\frac{Q_r}{L} = -2\pi r k \frac{dT}{dr}$$

$$= -(2)(\pi)(0.025)\,[m]\,(25)\left[\frac{W}{m\,{}^{\circ}C}\right](2 \times 10^4)\left[\frac{{}^{\circ}C}{m}\right]$$

$$= 0.75 \times 10^5 \frac{W}{m} \tag{7.2.5-14}$$

Example 7.2.5-2   Conduction through an insulated pipe

A 3-inch glass-lined steel pipe carries steam which keeps the inside surface of the glass at 350°F. The glass coating is 1/32-inch thick and the pipe is coated on the outside with 1-inch of magnesia insulation.

Calculate the heat loss per foot of pipe if the outside surface temperature of the insulation may be assumed to be 120°F.

The following properties may be used:

      Inside diameter of 3-in. pipe = 3 inches
      Outside diameter of 3-in. pipe = 3.5 inches
      k_{glass} = 0.5 Btu/(hr ft °F)
      k_{steel} = 26 Btu/(hr ft °F)
      $k_{magnesia}$ = 0.03 Btu/(hr ft °F)

Solution

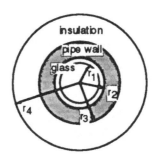

The relationship for conduction through a series of cylindrical resistances is

$$\dot{Q}_r = \frac{-1}{\sum_{i=A, B, C, \dots} \left[\frac{\Delta r}{k A_{lm}}\right]_i} \Delta T_{overall} \tag{7.2.5-15}$$

We are given that

$$\Delta T_{overall} = T_4 - T_1 = \left(120 - 350\right){}^{\circ}F \tag{7.2.5-16}$$

Since we are required to find the loss per foot of pipe, we take L = 1 ft. Remembering that we are given diameters, not radii for the pipe, we can compute

$$[\Delta r]_{glass} = \frac{(1)}{(32)} \text{ in} \frac{ft}{(12) \text{ in}} = 2.604 \times 10^{-3} \text{ ft}$$

$$[\Delta r]_{steel} = \frac{(3.5 - 3.0)}{(2)} \text{ in} \frac{ft}{(12) \text{ in}} = 2.083 \times 10^{-2} \text{ ft}$$

$$[\Delta r]_{magnesia} = (1) \text{ in} \frac{ft}{(12) \text{ in}} = 8.333 \times 10^{-2} \text{ ft} \tag{7.2.5-17}$$

and

$$
\begin{aligned}
[A_{lm}]_{glass} &= \frac{[\pi D L]_2 - [\pi D L]_1}{\ln \frac{[\pi D L]_2}{[\pi D L]_1}} \\
&= \frac{\pi (D_2 - D_1) L}{\ln \frac{D_2}{D_1}} = \frac{\pi \left[3.0 - \left(3.0 - 2 \times \frac{1}{32}\right)\right] \text{ in} (1) \text{ ft}}{\ln \left[\frac{3.0}{\left(3.0 - 2 \times \frac{1}{32}\right)}\right]} \frac{ft}{(12) \text{ in}} \\
&= 0.7772 \text{ ft}^2
\end{aligned}
\tag{7.2.5-18}
$$

$$\left[A_{lm}\right]_{steel} = \frac{\pi\left(D_3 - D_2\right)L}{\ln\frac{D_3}{D_2}} = \frac{\pi\left[3.5 - 3.0\right] \text{in} \left(1\right) \text{ft}}{\ln\left[\frac{\left(3.5\right)}{\left(3.0\right)}\right]} \frac{\text{ft}}{\left(12\right) \text{in}}$$

$$= 0.8492 \text{ ft}^2$$

(7.2.5-19)

$$\left[A_{lm}\right]_{magnesia} = \frac{\pi\left(D_4 - D_3\right)L}{\ln\frac{D_4}{D_3}} = \frac{\pi\left[\left(3.5 + 2\right) - 3.5\right] \text{in} \left(1\right) \text{ft}}{\ln\left[\frac{\left(3.5 + 2\right)}{3.5}\right]} \frac{\text{ft}}{\left(12\right) \text{in}}$$

$$= 1.158 \text{ ft}^2$$

(7.2.5-20)

which yields

$$\left[\frac{\Delta r}{k\,A_{lm}}\right]_{glass} = \frac{\left(2.604 \times 10^{-3}\right) \text{ft}}{\left(0.5\right)\frac{\text{Btu}}{\text{hr ft °F}}\left(0.7772\right) \text{ft}^2} = 6.701 \times 10^{-3} \frac{\text{hr °F}}{\text{Btu}}$$

$$\left[\frac{\Delta r}{k\,A_{lm}}\right]_{steel} = \frac{\left(2.083 \times 10^{-2}\right) \text{ft}}{\left(26\right)\frac{\text{Btu}}{\text{hr ft °F}}\left(0.8492\right) \text{ft}^2} = 9.434 \times 10^{-4} \frac{\text{hr °F}}{\text{Btu}}$$

$$\left[\frac{\Delta r}{k\,A_{lm}}\right]_{magnesia} = \frac{\left(8.333 \times 10^{-2}\right) \text{ft}}{\left(0.03\right)\frac{\text{Btu}}{\text{hr ft °F}}\left(1.158\right) \text{ft}^2} = 2.399 \frac{\text{hr °F}}{\text{Btu}}$$

(7.2.5-21)

Substituting in the original relationship

$$\dot{Q}_r = \frac{-\left(120 - 350\right)°\text{F}}{6.701 \times 10^{-3} \frac{\text{hr °F}}{\text{Btu}} + 9.434 \times 10^{-4} \frac{\text{hr °F}}{\text{Btu}} + 2.399 \frac{\text{hr °F}}{\text{Btu}}}$$

$$= 95.6 \frac{\text{Btu}}{\text{hr}} \text{ for each foot of length}$$

(7.2.5-22)

It is worth noting that

- The resistance of the magnesia dominates the resistance terms, as it should

- Even though the glass is thinner than the pipe wall, its thermal resistance is much larger
- It is probably not worth using the logarithmic mean area over the arithmetic mean for the glass
- In fact, the first two terms in the denominator could probably be neglected altogether - this is an example of a **controlling resistance**, specifically, the magnesia. This, of course, is the usual business of insulation.

7.2.6 One-dimensional steady-state conduction in spherical coordinates

If one repeats for radial conduction in spherical coordinates (see Figure 7.2.6-1) the approach previously used in the rectangular and cylindrical coordinate frames, one has from the Fourier equation

$$\left(\frac{\dot{Q}}{A}\right)_r = q_r = -k\frac{dT}{dr} \tag{7.2.6-1}$$

just as for cylindrical coordinates. An energy balance on a system between any two surfaces of constant r again tells us that $\dot{Q}$ is constant, since we have no thermal energy generation in the model under consideration.

However, now the area is **a different function of radius**:

$$A = 4\pi r^2 \tag{7.2.6-2}$$

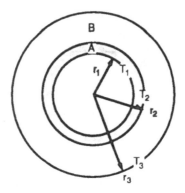

Figure 7.2.6-1 Radial conduction in spherical geometry

This changes the integration (we continue to assume constant thermal conductivity) to

$$\frac{\dot{Q}}{4\pi} \int_{r_1}^{r_2} \frac{dr}{r^2} = -k \int_{T_1}^{T_2} dT \tag{7.2.6-3}$$

or

$$\frac{\dot{Q}}{4\pi} \left(-\frac{1}{r_2} + \frac{1}{r_1} \right) = -k \left(T_2 - T_1 \right) \tag{7.2.6-4}$$

As before, we multiply and divide by $(r_2 - r_1)$ and get

$$\frac{\dot{Q}\left(r_2 - r_1\right)}{\left(\dfrac{4\pi\left(r_2 - r_1\right)}{\dfrac{1}{r_1} - \dfrac{1}{r_2}}\right)} = -k\left(T_2 - T_1\right) \tag{7.2.6-5}$$

but, rearranging the denominator of the left-hand side:

$$\left(\frac{4\pi\left(r_2-r_1\right)}{\dfrac{1}{r_1}-\dfrac{1}{r_2}}\right) = \left(\frac{4\pi\left(r_2-r_1\right)}{\dfrac{\left(r_2-r_1\right)}{r_2 r_1}}\right)$$

$$= 4\pi r_2 r_1$$

$$= \sqrt{\left(4\pi r_2^2\right)\left(4\pi r_1^2\right)}$$

$$= \sqrt{A_1 A_2} \tag{7.2.6-6}$$

and now the appropriate mean area is the *geometric* mean.

$$\boxed{A_{gm} = \sqrt{A_1 A_2}} \tag{7.2.6-7}$$

Then

$$\frac{\dot{Q}\left(r_2-r_1\right)}{A_{gm}} = -k\left(T_2-T_1\right) \tag{7.2.6-8}$$

in *spherical* coordinates, which corresponds to Equations (7.2.4-5) and (7.2.5-8) in the rectangular and cylindrical systems.

For a *composite* sphere we get, using the same procedure as before

$$\boxed{\dot{Q} = \frac{-1}{\displaystyle\sum_{i=A,B,C,\dots}\left[\frac{\Delta r}{k\,A_{gm}}\right]_i}\left(\Delta T\right)_{overall}} \tag{7.2.6-9}$$

Example 7.2.6-1   Conduction through shielding

We are faced with the problem of disposal of radioactive waste. It has been proposed to coat the waste with a new organic coating containing a shielding material, followed by disposal in underground cavern storage. At least a 3-inch thickness of coating is necessary to obtain the required shielding. One of many problems associated with this scheme is that the thermal energy generated by the waste must be transferred through the coating to the surroundings.

We are concerned about whether the minimum shielding thickness will satisfy the thermal energy removal requirement. (Obviously many other considerations such as mechanical strength, stability, cost, etc., will also have to be addressed sooner or later.)

To make a quick assessment of the practicality of such a coating, as an initial model we will assume that the waste is a 1-foot radius sphere which generates 440 Btu/hr. The proposed coating material degrades thermally above 400°F. For this preliminary calculation we will assume that the cavern environment will keep the outer surface of the coating at a maximum of 200°F. The thermal conductivity of the coating will be assumed to be 0.02 Btu/(hr ft °F) and to be independent of temperature.

What maximum thickness of coating could be used and still avoid thermal degradation of the interior surface of the coating?

Solution

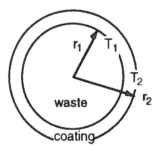

An energy balance using the coating as the system shows that Q is constant. We can express Q in terms of the temperatures using Fourier's law. We will obtain

$$\frac{\dot{Q}\left(r_2 - r_1\right)}{A_{gm}} = -k\left(T_2 - T_1\right)$$

$$(7.2.6\text{-}10)$$

where

$$A_{gm} = \sqrt{A_1 A_2}$$
$$= \sqrt{\left(4\pi r_1^2\right)\left(4\pi r_2^2\right)}$$

$$= \left[\left(4 \pi \ 1 \ ft^2 \right) \left(4 \pi \ r_2^2 \right) \right]^{0.5}$$
$$= \left(4 \pi \ r_2 \right) ft \tag{7.2.6-11}$$

Rearranging (a) and substituting known values

$$\left(r_2 - r_1 \right) = - \frac{k \, A_{gm} \left(T_2 - T_1 \right)}{Q}$$

$$= - \frac{(0.02) \dfrac{Btu}{hr \, ft \, {}^\circ F} \left(4 \pi \ r_2 \right) ft^2 \left(200 - 400 \right) {}^\circ F}{(440) \dfrac{Btu}{hr}} \tag{7.2.6-12}$$

$$\left(r_2 - 1 \right) = 0.1142 \, r_2$$
$$r_2 = 1.129 \, ft \tag{7.2.6-13}$$

which gives as the **maximum** coating thickness

$$\text{maximum thickness} = \left(r_2 - r_1 \right) = \left(1.129 - 1 \right) ft$$
$$= 0.129 \, ft = 1.55 \, in \tag{7.2.6-14}$$

We are faced with the usual design dilemma - two requirements are in conflict, i.e., they cannot both be satisfied simultaneously. If we use a sufficiently thick coating for adequate shielding, we will have thermal degradation; if we use a sufficiently thin coating to avoid thermal degradation, we will not have adequate shielding.

To reconcile such conflicts is the heart of engineering. The following list includes only a few of the design questions that should be explored next:

- Is the data used for shielding requirements and thermal degradation trustworthy?
- Is the assumed outer surface temperature realistic? If so, is there a way to lower it?
- Can we store the waste in a form with a higher surface area to volume ratio? (For example, as spheres of smaller diameter, or in sheet form)
- Can we modify the coating so as either to require less thickness for shielding or to improve its thermal stability?

• Is there another coating that could be used?
• Is there an alternative form of disposal - e.g., recycling?
• Can we prevent the generation of the waste from the outset?

The *finite element method* differs from the finite difference method in that the finite element method relies upon fitting a series of approximating functions to the **solution** of the differential equation, not to the equation itself as in the finite difference method.

7.2.7   Two-dimensional steady-state conduction

Two-dimensional steady-state conduction in a material with constant thermal conductivity in the presence of thermal energy generation can be described by a model obtained by deleting the time-dependent term in the microscopic energy balance, which leaves

$$k \left[\frac{\partial^2 T}{\partial x^2} + \frac{\partial^2 T}{\partial y^2} \right] + \gamma_e = 0 \qquad\qquad (7.2.7\text{-}1)$$

Without thermal energy generation , dividing by k leaves the two-dimensional Laplace equation

$$\nabla^2 T = \left[\frac{\partial^2 T}{\partial x^2} + \frac{\partial^2 T}{\partial y^2} \right] = 0 \qquad\qquad (7.2.7\text{-}2)$$

This elliptic[17] equation occurs in many contexts other than steady-state heat transfer; for example, we have already used it to describe potential flow, where the dimensionless temperature is replaced by the velocity potential. We shall later see its utility in describing molecular diffusion, where the dependent variable becomes some species concentration-related (chemical potential-related) quantity.

---

[17] Second-order partial differential equations are classified as elliptic, parabolic, or hyperbolic after writing the equation in the form

$$A \frac{\partial^2 u}{\partial x^2} + B \frac{\partial^2 u}{\partial x \partial y} + C \frac{\partial^2 T}{\partial y^2} + D \left(x, y, u, \frac{\partial u}{\partial x}, \frac{\partial u}{\partial y} \right) = 0$$

according to

$B^2 - 4AC < 0 \;\Rightarrow\;$ Elliptic
$B^2 - 4AC = 0 \;\Rightarrow\;$ Parabolic
$B^2 - 4AC > 0 \;\Rightarrow\;$ Hyperbolic

Here we will be concerned only with elliptic and parabolic equations.

The primary difficulty, as is more or less true in general for partial differential equations, is in satisfying the boundary conditions. Irregular boundaries tend to complicate analytic solutions immensely, finite difference solutions somewhat, and finite element solutions least.

Analytically, the Laplace equation is often solved for multi-dimensional problems with aid of the technique from the theory of complex variables[18] known as *conformal mapping*, which is basically a method of transforming a problem with more difficult geometry into one with simpler geometry.

Numerically, a solution method often adopted is the *finite difference method*, which replaces the differential equation by a difference (algebraic) equation based on representing the derivatives in the differential equation by differences developed using the Taylor series. Like all numerical solutions, the computation is subject to errors - **truncation** error from termination of infinite series at a finite number of terms, and **round-off** error from the fact that digital computers must operate with a finite number of digits to represent variables.[19] (In general, we also need to be concerned with the statistical concepts of **accuracy**, which has to do with whether the average of the data is correct, and **precision**, which is a measure of the scatter of data. Error can also **propagate** through calculations and thereby be magnified. Finally, there is error introduced through good old-fashioned **blunders and mistakes**.)

We very often work with the partial differential equations and boundary conditions in dimensionless form, e.g., by defining in the case of energy transfer the following dimensionless variables

$$\theta = \frac{T - T_0}{T_1 - T_0}$$
$$x^* = \frac{x}{L}$$
$$y^* = \frac{y}{W}$$

(7.2.7-3)

From the basic equation

---

[18] See any text on complex variables or complex analysis; e.g., Churchill, R. V. (1948). *Introduction to Complex Variables and Applications*. New York, NY, McGraw-Hill.

[19] See, for example, Nakamura, S. (1991). *Applied Numerical Methods with Software*. Englewood Cliffs, NJ, Prentice Hall, Chapter 1.

$$k\left[\frac{\partial^2 T}{\partial x^2} + \frac{\partial^2 T}{\partial y^2}\right] + \gamma_\theta = 0 \tag{7.2.7-4}$$

$$\frac{\partial^2 T}{\partial x^2} + \frac{\partial^2 T}{\partial y^2} + \frac{\gamma_\theta}{k} = \frac{\partial}{\partial x}\left[\frac{\partial T}{\partial x}\right] + \frac{\partial}{\partial y}\left[\frac{\partial T}{\partial y}\right] + \frac{\gamma_\theta}{k} = 0 \tag{7.2.7-5}$$

Applying the chain rule to transform to dimensionless variables

$$\frac{\partial x^*}{\partial x}\frac{\partial}{\partial x^*}\left[\frac{\partial x^*}{\partial x}\frac{\partial \theta}{\partial x^*}\frac{\partial T}{\partial \theta}\right] + \frac{\partial y^*}{\partial y}\frac{\partial}{\partial y^*}\left[\frac{\partial y^*}{\partial y}\frac{\partial \theta}{\partial y^*}\frac{\partial T}{\partial \theta}\right] + \frac{\gamma_\theta}{k} = 0 \tag{7.2.7-6}$$

Substituting using the definitions of the dimensionless variables

$$\frac{1}{L}\frac{\partial}{\partial x^*}\left[\frac{1}{L}\frac{\partial \theta}{\partial x^*}(T_1 - T_0)\right] + \frac{1}{W}\frac{\partial}{\partial y^*}\left[\frac{1}{W}\frac{\partial \theta}{\partial y^*}(T_1 - T_0)\right] + \frac{\gamma_\theta}{k} = 0 \tag{7.2.7-7}$$

and requiring that $L = W$

$$\frac{\partial^2 \theta}{\partial (x^*)^2} + \frac{\partial^2 \theta}{\partial (y^*)^2} + \frac{L^2 \gamma_\theta}{k(T_1 - T_0)} = 0 \tag{7.2.7-8}$$

In the absence of thermal energy generation this becomes

$$\frac{\partial^2 \theta}{\partial (x^*)^2} + \frac{\partial^2 \theta}{\partial (y^*)^2} = 0 = \nabla^2 \theta \tag{7.2.7-9}$$

Dimensionless Dirichlet boundary conditions for the typical problem we shall consider are shown in the following sketch

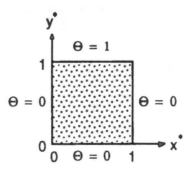

which correspond to dimensionless boundary conditions

$$\theta\left(0, y^*\right) = 0$$
$$\theta\left(1, y^*\right) = 0$$
$$\theta\left(x^*, 0\right) = 0 \qquad\qquad (7.2.7\text{-}10)$$
$$\theta\left(x^*, 1\right) = 1$$

Taylor series

The Taylor series with a remainder in two variables may be written as

$$
\begin{aligned}
f\left[x+h, y+k\right] = f\left[x, y\right] &+ \left(h \frac{\partial}{\partial x} + k \frac{\partial}{\partial y}\right) f\left[x, y\right] \\
&+ \frac{1}{2!}\left(h \frac{\partial}{\partial x} + k \frac{\partial}{\partial y}\right)^2 f\left[x, y\right] \\
&+ \frac{1}{3!}\left(h \frac{\partial}{\partial x} + k \frac{\partial}{\partial y}\right)^3 f\left[x, y\right] \qquad (7.2.7\text{-}11) \\
+ \dots &+ \frac{1}{(n-1)!}\left(h \frac{\partial}{\partial x} + k \frac{\partial}{\partial y}\right)^{n-1} f\left[x, y\right] + R_n
\end{aligned}
$$

with remainder

$$R_n = \frac{1}{(n)!}\left(h \frac{\partial}{\partial x} + k \frac{\partial}{\partial y}\right)^n f\left[x+\xi h, y+\zeta k\right]$$
$$0 \le \xi \le 1; 0 \le \zeta \le 1 \qquad\qquad (7.2.7\text{-}12)$$

which is a more general form than the (perhaps) more familiar one-variable form

$$
\begin{aligned}
f[x+h] = f[x] + \left(h\frac{d}{dx}\right) f[x] \\
+ \frac{1}{2!}\left(h\frac{d}{dx}\right)^2 f[x] \\
+ \frac{1}{3!}\left(h\frac{d}{dx}\right)^3 f[x] \\
+ \cdots + \frac{1}{(n-1)!}\left(h\frac{d}{dx}\right)^{n-1} f[x] + R_n
\end{aligned}
$$

(7.2.7-13)

with remainder

$$
R_n = \frac{1}{(n)!}\left(h\frac{d}{dx}\right)^n f[x+\xi h] \quad 0 \le \xi \le 1
$$

(7.2.7-14)

which is frequently written in another form as

$$
f(x) = f(x_0) + \sum_{n=1}^{N}\left[\frac{f^n(x_0)}{n!}(x-x_0)^n\right] + R_N
$$

(7.2.7-15)

with remainder

$$
R_N = \frac{f^{N+1}(\xi)}{(N+1)!}(x-x_0)^{N+1} \quad (x < \xi < x_0)
$$

(7.2.7-16)

sometimes written as

$$
f(x) = f(x_0) + \sum_{n=1}^{N}\left[\frac{f^n(x_0)}{n!}(x-x_0)^n\right] + O\left[(x-x_0)^{N+1}\right]
$$

(7.2.7-17)

where the last term is read as "terms of the order of $(x - x_0)^{N+1}$." Notice that this does **not** mean that the remainder is of **magnitude** $(x - x_0)^{N+1}$, because we have no general way of predicting the magnitude of the derivative in the remainder.[20] We are merely regarding the remainder as a function of the magnitude of $\Delta x = x - x_0$.

$$R_N \left(x - x_0 \right) = \frac{f^{N+1}\left(\xi\right)}{(N+1)!} \left(x - x_0\right)^{N+1} \tag{7.2.7-18}$$

The significance of the statement, therefore, is that if we cut the size of Δx by, for example, half, if the value of ξ did not change (which may or may not be true), the error (whatever it is) will be reduced by

$$\frac{R_N\left(\Delta x_2\right)}{R_N\left(\Delta x_1\right)} = \frac{\dfrac{f^{N+1}\left(\xi\right)}{(N+1)!}\left(\Delta x_2\right)^{N+1}}{\dfrac{f^{N+1}\left(\xi\right)}{(N+1)!}\left(\Delta x_1\right)^{N+1}} = \frac{\left(\dfrac{\Delta x_1}{2}\right)^{N+1}}{\left(\Delta x_1\right)^{N+1}} = \left(\tfrac{1}{2}\right)^{N+1} \tag{7.2.7-19}$$

Analytical solution

The heat conduction problem can often be solved by a series solution method for relatively regular geometries, i.e., rectangular parallelepipeds, cylinders, spheres, and shapes capable of being mapped into such regular geometries. The primary problem in solving partial differential equations lies usually in satisfying the boundary conditions. Since there are many different combinations of boundary conditions possible, historically there have been many techniques used for analytical solution of such equations.

In this text we will not attempt in any sense to be exhaustive in enumerating possible solution methods; instead we will attempt to illustrate common methods which give general insight into all methods as well as exemplifying a large class of solutions the reader is apt to encounter.

In the realm of two-dimensional steady-state conduction, we choose to illustrate using the method of separation of variables, which leads to a series solution - in the geometry chosen, a Fourier series. The typical problem we use

---

[20] Strictly speaking, the notation signifies [see Gerald, C. F. and P. O. Wheatley (1994). *Applied Numerical Analysis*. Reading, MA, Addison-Wesley]

Error $= O\left[(\Delta x)^n\right]$

if

$\lim\limits_{\Delta x \to 0}\left(\text{Error}\right) = K\,\Delta x^n$

for some constant $K \neq 0$

is the dimensionless two-dimensional steady-state conduction equation presented in Equations (7.2.7-9) and (7.2.7-10).[21]

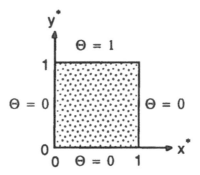

$$\nabla^2 \theta = 0 = \frac{\partial^2 \theta}{\partial (x^*)^2} + \frac{\partial^2 \theta}{\partial (y^*)^2} \qquad (7.2.7\text{-}20)$$

$$\theta(0, y^*) = 0$$
$$\theta(1, y^*) = 0$$
$$\theta(x^*, 0) = 0 \qquad (7.2.7\text{-}21)$$
$$\theta(x^*, 1) = 1$$

The method of separation of variables revolves about the assumption that the two-dimensional function that satisfies the partial differential equation can be written as the product of two functions, the first a function of only the first variable, and the second a function of only the second variable, e.g.,

$$\theta(x^*, y^*) = \Phi(x^*) \Psi(y^*) \qquad (7.2.7\text{-}22)$$

The validity of this assumption rises and falls on whether or not we can find, as a result of using the assumption, two functions which, when multiplied together, yield a resultant function which satisfies the partial differential equation and the boundary conditions. Such a resultant function is, by definition, a solution of the problem.

---

[21] The same differential equation and boundary conditions are also applicable to the appropriate problems in diffusion of mass.

We proceed by substituting the assumption into the partial differential equation

$$\frac{\partial^2 \theta \left(x^*, y^*\right)}{\partial \left(x^*\right)^2} + \frac{\partial^2 \left[\theta \left(x^*, y^*\right)\right]}{\partial \left(y^*\right)^2} = \frac{\partial^2 \left[\Phi \left(x^*\right) \Psi \left(y^*\right)\right]}{\partial \left(x^*\right)^2} + \frac{\partial^2 \left[\Phi \left(x^*\right) \Psi \left(y^*\right)\right]}{\partial \left(y^*\right)^2} = 0$$

$$(7.2.7\text{-}23)$$

Noting that we are now dealing with functions of one variable, and further, that functions of only x^* act as constants when differentiating with respect to y^* and vice versa

$$\Psi \left(y^*\right) \frac{d^2 \left[\Phi \left(x^*\right)\right]}{d \left(x^*\right)^2} + \Phi \left(x^*\right) \frac{d^2 \left[\Psi \left(y^*\right)\right]}{d \left(y^*\right)^2} = 0$$

$$(7.2.7\text{-}24)$$

which can be rearranged to

$$\frac{1}{\Phi \left(x^*\right)} \frac{d^2 \left[\Phi \left(x^*\right)\right]}{d \left(x^*\right)^2} = -\frac{1}{\Psi \left(y^*\right)} \frac{d^2 \left[\Psi \left(y^*\right)\right]}{d \left(y^*\right)^2}$$

$$(7.2.7\text{-}25)$$

Notice that we have an equation which says that a function of x^* only is equal to a function of y^* only.

$$f \left(x^*\right) = g \left(y^*\right)$$

$$(7.2.7\text{-}26)$$

If this is to be true for arbitrary values of x^* and y^*, the only way that this can occur is if each of the functions is equal to the same constant.

We choose this constant to be ($-c^2$) for purposes of later algebraic simplicity and write (now dropping the indication of independent variable in each function)

$$\frac{1}{\Phi} \frac{d^2 \Phi}{d \left(x^*\right)^2} = -\frac{1}{\Psi} \frac{d^2 \Psi}{d \left(y^*\right)^2} = -c^2$$

$$(7.2.7\text{-}27)$$

This relation now gives us two second-order linear homogeneous ordinary differential equations with constant coefficients

$$\frac{d^2\Phi}{d(x^*)^2} + c^2\,\Phi = 0 \tag{7.2.7-28}$$

$$\frac{d^2\Psi}{d(y^*)^2} - c^2\,\Psi = 0 \tag{7.2.7-29}$$

Solving these equations in order, we find the characteristic (or auxiliary) equation for the first to be[22]

$$\lambda^2 + c^2 = 0 \tag{7.2.7-30}$$

$$\lambda = \pm i\,c \tag{7.2.7-31}$$

This is the case of two complex conjugate roots, so the corresponding general solution is

$$\Phi = K_1'\,e^{icx^*} + K_2'\,e^{-icx^*} \tag{7.2.7-32}$$

Using the identity

$$e^{i\omega} = \cos\omega + i\sin\omega \tag{7.2.7-33}$$

we can write this result as

$$\Phi = K_1'\Big[\cos\big(c\,x^*\big) + i\sin\big(c\,x^*\big)\Big] \\ + K_2'\Big[\cos\big(-c\,x^*\big) + i\sin\big(-c\,x^*\big)\Big] \tag{7.2.7-34}$$

but using the identities

$$\cos\big(\omega\big) = \cos\big(-\omega\big) \tag{7.2.7-35}$$

$$\sin\big(\omega\big) = -\sin\big(-\omega\big) \tag{7.2.7-36}$$

we can write

---

[22] This method of solution can be found in most treatments of differential equations or advanced engineering mathematics, for example, Kreyszig, E. (1993). *Advanced Engineering Mathematics*. New York, NY, John Wiley and Sons, Inc., p. 628.

$$\Phi = K_1' \left[\cos\left(c\,x^* \right) + i \sin\left(c\,x^* \right) \right]$$
$$+ K_2 \left[\cos\left(c\,x^* \right) - i \sin\left(c\,x^* \right) \right] \qquad (7.2.7\text{-}37)$$

$$\Phi = \left(K_1' + K_2' \right) \cos\left(c\,x^* \right)$$
$$+ i \left(K_1' - K_2' \right) \sin\left(c\,x^* \right) \qquad (7.2.7\text{-}38)$$

Defining new constants for compactness of notation

$$K_1 = \left(K_1' + K_2' \right)$$
$$K_2 = i \left(K_1' - K_2' \right) \qquad (7.2.7\text{-}39)$$

we have

$$\Phi = K_1 \cos\left(c\,x^* \right) + K_2 \sin\left(c\,x^* \right) \qquad (7.2.7\text{-}40)$$

Next considering the equation for Ψ we have the characteristic equation

$$\lambda^2 - c^2 = 0 \qquad (7.2.7\text{-}41)$$
$$\lambda = \pm c \qquad (7.2.7\text{-}42)$$

This is the case of two real conjugate roots, so the corresponding general solution is

$$\Psi = K_3 e^{c\,y^*} + K_4 e^{-c\,y^*} \qquad (7.2.7\text{-}43)$$

Taking the product of the two solutions we have a function that satisfies the original partial differential equation

$$\theta = \left[K_1 \cos\left(c\,x^* \right) + K_2 \sin\left(c\,x^* \right) \right] \left[K_3 e^{c\,y^*} + K_4 e^{-c\,y^*} \right] \qquad (7.2.7\text{-}44)$$

We now attempt to make this solution satisfy the boundary conditions. We begin with the boundary condition

$$\theta\left(0, y^* \right) = 0 \qquad (7.2.7\text{-}45)$$

Substituting in our function

$$0 = \left[K_1 \cos(0) + K_2 \sin(0)\right]\left[K_3 e^{cy^*} + K_4 e^{-cy^*}\right]$$
$$0 = \left[K_1\right]\left[K_3 e^{cy^*} + K_4 e^{-cy^*}\right] \tag{7.2.7-46}$$

For this to vanish for arbitrary values of y* (we cannot make the second term in brackets equal to zero without losing the y* dependence of our solution)

$$K_1 = 0 \tag{7.2.7-47}$$

leaving

$$\theta = \left[K_2 \sin(c x^*)\right]\left[K_3 e^{cy^*} + K_4 e^{-cy^*}\right] \tag{7.2.7-48}$$

We now apply the boundary condition

$$\theta(x^*, 0) = 0 \tag{7.2.7-49}$$

giving

$$0 = \left[K_2 \sin(c x^*)\right]\left[K_3 + K_4\right] \tag{7.2.7-50}$$

which leads to the conclusion that

$$K_3 = -K_4 \tag{7.2.7-51}$$

so

$$\theta = \left[K_2 \sin(c x^*)\right]\left[K_3 e^{cy^*} - K_3 e^{-cy^*}\right]$$
$$\theta = \left[K_2 K_3 \sin(c x^*)\right]\left[e^{cy^*} - e^{-cy^*}\right] \tag{7.2.7-52}$$

Applying the third boundary condition

$$\theta(1, y^*) = 0 \tag{7.2.7-53}$$

we have

$$0 = \left[K_2 K_3 \sin (c) \right] \left[e^{cy^*} - e^{-cy^*} \right] \qquad (7.2.7-54)$$

For this to be satisfied without eliminating part of our solution

$$\sin (c) = 0 \qquad (7.2.7-55)$$

$$c = n\pi: \quad n \neq 0; n = 1, 2, 3, \dots \qquad (7.2.7-56)$$

The value $n = 0$ is excluded because this would yield

$$0 = \left[K_2 K_3 (0) \right] \left[1 - 1 \right] \qquad (7.2.7-57)$$

removing the dependence on y^*. Multiplying and dividing by 2, and defining a new constant K_n (dependent on n) which incorporates the 2 inserted in the numerator

$$\theta = K_n \sin \left(n\pi x^* \right) \left[\frac{e^{n\pi y^*} - e^{-n\pi y^*}}{2} \right]: \quad n \neq 0; n = 1, 2, 3, \dots$$

$$(7.2.7-58)$$

$$\theta = K_n \sin \left(n\pi x^* \right) \sinh \left(n\pi y^* \right): \quad n \neq 0; n = 1, 2, 3, \dots \qquad (7.2.7-59)$$

Now we do not have a single solution, but an infinity of solutions depending on the value of n. Fortunately, since we are dealing with a linear homogeneous partial differential equation these solutions may be superposed to give the most general solution[23]

$$\theta = \sum_{n=1}^{\infty} K_n \sin \left(n\pi x^* \right) \sinh \left(n\pi y^* \right) \qquad (7.2.7-60)$$

To incorporate the fourth boundary condition, we need to use the concept of *orthogonal functions*, so we will make a brief digression.

---

[23] Kreyszig, E. (1993). *Advanced Engineering Mathematics*. New York, NY, John Wiley and Sons, Inc., p. 646.

Orthogonal functions

The concept of orthogonal vectors has already been used in this work in the scalar (or dot) product

$$u \cdot v = 0 \tag{7.2.7-61}$$

in symbolic form, which indicates that two vectors (here, in 2-space) are at right angles to each other, as shown in the following illustration.

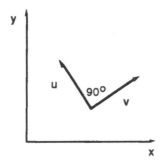

Similarly, we can represent any vector by a combination of a set of orthogonal vectors, or *basis*. In 2-space rectangular coordinates, the basis is the set of unit vectors along the x- and y- axes. Thus, we can represent the vector u as

$$u = u_x \, i + u_y \, j \tag{7.2.7-62}$$

The idea of orthogonality can be extended to functions, by defining that if

$$\int_a^b g_1(x) \, g_2(x) \, dx = 0 \tag{7.2.7-63}$$

the two functions of x, $g_1(x)$ and $g_2(x)$ are orthogonal *on the interval [a, b]*.

Any function of x can be represented in terms of a set of *orthogonal functions* $g_n(x)$

$$f(x) = \sum_{n=1}^{\infty} A_n \, g_n(x) \tag{7.2.7-64}$$

In order to compute with this formula, we must have a way to determine the coefficients A_n. We establish such a method by multiplying both sides of the equation by $g_m(x)$ and integrating from a to b

$$\int_a^b g_m(x) f(x) dx = \int_a^b \left[\sum_{n=1}^{\infty} A_n g_m(x) g_n(x) \right] dx \qquad (7.2.7\text{-}65)$$

Interchanging summation and integration

$$\int_a^b g_m(x) f(x) dx = \sum_{n=1}^{\infty} \left(\int_a^b \left[A_n g_m(x) g_n(x) \right] dx \right) \qquad (7.2.7\text{-}66)$$

Recognizing that A_n is not a function of x

$$\int_a^b g_m(x) f(x) dx = \sum_{n=1}^{\infty} \left(A_n \int_a^b \left[g_m(x) g_n(x) \right] dx \right) \qquad (7.2.7\text{-}67)$$

Notice that because of the orthogonality of the set of functions, all the integrals on the right are zero except for the one in which m is equal to n. We therefore write

$$\int_a^b g_n(x) f(x) dx = A_n \int_a^b \left[g_n(x) \right]^2 dx \qquad (7.2.7\text{-}68)$$

so

$$A_n = \frac{\int_a^b g_n(x) f(x) dx}{\int_a^b \left[g_n(x) \right]^2 dx} \qquad (7.2.7\text{-}69)$$

which is the formula required to evaluate A_n. This ends our digression.

We now return to attempting to make our solution

$$\theta = \sum_{n=1}^{\infty} K_n \sin\left(n \pi x^*\right) \sinh\left(n \pi y^*\right) \qquad (7.2.7\text{-}70)$$

satisfy the fourth boundary condition

$$\theta\left(x^{*}, 1\right) = 1 \qquad (7.2.7\text{-}71)$$

Substituting

$$1 = \sum_{n=1}^{\infty} K_{n} \sin\left(n \pi x^{*}\right) \sinh\left(n \pi\right) \qquad (7.2.7\text{-}72)$$

We now state without proof that the sine functions above (the sinh function is just a constant) are a form of Fourier series and are orthogonal on our interval of interest.[24]

Following the example of orthogonal functions, if we now multiply both sides of the equation by $\sin(m\pi x^{*})$ and integrate over [0, 1]

$$\int_{0}^{1} (1) \sin\left(m \pi x^{*}\right) dx^{*} = \int_{0}^{1} \left[\sum_{n=1}^{\infty} K_{n} \sin\left(m \pi x^{*}\right) \sin\left(n \pi x^{*}\right) \sinh\left(n \pi\right)\right] dx^{*}$$

$$(7.2.7\text{-}73)$$

Since

$$K_{n} \sinh\left(n \pi\right) \qquad (7.2.7\text{-}74)$$

is not a function of x^{*}, it may be moved outside the integral sign

$$\int_{0}^{1} (1) \sin\left(m \pi x^{*}\right) dx^{*} = \sum_{n=1}^{\infty} \left[K_{n} \sinh\left(n \pi\right) \int_{0}^{1} \sin\left(m \pi x^{*}\right) \sin\left(n \pi x^{*}\right) dx^{*}\right]$$

$$(7.2.7\text{-}75)$$

Again, the only terms not equal to zero are those for which m = n

$$\int_{0}^{1} (1) \sin\left(n \pi x^{*}\right) dx^{*} = K_{n} \sinh\left(n \pi\right) \int_{0}^{1} \sin^{2}\left(n \pi x^{*}\right) dx^{*} \qquad (7.2.7\text{-}76)$$

---

[24] Kreyszig, E. (1993). *Advanced Engineering Mathematics*. New York, NY, John Wiley and Sons, Inc., p. 250.

$$K_n \sinh\left(n\,\pi\right) = \frac{\int_0^1 (1)\sin\left(n\,\pi\,x^*\right)dx^*}{\int_0^1 \sin^2\left(n\,\pi\,x^*\right)dx^*} \qquad (7.2.7\text{-}77)$$

Comparing this with our earlier result

$$A_n = \frac{\int_a^b g_n\left(x\right)f\left(x\right)dx}{\int_a^b \left[g_n\left(x\right)\right]^2 dx} \qquad (7.2.7\text{-}78)$$

we see that we can identify

$$1 \Leftrightarrow f\left(x\right) \qquad\qquad (7.2.7\text{-}79)$$
$$\sin\left(n\,\pi\,x^*\right) \Leftrightarrow g_n\left(x\right) \qquad\qquad (7.2.7\text{-}80)$$
$$K_n \sinh\left(n\,\pi\right) \Leftrightarrow A_n \qquad\qquad (7.2.7\text{-}81)$$
$$0 \Leftrightarrow a \qquad\qquad (7.2.7\text{-}82)$$
$$1 \Leftrightarrow b \qquad\qquad (7.2.7\text{-}83)$$
$$x^* \Leftrightarrow x \qquad\qquad (7.2.7\text{-}84)$$

with the situation previously considered in our aside regarding orthogonal functions.

First considering the denominator

$$\begin{aligned}
\int_0^1 \sin^2\left(n\,\pi\,x^*\right)dx^* &= \int_0^1 \left(\tfrac{1}{2}\right)\left[1 - \cos\left(2\,n\,\pi\,x^*\right)\right]dx^* \\
&= \int_0^1 \left(\tfrac{1}{2}\right)dx^* - \tfrac{1}{2\,n\,\pi}\int_0^1 \left[\cos\left(2\,n\,\pi\,x^*\right)\right]2\,n\,\pi\,dx^* \\
&= \tfrac{x}{2}\Big|_0^1 - \tfrac{1}{2\,n\,\pi}\left[2\sin\left(n\,\pi\,x^*\right)\cos\left(n\,\pi\,x^*\right)\right]\Big|_0^1 \\
&= \tfrac{1}{2} \qquad\qquad (7.2.7\text{-}85)
\end{aligned}$$

Then the numerator

$$\int_0^1 \sin\left(n \pi x^*\right) dx^* = \left.\frac{-\cos\left(n \pi x^*\right)}{n \pi}\right|_0^1$$

$$= -\left[\frac{(-1)^n - 1}{n \pi}\right] = \frac{(-1)^{n+1} + 1}{n \pi} \qquad (7.2.7\text{-}86)$$

Therefore,

$$K_n \sinh\left(n \pi\right) = (2)\left[\frac{(-1)^{n+1} + 1}{n \pi}\right] \qquad (7.2.7\text{-}87)$$

$$K_n = \frac{(2)}{n \pi}\left[(-1)^{n+1} + 1\right]\frac{1}{\sinh\left(n \pi\right)} \qquad (7.2.7\text{-}88)$$

and substituting in

$$\theta = \sum_{n=1}^{\infty} K_n \sin\left(n \pi x^*\right) \sinh\left(n \pi y^*\right) \qquad (7.2.7\text{-}89)$$

we have

$$\theta = \sum_{n=1}^{\infty} \left(\frac{2}{n \pi}\right)\left[(-1)^{n+1} + 1\right] \sin\left(n \pi x^*\right) \frac{\sinh\left(n \pi y^*\right)}{\sinh\left(n \pi\right)} \qquad (7.2.7\text{-}90)$$

$$\theta = \frac{2}{\pi} \sum_{n=1}^{\infty} \left[\frac{(-1)^{n+1} + 1}{n}\right] \sin\left(n \pi x^*\right) \frac{\sinh\left(n \pi y^*\right)}{\sinh\left(n \pi\right)} \qquad (7.2.7\text{-}91)$$

This function satisfies both the partial differential equation and the boundary conditions and therefore constitutes a solution to the problem posed, thereby verifying the validity of the separation of variables originally postulated. One can recover the original dependent variable, temperature, by using the definition of θ.

Example 7.2.7-2   Convergence of steady-state rectangular coordinate solution

For the dimensionless temperature profile

$$\theta = \frac{2}{\pi} \sum_{n=1}^{\infty} \left[\frac{(-1)^{n+1} + 1}{n} \right] \sin\left(n \pi x^*\right) \frac{\sinh\left(n \pi y^*\right)}{\sinh\left(n \pi\right)} \qquad (7.2.7\text{-}92)$$

Investigate the speed of convergence at the following points

x^*	y^*
0.5	0.5
0.1	0.1
0.1	0.8
0.1	0.5

Solution

Notice that the factor

$$\left[\frac{(-1)^{n+1} + 1}{n} \right] \qquad (7.2.7\text{-}93)$$

is zero for even values of n, so the sum changes only every other term. We can easily calculate using a spreadsheet, other mathematical software, or by programming

n	x^*	y^*	θ
1	0.5	0.5	0.25371694
2	0.5	0.5	0.25371694
3	0.5	0.5	0.249904602
4	0.5	0.5	0.249904602
5	0.5	0.5	0.250003458
6	0.5	0.5	0.250003458
7	0.5	0.5	0.250000406
8	0.5	0.5	0.250000406
9	0.5	0.5	0.250000509
10	0.5	0.5	0.250000509

n	x^*	y^*	Θ
1	0.1	0.1	0.010880004
2	0.1	0.1	0.010880004
3	0.1	0.1	0.010940318
4	0.1	0.1	0.010940318
5	0.1	0.1	0.010940494
6	0.1	0.1	0.010940494
7	0.1	0.1	0.010940495
8	0.1	0.1	0.010940495
9	0.1	0.1	0.010940495
10	0.1	0.1	0.010940495

n	x^*	y^*	Θ
1	0.1	0.8	0.2089153
2	0.1	0.8	0.2089153
3	0.1	0.8	0.26104934
4	0.1	0.8	0.26104934
5	0.1	0.8	0.272053713
6	0.1	0.8	0.272053713
7	0.1	0.8	0.273863574
8	0.1	0.8	0.273863574
9	0.1	0.8	0.274016605
10	0.1	0.8	0.274016605

n	x^*	y^*	Θ
1	0.1	0.5	0.078402782
2	0.1	0.5	0.078402782
3	0.1	0.5	0.081487026
4	0.1	0.5	0.081487026
5	0.1	0.5	0.081585882
6	0.1	0.5	0.081585882
7	0.1	0.5	0.081588351
8	0.1	0.5	0.081588351
9	0.1	0.5	0.081588383
10	0.1	0.5	0.081588383

Notice that the series converges quite rapidly, with only a few terms (only half of which are non-zero) being more than adequate for most engineering calculations. This is also true for the corresponding solutions in cylindrical and spherical coordinates, although the series are in terms of Bessel functions and Legendre polynomials, respectively, rather than Fourier series as in this case.

Numerical solution

The most common approaches to numerical solution of both steady-state and unsteady-state conduction problems are

- finite difference solutions

- finite element solutions

Both of the methods involve covering the region of interest (the interior and boundary of the material within which the differential equation is to be satisfied) with a grid. The finite element method has the advantage of being able to utilize grids based on shapes such as triangles or irregular quadrilaterals, and therefore being able to follow irregular boundaries more closely. Because of this flexibility in grid compositions, finite element methods offer the advantage of being able to use a coarse grid where the solution is changing slowly, and a finer grid where the solution is changing rapidly, which offers considerable advantages in computation.

Finite difference solutions return the approximate value of the desired function at only the points of the grid, and therefore require a decision as to interpolation method if values at intermediate points are required. Finite element methods, on the other hand, return an approximating function valid throughout the region of solution - the interpolation is an intrinsic part of the method.

Finite element methods for two- and three-dimensional problems, although not difficult to comprehend, are complex and require a level of software dependence not required for finite difference methods. The fact that we used and will use only one-dimensional illustration for the finite element method should not, therefore, be construed by the reader to imply any lack of utility for application to higher-dimensional problems - the reason is purely pedagogical.

Finite difference method

The finite difference solution of differential equations (including those describing steady-state heat conduction) consists of a sequence of steps

- Covering the region of interest (the interior and boundary of the material within which the differential equation is to be satisfied) with a grid

• At the nodes (intersection points) of the grid, replacing the derivatives in the differential equation with an algebraic approximation

• Solving the resultant set of algebraic equations to obtain an estimate of the dependent variable (temperature) at the nodal points

We proceed to apply this series of steps to the steady-state conduction problem.

The following illustration is a typical set of four nodal points from such a grid.

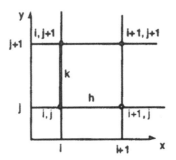

The algebraic approximation to the derivatives in the differential equation is obtained by truncating a series expansion about a grid node - here, a two-dimensional Taylor series expansion, which, in general, can be written using the grid spacing shown above as

$$
\begin{aligned}
f\left[x+h, y+k\right] = {} & f\left[x, y\right] + \left(h\frac{\partial}{\partial x} + k\frac{\partial}{\partial y}\right) f\left[x, y\right] \\
& + \frac{1}{2!}\left(h\frac{\partial}{\partial x} + k\frac{\partial}{\partial y}\right)^2 f\left[x, y\right] \\
& + \frac{1}{3!}\left(h\frac{\partial}{\partial x} + k\frac{\partial}{\partial y}\right)^3 f\left[x, y\right] \\
+ \ldots + {} & \frac{1}{(n-1)!}\left(h\frac{\partial}{\partial x} + k\frac{\partial}{\partial y}\right)^{n-1} f\left[x, y\right] + R_n
\end{aligned}
\tag{7.2.7-94}
$$

with remainder

$$R_n = \frac{1}{(n)!}\left(h\frac{\partial}{\partial x} + k\frac{\partial}{\partial y}\right)^n f\left[x + \xi h, y + \zeta k\right] \qquad 0 \le \xi \le 1; 0 \le \zeta \le 1$$

$$(7.2.7\text{-}95)$$

We first examine the x-direction approximations to the first partial derivative, then to the second partial derivative. The y-direction approximations follow directly by the same method and will therefore not be developed separately.

Forward difference approximation to the first derivative

Expanding in the forward x-direction from the point at (i, j)

$$h = +\Delta x$$
$$k = \Delta y = 0 \qquad\qquad (7.2.7\text{-}96)$$

Substituting in the Taylor series we obtain a form reminiscent of the one-dimensional Taylor series approximation, but incorporating partial rather than ordinary derivatives.

$$f_{i+1,j} = f_{i,j} + \Delta x\left[\frac{\partial f}{\partial x}\right]_{i,j} + \frac{(\Delta x)^2}{2!}\left[\frac{\partial^2 f}{\partial x^2}\right]_{i,j} + \frac{(\Delta x)^3}{3!}\left[\frac{\partial^3 f}{\partial x^3}\right]_{i,j} + \frac{(\Delta x)^4}{4!}\left[\frac{\partial^4 f}{\partial x^4}\right]_{i,j} + \ldots$$

$$(7.2.7\text{-}97)$$

Truncating the series after the linear term

$$f_{i+1,j} \cong f_{i,j} + \Delta x\left[\frac{\partial f}{\partial x}\right]_{i,j} \qquad\qquad (7.2.7\text{-}98)$$

and solving for the first partial derivative, we obtain the **forward difference** approximation

$$\left[\frac{\partial f}{\partial x}\right]_{i,j} = \frac{f_{i+1,j} - f_{i,j}}{\Delta x} + O\left[\Delta x\right] \qquad\qquad (7.2.7\text{-}99)$$

Backward difference approximation to the first derivative

Expanding in the backward x-direction from the point at (i, j)

$$h = -\Delta x$$
$$k = \Delta y = 0 \tag{7.2.7-100}$$

Substituting in the Taylor series we again obtain the usual one-dimensional series approximation

$$f_{i-1,j} = f_{i,j} - \Delta x \left[\frac{\partial f}{\partial x}\right]_{i,j} + \frac{(\Delta x)^2}{2!}\left[\frac{\partial^2 f}{\partial x^2}\right]_{i,j} - \frac{(\Delta x)^3}{3!}\left[\frac{\partial^3 f}{\partial x^3}\right]_{i,j} + \frac{(\Delta x)^4}{4!}\left[\frac{\partial^4 f}{\partial x^4}\right]_{i,j} + \cdots$$

$$\tag{7.2.7-101}$$

Truncating the series after the linear term

$$f_{i-1,j} \cong f_{i,j} - \Delta x \left[\frac{\partial f}{\partial x}\right]_{i,j} \tag{7.2.7-102}$$

and solving for the first partial derivative, we obtain the **backward difference** approximation

$$\left[\frac{\partial f}{\partial x}\right]_{i,j} = \frac{f_{i,j} - f_{i-1,j}}{\Delta x} + O\left[\Delta x\right] \tag{7.2.7-103}$$

Central difference approximation to the first derivative

Subtracting the backward difference series from the forward difference series, but now truncating each series after **three** terms rather than two, we have

$$f_{i+1,j} - f_{i-1,j} = \left[f_{i,j} + \Delta x \left[\frac{\partial f}{\partial x}\right]_{i,j} + \frac{(\Delta x)^2}{2!}\left[\frac{\partial^2 f}{\partial x^2}\right]_{i,j}\right]$$

$$- \left[f_{i,j} - \Delta x \left[\frac{\partial f}{\partial x}\right]_{i,j} + \frac{(\Delta x)^2}{2!}\left[\frac{\partial^2 f}{\partial x^2}\right]_{i,j}\right] + O\left[(\Delta x)^2\right]$$

$$\tag{7.2.7-104}$$

$$f_{i+1,j} - f_{i-1,j} = \Delta x \left[\frac{\partial f}{\partial x}\right]_{i,j} + \Delta x \left[\frac{\partial f}{\partial x}\right]_{i,j} + O\left[(\Delta x)^2\right] \tag{7.2.7-105}$$

Notice that, in the act of subtracting, the third terms negate each other and therefore, fortuitously, we obtain for the central difference a better order of approximation than that derived from either the forward or backward difference form.

$$\left[\frac{\partial f}{\partial x}\right]_{i,j} = \frac{f_{i+1,j} - f_{i-1,j}}{2\,\Delta x} + O\left[(\Delta x)^2\right] \tag{7.2.7-106}$$

Corresponding forms for forward, backward, and central difference approximations are found for the y-direction in identical fashion.

Approximation of second derivative

If, rather than subtracting the backward difference series from the forward difference series, we now **add** the two series, retaining truncation after three terms in each case, we can develop an approximation for the second partial derivative.

$$f_{i+1,j} + f_{i-1,j} = \left[f_{i,j} + \Delta x\left[\frac{\partial f}{\partial x}\right]_{i,j} + \frac{(\Delta x)^2}{2!}\left[\frac{\partial^2 f}{\partial x^2}\right]_{i,j}\right]$$

$$+ \left[f_{i,j} - \Delta x\left[\frac{\partial f}{\partial x}\right]_{i,j} + \frac{(\Delta x)^2}{2!}\left[\frac{\partial^2 f}{\partial x^2}\right]_{i,j}\right] + O\left[(\Delta x)^2\right]$$

$$\tag{7.2.7-107}$$

$$f_{i+1,j} + f_{i-1,j} = 2 f_{i,j} + (\Delta x)^2 \left[\frac{\partial^2 f}{\partial x^2}\right]_{i,j} + O\left[(\Delta x)^2\right] \tag{7.2.7-108}$$

$$\left[\frac{\partial^2 f}{\partial x^2}\right]_{i,j} = \frac{f_{i+1,j} - 2 f_{i,j} + f_{i-1,j}}{(\Delta x)^2} + O\left[(\Delta x)^2\right] \tag{7.2.7-109}$$

Again, a corresponding form for the approximation to the second partial derivative with respect to y is found by the identical method.

Finite difference approximation to the Laplace equation

For a rectangular (L is not necessarily equal to W, even though it has been so drawn) slab with Dirichlet boundary conditions as shown, let us dedimensionalize the partial differential equation and boundary conditions.

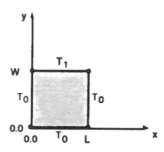

Solution

$$\nabla^2 T = 0 \tag{7.2.7-110}$$

Boundary conditions:

$$
\begin{aligned}
T(0, y) &= T_0 \\
T(L, y) &= T_0 \\
T(x, 0) &= T_0 \\
T(x, W) &= T_1
\end{aligned}
\tag{7.2.7-111}
$$

Dedimensionalize using

$$
\begin{aligned}
\theta &= \frac{T - T_0}{T_1 - T_0} \\
x^* &= \frac{x}{L} \\
y^* &= \frac{y}{W}
\end{aligned}
\tag{7.2.7-112}
$$

This maps the problem into

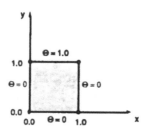

$$\nabla^2\theta = 0 = \frac{\partial^2\theta}{\partial(x^*)^2} + \frac{\partial^2\theta}{\partial(y^*)^2}$$

(7.2.7-113)

with boundary conditions

$$\theta(0, y^*) = 0$$
$$\theta(1, y^*) = 0$$
$$\theta(x^*, 0) = 0$$
$$\theta(x^*, 1) = 1$$

(7.2.7-114)

Combining the above expression for the x-direction with the corresponding approximation for the y-direction, we can develop a finite difference approximation to the Laplace equation (using the dimensionless form for purposes of illustration)

$$\nabla^2\theta = 0 = \frac{\partial^2\theta}{\partial(x^*)^2} + \frac{\partial^2\theta}{\partial(y^*)^2}$$
$$= \frac{\theta_{i+1,j} - 2\theta_{i,j} + \theta_{i-1,j}}{(\Delta x^*)^2} + \frac{\theta_{i,j+1} - 2\theta_{i,j} + \theta_{i,j-1}}{(\Delta y^*)^2} + O\left[(\Delta x)^2\right]$$

(7.2.7-115)

Requiring that $\Delta x^* = \Delta y^* = h^*$

$$\theta_{i+1,j} + \theta_{i,j+1} - 4\theta_{i,j} + \theta_{i-1,j} + \theta_{i,j-1} = 0$$

(7.2.7-116)

This is a linear algebraic equation that can be written for each interior node. The system of equations thus obtained can be solved to obtain the approximate value of Θ at each interior node.

Example 7.2.7-1   Determination of steady-state temperature distribution in a rectangular slab[25]

Apply the dedimensionalized form just obtained to calculation of the steady-state temperature distribution in the slab shown in the following figure.

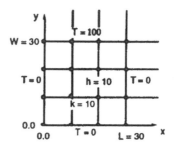

Mapping the problem into dimensionless coordinates

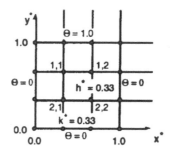

At interior points replace the dimensionless differential equation with

$$\frac{\partial^2 \theta}{\partial (x^*)^2} + \frac{\partial^2 \theta}{\partial (y^*)^2} = \frac{\theta_{i+1,j} - 2\theta_{i,j} + \theta_{i-1,j}}{(h^*)^2} + \frac{\theta_{i,j+1} - 2\theta_{i,j} + \theta_{i,j-1}}{(k^*)^2} = 0$$

(7.2.7-117)

Since h = k

---

[25] Note that this example applies equally well to diffusion

$$\nabla^2 c_A = 0$$

with corresponding boundary conditions and with the thermal conductivity replaced by the diffusivity by using

$$\theta = \frac{c_A - c_{A0}}{c_{A1} - c_{A0}}$$

$$\theta_{i+1,j} + \theta_{i,j+1} - 4\,\theta_{i,j} + \theta_{i-1,j} + \theta_{i,j-1} = 0 \qquad (7.2.7\text{-}118)$$

which can be written in terms of a computational molecule as

$$\left\{ \begin{array}{c} \boxed{\begin{array}{ccc} & 1 & \\ 1 & -4 & 1 \\ & 1 & \end{array}} \end{array} \right\} \theta_{i,j} = 0 \qquad (7.2.7\text{-}119)$$

Apply at interior points

$$
\begin{array}{ll}
\text{at } (1,1): & \theta_{1,2} + 1 \quad -4\,\theta_{1,1} + 0 \quad + \theta_{2,1} = 0 \\
\text{at } (1,2): & 0 \quad +1 \quad -4\,\theta_{1,2} + \theta_{1,1} + \theta_{2,2} = 0 \\
\text{at } (2,1): & \theta_{2,2} + \theta_{1,1} - 4\,\theta_{2,1} + 0 \quad +0 \quad = 0 \\
\text{at } (2,2): & 0 \quad +\theta_{1,2} - 4\,\theta_{2,2} + \theta_{2,1} + 0 \quad = 0
\end{array}
\qquad (7.2.7\text{-}120)
$$

$$
\begin{bmatrix}
-4 & 1 & 1 & 0 \\
1 & -4 & 0 & 1 \\
0 & 1 & -4 & 1 \\
0 & 1 & 1 & -4
\end{bmatrix}
\begin{bmatrix}
\theta_{1,1} \\ \theta_{1,2} \\ \theta_{2,1} \\ \theta_{2,2}
\end{bmatrix}
=
\begin{bmatrix}
-1 \\ -1 \\ 0 \\ 0
\end{bmatrix}
\qquad (7.2.7\text{-}121)
$$

This is a system of linear algebraic equations - here, four equations in four unknowns - which can readily be solved by the application of elementary matrix manipulation by

$$[A][x] = [b] \qquad (7.2.7\text{-}122)$$
$$[A][A]^{-1}[x] = [A]^{-1}[b] \qquad (7.2.7\text{-}123)$$
$$[x] = [A]^{-1}[b] \qquad (7.2.7\text{-}124)$$

This relationship can readily be solved numerically via computer by using software such as EXCEL, MATLAB, MATHCAD, MAPLE, MATHEMATICA, FORTRAN, C, PASCAL, SAS, IMSL, etc.

A				A⁻¹				b	A⁻¹*b
-4	1	1	0	-0.275	-0.1	-0.08	-0.045	-1	0.375
1	-4	0	1	-0.075	-0.3	-0.04	-0.085	-1	0.375
0	1	-4	1	-0.025	-0.1	-0.28	-0.095	0	0.125
0	1	1	-4	-0.025	-0.1	-0.08	-0.295	0	0.125

$$\begin{bmatrix} \theta_{1,1} \\ \theta_{1,2} \\ \theta_{2,1} \\ \theta_{2,2} \end{bmatrix} = \begin{bmatrix} 0.375 \\ 0.375 \\ 0.125 \\ 0.125 \end{bmatrix} \qquad\qquad (7.2.7\text{-}125)$$

One can then recover the nodal estimated temperatures

$$\theta_{1,1} = 0.375 = \frac{(T-0)}{(100-0)} \quad \text{at} \quad \frac{x}{30} = 0.33, \frac{y}{30} = 0.66$$

$$\Rightarrow \ T = 37.5 \qquad \text{at} \quad x = 10, \quad y = 20 \qquad (7.2.7\text{-}126)$$

$$\theta_{1,2} = 0.375 = \frac{(T-0)}{(100-0)} \quad \text{at} \quad \frac{x}{30} = 0.66, \frac{y}{30} = 0.66$$

$$\Rightarrow \ T = 37.5 \qquad \text{at} \quad x = 20, \quad y = 20 \qquad (7.2.7\text{-}127)$$

$$\theta_{2,1} = 0.125 = \frac{(T-0)}{(100-0)} \quad \text{at} \quad \frac{x}{30} = 0.33, \frac{y}{30} = 0.33$$

$$\Rightarrow \ T = 12.5 \qquad \text{at} \quad x = 10, \quad y = 10 \qquad (7.2.7\text{-}128)$$

$$\theta_{2,2} = 0.125 = \frac{(T-0)}{(100-0)} \quad \text{at} \quad \frac{x}{30} = 0.66, \frac{y}{30} = 0.33$$

$$\Rightarrow \ T = 12.5 \qquad \text{at} \quad x = 20, \quad y = 10 \qquad (7.2.7\text{-}129)$$

Irregular boundaries. Dirichlet boundary conditions[26]

One of the problems with the finite difference method is that, for irregularly shaped objects, it is not always possible to make the boundary intersect the grid points that are proximate to the boundary. Consider the problem of using the very coarse grid shown below to compute the solution to the Laplace equation in

---

[26] Neumann (derivative) types of boundary conditions are complicated to handle for irregular boundaries and will not be treated here.

the region indicated. The boundary of the region does not pass though most of the grid points adjacent to the boundary.

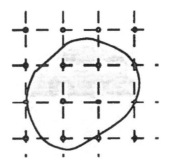

Admittedly, one can make the boundary approach the grid points more closely by using a finer mesh spacing; however, this comes at considerable increased computational time because as the number of nodes increases, so does the number of simultaneous equations that must be solved.

Even with a finer mesh, however, without some sort of mapping into a different space where the region assumes a more regular shape, although the mesh points will approach the boundary more and more closely, true coincidence will not be obtained. Consequently, we seek to develop a method to deal with irregular boundaries.

We shall simultaneously consider all four cases of irregular boundaries, that is, boundaries that pass between grid points rather than through them.

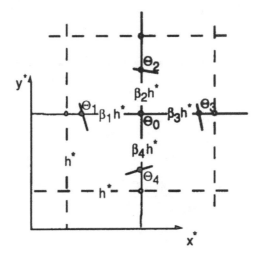

As shown, we consider boundaries that pass at distances of $\beta_1 h$, $\beta_2 h$ $\beta_3 h$ and $\beta_4 h$ ($0 < \beta < 1$) measured from an arbitrary grid point Θ_0. (Open circles denote regular grid points; darkened circles denote points where the boundary intersects the grid.)

We can approximate the first partial derivatives in the x-direction at Θ_0 by using a truncated Taylor expansion

$$\frac{\partial \theta}{\partial x^\bullet}\bigg|_{1 \to 0} = \frac{\theta_0 - \theta_1}{\beta_1 h^\bullet} + O[h^\bullet] \tag{7.2.7-130}$$

$$\frac{\partial \theta}{\partial x^\bullet}\bigg|_{0 \to 3} = \frac{\theta_3 - \theta_0}{\beta_3 h^\bullet} + O[h^\bullet] \tag{7.2.7-131}$$

and in the y-direction by

$$\frac{\partial \theta}{\partial y^\bullet}\bigg|_{4 \to 0} = \frac{\theta_0 - \theta_4}{\beta_4 h^\bullet} + O[h^\bullet] \tag{7.2.7-132}$$

$$\frac{\partial \theta}{\partial y^\bullet}\bigg|_{0 \to 2} = \frac{\theta_2 - \theta_0}{\beta_2 h^\bullet} + O[h^\bullet] \tag{7.2.7-133}$$

We can then use the definition of the second partial derivatives

$$\frac{\partial^2 \theta}{\partial x^{\bullet 2}} = \frac{\partial}{\partial x^\bullet}\left[\frac{\partial \theta}{\partial x^\bullet}\right] \tag{7.2.7-134}$$

$$\frac{\partial^2 \theta}{\partial y^{\bullet 2}} = \frac{\partial}{\partial y^\bullet}\left[\frac{\partial \theta}{\partial y^\bullet}\right] \tag{7.2.7-135}$$

to approximate the second partial derivatives using the differences between the first partial derivatives divided by the appropriate space interval. For the x-direction

$$\frac{\partial^2 \theta}{\partial x^{\bullet 2}} \approx \frac{\dfrac{\theta_3 - \theta_0}{\beta_3 h^\bullet} - \dfrac{\theta_0 - \theta_1}{\beta_1 h^\bullet}}{\frac{1}{2}\left(\beta_3 h^\bullet + \beta_1 h^\bullet\right)}$$

$$\approx \frac{2}{h^{\bullet 2}}\left[\frac{\theta_1 - \theta_0}{\beta_1 \left(\beta_1 + \beta_3\right)} + \frac{\theta_3 - \theta_0}{\beta_3 \left(\beta_1 + \beta_3\right)}\right] \tag{7.2.7-136}$$

The same procedure gives for the y-direction

$$\frac{\partial^2 \theta}{\partial y^{*2}} \approx \frac{2}{h^{*2}} \left[\frac{\theta_2 - \theta_0}{\beta_2 (\beta_2 + \beta_4)} + \frac{\theta_4 - \theta_0}{\beta_4 (\beta_2 + \beta_4)} \right] \qquad (7.2.7\text{-}137)$$

Each of these approximations has error $O[h^*]$. Applying them to the Laplace equation

$$\nabla^2 \theta = \frac{\partial^2 \theta}{\partial (x^*)^2} + \frac{\partial^2 \theta}{\partial (y^*)^2} = 0$$

$$\approx \frac{2}{h^{*2}} \left[\frac{\theta_1 - \theta_0}{\beta_1 (\beta_1 + \beta_3)} + \frac{\theta_3 - \theta_0}{\beta_3 (\beta_1 + \beta_3)} \right] + \frac{2}{h^{*2}} \left[\frac{\theta_2 - \theta_0}{\beta_2 (\beta_2 + \beta_4)} + \frac{\theta_4 - \theta_0}{\beta_4 (\beta_2 + \beta_4)} \right]$$

$$(7.2.7\text{-}138)$$

Rearranging

$$\left[\frac{1}{\beta_1 (\beta_1 + \beta_3)} \right] \theta_1 + \left[\frac{1}{\beta_2 (\beta_2 + \beta_4)} \right] \theta_2$$

$$+ \left[\frac{1}{\beta_3 (\beta_1 + \beta_3)} \right] \theta_3 + \left[\frac{1}{\beta_4 (\beta_2 + \beta_4)} \right] \theta_4 \qquad (7.2.7\text{-}139)$$

$$- \left[\frac{1}{\beta_1 \beta_3} + \frac{1}{\beta_2 \beta_4} \right] \theta_0 = 0$$

We apply this relationship at nodes adjacent to the boundary that do not lie on the boundary. One can think of the weighting factors in terms of a computational module, albeit more complicated than the one used for internal nodes.

$$\left\{ \frac{\left[\dfrac{1}{\beta_1(\beta_1+\beta_3)}\right] \left|\dfrac{\left[\dfrac{1}{\beta_2(\beta_2+\beta_4)}\right]}{-\left[\dfrac{1}{\beta_1\beta_3}+\dfrac{1}{\beta_2\beta_4}\right]}\right| \left[\dfrac{1}{\beta_3(\beta_1+\beta_3)}\right]}{\left[\dfrac{1}{\beta_4(\beta_2+\beta_4)}\right]} \right\} \theta_{i,j} = 0$$

$$(7.2.7\text{-}140)$$

Normal derivative (Neumann) boundary condition at nodal point

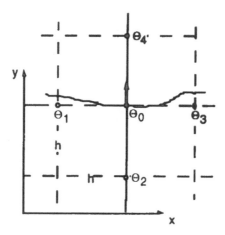

We will treat only boundary conditions that fix normal derivatives (Neumann conditions) at a nodal point on the boundary of our region. We will not consider irregular boundaries in the context of normal derivatives.

If the value of Θ at the boundary point Θ_0 is unknown, and instead a normal derivative C_0 is prescribed at Θ_0 (on the boundary of the region in question), we have an additional unknown, Θ_0, and therefore need an additional independent equation.

To circumvent this problem we introduce a fictitious point, Θ_4, and apply at Θ_0 the usual approximation for the Laplace equation.

$$\theta_3 + \theta_4 - 4\theta_0 + \theta_1 + \theta_2 = 0$$

$$(7.2.7\text{-}141)$$

At first glance this does not seem to have helped the situation much, because although we have written an additional independent equation, we have also introduced another unknown, Θ_4. However, we can use the central difference approximation to the derivative at Θ_0 to substitute for Θ_4.

$$\frac{\partial \theta}{\partial y^{\bullet}} = \text{constant} = C_0 = \frac{\theta_4 - \theta_2}{2 h^{\bullet}} + O\left[h^{\bullet 2}\right] \qquad (7.2.7\text{-}142)$$

$$\theta_2 - \theta_4 + 2 h^{\bullet} C_0 = 0 \qquad (7.2.7\text{-}143)$$

$$\theta_4 = \theta_2 + 2 h^{\bullet} C_0 \qquad (7.2.7\text{-}144)$$

Note that the central difference approximation is $O[h^2]$, the same order as the approximation we used for the second partial derivatives in the Laplace equation. Substituting the approximation at Θ_0 in the Laplace equation

$$\theta_3 + \left(\theta_2 + 2 h^{\bullet} C_0\right) - 4 \theta_0 + \theta_1 + \theta_2 = 0 \qquad (7.2.7\text{-}145)$$

$$\theta_3 + 2 \theta_2 - 4 \theta_0 + \theta_1 = -2 h^{\bullet} C_0 \qquad (7.2.7\text{-}146)$$

We simply append this equation to our set of linear algebraic equations and solve as before.

Generation terms

The Laplace equation describes conduction of thermal energy **without** thermal energy generation. As we saw earlier in this chapter, however, the form of the differential equation incorporating thermal energy generation is not greatly different.

$$\frac{\partial^2 T}{\partial x^2} + \frac{\partial^2 T}{\partial y^2} + \frac{\gamma_\theta}{k} = 0 \qquad (7.2.7\text{-}147)$$

The units of the last term on the left-hand side are

$$\frac{\gamma_\theta}{k} = \frac{\left[\dfrac{\text{energy}}{(\text{length})^3 \, \text{time}}\right]}{\left[\dfrac{\text{energy}}{\text{length time temperature}}\right]} = \left[\frac{\text{temperature}}{(\text{length})^2}\right] \qquad (7.2.7\text{-}148)$$

which matches, as it must, the units on the second partial derivatives of temperature.

$$\frac{\partial^2 T}{\partial x^2} = \frac{\partial}{\partial x}\left(\frac{\partial T}{\partial x}\right) \Rightarrow \left[\frac{1}{L}\right]\left[\frac{T}{L}\right] = \left[\frac{T}{(L)^2}\right] \qquad (7.2.7\text{-}149)$$

Inserting our approximations for $\Delta x = h = \Delta y$

$$\frac{\partial^2 T}{\partial x^2} + \frac{\partial^2 T}{\partial y^2} + \frac{\gamma_\theta}{k} = 0 = \frac{T_{i+1,j} - 2\,T_{i,j} + T_{i-1,j}}{(h)^2} + \frac{T_{i,j+1} - 2\,T_{i,j} + T_{i,j-1}}{(h)^2} + \frac{(\gamma_\theta)_{i,j}}{k}$$

$$(7.2.7\text{-}150)$$

$$0 = T_{i+1,j} - 2\,T_{i,j} + T_{i-1,j} + T_{i,j+1} - 2\,T_{i,j} + T_{i,j-1} + \frac{(h)^2\,(\gamma_\theta)_{i,j}}{k}$$

$$(7.2.7\text{-}151)$$

$$T_{i+1,j} + T_{i,j+1} - 4\,T_{i,j} + T_{i-1,j} + T_{i,j-1} + \frac{(h)^2\,(\gamma_\theta)_{i,j}}{k} = 0 \qquad (7.2.7\text{-}152)$$

Therefore, all that is necessary in order to incorporate thermal energy generation into our numerical solution is to add the specified term at any node where generation is occurring. Notice that the generation can be different at different nodes.

Remember that k in Equation (7.2.7-152) is the thermal conductivity, not Δy. We have assumed that k is constant in the differential equation model that we are using, so k carries no subscripts.[27]

One can, of course, work instead with the dimensionless form of the equation after performing the appropriate transformation. First dedimensionalizing the dependent variable

$$\left(T_1 - T_0\right)\frac{\partial^2\left[\dfrac{T - T_0}{T_1 - T_0}\right]}{\partial x^2} + \left(T_1 - T_0\right)\frac{\partial^2\left[\dfrac{T - T_0}{T_1 - T_0}\right]}{\partial y^2} + \frac{\gamma_\theta}{k} = 0 \qquad (7.2.7\text{-}153)$$

$$\frac{\partial^2\theta}{\partial x^2} + \frac{\partial^2\theta}{\partial y^2} + \frac{\gamma_\theta}{k\left(T_1 - T_0\right)} = 0 \qquad (7.2.7\text{-}154)$$

---

[27] Notice that the finite difference solution lends itself readily to space varying thermal conductivity and thermal energy generation, as well as initial temperature distributions and/or Dirichlet or Neumann boundary conditions that vary in space.

then applying the chain rule to dedimensionalize the independent variables

$$\frac{\partial x^*}{\partial x}\frac{\partial}{\partial x^*}\left[\frac{\partial x^*}{\partial x}\frac{\partial \theta}{\partial x^*}\right] + \frac{\partial y^*}{\partial y}\frac{\partial}{\partial y^*}\left[\frac{\partial y^*}{\partial y}\frac{\partial \theta}{\partial y^*}\right] + \frac{\gamma_\theta}{k(T_1 - T_0)} = 0 \quad (7.2.7\text{-}155)$$

$$\frac{1}{L}\frac{\partial}{\partial x^*}\left[\frac{1}{L}\frac{\partial \theta}{\partial x^*}\right] + \frac{1}{W}\frac{\partial}{\partial y^*}\left[\frac{1}{W}\frac{\partial \theta}{\partial y^*}\right] + \frac{\gamma_\theta}{k(T_1 - T_0)} = 0 \quad (7.2.7\text{-}156)$$

and assuming for purposes of illustration that L = W, we have the dimensionless heat conduction equation, including thermal energy generation

$$\frac{\partial^2 \theta}{\partial(x^*)^2} + \frac{\partial^2 \theta}{\partial(y^*)^2} + \frac{L^2 \gamma_\theta}{k(T_1 - T_0)} = 0 \quad (7.2.7\text{-}157)$$

Defining a dimensionless grid size

$$h^* = \frac{h}{L} \quad (7.2.7\text{-}158)$$

to relate the dimensionless grid to the dimensional grid, and introducing the approximations to the second partial derivatives, we obtain

$$\frac{\theta_{i+1,j} - 2\theta_{i,j} + \theta_{i-1,j}}{(h^*)^2} + \frac{\theta_{i,j+1} - 2\theta_{i,j} + \theta_{i,j-1}}{(h^*)^2} + \frac{L^2(\gamma_\theta)_{i,j}}{k(T_1 - T_0)} = 0 \quad (7.2.7\text{-}159)$$

or

$$\theta_{i+1,j} + \theta_{i,j+1} - 4\theta_{i,j} + \theta_{i-1,j} + \theta_{i,j-1} + \frac{(h^*)^2 L^2(\gamma_\theta)_{i,j}}{k(T_1 - T_0)} = 0 \quad (7.2.7\text{-}160)$$

Example 7.2.7-2   Finite difference solution of 2-D steady-state conduction

The irregular shape shown in (x, y) coordinates with the origin at the lower left-hand corner of the grid has dimensions of 10 mm in both the x- and y-directions, and is 1 mm thick.

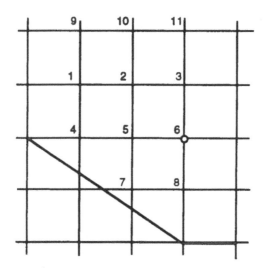

The left-hand vertical side of the object is maintained at 100 K, the right-hand side at 50 K, and the lower side and diagonal side at 75 K. The derivative of the temperature at the top side is maintained at 7 K/mm.

   • The thermal conductivity of the material is 0.6 W/(m K).

   • At point 6 a steady-state thermal energy source (microwave energy) is applied at the rate of 0.05 W/mm^3.

   Calculate the steady-state temperature distribution using the grid shown, assuming the origin to be at the lower left-hand corner of the grid. (In a real-world problem, it is, to be sure, highly unlikely that we would ever use such a coarse grid - here we are interested in the solution from a pedagogical point of view and therefore wish to keep the system of equations small.)

Solution

   We can solve this problem either in dimensional or dimensionless form. The dimensionless method is more useful (both because of its applicability to a more general class of problems as well as the mapping of certain boundary conditions into the convenient values of zero and one).

<u>Solution In Dimensionless Form</u>

   We define the variables

$$\theta = \frac{T-T_0}{T_1-T_0} = \frac{T-50}{100-50}$$

$$x^{\bullet} = \frac{x}{L} = \frac{x}{10}$$

$$y^{\bullet} = \frac{y}{W} = \frac{y}{10} \qquad (7.2.7\text{-}161)$$

$$h^{\bullet} = \frac{h}{10}$$

The differential equation transforms to [cf. Equation (7.2.7-157)]

$$\frac{\partial^2\theta}{\partial(x^{\bullet})^2} + \frac{\partial^2\theta}{\partial(y^{\bullet})^2} + \frac{L^2\gamma_\theta}{k\left(T_1-T_0\right)} = 0 \qquad (7.2.7\text{-}162)$$

Noting that

$$\text{at } T = 100, \theta = \frac{100-50}{100-50} = 1$$

$$\text{at } T = 75, \theta = \frac{75-50}{100-50} = 0.5 \qquad (7.2.7\text{-}163)$$

$$\text{at } T = 50, \theta = \frac{50-50}{100-50} = 0$$

$$\text{at } x = 0, x^{\bullet} = \frac{0}{10} = 0$$

$$\text{at } x = 10, x^{\bullet} = \frac{10}{10} = 1 \qquad (7.2.7\text{-}164)$$

$$\text{at } y = 0, y^{\bullet} = \frac{0}{10} = 0$$

$$\text{at } y = 10, y^{\bullet} = \frac{10}{10} = 1 \qquad (7.2.7\text{-}165)$$

$$h^{\bullet} = \frac{2.5}{10} = 0.25 \qquad (7.2.7\text{-}166)$$

$$\frac{\partial\theta}{\partial y^{\bullet}} = \frac{\partial y}{\partial y^{\bullet}}\frac{\partial\theta}{\partial T}\frac{\partial T}{\partial y} = W\frac{1}{\left(T_1-T_0\right)}\frac{\partial T}{\partial y}$$

$$= (10)\left(\frac{1}{100-50}\right)(7) = 1.4 \qquad (7.2.7\text{-}167)$$

in $(x^{\bullet}, y^{\bullet})$ coordinates with the origin at the lower left corner of the grid, the boundary conditions transform to

$$x^* = 0; 0.5 \le y^* \le 1 \quad \Rightarrow \quad \theta = 1$$
$$x^* = 1; 0 \le y^* \le 1 \quad \Rightarrow \quad \theta = 0$$
$$0 \le x^* \le 0.75; y^* = -\frac{2}{3}x^* + 0.5 \quad \Rightarrow \quad \theta = 0.5 \qquad (7.2.7\text{-}168)$$
$$0.75 \le x^* \le 1; y^* = 0 \quad \Rightarrow \quad \theta = 0.5$$
$$0 \le x^* \le 1; y^* = 1; \quad \Rightarrow \quad \frac{\partial \theta}{\partial y^*} = 1.4$$

so the problem maps into

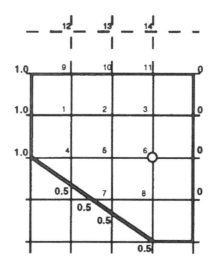

Here, h = 0.25 mm.

At points 1, 2, 3, 5, and 8 we can write the approximation to the Laplace equation using the form of Equation (7.2.7-160), omitting the generation term and substituting the known Dirichlet boundary conditions where appropriate.

$$1 + \theta_9 + \theta_2 + \theta_4 - 4\theta_1 = 0$$
$$\theta_1 + \theta_{10} + \theta_3 + \theta_5 - 4\theta_2 = 0$$
$$\theta_2 + \theta_{11} + 0 + \theta_6 - 4\theta_3 = 0 \qquad (7.2.7\text{-}169)$$
$$\theta_4 + \theta_2 + \theta_6 + \theta_7 - 4\theta_5 = 0$$
$$\theta_7 + \theta_6 + 0 + 0.5 - 4\theta_8 = 0$$

Rearranging

$$\theta_9 + \theta_2 + \theta_4 - 4\theta_1 = -1$$
$$\theta_1 + \theta_{10} + \theta_3 + \theta_5 - 4\theta_2 = 0$$
$$\theta_2 + \theta_{11} + \theta_6 - 4\theta_3 = 0 \qquad\qquad (7.2.7\text{-}170)$$
$$\theta_4 + \theta_2 + \theta_6 + \theta_7 - 4\theta_5 = 0$$
$$\theta_7 + \theta_6 - 4\theta_8 = -0.5$$

At point 6 the difference equation retains the generation term

$$-\frac{\left[h^{\bullet}\right]^2 L^2 \left[\gamma_\theta\right]_6}{k\left(T_1 - T_0\right)} = -\frac{(0.25)^2 (10)^2 \left[mm^2\right](0.05)\left[\frac{W}{mm^3}\right]}{(0.6)\left[\frac{W}{m\,K}\right]\left[\frac{m}{1000\,mm}\right](100-50)\left[K\right]} = -10.42$$

$$(7.2.7\text{-}171)$$

We apply Equation (7.2.7-160)

$$\theta_5 + \theta_3 - 4\theta_6 + 0 + \theta_8 = -10.42$$
$$\theta_5 + \theta_3 - 4\theta_6 + \theta_8 = -10.42 \qquad\qquad (7.2.7\text{-}172)$$

At point 4, $\beta_1 = \beta_2 = \beta_3 = 1$, $\beta_4 = 0.6667$

$$\left[\frac{1}{(1)(1+1)}\right](1) + \left[\frac{1}{(1)(1+0.6667)}\right](\theta_1)$$
$$+ \left[\frac{1}{(1)(1+1)}\right](\theta_5) + \left[\frac{1}{(0.6667)(1+0.6667)}\right](0.5)$$
$$- \left[\frac{1}{(1)(1)} + \frac{1}{(1)(0.6667)}\right]\theta_4 = 0$$

$$0.5 + 0.6\,\theta_1 + 0.5\,\theta_5 + 0.45 - 2.5\,\theta_4 = 0$$
$$\theta_5 - 5\,\theta_4 + 1.2\,\theta_1 = -0.95 \qquad\qquad (7.2.7\text{-}173)$$

At point 7, $\beta_1 = 1/2$, $\beta_2 = 1$, $\beta_3 = 1$, $\beta_4 = 1/3$

$$\left[\frac{1}{(0.5)(0.5+1)}\right](0.5) + \left[\frac{1}{(1)(1+0.3333)}\right](\theta_5)$$

$$+ \left[\frac{1}{(1)(0.5+1)}\right](\theta_8) + \left[\frac{1}{(0.3333)(1+0.3333)}\right](0.5)$$

$$- \left[\frac{1}{(0.5)(1)} + \frac{1}{(1)(0.3333)}\right]\theta_7 = 0$$

$$0.6667 + 0.75\,\theta_5 + 0.6667\,\theta_8 + 1.125 - 5\,\theta_7 = 0$$

$$0.75\,T_5 + 0.6667\,T_8 - 5\,T_7 = -1.792 \qquad\qquad (7.2.7\text{-}174)$$

By adding a fictitious row of points across the top, we generate the following three equations

$$1 + \theta_{12} + \theta_{10} + \theta_1 - 4\,\theta_9 = 0$$

$$\theta_9 + \theta_{13} + \theta_{11} + \theta_2 - 4\,\theta_{10} = 0 \qquad\qquad (7.2.7\text{-}175)$$

$$\theta_{10} + \theta_{14} + 0 + \theta_3 - 4\,\theta_{11} = 0$$

We introduce the Neumann boundary condition via

$$\left.\frac{\partial\theta}{\partial x^*}\right|_9 = \frac{\theta_{12} - \theta_1}{2\,h^*}$$

$$\left.\frac{\partial\theta}{\partial x^*}\right|_{10} = \frac{\theta_{13} - \theta_2}{2\,h^*} \qquad\qquad (7.2.7\text{-}176)$$

$$\left.\frac{\partial\theta}{\partial x^*}\right|_{11} = \frac{\theta_{14} - \theta_3}{2\,h^*}$$

which can be rearranged to

$$\theta_{12} = \theta_1 + 2\,h^* \left.\frac{\partial\theta}{\partial x^*}\right|_9 = \theta_1 + (2)(0.25)(1.4) = \theta_1 + 0.7$$

$$\theta_{13} = \theta_2 + 2\,h^* \left.\frac{\partial\theta}{\partial x^*}\right|_{10} = \theta_2 + (2)(0.25)(1.4) = \theta_2 + 0.7 \quad (7.2.7\text{-}177)$$

$$\theta_{14} = \theta_3 + 2\,h^* \left.\frac{\partial\theta}{\partial x^*}\right|_{11} = \theta_3 + (2)(0.25)(1.4) = \theta_3 + 0.7$$

Substituting gives

$$1 + \theta_1 + 0.7 + \theta_{10} + \theta_1 - 4\,\theta_9 = 0$$
$$\theta_9 + \theta_2 + 0.7 + \theta_{11} + \theta_2 - 4\,\theta_{10} = 0 \qquad (7.2.7\text{-}178)$$
$$\theta_{10} + \theta_3 + 0.7 + 0 + \theta_3 - 4\,\theta_{11} = 0$$

Collecting terms and rearranging

$$2\,\theta_1 + \theta_{10} - 4\,\theta_9 = -1.7$$
$$\theta_9 + 2\,\theta_2 + \theta_{11} - 4\,\theta_{10} = -0.7 \qquad (7.2.7\text{-}179)$$
$$\theta_{10} + 2\,\theta_3 - 4\,\theta_{11} = -0.7$$

Appending these three equations to the equations already developed for points 1, 2, 3, 5, and 8, and for points 4, 6, and 7, gives the set of 11 equations in 11 dimensionless unknowns

$$\theta_9 + \theta_2 + \theta_4 - 4\,\theta_1 = -1$$
$$\theta_1 + \theta_{10} + \theta_3 + \theta_5 - 4\,\theta_2 = 0$$
$$\theta_2 + \theta_{11} + \theta_6 - 4\,\theta_3 = 0$$
$$\theta_4 + \theta_2 + \theta_6 + \theta_7 - 4\,\theta_5 = 0$$
$$\theta_7 + \theta_6 - 4\,\theta_8 = -0.5$$
$$\theta_5 - 5\,\theta_4 + 1.2\,\theta_1 = -0.95$$
$$\theta_5 + \theta_3 - 4\,\theta_6 + \theta_8 = -10.42$$
$$0.75\,T_5 + 0.6667\,T_8 - 5\,T_7 = -1.792$$
$$2\,\theta_1 + \theta_{10} - 4\,\theta_9 = -1.7 \qquad (7.2.7\text{-}180)$$
$$\theta_9 + 2\,\theta_2 + \theta_{11} - 4\,\theta_{10} = -0.7$$
$$\theta_{10} + 2\,\theta_3 - 4\,\theta_{11} = -0.7$$

Writing the system in the form

$$A\,\theta = b \qquad (7.2.7\text{-}181)$$

note that the matrix **A** and its inverse are each unchanged from the dimensional solution.

The vector **b** is now

-1.000
0.000
0.000
0.000
-0.500
-0.950
-10.420
-1.792
-1.700
-0.700
-0.700

Applying

$$A^{-1} A \, \theta = A^{-1} b$$
$$\theta = A^{-1} b$$

(7.2.7-182)

yields the Θ vector as

1.22
1.60
1.71
0.83
1.75
3.79
0.79
1.27
1.46
1.70
1.46

Transforming Θ back to the original **T** coordinate we have for the vector **T**

111
130
136
92
138
239
90
113
123
135
123

7.2.8   One-dimensional unsteady-state conduction

One-dimensional unsteady-state conduction in a medium with constant thermal conductivity and without thermal energy generation is described by a differential equation in one space dimension and time

$$\rho \, C_p \frac{\partial T}{\partial t} = k \frac{\partial^2 T}{\partial x^2} \tag{7.2.8-1}$$

Using the definition of the thermal diffusivity, α

$$\alpha \equiv \frac{k}{\rho \, C_p} \tag{7.2.8-2}$$

gives

$$\frac{\partial T}{\partial t} = \alpha \frac{\partial^2 T}{\partial x^2} \tag{7.2.8-3}$$

We now have moved from a steady-state problem which is defined by an elliptic equation to an unsteady state problem defined by a parabolic equation (because of the presence of the first partial derivative with respect to time). The method of solution of this equation, like most partial differential equations, depends heavily on the type of boundary conditions involved. We will consider a series of examples which will illustrate a variety of methods of solution. For a given problem, more than one of these approaches may work.

We cover the example in rectangular coordinates in detail an example of both the analytical procedure of applying separation of variables techniques followed by series solution and the numerical procedures available. The choice both of ultimate solution method, i.e., analytical or numerical, and the particular implementation of the solution method chosen depend on the particular problem at hand, and, especially, the shape of the boundary and form of the boundary conditions.

Analytical methods for one-dimensional unsteady-state conduction

Semi-infinite slab

Consider a heat conduction in the x-direction in a semi-infinite (bounded only by one face) slab initially at a uniform temperature, T_i, whose face suddenly at time equal to zero is raised to and maintained at T_S. The temperature profile will "penetrate" into the slab with time in the general fashion shown in Figure 7.2.8-1.

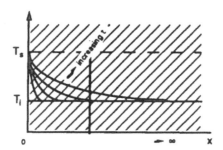

Figure 7.2.8-1 Semi-infinite slab with constant face temperature

The partial differential equation which is the microscopic energy balance

$$\frac{\partial T}{\partial t} = \alpha \frac{\partial^2 T}{\partial x^2}$$

(7.2.8-4)

has boundary conditions

t = 0: T = T_i, all x   (initial condition)
x = 0: T = T_s, all t > 0 (boundary condition)
x = ∞: T = T_i, all t > 0 (boundary condition)

$$(7.2.8\text{-}5)$$

Consider the dedimensionalization of the dependent variable in the equation using

$$T^\bullet = \frac{T - T_i}{T_s - T_i}$$

$$(7.2.8\text{-}6)$$

which will give, dedimensionalizing term by term

$$\frac{\partial T}{\partial t} = \frac{\partial T}{\partial T^\bullet} \frac{\partial T^\bullet}{\partial t} = \left(T_s - T_i\right) \frac{\partial T^\bullet}{\partial t}$$

$$(7.2.8\text{-}7)$$

$$\frac{\partial^2 T}{\partial x^2} = \frac{\partial}{\partial x}\left[\frac{\partial T}{\partial x}\right] = \frac{\partial}{\partial x}\left[\frac{\partial T}{\partial T^\bullet} \frac{\partial T^\bullet}{\partial x}\right] = \left(T_s - T_i\right) \frac{\partial^2 T^\bullet}{\partial x^2}$$

$$(7.2.8\text{-}8)$$

Substituting

$$\left(T_s - T_i\right) \frac{\partial T^\bullet}{\partial t} = \left(T_s - T_i\right) \frac{k}{\rho\, \hat{C}_p} \frac{\partial^2 T^\bullet}{\partial x^2}$$

$$(7.2.8\text{-}9)$$

yields the transformed partial differential equation

$$\frac{\partial T^\bullet}{\partial t} = \alpha \frac{\partial^2 T^\bullet}{\partial x^2}$$

$$(7.2.8\text{-}10)$$

with transformed boundary conditions

t = 0: $T^\bullet$ = 0, all x   (initial condition)
x = 0: $T^\bullet$ = 1, all t > 0 (boundary condition)
x = ∞: $T^\bullet$ = 0, all t > 0 (boundary condition)

$$(7.2.8\text{-}11)$$

This is the analogous problem to that of velocity profiles in a Newtonian fluid adjacent to a wall suddenly set in motion, and to that of certain binary diffusion in a semi-infinite medium. The dimensionless temperature, T^*, is analogous to the dimensionless velocity

$$v_y^{\bullet} = \frac{v_y - v_i}{V - v_i} = \frac{v_y - 0}{V - 0} = \frac{v_y}{V} \qquad (7.2.8\text{-}12)$$

(V being the velocity of the wall) or the dimensionless concentration

$$c_A^{\bullet} = \frac{c_A - c_{Ai}}{c_{As} - c_{Ai}} \qquad (7.2.8\text{-}13)$$

and the thermal diffusivity is analogous to the kinematic viscosity, v, or the binary diffusivity, $\mathcal{D}_{AB}$.

The solution to the differential equation

$$T^{\bullet} = f_1(x, t, \alpha) \qquad (7.2.8\text{-}14)$$

can be written as

$$f_2(T^{\bullet}, x, t, \alpha) = T^{\bullet} - f_1(x, t, \alpha) = 0 \qquad (7.2.8\text{-}15)$$

where we observe that $f_2(T^{\bullet}, x, t, \alpha)$ must be a dimensionally homogeneous function.

Applying the principles of dimensional analysis in a system of two fundamental dimensions (L, t) indicates that we should obtain two dimensionless groups. One of our variables, $T^{\bullet}$, is already dimensionless and therefore serves as the first member of the group of two. A systematic analysis such as we developed in Chapter 5 will give the other dimensionless group as the similarity variable

$$\eta = \frac{x}{\sqrt{4\alpha t}} \qquad (7.2.8\text{-}16)$$

which implies that

$$f_3(T^{\bullet}, \eta) = 0 \qquad (7.2.8\text{-}17)$$

or that

$$T^{\bullet} = \varphi(\eta) \qquad (7.2.8\text{-}18)$$

Transforming the terms in the differential equation we have

$$\frac{\partial T^{\bullet}}{\partial t} = \frac{dT^{\bullet}}{d\eta} \frac{\partial \eta}{\partial t} = \frac{d[\varphi(\eta)]}{d\eta} \frac{\partial\left[\frac{x}{\sqrt{4\alpha t}}\right]}{\partial t}$$

$$= \varphi'\left(\frac{x}{\sqrt{4\alpha}}\right)\left(-\frac{1}{2}t^{-\frac{3}{2}}\right) = \varphi'\left(\frac{x}{\sqrt{4\alpha t}}\right)\left(-\frac{1}{2t}\right)$$

$$= -\frac{1}{2}\frac{\eta}{t}\varphi' \qquad\qquad (7.2.8\text{-}19)$$

$$\frac{\partial^2 T^{\bullet}}{\partial x^2} = \frac{\partial}{\partial x} \frac{\partial T^{\bullet}}{\partial x} = \frac{\partial}{\partial x} \frac{\partial[\varphi(\eta)]}{\partial x} = \frac{\partial \eta}{\partial x} \frac{d}{d\eta}\left\{\frac{d[\varphi(\eta)]}{d\eta} \frac{\partial \eta}{\partial x}\right\}$$

$$= \frac{\partial\left[\frac{x}{\sqrt{4\alpha t}}\right]}{\partial x} \frac{d}{d\eta}\left\{\frac{d[\varphi(\eta)]}{d\eta} \frac{\partial\left[\frac{x}{\sqrt{4\alpha t}}\right]}{\partial x}\right\}$$

$$= \frac{1}{\sqrt{4\alpha t}} \frac{d}{d\eta}\left\{\frac{d[\varphi(\eta)]}{d\eta} \frac{1}{\sqrt{4\alpha t}}\right\}$$

$$= \frac{1}{\sqrt{4\alpha t}} \frac{1}{\sqrt{4\alpha t}} \frac{d}{d\eta}\left\{\frac{d[\varphi(\eta)]}{d\eta}\right\} = \frac{1}{4\alpha t}\varphi' \qquad (7.2.8\text{-}20)$$

Substituting in the differential equation

$$-\frac{1}{2}\frac{\eta}{t}\varphi' = \alpha\left(\frac{1}{4\alpha t}\varphi'\right) \qquad\qquad (7.2.8\text{-}21)$$

Rearranging

$$\varphi'' + 2\eta\varphi' = 0 \qquad\qquad (7.2.8\text{-}22)$$

which leaves us with an ordinary differential equation because of the combining of two independent variables into one via the similarity transformation.

The initial and boundary conditions transform as

$$t = 0 \Rightarrow \eta = \infty: \ T^{\bullet} = 0 \Rightarrow \varphi(\eta) = 0$$
$$x = 0 \Rightarrow \eta = 0: \ T^{\bullet} = 1 \Rightarrow \varphi(\eta) = 1 \qquad (7.2.8\text{-}23)$$
$$x = \infty \Rightarrow \eta = \infty: \ T^{\bullet} = 0 \Rightarrow \varphi(\eta) = 0$$

which gives the transformed boundary conditions as

$$\eta = \infty: \quad \varphi(\eta) = 0$$
$$\eta = 0: \quad \varphi(\eta) = 1 \tag{7.2.8-24}$$

Equation (7.2.8-22) is easily solved via the simple expedient of substituting ψ for ϕ' which renders it separable

$$\psi' + 2\eta\psi = 0 \tag{7.2.8-25}$$

$$\frac{d\psi}{d\eta} + 2\eta\psi = 0 \tag{7.2.8-26}$$

$$\int \frac{d\psi}{\psi} = -\int 2\eta\,d\eta \tag{7.2.8-27}$$

$$\ln\psi = -\eta^2 + \ln C_1 \tag{7.2.8-28}$$

$$\psi = C_1\,e^{-\eta^2} \tag{7.2.8-29}$$

Back substituting for ψ

$$\varphi' = C_1\,e^{-\eta^2}$$
$$\int \varphi'\,d\eta = C_1 \int e^{-\eta^2}d\eta \tag{7.2.8-30}$$
$$\varphi = C_1 \int_0^\eta e^{-\zeta^2}d\zeta + C_2$$

As we saw in the corresponding momentum transfer example in Chapter 6, the integral here is one that we cannot evaluate in closed form.[28]

Introducing the boundary conditions and substituting the appropriate value for the definite integrals[29]

---

[28] Engineers have a certain regrettable proclivity to use the same symbol for the dummy variable of integration that appears in the integrand and for the variable in the limits. This often does not cause confusion; here, however, it is clearer to make the distinction.

[29] The error function plays a key role in statistics. We do not have space here to show the details of the integration of the function to obtain the tables listed; for further details the reader is referred to any of the many texts in mathematical statistics.

$$1 = C_1 \int_0^0 e^{-\zeta^2} d\zeta + C_2 \quad \Rightarrow \quad C_2 = 1 - (C_1)(0) = 1$$

$$0 = C_1 \int_0^\infty e^{-\zeta^2} d\zeta + 1 \quad \Rightarrow \quad C_1 = -\frac{1}{\displaystyle\int_0^\infty e^{-\zeta^2} d\zeta} = -\frac{2}{\sqrt{\pi}} \qquad (7.2.8\text{-}31)$$

Substituting the constants in the equation and re-ordering the terms gives

$$\varphi = 1 - \frac{2}{\sqrt{\pi}} \int_0^\eta e^{-\zeta^2} d\zeta \qquad (7.2.8\text{-}32)$$

We defined the error function and the complementary error function in Chapter 6

$$\text{error function} = \text{erf}(\eta) \equiv \frac{2}{\sqrt{\pi}} \int_0^\eta e^{-\zeta^2} d\zeta$$

$$\text{complementary error function} = \text{erfc}(\eta) \equiv 1 - \text{erf}(\eta) \qquad (7.2.8\text{-}33)$$

allowing us to write

$$\varphi = 1 - \text{erf}[\eta]$$
$$= \text{erfc}[\eta] \qquad (7.2.8\text{-}34)$$

The solution of our problem is then

$$T^* = 1 - \text{erf}\left[\frac{x}{\sqrt{4\,\alpha\,t}}\right]$$
$$= \text{erfc}\left[\frac{x}{\sqrt{4\,\alpha\,t}}\right] \qquad (7.2.8\text{-}35)$$

A table of values of the error function is listed in Appendix B.

Example 7.2.8-1   Semi-infinite slab: conduction in a brick wall

A brick furnace wall, so thick as to be essentially infinite in the z-direction, is initially at a uniform temperature of 400°F

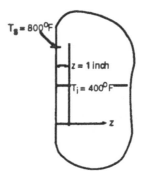

At time t = 0, a very high velocity jet of hot gas at a temperature of 800°F impinges upon the inside wall of the furnace. The heat transfer from the jet of hot gas to the wall is good enough that the convective resistance can be neglected, and the wall surface assumed to be maintained at 800°F.

The properties of the wall are listed in the following table.

Property	Value
Thermal conductivity[30]	1 Btu/(hr ft^2 °F)
Density[31]	110 lbmass/ft^3
Heat Capacity[32]	0.2 Btu/(lbmass °F)

What is the temperature one inch from the wall surface after five minutes?

Solution

The thermal diffusivity is

$$\alpha = \frac{k}{\rho \, \hat{C}_p}$$

[30] Perry, R. H. and D. W. Green, Eds. (1984). *Chemical Engineer's Handbook*. New York, McGraw-Hill, p. 3-146.

[31] Perry, R. H. and D. W. Green, Eds. (1984). *Chemical Engineer's Handbook*. New York, McGraw-Hill, p. 3-95.

[32] Perry, R. H. and D. W. Green, Eds. (1984). *Chemical Engineer's Handbook*. New York, McGraw-Hill, p. 3-128.

$$= \frac{(1)\left[\frac{Btu}{hr\ ft\ °F}\right]}{(110)\left[\frac{lbmass}{ft^3}\right](0.2)\left[\frac{Btu}{lbmass\ °F}\right]}\left[\frac{hr}{(3600)\ s}\right]\left[\frac{(144)\ in^2}{ft^2}\right]$$

$$= 1.818 \times 10^{-3}\ \frac{in^2}{s} \tag{7.2.8-36}$$

From the error function solution for a semi-infinite slab we have that

$$\theta = \frac{T - T_i}{T_s - T_i} = 1 - erf\left(\frac{z}{\sqrt{4\,\alpha\,t}}\right)$$

$$= 1 - erf\left(\frac{(1)\left[in\right]}{\sqrt[\cdot]{(4)\left(1.818 \times 10^{-3}\right)\left[\frac{in^2}{s}\right](300)\left[s\right]}}\right)$$

$$= 1 - erf\left(0.677\right) \tag{7.2.8-37}$$

Interpolating linearly from the error function

$$\frac{\left[erf\left(0.677\right) - erf\left(0.64\right)\right]}{\left(0.667 - 0.64\right)} = \frac{\left[erf\left(0.68\right) - erf\left(0.64\right)\right]}{\left(0.68 - 0.64\right)}$$

$$erf\left(0.677\right) = \frac{\left[erf\left(0.68\right) - erf\left(0.64\right)\right]}{\left(0.68 - 0.64\right)}\left(0.667 - 0.64\right) + erf\left(0.64\right)$$

$$erf\left(0.677\right) = \frac{\left[0.66378 - 0.63459\right]}{\left(0.68 - 0.64\right)}\left(0.667 - 0.64\right) + 0.63459$$

$$erf\left(0.677\right) = 0.654 \tag{7.2.8-38}$$

Substituting

$$\frac{T - T_i}{T_s - T_i} = \frac{T - 400}{800 - 400} = 1 - erf\left(0.677\right) = 1 - 0.654$$

$$T = 538°F \tag{7.2.8-39}$$

Finite slab

Now let us again consider solutions to the equation

$$\frac{\partial T}{\partial t} = \alpha \frac{\partial^2 T}{\partial x^2} \tag{7.2.8-40}$$

but with boundary conditions that describe conduction in a **finite** (rather than **semi-infinite**) slab initially at T_i which at time $t = 0$ has each face suddenly raised to and maintained at T_s, as shown in Figure 7.2.8-2.

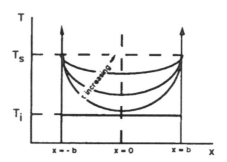

Figure 7.2.8-2 Finite slab with constant face temperatures

The appropriate initial and boundary conditions are

$$
\begin{array}{llll}
t = 0: & T = T_i, & -b \le x \le b & \text{(initial condition)} \\
x = -b: & T = T_s, & \text{all } t > 0 & \text{(boundary condition)} \\
x = b: & T = T_s, & \text{all } t > 0 & \text{(boundary condition)}
\end{array} \tag{7.2.8-41}
$$

We shall illustrate using the rectangular coordinate problem posed in Equations (7.2.8-40) and (7.2.8-41). The usual analytic solution of this problem relies upon a two step process: (a) separation of variables, followed by (b) power series solution.

First, we will dedimensionalize the differential equation and boundary conditions using the following dimensionless variables[33]

---

[33] Note that this is a different dimensionless temperature that we used in the case of the semi-infinite slab. However, the two dimensionless variables are simply related:

$$\frac{T - T_s}{T_i - T_s} = \frac{T - T_i + T_i - T_s}{T_i - T_s} = 1 - \frac{T - T_i}{T_s - T_i}$$

$$\theta = \frac{T - T_s}{T_i - T_s} \quad \Rightarrow \text{ dimensionless temperature}$$

$$\eta = \frac{x}{b} \quad \Rightarrow \begin{array}{l} \text{dimensionless distance} \\ \text{(not a similarity variable: contains} \\ \text{only one independent variable)} \end{array}$$

$$\tau = \frac{\alpha t}{b^2} \quad \Rightarrow \begin{array}{l} \text{dimensionless time} \\ \text{(also known as the} \\ \text{Fourier Number, Fo)} \end{array} \qquad (7.2.8\text{-}42)$$

Transforming to the dimensionless form of the derivatives

$$\frac{\partial T}{\partial t} = \frac{\partial \tau}{\partial t} \frac{\partial \theta}{\partial \tau} \frac{\partial T}{\partial \theta} = \frac{\alpha}{b^2} \frac{\partial \theta}{\partial \tau} (T_i - T_s) = \frac{\alpha (T_i - T_s)}{b^2} \frac{\partial \theta}{\partial \tau} \qquad (7.2.8\text{-}43)$$

$$\frac{\partial^2 T}{\partial x^2} = \frac{\partial}{\partial x} \left[\frac{\partial T}{\partial x} \right] = \frac{\partial \eta}{\partial x} \frac{\partial}{\partial \eta} \left[\frac{\partial \eta}{\partial x} \frac{\partial \theta}{\partial \eta} \frac{\partial T}{\partial \theta} \right]$$

$$= \frac{1}{b} \frac{\partial}{\partial \eta} \left[\frac{1}{b} \frac{\partial \theta}{\partial \eta} (T_i - T_s) \right] = \frac{(T_i - T_s)}{b^2} \frac{\partial^2 \theta}{\partial \eta^2} \qquad (7.2.8\text{-}44)$$

Substituting in the equation

$$\frac{\alpha (T_i - T_s)}{b^2} \frac{\partial \theta}{\partial \tau} = \alpha \frac{(T_i - T_s)}{b^2} \frac{\partial^2 \theta}{\partial \eta^2}$$

$$\frac{\partial \theta}{\partial \tau} = \frac{\partial^2 \theta}{\partial \eta^2} \qquad (7.2.8\text{-}45)$$

The dimensional initial and boundary conditions map into dimensionless conditions as

$$\left[t = 0 \Rightarrow \tau = 0 \right] \quad \left[T = T_i \Rightarrow \theta = 1 \right] \quad \left[-b \le x \le b \Rightarrow -1 \le \eta \le 1 \right]$$

$$\left[x = -b \Rightarrow \eta = -1 \right] \left[T = T_s \Rightarrow \theta = 0 \right] \left[\text{all } t > 0 \Rightarrow \text{all } \tau > 0 \right]$$

$$\left[x = b \Rightarrow \eta = 1 \right] \left[T = T_s \Rightarrow \theta = 0 \right] \left[\text{all } t > 0 \Rightarrow \text{all } \tau > 0 \right]$$

$$(7.2.8\text{-}46)$$

which gives the dimensionless initial and boundary conditions as

$$\tau = \quad 0: \quad \theta = 1, \quad -1 \le \eta \le 1$$
$$\eta = -1: \quad \theta = 0, \quad \text{all } \tau > 0$$
$$\eta = \quad 1: \quad \theta = 0, \quad \text{all } \tau > 0 \qquad (7.2.8\text{-}47)$$

We now assume that the variables can be separated (we must verify this assumption later); that is, that we can write[34]

$$\theta(\eta, \tau) = \zeta(\eta)\,\xi(\tau) \qquad (7.2.8\text{-}48)$$

where ζ and ξ are functions, respectively, of only η and τ. Applying this assumption to the differential equation, and noting that the partial derivative sign can be replaced by the ordinary derivative in the case of a function of only one variable (we write the independent variable in the first line to remind ourselves of this fact) yields

$$\frac{\partial\left[\zeta(\eta)\,\xi(\tau)\right]}{\partial\tau} = \frac{\partial^2\left[\zeta(\eta)\,\xi(\tau)\right]}{\partial\eta^2}$$

$$\zeta\frac{d\xi}{d\tau} = \xi\frac{d^2\zeta}{d\eta^2}$$

$$\frac{1}{\xi}\frac{d\xi}{d\tau} = \frac{1}{\zeta}\frac{d^2\zeta}{d\eta^2} \qquad (7.2.8\text{-}49)$$

But here we have the left-hand side, a function only of ζ, equal to the right-hand side, a function only of ξ. The only way that this can be true if ζ and ξ are free to take arbitrary values is if they are mutually equal to some constant. For convenience in later manipulation we choose to designate this constant as $(-c^2)$.

$$\frac{1}{\xi}\frac{d\xi}{d\tau} = \frac{1}{\zeta}\frac{d^2\zeta}{d\eta^2} = -c^2 \qquad (7.2.8\text{-}50)$$

This reduces our problem to the solution of two ordinary differential equations

---

[34] This is different from the case of a **separable equation** - that is, one where the variables can be separated, one to the left-hand side of the equation and the other to the right-hand side.

$$\frac{1}{\xi}\frac{d\xi}{d\tau} = -c^2$$

$$\frac{1}{\zeta}\frac{d^2\zeta}{d\eta^2} = -c^2 \qquad\qquad (7.2.8\text{-}51)$$

The initial condition applies to the first equation, the two boundary conditions to the second.

The first equation is separable and therefore easily integrated

$$\int \frac{d\xi}{\xi} = -c^2 \int d\tau$$

$$\ln\xi - \ln K_1 = -c^2\,\tau \qquad\qquad (7.2.8\text{-}52)$$

$$\xi = K_1\,e^{-c^2\tau}$$

The second equation is a linear, homogeneous, second-order ordinary differential equation with constant coefficients and likewise can be integrated easily[35]

$$\frac{d^2\zeta}{d\eta^2} + c^2\,\zeta = 0 \qquad\qquad (7.2.8\text{-}53)$$

The characteristic equation is

$$\lambda^2 + c^2 = 0 \qquad\qquad (7.2.8\text{-}54)$$

which has the complex conjugate roots

$$\lambda = \pm\,ic \qquad\qquad (7.2.8\text{-}55)$$

giving the solution

$$\zeta = K_2\cos\left(c\,\eta\right) + K_3\sin\left(c\,\eta\right) \qquad\qquad (7.2.8\text{-}56)$$

---

[35] For example, see Kreyszig, E. (1993). *Advanced Engineering Mathematics*. New York, NY, John Wiley and Sons, Inc., p. 74.

The problem is symmetric in space, which means that θ and therefore ζ must be an even[36] function of η. This implies that $K_3 = 0$, since the sine is an odd function. This leaves[37]

$$\zeta = K_2 \cos\left(c\,\eta\right) \qquad (7.2.8\text{-}57)$$

Application of the first boundary condition (the other could equally well be used) gives[38]

$$0 = K_2 \cos\left(c\right) \qquad (7.2.8\text{-}58)$$

We observe that $K_2 = 0$ is not an admissible solution because we would lose all variation of temperature in space. Therefore,

$$\begin{aligned} 0 &= \cos\left(c\right) \\ c &= \left(n+\tfrac{1}{2}\right)\pi \;\Leftarrow n \text{ an integer} \end{aligned} \qquad (7.2.8\text{-}59)$$

which gives as a particular solution to our original partial differential equation

$$\theta\left(n\right) = K_1\left(n\right) K_2\left(n\right) \exp\left[-\left(n+\tfrac{1}{2}\right)^2 \pi^2\, \tau\right] \cos\left[\left(n+\tfrac{1}{2}\right)\pi\,\eta\right] \qquad (7.2.8\text{-}60)$$

where the notation reminds us that K_1, K_2 and thereby θ may be functions of n.

If we now attempt to satisfy the initial condition using this solution we have, where we have replaced the product $(K_1\, K_2)$ by K_4

$$1 = K_4\left(n\right) \cos\left[\left(n+\tfrac{1}{2}\right)\pi\,\eta\right] \qquad (7.2.8\text{-}61)$$

---

[36] An even function of x, y(x), is one such that $f(-x) = f(x)$. Conversely, an odd function is one for which $f(-x) = -f(x)$.

[37] The recognition of ζ as an even function amounts to using one of the two boundary conditions.

[38] $\theta = 0$ implies that $\zeta = 0$, because the condition on θ holds for arbitrary values of $\tau > 0$, and we are not interested in the trivial case for which ξ is identically zero.

But since n is an integer and K_4 a constant, we cannot satisfy this relationship for arbitrary η. In essence we are attempting to represent an arbitrary function with only one term of a (Fourier) series.

Recognizing the fact that the sum of solutions is also a solution offers us a way around the difficulty. Adding all the solutions for $-\infty \le n \le \infty$

$$\theta = \sum_{n=-\infty}^{\infty} K_4(n) \exp\left[-\left(n+\tfrac{1}{2}\right)^2 \pi^2 \tau\right] \cos\left[\left(n+\tfrac{1}{2}\right)\pi\,\eta\right] \quad (7.2.8\text{-}62)$$

Using the fact that $\cos(-\beta) = \cos(\beta)$, and defining $K_5(n) = K_4(n) + K_4(-[n+1])$ we can change the range on the sum[39]

$$\theta = \sum_{n=0}^{\infty} K_5(n) \exp\left[-\left(n+\tfrac{1}{2}\right)^2 \pi^2 \tau\right] \cos\left[\left(n+\tfrac{1}{2}\right)\pi\,\eta\right] \quad (7.2.8\text{-}63)$$

We can then apply the initial condition to obtain (now ceasing to write the reminder that K_5 is a function of n)

$$1 = \sum_{n=0}^{\infty} K_5 \cos\left[\left(n+\tfrac{1}{2}\right)\pi\,\eta\right] \quad (7.2.8\text{-}64)$$

where we recognize this as the Fourier cosine series representation. We now need to determine the coefficients. To do this we use the orthogonality of the cosine function.

If we multiply both sides of the above relation by $\cos(m + 1/2)\,\pi\,\eta$ and integrate from $\eta = -1$ to $\eta = +1$

---

[39] In doing this we pair the terms

n	$-(n+1)$
0	-1
1	-2
2	-3
etc.	etc.

Note that since n is a dummy variable, we can continue to use the same symbol despite the change in range.

$$\int_{-1}^{1} \cos\left[\left(m+\tfrac{1}{2}\right)\pi\,\eta\right] d\eta$$

$$= \sum_{n=0}^{\infty} K_s \int_{-1}^{1} \cos\left[\left(m+\tfrac{1}{2}\right)\pi\,\eta\right] \cos\left[\left(n+\tfrac{1}{2}\right)\pi\,\eta\right] d\eta \qquad (7.2.8\text{-}65)$$

The left-hand side is in a form easily integrated

$$\frac{1}{\left(m+\tfrac{1}{2}\right)\pi} \int_{-1}^{1} \cos\left[\left(m+\tfrac{1}{2}\right)\pi\,\eta\right] d\left[\left(m+\tfrac{1}{2}\right)\pi\,\eta\right] = \frac{\sin\left[\left(m+\tfrac{1}{2}\right)\pi\,\eta\right]}{\left(m+\tfrac{1}{2}\right)\pi}\Bigg|_{\eta=-1}^{\eta=+1}$$

$$= \frac{(-1)^m - (-1)^{m+1}}{\left(m+\tfrac{1}{2}\right)\pi} = \frac{(-1)^m\left[1-(-1)\right]}{\left(m+\tfrac{1}{2}\right)\pi} = \frac{2(-1)^m}{\left(m+\tfrac{1}{2}\right)\pi}$$

$$(7.2.8\text{-}66)$$

Orthogonality of the cosine function over the interval (-1, 1) implies that all the integrals on the right-hand side vanish except for m = n

$$\frac{1}{\left(m+\tfrac{1}{2}\right)\pi} \int_{-1}^{1} \cos^2\left[\left(m+\tfrac{1}{2}\right)\pi\,\eta\right] d\left[\left(m+\tfrac{1}{2}\right)\pi\,\eta\right]$$

$$= \frac{1}{\left(m+\tfrac{1}{2}\right)\pi} \int_{-\left(m+\tfrac{1}{2}\right)\pi}^{+\left(m+\tfrac{1}{2}\right)\pi} \cos^2 x \, dx$$

$$= \frac{1}{\left(m+\tfrac{1}{2}\right)\pi} \left[\frac{x}{2} + \frac{\sin(2x)}{4}\right]_{-\left(m+\tfrac{1}{2}\right)\pi}^{+\left(m+\tfrac{1}{2}\right)\pi}$$

$$= 1 + \frac{1}{4\left(m+\tfrac{1}{2}\right)\pi} \left\{\sin\left[(2m+1)\pi\right] - \sin\left[-(2m+1)\pi\right]\right\}$$

$$= 1 \qquad\qquad (7.2.8\text{-}67)$$

Substituting

$$\frac{2(-1)^m}{\left(m+\frac{1}{2}\right)\pi} = \left(K_s\right)_m \tag{7.2.8-68}$$

and using this result in the expression for the solution

$$\theta = 2\sum_{n=0}^{\infty}\frac{(-1)^n}{\left(n+\frac{1}{2}\right)\pi} \exp\left[-\left(n+\frac{1}{2}\right)^2 \pi^2 \tau\right]\cos\left[\left(n+\frac{1}{2}\right)\pi\eta\right] \tag{7.2.8-69}$$

which in terms of the original variables is

$$\frac{T-T_s}{T_1-T_s} = 2\sum_{n=0}^{\infty}\frac{(-1)^n}{\left(n+\frac{1}{2}\right)\pi} \exp\left[-\left(n+\frac{1}{2}\right)^2 \pi^2 \frac{\alpha t}{b^2}\right]\cos\left[\left(n+\frac{1}{2}\right)\pi\frac{x}{b}\right]$$

$$\tag{7.2.8-70}$$

These solutions are plotted in Figure 7.2.8-3. Because of the large slope of the curves at short times, the graphical presentation is relatively of little use. However, the solution at very short times can be approximated quite well by the solution for the semi-infinite slab model for one-dimensional unsteady-state conductive heat transfer, since the profiles have not "penetrated" very far into the slab; hence, the effect of the opposite wall is quite small and the first wall in essence "sees" an infinite medium.

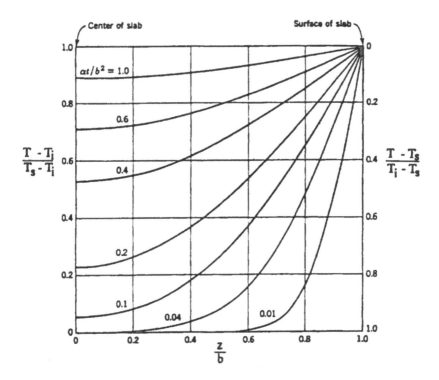

Figure 7.2.8-3 Unsteady-state heat transfer in a finite slab with uniform initial temperature and constant, equal surface temperatures[40]

Example 7.2.8-2   Finite slab model vs. semi-infinite slab model for one-dimensional unsteady-state conductive heat transfer

A brick wall 12 in. thick is initially at a uniform temperature of 100°F. At t = 0, a temperature of 500°F is applied and maintained at both surfaces.

[40] Carslaw, H. S. and J. C. Jaeger (1959). *Conduction of Heat in Solids*, Oxford, p. 200, as presented in Bird, R. B., W. E. Stewart, et al. (1960). *Transport Phenomena*. New York, Wiley, p. 357.

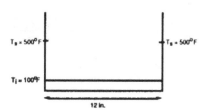

$T_s = 500°F$ $T_s = 500°F$

$T_i = 100°F$

12 in.

Properties of the brick may be taken as independent of temperature and with the following values

Property	Value
k Btu/(hr ft °F)	1
ρ lbm/ft^3	110
C_p Btu/(lbm °F)	0.2

a) Assuming that the wall is a **semi-infinite** slab, calculate and plot the temperature at 0.25-inch intervals for a total distance of 7 inches from the left-hand face at times t = 10 s, 180 s, 900 s, and 9000 s.

b) Treat the wall as a **finite** slab and using the **graphical** trigonometric series solution Figure 7.2.8-3 calculate the temperature at the following values of (t, z), where z is the distance from the left-hand face

(10, 0.25)   (180, 1.0)   (900, 2.0)   (9000, 6.0)

Plot on the same graph constructed in (a).

c) Calculate the value of temperature for the points (10, 0.25) and (9000, 6.0) using the **analytical** trigonometric series solution and plot the result vs. the number of terms used.

Solution

(a)

$$\alpha = \frac{k}{\rho\, \hat{C}_p} = (1)\left[\frac{Btu}{hr\,ft\,°F}\right]\left(\frac{1}{110}\right)\left[\frac{ft^3}{lbm}\right]\left(\frac{1}{0.2}\right)\left[\frac{lbm\,°F}{Btu}\right]$$
$$\times\left(\frac{1}{3600}\right)\left[\frac{hr}{s}\right](144)\left[\frac{in^2}{ft^2}\right]$$

$$\alpha = 1.818\times 10^{-3}\ \frac{in^2}{s} \tag{7.2.8-71}$$

$$\theta = \frac{T-T_i}{T_s-T_i} = 1-erf[\eta] = 1-erf\left[\frac{z}{\sqrt{4\,\alpha\, t}}\right] \tag{7.2.8-72}$$

$$T = (T_s-T_i)\left(1-erf\left[\frac{z}{\sqrt{4\,\alpha\, t}}\right]\right)+T_i$$

$$T = (500-100)\,[°F]\left(1-erf\left[\frac{z\,[in]}{\sqrt{(4)\left(1.818\times 10^{-3}\right)\left[\frac{in^2}{s}\right]t\,[s]}}\right]\right)+100\ °F \tag{7.2.8-73}$$

The calculations are readily carried out by a spreadsheet program. Results are plotted below.

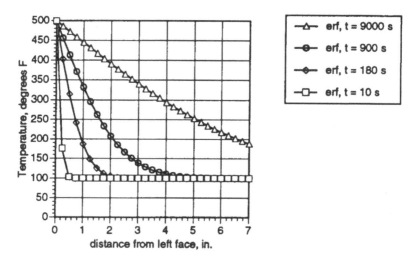

At longer and longer times this solution becomes less and less useful. In particular, because of symmetry we know that the derivative must go to zero at the centerline, which for this model it clearly does not at long times. At short times, since the influence of the other wall is minimal, the derivative goes essentially to zero by the time the center line is reached, so the model is valid.

(b)

For short times the graph (finite slab) is essentially useless because the slope of the solution is nearly vertical, making the temperature read extremely sensitive to the distance from the face. (The point at 10 s is for all practical purposes unreadable from the graph.) Fortunately, these are the times when the error function solution (semi-infinite slab) is the most useful.

Additional results are plotted below together with the results of part (a).

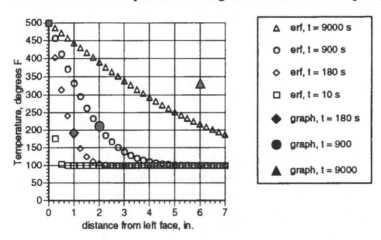

At intermediate times the graph (finite slab) and the error function (semi-infinite slab) give essentially the same answer.

For long times the graph (finite slab) gives a far superior answer because the error function solution (semi-infinite slab) does not take into account the influence of the other (right-hand) wall. This is reflected in the solution from the graph (finite slab) giving a higher temperature, since this model accounts for heat conducted into the slab from the right-hand wall as well as from the left-hand wall.

(c) Using the solution

$$\frac{T-T_s}{T_i-T_s} = 2 \sum_{n=0}^{\infty} \frac{(-1)^n}{\left(n+\frac{1}{2}\right)\pi} \exp\left[-\left(n+\frac{1}{2}\right)^2 \pi^2 \frac{\alpha t}{b^2}\right] \cos\left[\left(n+\frac{1}{2}\right)\pi \frac{x}{b}\right]$$

(7.2.8-74)

For t = 10 s

y	t	n	last term	sum
5.75	10	0	33.27	466.73
5.75	10	1	32.75	433.98
5.75	10	2	31.74	402.24
5.75	10	3	30.27	371.97
5.75	10	4	28.42	343.55
5.75	10	5	26.26	317.29
5.75	10	6	23.86	293.43
5.75	10	7	21.33	272.10
5.75	10	8	18.74	253.36
5.75	10	9	16.19	237.17
5.75	10	10	13.73	223.44
5.75	10	11	11.43	212.01
5.75	10	12	9.33	202.68
5.75	10	13	7.46	195.22
5.75	10	14	5.83	189.39
...	...	...	...	...
5.75	10	22	0.18	175.29
5.75	10	23	0.05	175.24
5.75	10	24	-0.03	175.27
5.75	10	25	-0.08	175.35
5.75	10	26	-0.09	175.44
5.75	10	27	-0.09	175.54
...	...	...	...	...
5.75	10	35	-0.01	175.90
5.75	10	36	-0.01	175.91
5.75	10	37	-0.01	175.92
5.75	10	38	0.00	175.92

For t = 9000 s

y	t	n	last term	sum
0	9000	0	165.94	334.06
0	9000	1	-0.01	334.07
0	9000	2	0.00	334.07
0	9000	3	0.00	334.07

Notice the speed of convergence varies greatly with the time - at 10 s, about 40 terms were required to achieve the same convergence as was obtained with only two terms at 9000 s.

We also can apply numerical methods to the solution of these equations.

Infinite cylinder and sphere

We can perform similar solutions for the case of surfaces with T = constant in cylindrical geometry and T = constant in spherical geometry. The difference is that following the separation of variables step in the cylindrical case, the second-order ordinary differential equation obtained is solved by a series in Bessel functions, and in the spherical case, by Legendre polynomials.

Bessel functions and Legendre polynomials are functions defined by infinite series just as the ordinary trigonometric functions (sines, cosines, etc.) can be defined by infinite series. They can be tabulated and plotted in the same manner as the trigonometric functions. The series forms are readily adaptable for use with numerical solutions via computers. We shall not go into the details of the series solutions here; rather, we present them in graphical form[41] in Figures 7.2.8-4 and 7.2.8-5.

---

[41] For details of various other power series solutions see, for example, Ozisik, M. N. (1968). *Boundary Value Problems of Heat Conduction*, International Textbook Company, Crank, J. (1975). *The Mathematics of Diffusion*, Oxford, or Carslaw, H. S. and J. C. Jaeger (1959). *Conduction of Heat in Solids*, Oxford.

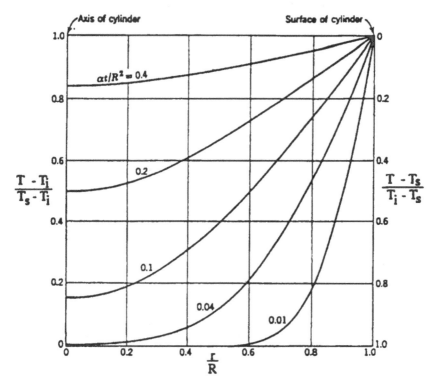

Figure 7.2.8-4 Unsteady-state heat transfer in an infinite
cylinder with uniform initial temperature and constant surface
temperature[42]

[42] Carslaw, H. S. and J. C. Jaeger (1959). *Conduction of Heat in Solids*, Oxford, p.
200, as presented in Bird, R. B., W. E. Stewart, et al. (1960). *Transport Phenomena*.
New York, Wiley, p. 357.

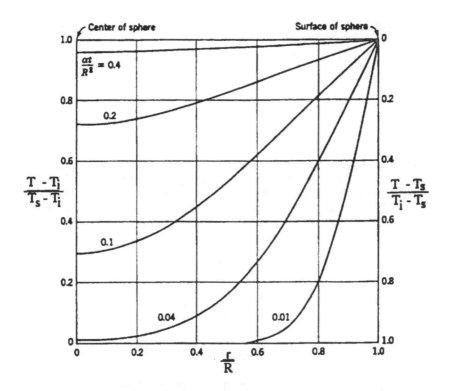

Figure 7.2.8-5 Unsteady-state heat transfer in a sphere with uniform initial temperature and constant surface temperature[43]

Numerical Methods for One-Dimensional Unsteady-State Conduction

The solution of unsteady-state heat conduction problems in one space dimension and one time dimension can be accomplished readily by numerical methods.

Consider the unsteady-state heat transfer problem in one space variable

$$\rho \, \bar{C}_p \frac{\partial T}{\partial t} = k \frac{\partial^2 T}{\partial x^2} \tag{7.2.8-75}$$

---

[43] Carslaw, H. S. and J. C. Jaeger (1959). *Conduction of Heat in Solids*, Oxford, p. 234, as presented in Bird, R. B., W. E. Stewart, et al. (1960). *Transport Phenomena*. New York, Wiley, p. 358.

$$\frac{\partial T}{\partial t} = \frac{k}{\rho\, C_p} \frac{\partial^2 T}{\partial x^2}$$

$$(7.2.8\text{-}76)$$

$$\frac{\partial T}{\partial t} = \alpha \frac{\partial^2 T}{\partial x^2}$$

$$(7.2.8\text{-}77)$$

Finite difference method

Finite difference explicit form

We approximate the conduction partial derivatives and the boundary conditions using the same technique based on Taylor series expansion as for steady-state conduction. Superscripts refer to the time step, subscripts to the space step. We now must define a grid in both space and time. We show the grid in Figure 7.2.8-6 as square; this need not necessarily be so.

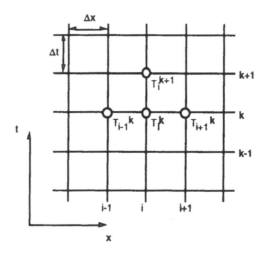

Figure 7.2.8-6 Finite difference grid for 1D unsteady-state conduction

Using the Taylor series approximations developed earlier with superscripts (k) to denote time steps (Δt) and subscripts (i) to denote space steps (Δz), using the forward difference at the ith space step for the time derivative, gives

$$\frac{\partial T}{\partial t} = \frac{T_i^{k+1} - T_i^k}{\Delta t} + O[\Delta t]$$

$$(7.2.8\text{-}78)$$

and the central difference at the kth time step for the space derivative

$$\frac{\partial^2 T}{\partial x^2} = \frac{T_{i+1}^k - 2\,T_i^k + T_{i-1}^k}{(\Delta x)^2} + O\!\left[(\Delta x)^2\right] \tag{7.2.8-79}$$

We have used different orders of approximation for t and x, but will receive certain benefits in the form obtained. Substituting in the equation gives

$$\frac{T_i^{k+1} - T_i^k}{\Delta t} = \alpha\,\frac{T_{i+1}^k - 2\,T_i^k + T_{i-1}^k}{(\Delta x)^2} \tag{7.2.8-80}$$

In the above representation, unknown quantity is at the (k + 1)st time step, known quantities at the kth time step. Solving for the unknown in terms of the knowns

$$T_i^{k+1} = \frac{\alpha\,\Delta t}{(\Delta x)^2}\,T_{i+1}^k + \left(1 - 2\,\frac{\alpha\,\Delta t}{(\Delta x)^2}\right) T_i^k + \frac{\alpha\,\Delta t}{(\Delta x)^2}\,T_{i-1}^k \tag{7.2.8-81}$$

To simplify we define

$$\boxed{\lambda \equiv \frac{\alpha\,\Delta t}{(\Delta x)^2}} \tag{7.2.8-82}$$

and write

$$T_i^{k+1} = \lambda\,T_{i+1}^k + (1 - 2\lambda)\,T_i^k + \lambda\,T_{i-1}^k \tag{7.2.8-83}$$

This is an explicit form for the temperature at the ith space coordinate, (k+1)st time coordinate, in terms of the three values at the kth time coordinate as shown in Figure 7.2.8-6. Using this form one can therefore calculate the values of temperature at the next time step one-by-one, using the three known temperatures at the previous time step as indicated.

If one knows the initial temperature distribution [(0)th time step], one can one-by-one calculate all the internal temperatures at the first time step. The left-most and the right-most (surface, boundary) temperatures are known for all time steps because they constitute the Dirichlet conditions. After determining all the temperatures at the first time step, one can repeat the process for the second time step, and so forth.

Unfortunately, the explicit method is both convergent and stable[44] only for

> Explicit form: both convergent and stable
> $$\lambda \leq \frac{1}{2}$$

(7.2.8-84)

and may oscillate, although the errors will not grow

> Explicit form: may oscillate with stable error
> $$\frac{1}{4} \leq \lambda \leq \frac{1}{2}$$

(7.2.8-85)

The practical implication of these restrictions is that one does not have simultaneous free choice of the fineness of both the time and space grids. One normally wants as fine a space grid as possible, in order to get precise temperature profiles as a function of time. However, the finer the grid in space, the finer the time step must be in order to preserve convergence and stability. If temperature profiles at long times are required, an inconveniently large number of steps may have to be taken.

Example 7.2.8-3   Unsteady-state heat transfer by explicit finite differences

Consider the case of a slab one foot in thickness, initially at a uniform temperature of 400°F, whose thermal diffusivity is 1.82×10^{-3} in^2/s. At time equal or greater than zero, the faces of the slab are raised to and maintained at 700°F.

> a) Use a value of $\Delta t = 30$ s and $\Delta x = 1$ inch in conjunction with the explicit method to plot the temperature history one inch from the face of the slab.

> b) Use a value of $\Delta t = 300$ s and $\Delta x = i$ inch in conjunction with the explicit method to plot the temperature history one inch from the face of the slab.

---

[44] For a more detailed discussion, see Chapra, S. C. and R. P. Canale (1988). *Numerical Methods for Engineers*. New York, McGraw-Hill. Gerald, C. F. and P. O. Wheatley (1994). *Applied Numerical Analysis*. Reading, MA, Addison-Wesley, and Carnahan, B., H. A. Luther, et al. (1969). *Applied Numerical Methods*. New York, John Wiley and Sons, Inc.

Solution

We can easily adapt the explicit method for spreadsheet calculation. We do not show the calculation, which is straightforward.

a) Note that for this case $\lambda = 0.05454$, well within the range of convergence and stability.

b) Note that for this case $\lambda = 0.5454$, outside of the range of convergence and stability.

Plots of the two cases are shown below for the first 1500 seconds. Note that the case $\Delta T = 100$ s is converging smoothly toward the steady-state value of 700°F, while the case $\Delta T = 300$ s is oscillating.

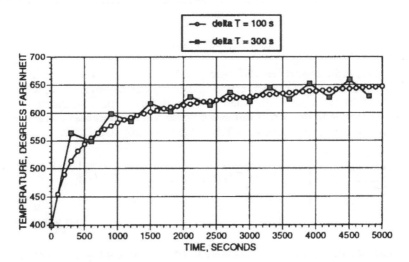

If we continue case (b) further in time, we see that it is indeed unstable, as shown in the following figure.

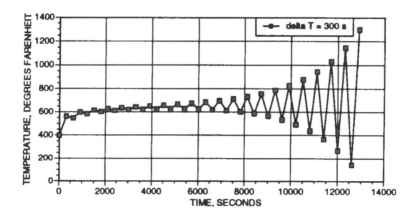

Finite difference implicit form

Implicit finite conduction difference methods can be used to overcome the stability problems of the explicit method. The Crank-Nicolson method, which is unconditionally stable for all values of λ, uses for the approximation of the second derivative in the partial differential equation the **arithmetic average** of the derivative approximations at the kth and (k+1)st time steps

$$\left[\frac{T_i^{k+1} - T_i^k}{\Delta t}\right] = \frac{\alpha}{2}\left[\frac{T_{i+1}^k - 2T_i^k + T_{i-1}^k}{(\Delta z)^2} + \frac{T_{i+1}^{k+1} - 2T_i^{k+1} + T_{i-1}^{k+1}}{(\Delta z)^2}\right]$$

$$(7.2.8\text{-}86)$$

It is therefore also second-order accurate in both space and time, as opposed to the explicit method, which was only first-order accurate in time.

Solving as before for the temperatures at the next time step

$$T_i^{k+1} - T_i^k = \frac{\lambda}{2}\left[T_{i+1}^k - 2T_i^k + T_{i-1}^k + T_{i+1}^{k+1} - 2T_i^{k+1} + T_{i-1}^{k+1}\right] \quad (7.2.8\text{-}87)$$

$$T_i^{k+1} + \lambda T_i^{k+1} = \frac{\lambda}{2}\left[T_{i+1}^k + T_{i-1}^k + T_{i+1}^{k+1} + T_{i-1}^{k+1}\right] + T_i^k - \lambda T_i^k \quad (7.2.8\text{-}88)$$

$$2(1+\lambda) T_i^{k+1} = \lambda\left[T_{i+1}^k + T_{i-1}^k + T_{i+1}^{k+1} + T_{i-1}^{k+1}\right] + 2(1-\lambda) T_i^k$$

$$(7.2.8\text{-}89)$$

$$-\lambda\, T_{i+1}^{k+1} + 2\left(1+\lambda\right) T_i^{k+1} - \lambda\, T_{i-1}^{k+1} \;=\; \lambda\, T_{i+1}^{k} + 2\left(1-\lambda\right) T_i^{k} + \lambda\, T_{i-1}^{k}$$

$$(7.2.8\text{-}90)$$

Notice that we now have three temperatures at the (k+1)st time step on the left-hand side of this equation.

Assume a problem with N interior points (i = 1, 2, ... N). For cases where the value of the index i denotes a point **not** immediately adjacent to the boundary (surface), i.e., i = (2, 3, ..., N-1), all three of these temperatures are unknown. For the two cases where the value of the index i denotes a point immediately adjacent to the boundary, i.e., i = 1 or i = N, one knows the temperature

$$T_0^{k+1}$$

$$(7.2.8\text{-}91)$$

in the equation at T_1 or the temperature

$$T_{N+1}^{k+1}$$

$$(7.2.8\text{-}92)$$

in the equation at T_N.

If we think of writing the Crank-Nicolson equation for points successively from left to right, this means that as we write the general equation

$$-\lambda\, T_{i+1}^{k+1} + 2\left(1+\lambda\right) T_i^{k+1} - \lambda\, T_{i-1}^{k+1} \;=\; \lambda\, T_{i+1}^{k} + 2\left(1-\lambda\right) T_i^{k} + \lambda\, T_{i-1}^{k}$$

$$(7.2.8\text{-}93)$$

for points interior to the slab, we generate for the point

$$T_1^{k+1}$$

$$(7.2.8\text{-}94)$$

only two unknowns

$$T_1^{k+1} \text{ and } T_2^{k+1}$$

$$(7.2.8\text{-}95)$$

because we know the value of

$$T_0^{k+1}$$

$$(7.2.8\text{-}96)$$

as a result of the Dirichlet boundary conditions.

As we continue to write equations for points not immediately adjacent to the boundary, each equation contains three unknowns, but two of these unknowns are common to previously written equations.

When we reach the equation for the point immediately adjacent to the right-hand boundary, we generate no additional unknowns, because the values of (assuming N interior points)

$$T_{N-1}^{k+1} \text{ and } T_N^{k+1} \qquad\qquad (7.2.8\text{-}97)$$

have already appeared as unknowns in the previous equation and the Dirichlet condition furnishes the value for

$$T_{N+1}^{k+1} \qquad\qquad (7.2.8\text{-}98)$$

Consequently, we generate a system of N linear equations in N unknowns at each time step, which system can be solved using a variety of well-known techniques. Our solution therefore proceeds by solution of a **system** of equations at each time step, rather than a series of solutions of **single** equations within each time step as was the case for the explicit method.

Writing the Crank-Nicolson algorithm for a general time step at each successive interior point, we have first for point 1

$$-\lambda\,T_2^{k+1} + 2\left(1+\lambda\right)T_1^{k+1} - \lambda\,T_0^{k+1} = \lambda\,T_2^k + 2\left(1-\lambda\right)T_1^k + \lambda\,T_0^k$$
$$(7.2.8\text{-}99)$$

Rearranging so that variables are on the left, constants on the right, we have at point 1

$$-\lambda\,T_2^{k+1} + 2\left(1+\lambda\right)T_1^{k+1} = \lambda\,T_2^k + 2\left(1-\lambda\right)T_1^k + \lambda\,T_0^k + \lambda\,T_0^{k+1}$$
$$(7.2.8\text{-}100)$$

At point 2 we do not need any rearrangement to separate the variables and constants to opposite sides of the equation

$$-\lambda\,T_3^{k+1} + 2\left(1+\lambda\right)T_2^{k+1} - \lambda\,T_1^{k+1} = \lambda\,T_3^k + 2\left(1-\lambda\right)T_2^k + \lambda\,T_1^k$$
$$(7.2.8\text{-}101)$$

nor do we at point N-1

$$-\lambda\, T_{N-2}^{k+1} + 2\left(1+\lambda\right) T_{N-1}^{k+1} - \lambda\, T_{N}^{k+1} = \lambda\, T_{N-2}^{k} + 2\left(1-\lambda\right) T_{N-1}^{k} + \lambda\, T_{N}^{k}$$

$$(7.2.8\text{-}102)$$

but at point N we again must rearrange because of the known boundary condition

$$2\left(1+\lambda\right) T_{N}^{k+1} - \lambda\, T_{N-1}^{k+1} = \lambda\, T_{N+1}^{k} + 2\left(1-\lambda\right) T_{N}^{k} + \lambda\, T_{N-1}^{k} + \lambda\, T_{N+1}^{k+1}$$

$$(7.2.8\text{-}103)$$

Assembling these equations, taking as an example the case of a grid of five elements in the x-direction (4 equations in 4 unknowns - $T_1{}^{k+1}$ ... $T_4{}^{k+1}$) gives in matrix form

$$\begin{bmatrix} 2\left(1+\lambda\right) & -\lambda & 0 & 0 \\ -\lambda & 2\left(1+\lambda\right) & -\lambda & 0 \\ 0 & -\lambda & 2\left(1+\lambda\right) & -\lambda \\ 0 & 0 & -\lambda & 2\left(1+\lambda\right) \end{bmatrix} \begin{bmatrix} T_1^{k+1} \\ T_2^{k+1} \\ T_3^{k+1} \\ T_4^{k+1} \end{bmatrix}$$

$$= \begin{bmatrix} 2\left(1-\lambda\right) & \lambda & 0 & 0 \\ \lambda & 2\left(1-\lambda\right) & \lambda & 0 \\ 0 & \lambda & 2\left(1-\lambda\right) & \lambda \\ 0 & 0 & \lambda & 2\left(1-\lambda\right) \end{bmatrix} \begin{bmatrix} T_1^{k} \\ T_2^{k} \\ T_3^{k} \\ T_4^{k} \end{bmatrix}$$

$$+ \begin{bmatrix} \lambda\left(T_0^{k} + T_0^{k+1}\right) \\ 0 \\ 0 \\ \lambda\left(T_5^{k} + T_5^{k+1}\right) \end{bmatrix} \qquad (7.2.8\text{-}104)$$

The matrices above are **tridiagonal**, a form suited to rapid machine computation.

This set of equations is of the form

$$[A] T^{k+1} = [C] T^k + b \qquad\qquad (7.2.8\text{-}105)$$

We can solve, for example, by multiplying by the inverse

$$T^{k+1} = [A]^{-1} [A] T^{k+1} = [A]^{-1} \left\{ [C] T^k + b \right\} \qquad (7.2.8\text{-}106)$$

Everything on the right-hand side is known at step k. Setting k = 0 permits calculation of all unknown temperatures for k = 1.

Repeating the procedure for the next time step does nothing more than increment the superscript of all temperatures by one. The temperatures at the first time step now become the knowns, and the temperatures at the second time step the unknowns.

No new inversion of the A matrix is necessary for different k, since the A matrix is independent of k, as are the C matrix and the b vector. Extension to further time steps should be obvious.

Example 7.2.8-4 Finite slab unsteady-state heat transfer by finite differences

Consider unsteady-state conduction in a 2-mm thick slab in which the initial temperature is 300 K, and both faces are instantly exposed to and maintained at a constant temperature of 500 K for times ≥ 0. Assume the thermal diffusivity to be 9.6 x 10^{-5} cm^2 / s, and use increments of $\Delta x = 0.4$ mm and $\Delta t = 10$ s to calculate the temperatures to t = 40 s. Plot the temperatures at each x grid point from x = 0 (left-hand face) to x = 5 vs. time.

Solution

$$\lambda = \frac{\left(9.6 \times 10^{-5}\right) \frac{cm^2}{s} (10)\, s}{(0.4)^2\, mm^2} \frac{(10)^2\, mm^2}{cm^2} = 0.6 \qquad (7.2.8\text{-}107)$$

Since $\lambda > 0.5$, we cannot use an explicit method because of instability; therefore, we apply Crank-Nicolson.

The general Crank-Nicolson algorithm is

$$-\lambda\,\theta_{k+1,i+1} + 2\left(1+\lambda\right)\theta_{k+1,i} - \lambda\,\theta_{k+1,i-1} = \lambda\,\theta_{k,i+1} + 2\left(1-\lambda\right)\theta_{k,i} + \lambda\,\theta_{k,i-1}$$

$$(7.2.8\text{-}108)$$

With $\lambda = 0.6$ (note this would give instability in an explicit solution)

$$-0.6\,\theta_{k+1,i+1} + 2\left(1+0.6\right)\theta_{k+1,i} - 0.6\,\theta_{k+1,i-1} =$$
$$0.6\,\theta_{k,i+1} + 2\left(1-0.6\right)\theta_{k,i} + 0.6\,\theta_{k,i-1} \qquad (7.2.8\text{-}109)$$

$$-0.6\,\theta_{k+1,i+1} + 3.2\,\theta_{k+1,i} - 0.6\,\theta_{k+1,i-1} =$$
$$0.6\,\theta_{k,i+1} + 0.8\,\theta_{k,i} + 0.6\,\theta_{k,i-1} \qquad (7.2.8\text{-}110)$$

Numbering the grid as shown below

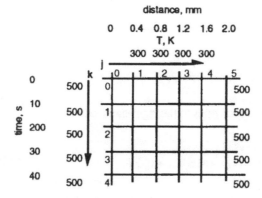

Applying the Crank-Nicolson relation at point $(0,1)$ - note that subscripts reflect [time, distance]

$$-0.6\,T_{12} + 3.2\,T_{1,1} - 0.6\,T_{1,0} =$$
$$0.6\,T_{02} + 0.8\,T_{0,1} + 0.6\,T_{00} \qquad (7.2.8\text{-}111)$$

Moving known quantities to the right-hand side

$$-0.6\,T_{12} + 3.2\,T_{1,1} =$$
$$0.6\,T_{02} + 0.8\,T_{0,1} + 0.6\,T_{0,0} + 0.6\,T_{1,0} \qquad (7.2.8\text{-}112)$$
$$-0.6\,T_{12} + 3.2\,T_{1,1} =$$
$$0.6\,T_{02} + 0.8\,T_{0,1} + 0.6\left(500\right) + 0.6\left(500\right) \qquad (7.2.8\text{-}113)$$

Notice that the last two terms remain constant for any value of the time subscript (Dirichlet boundary condition).

Applying the Crank-Nicolson relation at point $(0,2)$

$$-0.6\, T_{1,3} + 3.2\, T_{1,2} - 0.6\, T_{1,1} =$$
$$0.6\, T_{0,3} + 0.8\, T_{0,2} + 0.6\, T_{0,1} \qquad (7.2.8\text{-}114)$$

This time there are no known temperatures on the left-hand side.

Applying the Crank-Nicolson relation at point $(0,3)$

$$-0.6\, T_{1,4} + 3.2\, T_{1,3} - 0.6\, T_{1,2} =$$
$$0.6\, T_{0,4} + 0.8\, T_{0,3} + 0.6\, T_{0,2} \qquad (7.2.8\text{-}115)$$

Applying the Crank-Nicolson relation at point $(0,4)$

$$-0.6\, T_{1,5} + 3.2\, T_{1,4} - 0.6\, T_{1,3} =$$
$$0.6\, T_{0,5} + 0.8\, T_{0,4} + 0.6\, T_{0,3} \qquad (7.2.8\text{-}116)$$

Moving known quantities to the right-hand side

$$3.2\, T_{1,4} - 0.6\, T_{1,3} =$$
$$0.6\, T_{0,5} + 0.8\, T_{0,4} + 0.6\, T_{0,3} + 0.6\, T_{1,5} \qquad (7.2.8\text{-}117)$$

Again notice that two terms remain constant for any value of the time subscript (Dirichlet boundary condition)

$$3.2\, T_{1,4} - 0.6\, T_{1,3} =$$
$$0.6\,(500) + 0.8\, T_{0,4} + 0.6\, T_{0,3} + 0.6\,(500) \qquad (7.2.8\text{-}118)$$

Assembling these equations (4 equations in 4 unknowns: $T_{1,1}$ ... $T_{1,4}$) gives in matrix form

$$
\begin{bmatrix}
3.2 & -0.6 & 0 & 0 \\
-0.6 & 3.2 & -0.6 & 0 \\
0 & -0.6 & 3.2 & -0.6 \\
0 & 0 & -0.6 & 3.2
\end{bmatrix}
\begin{bmatrix}
T_{1,1} \\
T_{1,2} \\
T_{1,3} \\
T_{1,4}
\end{bmatrix}
$$

$$
=
\begin{bmatrix}
0.8 & 0.6 & 0 & 0 \\
0.6 & 0.8 & 0.6 & 0 \\
0 & 0.6 & 0.8 & 0.6 \\
0 & 0 & 0.6 & 0.8
\end{bmatrix}
\begin{bmatrix}
T_{0,1} \\
T_{0,2} \\
T_{0,3} \\
T_{0,4}
\end{bmatrix}
+
\begin{bmatrix}
(2)(0.6)(500) \\
0 \\
0 \\
(2)(0.6)(500)
\end{bmatrix}
\qquad (7.2.8\text{-}119)
$$

This relationship can be written for a general time step because of the constant boundary conditions

$$
\begin{bmatrix}
3.2 & -0.6 & 0 & 0 \\
-0.6 & 3.2 & -0.6 & 0 \\
0 & -0.6 & 3.2 & -0.6 \\
0 & 0 & -0.6 & 3.2
\end{bmatrix}
\begin{bmatrix}
T_{k+1,1} \\
T_{k+1,2} \\
T_{k+1,3} \\
T_{k+1,4}
\end{bmatrix}
$$

$$
=
\begin{bmatrix}
0.8 & 0.6 & 0 & 0 \\
0.6 & 0.8 & 0.6 & 0 \\
0 & 0.6 & 0.8 & 0.6 \\
0 & 0 & 0.6 & 0.8
\end{bmatrix}
\begin{bmatrix}
T_{k,1} \\
T_{k,2} \\
T_{k,3} \\
T_{k,4}
\end{bmatrix}
+
\begin{bmatrix}
(2)(0.6)(500) \\
0 \\
0 \\
(2)(0.6)(500)
\end{bmatrix}
\qquad (7.2.8\text{-}120)
$$

The relationship is of the form

$$
[A]x = [C]d + e \qquad (7.2.8\text{-}121)
$$

We can solve by multiplying by the inverse

$$
x = [A]^{-1}[A]x = [A]^{-1}\left\{[C]d + e\right\}[A]^{-1}[C]d + [A]^{-1}e
$$
$$
x = [A]^{-1}[C]d + [A]^{-1}e
$$
$$
x = [D]d + [E] \qquad (7.2.8\text{-}122)
$$

At each step everything on the right-hand side is known.

The inverse can be found using a spreadsheet program or other computer software, as can the results for the subsequent calculations.

A matrix

3.2	-0.6	0	0
-0.6	3.2	-0.6	0
0	-0.6	3.2	-0.6
0	0	-0.6	3.2

Inverse matrix

0.3243	0.0631	0.0123	0.0023
0.0631	0.3366	0.0654	0.0123
0.0123	0.0654	0.3366	0.0631
0.0023	0.0123	0.0631	0.3243

The results of this calculation are plotted in the following figure.

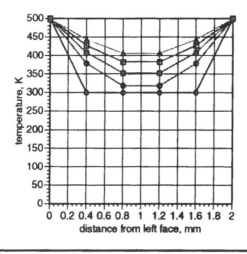 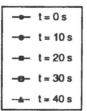

7.2.9 Multi-dimensional unsteady-state conduction

Analytical solution for regular geometries

We state without proof that the solution to multi-dimensional unsteady-state conduction problems can be obtained by the method of separation of variables, and, further, that this separation of variables leads to the possibility of expressing the solution to the multidimensional problem in regular geometries as the product of the one-dimensional solutions in each of the coordinate directions.

We define notation for the one-dimensional solutions as follows, where θ is the dimensionless temperature, T_i is the initial temperature, and T_∞ is the steady-state temperature (i.e., the surface temperature for Dirichlet boundary conditions)

- semi-infinite slab
$$\theta_s(z, t) = \frac{T(z, t) - T_\infty}{T_i - T_\infty} \qquad (7.2.9\text{-}1)$$

- finite slab
$$\theta_F(z, t) = \frac{T(z, t) - T_\infty}{T_i - T_\infty} \qquad (7.2.9\text{-}2)$$

- infinite cylinder
$$\theta_C(r, t) = \frac{T(r, t) - T_\infty}{T_i - T_\infty} \qquad (7.2.9\text{-}3)$$

We can then synthesize the solutions for a variety of regular solids by multiplying these solutions. All we have to decide for each coordinate direction is a) whether to use the rectangular or cylindrical form, and then b) if the rectangular form is called for, whether to use the semi-infinite or the finite form. For example, to obtain the solution for a semi-infinite cylinder, i.e., a cylinder bounded on one end, we observe that there are two coordinates of importance, r and z. In the r-coordinate direction we use the cylindrical form; in the z-coordinate the system rectangular, and further is semi-infinite; therefore, we obtain the solution by multiplying

$$\theta(z, r, t) = \theta_s(z, t)\, \theta_C(r, t) \qquad (7.2.9\text{-}4)$$

Figure 7.2.9-1 illustrates this procedure.

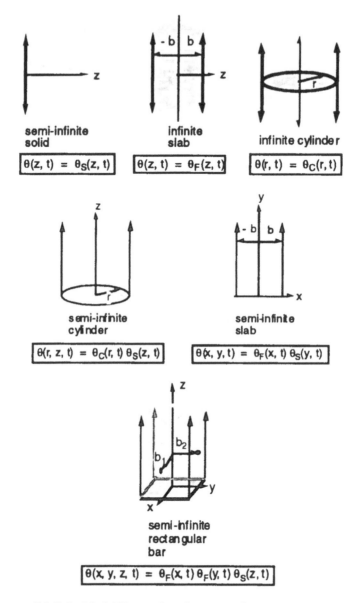

Figure 7.2.9-1 Multidimensional unsteady-state temperature profiles for conduction in regular geometries expressed as product of one-dimensional solutions

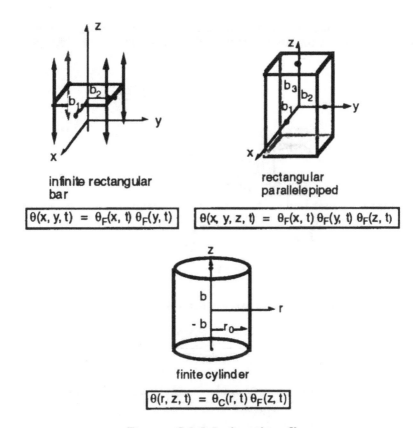

infinite rectangular
bar

$$\theta(x, y, t) = \theta_F(x, t)\,\theta_F(y, t)$$

rectangular
parallelepiped

$$\theta(x, y, z, t) = \theta_F(x, t)\,\theta_F(y, t)\,\theta_F(z, t)$$

finite cylinder

$$\theta(r, z, t) = \theta_C(r, t)\,\theta_F(z, t)$$

Figure 7.2.9-1 (continued)

Numerical solution of two-dimensional unsteady-state conduction

The solution of unsteady-state heat conduction problems in two space dimensions and one time dimension can be accomplished readily by numerical means. The form of the heat conduction equation is

$$\frac{\partial \theta}{\partial t^\bullet} = \alpha \left[\frac{\partial^2 \theta}{\partial x^{\bullet 2}} + \frac{\partial^2 \theta}{\partial y^{\bullet 2}} \right] \tag{7.2.9-5}$$

Finite difference method

A common method is the Alternating Direction Implicit (ADI) algorithm.

The ADI method breaks the time step in half, and alternates solution for unknown values of the dependent variable between the x and y directions as shown in Figure 7.2.9-2.

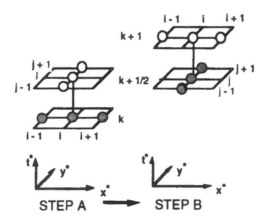

Figure 7.2.9-2 Alternating direction implicit method

For the first half of the time step (STEP A in Figure 7.2.9-2), using the subscripts i for the x-direction, j for the y-direction, and k for the time dimension

$$\frac{\theta_{i,j}^{k+1/2} - \theta_{i,j}^{k}}{\Delta t^* / 2} = \alpha \left[\frac{\theta_{i+1,j}^{k} - 2\theta_{i,j}^{k} + \theta_{i-1,j}^{k}}{(\Delta x^*)^2} + \frac{\theta_{i,j+1}^{k+1/2} - 2\theta_{i,j}^{k+1/2} + \theta_{i,j-1}^{k+1/2}}{(\Delta y^*)^2} \right]$$

$$(7.2.9\text{-}6)$$

If we require that $\Delta x^* = \Delta y^*$, and rearrange so that the unknowns are on the left-hand side and the knowns are on the right-hand side, defining

$$\lambda \equiv \frac{\alpha \Delta t^*}{(\Delta x^*)^2} \qquad (7.2.9\text{-}7)$$

we have

$$-\lambda \theta_{i,j+1}^{k+1/2} + 2(1+\lambda)\theta_{i,j}^{k+1/2} - \lambda \theta_{i,j-1}^{k+1/2} = \lambda \theta_{i+1,j}^{k} + 2(1-\lambda)\theta_{i,j}^{k} + \lambda \theta_{i-1,j}^{k}$$

$$(7.2.9\text{-}8)$$

As is evident, this is a tridiagonal system of equations, with the unknown values running in the x-direction shown in Figure 7.2.9-2 as solid dots, and the unknowns running in the y-direction as unfilled dots.

For the second half of the time step (STEP B in Figure 7.2.9-2), the algorithm becomes

$$\frac{\theta_{i,j}^{k+1} - \theta_{i,j}^{k+1/2}}{\Delta t^{*}/2} = \alpha \left[\frac{\theta_{i+1,j}^{k+1} - 2\theta_{i,j}^{k+1} + \theta_{i-1,j}^{k+1}}{\left(\Delta x^{*}\right)^{2}} + \frac{\theta_{i,j+1}^{k+1/2} - 2\theta_{i,j}^{k+1/2} + \theta_{i,j-1}^{k+1/2}}{\left(\Delta y^{*}\right)^{2}} \right]$$

$$(7.2.9-9)$$

which reduces to

$$-\lambda\,\theta_{i+1,j}^{k+1} + 2\left(1+\lambda\right)\theta_{i,j}^{k+1} - \lambda\,\theta_{i-1,j}^{k+1} = \lambda\,\theta_{i,j+1}^{k+1/2} + 2\left(1-\lambda\right)\theta_{i,j}^{k+1/2} + \lambda\,\theta_{i,j-1}^{k+1/2}$$

$$(7.2.9-10)$$

another tridiagonal system, with the unknown values (now known) from the previous step running in the y-direction, and a new set of unknowns running in the x-direction.

These alternate steps in calculation are followed to the desired time. Boundary conditions are treated in the usual manner. The tridiagonal character of the matrices makes them amenable to very efficient numerical solution.

Finite element method

We remark in passing that for certain problems the finite difference method is sometimes combined with the finite element approach. For example, in treatment of the two-space-dimension, one-time-dimension unsteady-state thermal energy equation with generation

$$\rho\,C_{p}\frac{\partial T}{\partial t} = k\left[\frac{\partial^{2}T}{\partial x^{2}} + \frac{\partial^{2}T}{\partial y^{2}}\right] + \gamma_{\theta}$$

$$(7.2.9-11)$$

one can approximate the time derivative with a finite difference and apply finite elements only to the spatial part of the problem (this is where irregular boundaries are most likely to give a problem). It is obviously less complicated to use an explicit formulation in time, if the step sizes required for stability permit doing so.

7.3   Convection Heat Transfer Models

In this section we discuss heat transfer by convection. Convective heat transfer is caused by *bulk fluid motion*. For example, the fluid adjacent to a heated horizontal surface is increased in internal energy by conduction from the surface to the fluid. Fluid motion induced by density gradients caused by the temperature change carries the heated fluid away from the wall. The colder fluid in the mainstream is heated by mixing with and conduction from the warmer fluid.

Heat is transferred in laminar flow mainly by molecular interaction between adjacent fluid "layers." In turbulent flow heat is transferred on a macroscopic level due to physical mixing of the fluids. We generally consider the fluid motion, and therefore the resulting heat transfer, as (1) *forced* or (2) *free* convection of the fluid. By *forced* convection we mean that the fluid is moved by an external force such as the water pump on a car engine supplying water to the radiator, or the fan on a hot air furnace; by *free* convection we mean the fluid is moved by gravity acting on density differences resulting from temperature or concentration variations within the fluid (e.g., the layer of cold air that tumbles down the interior side of the pane of a window in winter).

Convective heat transfer is a somewhat inappropriate term - it should perhaps be *convection with* heat transfer, because the truly convective part of the mechanism primarily **moves** material (even though it is relatively hotter or colder material) and there is little or no **heat transfer** (energy crossing the boundary of the system *not* associated with mass) during this process. Almost all the energy remains associated with mass in this step (there is, of course, some minor conduction loss). The **ultimate** transfer of heat is always by conduction or radiation, when the material that has convected interacts with the ambient material, usually by mixing with it. A better term would probably be *convective energy transfer*.

Most problems in convection are too complicated to permit the kind of detailed analytical solution we discussed for conduction, so for many convective problems we adopt an individual heat transfer coefficient approach combined with empirical correlations as exemplified by the Dittus-Boelter relation shown in Equation 7.3.2-78. There are, however, situations where more elaborate models, such as the boundary layer model or even full solution of the continuum equations can be applied.

7.3.1   The thermal boundary layer

If we consider steady laminar incompressible flow past a heated flat plate, and make microscopic (rather than macroscopic) balances of momentum and energy, we are led to differential equations; with suitable assumptions, these become the well-known boundary layer equations of Prandtl. In each case, the resulting differential equation in dimensionless form is

$$\frac{d^2\Omega}{d\eta^2} + \frac{K f(\eta)}{2} \frac{d\Omega}{d\eta} = 0 \qquad\qquad (7.3.1\text{-}1)$$

with boundary conditions

$$\begin{aligned}
&\text{at } \eta = 0,\, \Omega = 0 \\
&\text{at } \eta = \infty,\, \Omega = 1
\end{aligned} \qquad\qquad (7.3.1\text{-}2)$$

In this equation, Ω is a dimensionless velocity, temperature, or concentration depending on whether a momentum, energy, or mass balance was used to generate the equation. The coefficient K is dimensionless, and η (an example of a *similarity* variable) is a dimensionless distance[45] which incorporates *both* y, the distance from the plate surface, and x, the distance from the nose of the plate:

$$\eta = y\sqrt{\frac{v_o}{v\,x}} = \frac{y}{x}\sqrt{Re} \qquad\qquad (7.3.1\text{-}3)$$

where Re is the Reynolds number based on the distance from the front of the plate, x, and the free-stream velocity, v_0. The function $f(\eta)$ is a known function (which cannot be expressed in closed form but can be expressed as a series).

For momentum transfer one obtains

---

[45] One way to develop the variable η is use of dimensional analysis techniques. One writes the variables upon which the velocity depends, and combines these variables to obtain dimensionless groups, one of which is η. For a discussion see A. G. Hansen, *Similarity Analyses of Boundary Value Problems in Engineering*, Prentice-Hall, Inc., Englewood Cliffs, NJ, 1964.

$$\frac{d^2\left(\frac{v}{v_o}\right)}{d\eta^2} + \frac{f(\eta)}{2} \frac{d\left(\frac{v}{v_o}\right)}{d\eta} = 0 \tag{7.3.1-4}$$

For heat transfer one obtains

$$\frac{d^2\left[\frac{(T-T_s)}{(T_o-T_s)}\right]}{d\eta^2} + \Pr\frac{f(\eta)}{2} \frac{d\left[\frac{(T-T_s)}{(T_o-T_s)}\right]}{d\eta} = 0 \tag{7.3.1-5}$$

where the subscript 0 designates freestream conditions and S surface conditions.

As we shall see later, for mass transfer one obtains

$$\frac{d^2\left[\frac{(c_A-c_{As})}{(c_{Ao}-c_{As})}\right]}{d\eta^2} + \Sc\frac{f(\eta)}{2} \frac{d\left[\frac{(c_A-c_{As})}{(c_{Ao}-c_{As})}\right]}{d\eta} = 0 \tag{7.3.1-6}$$

Thus, $K = 1$ for momentum transfer, $K = \Pr$ for heat transfer, and $K = \Sc$ for mass transfer.

The boundary condition at $\eta = 0$ is at the surface of the plate, since $y = 0$ implies $\eta = 0$. Substituting in the appropriate form for Ω we see that this yields either $v = 0$ or $T = T_S$, as desired, depending on whether momentum or heat transfer is being modeled. The boundary condition at $\eta = \infty$ incorporates both $x = 0$ and $y = \infty$. Substitution gives the free-stream values $v = v_0$ and $T = T_0$ for the momentum transfer and the heat transfer case, respectively.

We can solve the above equation, and without going into details of the solution method we present a plot of the solution in Figure 7.3.1-1. The plot gives profiles of either velocity or temperature. This can be seen by considering a point in a given fluid at a given distance, x, from the nose of the plate. This means that everything is fixed in η except y, and so the curves give dimensionless velocity or temperature as a function of y - the profile in the boundary layer. This profile is plotted in a slightly different manner than the usual orientation in which velocity is the ordinate and distance from the surface of the plate the abscissa.

Notice that if Pr = 1, the dimensionless differential equation and boundary conditions are the same for both momentum and heat transfer. This means the solutions to the differential equation are identical and the profiles (and therefore the boundary layers) coincide.

Equations (7.3.1-4) to (7.3.1-6) also illustrate the fact that dimensionless groups obtained in the de-dimensionalization of a problem appear as the coefficients in the dimensionless differential equation describing the physical situation.

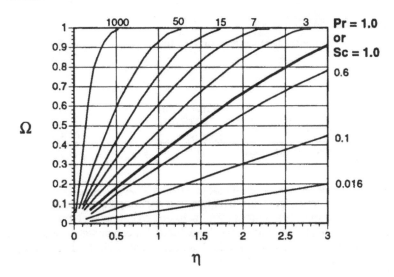

Figure 7.3.1-1 Solution of Equation (7.3.1-1) [Parameter = K]

7.3.2   Heat transfer coefficients

In this section we discuss heat transfer coefficients, their correlation and use. The model adopted is[46]:

$$\boxed{\text{heat flux } = \text{ coefficient} \times \text{driving force}}$$

$$(7.3.2\text{-}1)$$

---

[46] When written in the form

$$\dot{Q} = h\,A\,\Delta T$$

the definition of the heat transfer coefficient is sometimes designated Newton's "law" of cooling. This is obviously a **definition**, not a **law**.

This form of model equation is adopted for many types of heat transfer and the '"heat transfer coefficient" is calculated from this equation.

We can relate this model to an Ohm's law form of model by noting

$$\text{heat flow rate } = \frac{\text{driving force}}{\text{resistance}} = \text{conductance} \times \text{driving force} \quad (7.3.2\text{-}2)$$

and that

$$\begin{aligned} \text{heat flow rate } &= \text{ heat flux} \times \text{area} \\ &= \text{coefficient} \times \text{area} \times \text{driving force} \end{aligned} \quad (7.3.2\text{-}3)$$

so

$$\begin{aligned} \text{coefficient} \times \text{area} \times \text{driving force} &= \frac{\text{driving force}}{\text{resistance}} \\ \text{coefficient} = \frac{1}{\text{resistance} \times \text{area}} &= \frac{\text{conductance}}{\text{area}} \end{aligned} \quad (7.3.2\text{-}4)$$

Such transfer coefficients (here, *convective* heat transfer coefficients) are really conductance terms in a flux equation and are adopted for many kinds of mechanisms of transport, sometimes even including radiation heat transfer where the form of the model does not fit the fundamental mechanism. In the case of solids (conduction rather than convection) a coefficient could be assigned using this model, but this is usually not done.

A somewhat analogous relationship was used in defining the friction factor, which is proportional to a momentum transfer coefficient. (In the case of the friction factor as it is defined, f is a resistance rather than a conductance.)

This section discusses single-phase heat transfer coefficients based on the flux equation. We then introduce overall heat transfer coefficients, which permit us to treat heat transfer in terms of a driving force (temperature difference) over several phases. Finally, we present an extensive discussion of various graphical correlations and correlation equations which are used to estimate heat transfer coefficients for a variety of situations.

Single-phase heat transfer coefficients

The problem of theoretical prediction of convective heat transfer in both some laminar and all turbulent flows is very difficult. Since most commonly the fluid phase is bounded by solid walls, usually one adopts the model (depending on whether Q is to be positive or negative for the system or control volume chosen

$$Q = h A \left(T_s - T \right) \quad \text{or} \quad Q = h A \left(T - T_s \right) \qquad (7.3.2\text{-}5)$$

where T is defined in terms of the temperature field of the fluid (e.g., usually the freestream temperature for external flows; some average temperature such as the mixing cup temperature for internal flows) and T_S is the temperature at the interface (most commonly, a fluid/solid interface; however, liquid/gas and liquid/liquid interfaces are also important in some situations).

It is frequently assumed that the temperature of a solid surface and the fluid immediately adjacent to the surface are identical; that is, that there is no discontinuity in temperature, which would lead to the prediction of infinite flux by Fourier's law. This means we must be careful when defining fouling factors and fouling coefficients, which account for resistance at the solid/fluid interface caused by deposits.

We can adopt such a model with moderate confidence because:

a) we are quite certain that the heat flux depends linearly on the temperature difference - attempts[47] have been made to detect a higher order, e.g., quadratic dependence, but they have not been successful.

b) temperature *difference* does not depend on the level of other variables in the system - that is, that there is no interaction[48] between temperature difference and other system variables. We do not have such confidence for the variables in the system which depend on physical properties of the fluid: viscosity, thermal conductivity, density, heat capacity (these variables can, of course, be functions of temperature - sometimes we can neglect this effect and sometimes not); or the velocity, which depends on pressure and/or density differences in the system.

---

[47] Flumerfeld, R. V. and J. C. Slattery (1969). *AIChEJ* 15: 291.

[48] For example, see any text on statistics and/or experimental design, e.g., Box, G. E. P., W. G. Hunter, et al. (1978). *Statistics for Experimenters*. New York, John Wiley and Sons. Interaction gives terms in the model where the variable in question is multiplied by other variables of the system.

c) we are quite certain that the heat flux depends directly on the area by the very nature of its definition.

In essence, we lump all our ignorance into the coefficient, h.

The value of the heat transfer coefficient, h, for a specific physical situation may be found empirically from measurement of heat transfer rate, temperature drop, and associated heat transfer area for that situation at full scale under the identical flow field. This amounts to a direct solution of the differential equations that describe the situation. Even when the measurements are made on the full scale model, however, the questions of reproducibility and experimental error remain. Consequently, most methods of prediction of heat transfer coefficients rely upon correlation of data from scaled experimental models or solution of more or less idealized mathematical models for the temperature field.[49]

Two separate cases may be distinguished for the temperature T_f:

• internal flow
• external flow

An obvious temperature to use for external flow without phase change in the fluid is the freestream temperature. For internal flows without phase change in the fluid there are many possible choices, but we usually use the bulk temperature of the fluid, paralleling the way we treated momentum transfer. The major problem then is the prediction of h for any particular physical situation.

For convective heat transfer from the heated wall of a pipe to the fluid flowing inside, we use as T_f the mixing-cup average temperature $\langle T \rangle$:

$$\dot{Q} = h A \left(T_s - \langle T \rangle \right)$$

(7.3.2-6)

(The temperature difference is conventionally written so as to make $\dot{Q}$ positive.)

For external flow next to a heated surface we use the freestream temperature T_0

---

[49] The value of the heat transfer coefficient can be predicted from the solutions of theoretical models (which usually take the form of the temperature field as a function of space and time) by differentiating to obtain the temperature gradient in the fluid phase at a solid surface and calculating the flux assuming pure conduction in the fluid (since under most circumstances the fluid may be modeled as sticking to the surface, so heat transfer is by conduction only in the absence of radiation effects).

$$\dot{Q} = h A \left(T_s - T_0\right) \qquad (7.3.2\text{-}7)$$

T_0 is the temperature "far from the surface"' (at or beyond the edge of the thermal boundary layer).

The form of Equation (7.3.2-5) is also used for calculating heat transfer by boiling, condensation, and sometimes, even though the form is ill suited, radiation.

- The boiling coefficient is defined by[50]

$$\dot{Q} = h A \left(T_s - T_{sL}\right) \qquad (7.3.2\text{-}8)$$

The primary resistance to heat transfer in boiling is usually the resistance from the more or less continuous vapor film that forms immediately adjacent to the hot solid surface, and since this resistance is usually much larger than the other resistances, it is reasonable to use the temperature drop across this film as essentially equal to the overall temperature drop. The saturated liquid temperature T_{sL} exists on the liquid surface side of the film, and on the solid surface side, the temperature is that of the surface, T_s.

- The condensing coefficient is defined by

$$\dot{Q} = h A \left(T_{sv} - T_s\right) \qquad (7.3.2\text{-}9)$$

where T_{sv} is the saturated vapor temperature.

---

[50] Initially, boiling usually occurs by what is called the **nucleate boiling** mechanism, where individual nucleation sites give birth to a string of vapor bubbles which rise through the liquid. Bubbles are not normally nucleated in the liquid itself unless some small nucleation sites (small particles, for example) are present, because the surface force that a small bubble must overcome is tremendous. Instead, small pockets of gas trapped in pits and crevices of the solid usually serve as the nucleation sites in early stages of boiling. Initially these bubbles leave patches of the surface directly accessible to the liquid. At increased heat fluxes, however, the bubbles grow and coalesce with one another, giving a continuous film of vapor next to the surface; so-called **film boiling**. In film boiling, all the heat transferred to the liquid must first pass through the vapor film which slows the transfer rate markedly.

During condensation, the primary resistance to heat transfer usually is contributed by the liquid film that forms on the solid surface, and therefore we use the temperature drop across this film[51] (assuming condensation of a pure vapor). This may be invalid for cases of condensation of a liquid from a mixture of its vapor and a non-condensable gas, where diffusion of the vapor through the non-condensable gas to the liquid-gas interface may become controlling.

• One sometimes sees a radiation coefficient defined as

$$\dot{Q} = h_r A \left(T_1 - T_2 \right) \qquad (7.3.2\text{-}10)$$

where T_1 and T_2 are the temperatures of the two surfaces between which the radiation is transferred. Since radiation transfer is actually proportional to the difference of the fourth powers of the absolute temperatures, the radiation coefficient h_r defined as above is very sensitive to the temperature at which the process occurs.

Figure 7.3.2- 1[52] gives typical values of individual heat transfer coefficients for various situations. This figure gives the reader an idea of the relative magnitude of h under various conditions; it is not intended for use in calculations other than extremely approximate ones.

---

[51] Initially, condensation usually occurs by a dropwise mechanism, where individual drops form on the surface but leave patches of surface directly accessible to the vapor - so-called **dropwise condensation**. At higher transfer rates, however, these drops grow and coalesce to a continuous liquid film on the surface, so-called **filmwise condensation**. In an analogous fashion to film boiling, here all the heat transferred from the vapor must pass through the liquid film, which slows transfer rates.

[52] Data from McAdams, W. H. (1954). *Heat Transmission*. New York, NY, McGraw-Hill.

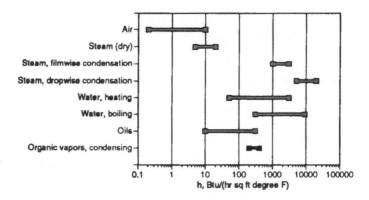

Figure 7.3.2-1 Single-phase heat transfer coefficients
1 Btu/(hr ft² °F) = 5.678 W/(m² K)
(for illustration only - not for design calculations)

Correlations for prediction of heat transfer

Heat transfer coefficients can be evaluated based on models at several levels of sophistication and complexity. At perhaps the simplest level, dimensional analysis may be applied to a set of variables identified through experience with a given process, producing in turn a set of dimensionless products that (hopefully) completely describe the process. The dimensionless groups (at least one of which will contain the heat transfer coefficient) are then varied systematically in the laboratory and the statistical correlation of these dimensionless groups which makes some function of the groups equal to zero is determined.

Note that when we speak of correlating the dimensionless groups, we mean performing a series of laboratory experiments which will give us the dependence of these groups on one another over a certain range of interest. What we are really doing is solving a physical model of the system rather than a mathematical model. Just as with the mathematical model, if the physical model (experimental setup) perfectly embodies the situation to be modeled, we will get the precise real-world answer - if we do our measurements without error - or will get the real-world answer on the average, if our measurements contain non-systematic error.

It is important to recognize that in the laboratory, **even without knowledge of the precise mathematical form of the differential equation(s) describing the process,** we still can obtain the **solution** to the differential equation(s) that describe our real-world process (the prototype). The solution is the data we measure in response to inputs to the prototypical physical situation (the full-scale, real-world model) or the laboratory physical model. The best solution is always the answer obtained from the prototype; however, if the physical modeling has been done properly, the laboratory model should also give the same answer.

For example, in a one-dimensional case, we might be measuring the response y as a function of changes in input x for our laboratory physical model (y might be a dimensionless temperature and x a dimensionless velocity.) The relationship between y and x might be a very complicated differential equation because of the shape of the solid, turbulent flow, etc., but measurements on the actual system give the solution to the differential equation and boundary conditions governing that system. If our laboratory physical model is chosen well, it will respond according to the same equation as the prototype physical situation; so, given the same dimensionless velocity, it will yield the same dimensionless temperature.

If one has somewhat more insight into physical mechanism, the (continuum) differential equation(s) which describes the process may be derived and the variables therein made dimensionless. In this way the fundamental variables are not arbitrarily selected, but enter in a natural way because of their appearance in the differential equation(s) describing the fundamental physical mechanism. The mathematical solution of the differential equation(s) then should give the same result as the real-world situation being modeled. The appropriate dimensionless groups will appear as coefficients in the differential equation(s) and boundary condition(s).

The differential equation(s) can then sometimes be solved analytically, which is probably the most fundamental way of modeling; if not, perhaps solved numerically. Even if the differential equation(s) cannot be solved, however, the dimensionless groups which are obtained as coefficients in the dimensionless differential equation(s) and boundary condition(s) can be correlated using physical models in the laboratory with increased confidence that they will completely describe the process, because the variables in these groups were obtained by a more rigorous procedure than arbitrary selection based on anecdotal experience.

Each of these levels of sophistication of models has its own particular advantages and disadvantages depending on the situation to be treated. Usually it is best to attempt to model the functional relation of the variables using a variety of these approaches (physical model solved in laboratory, mathematical model solved analytically, mathematical model solved numerically) before deciding how complex a model is needed for the accuracy and precision required in the answer.

Since we have shown there are analogies (models in common) among momentum, heat, and mass transfer, we may use physical models, for example, to solve a heat transfer problem by solving its corresponding physical momentum transfer model, or a mass transfer problem by solving a physical heat transfer model. The utility of this approach depends on the mechanism (the fundamental dimensionless differential equation) describing each of the two processes (e.g., mass transfer and heat transfer) being the same.

In this section, we are concerned primarily with application of correlating equations obtained over the years from physical modeling approaches. We elsewhere in the text emphasize solutions to mathematical models. For the engineer, however, applying models of **appropriate** degree of sophistication to practical problems is of primary importance.

To make the forms of our correlating equations (models) plausible, but certainly not to justify them in general, we consider an application involving the simplest level of model above - that is, the selection of a set of variables followed by dimensional analysis of this set to form dimensionless groups.

For heat transfer to a fluid in turbulent flow in a pipe, the pertinent variables are the pipe diameter, D; the thermal conductivity of the fluid, k; the velocity of the fluid, v; the density of the fluid, ρ; the viscosity of the fluid, μ, the specific heat of the fluid, C_p; and the heat transfer coefficient, h. Functionally

$$f_1 \left(D, k, \mu, \rho, \langle v \rangle, C_p, h \right) = 0 \qquad (7.3.2\text{-}11)$$

Forming dimensionless groups in the mass, length, temperature, time system gives the following groups (omitting the intermediate algebra):

$$\frac{hD}{k} = F_1 \left[\frac{D \langle v \rangle \rho}{\mu}, \frac{C_p \mu}{k} \right] \qquad (7.3.2\text{-}12)$$

$$Nu = F_1 \left[Re, Pr \right] \qquad (7.3.2\text{-}13)$$

where Nu is the Nusselt number (the dimensionless temperature gradient at the surface), Re is the Reynolds number (the ratio of inertial to viscous forces), and Pr is the Prandtl number (the ratio of momentum diffusivity to thermal diffusivity).

Similarly, for a free (natural) convection problem the variables are

$$f_2 \left(D, k, \mu, \rho, v_\infty, C_p, g, \beta, \Delta T, L, h \right) = 0 \qquad (7.3.2\text{-}14)$$

where the additional variables beyond those in Equation (7.3.2-11) are the acceleration due to gravity, g; the thermal coefficient of volume expansion, β[53]; the temperature difference between the heated surface and the fluid, ΔT; and the length of the heated section, L.

Dimensional analysis of the set of variables gives as a complete set of dimensionless groups

$$\frac{hD}{k} = F_2\left[\left(\frac{Dv\rho}{\mu}\right),\left(\frac{C_p\mu}{k}\right),\left(\frac{L}{D}\right),\right.$$
$$\left.\left(\frac{L^3\rho^2 g \Delta T \beta}{\mu^2}\right),\left(\frac{L^3\rho^2 g}{\mu^2}\right),\left(\frac{L^2\rho^2 k \Delta T}{\mu^3}\right)\right]$$

(7.3.2-15)

For problems of interest here, the last two groups can be neglected. For example for an infinite heated horizontal flat plate in the absence of forced convection, Re = 0, and results do not depend on location on the plate, so L/D is not a significant variable, giving

$$\frac{hL}{k} = F_2\left[\left(\frac{C_p\mu}{k}\right),\left(\frac{L^3\rho^2 g \Delta T \beta}{\mu^2}\right)\right]$$

(7.3.2-16)

or

$$Nu = F_2\,[Pr, Gr]$$

(7.3.2-17)

where Gr is the Grashof number (the ratio of buoyant to viscous forces).

---

[53] The thermal coefficient of volume expansion, β, is defined as

$$= \frac{1}{V}\left(\frac{\partial V}{\partial T}\right)_p = \frac{1}{\left(\frac{1}{\rho}\right)}\left(\frac{\partial\left(\frac{1}{\rho}\right)}{\partial T}\right)_p = -\frac{1}{\rho}\left(\frac{\partial\rho}{\partial T}\right)_p$$

For an ideal gas

$$\beta = \frac{1}{\left(\frac{RT}{P}\right)}\left[\frac{\partial\left(\frac{RT}{P}\right)}{\partial T}\right]_p = \frac{1}{\left(\frac{RT}{P}\right)}\frac{R}{P}\left[\frac{\partial T}{\partial T}\right]_p = \frac{1}{T}$$

We emphasize again that the design equations we will consider have been developed using various models, but all incorporate the groups obtainable by dimensional analysis. To show the model and solution method yielding each of the equations we present would be a long process indeed, so we shall simply point to the dimensional analysis as a rationale to make the form of the equations plausible to the reader, although we will not justify the specific form rigorously.

We first will consider a simple example which shows in slightly more detail how the form of certain design equations originates.

Many design equations have been developed by extension of a postulated analogy of momentum, heat, and mass transfer called the *Reynolds analogy* (model), developed for turbulent flow by the same Reynolds as was namesake of the Reynolds number.

The Reynolds analogy assumes that momentum, heat, and mass are transferred by the same mechanism. By this we mean that the transfer in each case is described by the same dimensionless mathematical model, or, in other words, that the dimensionless profiles (solutions) for velocity, temperature, and concentration coincide, as they did at one condition in the boundary layer example. As we shall see, this model is subject to some severe limitations.

We can arbitrarily extend our models of transport via molecular processes to include transport via convection processes by writing a so-called *eddy coefficient* to account for the convective part of the transport.

$$\tau_{xy} = -\left(\mu^t + \mu\right)\frac{dv_x}{dy} \tag{7.3.2-18}$$

$$q_y = -\left(k^t + k\right)\frac{dT}{dy} \tag{7.3.2-19}$$

$$J_{Ay}^\bullet = -\left(\mathcal{D}_{AB}^t + \mathcal{D}_{AB}\right)\frac{dc_A}{dy} \tag{7.3.2-20}$$

The quantities with superscript t (for turbulence) are the *eddy viscosity*, the *eddy thermal conductivity*, and the *eddy diffusivity*, respectively. Remember that viscosity, the thermal conductivity, and the diffusivity (the original coefficients) were properties of only the *fluid*. The eddy coefficients, by their very nature, are properties of the *flow field* as well.

We can take the ratio of any two of these equations. Since we are interested in heat transfer at the moment, let us divide Equation (7.3.2-18) by Equation (7.3.2-19)

$$\frac{\tau}{q} = \frac{(\mu^t + \mu)\dfrac{dv}{dy}}{(k^t + k)\dfrac{dT}{dy}} \tag{7.3.2-21}$$

Since we are interested in developing design equations, we consider Equation (7.3.2-11) as applied at a solid surface, e.g., the wall of a tube or the surface of a plate. We designate this surface temperature by T_s, and use a y-coordinate that has its origin at the surface.

We replace the flux at the wall in terms of our design coefficients h and f (or C_D), and for simplicity we restrict the discussion to fluids with constant density and heat capacity:

$$\frac{\frac{1}{2}\rho\langle v\rangle^2 f}{h(\langle T\rangle - T_s)} = \frac{\left[(\mu^t + \mu)\dfrac{dv}{dy}\right]_{y=0}}{\left[(k^t + k)\dfrac{dT}{dy}\right]_{y=0}} \tag{7.3.2-22}$$

(Here we have used f, <v>, and <T>, as we would for flow in a tube - C_D, v_o, and T_o would be used for external flows.)

The Reynolds assumption is that the *dimensionless* temperature and velocity profiles will coincide. So, dedimensionalizing, we take the difference between the quantity and its value at the wall divided by the difference between its bulk value and the wall value. (We could dedimensionalize y, the distance from the surface, but since it is the same in each derivative we would accomplish nothing):

$$\frac{dv}{dy} = (\langle v\rangle - 0)\frac{d\left[\dfrac{(v-0)}{(\langle v\rangle - 0)}\right]}{dy} \tag{7.3.2-23}$$

$$\frac{dT}{dy} = \left(\langle T \rangle - T_s\right)\frac{d\left[\frac{(T-T_s)}{(\langle T \rangle - T_s)}\right]}{dy}$$

(7.3.2-24)

Substituting

$$\frac{\left(\frac{1}{2}\right)\rho \langle v \rangle^2 f}{h\left(\langle T \rangle - T_s\right)} = \frac{\left\{\left(\mu^t + \mu\right)\langle v \rangle \frac{d\left(\frac{v}{\langle v \rangle}\right)}{dy}\right\}_{y=0}}{\left\{\left(k^t + k\right)\left(\langle T \rangle - T_s\right)\frac{d\left(\frac{T-T_s}{\langle T \rangle - T_s}\right)}{dy}\right\}_{y=0}}$$

(7.3.2-25)

or

$$\frac{\left(\frac{1}{2}\right)\rho \langle v \rangle f}{h} = \frac{\left(\mu^t + \mu\right)}{\left(k^t + k\right)}\frac{\left[\frac{d\left(\frac{v}{\langle v \rangle}\right)}{dy}\right]_{y=0}}{\left[\frac{d\left(\frac{T-T_s}{\langle T \rangle - T_s}\right)}{dy}\right]_{y=0}}$$

(7.3.2-26)

By the Reynolds assumption, the dimensionless velocity and temperature profiles coincide; therefore, it follows that the slopes at the wall (and everywhere else as well) must be the same, so that

$$\frac{\left[\frac{d\left(\frac{v}{\langle v \rangle}\right)}{dy}\right]_{y=0}}{\left[\frac{d\left(\frac{T-T_s}{\langle T \rangle - T_s}\right)}{dy}\right]_{y=0}} = 1.0$$

(7.3.2-27)

giving

$$\frac{\left(\frac{1}{2}\right)\rho\langle v\rangle f}{h} = \frac{\left(\mu^t + \mu\right)}{\left(k^t + k\right)} \tag{7.3.2-28}$$

We now examine the right-hand side of Equation (7.3.2-28) to see what conclusions we might draw.

Our dimensional analysis has shown that the dimensionless group involving the ratio of viscosity and thermal conductivity is the Prandtl number:

$$Pr = \frac{\mu C_p}{k} = \frac{\left(\frac{\mu}{\rho}\right)}{\left(\frac{k}{\rho C_p}\right)} = \frac{\nu}{\alpha} \tag{7.3.2-29}$$

where ν is the momentum diffusivity (kinematic viscosity) and α is the thermal diffusivity, so named because they represent the corresponding coefficients in the two differential equations whose solutions give, respectively, the velocity profile and the temperature profile as a function of time, and therefore describe the diffusion of momentum and thermal energy.[54]

We rewrite the right-hand side of Equation (7.3.2-28) as follows:

$$\frac{\left(\frac{1}{2}\right)\rho\langle v\rangle f}{h} = \frac{1}{C_p} \frac{\left(\frac{\mu^t}{\rho} + \frac{\mu}{\rho}\right)}{\left(\frac{k^t}{\rho C_p} + \frac{k}{\rho C_p}\right)} \tag{7.3.2-30}$$

---

[54] It is plausible that ν should measure the diffusion of momentum; for example, in a transient situation the more viscous a fluid is, the faster momentum transfers, but since the momentum involves both mass and velocity, a denser fluid will "soak up" more momentum rather than transferring it. A crude analogy is the diffusion of water in steel wool and a sponge: the sponge has a large capacity to absorb water and so it diffuses more slowly. The same comments hold for thermal energy: the lower the thermal conductivity the slower the diffusion; the denser and the higher the heat capacity, the more capacity for thermal energy.

However, if momentum and thermal energy are transported in the same flow field by the same mechanism, it is reasonable to expect ν^t to equal α^t. (This would not be reasonable if there were, for example, form drag, for which there is no parallel mechanism in thermal energy transfer.) If *in addition* $\nu = \alpha$ ($Pr = 1.0$), then Equation (7.3.2-30) becomes

$$\frac{\left(\frac{1}{2}\right) \rho \langle v \rangle f}{h} = \frac{1}{C_p} \qquad (7.3.2\text{-}31)$$

or

$$\frac{h}{C_p \rho \langle v \rangle} = \frac{f}{2} \qquad (7.3.2\text{-}32)$$

Note that this equation is greatly restricted by the assumptions.

Equation (7.3.2-32) is the Reynolds analogy. It is interesting not only because it suggests a way of correlating data but also because it shows that *under certain circumstances the heat transfer coefficient can be found from friction factor measurements* (and *vice versa*).

We can also take the ratio of momentum transfer to mass transfer or heat transfer to mass transfer and arrive at analogous equations to Equation (7.3.2-32). We emphasize again that **the transport mechanism must be the same for the above procedure to be valid.**

We define $h/(C_p \rho \langle v \rangle)$ as the Stanton number (St_H) for heat transfer

$$St_H \equiv \frac{h}{C_p \rho \langle v \rangle} = \frac{\left(\dfrac{h D}{k}\right)}{\left(\dfrac{D \langle v \rangle \rho}{\mu}\right)\left(\dfrac{\mu \rho C_p}{\rho k}\right)} = \frac{Nu}{Re \, Pr} \qquad (7.3.2\text{-}33)$$

The Reynolds analogy modified by multiplying by the 2/3 power of the Prandtl number gives the Chilton-Colburn analogy, often referred to as the j-factor

$$\frac{f}{2} = St_H \, Pr^{2/3} \equiv j_H \tag{7.3.2-34}$$

which is applicable roughly over the range $0.6 < Pr < 60$, as opposed to the strict requirement of the Reynolds analogy that the Prandtl number be one. (The requirement of a zero dimensionless pressure drop remains, but turbulent systems are relatively insensitive to this requirement.)

A corresponding development can be done for the case of mass transfer to yield a j-factor for mass transfer, j_M (see Chapter 8).

In the above paragraphs we showed how functional relations involving the variables associated with heat transfer can be developed for application to correlation of heat transfer data. Many useful design equations are determined by applying dimensional analysis and the Reynolds analogy to experimental data.

Some of the equations we will discuss have evolved from analytical solutions of the differential equations of change (microscopic balances) or from integral solutions of the boundary layer equations (simplified versions of the microscopic balances). Our emphasis at this point is on *use* of the design equations, to motivate the reader for further subsequent study of the development of these equations.

Average heat transfer coefficients

In many situations of interest, the local heat transfer coefficient will vary over the heat transfer area involved. For example, for flow over a flat plate, the local heat transfer coefficient may be described by

$$Nu_x = 0.332 \, Re_x^{1/2} \, Pr^{1/3} \tag{7.3.2-35}$$

where

x = distance from front of plate

which implies

$$\frac{h_x x}{k} = 0.332 \left(\frac{x \, v_\infty \, \rho}{\mu} \right)^{1/2} \left(\frac{\bar{C}_p \mu}{k} \right)^{1/3} \tag{7.3.2-36}$$

$$h_x = 0.332 \frac{k}{\sqrt{x}} \left(\frac{v_\infty \rho}{\mu}\right)^{1/2} \left(\frac{\hat{C}_p \mu}{k}\right)^{1/3}$$

(7.3.2-37)

and the driving force to be used with h to calculate local heat flux for a plate is

$$q_{local} = h_x \left(T_s - T_\infty\right)$$

(7.3.2-38)

where the temperatures are the surface and freestream temperatures, respectively.

If we wish to calculate heat transfer across the entire surface of a flat plate with constant surface and freestream temperatures, however, it would be more convenient to have an average heat transfer coefficient to use in an equation of the form, where we use h_L to designate this average coefficient

$$q_{average} = \frac{\dot{Q}}{A_{plate}} = h_{average} \left(T_s - T_\infty\right) = h_L \left(T_s - T_\infty\right)$$

(7.3.2-39)

We can obtain such a coefficient by requiring that the average coefficient give us the same answer as that obtained by integrating the local coefficient over the surface of the plate

$$\dot{Q} = h_L A_{plate} \left(T_s - T_\infty\right) = \int_0^{A_{plate}} h_x \left(T_s - T_\infty\right) dA$$

(7.3.2-40)

Since the temperature difference is constant

$$h_L A_{plate} = \int_0^{A_{plate}} h_x \, dA$$

(7.3.2-41)

and, using L for the length of the plate and W for the width

$$h_L = \frac{1}{LW} \int_0^{LW} h_x \, d(x\,W)$$
$$= \frac{W}{LW} \int_0^L h_x \, dx = \frac{1}{L} \int_0^L h_x \, dx$$

(7.3.2-42)

Substituting

$$h_L = \frac{1}{L}\int_0^L 0.332\frac{k}{\sqrt{x}}\left(\frac{v_\infty\rho}{\mu}\right)^{1/2}\left(\frac{\hat{C}_p\mu}{k}\right)^{1/3}dx \qquad (7.3.2\text{-}43)$$

$$h_L = \frac{1}{L}0.332\,k\left(\frac{v_\infty\rho}{\mu}\right)^{1/2}\left(\frac{\hat{C}_p\mu}{k}\right)^{1/3}\int_0^L x^{-1/2}\,dx \qquad (7.3.2\text{-}44)$$

$$h_L = \frac{1}{L}(0.332)\,k\left(\frac{v_\infty\rho}{\mu}\right)^{1/2}\left(\frac{\hat{C}_p\mu}{k}\right)^{1/3}(2)\,L^{1/2} \qquad (7.3.2\text{-}45)$$

$$h_L = (0.664)\frac{k}{L}\left(\frac{L\,v_\infty\rho}{\mu}\right)^{1/2}\left(\frac{\hat{C}_p\mu}{k}\right)^{1/3} \qquad (7.3.2\text{-}46)$$

which is the explicit expression for the average coefficient. Putting this expression into the usual dimensionless form

$$\frac{h_L L}{k} = 0.664\left(\frac{L\,v_\infty\rho}{\mu}\right)^{1/2}\left(\frac{\hat{C}_p\mu}{k}\right)^{1/3} \qquad (7.3.2\text{-}47)$$

or

$$Nu_L = 0.664\,Re_L^{1/2}\,Pr^{1/3} \qquad (7.3.2\text{-}48)$$

A similar approach can be adopted for other coefficients, although the result is not necessarily so neat.

Example 7.3.2-1   Average heat transfer coefficients for pipe flow

In order to calculate heat transfer for flow internal to a length of pipe it is more convenient to use average heat transfer coefficients than local coefficients. Determine the average heat transfer coefficient for a pipe with constant wall temperature based on the local coefficient and average temperature difference given by

$$\frac{\left(T_s - \langle T_{in}\rangle\right) + \left(T_s - \langle T_{out}\rangle\right)}{2} \qquad (7.3.2\text{-}49)$$

'where T_s is the temperature of the wall and $<T_{in}>$ and $<T_{out}>$ are bulk temperatures.

Solution

Defining the average coefficient to be used with the arithmetic average driving force as h_{avg}

$$Q = h_{avg} A \left[\frac{\left(T_s - \langle T_{in} \rangle\right) + \left(T_s - \langle T_{out} \rangle\right)}{2} \right]$$

$$= h_{avg} A \left(T_s - \langle T \rangle\right)_{avg} \qquad (7.3.2\text{-}50)$$

Writing an energy balance for heat transferred over a differential length, dx, of the pipe

$$\delta Q = h_x \left(T_s - \langle T \rangle\right) \left(2 \pi R\right) dx \qquad (7.3.2\text{-}51)$$

where h_x is the **local** coefficient. Integrating the right-hand side and summing the left-hand side, we obtain the total heat transfer rate as

$$Q = \int_0^L h_x \left(T_s - \langle T \rangle\right) \left(2 \pi R\right) dx \qquad (7.3.2\text{-}52)$$

and so we have

$$Q = 2 \pi R \int_0^L h_x \left(T_s - \langle T \rangle\right) dx$$

$$= h_{avg} \left(2 \pi R L\right) \left(T_s - \langle T \rangle\right)_{avg} \qquad (7.3.2\text{-}53)$$

or

$$h_{avg} = \frac{\int_0^L h_x \left(T_s - \langle T \rangle\right) dx}{L \left(T_s - \langle T \rangle\right)_{avg}} \qquad (7.3.2\text{-}54)$$

To obtain a numerical value for h_{avg} we must know h_x and $(T_s - <T>)$ as a function of x. In laminar flow, these functions can be determined from an analytical solution which gives T as a function of r and x as an infinite series, as outlined above. For turbulent flow, the functions would have to be determined from experimental measurement and the expression integrated numerically.

Design equations for convective heat transfer

To this point we have illustrated the utility of **using** heat transfer coefficients without concerning ourselves greatly with how a heat transfer coefficients may be **predicted**. In this section we will present some typical design equations for making such predictions. Once again, as in momentum transfer, it is convenient to distinguish two general classes of heat transfer problems: those involving flows (internal) **through** things and those involving flows (external) **around** things.

Ideally, one would like for a design equation to permit calculation of a single average coefficient which could be used to predict the heat transfer rate, much as a single friction factor could frequently be used to predict pressure drop in pipes. In heat transfer, however, several considerations intrude:

1. Fluid properties

   When thermal gradients are imposed on a liquid flow field, the assumption of constant physical properties frequently breaks down because viscosity is typically a fairly sensitive function of temperature. This, in turn, affects the velocity field. Usually less important than the viscosity change, but critical for certain problems, is the effect of temperature on heat capacity, thermal conductivity, and density (plus the associated induced natural convection).

2. Driving forces

   As opposed to momentum transfer in a pipe, where an unchanging velocity profile is frequently obtained for incompressible fluids, in the case of heat transfer the fluid temperature is continually increasing or decreasing except for cases of phase change such as boiling and condensation.

3. Entrance effects

We previously discussed an example in which the momentum and thermal boundary layers were assumed to coincide. This may or may not be the case, and also since the lengths of pipe or tube used for heat transfer are frequently relatively short, entrance effects may become very important.

These and other considerations for heat transfer make it far more difficult to obtain design equations that possess wide applicability than was the case for momentum transfer. They also make mathematical solution of the fundamental differential equations and boundary conditions quite difficult if not impossible in a preponderance of the real-world cases for which design calculations are required, forcing one to rely on the empirical results of careful experiments, argument using dimensional analysis, and various levels of approximation to obtain tractable fundamental equations and accompanying boundary conditions.

Dimensionless groups frequently used in correlating heat transfer data include

- the Prandtl number

$$\text{Pr} = \frac{\mu \hat{C}_p}{k} = \frac{\nu}{\alpha} = \frac{\text{momentum diffusivity}}{\text{thermal energy diffusivity}} \qquad (7.3.2\text{-}55)$$

- the Nusselt number

$$\text{Nu} = \frac{hL}{k} = \text{dimensionless temperature gradient @ surface} \qquad (7.3.2\text{-}56)$$

- the Grashof number for heat transfer

$$\text{Gr}_h = \frac{g\beta\rho^2 L^3 \Delta T}{\mu^2} = \frac{\text{temperature induced buoyancy forces}}{\text{viscous forces}} \qquad (7.3.2\text{-}57)$$

- the Biot number for heat transfer

$$\text{Bi}_h = \frac{hL}{k} = \frac{\text{internal resistance to conductive thermal energy transfer}}{\text{boundary layer resistance to convective thermal energy transfer}}$$
$$(7.3.2\text{-}58)$$

- the Fourier number for heat transfer

$$\text{Fo}_h = \frac{\alpha t}{L^2} = \frac{\text{rate of thermal energy conduction}}{\text{rate of thermal energy accumulation}} \qquad (7.3.2\text{-}59)$$

• the Peclet number for heat transfer (note the close relationship to the Graetz number following)

$$Pe_b = \frac{vL}{\alpha} = \left[\frac{vL\rho}{\mu}\right]\left[\frac{\mu C_p}{k}\right] = Re_L\, Pr$$

$$= \frac{vL\rho C_p}{k} = \frac{\text{bulk thermal energy transport}}{\text{rate of thermal energy conduction}} \qquad (7.3.2\text{-}60)$$

• the Graetz number

$$Gz = \frac{\langle v\rangle A\rho C_p}{kL} = Re_D\, Pr\, \frac{D}{L} = \frac{\text{bulk thermal energy transport}}{\text{rate of thermal energy conduction}}$$

$$(7.3.2\text{-}61)$$

• the Brinkman number

$$Br = \frac{\mu v^2}{k\,\Delta T} = \frac{\text{heating from viscous dissipation}}{\text{heating from temperature difference}} \qquad (7.3.2\text{-}62)$$

• the Eckert number

$$Ec = \frac{v^2}{C_p\left(T_s - T_\infty\right)} = \frac{Br}{Pr} \qquad (7.3.2\text{-}63)$$

• the Lewis number

$$Le = \frac{Sc}{Pr} = \frac{\text{thermal energy diffusivity}}{\text{mass diffusivity}} \qquad (7.3.2\text{-}64)$$

Forced convection in laminar flow

For laminar flow forced convection in tubes it is possible to construct quite accurate mathematical models (in this case, partial differential equations) which can be solved for certain boundary conditions and that give results useful in predicting heat transfer in real systems. These models usually use an assumed, constant velocity profile, constant physical properties, and a wall condition of either constant wall temperature or constant heat flux through the wall. All these features of the model may or may not correspond to the true situations for a given real-world problem.

The analytic solutions of the various differential equations by which the process is modeled are usually in the form of infinite series which, when evaluated, furnish a temperature profile in the tube at any given point.[55] The details of the mathematics should not be permitted to obscure the purpose of the calculation, which is to use design equations in the treatment of real problems.

For example, for heat transfer purposes one can model the velocity profile for steady-state laminar flow of a Newtonian fluid in a tube as being flat; that is, that the velocity is not a function of radius (even though we know from our momentum transfer studies that it is parabolic). Since, for a tube of uniform diameter, conservation of mass indicates that the velocity then is not a function of distance down the tube (z-direction), velocity will be constant. If we write Equation (3.2.3-16) for steady state by dropping the time-dependent term, we have

$$\rho \, C_p \, v_z \frac{\partial T}{\partial z} = k \left(\frac{\partial^2 T}{\partial r^2} + \frac{1}{r} \frac{\partial T}{\partial r} \right) \tag{7.3.2-65}$$

$$\frac{\rho \, C_p}{k} \, v_z \frac{\partial T}{\partial z} = \frac{\partial^2 T}{\partial r^2} + \frac{1}{r} \frac{\partial T}{\partial r} \tag{7.3.2-66}$$

$$\frac{\partial^2 T}{\partial r^2} + \frac{1}{r} \frac{\partial T}{\partial r} = \frac{v_z}{\alpha} \frac{\partial T}{\partial z} \tag{7.3.2-67}$$

Solving this partial differential equation with the boundary conditions of uniform entering temperature

$$v_z \left(r, z \right) = v_z \left(r, 0 \right) = T_0 \tag{7.3.2-68}$$

and uniform wall temperature

$$v_z \left(r, z \right) = v_z \left(R, z \right) = T_w \tag{7.3.2-69}$$

gives a solution in Bessel functions (tabulated functions which in cylindrical geometry play a role corresponding to that of the trigonometric functions in rectangular geometry)

---

[55] Graetz, L. (1883). *Ann. Phys.* **18**: 79; Graetz, L. (1885). *Ann. Phys.* **25**: 737; Drew, T. B. (1931). *Trans. AIChE* **26**: 26; Jacob, M. (1950). *Heat Transfer.* New York, John Wiley and Sons, Inc.; Sellars, J. R., M. Tribus, et al. (1956). *Trans. ASME* **78**: 441.

$$\frac{T_w - T}{T_w - T_0} = \sum_{n=1}^{\infty} \frac{2}{c_n J_1(c_n)} J_0\left(c_n \frac{r}{R}\right) \exp\left[\frac{-2 c_n^2 \frac{z}{R}}{Re\, Pr}\right]$$

(7.3.2-70)

where J_0 and J_1 are Bessel functions of the zeroth and first orders, respectively, and c_n is the nth root of

$$J_0(c_n) = 0$$

(7.3.2-71)

Such mathematical models may be thought of as "black boxes" which take as input the conditions of the problem and yield as output temperature profiles, just as Newton's second law takes an initial position, a force, and a mass, and upon integration yields position versus time (a "position profile" in time rather than space).

A temperature profile from a model can easily be converted to a heat transfer coefficient. Using the physical fact that fluid sticks to a wall, one concludes that heat transfer through the fluid just at the wall surface is by conduction only (radiation effects neglected). All the heat transferred, however, must pass through this stagnant layer. This means that the heat flux calculated using conduction in the fluid just at the wall surface must equal the flux calculated using a convective heat transfer coefficient: for example, consider flow in a tube:

$$h\left(\langle T \rangle - T_w\right) = q_r = -k \left[\frac{dT}{dr}\right]_{r=R}$$

(7.3.2-72)

or

$$\frac{h}{k} = -\frac{1}{\left(\langle T \rangle - T_w\right)} \left[\frac{dT}{dr}\right]_{r=R}$$

(7.3.2-73)

$$\left[\frac{hR}{k}\right] = \left[\frac{d\left(\frac{\langle T \rangle - T}{\langle T \rangle - T_w}\right)}{d\left(\frac{r}{R}\right)}\right]_{\frac{r}{R}=1}$$

(7.3.2-74)

$$Nu = \left[\frac{dT^*}{dr^*}\right]_{r^*=1}$$

(7.3.2-75)

so it can be seen that the Nusselt number is a dimensionless temperature profile at the surface.

The Nusselt number for laminar flow in ducts approaches a limiting value depending on the configuration of the enclosing walls and the boundary conditions applied. Selected values are given in Table 7.3.2-1.[56]

Table 7.3.2-1 Nusselt number limit for laminar flow in ducts with various cross-sections

Nusselt number limit for Gz > 4.0	Boundary condition	
Cross-section	Constant wall temperature	Constant wall heat flux
Circle	3.66	4.36
Equilateral triangle		3.00
Square	2.89	3.63
Rectangle, 1:2 aspect ratio	3.39	4.11
Rectangle, 1:3 aspect ratio		4.77
Parallel planes	7.60	8.24

The bulk temperature, the wall temperature, and the slope of the temperature profile at the wall can all be obtained in a straightforward manner from the temperature profile, so a knowledge of the temperature profile is equivalent to a knowledge of the local heat transfer coefficient.

The *average* coefficient is easier to use for many problems, and this is available for the constant wall-temperature case for use with both arithmetic mean and logarithmic mean temperature difference (mean difference in fluid temperature and wall temperature for the end conditions of the tube).

The viscosity frequently varies significantly, making the assumption of constant velocity profile a poor one. Sieder and Tate[57] empirically developed an equation incorporating a viscosity correction to account for this variation

---

[56] Perry, R. H. and D. W. Green, Eds. (1984). *Chemical Engineer's Handbook.* New York, McGraw-Hill, pp. 10-15.

[57] Sieder, E. N. and G. E. Tate (1936). *Ind. Eng. Chem.* **28**: 1429.

$$Nu = 1.86 \, Re^{1/3} \, Pr^{1/3} \left(\frac{D}{L}\right)^{1/3} \left(\frac{\mu}{\mu_0}\right)^{0.14} \tag{7.3.2-76}$$

In this equation

> a) all properties are evaluated at the arithmetic mean bulk temperature except μ_0, which is evaluated at the wall temperature, and

> b) the Nusselt number contains as h the *average* coefficient for a tube of length L.

Coefficients in short tubes may be one or two orders of magnitude higher than predicted by Equation (7.3.2-76) because of entrance effects. Fortunately the prediction is conservative - that is, it leads to overdesign (more area than is actually required) rather than a design that will not meet specifications - because resistance to heat transfer in the entrance region is smaller relative to that of the downstream section, so the same temperature gradient produces more transfer.

A plot of Nusselt number for the entrance region of tubes may be found in Greenkorn, R. A. and D. P. Kessler (1972) *Transfer Operations*, McGraw-Hill, p. 381. The Nusselt number becomes more or less constant beyond about

$$\frac{L_\bullet}{D} = 0.05 \, Re \, Pr \text{ to } 0.15 \, Re \, Pr \tag{7.3.2-77}$$

(This is the point beyond which h becomes relatively constant.)

Forced convection in turbulent flow[58]

The results of dimensional analysis of the variables for heat transfer to a fluid in *turbulent* flow have been used by different investigators to correlate heat transfer data. The resulting equations usually bear the name of the investigator.

In general, all the equations discussed are of the form of Equation (7.3.2-13). The differences in the equations usually are related to the temperature at which the fluid properties are evaluated and whether change in viscosity with temperature is important.

---

[58] For a critical review of modeling efforts in this regime, see Churchill, S. W. (1996). A Critique of Predictive and Correlative Models for Turbulent Flow and Convection. *Ind. Eng. Chem. Res.* 35: 3122-3140.

The Dittus-Boelter[59] equation for turbulent flow in tubes uses fluid properties evaluated at the arithmetic mean bulk temperature of the fluid, with the parameter a = 0.4 when the fluid is heated and a = 0.3 when the fluid is cooled. (For most gases Pr is approximately equal to 1 and so this variation of the exponent is of little significance for gases.)

$$\frac{hD}{k} = 0.023 \left(\frac{D\langle v\rangle \rho}{\mu}\right)^{0.8} \left(\frac{\hat{C}_p \mu}{k}\right)^{a}$$
(7.3.2-78)

The Dittus-Boelter equation resulted from an attempt to fit the available data for both liquids and gases with a set of two curves.

The Colburn equation[60] results from correlating data using fluid properties evaluated at the wall temperature, except for $\hat{C}_p$, which is evaluated at the arithmetic mean bulk fluid temperature

$$\frac{h}{\langle v\rangle \rho \hat{C}_p} \left(\frac{\hat{C}_p \mu}{k}\right)^{2/3} = 0.023 \left(\frac{D\langle v\rangle \rho}{\mu}\right)^{-0.2}$$
(7.3.2-79)

To show the parallel both to the Dittus-Boelter equation and to our dimensional analysis results, the Colburn equation may be rearranged as

$$\frac{hD}{k} = 0.023 \left(\frac{D\langle v\rangle \rho}{\mu}\right)^{0.8} \left(\frac{\hat{C}_p \mu}{k}\right)^{1/3}$$
(7.3.2-80)

or

$$Nu = 0.023\, Re^{0.8}\, Pr^{1/3}$$
(7.3.2-81)

The Sieder-Tate equation for turbulent flow takes into account variation in the viscosity of the fluid near the wall because of thermal gradients. This equation is valid up to Pr = 10^4 The properties of the fluids are evaluated at the arithmetic average bulk temperature except for μ_0, which is evaluated at the average wall temperature.

---

[59] Dittus, F. W. and L. M. E. Boelter (1930) Engineering Publication 2:443, University of California.
[60] Colburn, A. P. (1942) Purdue University.

$$\frac{h}{\langle v\rangle \rho \hat{C}_p}\left(\frac{\hat{C}_p \mu}{k}\right)^{2/3}\left(\frac{\mu_0}{\mu}\right)^{0.14} = 0.023\left(\frac{D\langle v\rangle \rho}{\mu}\right)^{-0.2}$$

(7.3.2-82)

Rearranged,

$$Nu = 0.023\, Re^{0.8}\, Pr^{1/3}\left(\frac{\mu}{\mu_0}\right)^{0.14}$$

(7.3.2-83)

Example 7.3.2-2   Comparison of the Dittus-Boelter, Colburn, and Sieder-Tate equations

Compare the Dittus-Boelter equation, the Colburn equation, and the Sieder-Tate equation prediction of the heat transfer coefficient for 1) dry air and 2) water flowing at 1 bar and at a bulk velocity of 15 ft/s in a tube 1 inch in diameter

a) being heated at an arithmetic mean bulk temperature of 80°F, wall temperature of 98°F

b) being cooled at an arithmetic mean bulk temperature of 80°F, wall temperature of 62°F

Solution

deg F	deg C	K
62	17	290
80	27	300
98	37	310

Substance	$\rho^{61, 62}$ lbmass/ft^3	Pr$^{63, 64}$	k$^{65, 66}$ * Btu/(hr ft °F) ** W/(m K)	$\mu^{67, 68}$ cP
HOH, 62°F	62.36	8.00	* 0.352	1.1
HOH, 80°F	62.22	5.69	* 0.357	0.9
HOH, 98°F	62.02	4.67	* 0.363	0.75
air, 62°F	0.0761	0.708	** 2.54 x 10^{-2}	0.0170
air, 80°F	0.0735	0.705	** 2.62 x 10^{-2}	0.0175
air, 98°F	0.0711	0.704	** 2.70 x 10^{-2}	0.0180

Substance	$\hat{C}_p$ Btu/(lbmass °F)	Re
HOH, 62°F	1.0	1.05 x 10^5
HOH, 80°F	1.02	1.29 x 10^5
HOH, 98°F	1.05	1.54 x 10^5
air, 62°F	0.25	8.33 x 10^3
air, 80°F	0.25	7.81 x 10^3
air, 98°F	0.25	7.35 x 10^3

Dittus Boelter:

$$\frac{h\,D}{k} = 0.023 \left(\frac{D\,\langle v \rangle\,\rho}{\mu} \right)^{0.8} \left(\frac{\hat{C}_p\,\mu}{k} \right)^{a}$$

(7.3.2-84)

[61] Perry, R. H. and D. W. Green, Eds. (1984). *Chemical Engineer's Handbook*. New York, McGraw-Hill, p. 3-237.

[62] Perry, R. H. and D. W. Green, Eds. (1984). *Chemical Engineer's Handbook*. New York, McGraw-Hill, p. 12-8.

[63] Perry, R. H. and D. W. Green, Eds. (1984). *Chemical Engineer's Handbook*. New York, McGraw-Hill, p. 3-255.

[64] Perry, R. H. and D. W. Green, Eds. (1984). *Chemical Engineer's Handbook*. New York, McGraw-Hill, p. 3-254.

[65] Perry, R. H. and D. W. Green, Eds. (1984). *Chemical Engineer's Handbook*. New York, McGraw-Hill, p. 3-253.

[66] Perry, R. H. and D. W. Green, Eds. (1984). *Chemical Engineer's Handbook*. New York, McGraw-Hill, p. 3-254.

[67] Perry, R. H. and D. W. Green, Eds. (1984). *Chemical Engineer's Handbook*. New York, McGraw-Hill, p. 3-252.

[68] Perry, R. H. and D. W. Green, Eds. (1984). *Chemical Engineer's Handbook*. New York, McGraw-Hill, p. 3-250.

Properties evaluated at arithmetic mean bulk temperature

$$h = 0.023 \frac{k}{D} \left(\frac{D \langle v \rangle \rho}{\mu} \right)^{0.8} \left(\frac{\hat{C}_p \mu}{k} \right)^a \qquad (7.3.2\text{-}85)$$

For water heating

$$h = (0.023) \frac{(0.357) \left[\dfrac{\text{Btu}}{\text{hr ft } ^\circ\text{F}} \right]}{\left(\dfrac{1}{12} \right) [\text{ft}]} \left(1.29 \times 10^5 \right)^{0.8} (5.69)^{0.4} = 2422 \frac{\text{Btu}}{\text{hr ft}^2 \, ^\circ\text{F}}$$

$$(7.3.2\text{-}86)$$

For water cooling

$$h = (0.023) \frac{(0.357) \left[\dfrac{\text{Btu}}{\text{hr ft } ^\circ\text{F}} \right]}{\left(\dfrac{1}{12} \right) [\text{ft}]} \left(1.29 \times 10^5 \right)^{0.8} (5.69)^{0.3} = 2035 \frac{\text{Btu}}{\text{hr ft}^2 \, ^\circ\text{F}}$$

$$(7.3.2\text{-}87)$$

For air heating

$$h = (0.023) \frac{\left(2.62 \times 10^{-2} \right) \left[\dfrac{\text{W}}{\text{m K}} \right] (0.578) \left[\dfrac{\text{Btu}}{\text{hr ft } ^\circ\text{F}} \dfrac{\text{m K}}{\text{W}} \right]}{\left(\dfrac{1}{12} \right) [\text{ft}]}$$

$$\times \left(7.81 \times 10^3 \right)^{0.8} (0.705)^{0.4}$$

$$= 4.73 \frac{\text{Btu}}{\text{hr ft}^2 \, ^\circ\text{F}} \qquad\qquad (7.3.2\text{-}88)$$

For air cooling

$$
h = (0.023) \frac{\left(2.62 \times 10^{-2}\right)\left[\frac{W}{m\,K}\right](0.578)\left[\frac{Btu}{hr\,ft\,°F}\frac{m\,K}{W}\right]}{\left(\frac{1}{12}\right)[ft]}
$$

$$
\times \left(7.81 \times 10^3\right)^{0.8} (0.705)^{0.3}
$$

$$
= 4.89 \frac{Btu}{hr\,ft^2\,°F} \tag{7.3.2-89}
$$

Colburn:

$$
\frac{h\,D}{k} = 0.023 \left(\frac{D\langle v\rangle \rho}{\mu}\right)^{0.8} \left(\frac{\hat{C}_p \mu}{k}\right)^{1/3} \tag{7.3.2-90}
$$

Properties evaluated at wall temperature except for heat capacity, which is evaluated at arithmetic mean bulk temperature

$$
h = 0.023 \frac{k}{D} \left(\frac{D\langle v\rangle \rho}{\mu}\right)^{0.8}_{wall} \left(\frac{\hat{C}_p \mu}{k}\right)^{1/3}_{wall} \left(\frac{[\hat{C}_p]_{mean\ bulk}}{[\hat{C}_p]_{wall}}\right)^{1/3} \tag{7.3.2-91}
$$

For water heating

$$
h = (0.023) \frac{(0.363)}{\left(\frac{1}{12}\right)} \left(1.54 \times 10^5\right)^{0.8} (4.67)^{1/3} \left(\frac{1.02}{1.05}\right)^{1/3}
$$

$$
= 2349 \frac{Btu}{hr\,ft^2\,°F} \tag{7.3.2-92}
$$

For water cooling

$$
h = (0.023) \frac{(0.352)}{\left(\frac{1}{12}\right)} \left(1.05 \times 10^5\right)^{0.8} (8.00)^{1/3} \left(\frac{1.02}{1.00}\right)^{1/3} = 2034 \frac{Btu}{hr\,ft^2\,°F}
$$

$$
\tag{7.3.2-93}
$$

For air heating

$$h = (0.023) \frac{\left(2.70 \times 10^{-2}\right)}{\left(\frac{1}{12}\right)} (0.578) \left(7.35 \times 10^{3}\right)^{0.8} (0.704)^{1/3} \left(\frac{0.25}{0.25}\right)^{1/3}$$

$$= 4.75 \frac{Btu}{hr\, ft^{2}\, °F} \qquad\qquad\qquad (7.3.2\text{-}94)$$

For air cooling

$$h = (0.023) \frac{\left(2.54 \times 10^{-2}\right)}{\left(\frac{1}{12}\right)} (0.578) \left(8.33 \times 10^{3}\right)^{0.8} (0.708)^{1/3} \left(\frac{0.25}{0.25}\right)^{1/3}$$

$$= 4.94 \frac{Btu}{hr\, ft^{2}\, °F} \qquad\qquad\qquad (7.3.2\text{-}95)$$

Sieder-Tate:

$$\frac{h}{\langle v \rangle \rho C_{p}} \left(\frac{\hat{C}_{p}\mu}{k}\right)^{2/3} \left(\frac{\mu_{o}}{\mu}\right)^{0.14} = 0.023 \left(\frac{D \langle v \rangle \rho}{\mu}\right)^{-0.2} \qquad (7.3.2\text{-}96)$$

Properties evaluated at arithmetic mean bulk temperature except for μ_{o}, which is evaluated at the average wall temperature

$$h = 0.023 \frac{k}{D} \left(\frac{D \langle v \rangle \rho}{\mu}\right)^{0.8} \left(\frac{\hat{C}_{p}\mu}{k}\right)^{1/3} \left(\frac{\mu}{\mu_{o}}\right)^{0.14} \qquad (7.3.2\text{-}97)$$

Note that this is the same as the Colburn equation except for a viscosity correction in place of the heat capacity correction.

For water heating

$$h = (0.023) \frac{(0.363)}{\left(\frac{1}{12}\right)} \left(1.54 \times 10^{5}\right)^{0.8} (4.67)^{1/3} \left(\frac{0.9}{0.75}\right)^{0.14} = 2433 \frac{Btu}{hr\, ft^{2}\, °F}$$

$$(7.3.2\text{-}98)$$

For water cooling

$$h = (0.023)\frac{(0.352)}{\left(\frac{1}{12}\right)}(1.05\times 10^5)^{0.8}(8.00)^{1/3}\left(\frac{0.9}{1.1}\right)^{0.14} = 1965\frac{Btu}{hr\ ft^2\ {}^{\circ}F}$$

$$(7.3.2\text{-}99)$$

For air heating

$$h = (0.023)\frac{(2.70\times 10^{-2})}{\left(\frac{1}{12}\right)}(0.578)(7.35\times 10^3)^{0.8}(0.704)^{1/3}\left(\frac{0.0175}{0.0180}\right)^{0.14}$$

$$= 4.73\frac{Btu}{hr\ ft^2\ {}^{\circ}F}$$

$$(7.3.2\text{-}100)$$

For air cooling

$$h = (0.023)\frac{(2.54\times 10^{-2})}{\left(\frac{1}{12}\right)}(0.578)(8.33\times 10^5)^{0.8}(0.708)^{1/3}\left(\frac{0.0175}{0.0170}\right)^{0.14}$$

$$= 4.96\frac{Btu}{hr\ ft^2\ {}^{\circ}F}$$

$$(7.3.2\text{-}101)$$

Tabulating the results to give a side-by-side comparison of the magnitude of the differences

	Dittus-Boelter h Btu/(hr ft² °F)	Colburn h Btu/(hr ft² °F)	Sieder-Tate h Btu/(hr ft² °F)
air heating	4.73	4.75	4.73
air cooling	4.89	4.94	4.96
water heating	2422	2349	2433
water cooling	2035	2034	1965

Heat transfer in non-circular conduits and annular flow

For flowing fluids heated in non-circular conduits or in annuli, the *equivalent diameter* (**four** times the hydraulic radius) is commonly used in place of the tube diameter in equations for estimating heat transfer coefficients.

For annular flow, where D_2 is the larger diameter,

$$D_{eq} = D_2 - D_1$$

(7.3.2-102)

An equation developed specifically for the *outer* wall of the *inside* pipe of an annulus $(12,000 < Re < 220,000)$ is[69]

$$Nu = 0.02\,Re^{0.8}\,Pr^{1/3}\left(\frac{D_2}{D_1}\right)^{0.45}$$

(7.3.2-103)

Fluid properties are evaluated at the arithmetic mean bulk temperature. For the *inner* wall of the *outside* pipe the heat transfer coefficient may be found from the Sieder-Tate equation, Equation (7.3.2-76).

External flows, natural and forced convection

Heat transfer coefficients for fluid flowing normal to a cylinder depend on the configuration of the boundary layer. The formation of a large wake changes the heat transfer characteristics. For *forced* convection

$$\frac{h\,D}{k} = b\left(\frac{D\,v\,\rho}{\mu}\right)^{n}$$

(7.3.2-104)

where values of b and n depend on the Reynolds number, as shown in Table 7.3.2-2. The equation retains the functional form derived from the earlier dimensional analysis.

---

[69] Monrad, C. C. and J. F. Pelton (1942). *Trans. ASME* **38**: 593.

Table 7.3.2-2 Values[70] of b and n for Equation (7.3.2-104)

$Re = \dfrac{D v \rho}{\mu}$	n	b
1 - 4	0.330	0.891
4 - 40	0.385	0.821
40 - 4000	0.466	0.615
4000 - 40,000	0.618	0.174
40,000 - 250,000	0.805	0.0239

For *natural* convection[71] with flow normal to a **horizontal surfaces**

$$Nu = a\left(Gr\, Pr\right)^{m}$$

$$(7.3.2\text{-}105)$$

where the characteristic length is L for flat surfaces, D for cylindrical ones, and the values of the parameters can be found in Table 7.3.2-3.

Table 7.3.2-3 Values of a and m for use with Equation (7.3.2-105)

Surface	Gr Pr	a	m
Horizontal cylinder, D < 8 in	$< 10^{-5}$	0.49	0
	$10^{-5} - 10^{-3}$	0.71	0.04
	$10^{-3} - 1.0$	1.09	0.1
	$1.0 - 10^{4}$	1.09	0.2
	$10^{4} - 10^{9}$	0.53	0.25
	$> 10^{9}$	0.13	0.333
Horizontal plane, face down	$3 \times 10^{5} - 3 \times 10^{10}$	0.27	0.25
Horizontal plane, face up	$10^{5} - 2 \times 10^{7}$	0.54	0.25
Horizontal plane, face up	$2 \times 10^{7} - 3 \times 10^{10}$	0.14	0.333

[70] Hilpert, R. (1933). *Forsch. Gebiete Ingenieur.* 4: 215.

[71] Perry, R. H. and D. W. Green, Eds. (1984). *Chemical Engineer's Handbook.* New York, McGraw-Hill, p. 10-13.

For *natural* convection on the surface of **vertical** plates or outside **vertical** tubes or cylinders,[72] $1 < Pr < 40$

$$Nu = 0.138 \, Gr^{0.36} \left(Pr^{0.175} - 0.55 \right) \qquad Gr > 10^9$$
$$Nu = 0.683 \, Gr^{0.25} \, Pr^{0.25} \left[\frac{Pr}{0.861 + Pr} \right]^{0.25} \qquad Gr < 10^9 \qquad (7.3.2\text{-}106)$$

For convection normal to the outside of pipes or tubes of diameter D **arranged in successive rows**, we need a modified equation since the heat transfer to or from downstream tubes is affected by the wake of tubes upstream:

$$\frac{h \, D}{k} = b \left(\frac{D \, G_{max}}{\mu} \right)^n \qquad (7.3.2\text{-}107)$$

where

- $G_{max} = r \, v'$
- v' is the velocity at the point of minimum flow cross-sectional area

and the values of b and n depend on the tube arrangement.

- For a square grid with center-to-center spacing of the tubes equal to twice the tube diameter, $b = 0.299$ and $n = 0.632$.
- For a pattern of equilateral triangles with center-to-center spacing equal to twice the tube diameter, $b = 0.482$ and $n = 0.556$

The equation above was developed for gases. To use Equation (7.3.2-107) for liquids multiply it by $(1.1) \, (Pr^{1/3})$.

Example 7.3.2-3   Heat transfer with flow normal to pipes

Water is being heated by passing it across a bank of rows of 1 in., 16 BWG gauge copper tubes 10 ft long, five tubes per row. The tubes are spaced on a square grid with center-to-center spacing of twice the tube diameter. Assume the space between the outside tube and the containing shell is one-half the tube diameter.

---

[72] Kato, Nishiwaki, et al. (1968). *Int. J. Heat Mass Transfer* 11: 1117.

The inside temperature of the tube walls is 230°F (neglect conduction resistance). If water flowing at 80 gpm enters the exchanger at 60°F and leaves at 140°F, find the mean heat transfer coefficient for the water.

Solution

Modifying Equation (7.3.2-107) as indicated previously for use with liquids

$$\frac{h\,D}{k} = 1.1\,b\left(\frac{D\,G_{max}}{\mu}\right)^{n} Pr^{1/3} \tag{7.3.2-108}$$

The minimum flow cross-sectional area is

$$
A_{min} = (4)\left[\text{openings}\right] \frac{\left(\frac{1}{12}\right)[\text{ft}]\,(10)\,[\text{ft}]}{[\text{opening}]}
$$

$$
+ (2)\left[\text{openings}\right] \frac{\left(\frac{1}{24}\right)[\text{ft}]\,(10)\,[\text{ft}]}{[\text{opening}]}
$$

$$
= 0.417\ \text{ft}^2 \tag{7.3.2-109}
$$

Note that we are neglecting end effects. The Reynolds number is

$$
Re = \left(\frac{D\,G_{max}}{\mu}\right) = 4{,}805
$$

$$
= \left\{ \left(\frac{1}{12}\right)[\text{ft}]\,(80)\left[\frac{\text{gal}}{\text{min}}\right]\left(\frac{1}{7.48}\right)\left[\frac{\text{ft}^3}{\text{gal}}\right](62.4)\left[\frac{\text{lbmass}}{\text{ft}^3}\right] \right.
$$

$$
\left. \times \left(\frac{1}{60}\right)\left[\frac{\text{min}}{\text{s}}\right]\left(\frac{1}{0.417}\right)\left[\frac{1}{\text{ft}^2}\right] \right\}
$$

$$
\div \left\{ (0.684)\,[\text{cP}]\left(6.72\times10^{-4}\right)\left[\frac{\text{lbmass}}{\text{ft s cP}}\right] \right\}
$$

$$
= 4805 \tag{7.3.2-110}
$$

and the Prandtl number is

$$\frac{\mu \, \hat{C}_p}{k} = \frac{(0.684)[cP]\left(6.72 \times 10^{-4}\right)\left[\dfrac{lbmass}{ft \ s \ cP}\right](0.999)\left[\dfrac{Btu}{lbmass \ ^\circ F}\right](3600)\left[\dfrac{s}{hr}\right]}{(0.363)\left[\dfrac{Btu}{hr \ ft \ ^\circ F}\right]}$$

$$= 4554 \tag{7.3.2-111}$$

Then

$$h = 1.1 \, b \left(\frac{k}{D}\right)\left(\frac{D \, G_{max}}{\mu}\right)^n Pr^{1/3}$$

$$= (1.1)(0.229)\frac{(0.363)\left[\dfrac{Btu}{hr \ ft \ ^\circ F}\right]}{\left(\dfrac{1}{12}\right)[ft]}(4805)^{0.632}(4554)^{1/3}$$

$$= 386 \frac{Btu}{hr \ ft^2 \ ^\circ F} \tag{7.3.2-112}$$

A useful design equation for air flowing outside and at right angles to a bank of **finned tubes** is[73]

$$h = 0.17\left(\frac{v^{0.6}}{D^{0.4}}\right)\left(\frac{L}{L-D}\right)^{0.6} \tag{7.3.2-113}$$

where v is the freestream velocity of the air in feet per minute, D is the outside diameter in inches of the bare tube (root diameter of fin), and L is the center-to-center spacing in inches of the tubes in a row. This equation is, unfortunately, dimensional, but the extreme complexity of the situation does not permit other than rough estimation from an overall design equation Obviously, to obtain more precise answers fin geometry must be considered, which further complicates the situation.

The correlation of Froessling is useful for determining the heat transfer coefficient for **spheres**. In this case, a constant is added to the functional form involving the Reynolds and Prandtl numbers

---

[73] Perry, R. H., C. H. Chilton, et al., Eds. (1963). *Chemical Engineer's Handbook.* New York, McGraw-Hill, p.10-24.

$$\frac{h\,D}{k} = 2 + 0.6 \left(\frac{D\,v_\infty\,\rho}{\mu}\right)^{1/2} \left(\frac{\tilde{C}_p\,\mu}{k}\right)^{1/3}$$

(7.3.2-114)

The constant term in Equation (7.3.2-114) accounts for the fact that heat transfer continues (by conduction) even when the Reynolds number goes to zero (for example, zero flow velocity).

The Froessling correlation is also used to correlate data for porous media, because one approach to modeling flow in a porous medium is to consider it as flow over a collection of submerged objects. For such a case, one replaces v_∞ with $(10.73)(v_{ap})$ where v_{ap} is the velocity of approach in the porous medium (see Section 6.9 on flow in porous media).

Heat transfer with phase change

The transfer of heat when phase change occurs, such as in boiling and condensation, is usually described by a heat transfer coefficient as defined in Equation (7.3.2-8) and Equation (7.3.2-9). Heat transfer which accompanies phase change usually occurs at a high rate. As we have seen in Table 7.3.2-1, coefficients for these cases can be very high.

Boiling - mechanism

Boiling can occur by more than one mechanism. The two main classes which are usually distinguished are

- nucleate boiling

- film boiling

The type of boiling one observes in heating water for a cup of tea or coffee, or in boiling an egg in a pan, is usually nucleate boiling. In this class of boiling, the vapor is evolved as a multitude of bubbles. As one increases heat flux, however, these bubbles grow large enough that they coalesce and form a continuous film of vapor at the surface, creating the regime known as film boiling.

An interesting illustration of film boiling is the Leidenfrost phenomenon. If one pours small amounts of water on the surface of a hot plate, the drops skitter about the surface and appear to bounce up and down. Leidenfrost explained this behavior by the mechanism that as the drop touches the hot surface, a film of vapor is formed between the drop and the surface which accelerates the drop in the upward direction. Since the film of vapor forms a relatively good insulator between the drop and the surface, the heat transfer rate then begins to decrease and the film thins until the drop again touches the surface, more water is evaporated, the drop shoots upward once more, etc., until all the water in the drop has been evaporated.

It is relatively obvious that, for heat transfer purposes, nucleate boiling is far preferable to film boiling because of the direct contact of much of the liquid phase with the heated surface as opposed to the separation of liquid from the surface by a gas film with poor heat transmission characteristics found in film boiling. Much work has been done on the formation of bubbles during boiling, and explanation of nucleation phenomena is far from trivial.

If one considers the problem of nucleating extremely small bubbles of vapor, one finds that the surface tension forces become very important since the ratio of surface area to volume of a sphere increases as the radius decreases. For this reason, it is easier to nucleate a bubble from a cavity which initially contains some trapped gas than to do so in the bulk of the liquid. The radius of curvature when the bubble begins to grow will then be relatively large, and less pressure will be required to overcome the surface tension force. There are motion pictures available which illustrate this type of bubble nucleation.[74]

We can illustrate the common mechanisms of boiling by considering the case of a platinum wire (immersed in water) through which an electric current is passed. If we conduct such an experiment, we will obtain results such as those shown in Figure 7.3.2-2, in which is plotted the heat flux versus the temperature of the wire. The heat flux as a function of wire temperature is readily obtained because

> (1) we can determine the surface area of the wire,
> (2) we can measure the electrical power input, all of which crosses the wire surface as heat, and
> (3) we can determine the temperature of the wire by optical pyrometry or by using the fact that wire temperature is a function of the wire resistance (which we can determine via the power input and the voltage drop)

---

[74] For example, see the PhD thesis of Roll (1962) The Effect of Surface Tension on Boiling Heat Transfer. School of Chemical Engineering, Purdue University, West Lafayette, IN (16 mm B/W, 12 min.)

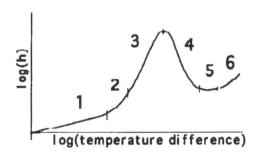

Figure 7.3.2-2 Boiling Curve[75]

In region 1 of Figure 7.3.2-2, superheated liquid rises to the surface of the bulk liquid, where evaporation takes place. The mechanism observed in regions 2 (where bubbles of vapor condense before reaching the bulk liquid surface) and 3 (where bubbles break the bulk liquid surface) is *nucleate* boiling. Bubbles of vapor form at discrete sites on the heated surface and detach almost immediately. As the temperature of the surface increases through region 4, bubbles coalesce and the surface becomes covered by a continuous film of vapor.

The continuous vapor film substantially prevents the liquid from reaching the surface. The mechanism then is called *film* boiling. In film boiling, vaporization takes place at the vapor-liquid interface. The film causes the heat transfer coefficient to drop precipitously - an unstable situation because decrease in heat transfer coefficient requires further increase in temperature drop to maintain the required power dissipation, which then causes further decrease in heat transfer coefficient. This positive feedback cycle continues until the film thickness and temperature drop reach a stable situation in region 5.

Region 4 is responsible for the phenomenon of "burnout" that can be experienced with resistance heaters if power input is maintained constant into region 4. The material in the resistance heater rod may melt before the stable film boiling region is reached. This can occur quite abruptly, leaving one with a heater that resembles a blown fuse - with a gap of missing material that looks as if it had "burned," even though it is more likely that it simply melted.

In region 6, radiation begins to contribute substantially to the heat transfer.

---

[75] Adapted by permission from Farber, E. A. and R. L. Scorah (1948). Heat Transfer to Water Boiling Under Pressure. *Transactions of the ASME* **70**:369.

Condensation - mechanism

Condensation, like boiling, occurs by two distinct mechanisms. The mechanisms are analogous to those for boiling. They are called *dropwise* condensation and *film* condensation. For vertical surfaces, if the condensing liquid does not wet the surface readily, drops will form, and as soon as the drop is large enough that its weight overcomes the surface tension forces holding it to the surface, the drop runs rapidly down the surface, clearing the surface for more condensation.

Large drops that run from the surface coalesce with small drops in their path and thus clear them from the surface. In the dropwise case, most of the heat transfer takes place directly between the vapor and the solid surface, and therefore the heat transfer coefficient is quite high.

If the condensing liquid wets the surface readily, however, one will obtain a continuous film of liquid over the heat transfer surface and heat transfer then becomes limited by conduction through this film of liquid. There has been a great deal of work done on additives or surface coatings which will promote dropwise rather than filmwise condensation, but process equipment usually runs in the film regime.

A third type of heat transfer with phase change is associated with freezing (or crystallization, adduction, inclusion, clathrate formation, etc.) processes, where a solid phase is formed directly from either a liquid phase or a gas phase. We normally distinguish only one mechanism for these processes (although there are probably many) because this simplistic view describes many processes adequately. This phase change regime, although extremely important for many processes, is less common than boiling and condensation, and so we will emphasize the latter two processes.

Boiling and condensation are both extremely interesting and extremely important processes in heat transfer; however, to discuss the theory necessary to develop these areas in detail would require a text in itself. Therefore, here we will simply present some of the more common design equations to illustrate their general form. In the solution of specific practical problems, one should always refer to the literature for data and theory, or to experiments which one develops oneself.

Boiling coefficients

For film boiling on the outer surface of a horizontal tube, one can predict a heat transfer coefficient using the equation

$$h = h_c + h_r \left(\frac{3}{4} + \frac{1}{4} \frac{h_r}{h_c} \frac{1}{\left[2.62 + \frac{h_r}{h_c} \right]} \right)^{1/4}$$

(7.3.2-115)

where the convective film coefficient and the radiative coefficient are given by the equations

$$h_c = 0.62 \left[\frac{g \rho_v \left(\rho_L - \rho_v \right) \lambda k_v^3}{D \mu_v \left(T_s - T_L \right)} \right]^{1/4}$$

(7.3.2-116)

and

$$h_r = \frac{\varepsilon \sigma \left(T_s^4 - T_L^4 \right)}{\left(T_s - T_L \right)}$$

(7.3.2-117)

For film boiling on the surface of a vertical tube, one modifies the above correlation by calculating a Reynolds number defined by

$$Re = \frac{4 w}{\pi D \mu_v}$$

(7.3.2-118)

and calculating the heat transfer coefficient from the equation[76]

$$h \left[\frac{\mu_v^2}{g \rho_v \left(\rho_L - \rho_v \right) \lambda k_v^3} \right]^{1/3} = 0.0020 \, Re^{0.6}$$

(7.3.2-119)

For nucleate boiling one can use an equation not written explicitly in terms of a heat transfer coefficient as follows[77]

---

[76] Bromley, A. (1950). *Chem. Eng. Prog.* **46**: 221.
[77] Hsu, S. T. (1963). *Engineering Heat Transfer*, D. Van Nostrand.

$$\frac{\hat{C}_{pL}\left(T_s - T_v\right)}{\lambda} = C\left[\frac{q}{\mu_L \lambda}\sqrt{\frac{4.17\times10^8\,\sigma}{g\left(\rho_L - \rho_v\right)}}\right]^{0.33} Pr_L^{1.7} \qquad (7.3.2\text{-}120)$$

See Table 7.3.2-3 for values of C.

Table 7.3.2-3 Values of C for nucleate boiling model[78]

Fluid/Surface	C
n-pentane/chromium	0.015
water/platinum	0.013
water/copper	0.013
carbon tetrachloride/copper	0.013
benzene/chromium	0.010
water/brass	0.006
n-butyl alcohol/copper	0.00305
50% K_2CO_3/copper	0.00275
ethyl alcohol/chromium	0.0027
isopropyl alcohol/copper	0.00225

Condensing coefficients

For condensation on vertical tubes, for laminar flow the mean heat transfer coefficient is given by

$$h_a\left(\frac{\nu^2}{k^3 g}\right)^{1/3} = 1.47\,Re_x^{-1/3} \qquad (7.3.2\text{-}121)$$

where

$$Re_x = \frac{4\,x_e\langle v\rangle\,\rho}{\mu} \qquad (7.3.2\text{-}122)$$

and x_e is the film thickness. For turbulent flow, $Re_x > 2,000$

---

[78] Rosenow, W. M. (1952). *Trans. ASME* **74**: 969.

$$h_a \left(\frac{v^2}{k^3 g} \right)^{1/3} = 0.0077 \, Re_x^{0.4}$$ (7.3.2-123)

For condensation on a row of horizontal tubes stacked vertically, where condensate flow is laminar, the mean coefficient for the entire bank of tubes is

$$h_a = 0.725 \left[\frac{k^3 \rho g \lambda}{v \, N \, D \left(T_{sv} - T_s \right)} \right]^{1/4}$$ (7.3.2-124)

where N is the number of tubes in a row.

The reader should be cautioned that condensing coefficients can be affected substantially by the presence of non-condensable gases. For example, if one has a heat exchanger in which ethanol condenses in the presence of nitrogen, a layer of nitrogen builds up at the condensing surface and changes the problem from one in heat transfer to one in mass transfer, since the limiting step becomes the diffusion of the ethanol across a stagnant nitrogen layer which accumulates next to the tube surface. This is, of course, an exaggeration - the gas will not remain in an undisturbed layer next to the tube wall but instead bubbles will rise to the top of the exchanger, eventually forming a large vapor space that may ultimately surround some of the tubes if it is not vented. For this reason, condensers on distillation columns are usually provided with some type of vent for non-condensable gases.

7.4 Conduction and Convection in Series

The general heat transfer problem involves both conduction and convection paths in both series and parallel.

In a series heat transfer path all the heat goes through every segment of the path, as shown in the two-segment path above. It is often the case that a certain segment of the series path in the problem will contain virtually all the resistance to heat transfer (in the illustration, segment 2, whose resistance is assumed far greater than the resistance of segment 1), which is another way of stating that essentially all of the temperature drop will be across this segment, much as most of the voltage drop in a series electrical circuit[79] occurs across a resistance which is much larger than any other resistance path in the circuit. In such a case we can often neglect the segment with the small resistance, which would amount to eliminating the part of the path shown enclosed in the dashed lines, thereby assuming that T_1 is applied directly to the left-hand side of the second segment.

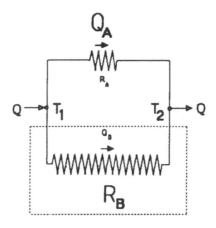

In the case of **parallel** paths (illustrated above for a two-segment path), we have a similar situation; however, now the controlling path is the one of **least** resistance, in the same fashion that in touching a "hot" wire while grounded we discover to our chagrin that most of the current goes through our body rather than the circuit intended. In parallel paths the temperature (voltage) drop is the same across all segments of the path, but the heat transfer (current) varies from segment to segment. If one segment is significantly lower in resistance than any of the others, virtually all the heat (current) will be transferred via this segment (segment A above). Low resistance in the controlling segment will require a relatively low temperature drop (voltage) to produce the heat flow (current). This same low temperature drop (voltage) applied across the much higher resistances of the remaining segments (here, segment B, which acts in effect as an open

---

[79] The reader is reminded that while the resistance of various segments is often considered to be a constant in the usual electrical circuit, in the case of heat transfer the thermal conductivity can be a function of temperature for conduction paths, and the heat transfer coefficient is frequently a strong function of the flow field for convective paths, so the analogy is not perfect.

circuit) will lead to negligible heat transfer (current) through these paths, so we can eliminate them (shown for this two-segment path by the dashed line).

We can frequently exploit such controlling steps to simplify our heat transfer calculation requirements. In some cases, we eliminate the necessity of calculation of entire segments of the heat transfer paths. In other cases, we reduce the order of the differential equations we are required to solve; often, we can reduce a problem requiring solution of partial differential equations to one of solution of ordinary differential equations. We shall consider some cases in the following section.

7.4.1 Lumped capacitance models

One-dimensional unsteady-state energy transfer thorough a solid (for simplicity in illustration, with constant thermal conductivity, although this restriction is not necessary) is modeled by the partial differential equation which describes the **microscopic** energy balance at a point

$$\rho\, C_p \frac{\partial T}{\partial t} = k \frac{\partial^2 T}{\partial x^2} \tag{7.4.1-1}$$

If we integrate the equation over the volume of the solid

$$\int_0^L \rho\, C_p \frac{\partial T}{\partial t} A\, dx = \int_0^L k \frac{\partial}{\partial x}\left(\frac{\partial T}{\partial x}\right) A\, dx \tag{7.4.1-2}$$

Sometimes we can model series heat transfer problems which include a solid segment by assuming that there is no spatial dependence of temperature (or density or heat capacity), for the illustration reducing the equation for the segment concerned to

$$A\, \rho\, C_p \frac{dT}{dt} \int_0^L dx = A\, \rho\, C_p \frac{dT}{dt} L = V\, \rho\, C_p \frac{dT}{dt}$$
$$= \int_0^L k \frac{\partial}{\partial x}(0)\, A\, dx = \text{constant} \tag{7.4.1-3}$$

$$V\, \rho\, C_p \frac{dT}{dt} = \text{constant} \tag{7.4.1-4}$$

Notice that this reduces the partial differential equation to an ordinary differential equation because under our assumption the temperature depends only on time. This is referred to as **lumping** the (thermal) capacitance because we have in essence "rolled up" the space dimensions into a single "lump" which contains no spatial variation. We have returned to the **macroscopic** viewpoint.

This is equivalent to assuming that thermal energy instantly distributes itself uniformly throughout the interior of the solid, so there are no temperature gradients in space. For thermal energy to move instantaneously implies infinite thermal conductivity. The constant is not dimensionless, but carries dimensions equivalent to those of the left-hand side, i.e.,

$$\left[L^3 \frac{M}{L^3} \left(\frac{M L^2}{t^2} \frac{1}{MT} \right) \frac{T}{t} \right] = \frac{M L^2}{t^3} \tag{7.4.1-5}$$

Although infinite thermal conductivity does not ordinarily occur, such a model furnishes a very good approximation to the case where the convective resistance to heat transfer at the surface of a solid is much greater than the conductive resistance within the material. The lumped model represents the two resistances in series

- the convective resistance at the surface of the object
- the conductive resistance within the object

with the first being larger and therefore the **controlling resistance**. As with all models, care must be taken that the model furnishes answers sufficiently close to the prototype.

For such models, to consider the problem in more than one space dimension we return to the macroscopic total energy balance to formulate the model equation. Consider a solid object of constant density at uniform initial temperature, T_i, to be instantly immersed in a colder liquid at constant temperature, T_∞. The macroscopic energy balance applied to a system consisting of the interior of the solid becomes, neglecting work terms and potential and kinetic energy terms, and recognizing that there are no convective terms (no inlets or outlets)

$$\frac{d}{dt} \int_V \hat{U} \rho \, dV = Q \tag{7.4.1-6}$$

Because of our assumptions that both density and temperature remain uniform in space within the solid, the internal energy will likewise be uniform in space.

$$\frac{d}{dt}\int_V \hat{U}\rho \, dV = \rho \frac{d}{dt}\left[\hat{U}\int_V dV\right] = \rho \frac{d}{dt}\left[\hat{U} \, V\right] \tag{7.4.1-7}$$

With no inlets and outlets and constant density, the macroscopic total mass balance shows that volume is constant in time

$$\rho \frac{d}{dt}\left[\hat{U} \, V\right] = \rho \, \hat{U}\frac{dV}{dt} + \rho \, V\frac{d\hat{U}}{dt} = \rho \, V\frac{d\hat{U}}{dt} = Q \tag{7.4.1-8}$$

But since

$$\frac{d\hat{U}}{dT} = \hat{C}_v \tag{7.4.1-9}$$

and for an incompressible material (which we will assume for our model)

$$\hat{C}_v = \hat{C}_p \tag{7.4.1-10}$$

we have

$$\rho \, V\frac{d\hat{U}}{dt} = \rho \, V\frac{d\hat{U}}{dT}\frac{dT}{dt} = \rho \, V\hat{C}_v\frac{dT}{dt} = \rho \, V\hat{C}_p\frac{dT}{dt} = Q \tag{7.4.1-11}$$

Q is the rate at which heat crosses the surface of the object, and can be expressed in terms of a convective heat transfer coefficient, the temperature of the surroundings, and the surface temperature of the object (which by assumption is the same uniform temperature as the interior). At this point we are at the same point as at the conclusion of the one-dimensional example, and we can see that Q is the constant in the one-dimensional example. As usual, we need to express Q in system parameters

$$Q = -h \, A\left(T - T_\infty\right) \tag{7.4.1-12}$$

where h, the convective heat transfer coefficient, has units of [energy/(area time)], and A is the surface area of the object. The negative sign occurs because $T > T_\infty$ by assumption, meaning that heat is flowing out of the system, so to maintain our sign convention of heat into the system being positive and out being negative, we require the negative sign. Therefore

$$\rho\, V\, C_p \frac{dT}{dt} = -h\, A\left(T - T_\infty\right) \tag{7.4.1-13}$$

Integrating

$$-\frac{\rho\, V\, C_p}{h\, A} \int_{T_i}^{T} \frac{dT}{\left(T - T_\infty\right)} = \int_{0}^{t} dt \tag{7.4.1-14}$$

$$-\left(\frac{\rho\, V\, C_p}{h\, A}\right) \ln\left[\frac{T - T_\infty}{T_i - T_\infty}\right] = t$$

$$\frac{T - T_\infty}{T_i - T_\infty} = \exp\left[-\left(\frac{h\, A}{\rho\, V\, C_p}\right) t\right] \tag{7.4.1-15}$$

The total temperature change (at infinite time) will be $(T_i - T_\infty)$, because as time approaches infinity, the temperature of the solid will approach T_∞. Therefore, the left-hand side of this relation can be defined as the (dimensionless) **unaccomplished temperature change, T^***.

Notice the units on the coefficient of time

$$\frac{h\, A}{\rho\, V\, C_p} \Rightarrow \frac{\left[\dfrac{FL}{L^2 tT}\right]\left[L^2\right]}{\left[\dfrac{M}{L^3}\right]\left[L^3\right]\left[\dfrac{FL}{MT}\right]} = \frac{1}{t} \tag{7.4.1-16}$$

so if we designate

$$\frac{\rho\, V\, C_p}{h\, A} \equiv t_c \tag{7.4.1-17}$$

we can see that t_c is the **time constant** of the process (when $t = t_c$, the unaccomplished temperature change will have a value of $1/e$). We can then define a **dimensionless time**[80] t^*

---

[80] This can be written in terms of dimensionless groups

$$t^* \equiv \frac{t}{t_c} \tag{7.4.1-18}$$

Equation (7.4.1-15) expresses the (dimensionless) unaccomplished temperature change, T^*, as a function of the dimensionless time, t^*.

$$T^* = e^{-t^*} \tag{7.4.1-19}$$

If we examine the components of t_c

$$\frac{1}{h A} \Rightarrow \frac{1}{\left[\frac{FL}{L^2 t T}\right][L^2]} = \frac{[T]}{\left[\frac{FL}{t}\right]} \tag{7.4.1-20}$$

$$\Rightarrow \frac{\text{driving potential}}{\text{energy flow rate}} = \text{thermal resistance}$$

$$\rho V C_p \Rightarrow \left[\frac{M}{L^3}\right][L^3]\left[\frac{FL}{MT}\right] = [FL] \tag{7.4.1-21}$$

$$\Rightarrow \text{energy} = \text{thermal capacity}$$

we see that t_c represents the product of the thermal resistance[81] and the (lumped) thermal capacity of the system,

The electrical analog of such a system is that of leakage of voltage charge on a capacitor through a resistor connected in parallel, as shown in Figure 7.4.1-1. The capacitor is initially charged by the battery; then the switch is opened and the charge allowed to decay through the resistor. Such a circuit could be used (and before the advent of digital computers frequently was used) to solve transient heat conduction models.

$$t^* = \frac{hAt}{\rho V C_p} = (h)\left(\frac{L^2}{L^3}\right)\left(\frac{t}{\rho C_p}\right) = \left(\frac{hL}{k}\right)\left(\frac{t}{\rho C_p L^2}\right) = \left(\frac{hL}{k}\right)\left(\frac{\alpha t}{L^2}\right)$$

$= \text{Bi Fo} \quad \Rightarrow \text{the product of two dimensionless groups, where}$

$\text{Bi} = \text{Biot number (see below) and}$

$\text{Fo} = \text{Fourier number} = \frac{\alpha t}{L^2}$

(another dimensionless time differing by the scale factor of the Biot number)

[81] For this lumped-capacitance model the resistance is only convective.

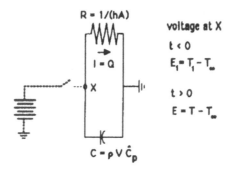

Figure 7.4.1-1 RC circuit analog of unsteady-state lumped-capacitance heat transfer

In the circuit shown in Figure 7.4.1-1, closing the switch the battery raises the voltage at point X to an initial voltage, E_i, which is analogous to the initial temperature difference $T_i - T_\infty$ (T_∞ is analogous to the ground potential of 0). At time $t = 0$ the switch is opened and the voltage (accumulated charge on the capacitor) allowed to decay by current leaking through the resistor. The resistor represents the convective resistance, the capacitor the thermal capacitance of the system. The electrical current is analogous to the heat flow.

Criteria for use of lumped capacitance models

The use of the lumped-capacitance assumption permitted us to reduce the partial differential equation describing unsteady-state heat transfer to an ordinary differential equation that was easily solved. Obviously we would like to make the lumped-capacitance assumption whenever possible to simplify computation.

In order to develop criteria for use of the assumption, for pedagogical reasons initially consider the case of one-dimensional **steady-state** conduction in a solid with convective resistance at the interface using rectangular coordinates (Figure 7.4.1-2). **Assume that the left-hand face of the solid is maintained at a constant temperature of T_1**, that the freestream temperature is T_∞ in the fluid to the right of the solid, and that the interfacial temperature at the conjunction of the right-hand face and the fluid is T_S.

A macroscopic energy balance about the conjunction of the right-hand face and the fluid (an infinitesimally thin system) shows only two terms remaining in the equation: the heat **conducted** to the face and the heat **convected** from the face

$$0 = Q_{conduction} + Q_{convection} \qquad\qquad (7.4.1\text{-}22)$$

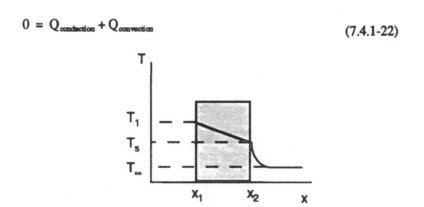

Figure 7.4.1-2 Steady-state conduction through solid with convection at interface

These two terms, when replaced with their numerical values, will have opposite signs because of our sign convention that heat into the system is positive.

Substituting using Fourier's law and the definition of the convective heat transfer coefficient

$$0 = -kA\frac{\left(T_s - T_1\right)}{\left(x_2 - x_1\right)} - hA\left(T_s - T_\infty\right) \qquad\qquad (7.4.1\text{-}23)$$

The negative sign preceding the first term on the right-hand side comes from Fourier's law (because the heat flow is in the opposite direction to the temperature gradient), not from the sign convention for heat. The negative sign preceding the second term on the right-hand side is there because of our sign convention regarding heat.

Since the gradient is negative for this example and k and A are both positive, the first term will have a net positive sign. Since the temperature difference is positive for this example and h and A are both positive, the second term will have a net negative sign.

Examining the ratio of the temperature drop in the solid to that in the fluid by rearranging Equation (7.4-23)

$$\frac{T_1 - T_s}{T_s - T_\infty} = \frac{h\, \Delta x}{k} = \frac{\left(\frac{\Delta x}{k}\right)}{\left(\frac{1}{h}\right)} = \frac{\text{conductive resistance}}{\text{convective resistance}} \qquad (7.4.1\text{-}24)$$

We define this ratio as the *Biot number*

$$\boxed{\text{Biot number} = \text{Bi} = \frac{h\,L}{k}} \qquad\qquad (7.4.1\text{-}25)$$

where L is an appropriate characteristic length for the particular problem. As reiterated below, it is common to define the characteristic length to be the ratio of the volume of the solid to its surface area.

> Notice that the Biot number, although nominally containing the same variables as the Nusselt number, *is not the same*, because the Biot number contains the thermal conductivity and space dimension of the solid phase and the heat transfer coefficient for the fluid phase, while the Nusselt number is based entirely on properties of the same (fluid) phase.

If the conductive resistance is small with respect to the convective resistance (therefore h is small relative to k, and the Biot number consequently small), the implication is that the temperature drop in the solid will be small with respect to that in the fluid; in other words, the temperature profile in the solid will be relatively constant, which suggests lumping all parts of the solid together as a single entity (this is analogous to the perfect mixing assumption for a fluid). This situation and its opposite extreme, very large h relative to k (large Biot number) are illustrated in Figure 7.4.1-3.

In the extreme case of very large Biot numbers, the surface temperature approaches that of the freestream, and the problem can be modeled as one with Dirichlet boundary conditions.

Now consider the a similar **unsteady**-state problem, still in one-dimensional rectangular coordinates, with a slab **initially at uniform temperature** being exposed to **identical** convective cooling at each of its faces. The derivative of the temperature with respect to x will be zero at the centerline, because otherwise there would have to be a thermal source or sink at the centerline.[82] Because of symmetry, we therefore need show only half the

---

[82] The problem is symmetric, so the derivative will be symmetric - a temperature gradient at the centerline would imply either thermal energy flowing into the center

slab. The temperature at the centerline (and throughout the slab) will continuously change until the slab reaches the temperature of the surroundings.

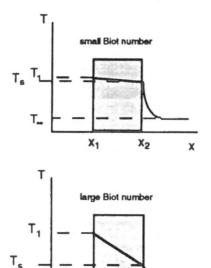

Figure 7.4.1-3 Steady-state conduction through solid with convection at interface, small vs. large Biot number

The temperature profile will no longer be linear because all the thermal energy transferred through a certain plane must be conducted also through any plane cut between that plane and the edge. There will be a steadily increasing thermal energy flux as one goes toward the edge of the slab (from the increasing thermal energy obtained from the cooling of the slab), so the temperature gradient must increase if the thermal conductivity and area are both constant. This is illustrated in Figure 7.4.1-4. The same general conclusions with respect to the Biot number still apply, however. We show the profile at some arbitrary time after the process starts.

---

plane from both directions and disappearing or flowing out of the center plane in both directions, implying a source.

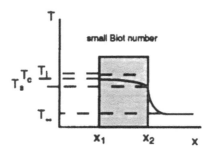

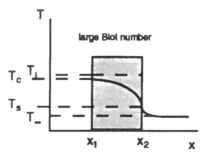

Figure 7.4.1-4 Unsteady-state conduction through solid with convection at interface, small vs. large Biot number, T_i = initial temperature, T_c = centerline temperature, T_s = surface temperature, T_∞ = freestream temperature

Although the appropriateness of a lumped-capacitance model depends on many factors, a rule of thumb often used is

$$\boxed{\text{Bi} < 0.1 \quad \Rightarrow \quad \text{lumping appropriate}} \qquad (7.4.1\text{-}26)$$

More conservative rules can, of course, be defined. Using the above constant for the rule of thumb, the characteristic length used (for both regular and irregular shapes) is typically the ratio of the volume of the solid to its surface area **through which transfer occurs**, which for typical regular shapes reduces to:

- For slabs with convection at both faces, the half-thickness
- For slabs with one insulated face, the thickness
- For long cylinders, half the radius
- For spheres, one-third the radius

Sometimes the more conservative value of L = radius is used for the cylinder and sphere.

Note that the Biot number should be computed before formulation or solution of the model whenever possible, because it may lead to a considerable reduction in computational labor by justifying a simpler model.

Example 7.4.1-1   Lumped capacitance models

A rectangular Pyrex wall at a uniform initial temperature of 100°C is cooled on each side by a gas at a freestream temperature of 20°C. Given the data below

- Is lumping appropriate to use in the calculation of the temperature of the wall with time?
- How long will it take the midpoint of the wall to cool to 40°C?

heat transfer coefficient	$10 \ W \ / \ (m^2 \ K)$

PYREX PROPERTIES[83]	
thermal conductivity	$1.4 \ W \ / \ (m \ K)$
density	$2250 \ kg \ / \ m^3$
heat capacity	$0.835 \ kJ \ / \ (kg \ K)$
thermal diffusivity	$0.74E\text{-}06 \ m^2 \ / \ s$
wall thickness	$20 \ mm$

Solution

$$Bi = \frac{hL}{k} = \frac{(10)\left[\frac{W}{m^2 K}\right](10)[mm]}{(1.4)\left[\frac{W}{m K}\right]} \frac{[m]}{(1000)[mm]} = 0.0714 < 0.1$$

$$(7.4.1\text{-}27)$$

Therefore, lumping is appropriate.

$$\frac{T - T_\infty}{T_i - T_\infty} = \exp\left[-\left(\frac{hA}{\rho V C_p}\right)t\right] = \exp\left[-\left(\frac{h 2 W H}{\rho L W H C_p}\right)t\right] \quad (7.4.1\text{-}28)$$

---

[83] Perry, R. H. and D. W. Green, Eds. (1984). *Chemical Engineer's Handbook*. New York, McGraw-Hill, p. 3-263.

$$\frac{T - T_\infty}{T_i - T_\infty} = \exp\left[-\left(\frac{2h}{\rho L \hat{C}_p}\right)t\right] \tag{7.4.1-29}$$

$$\left[\frac{40 - 20}{100 - 20}\right] =$$

$$\exp\left[-\left(\frac{(2)(10)\left[\frac{W}{m^2 K}\right]}{(2250)\left[\frac{kg}{m^3}\right](20)[mm](0.835)\left[\frac{kJ}{kg K}\right]}\right.\right. \tag{7.4.1-30}$$

$$\left.\left.\times \frac{(1000)[mm]}{[m]}\frac{[kJ]}{(1000)[J]}\left[\frac{W \cdot s}{J}\right]\right)(t)[s]\right]$$

$$[0.25] = \exp\left[-\left(5.32 \times 10^{-4}\right)(t)\right] \tag{7.4.1-31}$$

$$-\ln[0.25] = \left(5.32 \times 10^{-4}\right)(t)$$

$$t = 2606\,s \quad \Rightarrow \quad 43.4\,min \tag{7.4.1-32}$$

7.4.2   Distributed capacitance models

Thus far we have considered two unsteady-state situations for conduction in solids: (1) Dirichlet boundary conditions, where we had known surface temperatures, and (2) lumped parameter models, where we considered convective resistance at the surface to dominate and therefore the temperature gradient in the solid was insignificant.

We now wish to consider the case where significant gradients in temperature (i.e., thermal resistances) exist both at the surface and in the surrounding fluid. As we will see in Equation (7.4.2-5), this amounts to introducing a boundary condition of the third kind into our problem.

Starting with the microscopic thermal energy balance

$$\rho \, C_\varphi \left(\frac{\partial T}{\partial t} + v_x \frac{\partial T}{\partial x} + v_y \frac{\partial T}{\partial y} + v_z \frac{\partial T}{\partial z} \right) =$$

$$\left[\frac{\partial}{\partial x} \left(k \frac{\partial T}{\partial x} \right) + \frac{\partial}{\partial y} \left(k \frac{\partial T}{\partial y} \right) + \frac{\partial}{\partial z} \left(k \frac{\partial T}{\partial z} \right) \right] + \gamma_\theta \qquad (7.4.2\text{-}1)$$

for constant k, a stationary medium, equality of heat capacities at constant pressure and volume, and no thermal energy generation, we have as before

$$\rho \, \hat{C}_p \frac{\partial T}{\partial t} = k \frac{\partial^2 T}{\partial x^2} \qquad (7.4.2\text{-}2)$$

Our initial condition remains for the moment as before (we could let the initial temperature vary in space by using our numerical techniques, or by analytical techniques not covered in this text)

$$\text{at } t = 0: \quad \text{for} -L \le x \le L; \quad T = T_i \qquad (7.4.2\text{-}3)$$

By recognizing the fact that all the energy transferred to the surface of the slab must be removed by convection into the surrounding fluid (there is no generation term in the equation), the boundary conditions may be written as

$$\text{for } t > 0: \quad \text{at } x = \pm L; \quad -k \left[\frac{\partial T}{\partial x} \right]_{x \,=\, \pm L} = h \left(T_s - T_0 \right) \qquad (7.4.2\text{-}4)$$

Note that the quantities on the left-hand side of this equation pertain to the **solid**, those on the right to the **fluid**. If we rewrite the condition, we see that we have a **mixed** boundary condition - that is, one which incorporates both a Dirichlet and a Neumann condition

$$\text{for } t > 0: \quad \text{at } x = \pm L; \quad T_s = T_0 - \frac{k}{h} \left[\frac{\partial T}{\partial x} \right]_{x \,=\, \pm L} \qquad (7.4.2\text{-}5)$$

We can proceed to solve the equation using the techniques already introduced. We present the solution to this problem and to the corresponding problems in cylindrical and spherical coordinates in graphical form in Figures 7.4.2-1 through 7.4.2-6.

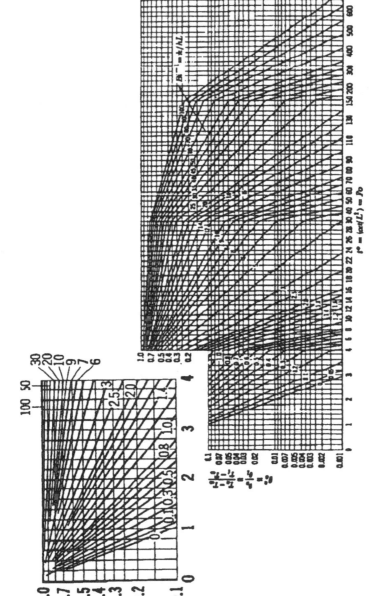

Figure 7.4.2-1   Mid-plane temperature for unsteady-state heat transfer in a slab of finite thickness 2L with uniform initial temperature and convective resistance at surfaces[84]

[84]Heisler, M. P. (1947). Temperature charts for induction and constant temperature heating. *Trans. ASME* 69:227-236.

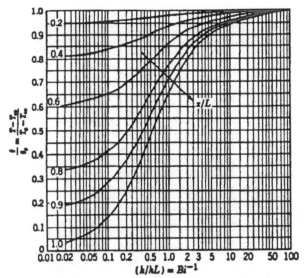

Figure 7.4.2-2 Temperature profile for unsteady-state heat transfer in a slab of finite thickness with uniform initial temperature and convective resistance at surfaces[85] x = distance from mid-plane; thickness = 2L

T_0 = centerline temperature

T_i = initial temperature

T_∞ = freestream temperature

$\theta = T - T_\infty$

$\theta_i = T_i - T_\infty$

$\theta_0 = T_0 - T_\infty$

$\theta_0^\cdot = \dfrac{\theta_0}{\theta_i} = \dfrac{T_0 - T_\infty}{T_i - T_\infty}$

$\dfrac{\theta}{\theta_0} = \dfrac{T - T_\infty}{T_0 - T_\infty}$

[85] Heisler, M. P. (1947). Temperature charts for induction and constant temperature heating. *Trans. ASME* **69**: 227-236.

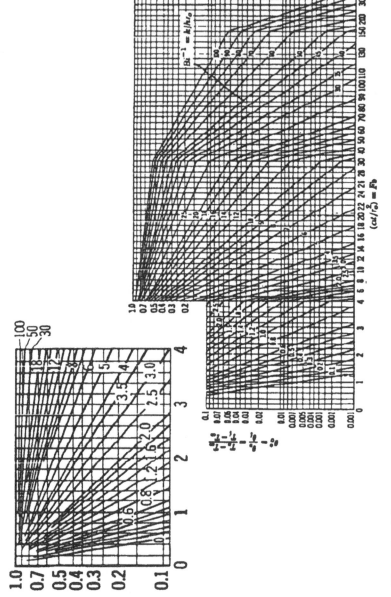

Figure 7.4.2-3   Centerline temperature for unsteady-state heat transfer in an infinite cylinder of radius r_o with uniform initial temperature and convective resistance at surfaces[86]

[86]Heisler, M. P. (1947). Temperature charts for induction and constant temperature heating. *Trans. ASME* 69:227-236.

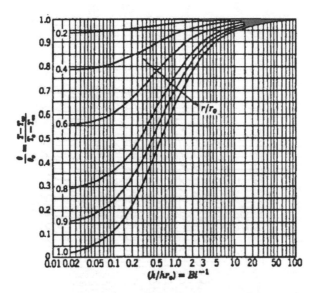

Figure 7.4.2-4 Temperature profile for unsteady-state heat transfer in an infinite cylinder of radius r_0 with uniform initial temperature and convective resistance at surface[87]

In Figures 7.4.2–3 and 7.4.2–4:

$$\boxed{\text{Bi} = \frac{h\,r_0}{k}}$$

as opposed to the lumped case where $\text{Bi} \equiv \dfrac{h\left(r_0/2\right)}{k}$

In Figures 7.4.2–5 and 7.4.2–6:

$$\boxed{\text{Bi} = \frac{h\,r_0}{k}}$$

as opposed to the lumped case where $\text{Bi} \equiv \dfrac{h\left(r_0/3\right)}{k}$

---

[87] Heisler, M. P. (1947). Temperature charts for induction and constant temperature heating. *Trans. ASME* **69**: 227-236.

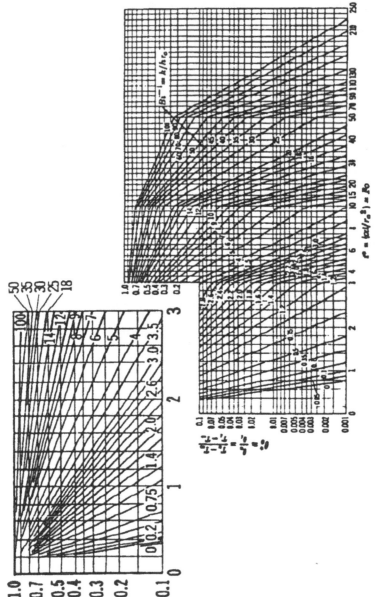

Figure 7.4.2-5 Center temperature for unsteady-state heat transfer in a sphere of radius r_o with uniform initial temperature and convective resistance at surface[88]

[88] Heisler, M. P. (1947). Temperature charts for induction and constant temperature heating. *Trans. ASME* 69:227-236.

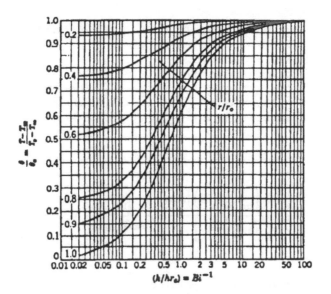

Figure 7.4.2-6 Temperature profile for unsteady-state heat transfer in a sphere of radius r_0 with uniform initial temperature and convective resistance at surface[89]

Example 7.4.2-1   Convective and conductive resistances in series

We now reconsider Example 7.4.1-1 under conditions such that lumping is no longer appropriate.

A rectangular Pyrex wall at a uniform initial temperature of 100°C is cooled on each side by water at a freestream temperature of 20°C. Given the data below

> • Is lumping appropriate to use in the calculation of the temperature of the wall with time?
> • How long will it take the midplane of the wall to cool to 40°C?
> • At this time what will be the temperature 4 mm from the face of the wall?

---

[89] Heisler, M. P. (1947). Temperature charts for induction and constant temperature heating. *Trans. ASME* **69**: 227-236.

heat transfer coefficient	300 W / (m² K)

PYREX PROPERTIES[90]

thermal conductivity	1.4 W / (m K)
density	2250 kg / m³
heat capacity	0.835 kJ / (kg K)
thermal diffusivity	0.74E-06 m² / s
wall thickness	20 mm

Solution

$$Bi = \frac{hL}{k} = \frac{(300)\left[\frac{W}{m^2\,K}\right](10)\left[mm\right]}{(1.4)\left[\frac{W}{m\,K}\right]} \frac{[m]}{(1000)\left[mm\right]} = 2.14 > 0.1$$

$$(7.4.2-6)$$

Therefore, lumping is no longer appropriate.

We calculate the ordinate of Figure 7.4.2-1

$$\theta_0^* = \frac{\theta_0}{\theta_i} = \frac{T_0 - T_\infty}{T_i - T_\infty} = \frac{40 - 20}{100 - 20} = 0.25 \qquad (7.4.2-7)$$

Using this together with the parameter on the graph

$$[Bi]^{-1} = \frac{1}{Bi} = \frac{1}{2.14} = 0.47 \qquad (7.4.2-8)$$

gives an abscissa value of

$$t^* = 1.35 = Fo = \frac{\alpha t}{L^2} = \frac{(0.74 \times 10^{-6})\left[\frac{m^2}{s}\right]}{(10)^2\left[mm\right]^2}\left[\frac{(1000)\,mm}{m}\right]^2 (t)$$

$$t = 182\,s$$

$$(7.4.2-9)$$

To find the temperature at 4 mm from the face, we use Figure 7.4.2-2. The abscissa is

---

[90] Perry, R. H. and D. W. Green, Eds. (1984). *Chemical Engineer's Handbook*. New York, McGraw-Hill, p. 3-263.

$$\left[Bi\right]^{-1} = 0.47 \tag{7.4.2-10}$$

and the parameter is

$$\frac{x}{L} = 0.47 = \frac{10-4}{10} = 0.6 \tag{7.4.2-11}$$

These two values yield

$$\frac{\theta}{\theta_0} = \frac{T-T_\infty}{T_0-T_\infty} = 0.79 = \frac{T-20}{40-20} \tag{7.4.2-12}$$

$$T = 36\ °F \tag{7.4.2-13}$$

7.5 Radiation Heat Transfer Models

Heat transfer by radiation occurs in a manner which does not parallel the mechanisms for transfer of mass or momentum, as was mentioned earlier. The transfer occurs without participation of an intervening medium (although at one time such a medium, the "ether," was postulated).

Thermal radiation - that is, the collective wavelengths of radiation that are the most important in heat transfer - is a relatively small part of the total electromagnetic spectrum as shown in Figure 7.5-1. The visible part of the electromagnetic spectrum contains a small part of what we classify as thermal radiation.[91]

---

[91] In an age of communications carried out in large part via electromagnetic waves, we can be grateful that our eyes are not sensitive to most of the spectrum - suppose that we had to see all the radio, television, microwave, etc. waves that surround us every day without having access to the *off* switch a radio or TV set possesses. There might be some dubious advantage in viewing police radar, but in exchange we would have to see (even if we chose not to hear) some of the more obnoxious forms of MTV. We would also have the questionable benefits of simultaneous visual bombardment by *Masterpiece Theater* and reruns of *Gilligan's Island*.

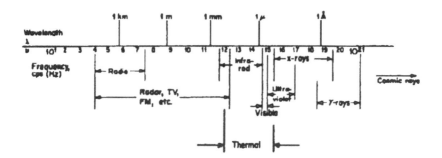

Figure 7.5-1 The electromagnetic spectrum[92]

7.5.1   Interaction of radiation and matter

Matter can reflect and/or transmit incident (incoming) radiation with or without a change in direction. It can also absorb incident radiation, and emit radiation in various directions.

The absorption and emission of radiation is accomplished by the transition of atoms and molecules among translational, rotational, vibrational, or electronic states which change the internal energy.

One important assumption that we shall use frequently in our models is that of a *diffuse surface:*

> **Diffuse surface: a surface with radiation properties independent of direction**

Geometric description of radiation

The geometry of radiation is described by solid angles measured in steradians.

> **Steradian: The ratio of the area of the spherical surface which subtends the angle to the *square* of the radius length, as illustrated in Figure 7.5.1-1**

---

[92] Greenkorn, R. A. and D. P. Kessler (1972). *Transfer Operations*, McGraw-Hill, p. 324.

$$\omega = \frac{A_{\text{subtending spherical surface}}}{r^2_{\text{sphere}}} \; [\text{steradians}] \qquad (7.5.1\text{-}1)$$

Since the area of the surface of a sphere is $4\pi r^2$, the surface of a sphere subtends 4π steradians

$$\omega_{\text{total surface of sphere}} = \frac{4 \pi r^2_{\text{sphere}}}{r^2_{\text{sphere}}} = 4 \pi \; [\text{steradians}] \qquad (7.5.1\text{-}2)$$

as opposed to the two-dimensional case, where a circle subtends only 2π radians.

To describe directions in space, the two angle coordinates from the spherical coordinate system suffice, as shown in Figure 7.5.1-1.

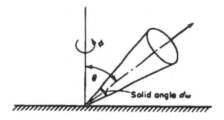

Figure 7.5.1-1 Directions in space[93]

Intensity of radiation

Radiation may be characterized by its **intensity**, which is defined as

the transmission rate of radiant energy

- of a particular wavelength,
- in a particular direction,
- per unit area of the surface **normal to this direction**,
- per unit solid angle about this direction, and
- per unit wavelength interval.

---

[93] Adapted from Sparrow, E. M. and R. D. Cess (1966). *Radiation Heat Transfer*, Brooks/Cole.

$$I_{\lambda,\theta,\varphi} = \frac{dQ_{\lambda,\theta,\varphi}}{dA_n \, d\omega \, d\lambda} = \frac{dQ_{\lambda,\theta,\varphi}}{dA_1 \cos\theta \, d\omega \, d\lambda}$$ (7.5.1-3)

where we have used the fact that the projected area normal to the (θ, φ)-direction of A_1 (with outward normal in the $\theta = 0$ direction) is $A_n = A_1 \cos\theta$.

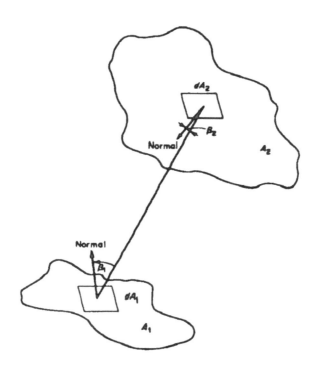

Figure 7.5.1-2 Radiation leaving A_1

We can see from Figure 7.5.1-2 that for A_2 normal to r, $\beta_2 = 0$, and $\beta_1 = \theta$

$$d\omega = \frac{dA_2}{r^2} = \frac{[r \, d\theta] \, [(r \sin\theta) \, d\varphi]}{r^2} = \sin\theta \, d\theta \, d\varphi$$ (7.5.1-4)

so the radiant energy flux **based on area A_1** is

$$\frac{d\dot{Q}_\lambda}{dA_1} = \left(dq_\lambda\right)_{A_1} = I_{\lambda,\theta,\varphi} \, \cos\theta \, \sin\theta \, d\theta \, d\varphi \tag{7.5.1-5}$$

where

$$dQ_\lambda = \frac{dQ}{d\lambda} \tag{7.5.1-6}$$

$$dq_\lambda = \frac{dq}{d\lambda} \tag{7.5.1-7}$$

Lumping of quantities used in modeling radiation

In general, all interactions of matter and radiation are functions of not only direction but also of wavelength. Furthermore, wavelength is a function of temperature. We frequently simplify our models through the use of **lumped** quantities, that is, quantities in which all of a particular class of radiation is grouped together by integration. We shall also assume diffuse surfaces.

The energy flux may be written for various types of radiation. Using a typical intensity designated as $I_{\lambda,\theta,\varphi}$, without specifying the nature of the radiation (incident, reflected, emitted etc.)

$$\left(dq_\lambda\right)_{A_1} = I_{\lambda,\theta,\varphi} \, \cos\theta \, \sin\theta \, d\theta \, d\varphi \tag{7.5.1-8}$$

Integrating the energy flux over a hemisphere

$$\iint_{hemi} dq_\lambda = \iint_{hemi} I_{\lambda,\theta,\varphi} \, \cos\theta \, \sin\theta \, d\theta \, d\varphi \tag{7.5.1-9}$$

$$\boxed{q_\lambda = \int_0^{2\pi} \int_0^{\frac{\pi}{2}} I_{\lambda,\theta,\varphi} \, \cos\theta \, \sin\theta \, d\theta \, d\varphi} \tag{7.5.1-10}$$

This is defined as the **spectral,** *hemispherical* energy flux.

Integrating over wavelength yields

$$\boxed{q = \int_0^\infty q_\lambda \, d\lambda} \tag{7.5.1-11}$$

(The lower limit on wavelength is zero since negative wavelengths are not physically realistic.) This is the **total, hemispherical** energy flux. **This is the lumping that we shall use for our introductory treatment.**

For a **diffuse surface**, $I_{\lambda,\theta,\phi}$ is not a function of θ or ϕ (so we add the subscript δ to remind ourselves that we have assumed diffuse behavior) and can therefore be factored out of the double integral in Equation (7.5.1-10).

$$
\begin{aligned}
q_{\lambda,\delta} &= \int_0^{2\pi} \int_0^{\frac{\pi}{2}} I_{\lambda,\delta} \cos\theta \, (\sin\theta \, d\theta) \, d\Phi \\
&= I_{\lambda,\delta} \int_0^{2\pi} \int_{-1}^0 \cos\theta \, d(-\cos\theta) \, d\Phi \\
&= I_{\lambda,\delta} \left[\int_0^{2\pi} d\phi \right] \left[\int_1^0 (-x) \, d(x) \right] \\
&= I_{\lambda,\delta} \left[\phi \right]_0^{2\pi} \left[\frac{-\left(x^2\right)}{2} \right]_1^0 \\
\end{aligned}
$$

$$\boxed{q_{\lambda,\delta} = \pi I_{\lambda,\delta}}$$

$$(7.5.1\text{-}12)$$

It then follows directly that

$$
\begin{aligned}
q_\delta &= \int_0^\infty q_{\lambda,\delta} \, d\lambda \;=\; \int_0^\infty \pi I_{\lambda,\delta} \, d\lambda \\
&= \pi \int_0^\infty I_{\lambda,\delta} \, d\lambda \\
\end{aligned}
$$

$$\boxed{q_\delta = \pi I_\delta}$$

$$(7.5.1\text{-}13)$$

Incident radiation

We will designate **incident radiation** by **G**, regardless of the source of that radiation (e.g., emission, reflection, transmission from another surface). In other words, we classify such radiation by its destination, not its origin.

We define the **irradiation** to include incident radiation from all directions. If the incident radiation is **diffuse** (independent of direction)

$$\boxed{G_{\lambda,\delta} = \pi\, I_{\lambda,\,\delta,\,\text{incident}}}$$ (7.5.1-14)

and

$$\boxed{G_{\delta} = \pi\, I_{\delta,\,\text{incident}}}$$ (7.5.1-15)

When radiation is incident on matter, several things may happen, the first two of which involve changes only in the direction of the radiation.

- First, it may pass through the substance unaltered except perhaps in direction (refracted); that is, it may be **transmitted**.

- Second, it may be **reflected** from the surface of the object. Reflection can occur in any combination of the two extreme cases known as **specular** and **diffuse** reflection. For specular reflection the angle of incidence is equal to the angle of reflection; for purely diffuse reflection the radiation is scattered with uniform intensity in all directions (see Figure 7.5.1-3).

- Third, the radiation may be **absorbed**, that is, converted into internal energy of the material. The absorption of photons can result in the direct change of electronic, vibrational, or rotational states of molecules, in the ionization of a molecule, or in the change in translational energy of a free electron.[94]

Absorption takes place near the surface for electrical conductors, but may take place up to several millimeters from the surface in non-conducting solids.[95] Liquids, solids, and many gases absorb significant amounts of thermal radiation. For example, monatomic and symmetric diatomic gases absorb little or no thermal radiation, but many polyatomic gases such as ammonia, carbon dioxide, water vapor, sulfur dioxide, long-chain hydrocarbons, etc., do absorb significant amounts of radiation.

---

[94] Sparrow, E. M. and R. D. Cess (1966). *Radiation Heat Transfer*, Brooks/Cole.

[95] Hsu, S. T. (1963). *Engineering Heat Transfer*, D. Van Nostrand, p. 426.

It should further be pointed out that gases and vapors do not absorb and emit radiation at all wavelengths. This characteristic is shared by liquids to a lesser extent. These materials are selective absorbers.

Models used herein neglect the effects of absorbing and/or radiating gases and vapors. Such models are useful but by no means universally applicable.

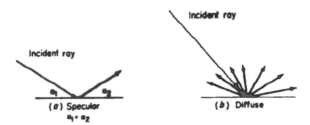

Figure 7.5.1-3 Extreme modes of reflection[96]

The **absorptivity**, α, the **reflectivity**, ρ, and the **transmittivity**, τ, describe the destination of incident radiation.

Absorptivity

The **total hemispherical absorptivity** is defined as

$$\alpha \equiv \frac{[G]_{absorbed}}{G} \qquad (7.5.1\text{-}16)$$

Reflectivity

Reflected radiation depends upon the direction of the incident ray. We will avoid this difficulty in our models by assuming diffuse reflection by our surfaces.

The **total hemispherical reflectivity** is

---

[96] Greenkorn, R. A. and D. P. Kessler (1972). *Transfer Operations*, McGraw-Hill, p. 325.

$$\boxed{\rho \equiv \frac{[G]_{\text{reflected}}}{G}}$$

(7.5.1-17)

Transmittivity

Bodies with a transmittivity of zero are called **opaque.** Transmittivity poses the same difficulty that reflectivity does, in that incident radiation can be transmitted at various angles. In this introductory treatment we restrict our attention to materials that are opaque, but we will remark that the following corresponding definitions can be made.

The **total hemispherical transmittivity** is defined as

$$\boxed{\tau \equiv \frac{[G]_{\text{transmitted}}}{G}}$$

(7.5.1-18)

Since absorption, reflection, and transmission account for all destinations of incident radiation, it follows that:

$$[G]_{\text{abs}} + [G]_{\text{refl}} + [G]_{\text{trans}} = G$$

(7.5.1-19)

$$\frac{[G]_{\text{abs}}}{G} + \frac{[G]_{\text{refl}}}{G} + \frac{[G]_{\text{trans}}}{G} = 1.0$$

(7.5.1-20)

$$\boxed{\alpha + \rho + \tau = 1.0}$$

(7.5.1-21)

Emitted radiation

Thus far we have discussed interaction of radiation and matter only via absorption, transmission, and reflection. Matter also *emits* radiation by making transitions between the same energy states as listed for absorption (but in the opposite direction).

The emission of radiation is quantified by defining the **emissive power** of a surface. We can make parallel definitions to those for incident radiation, but now using the emitted radiation.

We define the **spectral, hemispherical emissive power**, E_λ, [W/(m^2 μm)] as

$$E_\lambda = \int_0^{2\pi} \int_0^{\frac{\pi}{2}} \left[I_{\lambda,\theta,\varphi} \right]_{emitted} \cos\theta \sin\theta \, d\theta \, d\varphi$$

(7.5.1-22)

and the **total, hemispherical emissive power**, E_λ, [W/m^2] is defined as

$$E = \int_0^\infty E_\lambda \, d\lambda$$

(7.5.1-23)

If intensity of the emitted radiation is **diffuse** (independent of direction)

$$E_{\lambda,\delta} = \pi I_{\lambda,\delta,emitted}$$

(7.5.1-24)

and

$$E_\delta = \pi I_{\delta,emitted}$$

(7.5.1-25)

Blackbodies

A **blackbody** is an idealized body for which

$$\alpha = 1.0$$

(7.5.1-26)

for all wavelengths and directions - that is, all incident radiation is absorbed. In addition, blackbodies are diffuse emitters, and furthermore, no surface can emit more energy than a blackbody at a given temperature and wavelength.

In practice, true blackbodies do not exist, but the concept is valuable in theoretical considerations. Most substances which are thought of as very "black" - carbon black, black plush, black cats, tax collector's hearts, etc. - are really not very black by the standard of unit absorptivity.

In the laboratory the best approximation to a blackbody is an enclosure with a very small opening. Radiation entering this small opening bounces about from wall to wall, with little probability of being re-emitted by passing back through the entrance hole. At each reflection some of the radiation is absorbed, and so ultimately virtually all is absorbed.

Blackbody radiation

One of the early triumphs of quantum theory was the derivation of the spectral distribution of blackbody radiation[97]

$$
I_{\lambda, b} = \frac{2 h c_o^2 \lambda^{-5}}{e^{h c_o/(\lambda kT)} - 1}
$$

(7.5.1-27)

where T must be in absolute temperature, and

c_o = 2.998 x 10^8 [m / s] = speed of light in a vacuum
h = 6.6256 x 10^{-34} [J s] = Planck constant
k = 1.3805 x 10^{-23} [J / K] = Boltzmann constant

Because a blackbody is a diffuse emitter, its spectral emissive power is

$$
E_{\lambda, b} = \pi I_{\lambda, b} = \frac{C_1 \lambda^{-5}}{e^{C_2/(\lambda T)} - 1}
$$

(7.5.1-28)

where

C_1 = 3.742 x 10^8 [W μm^4 / m^2] = $2\pi h c_o^2$
C_2 = 1.439 x 10^4 [mm K] = $h c_o / k$

The **total emissive power of a blackbody** then is

$$
E_b = \int_0^\infty \frac{2 h c_o^2 \lambda^{-5}}{e^{h c_o/(\lambda kT)} - 1} \, d\lambda
$$

(7.5.1-29)

Evaluating the integral yields the **Stefan-Boltzmann law**.

---

[97] Planck, M. (1959). *The Theory of Heat Radiation.* New York, NY, Dover Publications.

$$\boxed{E_b = \sigma T^4}$$

(7.5.1-30)

where

$$\sigma = \text{Stefan–Boltzmann constant}$$

(7.5.1-31)

and

$$\sigma = 5.670 \times 10^{-8} \; \frac{W}{m^2 \, K^4}$$

(7.5.1-32)

$$\sigma = 0.1713 \times 10^{-8} \; \frac{Btu}{hr \, ft^2 \, R^4}$$

(7.5.1-33)

Emissivity

The *emissivity, ε,* is defined as the ratio of energy emitted at a given wavelength per unit area for the body in question to that of a blackbody at the same temperature. The total, hemispherical emissivity is defined as

$$\boxed{\varepsilon_T \equiv \frac{[E]_{emitted}}{E_{T,b}} = \frac{\int_0^\infty \varepsilon_{\lambda,T} E_{\lambda,T,b} \, d\lambda}{E_{T,b}}}$$

(7.5.1-34)

Radiosity

All the radiation leaving a surface, i.e., the reflection plus the emission, is called the **radiosity**, J.

For diffuse surfaces

$$\boxed{J_\lambda = \pi I_{\lambda,(e+r)}}$$

(7.5.1-35)

$$\boxed{J = \pi I_{(e+r)}}$$

(7.5.1-36)

7.5.2 Radiant heat exchange between two opaque bodies with no intervening medium

We now consider the problem of radiant heat exchange between surfaces. We will treat only fairly naive models analytically, but fortunately we can get fairly realistic approximations from such models.

Consider the case of two bodies:

 1. that radiate and absorb in the same spectral range (frequency)
 2. that are opaque
 3. that are situated so that all the radiation emitted from body I strikes body II and vice versa (e.g., infinite parallel planes)

Calculate the energy exchanged by radiation between the two bodies.

We can do this by considering a certain rate of emission of radiation by body I and subtracting from this amount

 (1) the portion which is reflected back and forth and ultimately reabsorbed by body I, plus
 (2) the radiation emitted from body II that is absorbed by body I.

Upon reaching body II, part of this radiation is absorbed and the remainder, since the body is opaque, is reflected back to body I. At body I, part of this reflected radiation is absorbed and the remainder reflected back again to body II, and so on. In Table 7.5.2-1 the reflected radiation from the previous step is enclosed each time in square brackets.

Table 7.5.2-1 History of radiation emitted

BDY	EMTTD	REFLECTED	ABSORBED
I	$\dot{Q}_I$		
II		$\dot{Q}_I\ (1\text{-}a_{II})$	$\dot{Q}_I\ a_{II}$
I		$[\dot{Q}_I\ (1\text{-}a_{II})]\ (1\text{-}a_I)$	$[\dot{Q}_I\ (1\text{-}a_{II})]\ a_I$
II		$[\dot{Q}_I\ (1\text{-}a_{II})\ (1\text{-}a_I)]\ (1\text{-}a_{II})$	$[\dot{Q}_I\ (1\text{-}a_{II})\ (1\text{-}a_I)]\ a_{II}$
I		$[\dot{Q}_I\ (1\text{-}a_{II})\ (1\text{-}a_I)$ $(1\text{-}a_{II})]\ (1\text{-}a_I)$	$[\dot{Q}_I\ (1\text{-}a_{II})\ (1\text{-}a_I)$ $(1\text{-}a_{II})]\ a_I$
II		$[\dot{Q}_I\ (1\text{-}a_{II})\ (1\text{-}a_I)$ $(1\text{-}a_{II})\ (1\text{-}a_I)]\ (1\text{-}a_{II})$	$[\dot{Q}_I\ (1\text{-}a_{II})\ (1\text{-}a_I)$ $(1\text{-}a_{II})\ (1\text{-}a_I)]\ a_{II}$
I		$[\dot{Q}_I\ (1\text{-}a_{II})\ (1\text{-}a_I)$ $(1\text{-}a_{II})\ (1\text{-}a_I)\ (1\text{-}a_{II})]\ (1\text{-}a_I)$	$[\dot{Q}_I\ (1\text{-}a_{II})\ (1\text{-}a_I)$ $(1\text{-}a_{II})\ (1\text{-}a_I)\ (1\text{-}a_{II})]\ a_I$
II		$[\dot{Q}_I\ (1\text{-}a_{II})\ (1\text{-}a_I)$ $(1\text{-}a_{II})\ (1\text{-}a_I)\ (1\text{-}a_{II})\ (1\text{-}a_I)]$ $(1\text{-}a_{II})$	$[\dot{Q}_I\ (1\text{-}a_{II})\ (1\text{-}a_I)$ $(1\text{-}a_{II})\ (1\text{-}a_I)\ (1\text{-}a_{II})\ (1\text{-}a_I)]$ a_{II}
etc.		etc.	etc.

If we continue the scheme shown, we can write that portion of radiation $\dot{Q}_I$ emitted by body I which is ultimately absorbed by body I as

Radiation ultimately absorbed by I =

$$\dot{Q}_I\,(1 + K + K^2 + K^3 + ...)\,(1 - \alpha_{II})\,\alpha_I \qquad (7.5.2\text{-}1)$$

where

$$K = \left(1 - \alpha_{II}\right)\left(1 - \alpha_I\right) \qquad (7.5.2\text{-}2)$$

but the sum of the infinite series is

$$1 + \sum_{i=1}^{n} K^i = \frac{1}{(1 - K)} \qquad (7.5.2\text{-}3)$$

and so Equation (7.5.2-1) reduces to

Radiation ultimately absorbed by I $= \dfrac{\dot{Q}_I\left(1-\alpha_{II}\right)\alpha_I}{\left(1-K\right)}$ (7.5.2-4)

If we follow the same procedure to calculate the portion of an emission from body II that is ultimately absorbed by body I we get

Emission from II ultimately absorbed by I $= \dfrac{\dot{Q}_{II}\,\alpha_I}{(1-K)}$ (7.5.2-5)

By subtracting (7.5.2-4) and (7.5.2-5) from $\dot{Q}_I$ we get the net heat transferred

$$\dot{Q} = \dot{Q}_I - \frac{\dot{Q}_I\left(1-\alpha_{II}\right)\alpha_I}{(1-K)} - \frac{\dot{Q}_{II}\,\alpha_I}{(1-K)}$$ (7.5.2-6)

Putting everything over a common denominator and substituting the definition of K yields

$$\dot{Q} = \frac{\dot{Q}_I\,\alpha_{II} - \dot{Q}_{II}\,\alpha_I}{\alpha_{\lambda I} + \alpha_{\lambda II} + \alpha_{\lambda I}\,\alpha_{II}}$$ (7.5.2-7)

This equation is true at non-equilibrium conditions as well as at equilibrium.

7.5.3 Kirchhoff's law

At thermal equilibrium

$$\dot{Q} = 0 \ \left(\text{net}\right)$$ (7.5.3-1)

Since the denominator of the right-hand side of Equation (7.5.2-7) cannot become infinite (the upper limit to absorptivity is 1.0) the numerator must vanish. Therefore

$$E_I\,\alpha_{II} - E_{II}\,\alpha_I = 0$$ (7.5.3-2)

or

$$\boxed{\frac{E_I}{\alpha_I} = \frac{E_{II}}{\alpha_{II}}}$$ (7.5.3-3)

This is *Kirchhoff's law*.

In Equation (7.5.3-3) there is nothing to say that one of the bodies cannot be a blackbody. Considering the special case for body II being black ($\alpha_{II} = 1.0$):

$$\frac{E_I}{E_{II,b}} = \alpha_I$$ (7.5.3-4)

But by definition

$$\frac{E_I}{E_{II,b}} = \varepsilon_I$$ (7.5.3-5)

Therefore, at *thermal* equilibrium

$$\boxed{\varepsilon_I = \alpha_I}$$ (7.5.3-6)

Figure 7.5.3-1 lists some total emissivities.

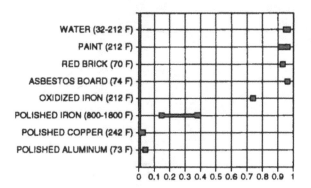

Figure 7.5.3-1 Total emissivity of some surfaces[98]

7.5.4   View factors

Consider radiant heat exchange between two arbitrarily oriented surfaces as shown in Figure 7.5.4-1. The radiation from surface 1 that is intercepted by surface 2 may be written as

$$d\dot{Q}_{1 \to 2} = I_1 \ dA_1 \cos \beta_1 \frac{dA_2 \cos \beta_2}{r^2}$$

$$(7.5.4-1)$$

where:

I_1 = total intensity for surface 1 (emitted plus reflected)
$dA_1 \cos \beta_1$ = projected area in r-direction
$dA_2 \cos \beta_2 / r^2$ = solid angle $d\omega$

A similar equation is obtained for radiation from surface 2 that is intercepted by surface 1.

$$d\dot{Q}_{2 \to 1} = I_2 \ dA_2 \cos \beta_2 \frac{dA_1 \cos \beta_1}{r^2}$$

$$(7.5.4-2)$$

Assuming each surface reflects and emits diffusely with uniform radiosity

---

[98] Data from H. C. Hottel as cited in McAdams, W. H. (1954). *Heat Transmission.* New York, NY, McGraw-Hill.

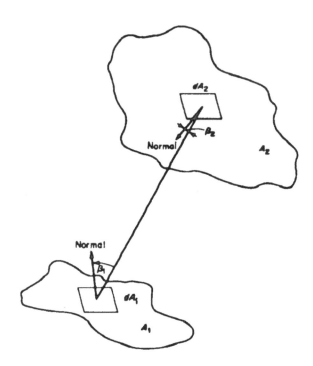

Figure 7.5.4-1 Radiation between surfaces[99]

$$J_1 = \pi I_{e+r}$$ (7.5.4-3)

Substituting and integrating

$$\dot{Q}_{1 \to 2} = J_1 \int_{A_1} \int_{A_2} \frac{\cos \beta_1 \cos \beta_2}{\pi r^2} dA_2 dA_1$$ (7.5.4-4)

We define the **view factor, F_{ij},** as the fraction of radiation leaving surface i that is intercepted by surface j. Notice that there exists a view factor between a surface and itself (i = j) because for concave surfaces radiation leaving one part of the surface can strike another part of the same surface.

[99] Greenkorn, R. A. and D. P. Kessler (1972). *Transfer Operations*, McGraw-Hill, p. 332.

For surfaces that reflect and emit diffusely, and have uniform radiosity, the view factor definition is expressed quantitatively as

$$F_{ij} \equiv \frac{1}{A_i} \int_{A_i} \int_{A_j} \frac{\cos\beta_i \cos\beta_j}{\pi r^2} \, dA_j \, dA_i \qquad (7.5.4\text{-}5)$$

Equation (7.5.4-5) must be integrated for each particular geometry. Since this is cumbersome, the results of this integration have been tabulated for common cases.

Then

$$\dot{Q}_{1\rightarrow 2} = J_1 \, A_1 F_{12} \qquad (7.5.4\text{-}6)$$

Developing a similar equation for radiant energy leaving surface 2 and intercepted by surface 1, we take the difference to obtain the net energy exchange as

$$\dot{Q}_{12} = J_1 \, A_1 F_{12} - J_2 \, A_2 F_{21} \qquad (7.5.4\text{-}7)$$

Some view factors can be evaluated intuitively. For example, no radiation leaving either a plane surface or a surface that is everywhere convex can strike the surface itself directly, so the view factor between such a surface and itself is zero. Conversely, the interior surface of an empty, hollow sphere has unity view factor with itself, since all radiation leaving the surface strikes (another point on) the same surface.

Reciprocity relation

From the definition of the view factor we can write

$$A_i \, F_{ij} = \int_{A_i} \int_{A_j} \frac{\cos\beta_i \cos\beta_j}{\pi r^2} \, dA_j \, dA_i \qquad (7.5.4\text{-}8)$$

and

$$A_j F_{ji} = \int_{A_j} \int_{A_i} \frac{\cos \beta_j \cos \beta_i}{\pi r^2} \, dA_i dA_j \qquad\qquad (7.5.4\text{-}9)$$

but by symmetry the integrals are equal, so

$$\boxed{A_i F_{ij} = A_j F_{ji}} \qquad\qquad (7.5.4\text{-}10)$$

which is the **reciprocity relation**.

Summation rule

If N surfaces form an *enclosure* - that is, all radiation is incident upon one of the surfaces (to use an incarceration analogy, none can "escape"), the definition of the view factor implies

$$\boxed{\sum_{j=1}^{N} F_{ij} = 1} \qquad\qquad (7.5.4\text{-}11)$$

This is the **summation rule**.

In an enclosure of N surfaces, there are N^2 view factors to be determined

$$\begin{bmatrix} F_{11} & F_{12} & \cdots & F_{1N} \\ F_{21} & F_{22} & \cdots & F_{2N} \\ \cdots & \cdots & \cdots & \cdots \\ F_{N1} & \cdots & \cdots & F_{NN} \end{bmatrix} \qquad\qquad (7.5.4\text{-}12)$$

However, N independent equations are supplied by the summation rule, leaving only $(N^2 - N)$ view factors required in order to compute all view factors. Furthermore, the reciprocity relation says that only half of these remaining view factors are independent, so there are in fact only

$$\frac{\left(N^2 - N\right)}{2} = \frac{N(N-1)}{2} \qquad\qquad (7.5.4\text{-}13)$$

independent view factors that need be determined in order to permit determination of all view factors.

The utility of the reciprocity relation and the summation rule is that they can at times be used to infer view factors that would otherwise require complicated integrations.

Example 7.5.4-1 Integration to obtain view factor

Using the definition, determine the view factor F_{12} by integration for radiant heat exchange between a small circular disk of area A_1 and a much larger, coaxial, parallel circular disk of area A_2. The smaller disk may be assumed to have a radius so small that it is essentially a point.

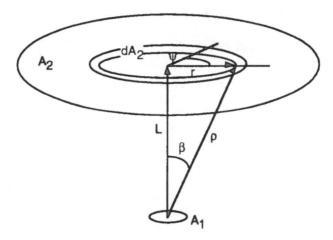

Solution

The definition of the view factor is

$$F_{ij} = \frac{1}{A_i} \int_{A_i} \int_{A_j} \frac{\cos \beta_i \cos \beta_j}{\pi \rho^2} \, dA_j \, dA_i \qquad (7.5.4\text{-}14)$$

The surfaces are normal to each other, so $\beta_1 = \beta_2 = \beta$. Since A_1 is so small, the integrand is constant over A_1. This leaves us with

$$F_{12} = \frac{1}{A_1} \int_{A_1} \int_{A_2} \frac{\cos^2 \beta}{\pi \rho^2} \, dA_2 \, dA_1 \qquad (7.5.4\text{-}15)$$

$$F_{12} = \frac{1}{A_1}\left[\int_{A_1} dA_1\right]\left[\int_{A_2} \frac{\cos^2 \beta}{\pi \rho^2} dA_2\right] \qquad (7.5.4\text{-}16)$$

$$F_{12} = \frac{A_1}{A_1}\left[\int_{A_2} \frac{\cos^2 \beta}{\pi \rho^2} dA_2\right] \qquad (7.5.4\text{-}17)$$

$$F_{12} = \int_{A_2} \frac{\cos^2 \beta}{\pi \rho^2} dA_2 \qquad (7.5.4\text{-}18)$$

We can write the differential area dA_2 as

$$dA_2 = [r\, d\psi][dr] = r\, dr\, d\psi \qquad (7.5.4\text{-}19)$$

Using the fact that $(L^2 + r^2) = \rho^2$ we can write the cosine as

$$\cos \beta = \frac{L}{\rho} = \frac{L}{\left(L^2 + r^2\right)^{1/2}} \qquad (7.5.4\text{-}20)$$

and the integral becomes

$$F_{12} = \int_0^{2\pi} \int_0^{r_2} \frac{\left[\dfrac{L}{\left(L^2+r^2\right)^{1/2}}\right]^2}{\pi\left(L^2+r^2\right)} r\, dr\, d\psi \qquad (7.5.4\text{-}21)$$

$$F_{12} = \left[\int_0^{2\pi} d\psi\right]\left[\int_0^{r_2} \frac{L^2}{\pi\left(L^2+r^2\right)^2} r\, dr\right] \qquad (7.5.4\text{-}22)$$

$$F_{12} = \left[2\pi\right]\left[\int_0^{r_2} \frac{L^2}{\pi\left(L^2+r^2\right)^2} r\, dr\right] \qquad (7.5.4\text{-}23)$$

$$F_{12} = \int_0^{r_2} \frac{L^2}{\left(L^2 + r^2\right)^2} 2\, r\, dr \tag{7.5.4-24}$$

$$F_{12} = L^2 \int_{L^2}^{L^2 + r_2^2} \frac{1}{\left(L^2 + r^2\right)^2} d\left(L^2 + r^2\right) \tag{7.5.4-25}$$

$$F_{12} = L^2 \left[\frac{-1}{\left(L^2 + r^2\right)} \right]_{L^2}^{L^2 + r_2^2} = L^2 \left[\left[\frac{-1}{\left(L^2 + r_2^2\right)} \right] - \left[\frac{-1}{L^2} \right] \right] \tag{7.5.4-26}$$

$$F_{12} = L^2 \left[\frac{-\left(L^2\right) + \left(L^2 + r_2^2\right)}{\left(L^2\right)\left(L^2 + r_2^2\right)} \right] = \frac{r_2^2}{\left(L^2 + r_2^2\right)} \tag{7.5.4-27}$$

Where r_2 is the radius of the upper disk.

View factors for more complicated geometries are obtained in a similar fashion (Table 7.5.4-1). View factors for many configurations are available in the literature.[100]

---

[100] For example, see Eckert, E. R. G. (1973). Radiation: Relations and Properties. *Handbook of Heat Transfer.* W. M. Rosenhow and J. P. Hartnett, eds. New York, NY, McGraw-Hill. Hamilton, D. C. and W. R. Morgan (1952). Radiant Interchange Configuration Factors, NACA. Howell, J. R. (1982). *A Catalog of Radiation Configuration Factors.* New York, NY, McGraw-Hill. Siegel, R. and J. R. Howell (1981). *Thermal Radiation Heat Transfer,* McGraw-Hill.

Table 7.5.4-1 View Factors

GEOMETRY	VIEW FACTOR
SEMI-INFINITE PARALLEL PLANES COMMON MIDPLANE Separation = L Width = w_k $w_k^* = \dfrac{w_k}{L}$	$F_{ij} = \left\{ \left[\left(w_i^* + w_j^* \right)^2 + 4 \right]^{1/2} \right.$ $- \left[\left(w_j^* - w_i^* \right)^2 + 4 \right]^{1/2} \Bigg\}$ $+ \left\{ 2\, w_i^* \right\}$
SEMI-INFINITE PLANES COMMON EDGE Common width = W Interior angle = α	$F_{ij} = 1 - \sin\left(\dfrac{\alpha}{2} \right)$
PERPENDICULAR SEMI-INFINITE PLANES COMMON EDGE Width = w_k	$F_{ij} = \dfrac{1 + \left(\dfrac{w_j}{w_i} \right) - \left[1 + \left(\dfrac{w_j}{w_i} \right)^2 \right]^{1/2}}{2}$
THREE-SIDED PLANAR SEMI-INFINITE ENCLOSURE Width = w_n	$F_{ij} = \dfrac{w_i + w_j - w_k}{2\, w_i}$
SEMI-INFINITE PARALLEL CIRCULAR CYLINDERS Axial separation = L+r_i+r_j $r^* = r_i / r_j,\ L^* = L / r_j,$ $M = 1 + r^* + L^*$ $C_1 = M^2 - (r^* + 1)^2$ $C_2 = M^2 - (r^* - 1)^2$ $C_3 = r^* / M - 1 / M$ $C_4 = r^* / M + 1 / M$	$F_{ij} = \dfrac{1}{2\pi} \left[\pi + \left(C_1 \right)^{1/2} - \left(C_2 \right)^{1/2} \right.$ $+ \left(r^* - 1 \right) \cos^{-1}\left(C_3 \right)$ $- \left(r^* + 1 \right) \cos^{-1}\left(C_4 \right) \Big]$

Example 7.5.4-2   Use of reciprocity relation and summation rule to infer view factor for concentric spheres

Determine the view factor F_{22} for the outermost of two concentric spheres. Designate the outer sphere with a subscript of 2 and the inner sphere with a subscript of 1.

Solution

By inspection.

$$F_{12} = 1.0$$
$$F_{11} = 0$$

$$(7.5.4-28)$$

Reciprocity gives

$$F_{21} = F_{12}\frac{A_1}{A_2} = \frac{A_1}{A_2}$$

$$(7.5.4-29)$$

and the summation rule gives

$$F_{21} + F_{22} = \frac{A_1}{A_2} + F_{22} = 1.0$$

$$F_{22} = 1.0 - \frac{A_1}{A_2} = 1.0 - \frac{4\pi r_1^2}{4\pi r_2^2} = 1.0 - \left(\frac{r_1}{r_2}\right)^2$$

$$(7.5.4-30)$$

Note that to obtain this result by integration is far from trivial.

7.5.5 Radiant heat exchange between blackbodies

For two blackbodies forming an enclosure, there is no reflection and the emissive power is known from the Stefan-Boltzmann law to be πI_b, so

$$\dot{Q}_{12} = J_1 A_1 F_{12} - J_2 A_2 F_{21}$$

$$(7.5.5-1)$$

becomes

$$\dot{Q}_{12} = [E_1]_b A_1 F_{12} - [E_2]_b A_2 F_{21}$$

$$(7.5.5-2)$$

or, using the Stefan-Boltzmann law and the reciprocity relation

$$\boxed{\dot{Q}_{12} = \sigma A_1 F_{12}\left(T_1^4 - T_2^4\right)}$$

$$(7.5.5-3)$$

which by symmetry is

$$\dot{Q}_{21} = \sigma A_2 F_{21} \left(T_2^4 - T_1^4 \right)$$

(7.5.5-4)

Example 7.5.5-1   Heat transfer by radiation - blackbody

Rather than applying conventional insulation directly to the outside of a furnace, it has been suggested that we use insulating material which is somewhat heat sensitive and build a double wall with an air gap as shown to prevent direct contact of the hot steel wall and the insulation. Given the data shown, what will be the inside temperature of the insulation? Neglect convection in the air gap.

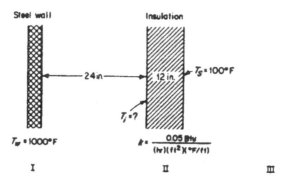

Solution

The heat transferred passes between I and II primarily by radiation (neglect natural convection), and through II by conduction. An energy balance shows

$$\dot{Q}_{I \to II} = \dot{Q}_{II}$$

(7.5.5-5)

Furthermore

$$\dot{Q}_{I \to II} = \sigma A F \left(T_w^4 - T_i^4 \right)$$

(7.5.5-6)

and

$$\dot{Q}_{II} = k A \left(\frac{T_i - T_s}{\Delta x_{II}}\right)$$

(7.5.5-7)

Substituting Equations (7.5.5-6) and (7.5.5-7) in Equation (7.5.5-5)

$$\sigma A F \left(T_w^4 - T_i^4\right) = k A \left(\frac{T_i - T_s}{\Delta x_{II}}\right)$$

(7.5.5-8)

Everything is known in the above relation except T_i

$$\left(0.1713 \times 10^{-8}\right) \left[\frac{Btu}{hr\ ft^2\ R^4}\right] (1.0) \left(1460^4 - T_i^4\right) \left[R^4\right]$$
$$= (0.05) \left[\frac{Btu}{hr\ ft\ R}\right] \left(\frac{T_i - 560}{1}\right) \left[\frac{R}{ft}\right]$$

(7.5.5-9)

$$0.1713 \times 10^{-8}\ T_i^4 + 0.05\ T_i - 7.82 \times 10^3 = 0$$

(7.5.5-10)

or, solving by Newton's method,[101] denoting

$$f\left(T_i\right) = 0.1713 \times 10^{-8}\ T_i^4 + 0.05\ T_i - 7.82 \times 10^3$$

(7.5.5-11)

$$f'\left(T_i\right) = 6.86 \times 10^{-9}\ T_i^3 + 0.05$$

(7.5.5-12)

$$\left(T_i\right)_{n+1} = \left(T_i\right)_n - \frac{f\left[\left(T_i\right)_n\right]}{f'\left[\left(T_i\right)_n\right]}$$

(7.5.5-13)

Newton's method converges very slowly for this problem.

$$T_i \cong 1458\ R = 998\ °F$$

(7.5.5-14)

Note that the air gap does little or no good - the inside temperature of the insulation remains almost the same as that of the steel wall.

---

[101] For example, see Gerald, C. F. and P. O. Wheatley (1994). *Applied Numerical Analysis*. Reading, MA, Addison-Wesley, p. 41.

Example 7.5.5-2   Use of view factor tables with blackbody radiation exchange

Two walls meet at a right angle as shown. The surfaces can be considered to be black. Surface temperatures are as shown.

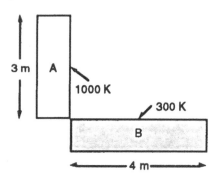

(a) Calculate the rate of radiant heat transfer from wall A to wall B per unit area of wall A.
(b) A third wall is constructed across the gap as shown. What is the net rate of heat transfer to wall B per square foot of B?

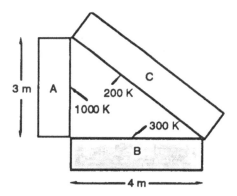

Solution

$$\dot{Q}_{ij} = \sigma A_i F_{ij} \left(T_i^4 - T_j^4 \right)$$

(7.5.5-15)

(a) The view factor relation from Table 7.5.4-1 is

$$F_{AB} = \frac{\left(1 + \frac{W_A}{W_B}\right) - \left(1 + \left[\frac{W_A}{W_B}\right]^2\right)^{1/2}}{2} \tag{7.5.5-16}$$

$$F_{AB} = \frac{\left(1 + \frac{3}{4}\right) - \left(1 + \left[\frac{3}{4}\right]^2\right)^{1/2}}{2} = 0.25 \tag{7.5.5-17}$$

Then

$$\frac{\dot{Q}_{AB}}{A_A} = \left(5.670 \times 10^{-8}\right)\left[\frac{W}{m^2 K^4}\right](0.25)\left(1000^4 - 300^4\right)\left[K^4\right] \tag{7.5.5-18}$$

$$\frac{\dot{Q}_{AB}}{A_A} = \left(1.406 \times 10^4\right)\left[\frac{W}{m^2}\right] \tag{7.5.5-19}$$

(b) The transfer to B is the sum of the transfer from A to B and from C to B. The appropriate view factor relation from Figure 7.5.4-1 is now

$$F_{ij} = \frac{W_i + W_j - W_k}{2 W_i} \tag{7.5.5-20}$$

Then

$$F_{BA} = \frac{W_B + W_A - W_C}{2 W_B} = \frac{4 + 3 - 5}{2(4)} = 0.25 \tag{7.5.5-21}$$

$$F_{BC} = \frac{W_B + W_C - W_A}{2 W_B} = \frac{4 + 5 - 3}{2(4)} = 0.75 \tag{7.5.5-22}$$

$$\frac{\dot{Q}_{BA}}{A_B} = \left(5.670 \times 10^{-8}\right)\left[\frac{W}{m^2 K^4}\right](0.25)\left(300^4 - 1000^4\right) K^4 \tag{7.5.5-23}$$

$$\frac{\dot{Q}_{BA}}{A_B} = -1.406 \times 10^4 \frac{W}{m^2} = -\frac{\dot{Q}_{AB}}{A_B} \tag{7.5.5-24}$$

$$\frac{\dot{Q}_{BC}}{A_B} = \left(5.670 \times 10^{-8}\right)\left[\frac{W}{m^2 K^4}\right](0.75)\left(300^4 - 200^4\right) K^4 \tag{7.5.5-25}$$

$$\frac{\dot{Q}_{BC}}{A_B} = 2.76 \times 10^2 \frac{W}{m^2} = -\frac{\dot{Q}_{CB}}{A_B} \tag{7.5.5-26}$$

Therefore, the net rate of transfer to B per unit area of B is

$$\frac{\dot{Q}_{net\,to\,B}}{A_B} = \frac{\dot{Q}_{AB}}{A_B} + \frac{\dot{Q}_{CB}}{A_B}$$

$$= 1.406 \times 10^4 \, \frac{W}{m^2} - 2.76 \times 10^2 \, \frac{W}{m^2}$$

$$= 1.378 \times 10^4 \, \frac{W}{m^2} \tag{7.5.5-27}$$

7.5.6   Radiative exchange between gray bodies

We can relax our restrictive assumption of blackbody behavior slightly by considering the case of *gray bodies*.

> A gray surface is one which absorbs a fixed fraction of incident radiation and emits the same fixed fraction that a blackbody emits at the same temperature.

We thus gain a degree of freedom in describing our physical situation but retain the independence of angle and wavelength we had with blackbody problems. This implies that gray bodies are **diffuse surfaces** - that emission and absorption do not depend upon direction.

By using our definition of emissivity we can therefore write for a *gray* body

$$E_g = \varepsilon \sigma T^4 \tag{7.5.6-1}$$

The **net** rate at which radiation leaves a gray surface, $\left[\dot{Q}_i \right]_g$, is the difference between the radiosity and irradiation (i.e., outgoing and incoming radiant energy)

$$\left[\dot{Q}_i \right]_g = A_i \left(J_i - G_i \right) \tag{7.5.6-2}$$

but the outgoing energy is made up of reflected energy plus emitted energy, so

$$J_i = \rho_i \, G_i + E_i \tag{7.5.6-3}$$

We can express emissive power for a gray body in terms of that of a blackbody, and since for a gray surface $[\rho_i]_g = 1 - [\alpha_i]_g = 1 - [\varepsilon_i]_g$, it follows that

$$J_i = \left(1 - [\varepsilon_i]_g\right) G_i + [\varepsilon_i]_g [E_i]_b \qquad (7.5.6\text{-}4)$$

which gives

$$G_i = \frac{J_i - [\varepsilon_i]_g [E_i]_b}{1 - [\varepsilon_i]_g} \qquad (7.5.6\text{-}5)$$

Substituting in Equation (7.5.6-2), the net rate at which radiation leaves a gray surface, i, may be written as (we now drop the subscript g since we are discussing only gray surfaces)

$$\dot{Q}_i = A_i \left(J_i - \frac{J_i - [\varepsilon_i] [E_i]_b}{1 - [\varepsilon_i]} \right) \qquad (7.5.6\text{-}6)$$

rearranging

$$\dot{Q}_i = \frac{\left([E_i]_b - J_i\right)}{\left(\dfrac{1 - [\varepsilon_i]_g}{[\varepsilon_i]_g A_i}\right)} \qquad (7.5.6\text{-}7)$$

This equation is in the form

$$\text{flow} = \frac{\text{potential difference}}{\text{resistance}} \qquad (7.5.6\text{-}8)$$

analogous to the electrical circuit equation

$$I = \frac{E}{R} \qquad (7.5.6\text{-}9)$$

The equation has the electrical circuit analog in Figure 7.5.6-1.

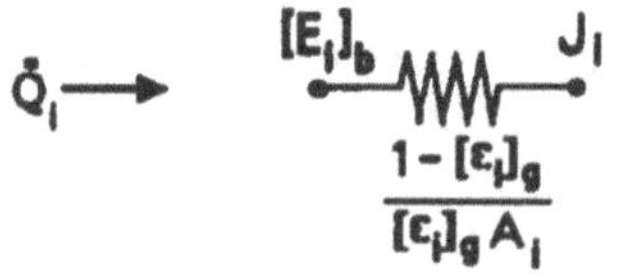

Figure 7.5.6-1 Electrical analog of net radiation from a gray surface

The total incident radiation to surface i comes from all the surfaces in the enclosure, including a possible contribution from the surface i itself. We can write this radiation as the sum of all these contributions by using the view factors as

$$A_i G_i = \sum_{j=1}^{N} F_{ji} A_j J_j \tag{7.5.6-10}$$

but using reciprocity

$$A_i G_i = \sum_{j=1}^{N} F_{ij} A_i J_j \tag{7.5.6-11}$$

so we can divide both sides by A_i, and substitute in our original expression for the net radiation leaving surface i

$$Q_i = A_i \left(J_i - \sum_{j=1}^{N} F_{ij} J_j \right) \tag{7.5.6-12}$$

Multiplying the first term on the right by 1.0 expressed as the summation rule gives (note the differing subscripts on J)

$$Q_i = A_i \left(\sum_{j=1}^{N} F_{ij} J_i - \sum_{j=1}^{N} F_{ij} J_j \right) = \sum_{j=1}^{N} A_i F_{ij} (J_i - J_j) = \sum_{j=1}^{N} Q_{ij} \tag{7.5.6-13}$$

Combining this relation with our expression for the electrical analog and rearranging the right-hand side to a parallel (driving force over resistance) form

$$\boxed{\frac{\left([E_i]_b - J_i\right)}{\left(\dfrac{1-[\varepsilon_i]_g}{[\varepsilon_i]_g A_i}\right)} = \sum_{j=1}^{N} \frac{\left(J_i - J_j\right)}{\left(A_i F_{ij}\right)^{-1}}}$$

$(7.5.6-14)$

This permits us to represent the exchange between surface i and the remaining surfaces in the enclosure as the network diagram in Figure 7.5.6-2. This diagram expresses the fact that the sum of the outflows to all surfaces from surface i is the outflow from surface i.

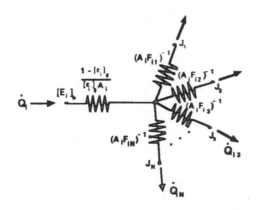

Figure 7.5.6-2 Network analog of radiation exchange with gray surfaces in an enclosure

Example 7.5.6-1   Two-gray-body exchange in enclosure

We wish to calculate from the general relationship the rate of radiative heat transfer between two gray surfaces that form an enclosure

$$Q_{ij} = \frac{\left([E_i]_b - J_i\right)}{\left(\dfrac{1-[\varepsilon_i]_g}{[\varepsilon_i]_g A_i}\right)} = \sum_{j=1}^{N} \frac{\left(J_i - J_j\right)}{\left(A_i F_{ij}\right)^{-1}}$$

$(7.5.6-15)$

This equation yields the following set of simultaneous equations (notice that one term on the right-hand side of each equation vanishes because the numerator becomes of the form $[J_i - J_i]$)

$$Q_{12} = Q_1 = \frac{[E_1]_b - J_1}{\dfrac{1-\varepsilon_1}{\varepsilon_1 A_1}} = \frac{J_1 - J_2}{\dfrac{1}{A_1 F_{12}}}$$

$$Q_{21} = Q_2 = \frac{[E_2]_b - J_2}{\dfrac{1-\varepsilon_2}{\varepsilon_2 A_2}} = \frac{J_2 - J_1}{\dfrac{1}{A_2 F_{21}}} \qquad (7.5.6\text{-}16)$$

The individual equations can be represented, respectively, by the equivalent circuit diagrams

Applying reciprocity in the second equation and multiplying both sides by (-1)

$$\frac{[E_1]_b - J_1}{\dfrac{1-\varepsilon_1}{\varepsilon_1 A_1}} = \frac{J_1 - J_2}{\dfrac{1}{A_1 F_{12}}}$$

$$-\frac{[E_2]_b - J_2}{\dfrac{1-\varepsilon_2}{\varepsilon_2 A_2}} = \frac{J_1 - J_2}{\dfrac{1}{A_1 F_{12}}} \qquad (7.5.6\text{-}17)$$

yields a set of equations which correspond to the circuit diagrams

But, since $\dot{Q}_1 = -\dot{Q}_2$, the two circuits are both represented in the single circuit

which from the electrical analog for a series circuit is modeled by

$$\text{Current} = \frac{\text{Potential difference}}{\Sigma\left[\text{Resistance}\right]}$$

$$\dot{Q}_1 = -\dot{Q}_2 = \frac{\left[E_1\right]_b - \left[E_2\right]_b}{\dfrac{1-\varepsilon_1}{\varepsilon_1 A_1} + \dfrac{1}{A_1 F_{12}} + \dfrac{1-\varepsilon_2}{\varepsilon_2 A_2}}$$

(7.5.6-18)

Example 7.5.6-2   Heat transfer by radiation - gray body

A 60° wedge-shaped surface, surface 1, is inserted into a long cylindrical duct, surface 2, of radius 15 mm as shown. The emissivity of the surface of the wedge is 0.73. The emissivity of the inside surface of the cylindrical section is 0.4.

A fluid is run through the wedge cross-section in such a fashion that surface (1) temperature is maintained at 900 K. The cylindrical surface (2) is maintained at 750 K.

a) Without using tables, determine F_{21}.
b) Calculate the heat transfer rate from the cylindrical surface to the surface of the wedge in W/mm² **based on the cylindrical surface.**

Solution

a) By inspection

$$F_{12} = 1.0 \tag{7.5.6-19}$$

By reciprocity

$$A_1 F_{12} = A_2 F_{21} \tag{7.5.6-20}$$

$$F_{21} = \frac{A_1}{A_2} F_{12} = \frac{(2)(15) \text{ mm} (L) \text{ mm}}{\left(\frac{300}{360}\right)(\pi)(30) \text{ mm} (L) \text{ mm}} (1.0)$$

$$F_{21} = 0.382 \tag{7.5.6-21}$$

b) This is a typical two-body enclosure. Therefore

$$Q_{12} = -Q_{21} = Q_1 = Q \neq Q_2 \tag{7.5.6-22}$$

where

$$Q = \frac{\sigma\left(T_1^4 - T_2^4\right)}{\dfrac{1-\varepsilon_1}{\varepsilon_1 A_1} + \dfrac{1}{A_1 F_{12}} + \dfrac{1-\varepsilon_2}{\varepsilon_2 A_2}} \tag{7.5.6-23}$$

Adjusting to the basis specified

$$\frac{-Q_{21}}{A_2} = \frac{Q}{A_2} = \frac{\sigma\left(T_1^4 - T_2^4\right)}{\dfrac{A_2\left(1-\varepsilon_1\right)}{\varepsilon_1 A_1} + \dfrac{A_2}{A_1 F_{12}} + \dfrac{1-\varepsilon_2}{\varepsilon_2}} \tag{7.5.6-24}$$

$$\frac{Q}{A_2} = \frac{5.67 \times 10^{-8} \dfrac{W}{m^2 K^4}\left(900^4 - 750^4\right) K^4}{\left| \dfrac{\left(\frac{300}{360}\right)(\pi)(30) \text{ mm} (L) \text{ mm} (1-0.73)}{(0.73)(2)(15) \text{ mm} (L) \text{ mm}} \right.}$$

$$\left. + \dfrac{\left(\frac{300}{360}\right)(\pi)(30) \text{ mm} (L) \text{ mm}}{(2)(15) \text{ mm} (L) \text{ mm} (1)} + \dfrac{(1-0.4)}{(0.4)} \right| \tag{7.5.6-25}$$

$$\frac{Q}{A_2} = 3790 \frac{W}{m^2}$$

$$\frac{Q_{21}}{A_2} = -3790\,\frac{W}{m^2}$$

(7.5.6-26)

7.6 Overall Heat Transfer Coefficients

We previously considered the problem of resistances in series during our consideration of **conductive** heat transfer. Now consider adding **convective** resistances to the path, such as for the case of the insulated pipe in Figure 7.6-1.

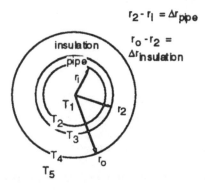

Figure 7.6-1 Insulated pipe

Let the heat transfer coefficient for the fluid inside the pipe (inside coefficient) be denoted as h_i, and the coefficient for the fluid outside of the insulation (outside coefficient) as h_o. We would like to develop an equation of the same **form** as that which we had for the **single-phase coefficients** [Equation (7.3.2-6) and Equation (7.3.2-7)], where $\dot{Q}$ is equal to an **overall** conductance (which we will call the *overall heat transfer coefficient*) times an appropriate area times the **overall** temperature drop. In cases where the area changes as we go in the direction of heat transfer (such as in cylindrical or spherical geometries with radial heat transfer) we must choose an area on which to base the coefficient.

We normally base the overall coefficient on either the inside heat transfer area or the outside heat transfer area- for example, for an uninsulated pipe with fluids on the inside and outside, we would normally use either the inside area of the pipe or the outside area of the pipe.

It is important to recognize that every overall heat transfer coefficient has associated with it a particular area. These two quantities (the overall heat transfer coefficient and its associated area) must be used together.

At steady state, the heat flow in the radial direction through the pipe wall is expressed by each of the following equations

$$\dot{Q}_r = h_i A_i \left(T_1 - T_2\right) = -h_i A_i \left(\Delta T\right) = -h_i A_i \left(T_2 - T_1\right)$$
(7.6-1)

$$\dot{Q}_r = \frac{k_P A_{P,\,lm}\left(T_2 - T_3\right)}{\Delta r_P}$$
(7.6-2)

$$\dot{Q}_r = \frac{k_I A_{I,\,lm}\left(T_3 - T_4\right)}{\Delta r_I}$$
(7.6-3)

$$\dot{Q}_r = h_o A_o \left(T_4 - T_5\right)$$
(7.6-4)

We could use any one of the above equations to calculate the heat flow, but the various temperature drops are usually not available. Only the overall temperature drop is ordinarily at our disposal.

To develop an equation in terms of overall temperature drop, we use exactly the same development as we did for a series of purely conductive resistances; that is, to determine the overall resistance we add all the individual temperature drops to get the overall temperature drop, and then substitute for the individual temperature drops using the heat flux equation

Writing Equation (7.6-1) to Equation (7.6-4) in terms of the appropriate temperature differences:

$$T_1 - T_2 = \frac{\dot{Q}_r}{h_i A_i}$$
(7.6-5)

$$T_2 - T_3 = \frac{\dot{Q}_r \Delta r_P}{k_P A_{P,\,lm}}$$
(7.6-6)

$$T_3 - T_4 = \frac{\dot{Q}_r \Delta r_I}{k_I A_{I,\,lm}}$$
(7.6-7)

$$T_4 - T_s = \frac{\dot{Q}_r}{h_o A_o} \qquad (7.6\text{-}8)$$

Adding Equation (7.6-5) through Equation (7.6-8) gives

$$\dot{Q}_r = \frac{T_1 - T_s}{\dfrac{1}{h_i A_i} + \dfrac{\Delta r_P}{k_P A_{P,\text{lm}}} + \dfrac{\Delta r_I}{k_I A_{I,\text{lm}}} + \dfrac{1}{h_o A_o}} = \frac{(\Delta T)_{\text{overall}}}{\Sigma R} \qquad (7.6\text{-}9)$$

where ΣR represents the sum of the thermal resistances.

First, let us develop the average coefficient associated with the **inside** area. If we multiply Equation (7.6-9) by the ratio of the inside area of the pipe to itself, A_i/A_i (multiplying by 1.0)

$$\dot{Q}_r = A_i (\Delta T)_{\text{overall}} \left[\frac{1}{\dfrac{1}{h_i} + \dfrac{\Delta r_P A_i}{k_P A_{P,\text{lm}}} + \dfrac{\Delta r_I A_i}{k_I A_{I,\text{lm}}} + \dfrac{A_i}{h_o A_o}} \right] \qquad (7.6\text{-}10)$$

The term in brackets in Equation (7.6-10) is called the **overall heat transfer coefficient based on the inside area, U_i**, and

$$\dot{Q}_r = U_i A_i (\Delta T)_{\text{overall}} \qquad (7.6\text{-}11)$$

In a similar fashion we can develop the overall coefficient associated with the **outside** area. If we multiply Equation (7.6-9) by the ratio of outside area of the insulation to itself, A_o/A_o

$$\dot{Q}_r = A_o (\Delta T)_{\text{overall}} \left[\frac{1}{\dfrac{A_o}{h_i A_i} + \dfrac{\Delta r_P A_o}{k_P A_{P,\text{lm}}} + \dfrac{\Delta r_I A_o}{k_I A_{I,\text{lm}}} + \dfrac{1}{h_o}} \right] \qquad (7.6\text{-}12)$$

The term in the brackets is the **overall heat transfer coefficient based on the outside area, U_0,** and

$$\boxed{Q_r = U_o A_o (\Delta T)_{\text{overall}}}$$

(7.6-13)

Typical values of U are shown in Figures 7.6.2a and b. They are only illustrative values and are not meant for calculation purposes other than for very crude approximation.

Heat transfer surfaces can increase in thermal resistance because of chemical transformations of the surface itself, such as rusting. Also, many fluids form deposits on heat transfer surfaces (e.g., by pyrolysis, precipitation, etc.) creating an additional thermal resistance. This resistance can become large enough that the exchanger must be periodically cleaned.

Cleaning of the surface may be done chemically, or, for deposits on the inner wall of tubes, by removing the exchanger head and passing through a rod with a scraper or brush attached which mechanically abrades the deposits. Some exchangers are designed to be self cleaning through incorporation of balls or other configurations that are trapped in baskets at the end of the tubes and may be forced through the tubes by reversing flow. Other exchangers, such as spiral exchangers, offer somewhat easier disassembly for cleaning. Obviously, when designing an exchanger, to facilitate cleaning one must consider where the deposits will form.

It is customary to assign a heat transfer coefficient to account for the resistance of deposits that progressively foul surfaces. This coefficient is called the **fouling** *coefficient,* h_f. (Another quantity, the **fouling** *factor* = $1000/h_f$, is sometimes used.)

The area associated with the fouling coefficient is taken as the area of the surface on which the material is deposited. The overall heat transfer coefficient based on the outside area of the insulation for an insulated pipe with fouling on the inside surface of the pipe is

$$U_o = \left[\cfrac{1}{\cfrac{A_o}{h_i A_i} + \cfrac{A_o}{h_f A_i} + \cfrac{\Delta r_P A_o}{k_P A_{P,\text{lm}}} + \cfrac{\Delta r_I A_o}{k_I A_{I,\text{lm}}} + \cfrac{1}{h_o}} \right]$$

(7.6-14)

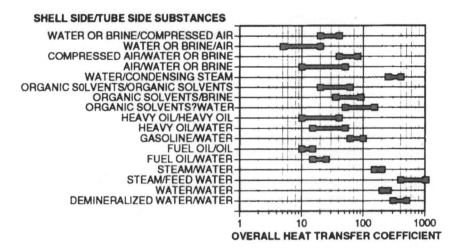

Figure 7.6-1a Overall heat transfer coefficients, Btu/(h ft² °F): tubular exchangers[102]

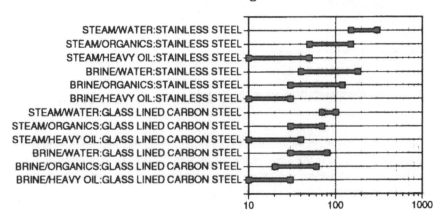

Figure 7.6-1b Overall heat transfer coefficients, Btu/(h ft² °F): jacketed vessels[103]

---

[102] Perry, R. H. and D. W. Green, Eds. (1984). *Chemical Engineer's Handbook*. New York, McGraw-Hill, p. 10-44.

[103] Perry, R. H. and D. W. Green, Eds. (1984). *Chemical Engineer's Handbook*. New York, McGraw-Hill, p. 10-46.

where h_f is the fouling coefficient. Fouling coefficients vary tremendously both in magnitude and time dependence, depending primarily on the particular fluids being used; typical values might be in the range 100-1000 Btu / (hr ft^2 °F).[104]

Sometimes the fluid used in heat exchange is not subject to change - for example, if it is one of the process streams. If, however, the fluid functions only as a heat exchange agent, one can select from a wide range of heat transfer fluids marketed commercially. Use of such fluids sometimes permits operation in more desirable temperature/pressure ranges. Many of the fluids do, however, present significant health or environmental hazards.

The procedure used above to develop overall heat transfer coefficients can be applied to a general problem with any number of conductive and convective resistances in series. The same form of result is obtained; that is, the overall heat transfer coefficient is the reciprocal of the sum of thermal resistances, which can be written for conduction resistances as

$$\boxed{\text{sum of conduction resistances} = \sum_{i=1}^{m} \frac{\Delta x_i \, A_j}{k_i \, A_{mean}}} \tag{7.6-15}$$

and for convection resistances as

$$\boxed{\text{sum of convection resistances} = \sum_{i=1}^{n} \frac{A_j}{h_i \, A_i}} \tag{7.6-16}$$

where Δx is the appropriate thickness coordinate, A_j is the area upon which U_j is based, and A_{mean} is the appropriate mean area for the conduction path, e.g., logarithmic for cylindrical geometry. In equation form, for m conductive and n convective resistances:

$$\boxed{U_j = \frac{1}{\displaystyle\sum_{i=1}^{m} \frac{\Delta x_i \, A_j}{k_i \, A_{mean}} + \sum_{i=1}^{n} \frac{A_j}{h_i \, A_i}}} \tag{7.6-17}$$

This equation is for resistances in **series**, which is the most common case.

---

[104] Perry, R. H. and D. W. Green, Eds. (1984). *Chemical Engineer's Handbook.* New York, McGraw-Hill, p. 10-43.

The above general approach can also easily be applied for parallel or series-parallel resistances.

Example 7.6-1   Controlling resistance for heat transfer resistances in series - spherical container of liquid oxygen

The stirred, insulated [k = 0.002 W/(m K)] spherical copper container of liquid oxygen at 87 K shown in the sketch is in ambient air at 298 K. The external heat transfer coefficient may be taken to be 18 W/(m² K). What is the leakage of heat into the container neglecting conduction of heat through any fittings passing through the insulation?

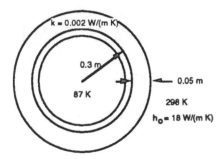

Solution

Since the problem is symmetrical, it is a one-dimensional series heat transfer problem in spherical coordinates (the heat travels in the radial direction only). For very highly insulated systems, the assumption of neglect of heat conducted through fittings that pierce the insulating layer should be checked carefully.

The path involves four resistances:

- the convective resistance in the internal fluid (liquid oxygen)
- the conductive resistance of the copper wall
- the conductive resistance of the insulation
- the convective resistance in the outer fluid (ambient air)

The internal forced convection heat transfer coefficient for the liquid oxygen should be much larger that the external free convection heat transfer coefficient for the gas, air; hence, the internal convective resistance should be much less than the external convective resistance and we shall neglect it.

The resistance terms for conduction through the vessel wall and the insulation are of the form

$$\frac{\Delta r}{k\,A_{gm}} \tag{7.6-18}$$

The sketch, even though not to scale, shows the thickness of the insulation to be the same magnitude as the thickness of the vessel wall, if not larger. The areas are of the same order of magnitude. The thermal conductivity of copper is of the order of 400 W/(m K)[105] at its maximum over the temperature range involved. Thus, the ratio of resistances is of the order of

$$\frac{\text{resistance of insulation}}{\text{resistance of vessel wall}} = \frac{\left[\dfrac{\Delta r}{k\,A_{gm}}\right]_{\text{insulation}}}{\left[\dfrac{\Delta r}{k\,A_{gm}}\right]_{\text{vessel wall}}} \approx \frac{\left[\dfrac{1}{0.002}\right]_{\text{insulation}}}{\left[\dfrac{1}{400}\right]_{\text{vessel wall}}} = 200{,}000 \tag{7.6-19}$$

so we will also neglect the thermal resistance of the vessel wall compared to that of the insulation.

$$T_i \; -\!\!\bigwedge\!\!\bigwedge\;-\!\!\bigwedge\!\!\bigwedge\;- \; T_o$$
$$[\Delta r/(kA_{gm})] \quad 1/(h_o A_o)$$
$$\longleftarrow Q_r$$

$$\dot{Q}_r = \frac{\Delta T_{\text{overall}}}{\dfrac{\Delta r_{\text{insulation}}}{k_{\text{insulation}}\,A_{\text{geometric mean}}} + \dfrac{1}{h_o\,A_o}} \tag{7.6-20}$$

$$\dot{Q}_r = \frac{T_o - T_i}{\dfrac{\left(r_{\text{outer, insulation}} - r_{\text{inner, insulation}}\right)}{k_{\text{insulation}}\left[\left(4\,\pi\,r^2_{\text{outer, insulation}}\right)\left(4\,\pi\,r^2_{\text{inner, insulation}}\right)\right]^{1/2}} + \dfrac{1}{h_o\,4\,\pi\,r^2_o}} \tag{7.6-21}$$

---

[105] Perry, R. H. and D. W. Green, Eds. (1984). *Chemical Engineer's Handbook*. New York, McGraw-Hill, p. 3-261.

$$\dot{Q}_r = \frac{T_o - T_i}{\dfrac{r_{outer,\ insulation} - r_{inner,\ insulation}}{4\pi k_{insulation}\, r_{outer,\ insulation}\, r_{inner,\ insulation}} + \dfrac{1}{h_o\, 4\pi r_o^2}} \tag{7.6-22}$$

$$\dot{Q}_r = \frac{T_o - T_i}{\dfrac{1}{4\pi k_{insulation}}\left[\dfrac{1}{r_i} - \dfrac{1}{r_o}\right] + \dfrac{1}{4\pi h_o r_o^2}} \tag{7.6-23}$$

$$\dot{Q}_r = \frac{(298 - 87)\,K}{\left\{\dfrac{1}{(4\pi)(0.002)\left[\dfrac{W}{m\,K}\right]}\left[\dfrac{1}{(0.35)\,m} - \dfrac{1}{(0.30)\,m}\right] \right.}$$

$$\left. + \dfrac{1}{(4\pi)(18)\left[\dfrac{W}{m^2\,K}\right](0.35)^2\,m^2}\right\} \tag{7.6-24}$$

$$\dot{Q}_r = \frac{(298 - 87)\,K}{(18.9) + (0.0361)} \tag{7.6-25}$$

$$\dot{Q}_r = 11.12\,W \tag{7.6-26}$$

Notice that the first term in the denominator, the conduction resistance of the insulation, is much larger than the free convection resistance at the outside of the sphere. This is what we would expect, because insulation by its very nature is designed to be the controlling resistance in a system

---

Example 7.6-2   Controlling resistance in replacement of section of wall of distillation column

A pilot plant distillation column is made of cylindrical sections of 10 in OD (outside diameter) glass 3/8 in. thick. It is proposed to replace one section of the column with an aluminum section with 0.37-in. wall thickness to facilitate machining to adapt instruments to the column. The question is to what extent the aluminum will increase heat loss from the column, thus condensing more liquid and generating additional internal reflux which could upset column operation. Insulating the aluminum section is an alternative, but will make changing the instrument taps more difficult as well as generally delaying the project and adding labor and some capital cost.

The inside heat transfer coefficient is estimated to be about 1000 Btu/(hr ft^2 °F) and the outside coefficient to be about 1 Btu/(hr ft^2 °F). Thermal conductivities of the glass and the aluminum are 7.5 Btu/(hr ft^2 [°F/in.]) and 1000 Btu/(hr ft^2 [°F/in.]), respectively.

Discuss the size of the change that results.

Solution

The equivalent resistance diagram is

Since the wall thickness is small compared to the radius, we will assume that

$$A_o = A_i = A_{lm} \tag{7.6-27}$$

Calculating the **relative** convection resistances (without the area, which is the same for each resistance under the above assumption)

$$\frac{1}{h_o} = \frac{1}{1.0} = 1 \frac{hr\ ft^2\ °F}{Btu} \tag{7.6-28}$$

$$\frac{1}{h_i} = \frac{1}{1000} = 0.001 \frac{hr\ ft^2\ °F}{Btu} \tag{7.6-29}$$

The **relative** conduction resistance for the glass wall is

$$\left[\frac{\Delta r}{k}\right]_{glass} = \frac{0.375\ in}{7.5\ \dfrac{Btu}{hr\ ft^2\ °F/in}} = 0.05 \frac{hr\ ft^2\ °F}{Btu} \tag{7.6-30}$$

and the **relative** conduction resistance for the aluminum wall is

$$\left[\frac{\Delta r}{k}\right]_{\text{aluminum}} = \frac{0.37\ \text{in}}{1000\ \dfrac{\text{Btu}}{\text{hr ft}^2\ ^\circ\text{F}/\text{in}}} = 0.0037\ \frac{\text{hr ft}^2\ ^\circ\text{F}}{\text{Btu}} \qquad (7.6\text{-}31)$$

Therefore, the percent change in heat transfer will be inversely proportional to the percent change in relative resistance

$$\frac{\left[\displaystyle\sum_{\text{glass}}\left(\text{relative resistances}\right) - \sum_{\text{aluminum}}\left(\text{relative resistances}\right)\right]}{\displaystyle\sum_{\text{glass}}\left(\text{relative resistances}\right)}(100)$$

$$= \frac{\left[\left(0.001 + 0.05 + 1\right) - \left(0.001 + 0.0037 + 1\right)\right]}{\left(0.001 + 0.05 + 1\right)}(100)$$

$$\approx 5\% \quad \text{(increase in heat transfer rate)} \qquad (7.6\text{-}32)$$

Whether this is significant or not depends upon the sensitivity of the particular separation being conducted in the column. For a pilot plant column, the convenience of the aluminum section probably would override the moderate upset in column operation, particularly since the aluminum section could always be insulated after the fact if a problem arose.

The external convection resistance is the dominant resistance in the transfer of heat, but not overwhelmingly controlling.

Example 7.6-3   Overall heat transfer coefficient with fouling

A multipass heat exchanger has two passes on the tube side and four passes on the shell side. The tubes are 1-in OD 16 BWG. Water on the shell side is used to cool oil on the tube side. Over time a fouling layer of scale has built up on the outside of the tubes from minerals in the cooling water.

Determine the overall heat transfer coefficient based on the outside tube area and neglecting the thermal resistance of the tube wall.

Heat transfer coefficients have been estimated as follows:

	h Btu/(hr ft^2 °F)
oil	48
water	170
scale	500

Solution

Heat exchanger tubes, like pipe, come in standard sizes. The wall thickness is usually specified in BWG (Birmingham Wire Gauge). For 1-in 16 BWG tubes

$$\frac{A_o}{A_i} = \frac{0.2618 \frac{ft^2}{\text{running ft}}}{0.2278 \frac{ft^2}{\text{running ft}}} = 1.15 \qquad (7.6-33)$$

The scale is on the water side of the tubes (outside) and so is associated with the outside area (this is not necessary, but is the logical area to employ). We assume that the inside surface, since it is in contact with oil, remains unfouled.

The overall heat transfer coefficient is then

$$U = \frac{1}{\frac{1}{h_o} + \frac{1}{h_f} + \frac{A_o}{A_i}\frac{1}{h_i}} \qquad (7.6-34)$$

$$U = \frac{1}{\left[\left(\frac{1}{170}\right) + \left(\frac{1}{500}\right) + (1.15)\left(\frac{1}{48}\right)\right]} \qquad (7.6-35)$$

$$U = 31.2 \frac{Btu}{hr\ ft^2} \qquad (7.6-36)$$

7.7 Heat Exchangers

Many, but by no means all, of the applications of heat transfer occur in exchanging heat between process streams via items equipment known (given the penchant of engineers for high literary creativity) as **heat exchangers**. These can be of many different configurations, the most common of which, however, is the shell-and-tube exchanger.

The shell-and-tube exchanger consists of a bundle of tubes with their axes parallel - much in the manner of soda straws in a carton - but supported at various points by baffles at right angles to the tube axes, which serve to keep the tubes fixed in space in a particular configuration, for example, with the axes spaced on equilateral triangles, or squares, etc. Fluid flows through groups of the tubes in parallel.

This tube bundle is encased in a shell, which confines the fluid which flows over the outside of the tubes in the tube bundle. The fluid which flows within the tubes (the tube-side fluid) may make a single pass through the exchanger or multiple passes, where the tubes are divided into groups via baffling at the ends of the exchangers and flow is reversed at the end of the exchanger and sent back and forth through different groups of tubes. Similarly, the fluid flowing external to the tubes (the shell-side fluid) may also make one or more passes across the tube bundle, depending upon the configuration of the baffles.

There are many practical considerations to heat exchanger design - thermal expansion and contraction stresses, removal of scale or other material that builds up on surfaces, etc. These details are usually best addressed by the vendor of the equipment. For the average engineer, it is necessary to be able to treat two generic problems

> • How to determine the area of an exchanger to be purchased for a particular heat exchange requirement - this is often referred to as the **design** problem

> • How to adapt an existing exchanger (that is, knowing the area) to operation with either altered flows or change in operating fluids - this is often referred to as the **retrofit** problem

These are the two problems we will emphasize here.

Before we begin, however, first we must address driving forces which vary through the heat exchange process.

7.7.1   Average overall temperature difference

So far we have applied the equation

$$\dot{Q} = U A (\Delta T)$$

$$(7.7.1-1)$$

for constant U and (ΔT) over area A.[106] In many cases of interest, however, ΔT varies as one moves about in the heat transfer area. It is convenient to retain this simple form of equation by defining appropriate average temperatures which permit the heat transfer to be written as

$$\dot{Q} = U A (\Delta T)_{average}$$

                                                                    (7.7.1-2)

The driving force for momentum transfer during fully developed flow of incompressible fluids in a uniform diameter pipe remains constant as we go down the pipe, because the velocity profile does not change. But in heat transfer, since the fluids involved change in internal energy, **the temperature profile changes, and therefore the driving force for heat transfer changes** - for example, the case of two fluids at different temperatures in steady concurrent flow, one within a tube and the other on the outside of the same tube, with neither changing phase.

As sensible heat is transferred from one fluid to the other, each will change in temperature as a result of first law considerations

$$\dot{Q} = w_i C_{pi} \Delta T_i$$

                                                                    (7.7.1-3)

and therefore the overall temperature difference, $(\Delta T)_{overall}$, between the two fluids in turn will vary. (As a second-order effect this also affects the properties of the fluid and thence the heat transfer coefficient, but we are neglecting these effects here.)

Consider the case where we use an overall heat transfer coefficient to describe heat transfer between a fluid, I, and another fluid, II. We would like to be able to describe the heat transfer by an equation of the form

$$\dot{Q} = U A \left(\Delta T \right)_{average}$$

                                                                    (7.7.1-4)

where $\dot{Q}$ is the total rate of heat transfer; U is the overall heat transfer coefficient based on a particular area; A is the prescribed area; and $(\Delta T)_{average}$ is a driving force calculated in a prescribed manner.

---

[106] Note that this implies that $\dot{Q}$ is positive for a positive temperature difference and vice versa. When applied to a thermal balance equation, this is the correct sign if the colder fluid is the system, but the negative of the proper quantity if $\dot{Q}$ is to be applied to the hotter fluid.

We restrict ourselves to cases where U may be assumed to be constant (frequently an excellent assumption) to avoid having to define **both** an average ΔT and an average U. Otherwise we would be confronted with an *embarras du choix* in which we could choose any definition for one of the averages and use the other to compensate accordingly. This problem is very similar to the one of defining an appropriate mean **area** to use with an overall ΔT in multilayer heat conduction problems.

We **define** this average overall temperature difference by requiring that it satisfy Equation (7.7.1-4), which implies the definition

$$(\Delta T)_{average} \equiv \frac{Q}{U\,A} \qquad\qquad (7.7.1\text{-}5)$$

In the development we assume

- The fluids flow at constant rate
- The fluids each have constant heat capacity
- U remains constant
- There is no phase change of either fluid (however, the case where a fluid absorbs or supplies heat **only** by phase change, i.e., its temperature remains constant, will also fit our ultimate definition)
- There is no heat exchange other than between the fluids - i.e., no heat losses from the exchanger to the surroundings
- There is no thermal energy source or sink in the system, or chemical reaction

We would now like to develop, for the two specific cases of strictly concurrent or strictly countercurrent flow (which, of course, includes such flows internal and external to pipes and tubes), an **average** overall temperature difference that can be used when (ΔT) changes rather than being constant as we go from point to point in the heat transfer area.

Our approach is to calculate the sum of the amounts of heat transferred across elemental areas, divide by UA, and write what remains in terms of temperature - this remnant, by definition, must be the required average ΔT.

We begin by writing $\dot{Q}$ in terms of the sum of amounts of heat transferred across elemental areas, of which dA is typical. The relative direction of flow of the fluids is unimportant to the final solution as we will develop it.

$$\dot{Q} = \lim_{\Delta A \to 0} \sum_A \delta \dot{Q} \qquad (7.7.1\text{-}6)$$

but

$$\delta \dot{Q} = U(\Delta T)(\Delta A) \qquad (7.7.1\text{-}7)$$

and so

$$\dot{Q} = \lim_{\Delta A \to 0} \sum_A U(\Delta T)(\Delta A) \qquad (7.7.1\text{-}8)$$

but this is

$$\dot{Q} = \int_A U(\Delta T) \, dA \qquad (7.7.1\text{-}9)$$

We now must evaluate the integral, but we face the problem of how ΔT varies with A (U is assumed constant and so causes no difficulty). We attack this problem by changing the independent variable in the integral.

We know that

1. We here can write $\dot{Q}$ as an exact differential (the difference between heat and work is an exact differential, and there is no work):

$$d\dot{Q} = U \Delta T \, dA \qquad (7.7.1\text{-}10)$$

$$\frac{dA}{d\dot{Q}} = \frac{1}{U \Delta T} \qquad (7.7.1\text{-}11)$$

This equation relates $\dot{Q}$ and A.

2. We also can relate $\dot{Q}$ and ΔT by observing that

$$d\dot{Q} = w_I \tilde{C}_{pI} \, dT_I \qquad (7.7.1\text{-}12)$$

$$-d\dot{Q} = w_{II} \tilde{C}_{pII} \, dT_{II} \qquad (7.7.1\text{-}13)$$

and

$$d\,(\Delta T\,) = d\,(T_I - T_{II}) = dT_I - dT_{II} \tag{7.7.1-14}$$

or, substituting,

$$d\,(\Delta T\,) = \frac{d\dot{Q}}{w_I\,\hat{C}_{pI}} + \frac{d\dot{Q}}{w_{II}\,\hat{C}_{pII}} = d\dot{Q}\left(\frac{1}{w_I\,\hat{C}_{pI}} + \frac{1}{w_{II}\,\hat{C}_{pII}}\right) \tag{7.7.1-15}$$

$$\frac{d\dot{Q}}{d\,(\Delta T)} = \frac{1}{\left(\dfrac{1}{w_I\,\hat{C}_{pI}} + \dfrac{1}{w_{II}\,\hat{C}_{pII}}\right)} \tag{7.7.1-16}$$

which relates $\dot{Q}$ and ΔT.

We change the independent variable by writing

$$\dot{Q} = U\int_A \Delta T\,dA = U\int_{\Delta T} \Delta T\,\frac{dA}{d\dot{Q}}\,\frac{d\dot{Q}}{d\,(\Delta T)}\,d\,(\Delta T) \tag{7.7.1-17}$$

$$\dot{Q} = U\int_{\Delta T_1}^{\Delta T_2} \Delta T\,\frac{1}{U\,\Delta T}\,\frac{1}{\left(\dfrac{1}{w_I\,\hat{C}_{pI}} + \dfrac{1}{w_{II}\,\hat{C}_{pII}}\right)}\,d\,(\Delta T) \tag{7.7.1-18}$$

$$\dot{Q} = \frac{1}{\left(\dfrac{1}{w_I\,\hat{C}_{pI}} + \dfrac{1}{w_{II}\,\hat{C}_{pII}}\right)}\int_{\Delta T_1}^{\Delta T_2} d\,(\Delta T) \tag{7.7.1-19}$$

$$\dot{Q} = \frac{\Delta T_2 - \Delta T_1}{\left(\dfrac{1}{w_I\,\hat{C}_{pI}} + \dfrac{1}{w_{II}\,\hat{C}_{pII}}\right)} \tag{7.7.1-20}$$

where $\Delta T1$ and $\Delta T2$ are the temperature differences at the ends of the tube. We have assumed heat capacity to be independent of T, and we have treated the heat capacity/flowrate products as constants since they do not change with ΔT as we go along the tube. To reduce the amount of writing required, we define

$$C \equiv w\hat{C}_p \tag{7.7.1-21}$$

We now substitute for $\dot{Q}$ in Equation (7.7.1-5) using Equation (7.7.1-20), and solve for $(\Delta T)_{avg}$ to obtain

$$(\Delta T)_{avg} = \frac{\Delta T_2 - \Delta T_1}{U A \left(\dfrac{1}{C_I} + \dfrac{1}{C_{II}}\right)}$$

(7.7.1-22)

This is not yet what we want because we have much on the right-hand side that is not temperature difference. Accordingly, we look for a way to eliminate those other things in terms of temperature difference. We observe that no single equation will do this, but a combination of Equations (7.7.1-10) and (7.7.1-15) to eliminate $d\dot{Q}$ will yield

$$d(\Delta T) = U (\Delta T) dA \left(\frac{1}{C_I} + \frac{1}{C_{II}}\right)$$

(7.7.1-23)

or

$$\int_{\Delta T_1}^{\Delta T_2} \frac{d(\Delta T)}{(\Delta T)} = U \left(\frac{1}{C_I} + \frac{1}{C_{II}}\right) \int_0^A dA$$

(7.7.1-24)

(where we merely indicate the area integration rather than writing it explicitly in radius and length). This yields

$$\ln\left[\frac{\Delta T_2}{\Delta T_1}\right] = U A \left(\frac{1}{C_I} + \frac{1}{C_{II}}\right)$$

(7.7.1-25)

and now we can substitute in Equation (7.7.1-22) to obtain

$$(\Delta T)_{average} = \frac{\Delta T_2 - \Delta T_1}{\ln\left[\dfrac{\Delta T_2}{\Delta T_1}\right]}$$

(7.7.1-26)

In other words, for heat exchange with constant U in tubes with fully developed flow of fluids with constant properties, Equation (7.7.1-4) is correct if used with the logarithmic mean temperature difference as the average temperature.

If $\Delta T_1 = \Delta T_2$, the logarithmic mean becomes indeterminate. A simple application of l'Hopital's rule rescues the situation, however:

$$\lim_{\left(\frac{\Delta T_2}{\Delta T_1}\right) \to 1} \left[\frac{\Delta T_2 - \Delta T_1}{\ln\left(\frac{\Delta T_2}{\Delta T_1}\right)}\right] = \lim_{\left(\frac{\Delta T_2}{\Delta T_1}\right) \to 1} \left[\frac{\left(\Delta T_2/\Delta T_1 - 1\right)\Delta T_1}{\ln\left(\frac{\Delta T_2}{\Delta T_1}\right)}\right]$$

(7.7.1-27)

$$\lim_{\left(\frac{\Delta T_2}{\Delta T_1}\right) \to 1} \left[\frac{\Delta T_2 - \Delta T_1}{\ln\left(\frac{\Delta T_2}{\Delta T_1}\right)}\right] = \lim_{\left(\frac{\Delta T_2}{\Delta T_1}\right) \to 1} \left[\frac{\frac{d}{d\left(\frac{\Delta T_2}{\Delta T_1}\right)}\left\{\left(\Delta T_2/\Delta T_1 - 1\right)\Delta T_1\right\}}{\frac{d}{d\left(\frac{\Delta T_2}{\Delta T_1}\right)}\left\{\ln\left(\frac{\Delta T_2}{\Delta T_1}\right)\right\}}\right]$$

(7.7.1-28)

$$\lim_{\left(\frac{\Delta T_2}{\Delta T_1}\right) \to 1} \left[\frac{\Delta T_2 - \Delta T_1}{\ln\left(\frac{\Delta T_2}{\Delta T_1}\right)}\right] = \lim_{\left(\frac{\Delta T_2}{\Delta T_1}\right) \to 1} \left[\frac{\Delta T_1}{\left(\frac{\Delta T_1}{\Delta T_2}\right)}\right] = \Delta T_2$$

(7.7.1-29)

but since $\Delta T_1 = \Delta T_2$, we can write for this case:

$$(\Delta T)_{average} = \Delta T$$

(7.7.1-30)

so for this case the appropriate form of Equation (7.7.1-2) is

$$Q = U A (\Delta T)_{\log \text{ mean}}$$

(7.7.1-31)

In the above development we assumed the heat transfer coefficient to be constant. If the heat transfer coefficient is not constant, there is an additional degree of freedom in the problem in that we can *arbitrarily* define $(\Delta T)_{avg}$ and define the average coefficient accordingly. In the following section we consider the design procedure for using average heat transfer coefficients based on a variety of driving forces.

Frequently the arithmetic mean is used in place of the logarithmic mean with sufficient accuracy. A discussion of the suitability of this approximation was included in Chapter 1 as an example.

7.7.2 Countercurrent vs. concurrent operation

One question that always faces the designer of an apparatus to exchange either energy or mass between two streams is whether to use **concurrent** (streams run in the same direction) or **countercurrent** (streams run in opposite directions) operation. Cross-flow and other varieties of operation are also possible; however, at the moment we wish merely to contrast concurrent and countercurrent flow.

Consider the exchange of sensible heat between two streams (for example, in a countercurrent double pipe heat exchanger). By **sensible** heat we mean that heat absorption results in a **change in temperature** of the stream (therefore, in principle the change can be **sensed**, as when one holds a hand in water issuing from a newly opened hot water tap as the water warms), as opposed to **latent** heat, which produces a change in phase at **constant temperature** (therefore, there is no change in temperature to be sensed, as when one holds a melting piece of ice in one's hand).

Designate the two streams with subscripts h for hot, c for cold. Further designate inlet conditions of a stream by i and outlet conditions by o. Let $\dot{Q}$ designate the amount of thermal energy transferred from the hot to the cold stream. If the cold stream is taken as the system, thermal energy will flow into the system and, therefore, by our sign convention $\dot{Q}$ will be positive; conversely, for the hot stream as the system, thermal energy will flow out of the system and, therefore, by our sign convention $\dot{Q}$ will be negative.

At steady state and in the absence of potential or kinetic energy changes or shaft work, the macroscopic energy balance (first law of thermodynamics) reduces to, for the hot stream as a system

$$\dot{H}_{ho} - \dot{H}_{hi} = \Delta \dot{H}_h = \left(w\, C_p \right)_h \left(T_{ho} - T_{hi} \right) = \dot{Q}_h \qquad (7.7.2\text{-}1)$$

and for the cold stream as a system

$$\dot{H}_{co} - \dot{H}_{ci} = \Delta \dot{H}_c = \left(w\, C_p \right)_c \left(T_{co} - T_{ci} \right) = \dot{Q}_c \qquad (7.7.2\text{-}2)$$

It follows that

$$Q_h = -Q_c$$
$$\Delta \dot{H}_h = -\Delta \dot{H}_c$$
$$\left(w\, C_p\right)_h \left(T_{ho} - T_{hi}\right) = -\left(w\, C_p\right)_c \left(T_{co} - T_{ci}\right) \qquad (7.7.2\text{-}3)$$

Only **differences** in enthalpy are significant in the above treatment, not the absolute value of enthalpies. This means that we can plot the temperature histories of the two streams on a T-H diagram without worrying about where we are along the H-axis, so long as the appropriate **difference** in enthalpy is preserved. We cannot, however, move along the T-axis with impunity, and, further, if we wish the plot to represent the relationship of the two streams in the exchanger, the appropriate points must line up vertically - for example, the inlet and outlet temperatures. Such a plot is shown in Figure 7.7.2-1 for two streams in countercurrent flow, each of which has heat capacity independent of temperature, which implies from the definition

$$w\, C_p \equiv \frac{dH}{dT} \qquad (7.7.2\text{-}4)$$

that the histories are straight lines.

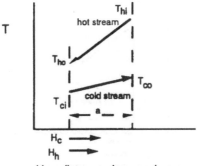

Figure 7.7.2-1 T-H diagram for countercurrent flow of two streams

We will align enthalpies according to their occurrence in space; that is, the enthalpy of the hot stream, H_h, at any point in the exchanger is to be plotted opposite the enthalpy of the cold stream, H_c, at that same point in the exchanger. If we do this, then the difference on the temperature scale, $T_h - T_c$, will represent the driving force for heat transfer at that point in the exchanger.

Although we shall continue to label the abscissa as H, because differences in enthalpy are directly related to the heat transfer, we need to remember that the abscissa really represents a space coordinate showing where we are in the exchanger, and that, therefore, in concurrent flow the scale for H will be reversed in direction for one of the fluids because as the space coordinate changes, both of the fluids are increasing in enthalpy (although the same interval on the scale will continue to represent the same difference in enthalpy for the hot and the cold fluid).

Notice that the absolute value of the distance **a** must be the same for both streams, because the change in enthalpy is the same for both except for sign. Figure 7.7.2-1 is for countercurrent flow, since the inlet for the hot stream is matched with the outlet for the cold stream and vice versa. The match of the streams at constant H corresponds to the match in space in the actual exchanger.

Concurrent operation is illustrated in Figure 7.7.2-2. Note that the direction of the arrows on the H-T curves is now the same. The absolute value of the distance **a** still must be the same for both streams. Again, the match of the streams at constant H corresponds to the match in space in the actual exchanger. Now, however, notice that the enthalpy scale for the cold fluid increases from right to left, while that for the hot fluid increases from left to right.

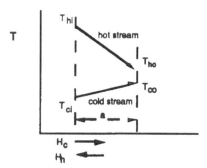

Figure 7.7.2-2 T-H diagram for concurrent flow of two streams

There is, however, a restriction on this plot imposed by the second law of thermodynamics, which requires that the hot stream to be always warmer than the cold stream in order for heat to flow from hot to cold. This presented no difficulty in the previous cases, because the hot stream was **everywhere** warmer than the cold stream. Consider, however, the slightly different case shown in Figure 7.7.2-3.

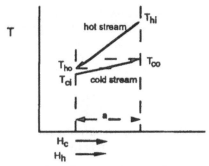

Figure 7.7.2-3 T-H diagram for countercurrent flow of two streams

Here we are heating a cold stream to an outlet temperature warmer than that of the cold stream in the previous illustration - the cold stream exit temperature is now above the hot stream exit temperature. As shown in Figure 7.7.2-3, we can still accomplish this with **countercurrent** operation without violating the second law of thermodynamics.

Now, however, let us consider attempting to accomplish the same task with **concurrent** operation, as shown in Figure 7.7.2-4.

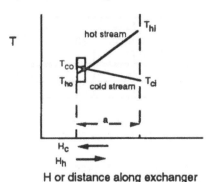

Figure 7.7.2-4 T-H diagram for concurrent flow of two streams

We have now tried to accomplish an impossibility - that portion of the cold stream and that portion of the hot stream that extend to the left beyond the point of intersection of the two streams **cannot exist**, because as soon as the cold stream and the hot stream reach the same temperature (**pinch** temperature), transfer of heat ceases. There is **enough** energy in the hot stream to heat the cold stream (first law), but the energy is at so low a temperature that the transfer cannot be accomplished. In fact, if we tried to operate in this manner, it would require a heat exchanger of infinite area simply to get the streams to the same temperature.

Temperature difference is the driving force for heat transfer. It can be seen that the **maximum** driving force obtainable is larger for concurrent flow, but this driving force quickly dissipates. The driving force for countercurrent flow, although smaller initially, decreases less rapidly.

Also notice that the **maximum temperature to which the cool stream can be heated** is larger for countercurrent than concurrent flow. Most heat exchangers are designed to operate in countercurrent flow.

Example 7.7.2-1   Concurrent vs. countercurrent flow in a concentric tube exchanger

We have two process liquid streams in our plant, one of which, stream A, is to be heated and the other of which, stream B, is to be cooled. No phase change is involved. Properties of the streams are listed in the following table. The minimum temperature of approach is to be 15°F (in order to keep the required area, and thus the capital cost, to a reasonable magnitude) and the streams are not to be split.

Stream	Flow rate (gpm)	Inlet Temperature (°F)	Density (lbmass/ft^3)	Heat Capacity [Btu/(lbmass °F)]
A (cold)	200	60	60	0.7
B (hot)	150	90	50	0.9

The overall heat transfer coefficient based on the outside area of the inner pipe may be taken to be constant at 100 Btu / (hr ft^2 °F)

Investigate using a double-pipe heat exchanger in both concurrent and countercurrent flow.

Solution

Calculating the slopes on an T-H diagram

$$\frac{dT}{dH} = \frac{1}{w\,C_p} \tag{7.7.2-5}$$

For the cold fluid

$$\left[\frac{dT}{dH}\right]_A = \frac{1}{\left[w\,C_p\right]_A}$$

$$= \left\{ (200)\left[\frac{gal}{min}\right]\left[\frac{ft^3}{(7.48)\,gal}\right](60)\left[\frac{lbmass}{ft^3}\right]\left[\frac{(0.7)\,Btu}{lbmass\,\circ F}\right] \right\}^{-1}$$

$$= 8.9 \times 10^{-4}\,\frac{\circ F}{Btu} \tag{7.7.2-6}$$

and for the hot fluid

$$\left[\frac{dT}{dH}\right]_B = \frac{1}{\left[w\,C_p\right]_B}$$

$$= \left\{ (150)\left[\frac{gal}{min}\right]\left[\frac{ft^3}{(7.48)\,gal}\right](50)\left[\frac{lbmass}{ft^3}\right]\left[\frac{(0.9)\,Btu}{lbmass\,\circ F}\right] \right\}^{-1}$$

$$= 1.11 \times 10^{-3}\,\frac{\circ F}{Btu} \tag{7.7.2-7}$$

CASE A: COUNTERCURRENT

A countercurrent exchanger will therefore have the pinch at the cold end, i.e., using a 15°F minimum temperature of approach $T_{hot,\ out} = 75°F$.

Calculating the heat duty from the known temperatures of the hot fluid

$$[\Delta H]_{hot} = [Q]_{hot} = \left[w\,C_p\,\Delta T\right]_{hot}$$

$$= (150)\left[\frac{gal}{min}\right]\left[\frac{ft^3}{(7.48)\,gal}\right](50)\left[\frac{lbmass}{ft^3}\right]\left[\frac{0.9\,Btu}{lbmass\,\circ F}\right](75°F - 90°F)$$

$$= -13{,}500\,\frac{Btu}{min} \tag{7.7.2-8}$$

Using the energy balance on the cold fluid

$$[\Delta H]_{cold} = [Q]_{cold} = [-Q]_{hot} = [w\,C_p\,\Delta T]_{cold}$$

$$13,500\,\frac{Btu}{min} = (200)\left[\frac{gal}{min}\right]\left[\frac{ft^3}{(7.48)\,gal}\right](60)\left[\frac{lbmass}{ft^3}\right]$$

$$\times\left[\frac{(0.7)\,Btu}{lbmass\,\,{}^\circ F}\right]\left(T_{cold,\,out}\,{}^\circ F - 60{}^\circ F\right)$$

$$T_{cold,\,out} = 72{}^\circ F$$

(7.7.2-9)

So we have

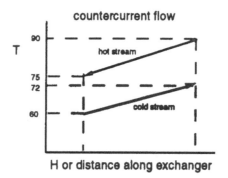

countercurrent flow

H or distance along exchanger

Using the log mean for the average temperature difference

$$(\Delta T)_{log\,mean} = \frac{\Delta T_2 - \Delta T_1}{\ln\left[\frac{\Delta T_2}{\Delta T_1}\right]} = \frac{(75-60){}^\circ F - (90-72){}^\circ F}{\ln\left[\frac{(75-60){}^\circ F}{(90-72){}^\circ F}\right]} = 16.5{}^\circ F$$

(7.7.2-10)

So we have

$$\dot{Q} = U_o\,A_o\,(\Delta T)_{log\,mean}$$

(7.7.2-11)

$$(13,500)\left[\frac{Btu}{min}\right](60)\left[\frac{min}{hr}\right] =$$
$$(100)\left[\frac{Btu}{hr\ ft^2\ °F}\right](A_o)\left[ft^2\right](16.5)°F$$
$$A_o = 491\ ft^2 \tag{7.7.2-12}$$

CASE B: CONCURRENT

$$[\Delta \hat{H}]_{cold} = [Q]_{cold} = [-Q]_{hot}$$
$$= \left[w\ \hat{C}_p\ \Delta T\right]_{cold} = -\left[w\ \hat{C}_p\ \Delta T\right]_{hot}(200)\left[\frac{gal}{min}\right]\left[\frac{ft^3}{(7.48)\ gal}\right]$$
$$\times(60)\left[\frac{lbmass}{ft^3}\right]\left[\frac{(0.7)\ Btu}{lbmass\ °F}\right]\left(T_{cold,\ out}°F - 60°F\right)$$
$$= -(150)\left[\frac{gal}{min}\right]\left[\frac{ft^3}{(7.48)\ gal}\right]$$
$$\times(50)\left[\frac{lbmass}{ft^3}\right]\left[\frac{(0.9)\ Btu}{lbmass\ °F}\right]\left(T_{hot,\ out}°F - 90°F\right)$$
$$\tag{7.7.2-13}$$

In order to maintain a 15 degree minimum approach temperature

$$T_{hot,\ out} = T_{cold,\ out} + 15 \tag{7.7.2-14}$$

Substituting

$$(1123)\left(T_{cold,\ out} - 60\right) = -902.4\left(T_{cold,\ out} + 15 - 90\right) \tag{7.7.2-15}$$
$$T_{cold,\ out} = 67$$
$$T_{hot,\ out} = 67 + 15 = 82 \tag{7.7.2-16}$$

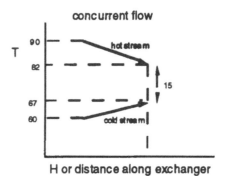

which means that we can transfer only

$$[Q]_{cold} = 1123\,(67-60) = 7860\,\frac{Btu}{min} \qquad (7.7.2\text{-}17)$$

CASE C: ALTERED TEMPERATURE OF APPROACH; CONCURRENT

Suppose we relax the 15 degree minimum approach temperature requirement to $(75 - 72) = 3$ degrees, which would permit us to transfer the same amount of energy as the countercurrent exchanger with a 15 degree approach temperature.

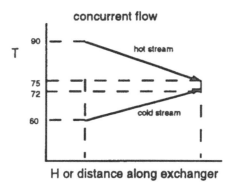

$$Q = U_o\,A_o\,(\Delta T)_{log\,mean} \qquad (7.7.2\text{-}18)$$

$$(\Delta T)_{log\,mean} = \frac{\left(\Delta T_2 - \Delta T_1\right)}{\ln\dfrac{\Delta T_2}{\Delta T_1}} = \frac{(75-72)\,(90-60)}{\ln\left[\dfrac{(75-72)}{(90-60)}\right]}$$

$$= 11.7\ ^\circ F \qquad (7.7.2\text{-}19)$$

$$Q = U_o A_o (\Delta T)_{\log\ mean}$$

$$(13,500)\left[\frac{Btu}{min}\right](60)\left[\frac{min}{hr}\right] = (100)\left[\frac{Btu}{hr\ ft^2\ {}_oF}\right](A_o)\left[ft^2\right](11.7)^oF$$

$$A_o = 692\ ft^2$$

(7.7.2-20)

CASE D: CORRESPONDING TEMPERATURE OF APPROACH; COUNTERCURRENT

A countercurrent exchanger with a 3 degree approach temperature would transfer, calculating the heat duty from the known temperatures of the hot fluid

$$\left[\Delta\hat{H}\right]_{hot} = Q_{hot} = \left[w\ C_p\ \Delta T\right]_{hot}$$

$$= (150)\left[\frac{gal}{min}\right]\left[\frac{ft^3}{(7.48)\ gal}\right](50)\left[\frac{lbmass}{ft^3}\right]$$

$$\times\left[\frac{(0.9)\ Btu}{lbmass\ {}_oF}\right](63^oF - 90^oF)$$

$$= -24,360\ \frac{Btu}{min}$$

(7.7.2-21)

Using the energy balance on the cold fluid

$$\left[\Delta\hat{H}\right]_{cold} = \left[Q\right]_{cold} = \left[-Q\right]_{hot} = \left[w\ C_p\ \Delta T\right]_{cold}$$

$$24,360\ \frac{Btu}{min} = (200)\left[\frac{gal}{min}\right]\left[\frac{ft^3}{(7.48)\ gal}\right](60)\left[\frac{lbmass}{ft^3}\right]$$

$$\times\left[\frac{(0.7)\ Btu}{lbmass\ {}_oF}\right]\left(T_{cold,\ out}\ {}^oF - 60^oF\right)$$

$$T_{cold,\ out} = 82^oF$$

(7.7.2-22)

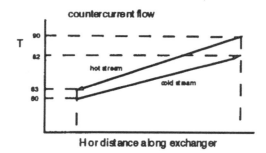

$$\dot{Q} = U_o A_o (\Delta T)_{log\,mean} \qquad\qquad\qquad (7.7.2\text{-}23)$$

$$(\Delta T)_{log\,mean} = \frac{\Delta T_2 - \Delta T_1}{\ln\left[\dfrac{\Delta T_2}{\Delta T_1}\right]} = \frac{(63-60)^\circ F - (90-82)^\circ F}{\ln\left[\dfrac{(63-60)^\circ F}{(90-82)^\circ F}\right]} = 5.1^\circ F$$

$$(7.7.2\text{-}24)$$

This would require an area of

$$Q = U_o A_o (\Delta T)_{log\,mean}$$

$$(24,360)\left[\frac{Btu}{min}\right](60)\left[\frac{min}{hr}\right] = (100)\left[\frac{Btu}{hr\,ft^2\,\circ F}\right](A_o)\left[ft^2\right](5.1)^\circ F$$

$$A_o = 2866\,ft^2$$

$$(7.7.2\text{-}25)$$

All the above have assumed unconstrained supplies of streams A and B.

7.7.3 NTU method for design of heat exchangers

In the examples in the previous section, we used streams for which the energy available (enthalpy change of hot stream) from the hot stream exactly equaled the energy required to heat the cold stream (enthalpy change of cold stream).

Consider now a particular example where we assume that we need to remove thermal energy from a hot stream with a high temperature, T_{hi} (inlet), by transferring the energy to a cold stream with a low temperature, T_{ci} (inlet). The first law of thermodynamics says that energy removed from the hot stream must be equal to that absorbed by the cold stream. The second law of thermodynamics requires that to transfer the energy from hot to cold, the hot stream must be at a higher temperature than the cold.

A typical situation is shown in Figures 7.7.3-1 and 7.7.3-2, for which the energy to take the hot stream from T_{hi} to T_{ho} and the energy required to take the cold stream from T_{ci} to T_{co} do not necessarily coincide. The slope of the individual curves depends on the heat capacity of the particular stream (we consider for the moment transfer of only sensible heat, not latent heat).

Note the differing direction in which the enthalpy (abscissa) scale for the cold stream increases on the two figures, which indicate countercurrent and concurrent flow, respectively. We now consider the transfer of sensible heat from the hot stream to the cold by putting the streams in either concurrent or countercurrent flow in a double pipe exchanger (either fluid may be considered to flow in the inner pipe, the other in the annular space).

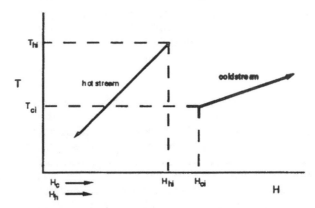

Figure 7.7.3-1 T-H diagram for two streams between which sensible heat is to be exchanged in countercurrent flow

Using the fact that only **differences** in enthalpy are important, not absolute values of enthalpy, we observe that we can slide either curve **horizontally** without affecting the first law balance, which involves only differences in enthalpy. (Moving the curve vertically, on the other hand, would represent a change in the temperatures of the stream.) The vertical distance between the two curves at any one point on the abscissa will then indicate the temperature difference of the streams (driving force for heat transfer) at that point in the exchanger.

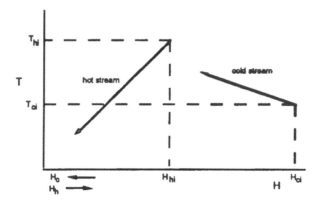

Figure 7.7.3-2 T-H diagram for two streams between which sensible heat is to be exchanged in concurrent flow

At any point where the curves intersect, the temperatures of the two streams become identical and the driving force for heat transfer becomes zero. Furthermore, to **reach** the point where the curves intersect (this point is called a **pinch**), we would need infinite area in our exchanger, so we obviously cannot go beyond the pinch.

For **concurrent** flow, the **inlets** of both streams are paired in space; for **countercurrent** flow the **inlet** of one stream is paired with the **outlet** of the other. Figure 7.7.3-3 and Figure 7.7.3-4 show the pinch situation for concurrent flow and countercurrent flow, respectively. Note that the horizontal distance a represents the enthalpy change for either stream (positive for the cold and negative for the hot), and, therefore, the heat transfer accomplished by the exchanger. The vertical distance, b, between the curves generally varies through the exchanger. The inlet driving force available for the concurrent case is larger than that at any point in the countercurrent exchanger.

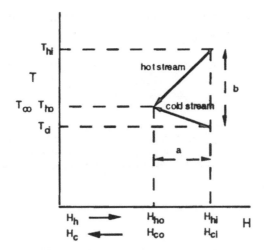

Figure 7.7.3-3 Pinch with concurrent operation

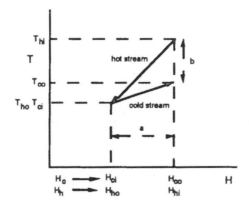

Figure 7.7.3-4 Pinch with countercurrent operation

Notice two things about these latter two figures:

1) The pinch conditions shown represent the **maximum heat transfer (infinite area)** in each case.

2) The maximum heat transfer attainable is larger for the countercurrent case than for the concurrent case. In the countercurrent case, the outlet temperature of the hot stream

becomes the inlet temperature of the cold stream and vice versa.

The conclusions of this ad hoc example may be extended to other cases. In particular, we will assume that countercurrent operation will always give a maximum heat transfer larger than concurrent.

Since the flow rate and heat capacity of each stream are each assumed to be constant here, and their product (which represents the thermal capacity of the stream - that is, the amount of energy that the stream can absorb per degree of temperature change) occurs repeatedly, we adopt the shorthand notation as we did in Section 7.3.2

$$C \equiv w \hat{C}_p \qquad\qquad (7.7.3\text{-}1)$$

We sort the possible cases for sensible heat exchange between two streams with constant flow rate and heat capacity by observing that for both the hot and the cold stream

$$\dot{H} - \dot{H}_i = \left(w\,\hat{C}_p\right)\left(T - T_i\right)$$
$$\dot{H} - \dot{H}_i = C\left(T - T_i\right)$$
$$\frac{dT}{d\dot{H}} = \frac{1}{C} \qquad\qquad (7.7.3\text{-}2)$$

so we can classify pairs of T-H curves by the reciprocal of their slopes as follows

1) $C_h > C_c$ $\qquad\qquad (7.7.3\text{-}3)$
2) $C_h < C_c$ $\qquad\qquad (7.7.3\text{-}4)$
3) $C_h = C_c$ $\qquad\qquad (7.7.3\text{-}5)$

The ad hoc example above was

$$\left(\text{slope}\right)_{hot} > \left(\text{slope}\right)_{cold} \qquad\qquad (7.7.3\text{-}6)$$
$$\left(\frac{dT}{d\dot{H}}\right)_{hot} > \left(\frac{dT}{d\dot{H}}\right)_{cold} \qquad\qquad (7.7.3\text{-}7)$$
$$\frac{1}{C_{hot}} > \frac{1}{C_{cold}} \qquad\qquad (7.7.3\text{-}8)$$
$$C_{hot} < C_{cold} \qquad\qquad (7.7.3\text{-}9)$$

therefore of case 2.

By extending the reasoning for the ad hoc case, we can write the maximum rate of heat transfer (which will be for countercurrent flow) as

$$Q_{maximum} = C_{minimum}\left(T_{hi} - T_{ci}\right) \tag{7.7.3-10}$$

We then can define an **effectiveness, ε,** as the ratio of the actual heat transfer rate to the maximum possible

$$\varepsilon \equiv \frac{Q}{Q_{max}} \tag{7.7.3-11}$$

so

$$Q = \varepsilon\, C_{minimum}\left(T_{hi} - T_{ci}\right) \tag{7.7.3-12}$$

If we define a quantity designated as NTU, the number of transfer units for heat transfer, as

$$\boxed{NTU \equiv \frac{U\,A}{C_{min}}} \tag{7.7.3-13}$$

we can then develop functional relationships of the form

$$f\left[\varepsilon, NTU, \frac{C_{min}}{C_{max}}\right] = 0 \tag{7.7.3-14}$$

for particular heat exchanger configurations. We define

$$C_r \equiv \frac{C_{min}}{C_{max}} \tag{7.7.3-15}$$

Tables 7.7.3-1a and b give such relationships for selected configurations. More extensive tabulations are available.[107]

---

[107] Kays, W. M. and A. L. London (1984). *Compact Heat Exchangers*. New York, NY, McGraw-Hill.

Table 7.7.3-1a Effectiveness/NTU relationships

EXCHANGER TYPE	FUNCTIONAL RELATION
Double pipe Concurrent flow	$\varepsilon = \dfrac{1 - \exp\left[-NTU\left(1 + C_r\right)\right]}{\left(1 + C_r\right)}$
Double pipe Countercurrent flow	$\varepsilon = \dfrac{1 - \exp\left[-NTU\left(1 - C_r\right)\right]}{1 - C_r \exp\left[-NTU\left(1 - C_r\right)\right]}$
Shell and tube One shell pass 2n tube passes	$\varepsilon = 2\left\{1 + C_r + \sqrt{\left(1 + C_r^2\right)}\right.$ $\left. \times \dfrac{1 + \exp\left[-NTU\sqrt{\left(1 + C_r^2\right)}\right]}{1 - \exp\left[-NTU\sqrt{\left(1 + C_r^2\right)}\right]}\right\}^{-1}$

Table 7.7.3-1b NTU/effectiveness relationships

EXCHANGER TYPE	FUNCTIONAL RELATION
Double pipe Concurrent flow	$NTU = \dfrac{\ln\left[1 - \varepsilon\left(1 + C_r\right)\right]}{\left(1 + C_r\right)}$
Double pipe Countercurrent flow	$NTU = -\dfrac{1}{C_r - 1}\ln\left[\dfrac{\varepsilon - 1}{\varepsilon\,C_r - 1}\right]$
Shell and tube One shell pass 2n tube passes	$NTU = -\dfrac{1}{\sqrt{\left(1 + C_r^2\right)}}\ln\left[\dfrac{E - 1}{E + 1}\right]$ $E = \dfrac{\frac{2}{\varepsilon} - \left(1 + C_r\right)}{\sqrt{\left(1 + C_r^2\right)}}$

Given such a relationship among ε, NTU, and the ratio of thermal capacities of the streams involved, one can readily find the heat transfer rate for the case where inlet temperatures of the streams are known, flow rates and heat capacities are known, and the area of the exchanger is known (one calculates U from the flow rate and type of exchanger). This is sometimes the situation when one is adapting an existing exchanger (therefore, the area is known) to use with different fluids and/or different entering temperatures (the **retrofit** problem). Such a situation can arise from incorporating used equipment in a process design, or from modifying a process to run with altered flow rates and/or fluids.

The NTU approach can also be adapted to the **design** problem; that of finding the area of exchanger required given, for example, the flow rate, inlet temperature and outlet temperature of one stream; and the inlet temperature and flow of the other stream.

Example 7.7.3-1   Determination of effectiveness for a concurrent flow exchanger

Consider a concurrent flow heat exchanger for which $C_{min} = C_{cold}$.

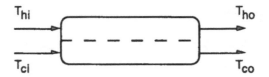

It follows that

$$\varepsilon = \frac{Q_c}{Q_{max}} = \frac{C_c \left(T_{co} - T_{ci}\right)}{C_{min} \left(T_{hi} - T_{ci}\right)} = \frac{C_c \left(T_{co} - T_{ci}\right)}{C_c \left(T_{hi} - T_{ci}\right)}$$

$$\varepsilon = \frac{\left(T_{co} - T_{ci}\right)}{\left(T_{hi} - T_{ci}\right)} \tag{7.7.3-16}$$

and, from energy balances on the hot and cold fluids[108]

$$C_c \left(T_{co} - T_{ci}\right) = \Delta H_c = Q_c$$
$$C_h \left(T_{ho} - T_{hi}\right) = \Delta H_h = Q_h \tag{7.7.3-17}$$
$$Q_h = -Q_c \tag{7.7.3-18}$$

which permits us to calculate

---

[108] We omit the caret overscore on the enthalpy in order to avoid the notational awkwardness of a double overscore. Enthalpy per unit mass is understood.

$$\frac{C_{min}}{C_{max}} = \frac{C_c}{C_h} = \frac{\dfrac{(\Delta\dot{H})_c}{(T_{co} - T_{ci})}}{\left[\dfrac{(\Delta\dot{H})_h}{(T_{ho} - T_{hi})}\right]} = \frac{\dfrac{(\Delta\dot{H})_c}{(T_{co} - T_{ci})}}{\left[\dfrac{-(\Delta\dot{H})_c}{(T_{ho} - T_{hi})}\right]}$$

$$\frac{C_{min}}{C_{max}} = \frac{(T_{hi} - T_{ho})}{(T_{co} - T_{ci})} \tag{7.7.3-19}$$

But

$$C_c\left(T_c - T_{ref}\right) = \Delta\dot{H}_c = Q_c \tag{7.7.3-20}$$

$$T_c = \frac{\dot{Q}_c}{C_c} + T_{ref} \tag{7.7.3-21}$$

$$dT_c = \frac{d\dot{Q}_c}{C_c} \tag{7.7.3-22}$$

and

$$C_h\left(T_h - T_{ref}\right) = \Delta\dot{H}_h = Q_h \tag{7.7.3-23}$$

$$T_h = \frac{Q_h}{C_h} + T_{ref} \tag{7.7.3-24}$$

$$dT_h = \frac{d\dot{Q}_h}{C_h} \tag{7.7.3-25}$$

so

$$\begin{aligned} d(\Delta T) = d\left(T_h - T_c\right) &= dT_h - dT_c \\ &= \frac{d\dot{Q}_h}{C_h} - \frac{d\dot{Q}_c}{C_c} \\ &= -\frac{d\dot{Q}_c}{C_h} - \frac{d\dot{Q}_c}{C_c} \end{aligned}$$

$$d(\Delta T) = -d\dot{Q}_c\left[\frac{1}{C_h} + \frac{1}{C_c}\right] \tag{7.7.3-26}$$

$$d\dot{Q}_c = \frac{-d(\Delta T)}{\left[\dfrac{1}{C_h} + \dfrac{1}{C_c}\right]} \tag{7.7.3-27}$$

But

$$dQ_c = U \, \Delta T \, dA \tag{7.7.3-28}$$

Substituting

$$\frac{-d(\Delta T)}{\left[\frac{1}{C_h} + \frac{1}{C_c}\right]} = U \, \Delta T \, dA \tag{7.7.3-29}$$

and integrating

$$\frac{d(\Delta T)}{\Delta T} = -U\left[\frac{1}{C_h} + \frac{1}{C_c}\right] dA \tag{7.7.3-30}$$

$$\int_{\Delta T_i}^{\Delta T_o} \frac{d(\Delta T)}{\Delta T} = -U\left[\frac{1}{C_h} + \frac{1}{C_c}\right]\int_0^A dA \tag{7.7.3-31}$$

$$\ln\frac{\Delta T_o}{\Delta T_i} = -U\left[\frac{1}{C_h} + \frac{1}{C_c}\right]A \tag{7.7.3-32}$$

$$\ln\left[\frac{T_{ho} - T_{co}}{T_{hi} - T_{ci}}\right] = -\left[\frac{UA}{C_c}\right]\left[\frac{C_c}{C_h} + 1\right] \tag{7.7.3-33}$$

Using the definition of NTU

$$\left(\frac{T_{ho} - T_{co}}{T_{hi} - T_{ci}}\right) = \exp\left\{-NTU\left[\frac{C_c}{C_h} + 1\right]\right\} \tag{7.7.3-34}$$

We can rewrite the left-hand side in terms of the effectiveness by adding and subtracting T_{hi} and T_{ci} in the numerator of the left-hand side

$$\frac{T_{ho} - T_{co}}{T_{hi} - T_{ci}} = \frac{T_{ho} + \left(T_{hi} - T_{hi}\right) + \left(T_{ci} - T_{ci}\right) - T_{co}}{T_{hi} - T_{ci}} \tag{7.7.3-35}$$

$$= \frac{\left(T_{ho} - T_{hi}\right) + \left(T_{hi} - T_{ci}\right) + \left(T_{ci} - T_{co}\right)}{T_{hi} - T_{ci}} \tag{7.7.3-36}$$

$$= \frac{\left(T_{ho} - T_{hi}\right)}{T_{hi} - T_{ci}} + 1 + \frac{\left(T_{ci} - T_{co}\right)}{T_{hi} - T_{ci}} \tag{7.7.3-37}$$

Using the definition of ε

$$= \frac{\left(T_{ho} - T_{hI}\right)}{\frac{1}{\varepsilon}\left(T_{\infty} - T_{cI}\right)} + 1 + \frac{\left(T_{cI} - T_{co}\right)}{\frac{1}{\varepsilon}\left(T_{\infty} - T_{cI}\right)} \qquad (7.7.3\text{-}38)$$

$$= \frac{\varepsilon\left(T_{ho} - T_{hI}\right)}{\left(T_{\infty} - T_{cI}\right)} + 1 - \varepsilon \qquad (7.7.3\text{-}39)$$

$$= -\varepsilon\frac{C_{min}}{C_{max}} + 1 - \varepsilon \qquad (7.7.3\text{-}40)$$

$$= -\varepsilon\, C_r + 1 - \varepsilon \qquad (7.7.3\text{-}41)$$

where

$$C_r = \frac{C_{min}}{C_{max}} \; . \qquad (7.7.3\text{-}42)$$

Then

$$\left(\frac{T_{ho} - T_{co}}{T_{hI} - T_{cI}}\right) = 1 - \varepsilon\left[1 + C_r\right] \qquad (7.7.3\text{-}43)$$

Substituting in Equation (7.7.3-34) and solving for ε

$$\varepsilon = \frac{1 - \exp\left\{-NTU\left[C_r + 1\right]\right\}}{\left[C_r + 1\right]} \qquad (7.7.3\text{-}44)$$

The same result is obtained starting with $\left(w\,\hat{C}_p\right)_{min} = \left(w\,\hat{C}_p\right)_{hot}$.

In a similar manner one can develop expressions for other types of heat exchangers.

Example 7.7.3-2   Calculation of area using NTU and ε for a concurrent flow exchanger

Four hundred pounds per minute of an oil is to be cooled from 250°F to 210°F in a concurrent double-pipe heat exchanger using 300 lbmass/min. of water that is available at 60°F. The overall heat transfer coefficient in the exchanger has been calculated to be 75 Btu/(hr ft^2 °F) and may be assumed to be constant. The following physical properties can be assumed to be constant.

Property	Water	Oil
Density, lbmass/ft^3	62.4	55
Heat Capacity, Btu/(lbmass °F)	1.0	0.7

Solution

We note that we do not have the water outlet temperature, so we first must calculate it to make sure that the second law is not violated. Since this is concurrent flow, the outlet temperature of the water must be below the outlet temperature of the oil.

Using the thermodynamic relationships for water

$$\left(w\, C_p \right)_c (T_{co} - T_{ci}) = \left(\Delta \hat{H} \right)_c \tag{7.7.3-45}$$

and for oil

$$\left(w\, C_p \right)_h (T_{ho} - T_{hi}) = \left(\Delta \hat{H} \right)_h \tag{7.7.3-46}$$

and the energy balances with water and oil as the respective systems

$$\left(\Delta \hat{H} \right)_c = \dot{Q}_c \tag{7.7.3-47}$$
$$\left(\Delta \hat{H} \right)_h = \dot{Q}_h \tag{7.7.3-48}$$

and noting that

$$\dot{Q}_c = -\dot{Q}_h \tag{7.7.3-49}$$

giving

$$C_c \left(T_{co} - T_{ci}\right) = -C_h \left(T_{ho} - T_{hi}\right) \qquad (7.7.3\text{-}50)$$

Substituting known values

$$\left(300 \frac{lbm}{min} \, 1 \, \frac{Btu}{lbm \, °F}\right)_c (T_{co} - 60) \, °F$$

$$= -\left(400 \frac{lbm}{min} \, 0.7 \, \frac{Btu}{lbm \, °F}\right)_h (210 - 250) \, °F$$

$$T_{co} = 97 \, °F \qquad (7.7.3\text{-}51)$$

Since 97°F (the outlet temperature of the water) is less than 210°F (the outlet temperature of the oil) we are in conformance with the second law.

We will obtain the area from NTU, which we will calculate using ε, so we first must determine C_{min} and C_{max}.

$$C_h = C_{oil} = (400) \frac{lbm}{min} (0.7) \frac{Btu}{lbm \, °F}$$

$$= 280 \frac{Btu}{min \, °F} = C_{min} \qquad (7.7.3\text{-}52)$$

$$C_c = C_{water} = (300) \frac{lbm}{min} (1) \frac{Btu}{lbm \, °F}$$

$$= 300 \frac{Btu}{min \, °F} = C_{max} \qquad (7.7.3\text{-}53)$$

Substituting in the relationship for ε derived in the previous example

$$\varepsilon = \frac{\left(T_{co} - T_{ci}\right)}{\left(T_{hi} - T_{ci}\right)} = \frac{(97 - 60)}{(250 - 60)} = 0.195 \qquad (7.7.3\text{-}54)$$

but from the previous example

$$\varepsilon = 0.195 = \frac{1 - \exp\left\{-NTU\left[C_r + 1\right]\right\}}{\left[C_r + 1\right]} \qquad (7.7.3\text{-}55)$$

$$= \frac{1 - \exp\left\{-NTU\left[\frac{0.7}{1.0} + 1\right]\right\}}{\left[\frac{0.7}{1.0} + 1\right]} \tag{7.7.3-56}$$

$$= \frac{1 - \exp\left\{-NTU\left[1.7\right]\right\}}{\left[1.7\right]} \tag{7.7.3-57}$$

Solving for NTU

$$NTU = -\frac{\ln\left\{1 - 0.195\left[1.7\right]\right\}}{1.7} = 0.237 \tag{7.7.3-58}$$

But using the definition of NTU

$$NTU \equiv \frac{U\,A}{\left(w\,\hat{C}_p\right)_{min}} \tag{7.7.3-59}$$

$$0.237 = \frac{(75)\dfrac{Btu}{hr\,ft^2\,°F}\,(A)\,ft^2}{400\dfrac{lbm}{min}\,0.7\dfrac{Btu}{lbm\,°F}}\,\dfrac{60\,min}{hr} \tag{7.7.3-60}$$

$$A = 53\,ft^2 \tag{7.7.3-61}$$

Example 7.7.3-3   Calculation of exit temperatures using NTU and e for a heat exchanger of known area

Five hundred pounds per minute of an organic liquid is available at 290°F and is to be cooled using 300 lbmass/min. of water that is available at 60°F.

An existing one shell pass, four tube pass heat exchanger with heat transfer area of 40 ft² is available in the salvage yard. The overall heat transfer coefficient in the exchanger is estimated to be 25 Btu/(hr ft² °F) with water on the shell side and the organic on the tube side, and may be assumed to be constant.

The following physical properties can be assumed to be constant.

Property	Water	Organic
Density, lbmass/ft^3	62.4	57
Heat Capacity, Btu/(lbmass °F)	1.0	0.8

Determine

- the heat that will be transferred using these two liquid streams with the existing exchanger and

- the outlet temperature of each stream.

Solution

We first calculate

$$C_h = C_{oil} = (500)\frac{lbm}{min}(0.8)\frac{Btu}{lbm\,°F}$$
$$= 400\frac{Btu}{min\,°F} = C_{max} \qquad\qquad (7.7.3\text{-}62)$$
$$C_c = C_{water} = (300)\frac{lbm}{min}(1)\frac{Btu}{lbm\,°F}$$
$$= 300\frac{Btu}{min\,°F} = C_{min} \qquad\qquad (7.7.3\text{-}63)$$

so

$$C_r = \frac{C_{min}}{C_{max}} = \frac{300}{400} = 0.75 \qquad\qquad (7.7.3\text{-}64)$$

and

$$NTU = \frac{UA}{\left(w\,C_p\right)_{min}} = \frac{(25)\frac{Btu}{hr\,ft^2\,°F}(40)\,ft^2}{(300)\frac{Btu}{hr\,°F}} = 3.33 \qquad\qquad (7.7.3\text{-}65)$$

From the table

$$\varepsilon = 2\left\{1 + C_r + \sqrt{\left(1 + C_r^2\right)}\right.$$
$$\left. \times \frac{1 + \exp\left[-NTU\sqrt{\left(1 + C_r^2\right)}\right]}{1 - \exp\left[-NTU\sqrt{\left(1 + C_r^2\right)}\right]}\right\}^{-1} \qquad (7.7.3\text{-}66)$$

$$\varepsilon = 2\left\{1 + 0.75 + \sqrt{\left(1 + (0.75)^2\right)}\right.$$
$$\left. \times \frac{1 + \exp\left[-3.33\sqrt{\left(1 + (0.75)^2\right)}\right]}{1 - \exp\left[-3.33\sqrt{\left(1 + (0.75)^2\right)}\right]}\right\}^{-1} \qquad (7.7.3\text{-}67)$$

$$\varepsilon = 0.658 \qquad (7.7.3\text{-}68)$$

So

$$\begin{aligned}
Q &= \varepsilon\, C_{min}\left(T_{hi} - T_{ci}\right) \\
&= (0.646)(300)\frac{Btu}{min\ °F}(250 - 60)\ °F \\
&= 36{,}822\,\frac{Btu}{hr}
\end{aligned} \qquad (7.7.3\text{-}69)$$

But

$$\begin{aligned}
Q_c &= C_c\left(T_{co} - T_{ci}\right) \\
36{,}822\,\frac{Btu}{min} &= (300)\frac{Btu}{hr\ °F}\left(T_{co} - 60\right)\ °F \\
T_{co} &= 183\ °F
\end{aligned} \qquad \begin{aligned}(7.7.3\text{-}70)\\[1.2em](7.7.3\text{-}71)\end{aligned}$$

and

$$\begin{aligned}
Q_h &= C_h\left(T_{ho} - T_{hi}\right) \\
-36{,}822\,\frac{Btu}{hr} &= (400)\frac{Btu}{hr\ °F}\left(T_{ho} - 290\right)\ °F \\
T_{ho} &= 198\ °F
\end{aligned} \qquad \begin{aligned}(7.7.3\text{-}72)\\[1.2em](7.7.3\text{-}73)\end{aligned}$$

This problem would have required an iterative solution if treated by the F-factor method that follows.

7.7.4   F-factor method for design of heat exchangers

Unfortunately, most heat exchangers have a rather more complicated flow pattern than strictly concurrent or strictly countercurrent flow, e.g., double coaxial pipes with one fluid flowing within the innermost pipe and another fluid flowing in the annular space. Instead, to give just one counterexample, shell and tube heat exchangers usually utilize substantial cross-flow components on the shell side.

Engineers, however, in their desire for simple models, apply the form of Equation (7.7.1-31) to shell and tube exchangers by introduction of a correction factor, F, to the log mean temperature difference

$$\dot{Q} = U A F (\Delta T)_{\text{log mean}}$$ (7.7.4-1)

The correction factor, which is (usually) presented in graphical form, is a result of integrating for the particular geometry and casting the result in the form of Equation (7.7.4-1).

In addition to the assumptions incorporated in the derivation of the log mean as the appropriate average, to wit

> • The fluids flow at constant rate
> • The fluids each have constant heat capacity
> • U remains constant
> • There is no phase change of either fluid (the case where a fluid absorbs or supplies heat **only** by phase change, i.e., its temperature remains constant, will also fit our ultimate definition, however)
> • There is no heat exchange other than between the fluids - i.e., no heat losses from the exchanger to the surroundings
> • There is no thermal energy source or sink in the system, e.g., chemical reaction

Two additional assumptions are used

      • The temperature of the shell-side fluid is uniform across any cross-section in any shell-side pass (perfect mixing on the shell side at any pass)
      • Each pass contains equal heat transfer area

The integrations to obtain the mean temperature difference are straightforward, although cumbersome. Considering the case of a two tube-side passes, one shell-side pass exchanger leads to the following expression[109] for the average temperature difference required for Equation (7.7.4-2)

$$(\Delta T)_{average} = \frac{\sqrt{(T_1 - T_2)^2 + (t_2 - t_1)^2}}{\ln\left[\dfrac{T_1 + T_2 - t_1 - t_2 + \sqrt{(T_1 - T_2)^2 + (t_2 - t_1)^2}}{T_1 + T_2 - t_1 - t_2 - \sqrt{(T_1 - T_2)^2 + (t_2 - t_1)^2}}\right]} \qquad (7.7.4\text{-}2)$$

Defining the dimensionless quantities

$$S = \frac{t_2 - t_1}{T_1 - t_1} \qquad\qquad (7.7.4\text{-}3)$$

$$R = \frac{T_1 - T_2}{t_2 - t_1} \qquad\qquad (7.7.4\text{-}4)$$

where the temperatures refer to

$$
\begin{aligned}
T_1 &= \text{shell side inlet} \\
T_2 &= \text{shell side outlet} \\
t_1 &= \text{tube side inlet} \\
t_2 &= \text{tube side outlet}
\end{aligned}
\qquad (7.7.4\text{-}5)
$$

and applying this mean temperature to the form of Equation (7.7.4-1)

$$(\Delta T)_{average} = F\,(\Delta T)_{lm} \qquad\qquad (7.7.4\text{-}6)$$

gives a correction factor

---

[109] Underwood, A. J. V. (1934). The calculation of the mean temperature difference in multipass heat exchangers. *Journal of Institute of Petroleum Technologists* **20**: 145.

$$F = \frac{\frac{\sqrt{R^2+1}}{R-1} \ln\left[\frac{1-S}{1-SR}\right]}{\ln\left[\frac{\frac{2}{S}-1-R+\sqrt{R^2+1}}{\frac{2}{S}-1-R-\sqrt{R^2+1}}\right]}$$

(7.7.4-7)

Figure 7.7.4-1 is a plot of this equation. This plot is also applicable (within the accuracy of the usual calculation) to one shell pass and 2n tube passes, n = 1, 2, 3, ... The indeterminate form for R = 1 can be removed by application of l'Hopital's rule to show that at this point

$$\lim_{z \to z_0}\left[\frac{f(z)}{g(z)}\right] = \frac{f'(z_0)}{g'(z_0)}$$

(7.7.4-8)

$$\lim_{R \to 1}\left[\frac{\ln\left[\frac{1-S}{1-SR}\right]}{R-1}\right] = \left[\frac{\frac{1}{\left[\frac{1-S}{1-SR}\right]}\frac{(1-S)S}{(1-SR)^2}}{-1}\right]_{R=1}$$

(7.7.4-9)

$$\frac{\frac{1}{\left[\frac{1-S}{1-S}\right]}\frac{(1-S)S}{(1-S)^2}}{-1} = \frac{S}{1-S}$$

(7.7.4-10)

The steep slope of the curves shows that this is not a very satisfying approach to exchanger design, because it is extremely sensitive to the ordinate value, S, over much of the ranges involved.

The plot in Figure 7.7.4-1 is merely typical. Corresponding figures for various configurations of exchangers (cross-flow, multiple shell passes, odd numbers of tube passes, etc.) are available.[110]

When applicable, the F correction factor technique is most convenient for **design** problems; that is, where the area of the exchanger is to be determined. It can, of course, be adapted to the **retrofit** problem - that is, adapting an existing exchanger to use with different fluids and/or different entering temperatures, but then requires an iterative solution.

---

[110] For example, see Perry, R. H. and D. W. Green, Eds. (1984). *Chemical Engineer's Handbook*. New York, McGraw-Hill, p. 10-27.

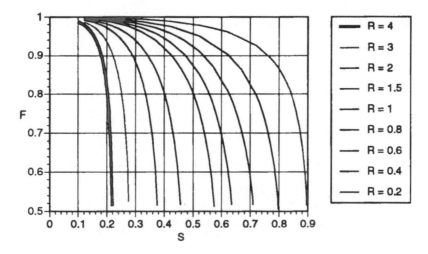

Figure 7.7.4-1 Correction factor to log mean temperature difference - one shell pass, 2^n tube passes

Example 7.7.4-1 Use of F Factor compared to effectiveness/NTU method

A countercurrent heat exchanger is required to cool 1000 gpm of oil from a temperature of 250°F to a temperature of 210°F. The coolant to be used is 250 gpm of water at a temperature of 80°F.

Oil property	Value
Specific gravity	0.8
Heat capacity, Btu/(lbmass °F)	0.7

The overall heat transfer coefficient has been calculated to be 150 Btu/(hr ft^2 °F)

Calculate the area required if an exchanger is used which has two tube passes and one shell pass, water on the tube side. Use both the F-factor method and the effectiveness/NTU method and compare.

Solution

A thermal energy balance on the oil gives the heat duty

$$Q = w_{oil} C_{p, oil} \left(T_{i, oil} - T_{o, oil} \right)$$

$$= (1000) \frac{gal}{min} (0.8) (8.33) \frac{lbmass}{gal} (0.7) \frac{Btu}{lbmass\ °F} (250 - 210)\ °F$$

$$= 1.87 \times 10^5 \frac{Btu}{min} \tag{7.7.4-11}$$

and an overall thermal energy balance (or one could use a balance on the water) gives the outlet temperature of the water

$$w_{oil} C_{p, oil} \left(T_{i, oil} - T_{o, oil} \right) = w_{HOH} C_{p, HOH} \left(T_{o, HOH} - T_{i, HOH} \right)$$

$$(1000) \frac{gal}{min} (0.8) (8.33) \frac{lbmass}{gal} (0.7) \frac{Btu}{lbmass\ °F} (250 - 210)\ °F$$

$$= (250) \frac{gal}{min} (8.33) \frac{lbmass}{gal} (1) \frac{Btu}{lbmass\ °F} \left(T_{o, HOH} - 80 \right)\ °F$$

$$T_{o, HOH} = 170\ °F \tag{7.7.4-12}$$

Then

$$(\Delta T)_{lm} = \frac{(210 - 80) - (250 - 170)}{\ln \left[\frac{(210 - 80)}{(250 - 170)} \right]} = 103\ °F \tag{7.7.4-13}$$

For the F-factor method

$$R = \frac{(T_1 - T_2)}{(t_2 - t_1)} = \frac{(250 - 210)}{(170 - 80)} = 0.44 \tag{7.7.4-14}$$

$$S = \frac{(t_2 - t_1)}{T_1 - t_1} = \frac{(170 - 80)}{(250 - 80)} = 0.53 \tag{7.7.4-15}$$

From Equation (7.7.4-7) and/or from Figure 7.7.4-1, F = 0.94

$$Q = U A F (\Delta T)_{lm} \tag{7.7.4-16}$$

$$A = \frac{Q}{U F (\Delta T)_{lm}}$$

$$A = \frac{\left(1.87 \times 10^5\right) \frac{Btu}{min}}{\left(150\right) \frac{Btu}{hr\ ft^2\ °F} \left(0.94\right) \left(103\right) °F} \frac{\left(60\right) min}{hr} = 773\ ft^2 \quad (7.7.4\text{-}17)$$

To do the same calculation using the effectiveness/NTU approach

Find C_{min} and C_{max}

$$C_{oil} = w_{oil} C_{p,oil}$$
$$= \left(1000\right) \frac{gal}{min} \left(0.8\right) \left(8.33\right) \frac{lbmass}{gal} \left(0.7\right) \frac{Btu}{lbmass\ °F}$$
$$= 4.66 \times 10^3 \frac{Btu}{min\ °F} \quad (7.7.4\text{-}18)$$

$$C_{HOH} = w_{HOH} C_{p HOH}$$
$$= \left(250\right) \frac{gal}{min} \left(8.33\right) \frac{lbmass}{gal} \left(1\right) \frac{Btu}{lbmass\ °F}$$
$$= 2.08 \times 10^3 \frac{Btu}{min\ °F} \quad (7.7.4\text{-}19)$$

Therefore $C_{min} = C_{HOH}$ and $C_{max} = C_{oil}$

$$C_r = \frac{C_{min}}{C_{max}} = \frac{2.08 \times 10^3}{4.66 \times 10^3} = 0.446 \quad (7.7.4\text{-}20)$$

The effectiveness is

$$\varepsilon = \frac{\dot{Q}}{\dot{Q}_{max}} = \frac{\dot{Q}}{C_{min} \left(T_{hi} - T_{ci}\right)}$$
$$= \frac{1.87 \times 10^5 \frac{Btu}{min}}{2.08 \times 10^3 \frac{Btu}{min\ °F} \left(250 - 80\right) °F}$$
$$= 0.529 \quad (7.7.4\text{-}21)$$

From Table 7.7.3-1, NTU = 1, from which

$$NTU = \frac{U A}{C_{min}} \quad (7.7.4\text{-}22)$$

$$A = \frac{C_{min}NTU}{U} = \frac{\left(2.08 \times 10^3\right)\frac{Btu}{min\ °F}\ (1)}{\left(150\right)\frac{Btu}{hr\ ft^2\ °F}}\ (60)\ \frac{min}{hr}$$

$$A = 832\ ft^2$$

(7.7.4-23)

Chapter 7 Problems

7.1 A composite furnace wall consists of 9 in. of firebrick [k = 0.96 Btu/(h ft °F)], 4 1/2 in. of insulating brick [k = 0. 183 Btu/(h ft °F)], and 4 1/2 in. of building brick [k = 0.40 Btu/(h ft °F)]

(a) Calculate the heat flux in Btu/(h ft²) when the inside surface temperature is 2400°F and the outside is 100°F.
(b) What are the temperatures at the interfaces of each layer of brick? Is the temperature of the insulating brick below the maximum allowable value of 2000°F?

7.2 A furnace wall consists of two layers of firebrick, k_1 = 0.8 Btu/(h ft °F) and k_2 = 0.1 Btu/(h ft °F), and a steel plate k_3 = 25 Btu/(h ft °F) as shown in the following illustration.

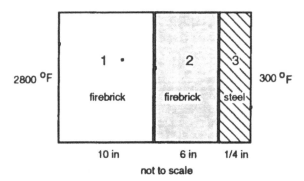

Calculate the heat flux [Btu/(h ft²)] and the interface temperatures between materials 1 and 2, 2 and 3.

7.3 A cylindrical conduit carries a gas which keeps the inside surface of the conduit at 600°F. The conduit is insulated with 4 in. of rock wool, and the outer surface of this insulation is at a temperature of 100°F. The thermal conductivity of the conduit is 0.88 Btu/(h ft °F).

The thermal conductivity of the rock wool is given by the empirical curve fit equation

$$k = 0.025 + 0.00005\,T$$

[T is in °F and k is in Btu/(h ft °F)]

If the conduit is 1/4-in. thick with a 3-1/2 in. ID, find the rate of heat loss per foot of conduit.

7.4 A standard 1 in. Schedule 40 iron pipe carries saturated steam at 250°F. The pipe is lagged (insulated) with a 2-in. layer of 85 percent magnesia, and outside this magnesia layer there is a 3-in. layer of cork.

The inside temperature of the pipe wall is 249°F, and the outside temperature of the cork is 90°F. Calculate

      (a) the rate of heat loss from 100 ft of pipe, in Btu/h, and
      (b) the temperatures at the boundary between metal and magnesia and between magnesia and cork.

7.5 The thermal insulation on a steam pipe (2 in. OD) consists of magnesia (85% magnesia) 2-in. thick with thermal conductivity $k = 0.038$ Btu/(h ft °F).

Calculate the temperature at the midpoint of the insulation when the inside surface is at 250°F and the outside surface is at 50°F.

7.6 A 6-in. nominal diameter (OD = 6.625 in.) pipe carrying vaporized heat transfer fluid at 750°F is covered with a 1.5-in. layer of high temperature insulation [$k = 0.09$ Btu/(h ft °F)] inside a 2-in. layer of 85% magnesia [$k = 0.038$ Btu/(h ft °F)].

If the outside surface of the outer layer has a temperature of 70°F, find

      (a) the heat loss in Btu/ h per linear foot of pipe, and
      (b) the temperature at the interface between the two insulating layers.
What assumptions are inherent in your model?

7.7 An empirical curve fit shows that the thermal conductivity of a solid plate in $\left[\dfrac{Btu}{h\,ft\,{}^\circ F}\right]$ varies with temperature in °F as

$$k\left(T\right) = 10\left(1 + 0.01\,T + 2.4 \times 10^{-5}\,T^2\right)$$

The large surfaces of the plate are at 300°F and 100°F, respectively. Assume 1-D heat transfer.

Find the heat flow rate through a section of plate 2 ft by 3 ft by 2 in. thick.

7.8 Consider the trapezoidal brick shown in the following figure.

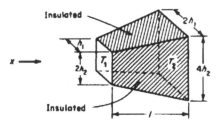

The brick is surrounded by insulation on its four lateral surfaces. The heat flow is from left to right in the figure, and we may consider the conduction of heat as being only in the x-direction to a good approximation. (The angle the sides make with the base of the trapezoid is smaller than is shown in the sketch.)

   (a) Find the temperature profile in the x- direction.
   (b) What is the heat flux as a function of x?
   (c) Starting with the Fourier equation, determine the correct
   mean area to use in the equation

$$\frac{Q_x}{A_m} = -k\left(T_2 - T_1\right)$$

7.9 The composite body shown in the following figure is 10 ft long in the direction perpendicular to the page.

$$k_I = 0.1 \frac{Btu}{h\ ft\ ^\circ F}$$

$$k_{II} = 0.8 \frac{Btu}{h\ ft\ ^\circ F}$$

$$A_1 = (1)\,[ft] \times (10)\,[ft] = 10\ ft^2$$

$$A_2 = A_3 = (0.5)\,[ft] \times (10)\,[ft] = 5\ ft^2$$

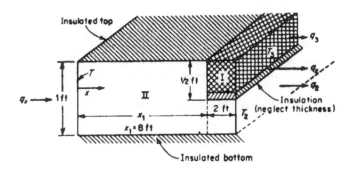

Considering heat flow to be only in the x-direction, find the temperature at x = x_1 and the heat fluxes q_2 and q_3 through the right side of the body.

7.10 Shown is a large scale model (model **larger** than prototype) of a high temperature fastener to be used to connect materials that are essentially insulators.

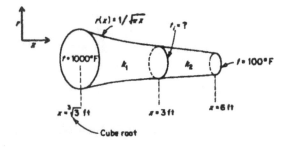

The model has a circular cross-section of radius, r, which varies with axial position, x, according to the relation

$$r = \frac{1}{\sqrt{\pi} \, x}$$

with

r = ft

x = ft

$$\left(\text{therefore, } \frac{1}{\sqrt{\pi}} \text{ has units of ft}^{3/2} \text{ to maintain dimensional homogeneity} \right)$$

The model is constructed of two materials having thermal conductivities k_1 and k_2. During a recent test of our scale model, the temperature of the large end was measured as 1000°F, while that of the small end was 100°F.

Assuming heat conduction only in the axial (x) direction, determine for these end temperatures:

> (a) The temperature profile in material 1.
> (b) The temperature T_j where the two materials join.
> (c) The heat flux across the end at the position x = 6 ft.
> (d) The heat flow in Btu/h across the junction at x = 3 ft.

7.11 A brass slab is cooled at steady state by water flow on the outside surface of its wall.

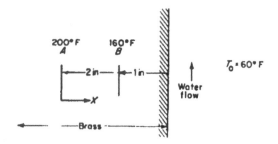

The brass has a thermal conductivity of 60 Btu/(h ft °F). The free stream water temperature is 60°F.

If the temperature in the brass at point A is 200°F and at point B is 160°F, what is the heat transfer coefficient on the outside surface?

7.12 A window consists of two layers of glass, each 1/4 in. thick, separated by a layer of dry, stagnant air, also 1/4 in. thick. If the outside surface to inside surface temperature drop over the composite system is 30°F, find the heat loss through such a window that is 10 ft wide and 4 ft high.

The window is proposed to replace a single glass pane, 3/8 in. thick, at a difference in cost of $120.

The thermal conductivity of the window glass is 0.5 Btu/(h ft °F).

Fuel costs $0.75 per million Btu combustion energy obtained. The furnace efficiency can be assumed to be 60 percent.

Find the number of days of operation under the specified temperature conditions to pay for the window. Determine the sensitivity of the number of days to the price of fuel.

Is the assumption of stagnant air a good one?

7.13 Calculate the direct radiant heat transfer between two parallel refracting black surfaces 12 by 12 ft spaced 12 ft apart, if the surface temperatures are 1300°F and 500°F, respectively.

What would be the radiant heat transfer if the two plane walls were connected by nonconducting but reradiating walls?

7.14 Two rectangular plates, both 6 in. by 3 in., are placed directly opposite each another with a plate spacing of 2 in. The larger plate is at 240°F, and has an emissivity of 0.5. The other plate is at 110°F, and has an emissivity of 0.33.

If there are no other surfaces present, what is the heat flow rate? If refractory walls now connect the plates, what is the heat flow rate?

7.15 Calculate the direct radiant heat transfer between two parallel refractory surfaces 10 by 10 ft spaced 10 ft apart. Assume that the temperatures of the surfaces are 6000°F and 1200°F, respectively.

What is the radiant heat transfer rate if the two surfaces are connected by nonconducting but reradiating walls?

7.16 Calculate the radiant heat loss from a 3-in. nominal diameter wrought iron oxidized pipe at 200°F which passes through a room at temperature 76°F.

What is the heat loss if the pipe is coated with aluminum paint?

The emissivities for the oxidized and painted surfaces are 0.8 and 0.3, respectively.

7.17 A large plane, perfectly insulated on one face and maintained at a fixed temperature, T, on the bare face, which has an emissivity of 0.90, loses 200 Btu/(h ft^2) when exposed to surroundings at absolute zero. A second plane has the same size as the first, and is also perfectly insulated on one face, but its bare face has emissivity of 0.45. When the bare face of the second plane is maintained at a fixed temperature, T, and exposed to surroundings at absolute zero, it loses 100 Btu/(h ft^2).

Let these planes be brought close together, so that the parallel bare faces are only 1 in. apart, and the heat supply to each be so adjusted that their respective temperatures remain unchanged. What will be the net heat flux between the planes, expressed in Btu/(h ft^2)?

7.18 Air is flowing steadily through a duct whose inner walls are at 800°F. A thermocouple housed in a 1-in. OD steel well with rusty exterior indicates a temperature of 450°F. The air mass velocity is 3600 lbmass/h ft^2. Estimate the true temperature of the air.

7.19 A thermocouple inserted in a long duct reads 200°F. The duct runs through a furnace that maintains the duct walls at 300°F. Inside the duct a nonabsorbing gas is flowing at such a velocity that the average convective heat transfer coefficient to the thermocouple is 10 Btu/(h ft^2 °F).

What is the gas temperature at the thermocouple in °F? (Assume that all surfaces are black and that no heat is conducted along the thermocouple leads.)

7.20 Calculate the heat loss per foot from a 3-in. Schedule 40 pipe [ID = 3.07 in, OD = 3.50 in.; k = 25 Btu/(h ft °F)] covered with a 1 1/4 in. thickness of insulation [k = 0.1 Btu/(h ft °F)].

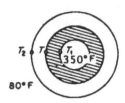

The pipe transports a fluid with mixing cup temperature of 350°F. Inside heat transfer coefficient, h_i, is 30 Btu/(h ft² °F). Outside heat transfer coefficient, h_o, is of 2 Btu/(h ft² °F). The pipe is exposed to air at an ambient temperature of 80°F.

     (a) Calculate the rate of heat loss per foot.
     (b) Calculate T_i, T_1, T_2.

7.21 Oil at a bulk temperature of 200°F is flowing in a 2-in. Schedule 40 pipe [ID = 2.067 in., OD = 2.375 in.; k = 26 Btu/(h ft °F)] at a bulk velocity of 4 ft/s. The outside of the pipe is covered with a 1/2-in. thick layer of insulation [k = 0.1 Btu/(h ft °F)]. The outside surface temperature of the insulation is 90°F. Find

     (a) The inside heat transfer coefficient.
     (b) The rate of heat loss from the pipe per foot.
     (c) The temperature of the inside pipe surface.

At 200°F for the oil

$$k = 0.074 \frac{Btu}{h\ ft^2\ °F}$$
$$Pr = 62$$
$$\hat{C}_p = 0.51 \frac{Btu}{lbmass\ °F}$$
$$\mu = 250 \times 10^{-5} \frac{lbmass}{ft\ s}$$
$$\rho = 54 \frac{lbmass}{ft^3}$$

7.22 If you were to be standing in a house with inside air temperature 80°F, would the inside wall temperature be the coldest:

(a) when the outside air temperature is 0°F and there is a 4 mph wind blowing parallel to the wall, or

(b) when the outside air temperature is 10°F and there is a 45 mph wind blowing parallel to the wall?

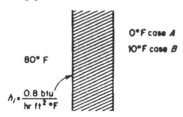

Thermal resistance of wall $= \dfrac{1.50}{\text{Area}} \left[\dfrac{\text{h ft}^2 \, {}^\circ\text{F}}{\text{Btu}} \right]$

$\text{Nu}_L = 0.086 \, \text{Re}^{0.8} \, \text{Pr}^{1/3}$

Properties of air $0^\circ \le T \le 10^\circ$

$\text{Pr} = 0.73$

$k = 0.0133 \dfrac{\text{Btu}}{\text{h ft } {}^\circ\text{F}}$

$\mu = 1.11 \times 10^{-5} \dfrac{\text{lbmass}}{\text{ft s}}$

$\rho = 0.086 \dfrac{\text{lbmass}}{\text{ft}^3}$

7.23 Consider the wall shown in the following sketch.

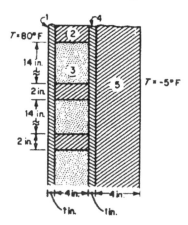

Properties:

Material	Thermal Conductivity Symbol	Thermal Conductivity Value, Btu/(h ft °F)
Wallboard	k_1	0.40
Wood	k_2	0.14
Insulation	k_3	0.016
Beaver Board	k_4	0.35
Brick	k_5	0.38

If $h_{inside} = 2$ Btu/(h ft^2 °F) and $h_{outside} = 12$ Btu/(h ft^2 °F), find

> (a) the rate of heat loss per unit area of the wall and
> (b) the inside wall temperature.

7.24 A single pane of glass [k = 0.50 Btu/(h ft °F)] 1/4-in. thick separates the air in a room which is at a mixing cup temperature of 70°F from the outdoors where the ambient atmospheric temperature is 0°F.

A thermocouple covered with a bit of insulation is attached to the inside surface of the glass. Assume that both the inside and the outside heat transfer coefficients are constant at 1 Btu/(h ft^2 °F). About what temperature should this thermocouple indicate?

7.25 A jacketed kettle is used for concentrating 1,000 lbmass/h of an aqueous salt solution using steam at 250°F. The concentrated solution boils at 220°F and has a latent heat of 965 Btu/lbmass. The vessel is made of steel [k = 35 Btu/(h ft °F)] and is 1/4 in. thick.

Determine the increase in capacity that might be expected if the kettle were replaced by one made of copper [k = 220 Btu/(h ft °F)].

7.26 Two thermometers are placed in an air stream flowing through a circular duct. One of these thermometers, which has been silvered on the outside, reads 100°F. The other is a plain mercury-in-glass thermometer and reads 150°F. The heat transfer coefficient for both thermometers is 2 Btu/(h ft^2 °F). The emissivity for the silvered surface is 0.02 and that of the glass surface 0.94.

Assuming that the readings are correct, calculate the temperature of the duct walls and that of the gas.

7.27 A hot fluid is to be cooled in a double-pipe heat exchanger from 245°F to 225°F. The cold fluid is heated from 135°F to 220°F.

Compare counterflow to parallel flow.

7.28 Calculate the heat loss per lineal foot from a 3-in. Schedule 40 steel pipe [3.07 in. ID, 3.500 in. OD, k = 25 Btu/(h ft °F)] covered with 1 in. thickness of insulation [k = 0.11 Btu/(h ft °F)]. The pipe transports a fluid at 300°F with an inner heat transfer coefficient of 40 Btu/(h ft °F), [h_{inside}], and is exposed to ambient air at 80°F with an outside heat transfer coefficient of 3 Btu/(h ft² °F), [$h_{outside}$].

7.29 Hot light oil is to be cooled by water from 250°F to 170°F in a double-pipe heat exchanger. The water is to be heated from 90°F to 150°F.

If the oil flow rate is 2.200 lbmass/h and $\left[C_p\right]_{oil}$ = 0.56 Btu/(lbmass °F), find the water flow rate.

If U_O = 45 Btu/(h ft² °F), compare countercurrent and concurrent heat exchangers.

Assuming a clean heat exchanger, which resistance - water side, metal, or oil side - would you expect to be the controlling resistance?

7.30 A hypothetical countercurrent heat exchanger having an infinite heat transfer area is used to cool 10,000 lbmass/h of water entering at 150°F. The coolant is 5,000 lbmass/h of water entering at 100°F.

    a) What are the exit temperatures of the streams?
    b) What is the rate of heat transferred?

7.31 A fluid is heated in a heat exchanger by condensing steam. Derive the log mean ΔT to use when

$$C_p = a + b\,T$$

7.32 Determine the heat loss from a bare, horizontal steam pipe, 150 ft long with 6 in. OD at 275°F. This pipe is located indoors where the temperature is 70°F.

Calculate the loss in the same room after this pipe is covered with a 3-in. layer of insulation [k = 0.11 Btu/(h ft °F)].

7.33 A composite wall is placed around a chemical reactor vessel. It is made up of concrete with steel beams centrally located for reinforcement and an outer wall of stainless steel. A typical cross-section is shown below.

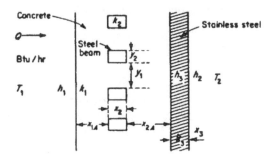

Assuming one-dimensional heat transfer throughout (i.e., vertical temperature uniformity), derive the following expression

$$ Q = \frac{\left(T_1 - T_2\right)\left(y_1 + y_2\right)}{\dfrac{1}{h_1} + \dfrac{x_1}{k_1} + \dfrac{x_3}{k_3} + \left(\dfrac{x_2}{k_1 y_1 + k_2 y_2}\right)\left(y_1 + y_2\right) + \dfrac{1}{h_2}} $$

where $x_1 = x_{1A} + x_{2A}$. Take a typical repeating section 1 ft wide (perpendicular to the plane of the sketch) and $(y_1 + y_2)$ high as the area.

Next, calculate $\dot{Q}$ based on this area for the following cases:

Case	T, °F	h, Btu/(h ft² °F)	k, Btu/(h ft² °F)	x, in.	y, in.
1	100	5	0.5	10	5
2	400	2	25	2	3
3	400	2	10	0.5	3

7.34 Insulation placed on a small-diameter pipe or electrical wire to prevent heat loss may actually increase the heat loss instead. This occurs because as the insulation radius r_2 increases, the insulation thermal resistance increases but the outside surface area is increased at the same time. The result is that the rate of heat loss will increase with increasing r_2 until a maximum heat loss is reached, after which further increase in r_2 lowers rate of heat loss.

If the inside surface temperature of the insulation, T_1, is known, and the surroundings are at T_2, find the value of r_2 for maximum rate of heat loss. Assume that T_1, T_2, k, h, and r_1 are constant.

7.35 Light oil enters the tube side of a heat exchanger (ID = 1.00 in.) at 150°F and exits at 300°F. The light oil flows at 1,200 gal/min. There are 60 tubes in the heat exchanger.

Property	150°F	300°F
ρ, lbmass/ft³	54.3	51.8
μ, lbmass/ft s	530 x 10⁻⁵	83 x 10⁻⁵
k, Btu/hr ft °F	0.075	0.073
Pr	122	22

Calculate the tube-side heat transfer coefficient.

7.36 Water is flowing in the shell side of a heat exchanger. The water may be modeled as flowing parallel to the tubes. The water enters at 80°F and leaves at 150°F. The inside diameter of the shell is 12 in., the outside tube diameter is 1.125 in., and there are 50 tubes in the heat exchanger. The water flow rate through the heat exchanger is 800 gal/min.

Property	80°F	150°F
ρ, lbmass/ft³	62.2	61.2
μ, lbmass/ft s	0.578 x 10⁻³	0.292 x 10⁻³
k, Btu/hr ft °F	0.353	0.384
Pr	5.89	2.74

(a) Calculate the equivalent diameter for the flow on the shell side of the heat exchanger.

(b) Calculate the average shell-side heat transfer coefficient.

7.37 Water is flowing in a thin-walled copper tube under conditions such that the coefficient of convective heat transfer from water to the tube is given by the obviously empirical equation

$$h_i\left[\frac{Btu}{h\,ft^2\,°F}\right] = 25\left[\frac{Btu\,s^{1/3}}{h\,ft^{7/3}\,°F}\right]\left\langle v\left[\frac{ft}{s}\right]\right\rangle^{1/3}$$

Heat from the water flows through the tube wall to the surrounding atmosphere at a rate such that the convection coefficient from the outer surface of the tube to the atmosphere is given by another empirical equation

$$h_o\left[\frac{Btu}{h\,ft^2\,°F}\right] = 0.5\left[\frac{Btu}{h\,ft^2\,°F^{5/4}}\right]\left(\Delta T\,[°F]\right)^{1/4}$$

Assume the heat transfer resistance of the tube wall to be negligible. Find the minimum water velocity necessary to keep ice from forming on the pipe wall when the bulk water temperature is 35°F and the atmosphere is at - 25°F.

7.38 The coefficient on the shell side of an exchanger having 1.0 in. OD tubes is 200 Btu/(h ft² °F) for water in turbulent flow. Estimate the coefficient for a similar exchanger constructed of 3/4-in. OD tubes under identical service conditions (i.e., same fluids, flow rates, and entering temperatures).

7.39 If you were designing a condenser or feed water heater that would heat incoming water on the inside of tubes and using condensing steam on the outside, which orientation (vertical or horizontal) of the tubes would give better heat transfer? Assume saturated steam at 7 psia, and a 1 in. OD by 6 ft long tube with an average wall temperature of 125°F.

7.40 A 1/2-in. OD by 6-ft-long tube is to be used to condense steam at 6 psia. Assume that the average tube wall temperature is 180°F. Estimate the mean heat transfer coefficient for this tube in both horizontal and vertical positions.

7.41 Water in turbulent flow is being heated in a concentric pipe exchanger by steam condensing in the annulus at 220°F. The water enters at 60°F and leaves at 80°F. If the water rate is doubled, estimate the new exit temperature.

7.42 In the production of salt from sea water, the sea water is being run into a large pond 100 ft long, 20 ft wide, and 2 ft deep. There the water is evaporated by exposure to the sun and dry air, leaving solid salt on the bottom of the pond.

At a certain point in this cycle of operations, the brine in the pond is 1 ft deep. On the evening of a clear day a workman finds that the sun has heated the brine to 90°F. During the following 10-hour period of the night, the average air temperature is 70°F, and the average effective blackbody temperature of the sky is -100°F.

Model the properties of the brine as those of water, and model the brine temperature as uniform from top to bottom. The coefficient of heat transfer from the surface of the pond to air by natural convection may be taken as

$$ h_c \left[\frac{Btu}{h\, ft^2\, °F} \right] = 0.38 \left[\frac{Btu}{h\, ft^2\, °F^{.54}} \right] \left\{ \left(T_{surface} - T_{air} \right) \left[°F \right] \right\}^{1/4} $$

Neglecting vaporization and/or crystallization effects, and assuming negligible heat transfer between the brine and the earth, estimate the temperature of the brine at the end of the 10-hour period.

7.43 A boiler is producing methanol vapor at 1 atm using hot oil flowing through a single 1-in. OD x 0.134-in. wall thickness 304 stainless tube as a source of heat. From one test of the equipment the following data are available

inlet oil temperature = 210°F   outlet oil temperature = 196°F
inside heat transfer coefficient = 100 Btu / (h ft² °F)

Determine the boiling heat transfer coefficient and the approximate heat flux.

8

MASS TRANSFER MODELS

8.1 The Nature of Mass Transfer

Mass transfer is of concern in many areas of engineering and science, but it is probably more important in conjunction with chemical reactions than in any other single area. In many cases, for implementation of chemical reactions on a commercial basis the capital investment in mass transfer equipment by far exceeds the capital investment associated with the reaction as such.

Why is this so? For several reasons:

• First, almost no chemical reaction is selective enough to produce only the desired product. Instead, a number of by-products are also produced. Separation of the major product from the by-products (which may also have considerable value) is a problem in mass transfer.

• Second, to implement a reaction the reacting molecules must somehow get together. Frequently the rate of reaction is exceedingly fast once the reactants "see" each other, and the limiting step reduces to one of transporting the reactants to the location where the reaction takes place (for example, a catalyst surface). In such cases the reaction is mass transfer limited rather than kinetically limited.

• Third, many chemicals are handled in solution (for convenience, temperature control, physical and chemical stability, etc.) and the creation of these solutions (mixing, dissolution) as well as the recovery of components from the solutions (absorption, desorption, distillation, extraction, precipitation, etc.) is a mass transfer problem.

• Fourth, much of the raw material for the chemical and pharmaceutical industry has its genesis in **naturally** occurring materials (air, natural gas, crude oil, coal, wood, various plant and animal products, etc.), and separation of desired constituents from these highly complex mixtures is again a mass transfer problem.

Many other examples could be listed, but it is to be hoped that the point has been established that a typical chemical engineer spends as much time moving and sorting molecules as he or she spends in assembling them into new structures.

Mass transfer is used here in a specialized sense. Mass is certainly transferred when **bulk flow** occurs, for example, when water flows in a pipe. This is **not**, however, what we mean by mass transfer. Mass transfer in the sense used here is:

> the migration of a component in a mixture either within the same phase or from phase to phase because of a displacement from equilibrium.

An example of the difference is furnished by the respiratory system. Air is inhaled and a gas of slightly different composition is exhaled, processes which are primarily bulk flow - analogous to the transport of air through a duct by a blower. These steps would not be called mass transfer by our definition.

At the surface of the lung, however, one component of the air, oxygen, transfers to the bloodstream because the blood there is not saturated with oxygen - the air and blood are displaced from physical/chemical equilibrium. Another mass transfer step simultaneously occurs - the transfer of CO_2 from the blood to the air, which is not saturated with CO_2.[1] Both of these are mass transfer by our criterion.

Other components in the air - for example, the nitrogen - are largely left behind. (There is, of course, a saturation concentration of nitrogen in the blood, which is dependent on pressure - to which any diver who has ever contracted "the bends" from too quick an ascent can readily attest. In the absence of variation in ambient pressure, however, the nitrogen reaches saturation and then does not transfer.)

---

[1] This analogy, like all analogies, is somewhat oversimplified as any physiologist well knows. We enter a standing apology to experts in disciplines external to engineering for all such excesses.

Mass transfer can be treated in much the same way as heat transfer, in that we are able to speak of driving forces, resistances, fluxes, boundary layers, etc. The important difference is that a transfer of mass, as opposed to transfer of heat, *always involves motion of the medium* (otherwise, no mass would be transferred). This means, for example, that the strict analog to heat conduction in a stagnant medium is found only by observing events from a coordinate frame *moving* in space rather than fixed in space. This complicates the mathematical manipulations somewhat, as we shall soon see.

8.2 Diffusive Mass Transfer Models

As a result of the second law of thermodynamics, systems displaced from equilibrium tend toward equilibrium (although perhaps very slowly) as time passes. The difference in the chemical potential of a component between one region in space and another is a measure of the displacement ("distance" in chemical potential coordinates) from equilibrium. There are a number of factors that can give rise to a difference in chemical potential:

- concentration differences,
- pressure differences,
- temperature differences, and
- differences in forces caused by external fields (gravity, magnetic, etc.).

By way of example, industrial application has been made of mass transfer obtained by applying a temperature gradient to a mixture - the Soret effect. (Conversely, we can obtain a flow of heat by imposing a concentration gradient - the Dufour effect.) The Soret effect, the foundation of thermal diffusion processes, requires discussion of irreversible thermodynamics for its systematic treatment. Here we do not develop the Onsager reciprocal relations which describe this and other coupled effects. We will, therefore, restrict ourselves in this text only to mass transfer which results from differences in concentration, both for the sake of brevity and because this is the effect by far most frequently used in practical problems.

8.2.1 Velocities of components in a mixture

A distinguishing characteristic of problems involving the transport of mass is that the flow velocities of the various molecular species in the mixture can differ. For example, consider the case of some pure diethyl ether placed in the bottom of a tall cylindrical vessel past the mouth of which is blown ether-free nitrogen (see Figure 8.2.1-1).

Figure 8.2.1-1 Diffusion of vapor from vessel

If we assume none of the ether to have evaporated initially, the initial concentration profile will look like curve a in Figure 8.2.1-2.

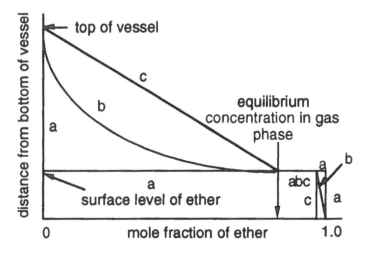

Figure 8.2.1-2 Evolution of concentration profile

Since initially the gas phase contains no ether and the liquid phase no nitrogen, equilibrium is not present. The chemical potential of ether differs between the gas phase and the liquid phase as does that of nitrogen. The nitrogen and ether, therefore, begin to move in such a way as to bring the system to its equilibrium state. Here the ether moves into the gas phase and the nitrogen (in a very small way, which is shown exaggerated in Figure 8.2.1-2) into the liquid phase.

At some intermediate time the profile may resemble curve b in Figure 8.2.1-2. Nitrogen will reach an equilibrium concentration in the gas phase at the liquid interface. Nitrogen is diffusing into the liquid ether, and ether is diffusing up the vessel and out its mouth. Notice, however, that the movement of nitrogen is halted by the bottom wall of the container.

When the liquid phase becomes saturated with nitrogen (i.e., the chemical potential in liquid and gas phases becomes the same), the concentration of nitrogen (and ether) in the liquid phase will be flat, and the nitrogen will cease to move because the driving force has disappeared. The ether, however, continues to be continuously removed by the steady stream of pure nitrogen past the mouth of the flask and the concentration of ether will reach a non-flat steady-state profile.

If we wait long enough, the concentration profile will reach a steady state as shown in curve c in Figure 8.2.1-2. (We assume that the supply of liquid is sufficiently large that the level does not change significantly during the unsteady-state portion of the process or that the ether liquid level is kept constant by addition of nitrogen-saturated liquid ether.) At this point the ether will continue to evaporate and transfer out the end of the vessel, but now at a steady rate; therefore, the vaporized ether will have some constant velocity (and concentration gradient) with respect to a coordinate system fixed in space. At steady state the nitrogen no longer moves relative to the vessel but rather is simply a stagnant medium (with respect to axes fixed in space) through which the ether diffuses. It is clear that the ether and the nitrogen have different velocities.

Until now we have left the concept of **velocity** of a fluid as something which, for a one-component system, was (it was hoped) intuitively obvious. With only one component it is not hard to replace the physical reality of lots of molecules banging into one another with the mathematical model describing a *continuum*. However, when we are faced with multicomponent systems, we must speak of velocities of **components** (not components of velocity). We face a more complicated situation, and to describe affairs we will introduce several types of velocity.

To begin at a level where physical intuition has not yet deserted us, let us consider the motion of a single molecule. At any one instant in time, if we ignore for the moment what we know of modern physics, an adequate model of the molecule is furnished by a small lump of matter with a given velocity vector **a** (see Figure 8.2.1-3).

Figure 8.2.1-3 Velocity of molecule

All is fine because we can clearly define a velocity for the molecule. Unfortunately, the molecule does not maintain this velocity for long, because just as it thinks it knows where it is going it runs into something (another molecule, perhaps free or perhaps fixed as part of a wall) which changes its course.

Consequently, if we wait just a little while we observe something like Figure 8.2.1-4, where our molecule has undergone several collisions with other molecules and therefore several changes in velocity. (We have not shown the velocity vectors of the other molecules for the sake of simplicity.) The molecule has traveled very fast, endured several collisions, and now has velocity b but really has not gone very far - only a distance $\Delta\lambda$. (We will not belabor the obvious parallel to life.)

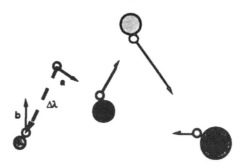

Figure 8.2.1-4 Changing velocity and displacement of single molecule[2] via collisions

For mass transfer purposes we are only really interested in how fast the molecule covered distance $\Delta\lambda$, not how fast it moved between collisions. The **apparent velocity** of the molecule is approximately

---

[2] Note that the solid arrows are **velocities**, the dashed arrow a **displacement**.

$$\boxed{v_{ij} \cong \frac{\Delta\lambda}{\Delta t}}$$

(8.2.1- 1)

where v_{ij} refers to the velocity of the jth molecule (that is, the 1st, 10th, 463rd, etc.) of the ith species (e.g., ethanol, oxygen, etc.).

Fortunately, the velocity at which a molecule hurtles about is so great in gases and liquids under ordinary conditions that it undergoes many collisions in any Δt large enough to be of interest to us. This is the same phenomenon that underlies our continuum assumption for the density.

Furthermore, since the many collisions are a stochastic process, $\Delta\lambda/\Delta t$ approaches a fairly constant value at "reasonable" values of Δt (not so small that individual collisions begin to make a difference, but not so large that the average velocity will change significantly if we are not at steady state). This, then, is the model of the velocity of a molecule.

We are not much interested in the velocity of a single molecule, but rather in the velocity of its chemical species. This we take to be the **number average velocity**, which weights the velocity of each individual molecule equally - below, by one (1)

$$v_i = \frac{\sum_{j=1}^{n} (1) v_{ij}}{\sum_{j=1}^{n} (1)} = \frac{v_{i1} + v_{i2} + v_{i3} + ... + v_{in}}{(1 + 1 + ... + 1)}$$
$$= \frac{v_{i1} + v_{i2} + v_{i3} + ... + v_{in}}{n}$$

(8.2.1-2)

where n is the number of molecules in some volume, ΔV. This is what is meant by **the velocity of a single species with respect to fixed co-ordinate axes.**

This average is simply the sum of the velocities of all the molecules of the ith species, divided by the total number of molecules of the ith species. Note that it parallels the general definition of average given earlier (with weighting factor equal to (1) except that the integrals are replaced by sums. To help illustrate what this velocity means physically, we will consider several idealized examples.

Example 8.2.1-1   Average velocity when individual particles have the same velocity

Consider the case of four molecules with identical velocity vectors as shown in Figure 8.2.1-5.

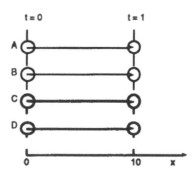

Figure 8.2.1-5 Identical molecules, identical velocities

Calculate the number average velocity.

Solution

Since molecules seldom behave this way, and since we wish to extend the analogy below, we might think of these molecules as students walking in the same direction at the same speed, say 10 ft/s. In this case the application of our definition gives an average velocity of 10 ft/s as follows

$$
v_{student} = \frac{\sum_{j=1}^{4} (1) \, v_{student \, j}}{\sum_{j=1}^{4} (1)} = \frac{v_{student \, 1} + v_{student \, 2} + v_{student \, 3} + v_{student \, 4}}{(1 + 1 + 1 + 1)}
$$

$$
= \frac{10 + 10 + 10 + 10}{4} = 10 \tag{8.2.1-3}
$$

(The 10s are vectors but are easily added without writing them in component form since they all have the same direction.) Notice that the students do not move relative to one another since they have the same velocity. Here the calculated velocity is intuitively satisfying because the average velocity multiplied by the time elapsed will give the distance traveled for each of the four students.

Example 8.2.1-2   Average velocity when individual particles have different velocities

Now consider a second case, shown in Figure 8.2.1-6 again for a unit elapsed time, where all the students again start along the y-axis, but now proceed with two different velocities.

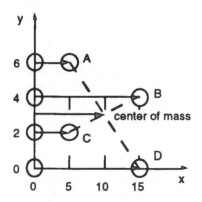

Figure 8.2.1-6 Identical molecules, differing velocities

Calculate the number average velocity.

Solution

Starting from the same initial configuration, the distance of one student from another no longer remains the same. Application of the definition of average velocity gives

$$v_{student} = \frac{\sum_{j=1}^{4} (1) \, v_{student \, j}}{\sum_{j=1}^{4} (1)} = \frac{v_{student \, 1} + v_{student \, 2} + v_{student \, 3} + v_{student \, 4}}{(1 + 1 + 1 + 1)}$$

$$= \frac{5 + 15 + 5 + 15}{4} = 10 \qquad (8.2.1\text{-}4)$$

The average velocity is the same, but notice that it does not correspond to the velocity of any one of the students. In this sense, it is a **fictitious** velocity. (The mistake of applying average characteristics to individual students is often committed in universities.)

Why are we interested in a fictitious velocity? Because it permits us to predict average **behavior**. Suppose the students each have a mass of 180 lbmass. The average velocity is the velocity at which the center of mass of the students moves.

The center of mass for several particles is found by multiplying the mass of each particle times its vector distance from an arbitrary origin, summing all these products (the *first moment* of the masses), and dividing this sum by the sum of all the individual masses. For our case here, a convenient coordinate system and origin to choose are the axes shown, which run parallel and perpendicular to the motion, respectively, with origin located to give one of the particles an initial position vector of zero. For the original configuration, denoting the position vector as **r**, the center of mass is calculated as

$$\mathbf{r}_{center\ of\ mass} = \frac{m_A\,\mathbf{r}_A + m_B\,\mathbf{r}_B + m_C\,\mathbf{r}_C + m_D\,\mathbf{r}_D}{(m_A + m_B + m_C + m_D)} \qquad (8.2.1\text{-}5)$$

$$\left(x_{center\ of\ mass}\,\mathbf{i} + y_{center\ of\ mass}\,\mathbf{j}\right)$$

$$= \frac{180\left(0\,\mathbf{i} + 6\,\mathbf{j}\right) + 180\left(0\,\mathbf{i} + 4\,\mathbf{j}\right) + 180\left(0\,\mathbf{i} + 2\,\mathbf{j}\right) + 180\left(0\,\mathbf{i} + 0\,\mathbf{j}\right)}{(180 + 180 + 180 + 180)}$$

$$(8.2.1\text{-}6)$$

Equating components and dividing numerator and denominator by 180

$$x_{center\ of\ mass} = 0$$

$$y_{center\ of\ mass} = \frac{6 + 4 + 2 + 0}{4} = 3 \qquad (8.2.1\text{-}7)$$

Thus, the original center of mass is as shown.

A similar calculation for the center of mass after a unit time interval shows that

$$\mathbf{r}_{center\ of\ mass} = \left(x_{center\ of\ mass}\,\mathbf{i} + y_{center\ of\ mass}\,\mathbf{j}\right)$$

$$= \frac{180\left(5\,\mathbf{i} + 6\,\mathbf{j}\right) + 180\left(15\,\mathbf{i} + 4\,\mathbf{j}\right) + 180\left(5\,\mathbf{i} + 2\,\mathbf{j}\right) + 180\left(15\,\mathbf{i} + 0\,\mathbf{j}\right)}{(180 + 180 + 180 + 180)}$$

$$(8.2.1\text{-}8)$$

Equating components and dividing numerator and denominator by 180

$$x_{center\ of\ mass} = \frac{5+15+5+15}{4} = 10$$
$$y_{center\ of\ mass} = \frac{6+4+2+0}{4} = 3 \qquad\qquad (8.2.1\text{-}9)$$

The change in position of the center of mass is shown in Figure 8.2.1-6. Notice that the distance the center of mass moves (10 units) is equal to the average velocity (10) times elapsed time (1).

The number average velocity we defined has been shown to correspond to the average velocity of the center of mass for a system of particles of equal mass. Instead of mass we could choose any property that all the objects possess equally, and our number average velocity will give us the velocity of the center of (first moment of) that property. The important thing is that each object possess the property to the same extent, so that the weighting factors in the numerator may all be made equal.

For example, we could define a property called "studentness." Assuming each student possesses "studentness" to the same extent (say 471.3), the number average velocity would give us the velocity of the "center of studentness." (It is obvious that the same procedure will not work with "scholarliness," which is possessed in widely varying degrees.)

In mass transfer we are interested in velocities of species, which by definition are collections sharing a common property (for example, chemical composition, molecular weight, etc.). For these cases the number average velocity gives us the rate of motion of the "center of" that property.

Example 8.2.1-3   Number average velocity, velocities in two dimensions

As a slightly more complicated example of number average velocity, consider oxygen molecules with velocities as shown in Figure 8.2.1-7.

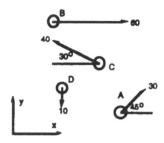

Figure 8.2.1-7 Number average velocity, velocities in two dimensions

Determine the number average velocity.

Solution

$$v_{O_2} = \frac{\sum_{j=1}^{n}(1)\,v_{O_2,j}}{\sum_{j=1}^{n}(1)} = \frac{\sum_{j=1}^{n}(1)\left(v_x\,i + v_y\,j\right)_{O_2,j}}{\sum_{j=1}^{n}(1)} \qquad (8.2.1\text{-}10)$$

$$v_{O_2} = \frac{\sum_{j=1}^{n}(1)\left[\left(v_x\right)i + \left(v_y\right)j\right]_{O_2,j}}{\sum_{j=1}^{n}(1)} \qquad (8.2.1\text{-}11)$$

$$= \frac{\left[\left(30\cos45°\right)i + \left(30\sin45°\right)j\right] + \left[(60)\,i + (0)\,j\right]}{(1+1+1+1)} \qquad (8.2.1\text{-}12)$$

$$- \frac{\left[\left(40\cos30°\right)i + \left(40\sin30°\right)j\right] + \left[(0)\,i + (10)\,j\right]}{(1+1+1+1)} \qquad (8.2.1\text{-}13)$$

$$v_{O_2} = 18.4\,i + 7.8\,j \qquad (8.2.1\text{-}14)$$

This is the velocity of the center of mass of the oxygen molecules. Or, in terms of speed and direction

$$\left|v_{O_2}\right| = \sqrt{(18.4)^2 + (7.8)^2} = 20 \qquad (8.2.1\text{-}15)$$

$$\theta = \arctan\left(\frac{7.8}{18.4}\right) = 23° \qquad (8.2.1\text{-}16)$$

We now need to define, for a fluid which is made up of more than one component, a **single** fluid velocity composed of the different velocities of the components. Suppose we think of our molecule as being one animal in a mixed herd made up of horses and steers rushing through a canyon. Assume that the horses are identical in weight and, being lighter and bred for more speed, move faster than the steers, which we also assume to be uniform in weight (but heavier). This difference in velocity leads to considerable jostling (collisions) and mixing, but, by applying our definition of species velocity, we could certainly compute a "horse" velocity and a "steer" velocity with respect to axes fixed in space (that is, as observed by a cowboy on a stationary horse). Someone might just ask us, however, what the "animal" velocity is.

One reply we could make to such a question is to ask in return, "Do you mean how *many* animals went past or what *mass* of animals went past?" The reason for such a reply is that if, for example, we are stationary (coordinates fixed in space) we certainly sense some sort of "animal" (fluid) velocity, but this velocity is generated by two different types of animals (molecules) - horses (component 1) and steers (component 2). Normally we wish to define the "animal" velocity only because we can use it to calculate certain **other** things (since the animals are not racing), for example, the *number* rate of animals passing or the *mass* rate of animals passing.

The same question faces us in mass transfer. In mass transfer, however, we are not usually counting numbers of molecules as we counted number of horses and steers in our example. Instead, we usually deal with concentrations in one form or another. We define the following symbols

$$c_i \equiv \text{molar concentration of ith species} \qquad \left[\frac{\text{mol } i}{\text{volume}}\right] \qquad (8.2.1\text{-}17)$$

$$c \equiv \text{total molar concentration} = \sum_{\text{all } i} c_i \qquad \left[\frac{\text{total mol}}{\text{volume}}\right] \qquad (8.2.1\text{-}18)$$

$$x_i \equiv \text{mole fraction of ith species} \qquad \left[\frac{\text{mol } i}{\text{total mol}}\right] \qquad (8.2.1\text{-}19)$$

$$\rho_i \equiv \text{mass concentration of ith species} \qquad \left[\frac{\text{mass } i}{\text{volume}}\right] \qquad (8.2.1\text{-}20)$$

$$\rho \equiv \text{total mass concentration (density)} = \sum_{\text{all } i} \rho_i \qquad \left[\frac{\text{total mass}}{\text{volume}}\right] \qquad (8.2.1\text{-}21)$$

$$\omega_i \equiv \text{mass fraction of ith species} \qquad \left[\frac{\text{mass } i}{\text{total mass}}\right] \qquad (8.2.1\text{-}22)$$

We are interested in the **number** of molecules of a particular species passing a certain point (the number of **moles**) as well as the **mass** of molecules of a particular species passing a certain point in space. Accordingly, we define *two* velocities with respect to axes fixed in space:

    (1) a molar average velocity, and
    (2) a mass average velocity.

The local[3] **molar average velocity** is defined as

$$v^* \equiv \frac{\sum_{i=1}^{n} c_i v_i}{\sum_{i=1}^{n} c_i} = \frac{\sum_{i=1}^{n} c_i v_i}{c}$$

(8.2.1-23)

This implies that

$$c\, v^* = v^* \sum_{i=1}^{n} c_i = \sum_{i=1}^{n} c_i v_i$$

(8.2.1-24)

The local **mass average velocity** is defined as

$$v \equiv \frac{\sum_{i=1}^{n} \rho_i v_i}{\sum_{i=1}^{n} \rho_i} = \frac{\sum_{i=1}^{n} \rho_i v_i}{\rho}$$

(8.2.1-25)

The parallel to the general definition of average should again be obvious. In fact, other average velocities may be defined - the *volume* average velocity, for example - but we need discuss only the two above for our purposes.

The above velocities are referred to axes *fixed in space*. We sometimes are interested in velocities relative to coordinate axes which themselves are moving. We will say more about this later.

From the above definitions we can see the usefulness of the molar and mass average velocities. The velocity v_i, when multiplied by the molar concentration c_i, and the resulting products summed over all components, yields the **total molar flux, N, with respect to axes fixed in space** since (the velocity of the coordinate frame being zero)

---

[3] Note that we are using a continuum model which assumes that the velocity exists at a point, even though we average over a finite number of molecules that would occupy a finite volume.

$$N = \sum_{i=1}^{n} c_i \left(v_i - 0 \right) = \sum_{i=1}^{n} c_i v_i = c v^*$$

(8.2.1-26)

is the sum of the molar fluxes of each individual species, (moles/ft^3) (ft/s) = [moles/(ft^2 s)].

Similarly, the velocity, v_i, when multiplied by the mass concentration, ρ_i, and the resulting products summed over all components, yields the **total mass flux, n, with respect to axes fixed in space**

$$n = \sum_{i=1}^{n} \rho_i \left(v_i - 0 \right) = \sum_{i=1}^{n} \rho_i v_i = \rho v$$

(8.2.1-27)

Our notation in discussing fluxes will be to use *lowercase* letters for *mass* fluxes and *uppercase* letters for *molar* fluxes. For fluxes referred to *stationary* coordinates we use the letter n (or N), while for fluxes referred to *moving* coordinates we use the letter j (or J). For non-stationary coordinate frames, the speed at which the coordinate frame is moving for cases we will consider will be either the *mass* average velocity or the *molar* average velocity. To distinguish these cases we append an asterisk superscript to fluxes referred to frames moving at the *molar* average velocity. These relations are summarized in Tables 8.2.1-1 and 8.2.1-2.

Table 8.2.1-1 Coordinate frame motion

Flux Units	Coordinate Frame Motion		
	Stationary	Mass Average Velocity	Molar Average Velocity
Mass	n_i	j_i	j_i^*
Moles	N_i	J_i	J_i^*

Table 8.2.1-2 Mass transfer relationships

$$\omega_i = \frac{\rho_i}{\rho}$$

$$x_i = \frac{c_i}{c}$$

$$c_i = \frac{\rho_i}{M_i}$$

$$v = \frac{\sum \rho_i v_i}{\sum \rho_i}$$

$$v^* = \frac{\sum c_i v_i}{\sum c_i}$$

$$n_i = \rho_i v_i$$

$$N_i = c_i v_i$$

$$j_i = \rho_i (v_i - v)$$

$$J_i^* = c_i (v_i - v^*)$$

8.2.2   Mechanisms of mass transfer

Mass transfer is carried out by two of the same mechanisms as transport of heat:

> • molecular diffusion, the analog of heat transfer by conduction
> • eddy diffusion, the analog of convective heat transfer.

The diffusive process is, in general, much slower than the convective process. We will first consider molecular diffusion and then convection.

8.2.3   Fick's law

The basic relation governing mass transfer by molecular diffusion in a binary mixture is called *Fick's law*.

$$\boxed{\mathbf{J}_A^{\bullet} = -c\, \mathcal{D}_{AB}\, \nabla x_A}$$

(8.2.3-1)

where

$J_A{}^{*}$ = molar flux of component A referred to axes moving at the molar average velocity

$\mathcal{D}_{AB}$ = constant of proportionality called the diffusivity of A through B

x_A = mole fraction of A

c = molar density of mixture

Fick's law is basically an empirical law, as is Fourier's law. As with Fourier's law, it is a constitutive relation which introduces the properties of a specific system into the more general balance equations. Remembering that a one-dimensional form of Fourier's law is

$$\frac{Q}{A} = -k\frac{dT}{dy}$$

(8.2.3-2)

the parallel of the one-dimensional form of Fick's law to Fourier's law is clear:

$$\boxed{J_{A_y}^{\bullet} = -c\, \mathcal{D}_{AB}\, \frac{dx_A}{dy}}$$

(8.2.3-3)

However, notice that the above flux is with respect to **moving** coordinate axes, while the flux in Fourier's law is with respect to stationary axes. Why should the law be written this way?

The reason is that use of a moving coordinate system makes the form of Fick's law - as introduced above - simpler. If the flux were instead rewritten with respect to coordinate axes fixed in space, one would have an **additional apparent flow** of magnitude Nx_A, where N is the *total* molar flux with respect to stationary coordinates.[4]

$$N_A = J_A^{\bullet} + N\, x_A$$

(8.2.3-4)

---

[4] Note that
$$N = N_A + N_B$$
by a mass balance.

or, using Fick's law

$$N_A = -c \, \mathcal{D}_{AB} \, \nabla x_A + N \, x_A \qquad\qquad (8.2.3\text{-}5)$$

When the diffusion equation is written as above there is an "extra term" which spoils the parallelism to the Fourier equation [Equation (8.2.3-2)]. This term also complicates the mathematics considerably except for special cases.

Table 8.2.3-1 gives equivalent forms of Fick's law.

Table 8.2.3-1 Equivalent forms of Fick's Law referred to coordinate systems in various motions

$$\mathbf{n}_A - \omega_A \left(\mathbf{n}_A + \mathbf{n}_B \right) = -\rho \, \mathcal{D}_{AB} \, \nabla \omega_A$$

$$\mathbf{N}_A - x_A \left(\mathbf{N}_A + \mathbf{N}_B \right) = -c \, \mathcal{D}_{AB} \, \nabla x_A$$

$$\mathbf{j}_A = -\rho \, \mathcal{D}_{AB} \, \nabla \omega_A$$

$$\mathbf{J}_A^* = -c \, \mathcal{D}_{AB} \, \nabla x_A$$

The fact that the flux of A in Equation (8.2.3-5) is made up of a flux from diffusion plus a flux from total molar flux (bulk flow) may perhaps be made clearer by considering an analogy. Suppose a sheet metal tray is constructed, 1 ft x 1 ft x 1 ft with shallow dimples in the bottom at regular intervals, and the tray dimples are filled with a regular pattern of 30 white (A) and 30 black (B) marbles (identical except for color) as shown in Figure 8.2.3-1.

The tray is then moved at some velocity, say 1 ft/s, past a fixed point in space. The heavy dotted line shows the center of number or mass in the direction of motion of all the marbles; the lighter dotted lines show the center of mass of the A and B groups of marbles, respectively.

During the one second that the tray takes to pass a fixed point, using the end of the tray as a basis area (a cross-section of 1 ft^2), an average *total flux* N of 60 marbles/(ft^2 s) with respect to the stationary coordinate system is observed (60 marbles passed through 1 ft^2 in one second). This flux corresponds to the flux in the last term in Equation (8.2.3-5). Multiplying this flux by the concentration of white marbles ($x_A = 1/2$) as indicated by the equation gives

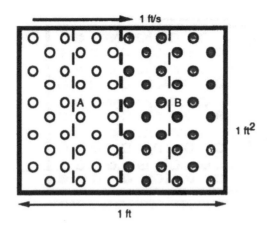

Figure 8.2.3-1 Flux of marbles without diffusion

$$N\,x_A = (60)\,(1/2) = 30\,\frac{\left(\text{white marbles}\right)}{(\text{ft}^2\,\text{s})} \tag{8.2.3-6}$$

(Although within Δt the flux of white marbles goes from 0 to 30 and back to 0, we are not interested in the *instantaneous* flux, just the *average* flux - this is an analogous problem to the one we had before with our *continuum* assumption.)

A fixed array of marbles produces no term corresponding to the *diffusion* term in Equation (8.2.3-5). Such a term could be produced if the tray were to be jiggled as it moves so that the marbles can trade sites.

Consider the case sketched in Figure 8.2.3-2.

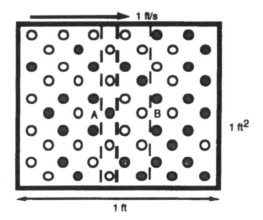

Figure 8.2.3-2 Flux of marbles with diffusion

The original heavy dotted line marking the center of mass or center of number for all of the marbles is still shown, and is unchanged. In addition, the center of mass or of number for the set of white marbles and the corresponding center for the set of black marbles is also given. (These facts can be verified by counting the number of marbles to the right and left of the respective line.)

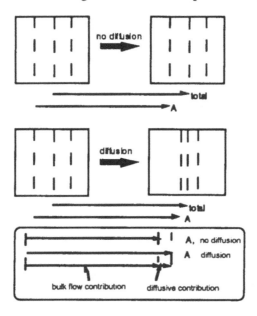

Figure 8.2.3-3 Fluxes compared

It can be seen that the center for the white marbles is relatively advanced compared to the non-diffusion case; that of the black marbles, retarded. This is illustrated in Figure 8.2.3-3.

In this second case the white marbles are farther along *on the average* - the difference is the flux from diffusion that gives the first term in Equation (8.2.3-5). The two cases are shown in simplified form, and the vectors for the white marbles (A) are compared below the cases. A similar comparison could be done for B.

8.2.4 Binary diffusivities

As one might expect from consideration of the mobility of the molecules, diffusivities are generally larger for liquids than for solids, and larger for gases than for liquids. The units of the diffusion coefficient are area per unit time, and most tabulations are in units of cm^2/s. To convert these units to ft^2/hr one multiplies by 3.87. There are a number of methods for predicting diffusivity a priori.[5] The diffusivity is a *property* and thus a function of thermodynamic variables - mainly of temperature and concentration, although pressure has a fairly pronounced effect on the diffusivity of gases.

Typical values for diffusion coefficients are listed in Figures 8.2.4-1, 8.2.4-2, and 8.2.4-3.[6]

[5] Bird, R. B., W. E. Stewart, et al. (1960). *Transport Phenomena*. New York, Wiley.

[6] Barrer, R. M. (1941). *Diffusion In and Through Solids*. New York, NY, The Macmillan Company.

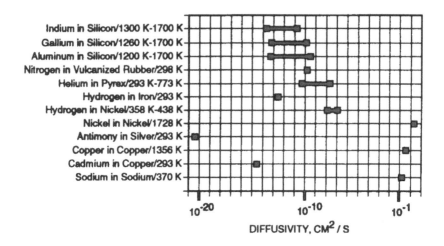

Figure 8.2.4-1 Diffusivities in solids

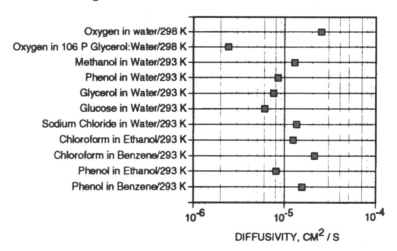

Figure 8.2.4-2 Diffusivities in liquids

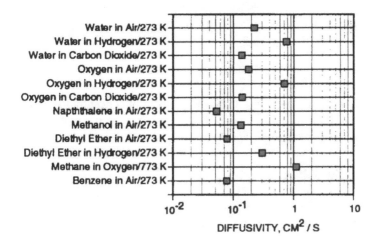

Figure 8.2.4-3 Diffusivities in gases

8.2.5 Solutions of the diffusion equation

The diffusion equation is a partial differential equation which derived from a balance on mass of a chemical species. Many analytic solutions for simple boundary shapes and various boundary conditions may be found in the literature.[7, 8] Many other solutions are possible by numerical techniques. We will consider one or two simple situations.

One-dimensional equimolar counterdiffusion in rectangular coordinates

By definition, equimolar counterdiffusion describes the case of one mole of A diffusing in a given direction for each mole of B diffusing in the opposite direction. Thus

$$N_A = -N_B \qquad\qquad (8.2.5-1)$$

Therefore, from Equation (8.2.3-6):

$$N = 0 \qquad\qquad (8.2.5-2)$$

For the one-dimensional case (using the y-direction) Fick's law reduces to

---

[7] Crank, J. (1975). *The Mathematics of Diffusion*, Oxford.
[8] Carslaw, H. S. and J. C. Jaeger (1959). *Conduction of Heat in Solids*, Oxford.

$$N_{Ay} = -c \, \mathcal{D}_{AB} \frac{dx_A}{dy} \tag{8.2.5-3}$$

(Remember that x_A is a concentration, *not* a coordinate.)

If we further assume c and $\mathcal{D}_{AB}$ to be independent of concentration and the temperature to be uniform (zero heat of mixing, etc.), Equations (8.2.5.1) to (8.2.5-3) may be integrated easily since the variables can be separated:

$$\int_0^L dy = -\frac{c \, \mathcal{D}_{AB}}{N_{Ay}} \int_{x_{A0}}^{x_{A1}} dx_A \tag{8.2.5-4}$$

where

        L = path length for diffusion
        x_{A0}, x_{A1} = bounding mole fractions

giving

$$L = -\frac{c \, \mathcal{D}_{AB}}{N_{Ay}} \left(x_{A1} - x_{A0} \right) \tag{8.2.5-5}$$

(Note that this type of behavior is observed in those binary distillation columns where a mole of B diffuses to the interface and condenses for every mole of A that evaporates and moves away from the interface; that is, those operating with constant molal overflow.)

We can rewrite Equation (8.2.5-5) as

$$N_{Ay} = -c \, \mathcal{D}_{AB} \frac{\left(x_{A1} - x_{A0} \right)}{L} \tag{8.2.5-6}$$

which is in the form of a flux equaling a conductance times a gradient (driving force).

For gases at moderate pressures we can use partial pressure as driving force instead of mole fraction and

$$N_{Ay} = -\frac{\mathcal{D}_{AB}}{RT} \frac{\left(p_{A1} - p_{A0} \right)}{L} \tag{8.2.5-7}$$

Example 8.2.5-1 Equimolar counterdiffusion

A mixture of benzene and toluene is supplied as vapor to the bottom of an insulated rectifying column. At one point in the column where the pressure is 1 atm, the vapor contains 80 mole % benzene and the interfacial liquid contains 70 mole % benzene. The temperature at this point is 89°C.

Assuming equilibrium at the interface, and assuming the diffusional resistance to transfer in the vapor phase to be equivalent to the diffusional resistance of a stagnant vapor layer 0.1 in. thick, calculate the rate of interchange of benzene and toluene between vapor and liquid.

The molal latent heats of vaporization of benzene and toluene may be assumed to be equal, and the system is close enough to ideal to use Raoult's law for the equilibrium relationship. The vapor pressure of benzene at 89°C is 958 mm Mg and the diffusivity for toluene/benzene may be assumed to be 0.198 ft^2/hr. (Note that one does not really know the equivalent film thickness; this would have to be determined by experimental measurement, but since it depends primarily on the fluid mechanics, it could be determined in the pilot plant in a different concentration range or even in a different chemical system so long as the appropriate similarity of the fluid flow were maintained.)

Solution

The liquid will be boiling and hence well mixed. In a rectifying column operating without heat loss, thermal energy from any toluene condensed will go toward vaporizing an equal number of moles of benzene, since the latent heats of vaporization per mole are assumed to be equal. At any point in the column, therefore, $N_A = -N_B$.

$$p_{Ao} = (0.7)(988)[mm\ Hg]\frac{[atm]}{(760)[mm\ Hg]} = 0.91\ atm$$

$$p_{A1} = 0.8\ atm$$

(8.2.5-8)

$$N_{benzene} = -\frac{\mathcal{D}_{AB}}{RT}\frac{(p_{A1} - p_{Ao})}{L}$$

$$= -\frac{0.198\ \frac{ft^2}{hr}}{0.728\ \frac{atm\ ft^3}{lbmol\ °R}\ 652.2\ °R}\frac{(0.91 - 0.8)\ atm}{\frac{0.1}{12}\ ft}$$

$$= 5.5 \times 10^{-3} \, \frac{\text{lbmol benzene}}{\text{hr ft}^2} \qquad (8.2.5\text{-}9)$$

One-dimensional diffusion of A through stagnant B observed in rectangular coordinates

If B does not move with respect to axes fixed in space we have

$$N_B = 0 \qquad (8.2.5\text{-}10)$$

and Fick's law reduces to

$$N_{Ay} = -c \, \mathcal{D}_{AB} \frac{dx_A}{dy} + N_{Ay} x_A \qquad (8.2.5\text{-}11)$$

or, after rearranging,

$$\int_0^L dy = -\frac{c \, \mathcal{D}_{AB}}{N_{Ay}} \int_{x_{A0}}^{x_{A1}} \frac{dx_A}{(1 - x_A)} \qquad (8.2.5\text{-}12)$$

Integrating and substituting limits:

$$N_{Ay} = -\frac{c \, \mathcal{D}_{AB}}{L} \ln \frac{(1 - x_{A0})}{(1 - x_{A1})} \qquad (8.2.5\text{-}13)$$

In principle this is as far as we need go, since we can calculate the flux from the above relation. Engineers, however, have a compulsion to use equations where they can identify flux, conductance, and gradient terms, and so to conform to convention we multiply and divide by $(x_{A1} - x_{A0})$, substitute x_B for $(1 - x_A)$, and use the definition of log mean to rewrite Equation (8.2.5-13) as

$$N_{Ay} = -\frac{c \, \mathcal{D}_{AB}}{(x_B)_{lm}} \frac{(x_{A1} - x_{A0})}{L} \qquad (8.2.5\text{-}14)$$

Note that for very dilute solutions $(x_B)_{lm}$ is approximately equal to 1.0. This is also an adequate model where B, although not completely stagnant, moves very slowly with respect to A.

Partial pressure may be used as the driving force where gases at moderate pressure are diffusing. Thus

$$N_{Ay} = -\frac{p\,\mathcal{D}_{AB}}{RT\,(p_B)_{lm}}\frac{(p_{A1} - p_{A0})}{L}$$

(8.2.5-15)

where

$$(p_B)_{lm} = \frac{(p_{B1} - p_{B0})}{\ln\left[\dfrac{p_{B1}}{p_{B0}}\right]}$$

(8.2.5-16)

Example 8.2.5-2   Diffusion of vapor through a stagnant gas

A fan is blowing 32°F air over a container filled to within 1 in. of the top with a 4-in. deep solution of 15% by weight ethyl alcohol in water. Assume the partial pressure of alcohol at the interface is 0.5 atm and the air flowing across the container does not contain alcohol. Also, assume that the mouth of the container is narrow enough that in effect we have 1 in. of stagnant gas.

$\mathcal{D}_{AB}$ for this system is about 0.4 ft^2/hr.

Will the alcohol content of the solution change significantly in one day?

Solution

We replace the real system by the model shown in Figure 8.2.5-1.

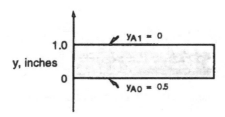

Figure 8.2.5-1 Diffusion through stagnant gas layer

$$p_{A0} = 0.5 \text{ atm}$$
$$p_{A1} = 0.0 \text{ atm}$$
$$p_{B0} = 0.5 \text{ atm}$$
$$p_{B1} = 1.0 \text{ atm}$$

$$\left(p_{\text{B}}\right)_{\text{lm}} = \frac{\left(p_{B1} - p_{B0}\right)}{\ln\left[\dfrac{p_{B1}}{p_{B0}}\right]} = \frac{\left(1 - 0.5\right)}{\ln\left[\dfrac{1}{0.5}\right]} = 0.72 \text{ atm} \tag{8.2.5-17}$$

Notice that there is little difference between the log mean and the arithmetic mean for this case, as compared to the general accuracy of the other assumptions involved.

$$\left(p_{\text{B}}\right)_{\text{arith}} = \frac{\left(p_{B1} + p_{B0}\right)}{2} = \frac{\left(1 + 0.5\right)}{2} = 0.75 \text{ atm} \tag{8.2.5-18}$$

Substituting

$$
\begin{aligned}
N_{Ay} &= -\frac{p\,\mathcal{D}_{AB}}{RT\left(p_{\text{B}}\right)_{\text{lm}}} \frac{\left(p_{A1} - p_{A0}\right)}{L} \\
&= -\frac{(1)(0.4)}{(0.728)(492)(0.72)} \frac{\left(0.5 - 0\right)}{\frac{1}{12}} \\
&= 9.3 \times 10^{-3} \frac{\text{lbmol}}{\text{hr ft}^2}
\end{aligned}
\tag{8.2.5-19}
$$

Model the density of the solution as that of water, and assume that the solution is perfectly mixed. Then in one day, one square foot of solution will lose **at the initial rate**

$$N_{Ay}\,A = \left(9.3 \times 10^{-3}\right) \frac{\text{lbmol}}{\text{hr ft}^2} (1)\, \text{ft}^2 (24)\, \text{hr} = 0.22 \text{ lbmol ethanol} \tag{8.2.5-20}$$

However, the solution beneath this square foot of surface contained initially only

$$\left(c_A\right)(A)(\text{depth}) = (0.15)(62.4)\frac{\text{lbmass}}{\text{ft}^3}\frac{1}{(46)}\frac{\text{lbmol}}{\text{lbmass}}(1)\,\text{ft}^2\left(\frac{4}{12}\right)\text{ft}$$

$$= 0.07 \text{ lbmol ethanol} \qquad (8.2.5\text{-}21)$$

The concentration of alcohol obviously will change significantly in one day (this model predicts that it would decrease so far as to become negative!), and, therefore, to assume the initial rate of evaporation of ethanol is maintained would not be a viable model of the process. (Such an assumption would, in general, give an upper bound to the amount of ethanol evaporated, since the driving force and therefore the rate of evaporation will decrease with time - in this case, however, the model says that **all** of the alcohol would evaporate, and we know that this should take an infinite amount of time.)

To calculate the liquid-phase concentration of ethanol after 24 hours in a proper fashion would require an unsteady-state model which incorporated a changing driving force for evaporation. In developing a revised model, the validity of the perfect mixing assumption of the liquid phase should be investigated also, because one could conceivably find the controlling resistance to be in the liquid phase, not the vapor. The assumption of a stagnant vapor phase is, of course, also questionable. This example shows that simple models are not always the best models.

One dimensional unsteady-state diffusion in a semi-infinite slab

Consider diffusion of component A in the x-direction in a semi-infinite (bounded only by one face) slab initially at a uniform concentration, c_{Ai}, whose face suddenly at time equal to zero is raised to and maintained at c_{As}. The concentration profile will "penetrate" into the slab with time in the general fashion shown in Figure 8.2.5-2.

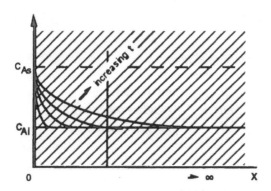

Figure 8.2.5-2 Semi-infinite slab with constant face concentration

The partial differential equation describing this situation springs from the microscopic mass balance

$$\nabla \cdot \rho_i v_i + \frac{\partial \rho_i}{\partial t} = r_i$$

$$(2.2.2\text{-}4)$$

For the case of no chemical reaction, one space dimension, and two components, A and B

$$\frac{\partial(\rho_A v_{Ax})}{\partial x} + \frac{\partial \rho_A}{\partial t} = 0$$

$$(8.2.5\text{-}22)$$

Dividing both sides by the molecular weight of A gives

$$\frac{\partial(c_A v_{Ax})}{\partial x} + \frac{\partial c_A}{\partial t} = \frac{\partial(N_{Ax})}{\partial x} + \frac{\partial c_A}{\partial t} = 0$$

$$(8.2.5\text{-}23)$$

Substituting using the appropriate form of Fick's Law

$$\frac{\partial\left(-\mathcal{D}_{AB}\frac{\partial c_A}{\partial x} + x_A\left[N_{Ax} + N_{Bx}\right]\right)}{\partial x} + \frac{\partial c_A}{\partial t} = 0$$

$$(8.2.5\text{-}24)$$

Assuming equimolar counterdiffusion yields the differential equation[9]

$$\frac{\partial c_A}{\partial t} = \mathcal{D}_{AB}\frac{\partial^2 c_A}{\partial x^2}$$

$$(8.2.5\text{-}25)$$

with initial and boundary conditions

$$
\begin{array}{llll}
t = 0: & c_A = c_{Ai}, & \text{all } x & \text{(initial condition)} \\
x = 0: & c_A = c_{As}, & \text{all } t > 0 & \text{(boundary condition)} \\
x = \infty: & c_A = c_{Ai}, & \text{all } t > 0 & \text{(boundary condition)}
\end{array}
$$

$$(8.2.5\text{-}26)$$

---

[9] For an example of how this same form of equation can arise for a *steady state* model, the case of absorption into a falling film, see Bird, R. B., W. E. Stewart, et al. (1960). *Transport Phenomena*. New York, Wiley, p. 537.

Consider the de-dimensionalization of the dependent variable in the equation using

$$c_A^* = \frac{c_A - c_{Ai}}{c_{As} - c_{Ai}} \tag{8.2.5-27}$$

Treating individual factors

$$\frac{\partial c_A}{\partial t} = \frac{\partial c_A}{\partial c_A^*} \frac{\partial c_A^*}{\partial t} = \left(c_{As} - c_{Ai} \right) \frac{\partial c_A^*}{\partial t} \tag{8.2.5-28}$$

$$\frac{\partial^2 c_A}{\partial x^2} = \frac{\partial}{\partial x} \left[\frac{\partial c_A}{\partial x} \right] = \frac{\partial}{\partial x} \left[\frac{\partial c_A}{\partial c_A^*} \frac{\partial c_A^*}{\partial x} \right] = \left(c_{As} - c_{Ai} \right) \frac{\partial^2 c_A^*}{\partial x^2} \tag{8.2.5-29}$$

Substituting in the differential equation

$$\left(c_{As} - c_{Ai} \right) \frac{\partial c_A^*}{\partial t} = \left(c_{As} - c_{Ai} \right) \mathcal{D}_{AB} \frac{\partial^2 c_A^*}{\partial x^2} \tag{8.2.5-30}$$

$$\frac{\partial c_A^*}{\partial t} = \mathcal{D}_{AB} \frac{\partial^2 c_A^*}{\partial x^2} \tag{8.2.5-31}$$

with transformed boundary conditions

$$
\begin{aligned}
t &= 0: \quad c_A^* = 0, \text{ all } x \quad &\text{(initial condition)} \\
x &= 0: \quad c_A^* = 1, \text{ all } t > 0 \quad &\text{(boundary condition)} \\
x &= \infty: \quad c_A^* = 0, \text{ all } t > 0 \quad &\text{(boundary condition)}
\end{aligned}
\tag{8.2.5-32}
$$

This is the analogous problem to that of velocity profiles in a Newtonian fluid adjacent to a wall suddenly set in motion, and to that of temperature profiles in a semi-infinite medium with one face abruptly raised to a higher temperature. The dimensionless concentration, c_A^* is analogous to the dimensionless velocity

$$v_y^* = \frac{v_y - v_i}{V - v_i} = \frac{v_y - 0}{V - 0} = \frac{v_y}{V} \tag{8.2.5-33}$$

(V being the velocity of the wall), or the dimensionless temperature

$$T^* = \frac{T - T_i}{T_s - T_i} \tag{8.2.5-34}$$

and the binary diffusivity $\mathcal{D}_{AB}$ is analogous to the kinematic viscosity, v, or the thermal diffusivity, α.

The solution to the differential equation is

$$c_A^\bullet = f_1(x, t, \mathcal{D}_{AB}) \tag{8.2.5-35}$$

and can be written as

$$f_2(c_A^\bullet, x, t, \mathcal{D}_{AB}) = c_A^\bullet - f_1(x, t, \mathcal{D}_{AB}) = 0 \tag{8.2.5-36}$$

where we observe that $f_2(c_A^\bullet, x, t, \mathcal{D}_{AB})$ must be a dimensionally homogeneous function.

Applying the principles of dimensional analysis in a system of two fundamental dimensions (L, t) indicates that we should obtain two dimensionless groups. One of our variables, $c_A^\bullet$, is already dimensionless and therefore serves as the first member of the group of two. A systematic analysis such as we developed in Chapter 5 will give the other dimensionless group as the similarity variable

$$\eta = \frac{x}{\sqrt{4 \, \mathcal{D}_{AB} \, t}} \tag{8.2.5-37}$$

which implies that

$$f_3(c_A^\bullet, \eta) = 0 \tag{8.2.5-38}$$

or that

$$c_A^\bullet = \varphi(\eta) \tag{8.2.5-39}$$

Transforming the terms in the differential equation we have

$$\frac{\partial c_A^\bullet}{\partial t} = \frac{dc_A^\bullet}{d\eta} \frac{\partial \eta}{\partial t} = \frac{d[\varphi(\eta)]}{d\eta} \frac{\partial\left[\frac{x}{\sqrt{4 \, \mathcal{D}_{AB} \, t}}\right]}{\partial t}$$

$$= \varphi'\left(\frac{x}{\sqrt{4 \, \mathcal{D}_{AB}}}\right)\left(-\frac{1}{2} t^{-\frac{3}{2}}\right) = \varphi'\left(\frac{x}{\sqrt{4 \, \mathcal{D}_{AB} \, t}}\right)\left(-\frac{1}{2t}\right)$$

$$= -\frac{1}{2}\frac{\eta}{t}\varphi' \tag{8.2.5-40}$$

$$\frac{\partial^2 c_A^*}{\partial x^2} = \frac{\partial}{\partial x}\frac{\partial c_A^*}{\partial x} = \frac{\partial}{\partial x}\frac{\partial[\phi(\eta)]}{\partial x} = \frac{\partial \eta}{\partial x}\frac{d}{d\eta}\left\{\frac{d[\phi(\eta)]}{d\eta}\frac{\partial \eta}{\partial x}\right\}$$

$$= \frac{\partial\left[\frac{x}{\sqrt{4\,\mathcal{D}_{AB}\,t}}\right]}{\partial x}\frac{d}{d\eta}\left\{\frac{d[\phi(\eta)]}{d\eta}\frac{\partial\left[\frac{x}{\sqrt{4\,\mathcal{D}_{AB}\,t}}\right]}{\partial x}\right\}$$

$$= \frac{1}{\sqrt{4\,\mathcal{D}_{AB}\,t}}\frac{d}{d\eta}\left\{\frac{d[\phi(\eta)]}{d\eta}\frac{1}{\sqrt{4\,\mathcal{D}_{AB}\,t}}\right\}$$

$$= \frac{1}{\sqrt{4\,\mathcal{D}_{AB}\,t}}\frac{1}{\sqrt{4\,\mathcal{D}_{AB}\,t}}\frac{d}{d\eta}\left\{\frac{d[\phi(\eta)]}{d\eta}\right\} = \frac{1}{4\,\mathcal{D}_{AB}\,t}\phi'' \tag{8.2.5-41}$$

Substituting in the differential equation

$$-\frac{1}{2}\frac{\eta}{t}\varphi' = \mathcal{D}_{AB}\left(\frac{1}{4\,\mathcal{D}_{AB}\,t}\right)\varphi'' \tag{8.2.5-42}$$

Rearranging

$$\varphi'' + 2\eta\,\varphi' = 0 \tag{8.2.5-43}$$

which leaves us with an ordinary differential equation because of the combining of two independent variables into one via the similarity transformation.

The initial and boundary conditions transform as

$$
\begin{aligned}
t = 0 &\Rightarrow \eta = \infty: \quad c_A^* = 0 \Rightarrow \varphi(\eta) = 0 \\
x = 0 &\Rightarrow \eta = 0: \quad c_A^* = 1 \Rightarrow \varphi(\eta) = 1 \\
x = \infty &\Rightarrow \eta = \infty: \quad c_A^* = 0 \Rightarrow \varphi(\eta) = 0
\end{aligned} \tag{8.2.5-44}
$$

which gives the transformed boundary conditions as

$$
\begin{aligned}
\eta = \infty: \quad \varphi(\eta) = 0 \\
\eta = 0: \quad \varphi(\eta) = 1
\end{aligned} \tag{8.2.5-45}
$$

This equation is easily solved via the simple expedient of substituting ψ for ϕ' because it is then separable

$$\psi' + 2\eta\psi = 0 \qquad\qquad\qquad (8.2.5\text{-}46)$$

$$\frac{d\psi}{d\eta} + 2\eta\psi = 0 \qquad\qquad\qquad (8.2.5\text{-}47)$$

$$\int \frac{d\psi}{\psi} = -\int 2\eta\,d\eta \qquad\qquad\qquad (8.2.5\text{-}48)$$

$$\ln\psi = -\eta^2 + C_1 \qquad\qquad\qquad (8.2.5\text{-}49)$$

$$\psi = C_1\,e^{-\eta^2} \qquad\qquad\qquad (8.2.5\text{-}50)$$

Back substituting for ψ

$$\phi' = C_1\,e^{-\eta^2} \qquad\qquad\qquad (8.2.5\text{-}51)$$

$$\int \phi'\,d\eta = C_1 \int e^{-\eta^2}d\eta \qquad\qquad\qquad (8.2.5\text{-}52)$$

$$\phi = C_1 \int_0^\eta e^{-\zeta^2}d\zeta + C_2 \qquad\qquad\qquad (8.2.5\text{-}53)$$

As we saw in the corresponding momentum transfer example in Chapter 6, the integral here is one that we cannot evaluate in closed form.

Introducing the boundary conditions and substituting the appropriate value for the definite integrals[10]

$$1 = C_1 \int_0^0 e^{-\zeta^2}d\zeta + C_2 \quad \Rightarrow \quad C_2 = 1 - C_1(0) = 1 \qquad (8.2.5\text{-}54)$$

$$0 = C_1 \int_0^\infty e^{-\zeta^2}d\zeta + 1 \quad \Rightarrow \quad C_1 = -\frac{1}{\displaystyle\int_0^\infty e^{-\zeta^2}d\zeta} = -\frac{2}{\sqrt{\pi}} \qquad (8.2.5\text{-}55)$$

Substituting the constants in the equation and re-ordering the terms gives

---

[10] The error function plays a key role in statistics. We do not have space here to show the details of the integration of the function to obtain the tables listed; for further details the reader is referred to any of the many texts in mathematical statistics.

$$\varphi = 1 - \frac{2}{\sqrt{\pi}} \int_0^\eta e^{-\zeta^2} d\zeta \tag{8.2.5-56}$$

We defined the error function and the complementary error function[11] in Chapter 6

$$\text{error function} = \text{erf}(\eta) \equiv \frac{2}{\sqrt{\pi}} \int_0^\eta e^{-\zeta^2} d\zeta$$

$$\text{complementary error function} = \text{erfc}(\eta) \equiv 1 - \text{erf}(\eta) \tag{8.2.5-57}$$

allowing us to write

$$\varphi = 1 - \text{erf}\left[\eta\right]$$
$$= \text{erfc}\left[\eta\right] \tag{8.2.5-58}$$

The solution of our problem is then

$$c_A^* = 1 - \text{erf}\left[\frac{x}{\sqrt{4\,\mathcal{D}_{AB}\,t}}\right]$$
$$= \text{erfc}\left[\frac{x}{\sqrt{4\,\mathcal{D}_{AB}\,t}}\right] \tag{8.2.5-59}$$

A table of values of the error function is listed in Appendix B.

8.2.6 Diffusion in porous solids

Many chemical systems utilize porous solids (for example, catalyst beds), and so we will consider briefly the effect of porous media on the diffusion coefficient. Diffusion in pores occurs by one or more of three mechanisms:

- ordinary diffusion
- Knudsen diffusion
- surface diffusion.

---

[11] Engineers have a certain sometimes regrettable proclivity to use the same symbol both for the dummy variable of integration that appears in the integrand and for the variable in the limits. This usually does not lead to confusion; here, however, it is clearer not to do so.

Although we can calculate diffusion in a single pore using the techniques already presented, we normally do not know the details of pore geometry. Therefore, we use the normal diffusion model but use an **effective** diffusion coefficient to account for the effect of the porous medium.

Ordinary diffusion takes place when the pores are large compared to the mean free path of the diffusing molecules. However, the effective diffusion coefficient is different from the ordinary diffusion coefficient because in porous media the whole cross-section is not available for flow, and, furthermore, the flow paths are tortuous. The ordinary coefficient must therefore be corrected both for the areal porosity (free cross-sectional area) and the tortuosity. The tortuosity is a factor that describes the relationship between the actual path length and the nominal length of the porous medium, taking into account the varying cross-section of the pores.

The effective diffusion coefficient in a porous medium is given by

$$\mathcal{D}_{eff} = \frac{\mathcal{D}\theta}{\tau} \tag{8.2.6-1}$$

where $\mathcal{D}$ is the ordinary diffusion coefficient, θ is the areal porosity, and τ is the tortuosity.

Knudsen diffusion takes place when the size of the pores is of the order of the mean free path of the diffusing molecule. In this situation a molecule collides with the wall as frequently or more frequently as with other molecules. From kinetic theory, for a straight cylindrical pore[12]

$$
\begin{aligned}
N &= \frac{\mathcal{D}_K}{x_0}\left(c_1 - c_2\right) \\
&= \frac{\mathcal{D}_K}{RT}\frac{\left(p_1 - p_2\right)}{x_0} \\
&= \frac{2\, r_e\, v}{3\, RT}\frac{\left(p_1 - p_2\right)}{x_0} \\
&= \frac{2\, r_e}{3\, RT}\left(\frac{8\, RT}{\rho M}\right)^{1/2}\frac{\left(p_1 - p_2\right)}{x_0}
\end{aligned}
\tag{8.2.6-2}
$$

where

---

[12] Satterfield, C. N. and T. K. Sherwood (1963). *The Role of Diffusion in Catalysis.* Reading, MA, Addison Wesley Publishing Co., Inc.

$$\mathcal{D}_K = 9700 \, r_e \sqrt{\frac{T}{M}} \qquad (8.2.6\text{-}3)$$

and r_e is the radius of the pore. Usually we correct Equation (8.2.6-3) empirically since pores, in general, are not cylindrical. Defining a mean pore radius

$$r_e = \frac{2V}{S} = \frac{2\theta}{S\rho} \qquad (8.2.6\text{-}4)$$

Thus

$$\begin{aligned}
\mathcal{D}_{K,eff} &= \frac{\mathcal{D}_K \theta}{\tau} = \frac{8\,\theta^2}{3\,\tau\,S\,\rho} \\
&= \sqrt{\frac{2RT}{\rho M}} \\
&= 19{,}400 \frac{\theta^2}{\tau\,S\,\rho} \sqrt{\frac{T}{M}}
\end{aligned} \qquad (8.2.6\text{-}5)$$

In the transition region between ordinary and Knudsen diffusion an effective diffusion coefficient is defined by

$$N_1 = -\tilde{\mathcal{D}}_{eff} \frac{dc}{dx} \qquad (8.2.6\text{-}6)$$

The expression for $\tilde{\mathcal{D}}_{eff}$ is found by integration of Fick's equation for specific boundary conditions.[13, 14]

Surface diffusion takes place when molecules adsorbed on solid surfaces are transported over the surface as a result of a two-dimensional concentration gradient on the surface.[15] Surface diffusion normally contributes little to overall transport of mass unless there is a large amount of adsorption.

---

[13] Scott, D. S. and F. A. L. Dullien (1962). *AIChEJ* 8:113.

[14] Rothfield, L. B. and C. C. Watson (1963). *AIChEJ* 9:19.

[15] Levich, V. G. (1962). *Physicochemical Hydrodynamics*. Englewood Cliffs, NJ, Prentice-Hall.

8.2.7   Dispersion

Dispersion is macroscopic mixing in a single phase caused by both diffusion and uneven flow. It is the result from the bulk motion of two or more components. Although dispersion is not primarily due to transport by molecular motion, this type of mixing of two or more components does result in mass transfer. Even though the primary mechanism is often not concentration gradient driven, we normally utilize a diffusion model to describe this macroscopic mixing by replacing the diffusion coefficient with an effective coefficient called a dispersion coefficient. (Normally, diffusion effects are included as a small contribution within the dispersion coefficient.[16])

We can represent macroscopic mixing of two identifiable species in fluid flow using a dispersion coefficient. For example, when fluid A displaces fluid B in a laminar flow in a cylindrical conduit, the fluid at the center of the conduit moves faster than the fluid near the wall, and, therefore, there is macroscopic mixing of the two species on the average, and this mixing can be described by a dispersion coefficient.

Likewise, when one fluid displaces another during flow through a porous medium the tortuous nature of the medium will cause the two species to mix. Again, the macroscopic mixing of the two species can be represented by a dispersion coefficient. As the flow velocity decreases there is less and less mixing by dispersion and relatively more mixing by molecular diffusion. For a porous medium one can plot the ratio of the dispersion coefficient to the diffusion coefficient versus a Peclet number as in Figure 8.2.7-1.

The Peclet[17] number, vL/D, represents the ratio of mass transport by bulk flow to that by dispersion. The Peclet number decreases as velocity decreases, and the ratio of dispersion coefficient to diffusion coefficient becomes constant since the dispersion coefficient becomes the diffusion coefficient as velocity goes to zero. There are several horizontal lines in Figure 8.2.7-1 which result from varying tortuosity of the medium. This means that at these low Peclet numbers mass transfer is dominated by diffusion.

---

[16] Whitaker (1967). Diffusion and Dispersion in Porous Media. *AIChEJ* 13(3): 420.

[17] When this dimensionless group incorporates the dispersion coefficient rather than the diffusivity, it is also called the Bodenstein number.

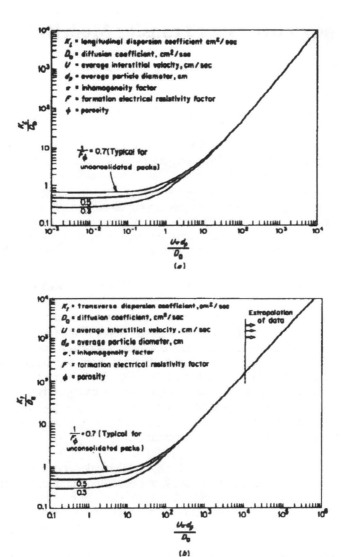

Figure 8.2.7-1 Dispersion and diffusion as a function of Peclet number[18]

---

[18] Pfannkuch, H. O. (1962). Contribution a l'etude des deplacement de fluides miscible dems on milleu-poreux. *Rev. Inst. Fr Petrol* **18**(2): 215.

8.3   Convective Mass Transfer Models

Until this point we have been discussing mass transfer

- *within* a single phase
- by the mechanism of *molecular diffusion.*

We now wish to extend our discussion to include *convective* mass transfer and, in addition, to consider transfer *between* phases.

Most problems in convection are too complicated to permit an analytical solution. There are many and various (and approximate) models used to attempt to predict mass transfer - film theory, boundary layer theory, and penetration-renewal theory - but for the moment we will confine ourselves only to investigating how we might make use of the predictions made via these theories.

8.3.1   The concentration boundary layer[19]

Consider the case of *mass* transfer to a flat plate in the same way that we considered momentum transfer and heat transfer. Suppose the flat plate is made of naphthalene (moth crystals, basically), which evaporates into a stream of nitrogen. If we plot concentration versus distance from the surface, we get something that looks like Figure 8.3.1-1 (a). This looks very little like our profiles in the cases of momentum and energy. We can change this, however, by plotting the *unaccomplished* change in concentration ($c_A - c_{AS}$), as in Figure 8.3.1-1 (b), and we see that we build up a concentration profile on the plate similar to those for temperature and momentum.

This procedure leads us to believe that by a suitable dedimensionalization of variables we should be able to write equations describing momentum transfer, heat transfer, and mass transfer in such a form that the equations would be identical for cases where the transport takes place by the same mechanism. As a particular example of this procedure, we can consider flow over a flat plate.

Flat plates do not exhibit form drag, so we do not have to worry about this particular form of momentum transport, which contains no analog in heat transfer or mass transfer. If we restrict ourselves to mass transfer rates sufficiently low that the mass transfer equations look the same written either in stationary coordinates or in coordinates moving at the molar average velocity, we

---

[19] For more discussion of this model see, for example, Kays, W. M. (1966). *Convective Heat and Mass Transfer.* New York, NY, McGraw-Hill.

will not have to worry about the bulk flow of the dissolving component affecting mainstream flow and complicating mass transfer results.

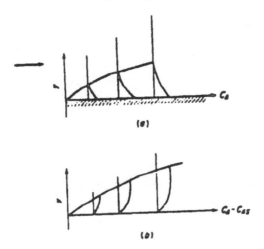

Figure 8.3.1-1 Concentration boundary layer

For such a situation, we can write a model which is an idealized set of differential equations involving several assumptions. In particular, for the case presented here, we will

> • consider only two-dimensional flow, that is, assume the flat plate is of infinite width normal to the flowing stream,
> • neglect transport by conductive or molecular diffusive mechanisms in all directions except at right angles to the plate surface, and
> • assume that the flow is at steady state.

Under these restrictions, we can write a single dimensionless differential equation which describes the velocity field, the mass transfer, and the heat transfer. The equation has the following dimensionless form

$$\frac{d^2\Omega}{d\eta^2} + K\frac{f(\eta)}{2}\frac{d\Omega}{d\eta} = 0 \qquad (8.3.1-1)$$

with boundary conditions

$$\eta = 0: \quad \Omega = 0$$
$$\eta = 1: \quad \Omega = 1 \tag{8.3.1-2}$$

In this equation Ω is a dimensionless velocity, temperature, or concentration depending on whether a momentum, energy, or mass balance was the source. The coefficient K is dimensionless and η is a dimensionless distance which incorporates both y, the distance perpendicular to the surface, and x, the distance along the surface from the nose of the plate (η is a *similarity* variable).

$$\eta = y\sqrt{\frac{v_0}{vx}} = \frac{y}{x}\sqrt{Re} \tag{8.3.1-3}$$

where Re is a Reynolds number based on distance from the nose, x, and freestream velocity, v_0. The function $f(\eta)$ is known but is expressed as an infinite series.

For mass transfer one thereby obtains

$$\frac{d^2\left[\frac{\rho_A - \rho_{AS}}{\rho_{A0} - \rho_{AS}}\right]}{d\eta^2} + Sc\frac{f(\eta)}{2}\frac{d\left[\frac{\rho_A - \rho_{AS}}{\rho_{A0} - \rho_{AS}}\right]}{d\eta} = 0 \tag{8.3.1-4}$$

Notice that Equation (8.3.1-3) is identical to the corresponding momentum and heat transfer equations except for the coefficient of the second term. If we require the Prandtl number to be equal to one in Equation (8.3.1-4) and the Schmidt number to be equal to one in Equation (8.3.1-3) the equations are identical, and since the dedimensionalized boundary conditions are the same, the identical solution is found for all three cases: momentum transfer, heat transfer, and mass transfer.

The solution to these equations can be obtained without great difficulty, and the solution is shown plotted in Figure 8.3.1-2. Notice that the ordinate is in each case the unaccomplished change in the dependent variable; for example,

> • v/v_0 is the unaccomplished velocity change (since v equals 0 at the surface of the plate),
> • $(T - T_s)/(T_0 - T_s)$ is the unaccomplished temperature change, and
> • $(\rho_A - \rho_{As})/(\rho_{A0} - \rho_{As})$ is the unaccomplished concentration change.

The abscissa in each case is simply the ratio of the distance from the surface of the plate to the boundary layer thickness at that point. The reason for the peculiar grouping of terms in the abscissa is that from exact solutions to the equations of motion we know that the boundary layer thickness is proportional to $(xt)^{1/2}$. (This comes from the solution to the problem of the suddenly accelerated flat plate.)

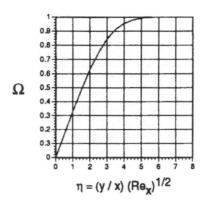

$$\eta = (y / x) (Re_x)^{1/2}$$

Figure 8.3.1-2 Boundary layer[20]

The appropriate time variable for our case is the distance from the front of the plate divided by the free-stream velocity. (This is approximately how long it takes a molecule to reach the point in question from the nose of the plate.) This gives

$$\frac{y}{\delta} = \frac{y}{\sqrt{\nu t}} = \frac{y}{\sqrt{\frac{\nu x}{v_o}}} \qquad (8.3.1\text{-}5)$$

The meaning of the abscissa can readily be seen by considering what the graph signifies at a constant value of x. At constant x, for a given fluid flowing at a given velocity,

$$\sqrt{\frac{v_o}{v_x}} = \text{constant} \qquad (8.3.1\text{-}6)$$

[20] Adapted from Schlichting, H. (1960). *Boundary Layer Theory*, New York, McGraw-Hill, p. 119.

   Therefore, the only thing that changes in the abscissa is y, so the graph gives us a profile of unaccomplished velocity, temperature, or concentration change as a function of y. This profile is not plotted in the way that you are accustomed to seeing it, because the ordinate and abscissa have been reversed. However, if you take this book page and hold it up to the light so that you can view the graph from the back of the page, by rotating the page 90° clockwise you will see the profile in the form in which you are accustomed to seeing it plotted.

   This is a very specific example, which is included only for illustration. The purpose is to point out that in cases where momentum, heat, and mass are transferred by the same mechanism, the same differential equation will describe all three processes, with the possible exception that one coefficient may be slightly different in each of the three cases. This coefficient, however, will involve the Prandtl number in the heat transfer case and the Schmidt number in the mass transfer case, and by choosing appropriate values for the Prandtl number and the Schmidt number (which are properties of the fluid only and not of the flow field), one can write the equation in exactly the same form.

   The physical meaning is that the boundary layers as described in dimensionless variables will coincide for such a case. One can also tabulate solutions for cases in which the Prandtl number and/or Schmidt number are not equal to unity, and such a solution is shown for the specific case of flow past the flat plate in Figure 8.3.1-3 (see Figure 7.3.1-1).

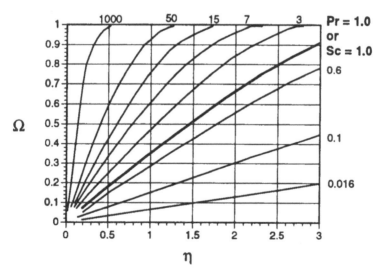

Figure 8.3.1-3 Boundary layer solution for a flat plate

What is the significance of this fact? The utility of these observations to the engineer is that it makes it possible for him or her to take data in a system other than the one for which he or she wishes to predict; for example, in cases where heat transfer and mass transfer occur by the same mechanism, *one can do heat transfer measurements to predict mass transfer, or vice versa.* In addition, for cases where the Schmidt number or Prandtl number is equal to one, *heat transfer or mass transfer measurements may be used to predict momentum transfer or vice versa* (remember that form drag cannot be present for this to be a valid procedure).

Such an approach is frequently useful. For example, it is often easier to make heat transfer measurements than to make mass transfer measurements because temperatures are, in general, easier to measure than concentrations. Again, it is emphasized that this is a general procedure despite the fact that we have chosen a particular example to illustrate this general procedure, namely, the flow over a flat plate.

For most cases of interest, the differential equation describing mass transfer cannot be solved (typically, for cases involving turbulent flow). However, we still can obtain the coefficients in the differential equation by dimensional analysis. Therefore, we can use the common model among momentum, heat, and mass transfer despite the fact that we cannot solve the problem analytically.

By making experimental measurements on our system we are, in fact, solving the differential equation. We are doing this by a procedure utilizing the best possible model (analog) of our system - that is, the system itself (the homologue). It must be emphasized that the important step in the procedure is the realization that momentum, heat, and mass are being transferred by the same mechanism and therefore have a common model (differential equation). Once the model has been determined, the method of solution of the model (by either analytical, numerical, or experimental means) depends on which method best satisfies the requirements of speed, cost, and accuracy for the particular problem.

8.3.2 Film theory and penetration-renewal theory[21]

Two other approaches to predicting mechanism of mass transfer across an interface are film theory and penetration-renewal theory. Film theory postulates that immediately adjacent to the interface there is a stagnant film which contains all the resistance to mass transfer. We know that most systems do not, in fact,

---

[21] See, for example, Davies, J. T. (1963). *Mass Transfer and Interfacial Phenomena.* New York, NY, Academic Press, Inc.

contain such a film, although the boundary layer can at times act in a very similar fashion. The difficulty with film theory models is that one is always faced with somehow predicting the fictitious film thickness.

Penetration-renewal theory models developed somewhat later in the theory of mass transfer. Penetration theory models are based on the assumption that there is an identifiable finite mass of fluid which is transported intact from the free stream to the interface without transferring mass en route, that this finite mass of fluid sits at the interface for a relatively short time with respect to the time required to saturate the mass of fluid with the transferring component, and that the mass of fluid, still intact, then transfers back to the free stream without transferring mass en route and then instantly mixes with the free-stream fluid.

This seems a somewhat idealized model, but under some circumstances is the sort of behavior one would expect of eddies that migrate to the interface. The difficulty in penetration-renewal theory models is calculating a rest time for the mass of fluid as it sits at the interface. Penetration theory models have also been combined with film theory models to form the film-penetration theory.

All the models that have been discussed, that is, boundary layer models, film models, and penetration-renewal models, as for any model, are useful only insofar as they give results which correspond to real mass transfer situations. The accuracy of most mass transfer data at present is perhaps plus or minus 30 percent, and so it is frequently virtually impossible to discriminate as to which of various models gives the best predictions. Most design of mass-transfer equipment is still done on the basis of highly empirical methods based on dimensional analysis. This is not to denigrate the importance of the theoretical approach, since it is only through some theoretical approach that we will achieve ultimate understanding; at the moment, however, most practical results are calculated from what is essentially empiricism.

8.4 The Mass Transfer Coefficient for a Single Phase

Most prediction of mass transfer is done in terms of a mass transfer coefficient, just as most prediction of heat transfer is done in terms of a heat transfer coefficient. This is true because most mass transfer situations of interest involve convective mass transfer for which analytic solutions are either impractical or not possible.

In this section we discuss single-phase mass transfer coefficients, their correlation, and their use. We later introduce the concept of the relationship between single-phase mass transfer coefficients and overall mass transfer coefficients, as we did for heat transfer.

Since we know that in many systems the amount of mass transferred is approximately proportional to the concentration difference, we can write

$$\text{Mass or molar flux} = \text{proportionality factor} \times \text{concentration driving force} \tag{8.4-1}$$

We call the proportionality factor in such a relation the *mass transfer coefficient*. This is exactly analogous to the way we defined the heat transfer coefficient, and again the usefulness of the coefficient so defined depends on its remaining fairly constant among similar mass transfer problems, and upon our being able to predict (or at least correlate) its performance based on physical parameters.

In contrast to the heat transfer coefficient, where the driving force was always expressed in terms of temperature, in the case of mass transfer there are a number of ways to express the driving force.

We must, therefore, define in a clear and unambiguous fashion precisely what we mean by the *driving force* for each particular mass transfer coefficient, because the value of the coefficient depends on the driving force with which it is to be used.

First let us consider the case of mass transfer *within* a single phase. For example, we can write[22]

$$N_A = k_z \left(z_{As} - z_A \right)$$
$$n_A = k_\omega \left(\omega_{As} - \omega_A \right) \tag{8.4-2}$$

where in the first equation z_{As} is the **mole** fraction of species A at the phase interface, and N_A is the **molar** flux with respect to fixed coordinates; in the second equation ω_{As} is the **mass** fraction of species A at the phase interface, and n_A is the **mass** flux with respect to fixed coordinates. Since we will often be concerned with molar fluxes in problems which simultaneously involve both gas

---

[22] In principle one **could** define a mass transfer coefficient which yielded a mass flux from a molar driving force, or vice versa. There is little motivation for such a course of action, and much to be said against it.

phases and liquid phases, for molar fluxes we replace the letter z by the letter x to denote liquid phase, and by the letter y to denote gas phase.

Mass transfer coefficients are also defined for driving forces of species mass density

$$n_A = k_\rho \left(\rho_{As} - \rho_A \right) \tag{8.4-3}$$

and, for gas phase coefficients, of partial pressure

$$N_A = k_G \left(p_{As} - p_A \right) \tag{8.4-4}$$

Note that the units of the various mass transfer coefficients differ depending on the driving force (see Example 8.4-2 and Example 8.4-3).

Note that this coefficient includes *all* the mass transferred (both by diffusion and convection). The order of terms in the driving force is usually taken so as to give a positive flux. This is *not* the only way to define the mass transfer coefficient. It is, however, a common way.

In general, the mass transfer coefficient depends on the mass transfer rate. Other mass transfer coefficients may also be defined with respect to axes *not* fixed in space - for example, moving at the molar average velocity. Since it is our purpose here to introduce the utility of mass transfer coefficients rather than to consider them in detail, we will limit our discussion to low mass transfer rates. In such problems the bulk flow term in Equation (8.1-9) approaches zero and, therefore, the mass transfer coefficient is the same for coordinate systems both stationary and moving at the molar average velocity. The reader is cautioned, however, that this assumption is not valid at high mass transfer rates.

For a pure solid such as iodine evaporating into a stream of nitrogen, where Equation (8.4-2) is applied to the gas phase, y_{AS} is the mole fraction of iodine in the gas phase at the solid surface (in the following discussion corresponding comments would hold for mass transfer coefficients based on mass fractions). This is not a quantity which is easily measured, but experiments have shown that in most cases one can assume that there is *no resistance to mass transfer at the interface*. This is equivalent to the assumption that *the phases immediately on either side of the interface are at equilibrium*. Therefore, we would evaluate Equation (8.4-2) using as y_{AS} the mole fraction of I_2 in a gas mixture at equilibrium with solid I_2 (saturated with N_2) at the pressure and temperature of the interface. Such an assumption is not, of course, valid in the case of a highly contaminated interface.

We must also pick a value to use for z_A. Two cases may be distinguished. First, for *external flows* such as nitrogen flowing past an evaporating sheet of ice, we use for z_A (here y_A) the *free stream concentration*. In this case, if the nitrogen initially contains no water vapor, $y_A = 0$. (This is the concentration outside the concentration boundary layer.) Second, for the case of *internal flows*, such as flow of liquid in a pipe where the wall is coated with a dissolving substance, z_A is the *bulk* or *mixing-cup* concentration. The bulk or mixing-cup concentration is entirely analogous to the bulk or mixing-cup temperature It is the concentration obtained if the stream is caught in a cup for a short while, then mixed perfectly, and it is defined as

$$\langle z_A \rangle = \frac{\int_A z_A \, \rho \left(v \cdot n \right) dA}{\int_A \rho \left(v \cdot n \right) dA} \qquad \text{(z in mass fraction)} \qquad (8.4\text{-}5)$$

$$\langle z_A \rangle = \frac{\int_A z_A \, c \left(v \cdot n \right) dA}{\int_A c \left(v \cdot n \right) dA} \qquad \text{(z in mole fraction)} \qquad (8.4\text{-}6)$$

We showed above that for certain simplified cases of molecular diffusion we could integrate an elementary differential equation and write the resulting expression for mass or molar flux in the form: conductance times driving force. That is, Equations (8.2.5-6) and (8.2.5-14) can be written as

$$N_{Ay} = \left[\frac{-c \, \mathcal{D}_{AB}}{L} \right] \left(x_{A1} - x_{A0} \right) \qquad (8.4\text{-}7)$$

and

$$N_{Ay} = \left[\frac{-c \, \mathcal{D}_{AB}}{L \left(x_B \right)_{lm}} \right] \left(x_{A1} - x_{A0} \right) \qquad (8.4\text{-}8)$$

The first equation applies to equimolar counterdiffusion and the second to diffusion of A through stagnant B.

The bracketed terms in Equations (8.4-4) and (8.4-5) are *exact* expressions for the mass transfer coefficients for these particular situations. In general, however, obtaining such an analytically derived expression is difficult and, in most cases, impossible. The same problems exist in attempting to treat convective heat transfer and momentum transfer; these problems are even more difficult in the case of mass transfer.

Example 8.4-1   Calculation of flux from a mass transfer coefficient

A thin film of water is flowing down a block of ice with air blowing countercurrent to the water. If the freestream air is dry and $k_y = 2$ mol/(hr ft^2), calculate N_A.

Solution

$$N_A = k_y \left(y_{As} - y_A \right)$$
(8.4-9)

The air is dry, so

$$y_A = 0$$
(8.4-10)

Assuming equilibrium at the interface, y_{As} is the water vapor concentration in equilibrium with liquid water saturated with air at 32°F. The system under these conditions will behave in a sufficiently ideal manner that we choose to use Raoult's law for our equilibrium model.

$$y_{As} p = x_{As} p'$$
(8.4-11)

where p is the system pressure (here 1 atm) and p' is the vapor pressure of the pure component (here water).

Since the solubility of air in water is slight under these conditions, we take x_{As} to be 1.0. The vapor pressure of water at 32°F is 0.0885 psi.

$$y_{As} = (1.0) \left[\frac{\text{mol HOH}}{\text{total mol}} \right] (0.0885) \left[\frac{\text{lbf}}{\text{in}^2} \right] \left(\frac{1}{14.7} \right) \left[\frac{\text{in}^2}{\text{lbf}} \right] = 0.00602 \frac{\text{mol HOH}}{\text{total mol}}$$
(8.4-12)

Substituting

$$N_{HOH} = (2.0)\left[\frac{\text{total mol}}{\text{hr ft}^2}\right](0.00602 - 0)\left[\frac{\text{mol HOH}}{\text{total mol}}\right] = 0.0120 \frac{\text{mol HOH}}{\text{hr ft}^2}$$

$$(8.4-13)$$

Note also that the resistance to mass transfer of water in the liquid phase is very nearly zero since the liquid phase everywhere is virtually pure water. We can see this by writing the flux equation for the **liquid** phase

$$N_A = k_x\left(x_A - x_{As}\right) \tag{8.4-14}$$

If we assume that the water evaporated is supplied by melting of the ice (maintaining a constant film thickness), N_A is finite. Also, x_A is almost identical with x_{As}, i.e., the driving force is almost zero. This implies that k_x is very large. Since k_x is a **conductance**, the **resistance** is therefore very small.

Example 8.4-2   Mass transfer using partial pressure as a driving force

Let k_G denote a mass transfer coefficient to be used to calculate gas phase molar fluxes but based on a driving force of **partial pressure**. Show that

$$k_y = k_G p \tag{8.4-15}$$

Solution

By definition

$$N_A = k_y\left(y_A - y_{As}\right) = k_G\left(p_A - p_{As}\right) \tag{8.4-16}$$

Using the definition of partial pressure

$$N_A = k_y\left(y_A - y_{As}\right) = k_G\left(y_A p - y_{As} p\right) \tag{8.4-17}$$

$$k_y = k_G p \tag{8.4-18}$$

Example 8.4-3   Mass transfer using species mass density as driving force

Let k_ρ denote a mass transfer coefficient based on a driving force of mass density to be used to calculate mass fluxes.[23]

Develop the functional relationship between this coefficient and k_z.

Solution

By definition

$$n_A = k_\rho \left(\rho_{As} - \rho_A \right) = N_A M_A = k_z M_A \left(z_{As} - z_A \right) \tag{8.4-19}$$

But

$$\left(\rho_A \right) \left[\frac{\text{mass A}}{\text{volume}} \right] = \left(z_A \right) \left[\frac{\text{mol A}}{\text{total mol}} \right] \left(M_A \right) \left[\frac{\text{mass A}}{\text{mol A}} \right] (c) \left[\frac{\text{total mol}}{\text{volume}} \right] \tag{8.4-20}$$

Substituting

$$k_\rho \left(z_{As} M_A c - z_A M_A c \right) = k_z M_A \left(z_{As} - z_A \right) \tag{8.4-21}$$

$$k_\rho = \frac{k_z}{c} \tag{8.4-22}$$

The units of k_z and k_ρ are not the same, despite their both being mass transfer coefficients.

$$k_z = \frac{N_A}{\left(z_A - z_{As} \right)} \quad \Rightarrow \quad \frac{\frac{\text{moles}}{\text{time area}}}{\text{mol fraction}} = \left[\frac{\text{mol}}{L^2 t} \right] \tag{8.4-23}$$

---

[23] We could similarly define a coefficient based on molar density to calculate molar fluxes by defining

$\tilde{\rho}$ = molar density, i.e., molar concentration, c

$$N_A = k_{\tilde{\rho}} \left(\tilde{\rho}_{As} - \tilde{\rho}_A \right) = k_{\tilde{\rho}} \left(c_{As} - c_A \right)$$

$$k_\rho = \frac{n_A}{(\rho_{A_s} - \rho_A)} \implies \frac{\frac{mass}{time\ area}}{\frac{mass}{volume}} = \left[\frac{L}{t}\right] \tag{8.4-24}$$

8.4.1 Design equations for single-phase mass transfer coefficients

Using the restrictions above, we have for the mass transfer version of the Reynolds analogy (see Chapter 7)

$$\frac{Sh}{Sc\ Re} \equiv St_M = \frac{f}{2} \tag{8.4.1-1}$$

One can regard the ratio of dimensionless numbers on the left-hand side as a Stanton number for mass transfer. As in the case of heat transfer, this model has severe restrictions.[24]

Flat plates

For *laminar* flow past a flat plate, an approximate expression for fluids with Sc > 0.6 is

$$Sh_x = 0.33\ Re_x^{1/2}\ Sc^{1/3} \tag{8.4.1-2}$$

where the characteristic length in Sh and Re is the distance from the nose of the plate. The Sherwood number is *local* in that it contains the local mass transfer coefficient and the distance from the nose of the plate.

Integrating over the surface of the plate yields an expression for the *mean* Sherwood number as

$$Sh_L = 0.66\ Re_L^{1/2}\ Sc^{1/3} \tag{8.4.1-3}$$

---

[24] One can in a similar manner to that suggested in Chapter 7 also obtain (via the Chilton-Colburn analogy) a broadly applicable j-factor for mass transfer as

$$\frac{f}{2} = \frac{Sh}{Re\ Sc}\ Sc^{2/3} = St_M\ Sc^{2/3} \equiv j_M$$

which is applicable over roughly the range 0.6 < Sc < 3000.

where the characteristic length is now the total length of the plate, and the mass transfer coefficient is a *mean* coefficient.

One local expression for *turbulent* flow past a flat plate is

$$Sh_x = 0.0202\, Re_x^{4/5} \qquad\qquad\qquad\qquad (8.4.1\text{-}4)$$

where the characteristic length is the distance from the nose of the plate, and the expression is valid only for fluids with Sc = 1.0.

Example 8.4.1-1   Average mass transfer coefficient from local coefficient

Given the model for local mass transfer coefficient for laminar flow over a flat plate of fluids with Sc > 0.6

$$Sh_x = 0.33\, Re_x^{0.5}\, Sc^{1/3} \qquad\qquad\qquad (8.4.1\text{-}5)$$

$$\left(\frac{k_y\, x}{c\, \mathcal{D}_{AB}}\right) = 0.33 \left(\frac{x\, v_0\, \rho}{\mu}\right)^{0.5} \left(\frac{\mu}{\rho\, \mathcal{D}_{AB}}\right)^{1/3} \qquad (8.4.1\text{-}6)$$

where x is the distance from the nose of the plate, develop an expression for the average coefficient. Use y as the normal distance from the surface of the plate.

Solution

By an **average coefficient of mass transfer**, $k_{y,avg}$ we mean a number that, when multiplied by the area of the plate and an appropriate driving force, gives the same mass transfer rate as the integral over the surface area of the plate of an integrand which is the product of the local coefficient, the local driving force, and a differential area

$$k_{y,avg}\left(y_{As} - y_A\right)_{avg} A_{plate} = \int_{A_{plate}} k_y\left(y_{As} - y_A\right) dA \qquad (8.4.1\text{-}7)$$

Notice that the definition of the average coefficient depends on what definition we choose for the average driving force - we must choose one of the pair for the other to be defined unambiguously.

If we have a **constant** driving force, as along a homogeneous plate with constant freestream conditions in the x-direction, this constant driving force is the obvious one to define as the average driving force, although this is not necessary. In the case where the driving force changes, as within a pipe with a dissolving wall where the bulk concentration changes in the axial direction, the choice is not so obvious. In practice, almost all average coefficients are based on either the arithmetic mean or the logarithmic mean driving force. Needless to say, to use an average coefficient correctly, one must know on what average driving force it is based.

For this model we assume that the conditions at the surface of the plate and in the freestream are constant; therefore, we have a constant driving force which we will use as the average. Substituting for the local coefficient and dividing both sides by the constant driving force

$$k_{y,\text{avg}} A_{\text{plate}} = \int_{A_{\text{plate}}} 0.33 \left(\frac{c\,\mathcal{D}_{AB}}{x}\right)\left(\frac{x\,v_0\,\rho}{\mu}\right)^{0.5}\left(\frac{\mu}{\rho\,\mathcal{D}_{AB}}\right)^{1/3} dA \qquad (8.4.1\text{-}8)$$

Letting the z-coordinate be in the direction of the width, W, of the plate, and letting L be the length of the plate in the x-direction

$$k_{y,\text{avg}} L\,W = \int_0^W \int_0^L 0.33 \left(\frac{c\,\mathcal{D}_{AB}}{x}\right)\left(\frac{x\,v_0\,\rho}{\mu}\right)^{0.5}\left(\frac{\mu}{\rho\,\mathcal{D}_{AB}}\right)^{1/3} dx\,dz$$

$$(8.4.1\text{-}9)$$

But none of the interior integral depends on z, so

$$k_{y,\text{avg}} L\,W = \left[\int_0^L 0.33 \left(\frac{c\,\mathcal{D}_{AB}}{x}\right)\left(\frac{x\,v_0\,\rho}{\mu}\right)^{0.5}\left(\frac{\mu}{\rho\,\mathcal{D}_{AB}}\right)^{1/3} dx\right]\int_0^W dz$$

$$(8.4.1\text{-}10)$$

$$k_{y,\text{avg}} L\,W = \left[\int_0^L 0.33 \left(\frac{c\,\mathcal{D}_{AB}}{x}\right)\left(\frac{x\,v_0\,\rho}{\mu}\right)^{0.5}\left(\frac{\mu}{\rho\,\mathcal{D}_{AB}}\right)^{1/3} dx\right][W]$$

$$(8.4.1\text{-}11)$$

Dividing both sides by W and moving constants outside the integral

$$k_{y,avg} L = 0.33 \, c \, \mathcal{D}_{AB} \left(\frac{v_0 \rho}{\mu}\right)^{0.5} \left(\frac{\mu}{\rho \, \mathcal{D}_{AB}}\right)^{1/3} \int_0^L \left(\frac{1}{x^{0.5}}\right) \, dx \qquad (8.4.1\text{-}12)$$

Integrating

$$k_{y,avg} L = 0.33 \, c \, \mathcal{D}_{AB} \left(\frac{v_0 \rho}{\mu}\right)^{0.5} \left(\frac{\mu}{\rho \, \mathcal{D}_{AB}}\right)^{1/3} \left(2 \, L^{0.5}\right) \qquad (8.4.1\text{-}13)$$

which can be rearranged to

$$\frac{k_{y,avg} L}{c \, \mathcal{D}_{AB}} = 0.66 \left(\frac{L \, v_0 \rho}{\mu}\right)^{0.5} \left(\frac{\mu}{\rho \, \mathcal{D}_{AB}}\right)^{1/3} \qquad (8.4.1\text{-}14)$$

$$Sh_L = 0.66 \, Re_L^{0.5} \, Sc^{1/3} \qquad (8.4.1\text{-}15)$$

where the subscript L denotes a quantity based on the length of the plate, and in the case of the Sherwood number also implies that it contains the average rather than the local coefficient.

Mass transfer in flow in pipes

Mass transfer for *laminar* flow with developed velocity distribution to or from the wall of a *pipe* may be treated using the corresponding heat transfer curves by substituting Sherwood number for Nusselt number and Schmidt number for Prandtl number.

For turbulent flow in pipes we will use the Reynolds analogy here, although several extensions of this treatment are available.

Mass transfer from spheres, drops, and bubbles

Mass transfer from spheres can be treated using the analogous equation to that for heat transfer

$$Sh = 2.0 + 0.6 \, Re^{1/2} \, Sc^{1/3} \qquad (8.4.1\text{-}16)$$

Mass transfer in drops and bubbles is a function of internal circulation, and solutions are available primarily for limiting cases. We will not treat this topic here because of its complexity, although it is one of the most important to engineers from a practical standpoint.

Example 8.4.1-2   Comparison of mass transfer coefficient models

For air at 50°F flowing at a bulk velocity of 50 ft/s in a 1-in. ID tube coated with a thin layer of naphthalene, compare the mass transfer coefficient predicted by the empirical correlation developed by Sherwood and Gilliland[25] for mass transfer

$$Sh = 0.023 \, Re^{0.83} \, Sc^{1/3}$$

(8.4.1-17)

to that predicted by the following model constructed by adapting to mass transfer the Dittus-Boelter equation for heat transfer in pipe flow

$$Sh = 0.023 \, Re^{0.8} \, Sc^{1/3}$$

(8.4.1-18)

Solution

For air at 50°F

r	0.078 lbmass/ft^3
m	1.2 x 10^{-5} lbmass/(ft s)
D_{AB}	0.2 ft^2/hr

We ignore the effect of the naphthalene on the properties of the air since the effect will be the same on both models.

$$Re = \frac{D \, v_0 \, \rho}{\mu}$$

(8.4.1-19)

$$Re = \frac{\frac{1}{(12)} \, [ft] \, (50) \left[\frac{ft}{s} \right] (0.078) \left[\frac{lbmass}{ft^3} \right]}{(1.2 \times 10^{-5}) \left[\frac{lbmass}{ft \, s} \right]} = 27,000$$

(8.4.1-20)

$$Sc = \frac{\nu}{D_{AB}}$$

(8.4.1-21)

---

[25] Sherwood, T. K. and F. A. L. Holloway (1940). *Trans AIChE* **30**: 39.

$$Sc = \frac{\left(1.2 \times 10^{-5}\right)\left[\frac{\text{lbmass}}{\text{ft s}}\right]}{\left(0.078\right)\left[\frac{\text{lbmass}}{\text{ft}^3}\right]\left(0.2\right)\left[\frac{\text{ft}^2}{\text{hr}}\right]\frac{1}{\left(3600\right)}\left[\frac{\text{hr}}{\text{s}}\right]} = 2.77 \qquad (8.4.1\text{-}22)$$

The Sherwood and Gilliland model gives

$$Sh = 0.023 \left(27,000\right)^{0.83} \left(2.77\right)^{1/3} = 154 \qquad (8.4.1\text{-}23)$$

while the model adapted from the Dittus-Boelter equation gives

$$Sh = 0.023 \left(27,000\right)^{0.8} \left(2.77\right)^{1/3} = 113 \qquad (8.4.1\text{-}24)$$

The mass transfer coefficients predicted will be in the same ratio as the Sherwood numbers.

$$\frac{154}{113} = 1.36 \qquad (8.4.1\text{-}25)$$

so for these conditions the Sherwood and Gilliland model gives a mass transfer coefficient almost 40% higher than the modified heat transfer model.

Note that the difference depends only on the ratio of the Reynolds numbers.

$$\frac{Re^{0.83}}{Re^{0.8}} = Re^{0.03} \qquad (8.4.1\text{-}26)$$

This means the difference is more pronounced at high Reynolds numbers.

Example 8.4.1-3   Mass transfer coefficient for dissolution of a sphere

Estimate the mass transfer coefficient for a 1/8-in. sphere of glucose dissolving in a water stream flowing with a freestream velocity of 0.5 ft/s. Temperature is uniform and constant at 25°C. Diffusivity of glucose in water at 25°C is 0.69 x 10^{-5} cm^2/s.

Properties of water at 25°C are

$$\rho = 62.2 \text{ lbmass/ft}^3$$
$$\mu = 0.9 \text{ cP}$$

Solution

We neglect the effect of the glucose on the physical properties of the water as a preliminary model. (Since concentrated glucose solutions are quite viscous compared to water, the validity of this assumption would have to be verified if we were interested in other than a rough estimate of mass transfer coefficient.)

We further assume that the mechanisms of mass transfer and heat transfer are similar. Note that we could not use an analogy with momentum transfer if our flow regime involves form drag, which has no counterpart in heat or mass transfer from a sphere. We adapt the Froessling correlation for heat transfer from a sphere (Equation 7.3.2-114) to the mass transfer situation by replacing Nusselt number with Sherwood number and Prandtl number with Schmidt number

$$\text{Sh} = 2.0 + 0.6 \, \text{Re}^{1/2} \, \text{Sc}^{1/3} \tag{8.4.1-27}$$

Then

$$\text{Re} = \frac{D \, v_0 \, \rho}{\mu} \tag{8.4.1-28}$$

$$\text{Re} = \frac{\frac{1}{(8)(12)} \, [\text{ft}] \, (0.5) \left[\frac{\text{ft}}{\text{s}}\right] (62.2) \left[\frac{\text{lbmass}}{\text{ft}^3}\right]}{(0.9) \, [\text{cP}] \left(6.72 \times 10^{-4}\right) \left[\frac{\text{lbmass}}{\text{ft s cP}}\right]} = 536 \tag{8.4.1-29}$$

$$\text{Sc} = \frac{\nu}{\mathcal{D}_{AB}} \tag{8.4.1-30}$$

$$\text{Sc} = \frac{\frac{(0.9) \, [\text{cP}] \left(6.72 \times 10^{-4}\right) \left[\frac{\text{lbmass}}{\text{ft s cP}}\right]}{(62.2) \left[\frac{\text{lbmass}}{\text{ft}^3}\right]}}{\left(0.69 \times 10^{-5}\right) \frac{\text{cm}^2}{\text{s}} \left[\frac{\text{in}}{(2.54) \, \text{cm}}\right]^2 \left[\frac{\text{ft}}{(12) \, \text{in}}\right]^2} = 1309 \tag{8.4.1-31}$$

Substituting

$$Sh = \frac{(k_{\omega})\left[\frac{ft}{hr}\right]\frac{1}{(8)(12)}[ft]}{\left(0.69 \times 10^{-5}\right)\left[\frac{cm^2}{s}\right](3.87)\left[\frac{ft^2\ s}{cm^2\ hr}\right]}$$

$$= 2.0 + 0.6\left(536\right)^{1/2}\left(1309\right)^{1/3} = 152 \tag{8.4.1-32}$$

$$k_{\omega} = 0.388\,\frac{ft}{hr} \tag{8.4.1-33}$$

Packed beds

Packed beds are frequently used as devices for contacting gas and liquid or liquid and liquid. Note that the mass transfer we will discuss with respect to packed beds is *between the flowing phases* and not between the packing in the bed and either of the phases (although this type of transfer is very important in some catalytic reactors, for example). Because of the fact that the interface is so ill-defined in a packed bed, making an interfacial area virtually impossible to determine, a slightly different approach to the mass transfer coefficient is taken. Since it is difficult to separate the interfacial area effect from other variables as a separate entity, it is usual to determine only the product of the interfacial area per unit volume and the mass transfer coefficient.

The interfacial area per unit volume of *empty* tower we denote as a:

$$a = \frac{A_i}{V_t} \tag{8.4.1-34}$$

where

> A_i = interfacial area
> V_t = volume of empty tower = (A_{xs}) x (height)
> A_{xs} = cross-sectional area of empty tower

even though we seldom can determine a. If, however, we rewrite our defining equation for the mass transfer coefficient (where W_A is the molar flow rate of species A)

$$N_A = \frac{W_A'}{A_i} = k_z\left(z_{As} - z_A\right) \tag{8.4.1-35}$$

Rearranging, and multiplying and dividing by the product $(A_{xs}\, h_t)$, where xs refers to the cross-section of the **empty** tower, and h_t to the height of the tower

$$\mathcal{W}_A = k_z \frac{A_i}{\left(A_{xs}\, h_t\right)} \left(A_{xs}\, h_t\right)\left(z_{AS} - z_A\right) \tag{8.4.1-36}$$

or

$$\mathcal{W}_A = k_z\, a\, A_{xs}\, h_t \left(z_{AS} - z_A\right) = k_z\, a\, V_t \left(z_{AS} - z_A\right) \tag{8.4.1-37}$$

where V_t is the volume of the empty tower.

This gives us an equation from which we can determine $k_z a$ (either $k_x a$ or $k_y a$). Using a specific packed bed, we can

- measure $\mathcal{W}_A$ from a mass balance by measuring flow and both inlet and outlet for either of the two fluids,
- obtain z_A from a bulk fluid sample,
- obtain V_t from the physical dimensions of the tower,

and from this information calculate an average $k_z a$ using Equation (8.4.1-37). As the height of tower over which we perform the measurements is decreased to a differential size, we approach calculation of the local $k_z a$.

Note: We cannot separate k_z and a unless in some manner we can determine interfacial area. This means that $k_z a$, in addition to varying with all the factors that influence the mass transfer coefficient, will also vary with factors which change the interfacial area per unit volume, such as packing method (for example, stacked, dry dumped, wet dumped, vibrated, etc.), packing configuration (rings, saddles, etc.), packing size, and so on.

Height of transfer unit models

There are two broad categories of mass transfer models for design of towers.

- The tower is modeled as a series of equilibrium contacts between the liquid and the gas streams, where in a succession of stages the streams are perfectly mixed and allowed to come to equilibrium, then separated, and the gas sent up the column to the next higher stage and the liquid down the column to the next lower stage. The number of these theoretical stages can then be calculated if the equilibrium data for the system is known. Departure of the mass transfer performance of the real tower from this idealized model is modeled by the introduction of a stage efficiency. The link between the model and the real-world column is made via a quantity usually called HETP: the *height equivalent to a theoretical plate*[26] - e.g., theoretical (equilibrium) stage. This approach is probably the most widely used in commercial design of both tray-type and packed columns.

• The model we shall employ here: one built around the concept of the mass transfer coefficient. This approach uses a model known as the HTU, the *height of a transfer unit*. This is less frequently used in design of tray-type columns, probably partly because of the appealing mental analog between equilibrium stages and a succession of trays. It does, however, despite the shortcomings of the mass transfer coefficient, furnish a more fundamental insight into mass transfer, which is our objective here. The reader concerned with design of commercial columns is referred to the many texts, computer design packages, and vendor publications available.

The HETP approach is probably more widely used than the more fundamental HTU model. For dilute systems with constant slope, m, of the equilibrium line, it can be shown[27] that the two models are united through the equation

$$\text{HETP} = \mathcal{H}_{\text{OG}} \frac{\ln\left[m\,\dfrac{G}{L}\right]}{\left(m\,\dfrac{G}{L} - 1\right)} \qquad (8.4.1\text{-}38)$$

where $\mathcal{H}_{\text{OG}}$ is the height of an overall transfer unit based on the gas phase (see Section 8.5).

---

[26] The term "plate" is both historical and colloquial in its origin. The shape of a tray in early distillation columns, which were (and still are) usually cylindrical, was that of a plate, and it usually contained an array of short slotted caps known as bubble caps designed to mix the liquid and gas. To keep the bubble caps submerged required that the tray retain a depth of liquid small in comparison to its diameter - e.g., a soup plate, and hence the terminology.

[27] For example, see Sherwood, T. K., R. L. Pigford, et al. (1975). *Mass Transfer*. New York, NY, McGraw-Hill.

To correlate mass transfer in packed beds one usually does not use $k_z a$ per se, but the **height of a transfer unit,** $\mathcal{H}$. This quantity combines superficial fluid rate with the mass transfer coefficient and interfacial area per unit volume. The reason for using this quantity, which has the units of length, is that this combination of variables arises very naturally in a ubiquitous integration when one designs packed columns. We consider this integration later; for now, we simply state the definition for $\mathcal{H}$ (in fact, several $\mathcal{H}$s)

The definitions are

$$\boxed{\mathcal{H}_G = \frac{G}{k_y a}}$$

(8.4.1-39)

$$\boxed{\mathcal{H}_L = \frac{L}{k_x a}}$$

(8.4.1-40)

where G and L are superficial molar velocities (moles per unit time per unit cross-section of empty tower), k_z is the single-phase mass transfer coefficient for gas or for liquid, and a is the interfacial area per unit tower volume.

We will encounter later (see Section 8.5.1) the corresponding heights that incorporate the overall transfer coefficient, K_z, based on the gas or the liquid phase.

$$\boxed{\mathcal{H}_{OG} = \frac{G}{K_y a}}$$

(8.4.1-41)

$$\boxed{\mathcal{H}_{OL} \equiv \frac{L}{K_x a}}$$

(8.4.1-42)

We turn to

- design equations
- rules of thumb
- experimental data

to obtain $\mathcal{H}_G$ and $\mathcal{H}_L$. Because of the widely variable predictions from various models for mass transfer in packed beds (basically prompted by the extremely complicated flow fields, which make theoretical constructs uncertain, further complicated by the scarcity and wide variability of data[28]), rules of thumb are almost as useful as mathematical models.

There are, however, HTU models that are useful, particularly for scaling within the same chemical system from one physical configuration to another, or from one chemical system to another with the same configuration. The following functional forms can be used either to scale or (less successfully) to design *ab initio*.

$$\mathcal{H}_G = \alpha\, \hat{G}^\beta\, \hat{L}^\gamma\, Sc_G^{0.5} \tag{8.4.1-43}$$

$$\mathcal{H}_L = \varphi \left(\frac{\hat{L}}{\mu}\right)^\eta Sc_L^{0.5} \tag{8.4.1-44}$$

Note: These are dimensional equations. $\hat{G}$ and $\hat{L}$ have units of lbmass/(hr ft^2) and μ has units of lbmass/(ft hr). $\mathcal{H}_G$ and $\mathcal{H}_L$ have units of ft.

The first of these relations evolved primarily from the data of Fellinger[29] for absorption of ammonia from air into water. The constants α, β, and γ are functions of packing type and range of $\hat{G}$ and $\hat{L}$; α is of the order of 10, β of the order of 0.5, and γ of the order of -0.5. The second is based on the data of Sherwood and Holloway.[30] The constants ϕ and η are functions of packing type and $\hat{G}$; ϕ is of the order of 0.001 to 0.01, η of 0.5.

---

[28] In determining data for single-phase mass transfer coefficients one would like to adapt the same approach taken in heat transfer: to make the resistance of that phase controlling. This is not so readily done in mass transfer experiments. It is possible to make the gas phase control by using a pure liquid and a gas that is relatively insoluble in the liquid - with no concentration gradient in the liquid, the gas phase clearly controls. Similarly, one could use a pure gas and a non-volatile liquid. These techniques are easier in the description than the execution. Alternatively, one can adjust for the resistance of one of the phases - an even more difficult task.

[29] Perry, R. H., C. H. Chilton, et al., Eds. (1963). *Chemical Engineer's Handbook*. New York, NY, McGraw-Hill. Sherwood, T. K. and R. L. Pigford (1952). *Absorption and Extraction*. New York, NY, McGraw-Hill. Treybal, R. E. (1955). *Mass Transfer Operations*. New York, NY, McGraw-Hill. cited with values of constants in Geankoplis, C. J. (1972). *Mass Transport Phenomena*, Holt, Rinehart, and Winston, p. 396.

[30] Sherwood, T. K. and F. A. L. Holloway (1940). *Trans AIChE* 30: 39. Treybal, R. E. (1955). *Mass Transfer Operations*. New York, NY, McGraw-Hill. cited with values of constants in Geankoplis, C. J. (1972). *Mass Transport Phenomena*, Holt, Rinehart, and Winston, p. 398.

For additional discussion of these and related equations, as well as more precise values of the constants see, for example:

- Bird, R. B., W. E. Stewart, et al. (1960). *Transport Phenomena*. New York, Wiley.
- Barrer, R. M. (1941). *Diffusion In and Through Solids*. New York, NY, The Macmillan Company.
- Scott, D. S. and F. A. L. Dullien (1962). *AIChEJ* **8:113**.
- Rothfield, L. B. and C. C. Watson (1963). *AIChEJ* **9:19**.
- Levich, V. G. (1962). *Physicochemical Hydrodynamics*. Englewood Cliffs, NJ, Prentice-Hall.
- Sherwood, T. K. and F. A. L. Holloway (1940). *Trans AIChE* **30**: 39.
- Sherwood, T. K., R. L. Pigford, et al. (1975). *Mass Transfer*. New York, NY, McGraw-Hill.
- Sherwood, T. K. and R. L. Pigford (1952). *Absorption and Extraction*. New York, NY, McGraw-Hill.
- Perry, R. H., C. H. Chilton, et al., Eds. (1963). *Chemical Engineer's Handbook*. New York, NY, McGraw-Hill.
- Treybal, R. E. (1955). *Mass Transfer Operations*. New York, NY, McGraw-Hill.
- Geankoplis, C. J. (1972). *Mass Transport Phenomena*. Holt, Rinehart, and Winston.
- Whitaker (1967). Diffusion and Dispersion in Porous Media. *AIChEJ* **13**(3): 420.
- Silvey, F. C. and G. J. Keller (1966). *Chemical Engineering Progress* **62**(1).
- Colburn, A. P. (1934). *Trans. AIChE* **35**: 211.
- Peters, M. S. and K. D. Timmerhaus (1991). *Plant Design and Economics for Chemical Engineers*. New York, McGraw-Hill, Inc.
- Friedlander, S. K. and M. Litt (1958). Diffusion controlled reaction in a laminar boundary layer. *Chem. Eng. Sci.* **7**: 229.
- Incropera, F. P. and D. P. DeWitt (1990). *Fundamentals of Heat and Mass Transfer*. New York, John Wiley and Sons.

8.4.2 Dimensional analysis of mass transfer by convection

We now show how to calculate the mass transfer coefficients we have discussed above. We do this, just as for heat transfer coefficients, in terms of design equations which are usually written in dimensionless terms. To introduce these design equations, we will first examine the significant dimensionless groups involved with mass transfer.

For mass transfer by molecular diffusion and *forced* convection, the pertinent variables are a characteristic length, D; the diffusivity, $\mathcal{D}_{AB}$; the velocity of the fluid, v; the mass density of the fluid, ρ; the viscosity of the fluid, μ; and the mass transfer coefficient, k_ρ. For the case of *natural* convection one also must include the acceleration due to gravity, g; and the mass density difference, $\Delta\rho_A$. If we carry out a dimensional analysis we find the significant dimensionless groups to be[31]

$$Sh \equiv \frac{k_\rho D}{\mathcal{D}_{AB}} = \frac{k_{\bar\rho} D}{\mathcal{D}_{AB}} = \frac{k_z D}{c\,\mathcal{D}_{AB}} \equiv Sherwood\,Number \qquad (8.4.2\text{-}1)$$

$$Sc \equiv \frac{\mu}{\rho\,\mathcal{D}_{AB}} \equiv Schmidt\,number \qquad (8.4.2\text{-}2)$$

$$Re \equiv \frac{D\,v\,\rho}{\mu} \equiv Reynolds\,number \qquad (8.4.2\text{-}3)$$

$$Fr \equiv \frac{v^2}{g\,D} \equiv Froude\,number \qquad (8.4.2\text{-}4)$$

$$Gr \equiv \frac{g\,D^3\,\rho\,\Delta\rho_A}{\mu^2} \equiv Grashof\,number\,for\,mass\,transfer \qquad (8.4.2\text{-}5)$$

The usual functional relationships used to correlate data are

$$Sh = f\left(Re,\,Sc\right) \qquad (8.4.2\text{-}6)$$

for forced convection and

$$Sh = f\left(Gr,\,Sc\right) \qquad (8.4.2\text{-}7)$$

for natural convection. The Froude number is normally unimportant in mass transfer correlations.

---

[31] Note that

$$n_A = k_\rho\left(\rho_{AS}-\rho_A\right) = N_A\,M_A = k_{\bar\rho}\left(\bar\rho_{AS}-\bar\rho_A\right)M_A$$

But

$$k_\rho\left(\rho_{AS}-\rho_A\right) = k_\rho\left(\bar\rho_{AS}-\bar\rho_A\right)M_A$$

so

$$k_\rho = k_{\bar\rho}$$

If we assume that mass and heat are transferred by the *same mechanism* (by the **same mechanism** we mean that the dimensionless differential equations for velocity, temperature, and concentration are identical in form), we see that the above amounts to a replacement of the Nusselt number by the Sherwood number and of the Prandtl number by the Schmidt number. The Sherwood number, analogous to the Nusselt number, is the ratio of total mass transferred to mass transferred by molecular diffusion, and the Schmidt number, analogous to the Prandtl number, represents the molecular diffusion of momentum to that of mass.

This all suggests that if the mechanism of mass and heat transfer is the same, we can use our correlations for heat transfer to predict mass transfer by the simple expedient of substituting Sh and Sc for Nu and Pr, respectively, and vice versa. We can do this (for both laminar and turbulent flow) only under fairly restrictive assumptions. We can even extend the technique to using momentum transfer data to predict mass or heat transfer, and other permutations of the same, but only under much more stringent restrictions because form drag has no counterpart in mass or heat transfer.

We have thus far avoided a very distasteful issue - namely, that our mass transfer coefficient as defined is a function of some things not in the list of variables we used to perform our dimensional analysis. For example, we showed in our treatment of molecular diffusion that the mass transfer coefficient depends on whether we have equimolar counterdiffusion or diffusion of A through stagnant B. In addition, the general coefficient we defined includes both molecular diffusion and bulk flow, and the bulk flow term is very sensitive to the flow field, i.e., the momentum transfer.

In fact, for a rigorous treatment we have to define several mass transfer coefficients, each appropriate for a different situation. The mass transfer coefficient as we have defined it holds for a specific mass transfer situation. It suffers, however, from the difficulty that it depends on mass transfer rate, because increasing rates of mass transfer distort velocity and concentration profiles. These effects are described by non-gradient terms in the differential equations, that is, terms that do not contain concentration derivatives (driving forces). These terms cannot be described well by gradient models

For this reason mass transfer coefficients sometimes are defined (a) which describe only the gradient contribution to transfer (as opposed to bulk flow) or perhaps (b) total transfer only for specified bulk flow situations such as equimolar counterdiffusion. These coefficients usually are designated by some type of superscript.[32] Our objective here, however, is to introduce the reader in a

---

[32] Bird, R. B., W. E. Stewart, et al. (1960). *Transport Phenomena*. New York, Wiley.

general way to the utility and prediction of mass transfer coefficients, so we will limit our discussion to situations which satisfy the following criteria:

1. constant physical properties
2. very small net bulk flow at the interface
3. no chemical reactions in the fluid
4. no viscous dissipation (or negligible temperature rise produced by viscous dissipation)
5. no radiant energy interchange
6. no pressure, thermal, or forced diffusion

Under these restrictions the several mass transfer coefficients we must define for a deeper discussion all reduce to the same number, and so we will be *consistent* and *correct* in what is to follow, but not *general*.

The above set of restrictions, although severe, also frequently gives useful approximations for real problems since the precision involved in mass transfer calculations is normally not fine. In exchange for the restrictions we gain vastly simplified notation and discussion.

8.5   Overall Mass Transfer Coefficients

Much of the time we are interested in mass transfer between two *fluid* phases (see Figure 8.5-1). In heat transfer and momentum transfer there is almost always some solid boundary involved in the transfer path, e.g., a pipe wall (excepting such cases, of course, as momentum and/or heat transfer with drops and bubbles). In mass transfer, however, we deal with cases such as transferring a component from a gas and/or a liquid to another liquid without any solid barrier between the two phases. Not only does this mean that the boundary can become severely distorted from flow forces, but also the precise location of the boundary is not always known.

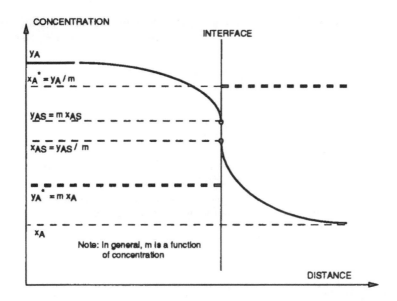

Figure 8.5-1 Mass transfer concentrations

In considering the mass transfer problem we are also confronted with a profound difference from the heat transfer situation in that *the variable of interest usually is no longer continuous across the interface.*

For heat transfer, equilibrium is usually assumed to exist immediately on either side of an interface. Thermal equilibrium implies equality of temperatures, so equilibrium at the interface requires that the temperature in the phases immediately adjacent to the interface be equal, i.e., that the temperature distribution be continuous.

On the other hand, for mass transfer, even if we assume equilibrium across an interface, we know from thermodynamics that two phases may be in equilibrium and possess *different concentrations.* At equilibrium the partial molar free energies (chemical potentials) of the given component in each phase must be the same, which is not necessarily the same (in fact, is seldom the same) as the concentrations being identical. In fact, a good part of our study of physical equilibrium in thermodynamics is aimed at predicting m, a function in general of temperature, pressure, and concentration, in

$$[y_A]_{eq} = m[x_A]_{eq}$$

(8.5-1)

Henry's law is an example of efforts in this direction, where m is a particularly simple form - a function of pressure only. In general, *m is not constant*. Henry's law

$$[p_A]_{eq} = [y_A]_{eq} p = H[x_A]_{eq} \tag{8.5-2}$$

yields

$$m = \frac{H}{p} \tag{8.5-3}$$

Raoult's law is another simple model, where m is a function of vapor pressure, which, in turn, is a function of temperature and pressure

$$[p_A]_{eq} = [y_A]_{eq} p = [x_A]_{eq} p' \tag{8.5-4}$$

$$[y_A]_{eq} = \frac{p'}{p}[x_A]_{eq} \tag{8.5-5}$$

which implies

$$m = \frac{p'}{p} \tag{8.5-6}$$

We would now like to define an *overall* coefficient for *mass* transfer in much the same way that we defined an overall coefficient of heat transfer, that is, in terms of the *overall driving force*.

The immediately obvious overall driving force to use for mass transfer would be the difference in bulk concentrations, $(y_A - x_A)$, by analogy to the heat transfer case, where the bulk temperatures on either side of the interface were used. However, the difference in bulk concentration is a function of m in the case of mass transfer. We can see this by adding and subtracting y_{AS} and x_{AS} to the driving force based on bulk concentrations

$$y_A - x_A = \left(y_A - y_{AS}\right) + \left(y_{AS} - x_{AS}\right) + \left(x_{AS} - x_A\right) \tag{8.5-7}$$

and then, assuming equilibrium at the interface, using Equation (8.5-1) to substitute for y_{AS} in terms of x_{AS} (we could equally well substitute for x_{AS} in terms of y_{AS})

$$\left(y_{AS} - x_{AS}\right) = \left(m\, x_{AS} - x_{AS}\right) = x_{AS}\left(m - 1\right) \tag{8.5-8}$$

Substituting in Equation (8.5-7)

$$y_A - x_A = \left(y_A - y_{AS}\right) + x_{AS}\left(m - 1\right) + \left(x_{AS} - x_A\right) \tag{8.5-9}$$

Note that, as the two phases approach equilibrium, y_A and x_A approach y_{AS} and x_{AS}, respectively, and the driving force defined in Equation (8.5-7) based on the difference in bulk concentrations approaches [x_{AS} (m - 1)] as shown by Equation (8.5-8)

$$\left(y_A - x_A\right) \Rightarrow \left(y_A - x_A\right)_{eq} = \left(y_{AS} - x_{AS}\right) = x_{AS}\left(m - 1\right) \tag{8.5-10}$$

This is not at all the behavior we wish for a driving force - we want driving forces to go to *zero* at equilibrium, not some finite value. This leads us instead to define one of two quantities for our driving force

$$\boxed{\text{Driving force} \equiv \left\{ \begin{array}{l} \text{either } x_A - x_A^{\bullet} \\ \text{or } \quad y_A - y_A^{\bullet} \end{array} \right\}} \tag{8.5-11}$$

> Quantities designated by asterisks are defined as those values of x or y that would be in equilibrium with the bulk fluid of the other phase.

Or, in equation form

$$\boxed{x_A^{\bullet} \equiv \frac{y_A}{m}} \tag{8.5-12}$$

$$\boxed{y_A^{\bullet} \equiv m\, x_A} \tag{8.5-13}$$

Note the driving force defined in this fashion goes to zero at equilibrium.

The concentrations x_A^* and y_A^* are not normally present anywhere in the real system, for at such a point the driving force for mass transfer would be zero; they are fictitious concentrations introduced for convenience. These concentrations are indicated by the heavy dashed lines in Figure 8.5-1. Note that for the case illustrated, y_A^* is below any concentration that exists in the y-phase; conversely, x_A^* is above any concentration that exists in the x-phase.

This driving force leads us to define an *overall* mass transfer coefficient

$$\boxed{N_A = K_z \left(z_A - z_A^* \right)}$$

(8.5-14)

where we substitute x for z to treat liquids and y for z to treat gases.

We now ask how to find x_A^*, y_A^*, x_{AS} and y_{AS}, given two bulk concentrations x_A and y_A. It is not difficult to locate x_A^* and y_A^* immediately - we simply apply the definition and the equilibrium relationship - for example, by using the definition that x_A^* is the composition that would be in equilibrium with y_A in conjunction with the appropriate form of m. The equilibrium relationship gives x_A^* when supplied with y_A. The value of y_A^* is found similarly using x_A.

To locate the interfacial compositions x_{AS}, y_{AS}, we use the fact that we must obtain the same value for N_A regardless of whether we calculate using k_x or k_y.

$$N_A = k_x \left(x_A - x_{AS} \right) = k_y \left(y_{AS} - y_A \right)$$

(8.5-15)

Rearranging

$$-\frac{k_x}{k_y} = \frac{\left(y_A - y_{AS} \right)}{\left(x_A - x_{AS} \right)}$$

(8.5-16)

This is the equation of a straight line through (x_A, y_A) with slope $(-k_x/k_y)$. The line also passes through (x_{AS}, y_{AS}), and this point, assuming equilibrium at the interface, must also lie on the equilibrium curve. We may, therefore, locate (x_{AS}, y_{AS})

- graphically, by drawing a line through (x_A, y_A) with slope $(- k_x/k_y)$ and taking its intersection with the equilibrium curve, or

- algebraically, by solving Equation (8.5-16) and the algebraic form of the equilibrium relationship simultaneously.

We illustrate the graphical method in Figure 8.5-2, where y denotes the concentration in the gas phase and x the concentration in the liquid phase. Since, in general, m is not constant, for pedagogical purposes we here plot the equilibrium line [Equation (8.5-1)] as the curve $(y_{AS} = 1.5 x_{AS}^2 + 0.4 x_{AS})$.

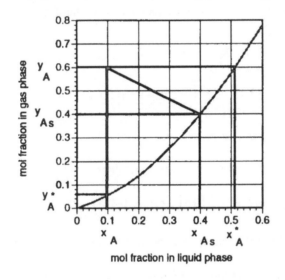

Figure 8.5-2 Interface conditions

Shown in Figure 8.5-2 is the graphical construction to find the interfacial concentrations corresponding to the bulk concentrations $(x_A, y_A) = (0.1, 0.6)$ for the case where the mass transfer coefficients are in the ratio

$$-\frac{k_x}{k_y} = -\frac{2}{3} \tag{8.5-17}$$

The same problem can be solved analytically by simultaneous solution of the equations

$$\frac{\left(0.6 - y_{AS}\right)}{\left(0.1 - x_{AS}\right)} = -\frac{2}{3}$$

$$y_{AS} = 1.5\, x_{AS}^2 + 0.4\, x_{AS}$$

(8.5-18)

Solving the first equation for y_{AS}

$$y_{AS} = -\frac{2}{3} x_{AS} + \frac{2.0}{3}$$

(8.5-19)

Substituting this result in the second equation and solving for the positive root

$$-\frac{2}{3} x_{AS} + \frac{2.0}{3} = 1.5\, x_{AS}^2 + 0.4\, x_{AS}$$

$$4.5\, x_{AS}^2 + 3.2\, x_{AS} - 2.0 = 0$$

$$x_{AS} = \frac{-3.2 \pm \sqrt{(3.2)^2 - (4)(4.5)(-2.0)}}{(2)(4.5)}$$

(8.5-20)

$$x_{AS} = 0.4$$

Substituting in the solved form of the first equation to obtain y_{AS}

$$y_{AS} = -\frac{2}{3} x_{AS} + \frac{2.0}{3} = -\frac{2}{3}(0.4) + \frac{2.0}{3} = 0.4$$

(8.5-21)

Which (necessarily) is the same result obtained by the graphical computation.

Example 8.5-1   Calculation of interface composition

A solute A is being absorbed from a gas mixture in a cylindrical wetted-wall column with the liquid in down flow on the walls and the gas in up flow. At a point in the column where the bulk liquid concentration is $x_A = 0.1$, the bulk gas concentration is $y_A = 0.4$.

The mass transfer coefficient for species A in the gas phase is 1.0 lbmol/(hr ft^2) based on concentration difference in mole fraction. The mass transfer coefficient for species A in the liquid phase is 1.5 lbmol/(hr ft^2) based on concentration difference in mole fraction.

Equilibrium data for the system is given in the following figure.

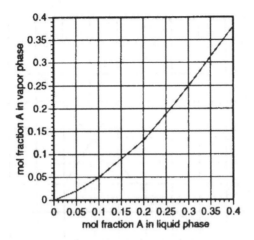

a) What is the interfacial concentration of A in both the gas and the liquid phases?

b) What is the flux of A?

Solution

a) We know the ratio of mass transfer coefficients which gives us a slope of a straight line

$$-\frac{k_x}{k_y} = -\frac{1.5}{1.0} = \frac{\left(y_A - y_{As}\right)}{\left(x_A - x_{As}\right)}$$

(8.5-22)

and we also know a point on the line, $(x_A, y_A) = (0.1, 0.4)$.

The interfacial concentrations are determined by constructing this straight line and taking its intersection with the equilibrium curve as shown in the following figure

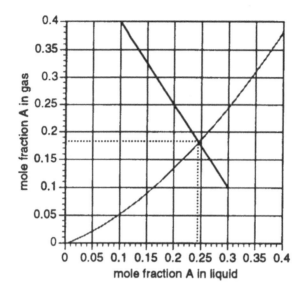

From the intersection we see that $(x_{AS}, y_{AS}) = (0.245, 0.18)$

b) To calculate the flux we use either

$$N_A = k_x \left(x_A - x_{AS} \right) = (1.5)\left[\frac{mol}{hr\,ft^2}\right](0.1 - 0.245) = -0.218\frac{mol}{hr\,ft^2}$$

(8.5-23)

or

$$N_A = k_y \left(y_{AS} - y_A \right) = (1.0)\left[\frac{mol}{hr\,ft^2}\right](0.18 - 0.4) = -0.22\frac{mol}{hr\,ft^2}$$

(8.5-24)

The two calculations agree within the accuracy of reading the graph.

The negative sign for the flux reflects our choice of order in the two terms in each of the driving forces

$$\left(x_A - x_{AS} \right)$$

(8.5-25)

and

$$\left(y_{AS} - y_A\right) \tag{8.5-26}$$

With this definition, mass transfer from the liquid to the gas gives a positive flux, since in such a case the concentration in the bulk liquid is greater than the liquid concentration at the interface, and the gas concentration at the interface is greater than the bulk gas concentration - in our example, however, the transfer was from the gas to the liquid, so we obtained a negative flux. If we had chosen instead to define the driving forces as

$$\left(x_{AS} - x_A\right) \tag{8.5-27}$$

$$\left(y_A - y_{AS}\right) \tag{8.5-28}$$

we would have obtained positive fluxes. The choice of order is arbitrary, but must be consistent.

8.5.1 Incorporation of overall mass transfer coefficient into height of transfer unit model

As was noted previously, heights of transfer units can also be defined that incorporate the overall mass transfer coefficients

$$\mathcal{H}_{OG} \equiv \frac{G}{K_y a} \tag{8.5.1-1}$$

$$\mathcal{H}_{OL} \equiv \frac{L}{K_x a} \tag{8.5.1-2}$$

Example 8.5.1-1   Overall transfer units

Derive the relationship among H_G, H_L, and H_{OL}.

Solution

From the definitions

$$\frac{N_A}{K_x a} = \left(x_A - x_A^*\right) \tag{8.5.1-3}$$

$$\frac{N_A}{K_x a} = \left(x_A - \frac{y_A}{m} \right)$$

(8.5.1-4)

Expanding the driving force

$$\frac{N_A}{K_x a} = \left(x_A - x_{As} + x_{As} - \frac{y_A}{m} \right)$$

(8.5.1-5)

Assuming m is constant

$$\frac{N_A}{K_x a} = x_A - x_{As} + \frac{y_{As}}{m} - \frac{y_A}{m} = \left(x_A - x_{As} \right) + \frac{1}{m} \left(y_{As} - y_A \right)$$

(8.5.1-6)

From the definitions

$$\frac{N_A}{k_x a} = \left(x_A - x_{As} \right)$$

(8.5.1-7)

$$\frac{N_A}{k_y a} = \left(y_{As} - y_A \right)$$

(8.5.1-8)

Substituting

$$\frac{N_A}{K_x a} = \left(\frac{N_A}{k_x a} + \frac{1}{m} \frac{N_A}{k_y a} \right)$$

(8.5.1-9)

$$\frac{1}{K_x a} = \left(\frac{1}{k_x a} + \frac{1}{m k_y a} \right)$$

(8.5.1-10)

$$\frac{L}{K_x a} = \left(\frac{L}{k_x a} + \frac{L}{m} \frac{G}{G k_y a} \right)$$

(8.5.1-11)

$$\mathcal{H}_{OL} = \left(\mathcal{H}_L + \frac{1}{m} \frac{L}{G} \mathcal{H}_G \right)$$

(8.5.1-12)

8.6 Relationship of Overall and Single-Phase Mass Transfer Coefficients

By using exactly the same procedure used to develop overall heat transfer coefficients from single-phase coefficients, one can sometimes develop similar relations between the overall and single-phase mass transfer coefficients. The procedure is first to *write the overall driving force as the sum of individual driving forces*. Let us use the liquid phase for our basis as an example. Substituting for x^*, remembering that, in general, m is a function of y_A

$$\left[x_A - x^*\right] = \left[x_A - \frac{y_A}{m(y_A)}\right] \tag{8.6-1}$$

Adding and subtracting x_{AS} from the right-hand side and substituting y_{AS}/m for one of the x_{AS}'s, again remembering that, in general, m is a function of y_A, i.e., $m = m(y_A)$

$$\left[x_A - x^*\right] = \left[x_A - x_{AS} + x_{AS} - \frac{y_A}{m(y_A)}\right] \tag{8.6-2}$$

$$\left[x_A - x^*\right] = \left[x_A - x_{AS} + \frac{y_{AS}}{m(y_{AS})} - \frac{y_A}{m(y_A)}\right] \tag{8.6-3}$$

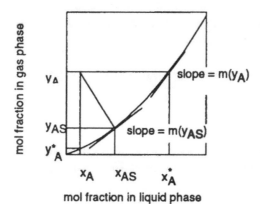

Figure 8.6-1 Assumption necessary to utilize overall mass transfer coefficient

Figure 8.6-1 illustrates the general case. If we now

> **assume that the m evaluated at y_A and the m evaluated at y_{AS} are the same**

e.g., for an essentially *straight* equilibrium line (although this could theoretically also happen fortuitously for the two points involved with a peculiarly curved equilibrium line, but with vanishing probability),

$$\left(x_A - x^*\right) = \left(x_A - x_{AS} + \frac{y_{AS}}{m} - \frac{y_A}{m}\right) \tag{8.6-4}$$

$$\left(x_A - x^*\right) = \left(x_A - x_{AS}\right) + \frac{1}{m}\left(y_{AS} - y_A\right) \tag{8.6-5}$$

Substituting using the defining expressions for the mass transfer coefficients for individual phases and for the overall mass transfer coefficient based on the liquid phase

$$\frac{N_A}{K_x} = \frac{N_A}{k_x} + \frac{N_A}{m\,k_y} \tag{8.6-6}$$

$$\frac{1}{K_x} = \frac{1}{k_x} + \frac{1}{m\,k_y} \tag{8.6-7}$$

These relations are true *only* for a straight equilibrium line because we have assumed that m is the same at y_{AS} and y_A. If the equilibrium line is *not* straight, this procedure is invalid. In other words, for significantly curved equilibrium lines one cannot develop a simple overall mass transfer coefficient in terms of the individual coefficients.

Similarly, starting with $(y_A^* - y_A)$ we can show that

$$\frac{1}{K_y} = \frac{m}{k_x} + \frac{1}{k_y} \tag{8.6-8}$$

Note that if either $1/k_y \ll m/k_x$ or vice versa, one phase controls the process just as in heat transfer. For example, if $1/k_y \rightarrow 0$ $(k_y \rightarrow \infty)$ the liquid phase controls.

Nowhere in any of this material have we shown how to calculate k or K from the physical structure (fluid mechanics, etc.). This will be our topic in the next section.

Example 8.6-1   Controlling resistance for mass transfer

In a system for which m = 300, k_x = 30, and k_y = 3, is the controlling resistance the liquid phase or the gas phase?

Solution

Using the overall gas phase coefficient

$$\frac{1}{K_y} = \frac{m}{k_x} + \frac{1}{k_y} \tag{8.6-9}$$

$$\frac{1}{K_y} = \frac{300}{30} + \frac{1}{2} = 10 + 0.5 \tag{8.6-10}$$

Therefore, the liquid phase controls. A similar result would be obtained using K_x.

8.7 Design of Mass Transfer Columns

Many mass transfer operations are concerned with the transfer of a component between two essentially immiscible phases. A classic illustration is gas absorption and desorption in dilute systems. For example, small amounts of SO_2 may be removed from air by absorption at nearly atmospheric pressure in an amine of low vapor pressure. The low vapor pressure of the amine means that little amine is transferred from liquid to gas; the low operating pressure ensures that little air dissolves in the amine; therefore, we transfer essentially only SO_2.

This sort of operation is frequently carried out in a column or tower, with the gas entering at the bottom and flowing countercurrent to the liquid, which enters at the top. If mass transfer rate is the only consideration (that is, assuming that there is not some sort of chemical reaction going on which liberates large amounts of heat which must be removed from the system), some sort of device is usually placed internal to the tower to increase the interfacial area between the gas and liquid and, therefore, to promote better mass transfer.

Such devices can take the form of trays of various types: bubble cap, sieve, jet, grid, etc. Alternatively, the tower may be packed with porous solids of varying configurations: cylinders, spheres, saddles, coils, mesh, etc. Such packings may be randomly dumped into the tower or structured either by placing the packing piece by piece (stacked packing) or pre-fabricating the packing in layers in a sort of hybrid between packing and trays.

Packings can be made of a variety of materials of construction - e.g., ceramics, metals, polymers, etc., and in a variety of configurations. "Loose" packings (those consisting of units small in dimension compared to the tower diameter) can be based on spheres, cylinders, saddles, or a variety of other geometrical shapes; the end objective simply being to create the maximum interfacial area between liquid and gas with the minimum pressure drop. Packing design requires a considerable amount of art as well as science at its present stage of development. An excellent discussion of packing design and summary of packing configurations including their historical evolution accompanied by many illustrations may be found in Kister, H. Z. (1992). *Distillation Design.* New York, NY, McGraw-Hill, Inc., p. 421.

The choice of type of device to use for a tower internal may be governed by a number of considerations. For example, if the pressure drop through the tower is large enough that compression cost of the gas phase becomes significant, a packed column may be preferable to a tray-type column because packed columns tend to have lower pressure drops. Many packed columns, however, have a tendency to channel,[33] and also suffer from the disadvantage that (if they are used with liquids that form deposits not readily removable by chemical cleaning) manual cleaning is quite difficult. The weight of the packing for a packed column can become a limiting consideration - sometimes it is large enough to crush the packing at the bottom unless intermediate supports are placed in the column. If reasonably rapid response for control purposes is required, packed towers usually have the advantage of less liquid holdup than plate-type towers. Obviously, the cost of the tower internals is also a consideration.

Many of the decisions regarding which type of tower internals to use are based on qualitative or rule-of-thumb criteria. Here we wish to stress the most important steps in tower design rather than confusing the reader with the myriad of details in the complete process, and so in what follows we will regard the choice of **type** of tower internals as having already been determined.

Design problems for mass transfer equipment come in many different forms; however, they all embody the same essential features. We will now introduce the method of calculation by considering an example.

A typical continuous contacting device is sketched in Figure 8.7-1. We will discuss countercurrent gas absorption (that is, transfer from gas to liquid - also

---

[33] Channeling refers to the existence of low frictional resistance passages in the column which permit liquid and/or gas to by-pass the mixing process.

sometimes called *scrubbing*) as our example. Treatment for devices that desorb a component from a liquid into a gas stream is analogous. The equations developed here may be applied to continuous contacting of immiscible phases (including liquid/liquid contact) in a spray column, packed column, plate column, wetted-wall column, or any other type of device which lends itself to the height of transfer unit/number of transfer units approach to handling the question of interfacial area per unit volume.

The simplified form of the usual gas absorption problem might be as follows.

Given:

1. A gas stream entering at a given rate, W_G, containing some given fraction of component A, y_{A0}.
2. A specified exit gas concentration of component A, y_{A1} (alternatively, the percent or fraction to be removed may be given).
3. A specified inlet fraction of A in the scrubbing liquid, x_{A1}.
4. A specified size and type of packing.

Calculate the height of packing and diameter of tower required to achieve the separation.[34] It is usually also the task of the engineer to acquire the equilibrium data for the system from the literature, via calculation, or by obtaining laboratory data.

The gas enters at a molar rate of W_G and contains y_{A0} mole fraction component A. This mole fraction is assumed to be small so that our assumptions with regard to the mass transfer coefficient will be satisfied. The liquid enters at an unknown molar velocity of L (moles per hour per square foot of empty tower) and contains x_{A1} mole fraction component A. The carrier gas is assumed insoluble in the liquid and the liquid is assumed to have a negligible vapor pressure. The cross-sectional area of the empty tower is denoted by A_{XS}.

---

[34] In practice, of course, it is also necessary to select the best type of packing, to decide on the allowable exit concentration, to optimize the economics, to insert any required intermediate packing supports, to design liquid and/or gas distributors, to determine disengagement space/method, etc. We are not interested in this much detail for our objective here, which is to consider the elemental mass transfer aspects of the problem.

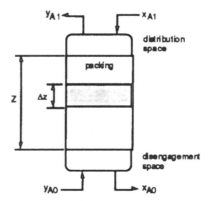

Figure 8.7-1 Typical countercurrent gas absorber

We have two separate tasks before us: first, to determine the height of the tower packing, and second, to determine the tower diameter. We will consider these tasks in the reverse of the order listed. Design of liquid and gas distributors and the disengagement space are topics we defer to more advanced treatments.

8.7.1 Determination of liquid-to-gas ratio

The problem of sizing the tower diameter is primarily one of momentum transfer rather than mass transfer; crudely stated: it is necessary to have sufficient room for the liquid and gas. To solve this fluid mechanics problem we need to know both the gas and the liquid rates. In our problem the gas rate is given; we must determine the liquid rate.

We can visualize the situation somewhat more easily by using a graph. In Figure 8.7.1-1 we show the equilibrium relationship for our typical system. (This must be given or obtained through literature values, experiment, or calculation, a problem not in transfer operations, but in thermodynamics.[35]) Points on this curve, as the name implies, are phases in equilibrium. We are concerned here with the dilute case and so we show only the extreme left-most portion of the curve.

---

[35] See, for example, Denbigh, K. (1966). *The Principles of Chemical Equilibrium.* New York, NY, Cambridge University Press.

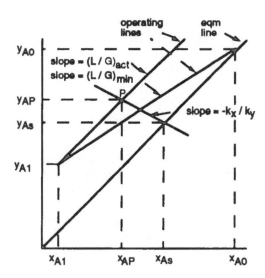

Figure 8.7.1-1 Gas absorption

Note that we can plot from the given information a point (x_{A1}, y_{A1}) representing the conditions at the top of the tower. Since in our example we are transferring from the gas to the liquid, for the driving force to be in the correct direction the concentration in the gas phase must be *larger* than the concentration that would be in equilibrium with the liquid and that in the liquid must be *smaller* than the concentration that would be in equilibrium with the gas. This is true not only for the top end of the tower, but for the gas and liquid phases at *any* point in the tower. (If we were transferring material from the liquid to the gas, our problem would be one in *desorption or stripping of a liquid*. In this case the point would lie below the equilibrium line.) The driving force is represented by the distance between the operating and equilibrium lines.

Now let us ask ourselves *how much liquid* is required to absorb the A which is transferred. Obviously this is not a unique amount, because we can absorb the A in a little or a lot of liquid. There is, however, a *lower limit* to the amount of liquid we can use. (Obviously if the proposed amount of liquid could not hold the required amount of A even when saturated with A, such an amount would be below the lower limit.)

We can find this lower limit by looking at the driving force for mass transfer. At any point in the tower, we find the interfacial composition by drawing a straight line through (x_A, y_A) with slope $(-k_X/k_y)$ and taking its intersection with the equilibrium curve, as shown in the figure for an arbitrary point, P. The length between the operating line and equilibrium line of this diagonal line is proportional to the driving force, since, for any point in the tower we choose, the length of a corresponding line is proportional to both (y_A - y_{As}) and (x_{As} - x_A), the driving forces in the system.

We can never *cross* the equilibrium line simply by mixing two phases (as in a flask), because when we *reach* the equilibrium line the driving force goes to zero and mass transfer stops. (Our process has no "inertia" to carry it past equilibrium.) Of course, as the driving force becomes smaller and smaller, the mass flux also becomes smaller and smaller, and so more and more surface area is required to transfer mass at the same rate (or more time is required to transfer a given amount of mass).

What does this say about our minimum liquid rate? It says that *even if we had a tower containing infinite interfacial area*, i.e., infinite height, *if we expect to maintain the required outlet gas concentration* we can at *best* reduce our driving force to something which *approaches* zero at one or more points in the tower, but we *cannot* cross the equilibrium line. If we attempt to cross the equilibrium curve, that is, if we reach it, mass transfer stops at that point.

How does this affect the rate of liquid we use? Assuming constant L and G, an overall balance on component A for the tower shows that

$$L\,A_{xs}\,x_{A1} + G\,A_{xs}\,y_{A0} = L\,A_{xs}\,x_{A0} + G\,A_{xs}\,y_{A1} \qquad (8.7.1\text{-}1)$$

For constant cross-sectional area (the usual case):

$$G\left(y_{A0} - y_{A1}\right) = L\left(x_{A0} - x_{A1}\right) \qquad (8.7.1\text{-}2)$$

Since $GA_{xs} = \mathcal{W}_G$, and y_{A0}, y_{A1}, and x_{A1} are specified, to make L as small as possible means that we want (x_{A0} - x_{A1}) as large as possible, that is, x_{A0} as large as possible. This corresponds to approaching equilibrium. In other words, the minimum liquid rate will be found where we *first* reach equilibrium at *some* point in a tower of infinite height as we decrease liquid rate, *assuming we continue to transfer the same amount of mass*.

This means that we need the locus of composition of the liquid phase for any given liquid rate. We can find this locus by writing a material balance for a system boundary around the top of the tower and through an arbitrary intermediate point at height z. Neglecting radial composition variation:

$$L\,x_{A1} + G\,y_A = L\,x_A + G\,y_{A1} \qquad\qquad (8.7.1\text{-}3)$$

or

$$y_A = \tfrac{L}{G}\,x_A + \left(y_{A1} - \tfrac{L}{G}\,x_{A1}\right) \qquad\qquad (8.7.1\text{-}4)$$

This is the equation of a line (called the *operating* line) which gives the locus of liquid phase compositions as a function of gas phase composition or vice versa. In other words, it relates the compositions of passing phases. The line passes through (x_{A1}, y_{A1}). It also passes through (x_{A0}, y_{A0}) as is easily demonstrated. For L/G constant, the line is straight. For our assumptions of a) immiscibility of carrier phases and b) dilute systems, L and G will be very nearly constant and we will take the line to be straight. For the moment we have assumed that $\mathcal{W}_G$, x_{A1}, and y_{A1} are given.

For any value of $\mathcal{W}_L$, therefore, we can plot the operating line by drawing it through (x_{A1}, y_{A1}) with slope

$$\frac{\mathcal{W}_L}{\mathcal{W}_G} = \frac{L\,A_{xs}}{G\,A_{xs}} = \frac{L}{G} \qquad\qquad (8.7.1\text{-}5)$$

Here we consider only problems with L and G essentially constant. If L and G vary through the tower, we must write Equation (8.7.1-1) as

$$L_1\,A_{xs}\,x_{A1} + G_0\,A_{xs}\,y_{A0} = L_0\,A_{xs}\,x_{A0} + G_1\,A_{xs}\,y_{A1} \qquad\qquad (8.7.1\text{-}6)$$

The criterion for minimum liquid rate is the same, but it is not so easily seen as for the case of constant L and G. The operating line, Equation (8.7.1-4), would become

$$y_A = \tfrac{L}{G}\,x_A + \left(\frac{G_1}{G}\,y_{A1} - \frac{L_1}{G}\,x_{A1}\right) \qquad\qquad (8.7.1\text{-}7)$$

and its slope would vary throughout the tower.

We now ask: "If we start with a straight line of large slope through (x_{A1}, y_{A1}) and decrease the slope (that is, decrease L), when will we first reach an equilibrium condition within the tower?" Since the tower is bounded by concentrations y_{A0} and y_{A1}, obviously the line labeled $(L/G)_{min}$ is the first to satisfy this requirement. For this line x_{A0} will be the concentration in equilibrium with y_{A0}. We can measure the slope of this line on the graph and calculate $\left(\mathcal{M}_L\right)_{min}$ as

$$\left(L\,A_{xs}\right)_{min} = \left(\mathcal{M}_L\right)_{min} = G\,A_{xs}\left(\frac{L}{G}\right)_{min} = \mathcal{M}_G\left(\frac{L}{G}\right)_{min} \tag{8.7.1-8}$$

Note that for curved equilibrium or operating lines it is not necessary that the first intersection or tangency be at the *end* of the tower.

Since to operate the tower with this liquid rate and satisfy y_{A1} would require an infinite area for mass transfer, we must actually operate at a somewhat higher liquid rate. The actual liquid rate is obtained by balancing the capital and operating costs involved[36]; here, since process economics is not our primary focus, we will, in general, assume that we will use

$$\mathcal{M}_L = 1.5\left(\mathcal{M}_L\right)_{min} \tag{8.7.1-9}$$

a not untypical figure.

We then would obtain an actual operating line as shown in Figure 8.7.1-2. Since the operating line equation is valid at the ends of the column as well as within the column, x_{A0} is located as shown. (The value of x_{A0} could also be obtained analytically via a mass balance on component A around the entire tower.)

We now have determined both liquid and gas rates and can proceed to size the tower cross-sectional area, A_{xs}. If our problem were one of desorption rather than absorption, all would proceed as above except the operating line would fall *below* the equilibrium curve and we would look for $(L/G)_{max}$, not $(L/G)_{min}$.

8.7.2 Calculation of tower diameter

As we mentioned earlier, the tower must be of sufficient diameter to accommodate the gas and liquid. Let us now be more precise in our statements.

---

[36] Lower liquid rates, while requiring less expenditure in terms of both investment in liquid and pumping and recovery costs would require a larger tower to achieve the same separation; finding the optimum is not a trivial problem.

If we have a packed column operating at a given liquid rate and we gradually increase the gas rate, we find that after a certain point the gas rate is so high that the drag of the gas on the liquid is sufficient to keep the liquid from flowing freely down the column. Liquid begins to accumulate and tends to block the entire cross-section for flow (so-called *loading*). This, of course, both increases the pressure drop and prevents the packing from mixing the gas and liquid effectively, and ultimately some liquid is even carried back up the column. This undesirable condition, known as *flooding*[37] occurs fairly abruptly, and the *superficial gas* velocity at which it occurs is called the *flooding velocity*.

We wish to design towers to avoid reaching the flooding velocity, and so we usually design for operation at some fraction of the flooding velocity (again, this is an economic judgment balancing pumping and capital costs) by using a larger tower diameter.

General models for calculation of pressure drop and/or flooding in packed towers continue to be a source of some controversy.[38] We will not concern ourselves with the details of this debate. Most of the models are some form of the GPDC (Generalized Pressure Drop Correlation) of Eckert,[39] which is based on earlier works by Leva and by Sherwood. This correlation is usually presented as a log-log plot with abscissa

$$X = \frac{L}{G} \sqrt{\frac{\rho_G}{\rho_L}} \qquad (8.7.2\text{-}1)$$

and ordinate some form similar to

---

[37] In fact, there are many different definitions of flooding that have been used in the literature, and so this is not so clean a term as implied by this statement - see, for example, Silvey, F. C. and G. J. Keller (1966). *Chemical Engineering Progress* 62(1).

[38] For example, see Kister, H. Z. (1992). *Distillation Design.* New York, NY, McGraw-Hill, Inc. Leva, M. (1992). Reconsider Packed-Tower Pressure-Drop Correlations. *Chemical Engineering Progress* January, p. 65. Hanley, B., B. Dunbobbin, et al. (1994). A Unified Model for Countercurrent Vapor/Liquid Packed Columns. 1. Pressure Drop. *Ind. Eng. Chem. Res.* 33, pp. 1208-1221. Kaiser, V. (1994). Correlate the Flooding of Packed Columns a New Way. *Chemical Engineering Progress* June, p. 55. Robbins, L. A. (1991). Improve Pressure-Drop Prediction with a New Correlation. *Chemical Engineering Progress* May, p. 87. Hanley, B., B. Dunbobbin, et al. (1994). A Unified Model for Countercurrent Vapor/Liquid Packed Columns. 2. Equations for the Mass-Transfer Coefficients, Mass-Transfer Area, the HETP, and the Dynamic Liquid Holdup. *Ind. Eng. Chem. Res.* 33, pp. 1222-1230.

[39] For example, see Strigle, R. F. (1987). *Random Packings and Packed Towers.* Houston, TX, Gulf Publishing.

$$Y = \frac{G^2 F \left[\frac{\rho_{HOH}}{\rho_L}\right] \mu^\alpha}{\rho_G \rho_L g_c}$$

(8.7.2-2)

with parametric lines for pressure drop per unit height of packing. X is usually termed the **flow parameter** and Y the **capacity parameter**. Care with units must be taken in using the plots, because the variables are not dimensionless.[40] For example, the units of viscosity are usually cP. F is a factor specific to the particular packing being used.

For purposes of illustration we will utilize a variant of this generic correlation type which is recommended by the Norton Chemical Process Products Corporation for the specific packing shown in Figure 8.7.2-1.

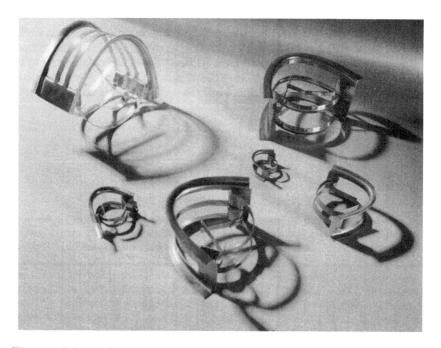

Figure 8.7.2-1 Norton Chemical Process Products Corporation Intalox® IMTP® Packing[41]

---

[40] For a graphic illustration of the problem see Leva, M. (1992). Letter to the Editor. *Chemical Engineering Progress, December*, p. 8.

[41] Norton Chemical Process Products Corporation, Akron, OH.

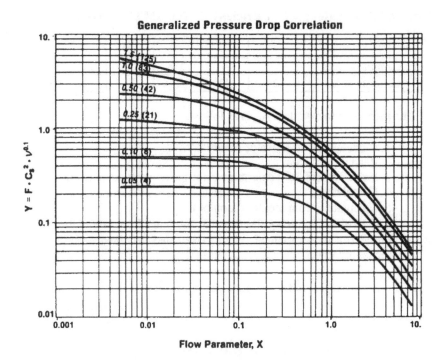

Figure 8.7.2-2 IMTP® packing pressure drop[42]

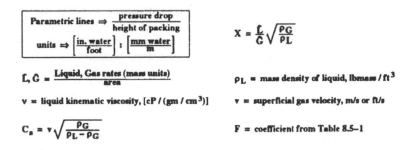

---

[42] Norton Chemical Process Products Corporation, Akron, OH.

Table 8.7.2-1 Values of coefficient F for IMTP® packings

IMTP® Size No.	15	25	40	50	70
F: C_s units = m/s	549	441	258	194	129
F: C_s units = ft/s	51	41	24	18	12

The procedure to calculate A_{xs} is as follows:

1. Calculate the abscissa of Figure 8.7.2-2.
2. Read the ordinate
3. From Table 8.7.2-1 determine F for the appropriate packing.
4. Calculate C_s
5. Calculate v from the definition of C_s
6. Calculate A_{xs} using the fact that the mass flow rate is the product of the superficial velocity, the cross-sectional area of the empty tower, and the mass density, so

$$A_{xs} = \frac{w_G}{v \, \rho_G}$$

(8.7.2-12)

8.7.3   Calculation of packing height

One approach to calculation of tower height using mass transfer coefficients (as opposed to use of equilibrium stages modified by a stage efficiency) is via H_G, H_L, H_{OG}, or H_{OL}; the procedures are similar for each of the four definitions. We will use H_G as an example. We must furnish sufficient interfacial area in the tower for the required amount of mass to be transferred with the existing driving force. For a given interfacial area per unit volume and a given tower cross-section, this is equivalent to specifying the height of the packing, since

$$A_{interfacial} = a \, A_{xs} \, Z$$

(8.7.3-1)

Consider a material balance for component A **on the gas phase only** for a disk of some arbitrary thickness Δz as shown in Figure 8.7-1. At steady state with no chemical reaction, mass is transferred into and out of the gas phase in this disk by convection, and through the interface by a mechanism that can be modeled using a mass transfer coefficient.

Output of A from gas phase by convection
  + Output of A from gas phase by mass transfer across gas/liquid interface[43]
  = Input of A to gas phase by convection

$$\left(A_{xs}\, G\, y_A\right)\Big|_{z+\Delta z} + k_y\, a\, A_{xs}\, \Delta z\left(y_A - y_{As}\right) = \left(A_{xs}\, G\, y_A\right)\Big|_{z} \qquad (8.7.3\text{-}2)$$

Rearranging

$$\frac{\left(G\, y_A\right)\Big|_{z+\Delta z} - \left(G\, y_A\right)\Big|_{z}}{\Delta z} = -k_y\, a\left(y_A - y_{As}\right) \qquad (8.7.3\text{-}3)$$

Taking the limit as $\Delta z \to 0$

$$\frac{d\left(G\, y_A\right)}{dz} = -k_y\, a\left(y_A - y_{As}\right) \qquad (8.7.3\text{-}4)$$

But since G is constant under our assumptions

$$-\frac{dy_A}{dz} = \frac{k_y\, a}{G}\left(y_A - y_{As}\right) = \frac{\left(y_A - y_{As}\right)}{\mathcal{H}_G} \qquad (8.7.3\text{-}5)$$

We observe that $\mathcal{H}_G$ is proportional to G, L, and Sc, all of which are constant for cases we consider here. We can write, therefore, since $\mathcal{H}_G$ is a constant

$$\int_0^z dz = -\mathcal{H}_G \int_{y_{A0}}^{y_{A1}} \frac{dy_A}{\left(y_A - y_{As}\right)} = \mathcal{H}_G \int_{y_{A1}}^{y_{A0}} \frac{dy_A}{\left(y_A - y_{As}\right)} \qquad (8.7.3\text{-}6)$$

$$Z = \mathcal{H}_G \int_{y_{A1}}^{y_{A0}} \frac{dy_A}{\left(y_A - y_{As}\right)} = \mathcal{H}_G\, n_G \qquad (8.7.3\text{-}7)$$

---

[43] We are using a gas absorber as our model system: however, the same equation will serve for a desorber, since the sign of the driving force, $(y_A - y_{As})$, will change and thus make the term an input (positive if moved to the right-hand side) instead of an output.

where Z is the tower packing height and the integral defines n_G, the *number of gas phase transfer units*.[44]

The integral may be evaluated graphically[45] by calculating $\mathcal{H}_G$ and $\mathcal{H}_L$, observing that

$$\frac{\mathcal{H}_G}{\mathcal{H}_L} = \frac{G}{k_y a} \frac{k_x a}{L} = \frac{k_x}{k_y} \frac{G}{L} \tag{8.7.3-8}$$

But we know L and G, so

$$\frac{k_x}{k_y} = \frac{L}{G} \frac{\mathcal{H}_G}{\mathcal{H}_L} \tag{8.7.3-9}$$

Selecting some arbitrary values of y_A, we can determine corresponding values of y_{AS} as outlined earlier. We then can plot $1/(y_A - y_{AS})$ versus y_A and determine the area under the curve between y_{A1} and y_{A0}. The value of this integral is the number of gas phase transfer units, n_G. We then multiply n_G by $\mathcal{H}_G$ to obtain the height of the tower. (Note that n_G is dimensionless.)

We can do a similar derivation using other combinations of height and number of transfer units to obtain equations of the form of Equation (8.7.3-2) for gas absorption, as shown in Table 8.7.3-1. For example, if we had used $\mathcal{H}_L$, the corresponding integral would define n_L; $\mathcal{H}_{OG}$, n_{OG}; $\mathcal{H}_{OL}$, n_{OL}. The latter two are called the number of overall gas and liquid transfer units, respectively. We can apply these same formulae to the case of **desorption**.

---

[44] Notice that this equation also will serve for the case of desorption, where the concentrated liquid enters and the concentrated gas leaves at the top of the tower. In this case $y_{AS} > y_A$, so the denominator is negative; however, $y_{A1} > y_{A0}$, so writing

$$z = \mathcal{H}_G \int_{y_{A1}}^{y_{A0}} \left[\frac{dy_A}{(y_A - y_{AS})} \right] = \mathcal{H}_G \int_{y_{A0}}^{y_{A1}} \left[\frac{-dy_A}{(y_A - y_{AS})} \right] = \mathcal{H}_G n_G$$

demonstrates that n_G remains a positive quantity.

[45] We can also sometimes more conveniently perform the integral analytically, as is shown in Example 8.7-2, or numerically.

Table 8.7.3-1 $\mathcal{H}$, n integrals

Basis	Height of tower
$\mathcal{H}_G$, n_G	$Z = \mathcal{H}_G \displaystyle\int_{y_{A1}}^{y_{A0}} \frac{dy_A}{\left(y_A - y_{AS}\right)} = \mathcal{H}_G\, n_G$
$\mathcal{H}_{OG}$, n_{OG}	$Z = \mathcal{H}_{OG} \displaystyle\int_{y_{A1}}^{y_{A0}} \frac{dy_A}{\left(y_A - y_A^*\right)} = \mathcal{H}_{OG}\, n_{OG}$
$\mathcal{H}_L$, n_L	$Z = \mathcal{H}_L \displaystyle\int_{x_{A1}}^{x_{A0}} \frac{dx_A}{\left(x_{As} - x_A\right)} = \mathcal{H}_L\, n_L$
$\mathcal{H}_{OL}$, n_{OL}	$Z = \mathcal{H}_{OL} \displaystyle\int_{x_{A1}}^{x_{A0}} \frac{dx_A}{\left(x_A^* - x_A\right)} = \mathcal{H}_{OL}\, n_{OL}$

8.7.4   Applications

Example 8.7.4-1   Analytical calculation of interfacial concentration[46]

A 1-ft inside diameter, 12-ft tall countercurrent gas absorption column operating at 1 atm and packed with 1-in. Raschig rings is absorbing methane into water from air. The ratio of water flow rate to air flow rate (both in moles/hr) is 20.

Previous experiments on the column have shown $\mathcal{H}_L$ = 10 ft and $\mathcal{H}_G$ = 4 ft. The Henry's law constant at the temperature of operation is 500 atm/(mole fraction).

A liquid sample withdrawn at a point 4 ft from the bottom of the column upon analysis shows a mole fraction of methane equal to 2×10^{-4}. A gas sample withdrawn at the same time at the same level shows a mole fraction of methane equal to 0.5.

---

[46] See Example 8.4-1 for the graphical method of determining interfacial concentration.

Calculate the concentration in the gas phase at the liquid/gas interface at this point.

Solution

The equilibrium curve may be determined by

$$p_A = y_A p = H x_A \tag{8.7.4-1}$$
$$y_A = \frac{H}{p} x_A$$

$$= \frac{(500) \text{ atm}}{(1) \text{ atm}} x_A$$

$$= 500 x_A \tag{8.7.4-2}$$

We need to find the intersection of the equilibrium curve with a line drawn through $(x_A, y_A) = (2 \times 10^{-4}, 0.5)$ having slope $- k_x/k_y$. To find this slope we use the definitions of the heights of the transfer units

$$\mathcal{H}_G = \frac{G}{k_y a} \tag{8.7.4-3}$$

$$\mathcal{H}_L = \frac{L}{k_x a} \tag{8.7.4-4}$$

$$\frac{k_x}{k_y} = -\frac{\left[\frac{L}{\mathcal{H}_L a}\right]}{\left[\frac{G}{\mathcal{H}_G a}\right]} = -\frac{L}{G}\frac{\mathcal{H}_G}{\mathcal{H}_L} = -\frac{L}{G}\frac{(4)\text{ ft}}{(10)\text{ ft}} \tag{8.7.4-5}$$

But

$$\frac{(\mathcal{W}_L)\frac{\text{moles}}{\text{hr}}}{(\mathcal{W}_G)\frac{\text{moles}}{\text{hr}}} = \frac{(L)\frac{\text{moles}}{\text{hr ft}^2}(A_{xs})\text{ ft}^2}{(G)\frac{\text{moles}}{\text{hr ft}^2}(A_{xs})\text{ ft}^2} = \frac{L}{G} = 20 \tag{8.7.4-6}$$

Therefore

$$-\frac{k_x}{k_y} = -20\frac{4}{10} = -8 = \frac{(y_A - y_{As})}{(x_A - x_{As})} \tag{8.7.4-7}$$

We need the intersection of the following two straight lines

$$y_A = 500 \, x_A \tag{8.7.4-8}$$

$$\frac{\left(y_A - 0.5\right)}{\left(x_A - 2 \times 10^{-4}\right)} = -8 \tag{8.7.4-9}$$

Substituting the first equation into the second

$$\frac{\left(500 \, x_A - 0.5\right)}{\left(x_A - 2 \times 10^{-4}\right)} = -8 \tag{8.7.4-10}$$

$$500 \, x_A - 0.5 = -8 \left(x_A - 2 \times 10^{-4}\right) = -8 \, x_A + 1.6 \times 10^{-3} \tag{8.7.4-11}$$

$$x_A = 9.87 \times 10^{-4} \tag{8.7.4-12}$$

Using the first equation

$$y_A = 500 \, x_A = \left(500\right) \left(9.87 \times 10^{-4}\right)$$
$$y_A = 0.494 \tag{8.7.4-13}$$

which is, as it must be, the same result as would have been found from the graphical method.

Example 8.7.4-2   Analytical determination of number of transfer units: straight operating and equilibrium lines

Until this point we have evaluated the integral which determines the number of transfer units by graphical means. If we can obtain analytical equations which describe the operating and equilibrium lines, and if these equations are sufficiently well behaved, we can perform the integration analytically. (If they are not sufficiently well behaved, we could, of course, integrate the equations numerically, which is equivalent to doing them graphically.)

We illustrate in the following example the development of an explicit equation for calculation of the number of transfer units for a particularly simple case: for both a straight operating line and a straight equilibrium line.

The operating line will sometimes approach straight-line behavior in practical cases because in very dilute systems, which are generally the conditions we treat in this text, little enough mass is transferred that L and G will remain essentially constant throughout the column; therefore, the operating line will be straight.

If, in addition, the solutions with which we deal are sufficiently ideal, we know that an equilibrium relationship of the form of Henry's law or Raoult's law will hold; these are straight-line equilibrium relationships. This may or may not be the case even in dilute solutions depending on the interaction of the solute and solvent molecules.

We now wish to develop an analytical equation for the number of transfer units that is explicit in the terminal conditions for the specific case of applying the number of overall gas transfer units, n_{OG}, to straight operating and equilibrium lines. (We could equally well choose to apply n_{OL}, n_G, or n_L.)

Solution

Our objective is to develop an analytical expression for the integral in the equation for packing height

$$Z = \mathcal{H}_{OG} \int_{y_{A1}}^{y_{A0}} \frac{dy_A}{\left(y_A - y_A^*\right)} = \mathcal{H}_{OG}\, n_{OG} \tag{8.7.4-14}$$

We do this by changing the variable of integration from y_A to $(y_A - y_A^*)$

$$n_{OG} = \int_{y_{A1}}^{y_{A0}} \frac{1}{y_A - y_A^*}\, dy_A$$

$$= \int_{y_{A1}-y_{A1}^*}^{y_{A0}-y_{A0}^*} \frac{1}{\left(y_A - y_A^*\right)} \frac{dy_A}{d\left(y_A - y_A^*\right)}\, d\left(y_A - y_A^*\right) \tag{8.7.4-15}$$

Now y_A (obviously) is linear in y_A; in addition, y_A^* is linear in x_A (the operating line) and x_A is linear in y_A (by assumption). Hence $(y_A - y_A^*)$ is the difference between two linear functions of y_A and therefore itself a linear function of y_A. To illustrate this fact:

a) y_A is linear in y_A

$$y_A = m_1 y_A + b_1 \qquad (8.7.4\text{-}16)$$

where

$$m_1 = 1$$
$$b_1 = 0 \qquad (8.7.4\text{-}17)$$

which is the trivial relationship

$$y_A = y_A \qquad (8.7.4\text{-}18)$$

b) Furthermore, y_A^* is linear in x_A (the equilibrium relationship)

$$y_A^* = m_2 x_A + b_2 \qquad (8.7.4\text{-}19)$$
$$b_2 = 0 \qquad (8.7.4\text{-}20)$$

or

$$\frac{y_A^*}{m_2} = x_A \qquad (8.7.4\text{-}21)$$

c) But x_A is linear in y_A (the operating line)

$$y_A = m_3 x_A + b_3 \qquad (8.7.4\text{-}22)$$

so

$$y_A = m_3 \frac{y_A^*}{m_2} + b_3 \qquad (8.7.4\text{-}23)$$

$$y_A^* = \frac{m_2}{m_3}\left(y_A - b_3\right) \qquad (8.7.4\text{-}24)$$

Defining

$$m_4 = \frac{m_2}{m_3} \qquad (8.7.4\text{-}25)$$

$$b_4 = \frac{m_2}{m_3} b_3 \qquad (8.7.4\text{-}26)$$

$$y_A^* = m_4 y_A + b_4 \qquad (8.7.4\text{-}27)$$

d) Taking the difference, and defining $m_5 = (1 - m_4)$

$$
\begin{aligned}
y_A - y_A^* &= y_A - m_4 y_A + b_4 \\
&= (1 - m_4) y_A + b_4 \\
&= m_5 y_A + b_4
\end{aligned}
\tag{8.7.4-28}
$$

Thus

$$
\frac{dy_A}{d(y_A - y_A^*)} = \frac{1}{m_5} = \text{constant} = \frac{\Delta y_A}{\Delta(y_A - y_A^*)}
\tag{8.7.4-29}
$$

$$
\frac{\Delta y_A}{\Delta(y_A - y_A^*)} = \frac{(y_{A0} - y_{A1})}{(y_{A0} - y_{A0}^*) - (y_{A1} - y_{A1}^*)}
\tag{8.7.4-30}
$$

Substituting in the integral

$$
\int_{y_{A1}}^{y_{A0}} \frac{1}{(y_A - y_A^*)} \, dy_A
$$

$$
= \int_{(y_{A1} - y_{A1}^*)}^{(y_{A0} - y_{A0}^*)} \frac{1}{(y_A - y_A^*)} \left[\frac{(y_{A0} - y_{A1})}{(y_{A0} - y_{A0}^*) - (y_{A1} - y_{A1}^*)} \right] d(y_A - y_A^*)
\tag{8.7.4-31}
$$

$$
= \left[\frac{(y_{A0} - y_{A1})}{(y_{A0} - y_{A0}^*) - (y_{A1} - y_{A1}^*)} \right] \int_{(y_{A1} - y_{A1}^*)}^{(y_{A0} - y_{A0}^*)} \frac{1}{(y_A - y_A^*)} d(y_A - y_A^*)
\tag{8.7.4-32}
$$

$$
= \left[\frac{(y_{A0} - y_{A1})}{(y_{A0} - y_{A0}^*) - (y_{A1} - y_{A1}^*)} \right] \ln \frac{(y_{A0} - y_{A0}^*)}{(y_{A1} - y_{A1}^*)}
\tag{8.7.4-33}
$$

$$
\int_{y_{A1}}^{y_{A0}} \frac{1}{(y_A - y_A^*)} \, dy_A = \left[\frac{(y_{A0} - y_{A1})}{(y_A - y_A^*)_{\text{log mean}}} \right] = n_{OG}
\tag{8.7.4-34}
$$

There is another approach to the above derivation; rather than substituting for the independent variable in the integration, substitute for y_A in terms of the equilibrium relationship and the operating line equation. This procedure, although it (as it must) leads to the same results, is somewhat more cumbersome from an algebraic point of view.[47]

This is only one example of a number of cases which can be in integrated analytically. For example, there are many theoretical thermodynamically based equations describing curved equilibrium lines which lead to expressions that can be evaluated analytically. We also can have instances in which the operating line is curved and can yet be integrated analytically. Some curved equilibrium data or operating lines can also be satisfactorily modeled over the entire particular range of operation with a simple functional form (e.g., a polynomial), which can then be integrated.

Our purpose here is not to provide an exhaustive tabulation of all the possible analytical solutions, but rather to point out the fact that one should not go to great effort integrating numerically (except, perhaps, in writing computer models to handle very general curved operating and equilibrium lines) without first checking to see if simpler analytical integration is possible in some form. Availability of general user-friendly computer models and the requisite hardware is becoming the method of choice in these calculations; however, the analytical approach will probably always have its uses.

Example 8.7.4-3   Effect of change of L/G on outlet composition

We have thus far considered the problem only of column **design**; that is, how one constructs a **new** column to perform a given separation. Perhaps the more frequent problem which concerns the engineer in a production or technical service function in industry is the effect of a changed set of operating conditions on performance of an **existing** column. This problem arises, for example, in cases where an existing column is to be converted for use in production of a different product, or the same product at a different rate.

In this example, we attempt to give some insight into the effect this has on the calculations. We also illustrate the applicability of our equation for the number of transfer units to desorption.

---

[47] Colburn, A. P. (1934). *Trans. AIChE* **35**: 211.

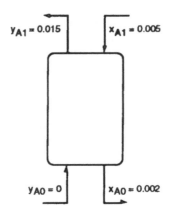

Assume that we have a packed column of constant cross-section being used as shown above to desorb component A into a stream of carrier gas from low concentration in a liquid. By utilizing a mole balance[48] on A around the entire column, we can calculate the liquid-to-gas molar flow ratio at which the column is operating

$$\text{Input} = \text{Output} \tag{8.7.4-35}$$

$$x_{A1}\,L + y_{A0}\,G = x_{A0}\,L + y_{A1}\,G \tag{8.7.4-36}$$

$$y_{A1} = x_{A1}\,\frac{L}{G} - x_{A0}\,\frac{L}{G} + y_{A0} \tag{8.7.4-37}$$

$$0.015 = 0.005\,\frac{L}{G} - 0.002\,\frac{L}{G} + 0 \tag{8.7.4-38}$$

$$\left(\frac{L}{G}\right) = 5 \tag{8.7.4-39}$$

We will model the interface as at equilibrium. Assume that equilibrium data for the system can be modeled by the straight-line relationship

$$y_A = 10\,x_A \tag{8.7.4-40}$$

Assume that the column operating condition is now changed to

$$\left(\frac{L}{G}\right) = 4 \tag{8.7.4-41}$$

---

[48] Here, since there is no chemical reaction, moles are conserved and a mole balance is equivalent to a mass balance.

with the same input rate of liquid, and the same liquid and gas inlet concentrations. What are the new outlet concentrations? Assume that the change in gas rate is not sufficient to alter $\mathcal{H}_G$ or $\mathcal{H}_L$ significantly.

Solution

If the heights of the liquid and gas transfer units are not changed significantly, and if the height of the packing remains the same, the number of transfer units should remain the same. We can calculate the number of transfer units under the original conditions from the given information

$$n_{OG} = \frac{(y_{A0} - y_{A1})}{(y_A - y_A^*)_{\text{log mean}}} \tag{8.7.4-42}$$

$$n_{OG} = \frac{(y_{A0} - y_{A1})}{\dfrac{(y_A - y_A^*)_0 - (y_A - y_A^*)_1}{\ln\left\{\dfrac{(y_A - y_A^*)_0}{(y_A - y_A^*)_1}\right\}}} \tag{8.7.4-43}$$

Substituting known values

$$n_{OG} = \frac{(0 - 0.015)}{\dfrac{(0 - 10[0.002]) - (0.015 - 10[0.005])}{\ln\left\{\dfrac{(0 - 10[0.002])}{(0.015 - 10[0.005])}\right\}}}$$

$$= -\ln\left[\frac{0.02}{0.035}\right] \tag{8.7.4-44}$$

$$n_{OG} = 0.56 \tag{8.7.4-45}$$

For the modified column operation

$$y_{A1} = x_{A1}\frac{L}{G} - x_{A0}\frac{L}{G} + y_{A0} \tag{8.7.4-46}$$

$$y_{A1} = 0.005\,(4) - x_{A0}\,(4) + 0 \tag{8.7.4-47}$$

$$y_{A1} = 4\left(0.005 - x_{A0}\right) \tag{8.7.4-48}$$

This relationship must be satisfied simultaneously with

$$0.56 = \cfrac{\left(0 - y_{A1}\right)}{\cfrac{\left(0 - 10\,x_{A0}\right) - \left(y_{A1} - 10\left[0.005\right]\right)}{\ln\left\{\cfrac{\left(0 - 10\,x_{A0}\right)}{\left(y_{A1} - 10\left[0.005\right]\right)}\right\}}} \tag{8.7.4-49}$$

$$0.56 = \cfrac{-y_{A1}}{\cfrac{\left(-10\,x_{A0}\right) - \left(y_{A1} - 0.05\right)}{\ln\left\{\cfrac{\left(-10\,x_{A0}\right)}{\left(y_{A1} - 0.05\right)}\right\}}} \tag{8.7.4-50}$$

Substituting the former relationship into the latter

$$0.56 = \cfrac{-4\left(0.005 - x_{A0}\right)}{\cfrac{\left(-10\,x_{A0}\right) - \left(4\left[0.005 - x_{A0}\right] - 0.05\right)}{\ln\left\{\cfrac{\left(-10\,x_{A0}\right)}{\left(4\left[0.005 - x_{A0}\right] - 0.05\right)}\right\}}} \tag{8.7.4-51}$$

$$0.56 = -\frac{\left(0.02 - 4\,x_{A0}\right)}{\left(0.03 - 6\,x_{A0}\right)}\ln\left\{\frac{10\,x_{A0}}{4\,x_{A0} + 0.03}\right\} \tag{8.7.4-52}$$

$$f\left(x_{A0}\right) = \frac{\left(0.02 - 4\,x_{A0}\right)}{\left(0.03 - 6\,x_{A0}\right)}\ln\left\{\frac{10\,x_{A0}}{4\,x_{A0} + 0.03}\right\} + 0.56 = 0 \tag{8.7.4-53}$$

This can readily be solved by trial, bracketing the root using a spreadsheet program. We know that the root is not too far from the solution for the original operating conditions, because the change in L/G was only about 20%. Since L/G has decreased, we are using more gas per unit of liquid; therefore, the concentration of the exiting liquid should go down. If we make the original condition the starting estimate, we have

x_{A0}	$f(x_{A0})$
0.002	0.13209741
0.001	-0.2558503
0.0015	-0.0236458
0.0016	0.0120133
0.00155	-0.0054794
0.00157	0.00159606
0.00156	-0.0019284
0.001565	-0.0001629

where we have carried out the solution to more significant figures than justified to show the speed of convergence. Using the new exit liquid concentration

$$x_{A0} = 0.00157 \tag{8.7.4-54}$$

$$y_{A1} = x_{A1}\frac{L}{G} - x_{A0}\frac{L}{G} + y_{A0} \tag{8.7.4-55}$$

$$y_{A1} = (0.005)(4) - (0.00157)(4) + 0 \tag{8.7.4-56}$$

$$y_{A1} = 0.0137 \tag{8.7.4-57}$$

So the new exit concentrations will be

$$x_{A0} = 0.00157 \tag{8.7.4-58}$$

$$y_{A1} = 0.0137 \tag{8.7.4-59}$$

As is illustrated by this procedure, calculation of the performance of an existing column under a new set of operating conditions is somewhat more complicated than that of designing a new column because an iterative procedure may be involved.

In general, if we consider the problem of changing L/G on an existing tower we also must account for the change in height of a transfer unit, which complicates the problem even further.

Tower operation can also be changed in such a way as to change the equilibrium relationship, for example, in decreasing the tower pressure - perhaps because of corrosion with age. (One would seldom be able to increase the tower pressure except in the case of extreme inadvertent overdesign of the initial unit.) The purpose here is to apply the mass transfer principles discussed earlier rather than equilibrium considerations, so calculation of such a modification is outside the scope of this text, although certainly within the scope of the models discussed.

Example 8.7.4-4   Design of absorber

We need to recover by absorption 90% of the C_3 hydrocarbons present in a vapor stream under a pressure of 100 psia. We have available an absorber oil that is free of the C_3 fraction and approximately a kerosene in composition.

The absorber must treat 11,000 mol/hr of entering gas that is 0.005 mol fraction C_3.

Although in general the selection of type and size of packing is part of the design problem, assume for purposes of this illustration that No. 40 random dumped IMTP® packing is to be used.

Design for 0.25 in water pressure drop per foot of packing depth. This also is a number subject to considerable uncertainty because of our limited knowledge about flooding. This value, however, lies within the rule-of-thumb range of values cited in Barrer, R. M. (1941). *Diffusion In and Through Solids*. New York, NY, The Macmillan Company.

Design the diameter and height of packing for an absorber to perform this separation assuming that

$$\left(\frac{L}{G}\right)_{act} = 1.5 \left(\frac{L}{G}\right)_{min}$$

(8.7.4-60)

Since the C_3 hydrocarbons are a relatively minor constituent of the input stream, we may apply our models for dilute systems.

Let A designate the C_3 fraction. Equilibrium data may be modeled by

$$y_{As} = 1.5 \, x_{As}$$

(8.7.4-61)

Property	Numerical value
Absorption oil density, lbm/ft^3	52
Gas phase density, lbm/ft^3	0.67
Effective molecular weight, oil	135
Effective molecular weight, gas	16.8
Oil viscosity, cP	20
Gas phase viscosity, cP	0.007
Oil phase diffusivity, cm^2/s	0.5×10^{-5}
Gas phase diffusivity, cm^2/s	0.14

Solution

The 90% recovery of C_3 implies the part not recovered is 10%, so

$$\frac{G\,y_{A1}}{G\,y_{A0}} = 0.10 \tag{8.7.4-62}$$

$$\frac{y_{A1}}{0.005} = 0.10 \tag{8.7.4-63}$$

$$y_{A1} = 5 \times 10^{-4} \tag{8.7.4-64}$$

The situation is shown in the following illustration:

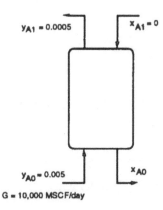

$y_{A1} = 0.0005$ $x_{A1} = 0$

$y_{A0} = 0.005$ x_{A0}

$G = 10,000$ MSCF/day

We have the equation of the equilibrium line

$$y_A = 1.5\,x_A \tag{8.7.4-65}$$

and we also know that the operating line must go through the point located by the compositions of the passing streams at the top (dilute end) of the column, both of which we know

$$\left(x_{A1}, y_{A1}\right) = \left(0, 0.0005\right) \tag{8.7.4-66}$$

We further know the gas concentration at the bottom (concentrated end) of the column to be

$$y_{A0} = 0.005 \tag{8.7.4-67}$$

and, therefore, we know that the point on the operating line representing the passing streams at the bottom of the column must lie along this y-coordinate. For the minimum L/G with straight operating and equilibrium lines, we know this point will lie at the intersection of this y-value and the equilibrium line.

Substituting the y-value into the equilibrium line equation we have

$$0.005 = 1.5 \, x_A \tag{8.7.4-68}$$
$$x_A = 0.00333 \tag{8.7.4-69}$$

Therefore, the minimum L/G is

$$\begin{aligned}
\left(\frac{L}{G}\right)_{min} &= [\text{slope}]_{min} = \left[\frac{\Delta y_A}{\Delta x_A}\right]_{min} \\
&= \left[\frac{y_{A0} - y_{A1}}{x_{A0} - x_{A1}}\right]_{min} = \frac{0.005 - 0.0005}{0.00333 - 0} = 1.35
\end{aligned} \tag{8.7.4-70}$$

The same slope may be found graphically by finding the slope of a straight line through

$$\left(x_{A1}, y_{A1}\right) = \left(0, 0.0005\right) \tag{8.7.4-71}$$

and the point of intersection of the equilibrium curve with the ordinate value

$$y_{A0} = 0.005 \tag{8.7.4-72}$$

as shown in the following figure.

We next can determine

$$\left(\frac{L}{G}\right)_{act} = 1.5 \left(\frac{L}{G}\right)_{min} = 1.5 \left(1.35\right) = 2.03 \tag{8.7.4-73}$$

This determines the actual operating line to be

$$y_A = 2.03 \, x_A + 0.0005 \tag{8.7.4-74}$$

We can find the actual exit concentration of the liquid by solving the equation of this line with the equation of the line

$$y_A = 0.005 \qquad (8.7.4\text{-}75)$$

which gives

$$x_{A0} = 0.00222 \qquad (8.7.4\text{-}76)$$

Once again the solution can also be done graphically: by drawing a line with slope 2.03 through the point

$$\left(x_{A1}, y_{A1}\right) = \left(0, 0.0005\right) \qquad (8.7.4\text{-}77)$$

as shown in the figure.

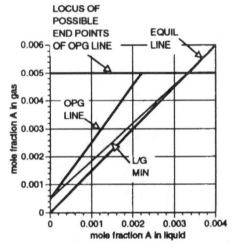

The number of transfer units is obtained from

$$n_{OG} = \frac{\left(y_{A0} - y_{A1}\right)}{\left(y_A - y_A^{*}\right)_{\text{log mean}}} \qquad (8.7.4\text{-}78)$$

$$n_{OG} = \frac{\left(y_{A0} - y_{A1}\right)}{\dfrac{\left(y_A - y_A^{*}\right)_0 - \left(y_A - y_A^{*}\right)_1}{\ln\left\{\dfrac{\left(y_A - y_A^{*}\right)_0}{\left(y_A - y_A^{*}\right)_1}\right\}}} \qquad (8.7.4\text{-}79)$$

Substituting known values

$$n_{OG} = \cfrac{\left(0.005 - 0.0005\right)}{\cfrac{\left(0.005 - 1.5\left[0.00222\right]\right) - \left(0.0005 - 1.5\left[0\right]\right)}{\ln\left\{\cfrac{\left(0.005 - 1.5\left[0.00222\right]\right)}{\left(0.0005 - 1.5\left[0\right]\right)}\right\}}}$$

$$= 3.85 \ln\left[3.34\right] \tag{8.7.4-80}$$

$$n_{OG} = 4.64 \tag{8.7.4-81}$$

We now determine $\mathcal{H}_{OG}$.

$$\mathcal{H}_{OG} = \frac{G}{K_y a} \tag{8.7.4-82}$$

$$\mathcal{H}_{OG} = \cfrac{G}{\left[\cfrac{1}{\left(\cfrac{1}{k_y} + \cfrac{m}{k_x}\right)}\right]a} = \cfrac{G}{\left[\cfrac{1}{\left(\cfrac{1}{k_y a} + \cfrac{m}{k_x a}\right)}\right]} \tag{8.7.4-83}$$

$$\mathcal{H}_{OG} = \cfrac{1}{\left[\cfrac{1}{\left(\cfrac{G}{k_y a} + m\cfrac{G}{L}\cfrac{L}{k_x a}\right)}\right]} = \left(\cfrac{G}{k_y a} + m\cfrac{G}{L}\cfrac{L}{k_x a}\right) \tag{8.7.4-84}$$

$$\mathcal{H}_{OG} = \mathcal{H}_G + m\frac{G}{L}\mathcal{H}_L \tag{8.7.4-85}$$

In order to calculate $\mathcal{H}_G$ and $\mathcal{H}_L$ we need information about $\hat{G}$ and $\hat{L}$. To obtain this information we now size the tower diameter. Calculating the abscissa of the pressure drop chart, Figure 8.7.2-2

$$\frac{\hat{L}}{\hat{G}} = \frac{L M_L}{G M_G} = 2.03 \frac{135\left(\frac{lbm}{mol}\right)_L}{16.8\left(\frac{lbm}{mol}\right)_G} = 16.3 \tag{8.7.4-86}$$

$$\sqrt{\frac{\rho_G}{\rho_L}} = \left[\frac{0.67 \frac{\text{lbmass}}{\text{ft}^3}}{52 \frac{\text{lbmass}}{\text{ft}^3}}\right]^{1/2} = 0.114 \tag{8.7.4-87}$$

$$X = \frac{L}{G}\sqrt{\frac{\rho_G}{\rho_L}} = (16.3)(0.114) = 1.85 \tag{8.7.4-88}$$

Reading the corresponding ordinate value for a pressure drop of 0.25 in. water/ft packing depth

$$Y = 0.17 = F C_s^2 \left(\frac{\mu_L}{\rho_L}\right)^{0.1} \tag{8.7.4-89}$$

From Table 8.7.2-1 using No. 40 IMTP® packing

$$F = 24 \tag{8.7.4-90}$$

$$v = \frac{Q}{A_{xs}} = 11,000 \frac{\text{mol}}{\text{hr}} \, 16.8 \frac{\text{lbm}}{\text{mol}} \, 0.67 \frac{\text{ft}^3}{\text{lbm}} \, \frac{1}{3600} \frac{\text{hr}}{\text{s}} \, \frac{1}{A_{xs} \, \text{ft}^2}$$

$$v = \frac{34.4}{A_{xs}} \tag{8.7.4-91}$$

$$C_s = v\sqrt{\frac{\rho_G}{\rho_L - \rho_G}} = \frac{34.4}{A_{xs}}\sqrt{\frac{0.67}{52 - 0.67}}$$

$$C_s = \frac{3.93}{A_{xs}} \tag{8.7.4-92}$$

$$\left(\frac{\mu_L}{\rho_L}\right)^{0.1} = \left(\frac{20 \text{ cP}}{52 \frac{\text{lbmass}}{\text{ft}^3} \, \frac{454 \text{ gm}}{\text{lbmass}} \, \frac{\text{ft}^3}{2.83 \times 10^4 \text{ cm}^3}}\right)^{0.1} = 1.37 \tag{8.7.4-93}$$

Substituting in the equation for Y

$$0.17 = F C_s^2 \left(\frac{\mu_L}{\rho_L}\right)^{0.1} = (24)\left(\frac{3.93}{A_{xs}}\right)^2 (1.37) \tag{8.7.4-94}$$

$$A_{xs} = 54.7 \text{ ft}^2 = \pi \frac{D^2}{4} \tag{8.7.4-95}$$

$$D = 8.34 \text{ ft} \tag{8.7.4-96}$$

We now determine $\hat{G}$ and $\hat{L}$.

$$G = \frac{11{,}000 \frac{mol}{hr} 16.8 \frac{lbm}{mol}}{54.7 \text{ ft}^2} = 3378 \frac{lbm}{hr \text{ ft}^2} \qquad (8.7.4\text{-}97)$$

$$L = \frac{11{,}000 \frac{mol \; gas}{hr} 2.03 \frac{mol \; liquid}{mol \; gas} 135 \frac{lbmass}{mol \; liquid}}{54.7 \text{ ft}^3}$$

$$= 55{,}110 \frac{lbmass}{hr \text{ ft}^2} \qquad (8.7.4\text{-}98)$$

In order ensure that we have good mass transfer data for $\mathcal{H}_G$ and $\mathcal{H}_L$, we decide to take data in our own laboratory. We request that the lab use the same packing that we intend for the tower and run at the above values of $\hat{G}$, and $\hat{L}$. The laboratory replies that they can accommodate these requests,[49] but their equipment is not explosion proof and, therefore, they can run only air/water with ammonia as the transferring component.

Using a dilute air/ammonia/water system at room temperature with the same packing, $\hat{G}$, and $\hat{L}$, the laboratory determines

$$\left[\mathcal{H}_G\right]_{lab} = 4 \text{ ft} \qquad\qquad (8.7.4\text{-}99)$$

$$\left[\mathcal{H}_L\right]_{lab} = 2 \text{ ft} \qquad\qquad (8.7.4\text{-}100)$$

We now adjust $\mathcal{H}_L$ and $\mathcal{H}_G$ from the laboratory data to our conditions.

We use the following values for the laboratory data

---

[49] Notice that the laboratory probably does not have a 7-8 ft diameter column to use for the experiment. If they run in a smaller diameter column, it is necessary that the packing diameter/column diameter ratio not be too large, e.g., perhaps not greater than 1/20, or the column will tend to have too large an area next to the wall, where porosity is less, compared to the total column cross-section, leading to a different flow pattern because of the short-circuiting next to the wall. We ignore such questions for the purpose of this example.

Property	Numerical value
Water density, lbm/ft^3	62.4
Air phase density, lbm/ft^3	0.0806
Water phase viscosity, cP	1.0
Air phase viscosity, cP	0.018
Water phase diffusivity, cm^2/s	1.3 x 10^{-5}
Air phase diffusivity, cm^2/s	0.23

$\mathcal{H}_G$ [Equation (8.4.1-43)] should vary only with the Schmidt number of the gas phase, since the remaining constants are functions of packing type, $\hat{G}$, and $\hat{L}$, which were maintained constant between the lab data and the proposed design.

$$\frac{[\mathcal{H}_G]_{act}}{[\mathcal{H}_G]_{lab}} = \frac{\left[\alpha\, G^\beta\, L^\gamma\, Sc_G^{0.5}\right]_{act}}{\left[\alpha\, G^\beta\, L^\gamma\, Sc_G^{0.5}\right]_{lab}} = \frac{\left[Sc_G^{0.5}\right]_{act}}{\left[Sc_G^{0.5}\right]_{lab}} = \frac{\left[\left(\frac{\mu}{\rho\, \mathcal{D}_{AB}}\right)_G^{0.5}\right]_{act}}{\left[\left(\frac{\mu}{\rho\, \mathcal{D}_{AB}}\right)_G^{0.5}\right]_{lab}} \quad (8.7.4\text{-}101)$$

$$[\mathcal{H}_G]_{act} = [\mathcal{H}_G]_{lab} \frac{\left[\left(\frac{\mu}{\rho\, \mathcal{D}_{AB}}\right)_G^{0.5}\right]_{act}}{\left[\left(\frac{\mu}{\rho\, \mathcal{D}_{AB}}\right)_G^{0.5}\right]_{lab}} = [4]\frac{\left[\frac{0.007}{(0.67)(0.14)}\right]^{0.5}}{\left[\frac{0.018}{(0.0806)(0.23)}\right]^{0.5}} \quad (8.7.4\text{-}102)$$

$$[\mathcal{H}_G]_{act} = 1.11 \text{ ft} \quad (8.7.4\text{-}103)$$

$\mathcal{H}_L$ [Equation (8.4.1-44)] should vary only with the Schmidt number of the liquid phase, since the remaining constants are functions of packing type and $\hat{L}$, which were maintained constant between the lab data and the proposed design.

$$\frac{[\mathcal{H}_L]_{act}}{[\mathcal{H}_L]_{lab}} = \frac{\left[\varphi\left(\frac{L}{\mu}\right)^\eta Sc_L^{0.5}\right]_{act}}{\left[\varphi\left(\frac{L}{\mu}\right)^\eta Sc_L^{0.5}\right]_{lab}} = \frac{\left[Sc_L^{0.5}\right]_{act}}{\left[Sc_L^{0.5}\right]_{lab}} = \frac{\left[\left(\frac{\mu}{\rho\, \mathcal{D}_{AB}}\right)_L^{0.5}\right]_{act}}{\left[\left(\frac{\mu}{\rho\, \mathcal{D}_{AB}}\right)_L^{0.5}\right]_{lab}} \quad (8.7.4\text{-}104)$$

$$\left[\mathcal{H}_L\right]_{act} = \left[\mathcal{H}_L\right]_{lab} \frac{\left[\left(\frac{\mu}{\rho \, \mathcal{D}_{AB}}\right)_L^{0.5}\right]_{act}}{\left[\left(\frac{\mu}{\rho \, \mathcal{D}_{AB}}\right)_L^{0.5}\right]_{lab}} = [2] \frac{\left[\dfrac{20}{(52)\left(0.5 \times 10^{-5}\right)}\right]^{0.5}}{\left[\dfrac{1.0}{(62.4)\left(1.3 \times 10^{-5}\right)}\right]^{0.5}}$$

(8.7.4-105)

$$\left[\mathcal{H}_L\right]_{act} = 15.8 \text{ ft}$$

(8.7.4-106)

We now can calculate $\mathcal{H}_{OG}$

$$\mathcal{H}_{OG} = \mathcal{H}_G + m \frac{G}{L} \mathcal{H}_L$$

(8.7.4-107)

$$\mathcal{H}_{OG} = \left(1.11\right) \text{ft} + \left(1.5\right)\left(\frac{1}{2.03}\right)\left(15.8\right) \text{ft}$$

$$\mathcal{H}_{OG} = 12.8 \text{ ft}$$

(8.7.4-108)

Therefore, the column height is

$$Z = \mathcal{H}_{OG} \, n_{OG} = \left(12.8\right)\left(4.64\right) = 59.3 \text{ ft}$$

(8.7.4-109)

This height, of course, is the depth of the packing and does not include space for a liquid feed pipe(s) and distributor at the top or room at the bottom for gas and liquid to disengage. Because of the weight of the packing, it is also sometimes necessary to furnish intermediate support plates. Bed limiting plates are also sometimes necessary at the top of the packing to restrict packing being carried out the top of the unit or into a position to plug piping. Intermediate liquid redistribution is also sometimes required.

Example 8.7.4-5   Economic optimization of an absorber

The problem of optimizing absorber design is one of minimizing an objective (cost) function which includes all costs associated with the unit. The interaction of all the many components of total product cost makes true optimal design a matter for a large computer. In economics, however, as with other types of problems, we try to isolate limiting steps and concentrate our attention on them so as to reduce the number of variables to be handled.

Total product cost is usually considered to be composed of

- General expense
- Manufacturing cost

Process design engineers usually have influence over only part of the manufacturing cost segment of total product cost. General expenses such as marketing, research, and administration are usually determined by decisions at a different (usually higher) level in the organization.

Manufacturing cost includes

- Direct cost
- Fixed charges
- Plant overhead

Process design probably has its most substantial effect on the raw material aspect (via yield) of direct costs and the depreciation component of fixed charges.[50]

For purposes of illustration, we consider, using data from the previous example, optimizing the relative costs of

- recovering propane by absorption
- unrecovered propane and its disposal.

Assume that unrecovered propane and its disposal will entail a net cost of $0.01/lbmass, while the installed cost of the absorption tower is about $10,000 per foot, including capitalized cost of absorbing liquid and other utilities.[51]

---

[50] Of course, the design of a process indirectly influences many other costs; however, for pedagogical reasons we wish to keep our objective function simple. For a detailed look at plant design and process economics, see Peters, M. S. and K. D. Timmerhaus (1991). *Plant Design and Economics for Chemical Engineers*. New York, McGraw-Hill, Inc.

[51] This figure is purely for illustrative purposes. Economic data change with time because of inflation, exchange rates, competitive situation, etc. In any study other than very rough approximation (sometimes used for crude screening of multiple processes or because of deadline pressure), one would consider separately the cost of the shell, the packing, and auxiliaries, including column configuration and material of construction. Also, it is not customary to capitalize the cost of the absorbing oil and other utilities. The point of this example is to emphasize that design involves optimization, and a simple objective function was desirable. Many texts on process economics are available, some of which include design; one which combines the (continued on following page)

Assuming that $\mathcal{H}_{OG}$ will not change drastically, that the life of the equipment is 10 years with no salvage value, and that straight-line depreciation is used, what recovery of propane should we specify, as opposed to the 90 percent recovery arbitrarily chosen in Example 8.7.4-4? Use 340 operating days per year and L/G = 1.5 (L/G)$_{min}$.

Solution

Our purpose is to minimize the following annual cost

$$\text{Annual cost} = C_{annual} = C_{propane} + C_{depr} \qquad (8.7.4\text{-}110)$$

Defining

$$F \equiv \text{Fraction of propane recovered}$$

$$= \frac{\left(w_A\right)_{\text{gas phase, in}} - \left(w_A\right)_{\text{gas phase, out}}}{\left(w_A\right)_{\text{gas phase, in}}} \qquad (8.7.4\text{-}111)$$

The cost of lost propane is

$$C_{propane} = (11{,}000)\left[\frac{\text{total lbmol}}{\text{hr}}\right](0.005)\left[\frac{\text{lbmol propane}}{\text{total lbmol}}\right]$$
$$\times (44.1)\left[\frac{\text{lbmass propane}}{\text{lbmol propane}}\right](24)\left[\frac{\text{hr}}{\text{day}}\right](340)\left[\frac{\text{days}}{\text{yr}}\right]$$
$$\times (1-F)(0.01)\left[\frac{\$}{\text{lbmass propane}}\right] \qquad (8.7.4\text{-}112)$$

$$C_{propane} = \frac{\$197{,}900\,(1-F)}{\text{yr}} \qquad (8.7.4\text{-}113)$$

The (straight line) depreciation cost[52] is

---

economics with a more detailed look at packed tower design is Peters, M. S. and K. D. Timmerhaus (1991). *Plant Design and Economics for Chemical Engineers*. New York, McGraw-Hill, Inc., p. 768.

[52] This is not the true out-of-pocket cost, but relates to tax accounting. If the reader understands this distinction, he/she already has knowledge of process economics beyond the scope of this text and should skip this oversimplified example.

$$C_{depr} = \frac{\text{installed cost} - \text{salvage value}}{\text{service life}}$$

$$= \frac{(Z)\,[\text{ft}]\,(10{,}000)\left[\frac{\$}{\text{ft}}\right] - (0)}{(10)\,[\text{yr}]} \tag{8.7.4-114}$$

$$C_{depr} = \frac{\$1000\,(Z)}{\text{yr}} \tag{8.7.4-115}$$

This gives the annual cost as

$$C_{annual} = \frac{\$197{,}900\,(1-F)}{\text{yr}} + \frac{(\$1000)\,(Z)}{\text{yr}} \tag{8.7.4-116}$$

$$C_{annual} = \frac{\$197{,}900\,(1-F)}{\text{yr}} + \frac{(\$1000)\,\left(H_{OG}\,n_{OG}\right)}{\text{yr}} \tag{8.7.4-117}$$

$$C_{annual} = \frac{\$197{,}900\,(1-F)}{\text{yr}} + \frac{(\$1000)\,\left(12.8\,n_{OG}\right)}{\text{yr}} \tag{8.7.4-118}$$

$$C_{annual} = 197{,}900\,(1-F) + (12{,}800)\,n_{OG} \tag{8.7.4-119}$$

To minimize this function with respect to F, we must relate n_{OG} and F. We can do this via y_{A1}, since both n_{OG} and F are functions of y_{A1}.

From the definition of recovery

$$F = \frac{\left(\frac{W_A}{M_A}\right)_{\text{gas phase, in}} - \left(\frac{W_A}{M_A}\right)_{\text{gas phase, out}}}{\left(\frac{W_A}{M_A}\right)_{\text{gas phase, in}}} \tag{8.7.4-120}$$

$$= \frac{\mathcal{W}_G\,y_{A0} - \mathcal{W}_G\,y_{A1}}{\mathcal{W}_G\,y_{A0}} \tag{8.7.4-121}$$

$$F = \frac{y_{A0} - y_{A1}}{y_{A0}} = \frac{0.005 - y_{A1}}{0.005} \tag{8.7.4-122}$$

We also know that

$$n_{OG} = \frac{\left(y_{A0} - y_{A1}\right)}{\left(y_A - y_A^*\right)_{\text{log mean}}}$$

$$n_{OG} = \frac{\left(y_{A0} - y_{A1}\right)}{\dfrac{\left(y_{A0} - y_{A0}^{*}\right) - \left(y_{A1} - y_{A1}^{*}\right)}{\ln\left\{\dfrac{\left(y_{A0} - y_{A0}^{*}\right)}{\left(y_{A1} - y_{s1}^{*}\right)}\right\}}} \qquad (8.7.4\text{-}123)$$

We have the equilibrium relationship

$$y_A^{*} = m\,x_A = 1.5\,x_A \qquad (8.7.4\text{-}124)$$

We know that y_{A1}^{*} is zero since x_{A1} is zero; we replace y_{A0}^{*} in terms of x_{A0}

$$n_{OG} = \frac{\left(y_{A0} - y_{A1}\right)}{\dfrac{\left(y_{A0} - m\,x_{A0}\right) - \left(y_{A1}\right)}{\ln\left\{\dfrac{\left(y_{A0} - m\,x_{A0}\right)}{\left(y_{A1}\right)}\right\}}} = \frac{\left(0.005 - y_{A1}\right)}{\dfrac{\left(0.005 - 1.5\,x_{A0}\right) - \left(y_{A1}\right)}{\ln\left\{\dfrac{\left(0.005 - 1.5\,x_{A0}\right)}{\left(y_{A1}\right)}\right\}}} \qquad (8.7.4\text{-}125)$$

We can use the operating line in the following form to find x_{A0}

$$\frac{L}{G} = \frac{y_{A0} - y_{A1}}{x_{A0} - x_{A1}} = \frac{y_{A0} - y_{A1}}{x_{A0}} \qquad (8.7.4\text{-}126)$$

$$x_{A0} = \frac{G}{L}\left(y_{A0} - y_{A1}\right) = \frac{G}{L}\left(0.005 - y_{A1}\right) \qquad (8.7.4\text{-}127)$$

G/L can be determined as a function of y_{A1}. This is done in the usual manner: by first determining the minimum L/G and then applying the appropriate factor, here,[53] 1.5.

The minimum L/G is determined by using the fact that maximum value of x_{A0} is the concentration that would be in equilibrium with y_{A0} and again noting

$$x_{A1} = 0 \qquad (8.7.4\text{-}128)$$

Then

---

[53] The choice in this example of 1.5 both as the constant m in the equilibrium relationship and as the multiple of the minimum L/G is unfortunate; this 1.5 is not m, but the ratio of actual to minimum L/G.

$$
\begin{aligned}
\left(\frac{L}{G}\right)_{min} &= \left[\text{slope}\right]_{min} = \left[\frac{\Delta y_A}{\Delta x_A}\right]_{min} \\
&= \left[\frac{y_{A0} - y_{A1}}{x_{A0} - x_{A1}}\right]_{min} = \left[\frac{y_{A0} - y_{A1}}{\frac{y_{A0}}{1.5}}\right]_{min} \\
&= \left[\frac{0.005 - y_{A1}}{\frac{0.005}{1.5}}\right] = \left[\frac{0.005 - y_{A1}}{0.00333}\right]
\end{aligned}
\tag{8.7.4-129}
$$

and

$$
\left(\frac{L}{G}\right)_{act} = 1.5 \left(\frac{L}{G}\right)_{min}
\tag{8.7.4-130}
$$

The set of equations above could be solved analytically to yield annual cost as a function of F alone, and then the optimum F obtained by differentiating and setting the derivative equal to zero. A simpler method is to use a spreadsheet program, assuming a variety of values of y_{A1}, and solve for the annual cost.

1. Calculate $(L/G)_{min}$ from equation (i) using the assumed y_{A1}
2. Calculate $(L/G)_{act}$ using equation (j)
3. Calculate x_{A0} from equation (h)
4. Calculate n_{OG} from equation (g)
5. Calculate F from equation (f)
6. Calculate C_{annual} from equation (e)
7. Plot C_{annual} vs. F

The following table shows the result of such a calculation.

y_{A1}	$(L/G)_{min}$	$(L/G)_{act}$	x_{A0}	n_{OG}	F	C_{annual}
1.00E-06	1.501	2.252	2.22E-03	22.23	1.000	$284,535
2.00E-06	1.501	2.251	2.22E-03	20.16	1.000	$258,103
5.00E-06	1.500	2.250	2.22E-03	17.43	0.999	$223,346
7.50E-06	1.499	2.249	2.22E-03	16.23	0.999	$208,083
1.00E-05	1.498	2.248	2.22E-03	15.38	0.998	$197,321
2.00E-05	1.495	2.243	2.22E-03	13.35	0.996	$171,735
5.00E-05	1.486	2.230	2.22E-03	10.72	0.990	$139,203
7.50E-05	1.479	2.218	2.22E-03	9.58	0.985	$125,614
1.00E-04	1.471	2.207	2.22E-03	8.79	0.980	$116,431
2.00E-04	1.441	2.162	2.22E-03	6.93	0.960	$96,618
5.00E-04	1.351	2.027	2.22E-03	4.64	0.900	$79,161
7.50E-04	1.276	1.914	2.22E-03	3.70	0.850	$77,019
1.00E-03	1.201	1.802	2.22E-03	3.06	0.800	$78,769
2.00E-03	0.901	1.351	2.22E-03	1.64	0.600	$100,143
4.50E-03	0.150	0.225	2.22E-03	0.18	0.100	$180,352

The results are plotted in the following figure.

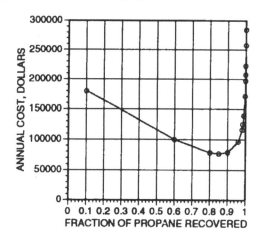

We see that our recovery of 0.9 was not too far from the optimum (minimum cost) design point - certainly probably adequate within the accuracy to which the data would be known. Also note the sharply rising costs as fractional recoveries closer and closer to 1.0 are used, requiring rapidly escalating capital investment in equipment, the depreciation of which dominates the total annual cost. This is a general result: costs increase dramatically as components must be virtually completely recovered - for example, for environmental reasons.

8.8 Mass Transfer with Chemical Reaction

Many applications of mass transfer are found in conjunction with chemical reactions. In such situations, either the mass transfer rate or the chemical reaction rate can be the controlling factor, but frequently neither is completely dominant. In this section we consider briefly the effects of chemical reaction on the apparent mass transfer rate. In general, treatment of simultaneous mass transfer and chemical reaction is quite complicated, and so we will attempt only to give a qualitative insight into such effects.

As an example of how chemical reaction can affect mass transfer rates, consider the situation shown in Figure 8.8-1.

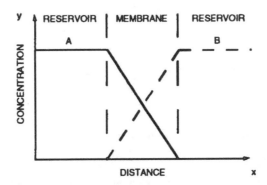

Figure 8.8-1 Diffusion in a membrane

On the left-hand side of a porous membrane infinite in the y- and z-direction, we maintain a large reservoir of pure component A. On the right-hand side of the membrane, we maintain a corresponding reservoir, but of pure B. We assume both these reservoirs to be perfectly stirred, so no concentration gradients can appear in the reservoirs, and that A and B have the same diffusivities.

Steady-state concentration profiles in the porous membrane will look the same as if the membrane were a stagnant film of fluid. We will (for one-dimensional diffusion) get concentration profiles which appear somewhat as shown. A will continuously diffuse through the membrane in one direction, and B will diffuse in the opposite direction. Since the cross-sectional area normal to the diffusion path remains constant, and we are assuming that densities, diffusion coefficients, etc., do not vary, the profiles will be linear.

Now let us consider the case where A and B react according to the stoichiometry

$$A + B \rightarrow AB \tag{8.8-1}$$

and let us assume that this reaction is instantaneous and irreversible. This means that as soon as a molecules of A and B "see" one another, they immediately react.

The concentration profiles in this instance will look something like those shown in Figure 8.8-2.

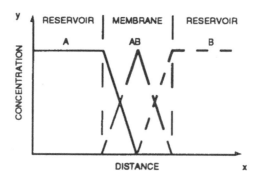

Figure 8.8-2 Diffusion with instantaneous irreversible reaction in a membrane

A will still diffuse into the porous membrane and B will diffuse from the opposite side, but product AB is formed immediately whenever A and B encounter each other. This product must now diffuse out of the porous membrane into the bulk fluids.

A diffuses from left to right, B diffuses from right to left, and the product formed by their reaction diffuses both to the left and to the right to the respective bulk solution. The important thing to notice is that the net effect of the chemical reaction has been to steepen the concentration gradient for both component A and component B. Since everything else remains constant, this means that for this case, diffusion rates (mass transfer rates) will be increased in the presence of the chemical reaction.

It is possible to solve the case analytically where component A is in the form of a solid flat plate and component B is the fluid flowing past this plate, with rate of reaction very fast compared to mass transfer rate. We can compare the Sherwood number for this situation with the Sherwood number obtained from our correlation for mass transfer with flow over a flat plate.

For the case where the Schmidt number of component A and the Schmidt number of component B are identical, the solution leads to a mean Sherwood number of[54]

$$\text{Sh} = \left(0.664\right)\left(1 + \frac{c_B}{c_{As}}\right)\left(\text{Sc}_A\right)^{\frac{1}{3}}\left(\text{Re}_L\right)^{\frac{1}{2}} \tag{8.8-2}$$

where c_{As} is the concentration of A at the interface in moles per unit volume of the fluid, and c_B is the freestream concentration of B in the same units.

Notice that this is the mean Sherwood number for laminar flow over a flat plate without reaction, multiplied by $(1 + c_B/c_{As})$. Since this multiplying factor is always greater than 1 because the concentration ratio cannot be negative, for this case chemical reaction **always** speeds up mass transfer. The treatment can be extended to differing Schmidt numbers, different stoichiometry, etc.

Example 8.8-1   Acidization of an oil well

To increase the porosity of an oil-bearing formation and thus the crude oil production rate of wells, several stimulation techniques are used. In limestone reservoirs, which usually contain natural fractures, one technique is to pump acid into the formation to leach the limestone and thus increase porosity.

Assume 0.2 mole fraction aqueous hydrochloric acid is pumped into a fractured limestone formation, and further assume the fractures are short and wide so that the model for flow developing over a flat plate is a reasonable model. The velocity in the fractures is estimated to be about 1 ft/s. Discuss the mass transfer.

Solution

To include the effect of stoichiometry of the essentially instantaneous chemical reaction, Equation (8.8-2) can be modified as follows[55]

---

[54] We omit the development. See Friedlander, S. K. and M. Litt (1958). Diffusion controlled reaction in a laminar boundary layer. *Chem. Eng. Sci.* **7**: 229.

[55] Friedlander, S. K. and M. Litt (1958). Diffusion controlled reaction in a laminar boundary layer. *Chem. Eng. Sci.* **7**: 229. This solution assumes equal Schmidt numbers for A and B, which is not correct, but for the purposes of this illustration the difference is unimportant.

$$Sh = (0.664)\left(1 + \frac{a\,c_B}{b\,c_{As}}\right)(Sc_A)^{\frac{1}{3}}(Re_L)^{\frac{1}{2}}$$

$$(8.8-3)$$

where a and b are the stoichiometric coefficients in the reaction

$$a\,A + b\,B \rightarrow products$$

$$(8.8-4)$$

Here our reaction is

$$CaCO_3 + 2\,HCl \rightarrow CaCl_2 + H_2O + CO_2$$

$$(8.8-5)$$

so $a = 1$, $b = 2$.

We next evaluate the factor $\left(1 + a\,c_B/b\,c_{As}\right)$. The solubility of $CaCO_3$ in water is about 4×10^{-6} mole fraction. We can relate mole fraction and molar density by

$$c_i = x_i\,c$$

$$(8.8-6)$$

where x is mole fraction and c is total molar density If we assume the total molar density to be the same at the interface as under freestream conditions, we can write

$$1 + \frac{a\,c_B}{b\,c_{As}} = 1 + \frac{a\,x_B\,c}{b\,x_{As}\,c} = 1 + \frac{a\,x_B}{b\,x_{As}}$$

$$1 + \frac{a\,c_B}{b\,c_{As}} = 1 + \frac{(1)\,(0.2)}{(2)\,\left(4 \times 10^{-6}\right)} = 2.5 \times 10^4$$

$$(8.8-7)$$

This tremendous enhancement of the mass transfer rate is logical, because in the absence of the reaction we would simply be trying to dissolve $CaCO_3$ in water. Since the solubility is so low, there would be an extremely small concentration gradient (between surface and freestream conditions) to drive the mass transfer. With chemical reaction, the driving force is still low, but it is applied over a much shorter distance, since the reaction takes place very close to the surface. The gradient, therefore, is much larger.

The elementary Example 8.8-1, selected from the vast class of problems which involve simultaneous mass transfer and chemical reaction, assumes instantaneous irreversible chemical reaction, which implies that A and B are never simultaneously present in solution at the same point. With slower reaction or with reversible reaction, one would not have an abrupt plane in which the chemical reaction takes place; rather, this plane would expand to a zone in space as indicated in Figure 8.8-3 for the aforementioned case of a porous membrane. Change in the relative diffusivities of A and B will move this zone toward one side of the membrane or the other. The stoichiometric coefficients of A and B in the chemical reaction also affect the profiles.

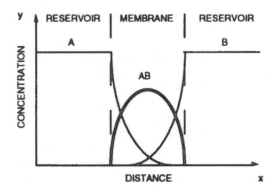

Figure 8.8-3 Mass transfer with slow or reversible chemical reaction

If chemical reaction is not instantaneous, we must model the rate in order to make quantitative predictions about mass transfer. In general, for a reversible chemical reaction of the form

$$\sigma_1 s_1 + \sigma_2 s_2 + \sigma_3 s_3 + \cdots + \sigma_R s_R \leftrightarrow$$
$$\sigma_{R+1} s_{R+1} + \sigma_{R+2} s_{R+2} + \sigma_{R+3} s_{R+3} + \cdots + \sigma_S s_S$$

$$\sum_{q=1}^{R} \sigma_q s_q \leftrightarrow \sum_{q=R+1}^{S} \sigma_q s_q \tag{8.8-8}$$

where

- σ_q is the stoichiometric coefficient of the qth species
- s_q is the molecular formula of the qth species
- R is the number of reactant species (**not** the rate of reaction)
- S is the total number of species

The rate is usually modeled with an expression of the form

$$\text{Rate} = \left[k_f c_{s_1}^{n_1} c_{s_2}^{n_2} c_{s_3}^{n_3} \cdots c_{s_R}^{n_R} \right] - \left[k_r c_{s_{R+1}}^{n_{R+1}} c_{s_{R+2}}^{n_{R+2}} c_{s_{R+3}}^{n_{R+3}} \cdots c_{s_S}^{n_S} \right]$$

(8.8-9)

where

> • k_f and k_r are the forward and reverse rate constants, respectively
> • c is concentration (although in a strict thermodynamic sense should be chemical potential)

Chemical reactions can be classified into the broad categories of homogeneous and heterogeneous.

> • **Homogeneous** reactions occur throughout a control volume. For example, in the neutralization of a tank of hydrochloric acid with sodium hydroxide, the reaction occurs throughout the fluid in the tank. This gives rise to a generation term **in the differential equation(s)** [species mass balances] describing the process.
>
> Since the effect of homogeneous reactions appears as a generation term in the differential equation, the rate constant is usually selected to give rates with units of mol/(time volume). We designate homogeneous rate constants as $k^{\sim}$ to remind ourselves of this volume proportionality. For example, for an irreversible second-order reaction (for this example we have chosen a reaction order that follows the stoichiometry - a situation not true, in general)
>
> $$2\,A \rightarrow C + 3D$$
>
> $$R_A \left[\frac{\text{mol}}{t\,L^3} \right] = k^{\sim} \left[\frac{L^3}{t\,\text{mol}} \right] c_A^2 \left[\frac{\text{mol}}{L^3} \right]^2 \qquad (8.8\text{-}10)$$

• **Heterogeneous** reactions, on the other hand, occur external to the control volume. For example, in many catalytic reactions the catalyst is a solid and reaction occurs on the solid surface. Mass transfer occurs in the fluid phase adjoining the surface, and reaction occurs on the surface.[56] The effect of reaction, therefore, is incorporated in the **boundary conditions** of the differential equation(s) rather than the equation(s) itself (themselves).

Since the effect of heterogeneous reactions appears in the boundary conditions, the rate constant is usually selected to give rates with units of mol/(time area). We designate heterogeneous rate constants as $k^{''}$ to remind ourselves that the proportionality is to area. Using as an example the same second-order irreversible reaction, we define the rate constant as

$$N_A \left[\frac{mol}{t\,L^2} \right] = k^{''} \left[\frac{L^4}{t\,mol} \right] c_A^2 \left[\frac{mol}{L^3} \right]^2 \qquad (8.8\text{-}11)$$

Note that the units of the rate constant depend both on the order of the reaction and on the classification of the reaction as homogeneous or heterogeneous.

Finally, most reactions either absorb or give off energy, and the mass transfer problem with chemical reaction thereby, in addition, can be coupled to the heat transfer problem.

In general, the reader should remember that mass transfer will usually be enhanced by the presence of chemical reaction because of the effect of steepening of concentration gradients. Chemical reaction also furnishes a further benefit in operations such as absorption of a component of a gas phase into a liquid phase, in that the chemical reaction permits the liquid phase to assimilate more of the transferring component than would be possible by solubility alone.

Example 8.8-2   Mass transfer with heterogeneous reaction

Consider a membrane coating a catalytic surface as shown in the following figure, with the indicated z-coordinate ordinate and with concentration plotted along the abscissa. Assume that a solution of A is flowing past the top surface of the membrane fast enough to sweep away any B as it reaches the freestream. We ignore the effect of any solvent carrying A since the solvent does not move in the z-direction at steady state.

---

[56] It is, of course, also possible to have concentration gradients on the surface and two-dimensional diffusion. We exclude such cases from the example.

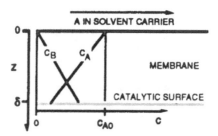

Let the reaction occurring at the catalytic surface be

$$2\,A \rightarrow B$$

(8.8-12)

and the reaction is irreversible and first order in A so the rate at the catalyst surface can be described as

$$N_{Az}\left[\frac{mol}{t\,L^2}\right] = k''\left[\frac{L}{t}\right]c_A\left[\frac{mol\,A}{L^3}\right] = k''\left[\frac{L}{t}\right]c\left[\frac{tot\,mol}{L^3}\right]x_A\,\frac{mol\,A}{tot\,mol}$$

(8.8-13)

Develop an expression for the flux of A at steady state.

Solution

Since there is no generation within the membrane and the area normal to flow remains constant, the microscopic species mass balance shows that

$$\frac{dN_A}{dz} = 0$$

(8.8-14)

The constitutive relationship is Fick's law in the form

$$N_{Az} - x_A\left(N_{Az} + N_{Bz}\right) = -c\,\mathcal{D}_{AB}\,\frac{dx_A}{dz}$$

(8.8-15)

However, the stoichiometry shows us that

$$N_{Az} = -2\,N_{Bz}$$

(8.8-16)

so

$$N_{Az} - x_A \left(N_{Az} - \frac{1}{2} N_{Az} \right) = -c \, \mathcal{D}_{AB} \frac{dx_A}{dz} \tag{8.8-17}$$

$$N_{Az} \left(1 - \frac{1}{2} x_A \right) = -c \, \mathcal{D}_{AB} \frac{dx_A}{dz} \tag{8.8-18}$$

$$N_{Az} = \frac{-c \, \mathcal{D}_{AB}}{\left(1 - \frac{1}{2} x_A \right)} \frac{dx_A}{dz} \tag{8.8-19}$$

Substituting in the microscopic species mass balance gives the differential equation

$$-\frac{d}{dz} \left[\frac{c \, \mathcal{D}_{AB}}{\left(1 - \frac{1}{2} x_A \right)} \frac{dx_A}{dz} \right] = 0 \tag{8.8-20}$$

$$\frac{d}{dz} \left[\frac{1}{\left(1 - \frac{1}{2} x_A \right)} \frac{dx_A}{dz} \right] = 0 \tag{8.8-21}$$

with boundary conditions

$$z = 0: \quad x_A = x_{A0}$$
$$z = \delta: \quad x_A = \frac{N_{Az}}{c \, k} \tag{8.8-22}$$

Integrating once

$$\int d \left[\frac{1}{\left(1 - \frac{1}{2} x_A \right)} \frac{dx_A}{dz} \right] = \int 0 \, dz \tag{8.8-23}$$

$$\frac{1}{\left(1 - \frac{1}{2} x_A \right)} \frac{dx_A}{dz} = C_1 \tag{8.8-24}$$

and a second time

$$\int \frac{dx_A}{\left(1 - \frac{1}{2} x_A\right)} = \int C_1 \, dz \tag{8.8-25}$$

$$-2 \ln \left(1 - \frac{1}{2} x_A\right) = C_1 z + C_2 \tag{8.8-26}$$

Introducing the boundary conditions

$$z = 0: \quad x_A = x_{A0} \tag{8.8-27}$$

$$-2 \ln \left(1 - \frac{1}{2} x_{A0}\right) = C_2 \tag{8.8-28}$$

$$z = \delta: \quad x_A = \frac{N_{Az}}{ck} \tag{8.8-29}$$

$$-2 \ln \left(1 - \frac{1}{2} \frac{N_{Az}}{ck}\right) = C_1 \delta - 2 \ln \left(1 - \frac{1}{2} x_{A0}\right) \tag{8.8-30}$$

$$C_1 = \frac{2 \ln \left(1 - \frac{1}{2} x_{A0}\right) - 2 \ln \left(1 - \frac{1}{2} \frac{N_{Az}}{ck}\right)}{\delta} \tag{8.8-31}$$

Substituting the values of the constants of integration into the solution

$$-2 \ln \left(1 - \frac{1}{2} x_A\right) = \frac{2 \ln \left(1 - \frac{1}{2} x_{A0}\right) - 2 \ln \left(1 - \frac{1}{2} \frac{N_{Az}}{ck}\right)}{\delta} z \\ - 2 \ln \left(1 - \frac{1}{2} x_{A0}\right) \tag{8.8-32}$$

$$-\ln \left(1 - \frac{1}{2} x_A\right) = \left[\ln \left(1 - \frac{1}{2} x_{A0}\right) - \ln \left(1 - \frac{1}{2} \frac{N_{Az}}{ck}\right)\right] \frac{z}{\delta} \\ - \ln \left(1 - \frac{1}{2} x_{A0}\right) \tag{8.8-33}$$

$$\ln \left(1 - \frac{1}{2} x_A\right) = \frac{z}{\delta} \ln \left(1 - \frac{1}{2} \frac{N_{Az}}{ck}\right) + \left(1 - \frac{z}{\delta}\right) \ln \left(1 - \frac{1}{2} x_{A0}\right) \tag{8.8-34}$$

$$\ln \left(1 - \frac{1}{2} x_A\right) = \ln \left(1 - \frac{1}{2} \frac{N_{Az}}{ck}\right)^{\frac{z}{\delta}} + \ln \left(1 - \frac{1}{2} x_{A0}\right)^{\left(1 - \frac{z}{\delta}\right)} \tag{8.8-35}$$

$$\ln\left(1-\frac{1}{2}x_A\right) = \ln\left[\left(1-\frac{1}{2}\frac{N_{Az}}{ck''}\right)^{\frac{z}{\delta}}\left(1-\frac{1}{2}x_{A0}\right)^{\left(1-\frac{z}{\delta}\right)}\right] \tag{8.8-36}$$

$$\left(1-\frac{1}{2}x_A\right) = \left(1-\frac{1}{2}\frac{N_{Az}}{ck''}\right)^{\frac{z}{\delta}}\left(1-\frac{1}{2}x_{A0}\right)^{\left(1-\frac{z}{\delta}\right)} \tag{8.8-37}$$

This is the mole fraction profile in the membrane. It is plotted in the following figure for the case of $x_{A0} = 1.0$.

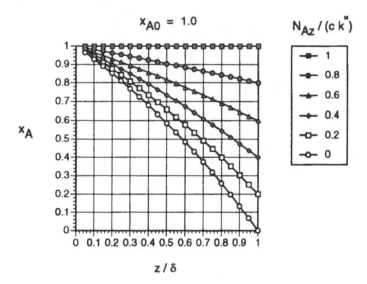

Before computing the flux, we can observe that for the case of instantaneous reaction (infinite reaction rate constant)

$$\left(1-\frac{1}{2}x_A\right) = \left(1-\frac{1}{2}x_{A0}\right)^{\left(1-\frac{z}{\delta}\right)} \tag{8.8-38}$$

We can compute the flux for the general case from the Equation (8.8-19) we obtained from Fick's Law

$$N_{Az} = \frac{-c\,\mathcal{D}_{AB}}{\left(1-\frac{1}{2}x_A\right)}\frac{dx_A}{dz} \tag{8.8-39}$$

The denominator of the first term on the right-hand side we get from Equation (8.8-37). From the same relation we need to develop the derivative in Equation (8.8-39). To simplify the amount of writing required, we define

$$z^{\bullet} = \frac{z}{\delta}$$

$$u = \left(1 - \frac{1}{2}\frac{N_{Az}}{c\,k_z}\right)$$

$$v = \left(1 - \frac{1}{2}x_{A0}\right)$$

(8.8-40)

which transforms Equation (8.8-37) into

$$\left(1 - \frac{1}{2}x_A\right) = u^{z^{\bullet}}\,v^{(1-z^{\bullet})}$$

(8.8-41)

Solving for x_A

$$x_A = 2\left[1 - u^{z^{\bullet}}\,v^{(1-z^{\bullet})}\right]$$

(8.8-42)

Applying the chain rule

$$\frac{dx_A}{dz} = \frac{dx_A}{dz^{\bullet}}\frac{dz^{\bullet}}{dz} = \frac{1}{\delta}\frac{dx_A}{dz^{\bullet}}$$

(8.8-43)

From Equation (8.8-42)

$$\frac{dx_A}{dz^{\bullet}} = -2\left[u^{z^{\bullet}}\,\frac{d\left(v^{(1-z^{\bullet})}\right)}{dz^{\bullet}} + v^{(1-z^{\bullet})}\frac{d\left(u^{z^{\bullet}}\right)}{dz^{\bullet}}\right]$$

(8.8-44)

Recalling that

$$\frac{d\left(a^x\right)}{dx} = a^x \ln\left[a\right]$$

(8.8-45)

$$\frac{dx_A}{dz^{\bullet}} = 2\left[u^{z^{\bullet}}\,v^{(1-z^{\bullet})}\ln\left(v\right) - v^{(1-z^{\bullet})}\,u^{z^{\bullet}}\ln\left(u\right)\right]$$

(8.8-46)

$$\frac{dx_A}{dz^{\bullet}} = 2\,u^{z^{\bullet}}\,v^{(1-z^{\bullet})}\ln\left(\frac{v}{u}\right)$$

(8.8-47)

Substituting Equation (8.8-47) in Equation (8.8-43), the result and Equation (8.8-41) in Equation (8.8-39) yields

$$N_{Az} = \frac{-c \, \mathcal{D}_{AB}}{u^{z^{\bullet}} \, v^{\left(1-z^{\bullet}\right)}} \, \frac{2 \, u^{z^{\bullet}} \, v^{\left(1-z^{\bullet}\right)}}{\delta} \, \ln\left(\frac{v}{u}\right) \qquad (8.8\text{-}48)$$

$$N_{Az} = \frac{2 \, c \, \mathcal{D}_{AB}}{\delta} \, \ln\left(\frac{u}{v}\right) \qquad (8.8\text{-}49)$$

Replacing u and v by their definitions

$$N_{Az} = \frac{2 \, c \, \mathcal{D}_{AB}}{\delta} \, \ln\left[\frac{1 - \frac{1}{2} \dfrac{N_{Az}}{c \, k^{*}}}{1 - \frac{1}{2} x_{A0}}\right] \qquad (8.8\text{-}50)$$

This relationship is not explicit in the flux, which appears on both sides of the equation. We can solve numerically for the flux given the parameters of the system.

Example 8.8-3   Mass transfer with homogeneous reaction

Consider the case shown in the following figure of a perfectly mixed carrier gas, G, containing a very dilute concentration, c_{A0}, of a component A, which is being absorbed into a film of thickness, δ, of an initially pure liquid B flowing on an impermeable surface.

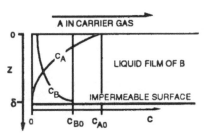

Within the film the following irreversible chemical reaction occurs

$$A + B \rightarrow C \qquad (8.8\text{-}51)$$

B may be considered to have negligible vapor pressure and G may be considered insoluble in B. The concentration of C is not sufficiently large to affect D_{AB} significantly.

Develop the expressions for the concentration profile and the flux of A at the film surface at steady state.

Solution

This is a case of homogeneous chemical reaction; therefore, the effect will appear as a generation term within the microscopic species mass balance.

$$\frac{dN_A}{dz} + k'' c_A = 0 \tag{8.8-52}$$

Because of its insolubility and the fact that it is perfectly mixed, carrier gas G will not move in the z-direction. Similarly, B will not move in the z-direction since it is not evaporating into G, although a concentration profile will develop in B because of the chemical reaction with A as A diffuses into the film.

The flux relationship (Fick's law)

$$N_{Az} - x_A \left(N_{Az} + N_{Bz} \right) = -\mathcal{D}_{AB} \frac{dc_A}{dz} \tag{8.8-53}$$

can be simplified by observing

$$N_{Bz} = 0 \tag{8.8-54}$$

which gives

$$N_{Az} \left(1 - x_A \right) = -\mathcal{D}_{AB} \frac{dc_A}{dz} \tag{8.8-55}$$

We assume the concentration of A to be sufficiently small that we may further model

$$\left(1 - x_A \right) \approx 1 \tag{8.8-56}$$

which leaves the flux model as

$$N_{Az} = -\mathcal{D}_{AB} \frac{dc_A}{dz} \tag{8.8-57}$$

Substituting the flux relationship into the microscopic species mass balance yields as our final process model

$$\frac{d\left[-\mathcal{D}_{AB} \frac{dc_A}{dz}\right]}{dz} + k'' c_A = 0 \tag{8.8-58}$$

$$-\mathcal{D}_{AB} \frac{d^2c_A}{dz^2} + k'' c_A = 0 \tag{8.8-59}$$

$$\frac{d^2c_A}{dz^2} - \frac{k''' c_A}{\mathcal{D}_{AB}} = 0 \tag{8.8-60}$$

which is a linear, second-order, ordinary differential equation with constant coefficients. The boundary conditions are

$$z = 0: \quad c_A = c_{A0}$$
$$z = \delta: \quad N_{Az} = 0, \text{ which implies that } \frac{dc_A}{dz} = 0 \tag{8.8-61}$$

To make the notation more compact we define

$$\alpha = \frac{k'''}{\mathcal{D}_{AB}} \tag{8.8-62}$$

The characteristic equation is then

$$\lambda^2 - \alpha = 0 \tag{8.8-63}$$

with roots

$$\lambda = \pm \sqrt{\alpha} \tag{8.8-64}$$

which gives as the solution

$$c_A = C_1 e^{\sqrt{\alpha} z} + C_2 e^{-\sqrt{\alpha} z} \tag{8.8-65}$$

Application of the first boundary conditions gives

$$c_{A0} = C_1 e^{\sqrt{\alpha} \, 0} + C_2 e^{-\sqrt{\alpha} \, 0} = C_1 + C_2 \tag{8.8-66}$$

and of the second

$$\frac{dc_A}{dz}\bigg|_{z=\delta} = 0 = \frac{d}{dz}\Big[C_1 e^{\sqrt{\alpha} z} + C_2 e^{-\sqrt{\alpha} z}\Big]\bigg|_{z=\delta} \tag{8.8-67}$$

$$\Big[C_1 \sqrt{\alpha} \, e^{\sqrt{\alpha} z} - C_2 \sqrt{\alpha} \, e^{-\sqrt{\alpha} z}\Big]\bigg|_{z=\delta} = C_1 \sqrt{\alpha} \, e^{\sqrt{\alpha} \delta} - C_2 \sqrt{\alpha} \, e^{-\sqrt{\alpha} \delta} = 0 \tag{8.8-68}$$

    Solving the equations

$$c_{A0} = C_1 + C_2 \tag{8.8-69}$$
$$0 = C_1 \sqrt{\alpha} \, e^{\sqrt{\alpha} \delta} - C_2 \sqrt{\alpha} \, e^{-\sqrt{\alpha} \delta} \tag{8.8-70}$$

simultaneously gives

$$C_1 = c_{A0}\left(1 - \frac{e^{\sqrt{\alpha} \delta}}{e^{\sqrt{\alpha} \delta} + e^{-\sqrt{\alpha} \delta}}\right) \tag{8.8-71}$$

$$C_2 = c_{A0}\left(\frac{e^{\sqrt{\alpha} \delta}}{e^{\sqrt{\alpha} \delta} + e^{-\sqrt{\alpha} \delta}}\right) \tag{8.8-72}$$

Using

$$\cosh(\zeta) = \frac{e^{\zeta} + e^{-\zeta}}{2} \tag{8.8-73}$$

yields

$$C_1 = c_{A0}\left(1 - \frac{e^{\sqrt{\alpha} \delta}}{2 \cosh(\sqrt{\alpha} \, \delta)}\right) \tag{8.8-74}$$

$$C_2 = c_{A0}\left(\frac{e^{\sqrt{\alpha} \delta}}{2 \cosh(\sqrt{\alpha} \, \delta)}\right) \tag{8.8-75}$$

    Substituting the values of the constants into the solution

$$c_A = c_{A0}\left(1 - \frac{e^{\sqrt{\alpha}\,\delta}}{2\cosh(\sqrt{\alpha}\,\delta)}\right)e^{\sqrt{\alpha}\,z} + c_{A0}\left(\frac{e^{\sqrt{\alpha}\,\delta}}{2\cosh(\sqrt{\alpha}\,\delta)}\right)e^{-\sqrt{\alpha}\,z} \qquad (8.8\text{-}76)$$

$$\frac{c_A}{c_{A0}} = e^{\sqrt{\alpha}\,z} - \frac{e^{\sqrt{\alpha}\,z}\,e^{\sqrt{\alpha}\,\delta}}{2\cosh(\sqrt{\alpha}\,\delta)} + \frac{e^{-\sqrt{\alpha}\,z}\,e^{\sqrt{\alpha}\,\delta}}{2\cosh(\sqrt{\alpha}\,\delta)} \qquad (8.8\text{-}77)$$

$$\frac{c_A}{c_{A0}} = \frac{1}{2\cosh(\sqrt{\alpha}\,\delta)}\left[e^{\sqrt{\alpha}\,z}\,2\cosh(\sqrt{\alpha}\,\delta) - e^{\sqrt{\alpha}\,z}\,e^{\sqrt{\alpha}\,\delta} + e^{-\sqrt{\alpha}\,z}\,e^{\sqrt{\alpha}\,\delta}\right]$$

$$(8.8\text{-}78)$$

$$\frac{c_A}{c_{A0}} = \frac{1}{2\cosh(\sqrt{\alpha}\,\delta)}\left[e^{\sqrt{\alpha}\,z}\left(e^{\sqrt{\alpha}\,\delta} + e^{\sqrt{\alpha}\,\delta}\right) - e^{\sqrt{\alpha}\,z}\,e^{\sqrt{\alpha}\,\delta} + e^{-\sqrt{\alpha}\,z}\,e^{\sqrt{\alpha}\,\delta}\right]$$

$$(8.8\text{-}79)$$

$$\frac{c_A}{c_{A0}} = \frac{1}{2\cosh(\sqrt{\alpha}\,\delta)}\left[e^{\sqrt{\alpha}\,z}\,e^{-\sqrt{\alpha}\,\delta} + e^{-\sqrt{\alpha}\,z}\,e^{\sqrt{\alpha}\,\delta}\right] \qquad (8.8\text{-}80)$$

$$\frac{c_A}{c_{A0}} = \frac{1}{2\cosh(\sqrt{\alpha}\,\delta)}\left[e^{\sqrt{\alpha}\,(\delta-z)} + e^{-\sqrt{\alpha}\,(\delta-z)}\right] \qquad (8.8\text{-}81)$$

$$\frac{c_A}{c_{A0}} = \frac{1}{2\cosh(\sqrt{\alpha}\,\delta)}\left[2\cosh\left[\sqrt{\alpha}\,(\delta-z)\right]\right] \qquad (8.8\text{-}82)$$

$$\frac{c_A}{c_{A0}} = \frac{\cosh\left[\sqrt{\alpha}\,(\delta-z)\right]}{\cosh(\sqrt{\alpha}\,\delta)} \qquad (8.8\text{-}83)$$

Substituting the value of α

$$\frac{c_A}{c_{A0}} = \frac{\cosh\left[\sqrt{\dfrac{k'''}{\mathcal{D}_{AB}}}\,(\delta-z)\right]}{\cosh\left[\sqrt{\dfrac{k'''}{\mathcal{D}_{AB}}}\,\delta\right]} \qquad (8.8\text{-}84)$$

$$\frac{c_A}{c_{A0}} = \frac{\cosh\left[\sqrt{\dfrac{k'''\,\delta^2}{\mathcal{D}_{AB}}}\,(\delta-z)\right]}{\cosh\left[\sqrt{\dfrac{k'''\,\delta^2}{\mathcal{D}_{AB}}}\,\delta\right]} \qquad (8.8\text{-}85)$$

Writing the z-coordinate in dimensionless form

$$\frac{c_A}{c_{A0}} = \frac{\cosh\left[\sqrt{\frac{k'' \delta^2}{\mathcal{D}_{AB}}}\left(1 - \frac{z}{\delta}\right)\right]}{\cosh\left[\sqrt{\frac{k'' \delta^2}{\mathcal{D}_{AB}}}\right]} \tag{8.8-86}$$

Defining

$$\beta = \sqrt{\frac{k'' \delta^2}{\mathcal{D}_{AB}}} \tag{8.8-87}$$

$$\frac{c_A}{c_{A0}} = \frac{\cosh\left[\beta\left(1 - \frac{z}{\delta}\right)\right]}{\cosh[\beta]} \tag{8.8-88}$$

A plot of the resultant profiles is shown in the following figure.

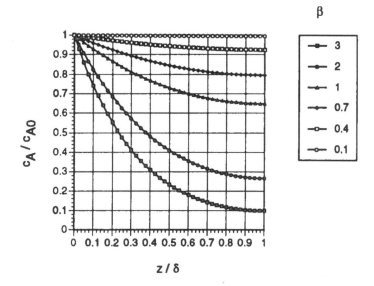

We obtain the flux at the surface by observing

$$N_{Az}\Big|_{z=0} = -\mathcal{D}_{AB} \frac{dc_A}{dz}\Big|_{z=0} = -\mathcal{D}_{AB} \frac{d}{dz}\left[c_{A0} \frac{\cosh\left[\beta\left(1-\frac{z}{\delta}\right)\right]}{\cosh[\beta]}\right]\Bigg|_{z=0}$$

(8.8-89)

$$N_{Az}\Big|_{z=0} = -\frac{\mathcal{D}_{AB}\,c_{A0}}{\cosh[\beta]} \frac{d}{dz}\left[\cosh\left[\beta\left(1-\frac{z}{\delta}\right)\right]\right]\Bigg|_{z=0}$$

(8.8-90)

$$N_{Az}\Big|_{z=0} = -\frac{\mathcal{D}_{AB}\,c_{A0}}{\cosh[\beta]} \sinh\left[\beta\left(1-\frac{z}{\delta}\right)\right]\left[\beta\left(1-\frac{z}{\delta}\right)\right]\left[-\frac{1}{\delta}\right]\Bigg|_{z=0}$$

(8.8-91)

$$N_{Az}\Big|_{z=0} = -\frac{\mathcal{D}_{AB}\,c_{A0}}{\cosh[\beta]} \sinh[\beta]\,[\beta]\left[-\frac{1}{\delta}\right]$$

(8.8-92)

$$N_{Az}\Big|_{z=0} = \frac{\mathcal{D}_{AB}\,c_{A0}}{\delta}\,[\beta]\,\tanh[\beta]$$

(8.8-93)

These examples are obviously extremely oversimplified, but were chosen to give a rudimentary feeling for the treatment of mass transfer with chemical reaction. The interested reader is referred to the literature for details on heat effects, unequal Schmidt numbers, etc.

Chapter 8   Problems

8.1 Compare Fick's law of diffusion, Newton's law of viscosity and Fourier's law of thermal conduction. In each case:

(a) What is being transported?
(b) What is the driving force?
(c) Express the above equations in a form such that the diffusivities all have the same units.

8.2 Rewrite Fick's law in terms of each of the fluxes in Table 8.2.1-1.

8.3 Consider the beaker shown below with CCl_4 in the bottom.

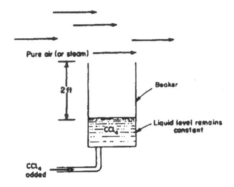

The CCl₄ evaporates into pure air. Use a steady-state model. Model gases as
ideal. Assume equilibrium at the interface. Raoult's law can be used for the
equilibrium model. The vapor pressure of CCl₄ is 50 mm Hg at the temperature
of the system. The total pressure is 1 atm. The diffusion coefficient of CCl_4 in
air is 0.1 cm²/s.

Find the rate of evaporation.

8.4 Consider the same circumstances as in Problem 1 but with steam instead of
air. Use a steady-state model. Model gases as ideal. The steam condenses and
increases the evaporation rate of CCl_4. The latent heat of steam is twice that of
CCl_4. Model the diffusion coefficient as the same as with air.

Find the rate of evaporation with steam.

8.5 For the diffusion of a component A through stagnant B where the cross-
sectional area is constant,

$$N_{Ay} = \frac{-c\,\mathcal{D}_{AB}}{L} \frac{\left(x_{A0} - x_{A1}\right)}{\left(x_B\right)_{lm}}$$

which was derived from

$$N_{Ay} = -c\,\mathcal{D}_{AB}\frac{dx_A}{dy}\left(N_{Ay} + N_{By}\right)x_A$$

Derive an equation in terms of the partial pressures $\bar{p}_A$ and $\bar{p}_B$, where

$$p = \bar{p}_A + \bar{p}_B$$

which is analogous to (a), that is, with a driving force of

$$\left(\bar{p}_{A0} - \bar{p}_{A1} \right)$$

8.6 Derive an expression for counterdiffusion at equal mass rates.

8.7 Two grams of naphthalene are solidified in the bottom of a 1.5-in. long, 1/8-in. diameter glass tube. The air in the tube is assumed to be saturated at the interface and the room air has no naphthalene in it. How long will it take the solid to disappear at 30°C?

8.8 Calculate the diffusion rate of water vapor from the bottom of a 10 ft pipe, if dry air is blowing over the top of the pipe. The entire system is at 70°F and 1 atm pressure ($\mathcal{D}_{AB} = 1$ ft²/h).

8.9 If k_G denotes a mass transfer coefficient based on a driving force of partial pressure

$$N_A = k_G \left(\bar{p}_A - \bar{p}_{A_s} \right)$$

show that

$$k_y = k_G p$$

8.10 Develop the flux equation for mass transfer using a driving force based on mass concentration and a coefficient based on molar concentration.

8.11 The equilibrium equation for acetone distributed between air and water is

$$y_A = 1.13 \, x_A$$

If the mole fraction of acetone in a liquid is 0.015 and in the air is 0.0 and if

$$k_x = 30 \left[\frac{lbmol}{h\,ft^2} \right]$$

$$k_y = 15 \left[\frac{lbmol}{h\,ft^2} \right]$$

find y_{As}, x_{As}, y_{As}^*, and K_y graphically; then find N_A.

8.12 Solve the preceding problem algebraically.

8.13 In a particular dilute system the equilibrium data can be modeled by

$$y_A = 15.3\,x_A$$

If

$$k_x = 5 \left[\frac{lbmol}{h\,ft^2} \right]$$

$$k_y = 1 \left[\frac{lbmol}{h\,ft^2} \right]$$

what is the controlling resistance?

8.14 Mass is transferred from a cylindrical naphthalene bar with diameter of 1/2 in. and length of 6 in. suspended in a 50°F air stream moving with a velocity of 40 ft/s. Estimate the initial mass transfer coefficient.

8.15 Air is blowing at a bulk velocity of 50 ft/s through a 4-in. ID pipe whose walls are wetted with a continuous thin film of water. The temperature is uniform at 70°F. The diffusivity of water in air may be taken as 0.26 cm²/s. At a point where $y_A = 0.001$, calculate the flux of water being evaporated from the pipe wall.

8.16 An aqueous solution containing 2 mole percent of a volatile organic is to be stripped with steam. This aqueous solution is fed to the top of a packed column and steam is fed to the bottom. Ninety-five percent of the entering organic is to be stripped from the water solution by the steam. The equilibrium relation for this situation is given by

$$y_A = 0.1 \sqrt{x_A}$$

In actual operation L/G remains approximately constant

(a) Plot the equilibrium line. Determine the maximum value of L/G which is possible for column operation.

(b) Using the actual operating value of L/G which is equal to 0.5, plot the operating line. Find the outlet concentration y_1 for this case. Write the algebraic equation for the operating line.

(c) Find the number of overall liquid phase transfer units based on the mole fraction. Can you find this analytically?

(d) Given that K_x has been found experimentally to be 20 lbmol/h ft^3 for the above conditions, find n_{OL} for a liquid rate of 500 lbmol/h ft^2 and determine the height of the packed section.

8.17 Graphically calculate

(a) the interface composition in the tower shown below at the point where $y_A = 0.002$.

(b) y_A^* at the point where $x_A = 0.007$.

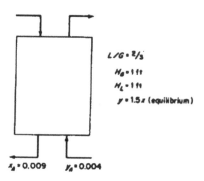

$L/G = \frac{2}{3}$

$N_G = 1 \text{ ft}$

$N_L = 1 \text{ ft}$

$y = 1.5x$ (equilibrium)

$x_A = 0.009$ $y_A = 0.004$

What is the maximum concentration that the exiting gas can have for the tower operating as shown?

8.18   A packed column is to be used to desorb component A from low concentration in a liquid into a stream of carrier gas. The column is operated in concurrent flow.

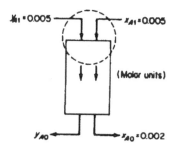

$x_{A1} = 0.005$ $x_{A1} = 0.005$

(Molar units)

y_{A0} $x_{A0} = 0.002$

Equilibrium data

$y_A = 3.0\, x_A$

(a) Develop the general equation of the operating line for concurrent flow in the form $y = mx + b$, assuming constant L and G.

(b) What is the limiting L/G in molar units?

(c) Is the above L/G_{max} or L/G_{min}?

8.19 The column in the preceding problem is now operated countercurrently as shown in the following illustration.

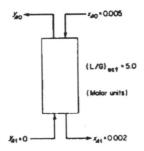

At the point in the tower where $y_A = 0.0005$, what x_{A_s}? For this mode of operation, $H_G = 2$ ft and $H_L = 1$ ft.

8.20 A tower packed with 1-in. ceramic Raschig rings treats 50,000 ft³ of entering gas per hour. The gas contains 2 mole percent SO_2 and is scrubbed with water initially SO_2 free, in such a manner that 90 percent of the SO_2 is removed. The tower pressure is 1 atm and temperature is 70°F. If the actual L/G rate is to be 25 percent greater than the minimum, and the gas velocity is to be 50 percent of the flooding velocity, what should the tower diameter be? The equilibrium curve is

$$y_A = 1.1 \, x_A$$

For I in Raschig rings $a/\varepsilon^3 = 159$. Also, $f = 1.0$.

8.21 It is desired to reduce the mole fraction of NH_3 in an air stream from $y = 0.01$ to $y = 0.001$ using pure water in counterflow in a packed column.

The equilibrium relation is

$$y_{NH_3} = 1.66 \, x_{NH_3}$$

      (a) Sketch the equilibrium line.
      (b) Sketch in the values $y = 0.01$ and $y = 0.001$.
      (c) Indicate on your sketch the operating line point that corresponds to the top of the tower.

(d) Indicate on your sketch the operating line point that corresponds to the bottom of the tower if minimum L/G is used.

(e) Sketch the operating line if $L/G = 2 \, (L/G)_{min}$.

(f) Sketch and calculate the concentration of the outlet water for (e).

(g) If H_{OG} is 1.6 ft, what will be the required height of packing?

APPENDIX A:

VECTOR AND TENSOR OPERATIONS

Scalars and **vectors**, although given special names, are merely subsets of the more general class, **tensors** (a scalar is a zero-order tensor, a vector is a first-order tensor). We can express tensors with either *symbolic* or *index* notation. Symbolic notation regards a tensor as an entity without reference to a coordinate system. Index notation, on the other hand, requires reference to a particular coordinate system.

A.1 Symbolic Notation

Symbolic notation permits us to write equations in terms of symbols (e.g., $\left[\mathbf{v} \times \mathbf{w} \right]$, $\left[\nabla \cdot \mathbf{v} \right]$) that are **independent** of coordinate system, but requires that we know the particular rules by which the operation (e.g., $[\nabla]$, $[\cdot]$) must be implemented in the particular coordinate system.

Table A.1 Operational properties of the del operator in different coordinate frames[1]

Coordinate Frame	$\nabla^2 s \quad \Leftarrow \quad$ where s is a scalar
Rectangular	$\dfrac{\partial^2 s}{\partial x^2} + \dfrac{\partial^2 s}{\partial y^2} + \dfrac{\partial^2 s}{\partial z^2}$
Cylindrical	$\dfrac{1}{r}\dfrac{\partial}{\partial r}\left(r\dfrac{\partial s}{\partial r}\right) + \dfrac{1}{r^2}\dfrac{\partial^2 s}{\partial \theta^2} + \dfrac{\partial^2 s}{\partial z^2}$
Spherical	$\dfrac{1}{r^2}\dfrac{\partial}{\partial r}\left[r^2\dfrac{\partial s}{\partial r}\right] + \dfrac{1}{r^2 \sin(\theta)}\dfrac{\partial}{\partial \theta}\left[\sin(\theta)\dfrac{\partial s}{\partial \theta}\right]$ $+ \dfrac{1}{r^2 \sin^2(\theta)}\dfrac{\partial^2 s}{\partial \varphi^2}$

Coordinate Frame	$\nabla s \quad \Leftarrow \quad$ where s is a scalar
Rectangular	$(\nabla s)_x = \dfrac{\partial s}{\partial x}\quad (\nabla s)_y = \dfrac{\partial s}{\partial y}\quad (\nabla s)_z = \dfrac{\partial s}{\partial z}$
Cylindrical	$(\nabla s)_r = \dfrac{\partial s}{\partial r}\quad (\nabla s)_\theta = \dfrac{1}{r}\dfrac{\partial s}{\partial \theta}\quad (\nabla s)_z = \dfrac{\partial s}{\partial z}$
Spherical	$(\nabla s)_r = \dfrac{\partial s}{\partial r}\quad (\nabla s)_\theta = \dfrac{1}{r}\dfrac{\partial s}{\partial \theta}\quad (\nabla s)_\varphi = \dfrac{1}{r \sin(\theta)}\dfrac{\partial s}{\partial \varphi}$

Coordinate Frame	$\nabla \cdot v \quad \Leftarrow \quad$ where v is a vector (first–order tensor)
Rectangular	$\dfrac{\partial v_x}{\partial x} + \dfrac{\partial v_y}{\partial y} + \dfrac{\partial v_z}{\partial z}$
Cylindrical	$\dfrac{1}{r}\dfrac{\partial}{\partial r}\left(r v_r\right) + \dfrac{1}{r}\dfrac{\partial v_\theta}{\partial \theta} + \dfrac{\partial v_z}{\partial z}$
Spherical	$\dfrac{1}{r^2}\dfrac{\partial}{\partial r}\left[r^2 v_r\right] + \dfrac{1}{r \sin(\theta)}\dfrac{\partial}{\partial \theta}\left[v_\theta \sin(\theta)\right] + \dfrac{1}{r \sin(\theta)}\dfrac{\partial v_\varphi}{\partial \varphi}$

---

[1] These tables encompass the operations most commonly used in this book; a few of the equations, however, list tensor operations not listed. For these a compact listing is provided by Bird, R. B., W. E. Stewart, et al. (1960). *Transport Phenomena*. New York, Wiley, p. 738 ff.

Table A.1 (continued)

Coordinate Frame	$\nabla \times \mathbf{v} \quad \Leftarrow \quad \mathbf{v}$ a vector (first order tensor)
Rectangular	$$\left(\nabla \times \mathbf{v}\right)_x = \frac{\partial v_z}{\partial y} - \frac{\partial v_y}{\partial z}$$ $$\left(\nabla \times \mathbf{v}\right)_y = \frac{\partial v_x}{\partial z} - \frac{\partial v_z}{\partial x}$$ $$\left(\nabla \times \mathbf{v}\right)_z = \frac{\partial v_y}{\partial x} - \frac{\partial v_x}{\partial y}$$
Cylindrical	$$\left(\nabla \times \mathbf{v}\right)_r = \frac{1}{r}\frac{\partial v_z}{\partial \theta} - \frac{\partial v_\theta}{\partial z}$$ $$\left(\nabla \times \mathbf{v}\right)_\theta = \frac{\partial v_r}{\partial z} - \frac{\partial v_z}{\partial r}$$ $$\left(\nabla \times \mathbf{v}\right)_z = \frac{1}{r}\frac{\partial}{\partial r}\left(r\, v_\theta\right) - \frac{1}{r}\frac{\partial v_r}{\partial \theta}$$
Spherical	$$\left(\nabla \times \mathbf{v}\right)_r = \frac{1}{r \sin\left(\theta\right)}\frac{\partial}{\partial \theta}\left[v_\varphi\, \sin\left(\theta\right)\right] - \frac{1}{r \sin\left(\theta\right)}\frac{\partial v_\theta}{\partial \varphi}$$ $$\left(\nabla \times \mathbf{v}\right)_\theta = \frac{1}{r \sin\left(\theta\right)}\frac{\partial v_r}{\partial \varphi} - \frac{1}{r}\frac{\partial}{\partial r}\left[r\, v_\varphi\right]$$ $$\left(\nabla \times \mathbf{v}\right)_z = \frac{1}{r}\frac{\partial}{\partial r}\left(r\, v_\theta\right) - \frac{1}{r}\frac{\partial v_r}{\partial \theta}$$

A.2 Index Notation

Index notation represents a tensor by its components, which are scalars. **We herein restrict our use of index notation to rectangular Cartesian coordinate systems only, and any equations written in index notation, therefore, should be applied only in such systems.** Index notation can be extended to curvilinear coordinate systems by use of superscripts as well as subscripts to distinguish covariant and contravariant components; however, this would introduce more complexity than is warranted for the level of treatment here.

Vector and tensor expressions can be written in index notation in terms of the algebra of the components; however, this leads to cumbersome expressions containing multiple summation signs.

Since for our problems summations will always be over the number of coordinate axes (two or three) we define the *summation convention*:[2]:

> **Any repeated index in a monomial implies the sum over the range of the number of coordinate axes**[3]

A.2.1 The unit tensor

The unit vectors along the coordinate axes of an orthogonal Cartesian coordinate system have the following properties (because the vectors are of unit magnitude and the cosine of the angle between the vectors is either 1 - the vector with itself - or 0 - the vector with either of the other two vectors):

$$\delta_i \cdot \delta_j \begin{Bmatrix} = 0 & \text{if } i \neq j \\ = 1 & \text{if } i = j \end{Bmatrix} \tag{A.2.1-1}$$

This is the same property as the *Kronecker delta*. The Kronecker delta represents the *unit tensor*

$$\delta_{ij} = \begin{bmatrix} 1 & 0 & 0 \\ 0 & 1 & 0 \\ 0 & 0 & 1 \end{bmatrix} \Rightarrow \begin{Bmatrix} = 0 & \text{if } i \neq j \\ = 1 & \text{if } i = j \end{Bmatrix} \tag{A.2.1-2}$$

A.2.2 The alternating tensor or permutation symbol

We define the *alternating unit tensor* (also called the *permutation symbol*) as

$$\varepsilon_{ijk} = \begin{cases} 1 & \text{if } \{i,j,k\} \text{ constitutes an even permutation of } \{1,2,3\} \\ -1 & \text{if } \{i,j,k\} \text{ constitutes an odd permutation of } \{1,2,3\} \\ 0 & \text{if } \{i,j,k\} \text{ does not constitute a permutation of } \{1,2,3\} \end{cases} \tag{A.2.2-1}$$

---

[2] Sometimes referred to as *Einstein convention*. Kyrala, A. (1967). *Theoretical Physics: Applications of Vectors, Matrices, Tensors, and Quaternions*. Philadelphia, PA, W. B. Saunders Company, p. 137.

[3] We continue to write many of our expressions in both index and symbolic notation. The reader is again cautioned that the index notation used is true only in rectangular Cartesian coordinate frames.

Some useful identities involving the alternating unit tensor are

a) $\varepsilon_{pqr}\varepsilon_{pqr} = 6$

b) $\varepsilon_{pqi}\varepsilon_{pqj} = 2\delta_{ij}$ (A.2.2-2)

c) $\varepsilon_{pij}\varepsilon_{pkl} = \delta_{ik}\delta_{jl} - \delta_{il}\delta_{jk}$

APPENDIX B:
ERROR FUNCTION

Table B-1 Gauss error function[1]

w	erf w	w	erf w	w	erf w
0.00	0.00000	0.36	0.38933	1.04	0.85865
0.02	0.02256	0.38	0.40901	1.08	0.87333
0.04	0.04511	0.40	0.42839	1.12	0.88679
0.06	0.06762	0.44	0.46622	1.16	0.89910
0.08	0.09008	0.48	0.50275	1.20	0.91031
0.10	0.11246	0.52	0.53790	1.30	0.93401
0.12	0.13476	0.56	0.57162	1.40	0.95228
0.14	0.15695	0.60	0.60386	1.50	0.96611
0.16	0.17901	0.64	0.63459	1.60	0.97635
0.18	0.20094	0.68	0.66378	1.70	0.98379
0.20	0.22270	0.72	0.69143	1.80	0.98909
0.22	0.24430	0.76	0.71754	1.90	0.99279
0.24	0.26570	0.80	0.74210	2.00	0.99532
0.26	0.28690	0.84	0.76514	2.20	0.99814
0.28	0.30788	0.88	0.78669	2.40	0.99931
0.30	0.32863	0.92	0.80677	2.60	0.99976
0.32	0.34913	0.96	0.82542	2.80	0.99992
0.34	0.36936	1.00	0.84270	3.00	0.99998

The Gauss error function is defined as

$$\text{erfc } w \equiv \frac{2}{\sqrt{\pi}} \int_0^w e^{-v^2} \, dv \tag{B-1}$$

The complementary error function is defined as

$$\text{erfc } w \equiv 1 - \text{erf } w \tag{B-2}$$

---

[1] Incropera, F. P. and D. P. DeWitt (1990). *Fundamentals of Heat and Mass Transfer*. New York, John Wiley and Sons, p. B3.

These values are shown plotted in Figure B-1.

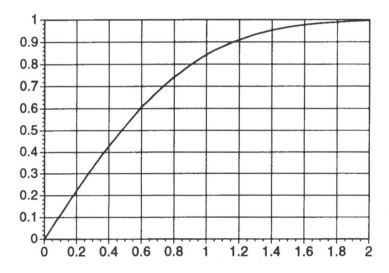

Figure B-1 Error function

APPENDIX C:
NOMENCLATURE

Symbol	Definition	Units in MLtT mol system

ENGLISH

Symbol	Definition	Units in MLtT mol system
A	area	L^2
A_{lm}	logarithmic mean area: $\left(A_2 - A_1\right) + \ln\left(A_2 / A_1\right)$	L^2
A_{gm}	geometric mean area: $\sqrt{A_1 A_2}$	L^2
A_{xs}	cross-sectional area of empty column	L^2
a	interfacial area per unit volume of empty vessel	L^{-1}
Bi_h	Biot number for heat transfer	
Bi_m	Biot number for mass transfer	
Br	Brinkman number	
c	total molar concentration	mol/L^3
c_i	molar concentration of species i	mol/L^3
C	discharge coefficient	
C_D	drag coefficient	
C_p	pitot tube coefficient	
$\hat{C}_p$	heat capacity per unit mass at constant pressure	$L^2/(t^2\,T)$
$\hat{C}_v$	heat capacity per unit mass at constant volume	$L^2/(t^2\,T)$
D	diameter	L
D_e	equivalent diameter	L
$\mathcal{D}_{ij}$	binary diffusivity for pair i, j	L^2/t
$\mathcal{D}_{ij}$	dispersion coefficient for pair i, j	L^2/t
$\bar{D}_s$	surface mean diameter	L

e	base of natural logarithms, 2.71828...	
E	emissive power of surface: rate at which radiant energy is emitted from a surface	M/t^2
Ec	Eckert number	
Eu	Euler number	
F	force	$F = M\,L/t^2$
Fo_h	Fourier number for heat transfer	
Fo_m	Fourier number for mass transfer	
Fr	Froude number	
F_{12}	view factor: fraction of radiation leaving surface 1 intercepted by surface 2	
f	friction factor (if no subscript, the Fanning friction factor)	
gmm	gram mass	M
gmmass	gram mass	M
gmmol	gram mol	mol
G	rate at which radiant energy is incident on a surface per unit area; irradiation	M/t^2
G	gas molar rate per unit area	$mol/(L^2\,t)$
$\bar{G}$	gas molar rate per unit area	$mol/(L^2\,t)$
$\hat{G}$	gas mass rate per unit area	$M/(L^2\,t)$
Gr_h	Grashof number for heat transfer	
Gr_m	Grashof number for mass transfer	
g	acceleration of gravity	L/t^2
g_c	dimensionless conversion factor necessary in FMLt systems (four rather than three fundamental dimensions)	$M\,L/(F\,t^2)$
H	enthalpy	$m\,L^2/t^2$
$\mathcal{H}$	height of transfer unit	L
h	heat transfer coefficient	$M/(t^3\,T)$
h	Planck's constant	$M\,L^2/t$
h	height above datum level (elevation)	L
h	hour	t
hr	hour	t
hemi	hemisphere	$(2\,\pi)$ steradians

$I_{\lambda,\theta,\varphi}$	the rate at which radiant energy of a particular wavelength is transmitted in a particular direction, per unit area of the surface normal to this direction, per unit solid angle about this direction, and per unit wavelength interval	
i	$\sqrt{-1}$	
J	rate at which radiant energy leaves a surface per unit area, both by emission and reflection; radiosity	M/t^2
J_i	molar flux of species i with respect to coordinate system moving at the mass average velocity	$mol/(L^2 t)$
J_i^*	molar flux of species i with respect to coordinate system moving at the molar average velocity	$mol/(L^2 t)$
j_i	mass flux of species i with respect to coordinate system moving at the mass average velocity	$M/(L^2 t)$
j_i^*	molar flux of species i with respect to coordinate system moving at the molar average velocity	$M/(L^2 t)$
j_D	Chilton-Colburn factor for mass transfer	
j_H	Chilton-Colburn factor for heat transfer	
K	kinetic energy	$M L^2/t^2$
k	roughness	L
k	shape factor	
k	thermal conductivity	$M L/(t^3 T)$
k	homogeneous nth order chemical reaction rate constant	$mol^{(1-n)}/[L^{(3-3n)} t]$
k_G	mass transfer coefficient based on driving force of partial pressure (to give molar flux)	$mol\ t/(M L)$
k_x	mass transfer coefficient based on driving force of liquid phase mole fraction (to give molar flux)	$mol/(L^2 t)$
k_y	mass transfer coefficient based on driving force of gas phase mole fraction (to give molar flux)	$mol/(L^2 t)$
k_z	mass transfer coefficient based on driving force of either gas phase mole fraction or liquid phase mole fraction (to give molar flux)	$mol/(L^2 t)$

k_ρ	mass transfer coefficient based on driving force of mass density (to give mass flux)	L/t
k_ω	mass transfer coefficient based on driving force of mass fraction (to give mass flux)	$M/(L^2 t)$
K_x	overall mass transfer coefficient based on driving force of gas phase concentration and equilibrium gas phase concentration to liquid phase	$mol/(L^2 t)$
K_y	overall mass transfer coefficient based on driving force of liquid phase concentration and equilibrium liquid phase concentration to gas phase	$mol/(L^2 t)$
kgm	kilogram mass	M
kgmass	kilogram mass	M
kgmol	kilogram mol	mol
lbm	pounds mass	M
lbmass	pounds mass	M
lbmol	pound mol	mol
L	length	L
L	liquid molar rate per unit area	$mol/(L^2 t)$
$\bar{L}$	liquid molar rate per unit area	$mol/(L^2 t)$
$\hat{L}$	liquid mass rate per unit area	$M/(L^2 t)$
Le	Lewis number	
L_e	equivalent length	L
M_i	molecular weight of species i	M/mol
Ma	Mach number	
m	mass of system	M
m_i	mass of component i	M
N	rotational speed, e.g., rpm	t^{-1}
N	total molar flux with respect to coordinates stationary in space	$mol/(L^2 t)$
N_i	molar flux of species i with respect to coordinates stationary in space	$mol/(L^2 t)$
Nu	Nusselt number	
n	number of moles	mol
n	total mass flux with respect to coordinates stationary in space	$M/(L^2 t)$
n_i	number of moles of species i	mol
n_i	mass flux of species i with respect to coordinates stationary in space	$M/(L^2 t)$

n_i	outward unit normal component in ith coordinate direction	
n	outward unit normal	
P	$p + \rho g h$ (ρ, g constant)	$M/(L\ t^2)$
p	fluid pressure	$M/(L\ t^2)$
p_i	vapor pressure of substance i	$M/(L\ t^2)$
$\bar{p}_i$	partial pressure of component i	$M/(L\ t^2)$
p	linear momentum of body	$M\ L/t$
P	angular momentum of body	$M\ L/t$
Pe_h	Peclet number for heat transfer	
Pe_m	Peclet number for mass transfer	
Pr	Prandtl number	
Q	volumetric flow rate	L^3/t
Q	amount of heat transfer across surface	$M\ L^2/t^2$
$\dot{Q}$	rate of heat transfer across surface	$M\ L^2/t^3$
Q_i	independent variables in dimensional analysis	depends on variable being considered
$\dot{Q}_1$	rate of radiant energy flow leaving surface 1	$M\ L^2/t^3$
$\dot{Q}_{12}$	net rate of radiant energy flow from surface 1 to surface 2	$M\ L^2/t^3$
$\dot{Q}_{1 \to 2}$	rate of radiant energy flow from surface 1 to surface 2	$M\ L^2/t^3$
q	energy flux relative to designated surface	M/t^3
R	universal gas constant	$M\ L^2/(t^2\ T)$
R	radius	L
R	$\dfrac{T_1 - T_2}{t_2 - t_1}$ where $\begin{aligned} T_1 &= \text{shell side inlet} \\ T_2 &= \text{shell side outlet} \\ t_1 &= \text{tube side inlet} \\ t_2 &= \text{tube side outlet} \end{aligned}$	
R_h	hydraulic radius	L
R_i	rate of generation per unit volume of moles of species i by chemical reaction	$mol/(L^3\ t)$
Re	Reynolds number	
Re_L, Re_D, Re_p	Reynolds number based on: distance from front of plate, diameter, particle diameter	

Re_m	modified Reynolds number for power law fluid $= \dfrac{D^a \langle v \rangle^{(2-a)} \rho}{\gamma}$	
r	radial distance	L
r_i	rate of generation of mass of species i per unit volume by chemical reaction	$M/(L^3\,t)$
S	surface area	L^2
S	$\dfrac{t_2 - t_1}{T_1 - t_1}$ where $\begin{aligned} T_1 &= \text{shell side inlet} \\ T_2 &= \text{shell side outlet} \\ t_1 &= \text{tube side inlet} \\ t_2 &= \text{tube side outlet} \end{aligned}$	
Sc	Schmidt number	
Sh	Sherwood number	
St_H	Stanton number for heat transfer: Nu/(Re Pr)	
St_M	Stanton number for mass transfer: Sh/(Re Sc)	
T	torque	ML^2/t^2
T	temperature	T
T, T_{ij}	total stress tensor	$M/(L\,t^2)$
t	time	t
U	internal energy	$M\,L^2/t^2$
U_i	overall heat transfer coefficient based on surface i	$M/(t^3\,T)$
V	volume, instantaneous velocity of flat plate	L^3
v	mass average velocity	L/t
v_i	number average velocity of species i	L/t
v^*	molar average velocity	L/t
W	quantity of work transferred across surface exclusive of flow work	$M\,L^2/t^2$
$\dot{W}$	rate of work transfer across surface	$M\,L^2/t^3$
W'	quantity of work transferred across surface, including flow work	$M\,L^2/t^2$
We	Weber number	
$\mathcal{W}$	total molar flow rate across surface	mol/t
$\mathcal{W}_i$	molar flow rate of species i across surface	mol/t
w	total mass flow rate across surface	M/t
w_i	mass flow rate of species i across surface	M/t
x	rectangular coordinate	L
x_i	mole fraction of species i in liquid or solid phase	

X_i	liquid or solid phase concentration on a basis free of substance i - e.g., if i is water, concentration on a dry basis	
$\bar{X}_i$	molar concentration of liquid or solid phase on a basis free of substance i, e.g., moles of i per moles of all remaining species	
$\hat{X}_i$	mass concentration of liquid or solid phase on a basis free of substance i, e.g., mass of i per mass of all remaining species	
y	rectangular coordinate	L
y_i	mole fraction of species i in gas phase	
z	rectangular coordinate	L
z	complex variable, $z = x + iy$	
$\bar{z}$	complex conjugate, $z = x - iy$	
Z	height of packing in tower	L

GREEK

α	thermal diffusivity, $k/(\rho\, \bar{C}_p)$	L^2/t
α	angle	radians (dimensionless)
α	absorptivity	
β	thermal coefficient of volume expansion $= \frac{1}{V}\left(\frac{\partial V}{\partial T}\right)_p = \frac{1}{\left(\frac{1}{\rho}\right)}\left(\frac{\partial\left[\frac{1}{\rho}\right]}{\partial T}\right)_p = -\frac{1}{\rho}\left(\frac{\partial \rho}{\partial T}\right)_p$ for an ideal gas $\beta = \frac{1}{\left(\frac{RT}{P}\right)}\left[\frac{\partial\left(\frac{RT}{P}\right)}{\partial T}\right]_p = \frac{1}{\left(\frac{RT}{P}\right)}\frac{R}{P}\left[\frac{\partial T}{\partial T}\right]_p = \frac{1}{T}$	T^{-1}
β	angle	radians (dimensionless)
γ	reflectivity	
γ	heat capacity ratio: C_p / C_v	
γ_θ	rate of generation per unit volume of thermal energy	$M/(L\, t^3)$
$\Delta(\ )$	operator designating state 2 minus state 1 of argument	dimensions of argument
δ	diffuse	
δ	thickness; boundary layer thickness	L
δ^*	displacement thickness (boundary layer)	L

δ	unit tensor	
δ_i	unit vector in ith direction in rectangular Cartesian coordinates	L
δ_{ij}	unit tensor; Kronecker delta	
ε	emissivity	
ε	void fraction	
$\varepsilon_{ijk}, \boldsymbol{\varepsilon}$	alternating tensor	
ζ	concentration coefficient of volume expansion $= \frac{1}{V}\left(\frac{\partial V}{\partial x_A}\right)_{p,T,x_i; i \ne A}$	
η	similarity variable	
η	non-Newtonian viscosity	M/(L t)
θ	angle	radians (dimensionless)
Θ	dimensionless temperature, concentration, velocity, etc.	
θ	momentum thickness (boundary layer)	L
λ	wavelength	L
μ	Newtonian viscosity	M/(L t)
ν	frequency	t^{-1}
ν	kinematic viscosity: μ/ρ	L^2/t
π	3.14159...	
π	pressure tensor	$M/(L\ t^2)$
ρ	mass density	M/L^3
$\tilde{\rho}$	molar density, c	mol/L^3
ρ	reflectivity	
ρ_i	mass concentration of species i	M/L^3
$\tilde{\rho}_i$	molar concentration of species i, c_i	mol/L^3
σ	Stefan-Boltzmann constant	$M/(t^3\ T^4)$
σ	total stress	$M/(L\ t^2)$
σ, σ_{ij}	stress tensor	$M/(L\ t^2)$
τ	transmittivity	
τ	shear stress	$M/(L\ t^2)$
τ, τ_{ij}	shear stress tensor	$M/(L\ t^2)$
ϕ	angle	radians (dimensionless)
Φ	potential energy	$M\ L^2/t^2$
Φ	velocity potential	depends on coordinate system

Ψ	stream function	depends on coordinate system
Ω	complex potential function: $\Phi + i\Psi$	depends on coordinate system
$\Omega_{ijk...}$	non-specific Cartesian tensor	
ω	angle	radians (dimensionless)
ω_i	mass fraction of species i	

OVERSCORES

$\sim$	per mole; in the case of dimensionless ratios, molar units in the ratio
$\sim$	in finite element method: approximating function
$\wedge$	mass units (e.g., as opposed to molar units); in the case of dimensionless ratios, mass units in the ratio
$\cdot$	rate: per unit time
$-$	time- or space-average (with real variables) complex conjugate (with complex variables)

BRACKETS

( )	denotes symmetric part of second-order tensor when placed around the subscripts of the parent tensor, e.g., $T_{(i,j)}$
[ ]	denotes antisymmetric part of second-order tensor when placed around the subscripts of the parent tensor, e.g., $T_{[i,j]}$
< >	area average

SUPERSCRIPTS

*	with respect to axes moving at a specified velocity; momentum or displacement thickness (turbulent flow)
*	estimate of quantity at end of interval using derivative at beginning of interval
**	estimate of quantity at end of interval using arithmetic average of derivatives at beginning and end of interval

*, **	dimensionless quantity
′	fluctuation from time-average value
+	quantity (e.g., distance from pipe wall, velocity) made dimensionless using the friction velocity
k	time step number, i.e., time-direction node number (in numerical approximation of differential equations)
ℓ	laminar
t	turbulent

SUBSCRIPTS

*	friction velocity
A, B, C...	species
act	actual
ap	approach (superficial, based on empty cross-section)
av	average
b	bulk or mixing-cup average; black body
cr	critical - at transition point
eq	equivalent, e.g., equivalent diameter D_{eq}
g	gray body
gm	geometric mean
G	gas
H	hydraulic, e.g., hydraulic radius R_H
H	for heat transfer, e.g., St_H
i	initial
i, j, k, ...	with mass, moles denote species i, j, k, ...; with tensors and vectors denote components 1, 2, 3, ...
i, j, k	x-, y-, z-direction node number (in numerical approximation of differential equations)
int	interstitial
lm	logarithmic mean
L	liquid
loc	local
M	for mass transfer, e.g., St_M
m	mean
p	particle, pipe
tot	total

r	reference temperature, e.g., for enthalpy
s	at the surface
t	thermal
xs	cross-section, excess
$0, \infty$	away from surface, freestream
1, 2, 3...	evaluated at point 1, 2, 3, ...

INDEX

TEMPERATURE CONVERSION

$$^\circ C = \frac{5}{9}\left(^\circ F - 32\right) \qquad ^\circ K = ^\circ C + 273.15$$

$$^\circ F = \frac{9}{5}\,^\circ C + 32 \qquad ^\circ R = ^\circ F + 459.67$$

°F	°C	°F	°C	°F	°C	°F	°C
-100	-73	85	29	270	132	455	235
-95	-71	90	32	275	135	460	238
-90	-68	95	35	280	138	465	241
-85	-65	100	38	285	141	470	243
-80	-62	105	41	290	143	475	246
-75	-59	110	43	295	146	480	249
-70	-57	115	46	300	149	485	252
-65	-54	120	49	305	152	490	254
-60	-51	125	52	310	154	495	257
-55	-48	130	54	315	157	500	260
-50	-46	135	57	320	160	505	263
-45	-43	140	60	325	163	510	266
-40	-40	145	63	330	166	515	268
-35	-37	150	66	335	168	520	271
-30	-34	155	68	340	171	525	274
-25	-32	160	71	345	174	530	277
-20	-29	165	74	350	177	535	279
-15	-26	170	77	355	179	540	282
-10	-23	175	79	360	182	545	285
-5	-21	180	82	365	185	550	288
0	-18	185	85	370	188	555	291
5	-15	190	88	375	191	560	293
10	-12	195	91	380	193	565	296
15	-9	200	93	385	196	570	299
20	-7	205	96	390	199	575	302
25	-4	210	99	395	202	580	304
30	-1	215	102	400	204	585	307
35	2	220	104	405	207	590	310
40	4	225	107	410	210	595	313
45	7	230	110	415	213	600	316
50	10	235	113	420	216	605	318
55	13	240	116	425	218	610	321
60	16	245	118	430	221	615	324
65	18	250	121	435	224	620	327
70	21	255	124	440	227	625	329
75	24	260	127	445	229	630	332
80	27	265	129	450	232	635	335

°F	°C	°F	°C	°F	°C	°F	°C
640	338	850	454	1060	571	1270	688
645	341	855	457	1065	574	1275	691
650	343	860	460	1070	577	1280	693
655	346	865	463	1075	579	1285	696
660	349	870	466	1080	582	1290	699
665	352	875	468	1085	585	1295	702
670	354	880	471	1090	588	1300	704
675	357	885	474	1095	591	1305	707
680	360	890	477	1100	593	1310	710
685	363	895	479	1105	596	1315	713
690	366	900	482	1110	599	1320	716
695	368	905	485	1115	602	1325	718
700	371	910	488	1120	604	1330	721
705	374	915	491	1125	607	1335	724
710	377	920	493	1130	610	1340	727
715	379	925	496	1135	613	1345	729
720	382	930	499	1140	616	1350	732
725	385	935	502	1145	618	1355	735
730	388	940	504	1150	621	1360	738
735	391	945	507	1155	624	1365	741
740	393	950	510	1160	627	1370	743
745	396	955	513	1165	629	1375	746
750	399	960	516	1170	632	1380	749
755	402	965	518	1175	635	1385	752
760	404	970	521	1180	638	1390	754
765	407	975	524	1185	641	1395	757
770	410	980	527	1190	643	1400	760
775	413	985	529	1195	646	1405	763
780	416	990	532	1200	649	1410	766
785	418	995	535	1205	652	1415	768
790	421	1000	538	1210	654	1420	771
795	424	1005	541	1215	657	1425	774
800	427	1010	543	1220	660	1430	777
805	429	1015	546	1225	663	1435	779
810	432	1020	549	1230	666	1440	782
815	435	1025	552	1235	668	1445	785
820	438	1030	554	1240	671	1450	788
825	441	1035	557	1245	674	1455	791
830	443	1040	560	1250	677	1460	793
835	446	1045	563	1255	679	1465	796
840	449	1050	566	1260	682	1470	799
845	452	1055	568	1265	685	1475	802

T - #0205 - 101024 - C0 - 234/156/56 [58] - CB - 9780824719722 - Gloss Lamination